52nd Conference on Engineering Mechanical Design and Stress Analysis (AIAS 2023)

IOP Conference Series: Materials Science and Engineering
Volume 1306

Genova, Italy
6-9 September 2023

ISBN: 978-1-7138-9813-9
ISSN: 1757-8981

Printed from e-media with permission by:

Curran Associates, Inc.
57 Morehouse Lane
Red Hook, NY 12571

Institute of Physics
Dirac House, Temple Back
Bristol BS1 6BE UK

Phone: 44 1 17 929 7481
Fax: 44 1 17 920 0979

techtracking@iop.org

Additional copies of this publication are available from:

Curran Associates, Inc.
57 Morehouse Lane
Red Hook, NY 12571 USA
Phone: 845-758-0400
Fax: 845-758-2633
Email: curran@proceedings.com
Web: www.proceedings.com

TABLE OF CONTENTS

IOP Conf. Series: Materials Science and Engineering 1306 (2024) 011001 doi:10.1088/1757-899X/1306/1/011001

Editorial

N. Bonora[1], U. Galietti[2], G. Arcidiacono[3], M. Avalle[4], M. Monti[4], F. Libonati[4], P. Livieri[5]

1. University of Cassino and Southern Lazio, Cassino I-03043, Italy
2. Polytechnic University of Bari, Bari, I-70126, Italy
3. University Guglielmo Marconi, Rome, I-00193, Italy
4. University of Genoa, Genoa, I-16126, Italy
5. University of Ferrara, Ferrara, I-44121, Italy

bonora@unicas.it

The Italian Association for Stress Analysis (AIAS) was founded in 1971 by researchers from academia, research centers and industry. AIAS was intended as a community where to discuss, to share and to develop scientific knowledge related to all technical aspects of stress analysis. In the years, from an initial focus on experimental techniques, AIAS contributed considerably to the development of modern numerical methods and computational techniques for the mechanical engineering design. In 2015, AIAS turned in the Italian Scientific Society of Mechanical Engineering Design.

Today, AIAS is an institutional partner that supports the instances from academia in subject area of the mechanical engineering design. Every year, AIAS organizes a technical conference offering the possibility to present research updates, share new ideas and foster collaborations. The AIAS conference has become a fundamental event for all those interested in current developments in mechanical engineering design and stress analysis, where to meet researchers, testing equipment and software developers.

The 52nd AIAS Conference edition was held in Genoa, Italy. The response of researchers and students has been outstanding: over 200 oral contributions have been presented during the three days of the conference, with four parallel sessions. In addition to the thematic sessions on AIAS traditional subjects, special sessions on *additive manufacturing, energy methods for structural analysis, circular design in mechanical engineering, mechanical behavior under extreme conditions*, have been successfully organized with the contribution of the AIAS technical committees.

Among all contributions presented at the conference, 44 have been selected to be published after peer review, in this volume. This was made possible thanks to the active participation of all AIAS members, to the work of the AIAS Scientific Committee panel (Profs Umberto Galietti, AIAS Scientific Coordinator, Gabriele Arcidiacono AIAS Scientific coordinator for external relationships, Enrico Armentani, Davide Castagnetti, Virgilio Fontanari and Aurelio Somà) and Conference Papers Reviewers. Their outstanding contribution is gratefully acknowledged.

Disclaimer

AIAS2023 Conference was held in presence on Sept. 6-9, 2023. Presentations were arranged on four parallel sessions with a time slot of 15 min assigned to each presenter. Approximately 280 participants

IOP Conf. Series: Materials Science and Engineering 1306 (2024) 011001 doi:10.1088/1757-899X/1306/1/011001

attended the conference during the three days event. All sessions have been continuously monitored by the organizers to provide technical support. The conference ran smoothly, and participants feedback was very positive.

IOP Conf. Series: Materials Science and Engineering 1306 (2024) 011002 doi:10.1088/1757-899X/1306/1/011002

Peer Review Statement

All papers published in this volume have been reviewed through processes administered by the Editors. Reviews were conducted by expert referees to the professional and scientific standards expected of a proceedings journal published by IOP Publishing.

- **Type of peer review:** Single Anonymous
- **Conference submission management system:** Morressier
- **Number of submissions received:** 44
- **Number of submissions sent for review:** 44
- **Number of submissions accepted:** 44
- **Acceptance Rate (Submissions Accepted / Submissions Received × 100):** 100
- **Average number of reviews per paper:** 2.090909090909091
- **Total number of reviewers involved:** 36
- **Contact person for queries:**
  Name: Giovanni Meneghetti
  Email: giovanni.meneghetti@unipd.it
  Affiliation: Universita degli Studi di Padova - Universita degli Studi di Padova Scuola di Ingegneria

IOP Conf. Series: Materials Science and Engineering 1306 (2024) 012001 doi:10.1088/1757-899X/1306/1/012001

An experimental approach for in-situ characterization of dynamic dissipative properties of road pavements

G Loi[1], G Marongiu[2], J Rombi[1], M Coni[1], M C Porcu[1], F Aymerich[2]

[1] Department of Civil, Environmental Engineering and Architecture, University of Cagliari, Piazza d'Armi, 09123 Cagliari, Italy
[2] Department of Mechanical, Chemical and Materials Engineering, University of Cagliari, Piazza d'Armi, 09123 Cagliari, Italy

Email address: gabriela.loi@unica.it

Abstract. The dissipative properties of road pavements may have beneficial effects to reduce vehicle vibrations, traffic noise, vehicles-structure dynamic interaction, and degradation of pavement materials. Assessing the dissipative capacity and the damping properties of road pavements is, therefore, of critical importance. Such assessment has been mainly conducted in recent years by laboratory-scale dynamic experiments, while little effort has been devoted to in-situ tests. The latter are, in fact, cumbersome for practical reasons and typically require a more advanced data analysis when highly coupled modes of vibration are involved. Due to the heterogeneity of the road structure, classical methods are not capable of accurately estimating the road damping properties. The present study proposes an alternative experimental approach based on recording signals from accelerometers embedded in the road, which is impacted by an instrumented hammer. The data are analyzed both in the frequency and in the time domains through the combined use of stabilization diagrams and energy decay tools. Multi-mode fitting algorithms are employed to construct stabilization diagrams for the identification of resonance frequencies, while energy decay curves allow for a robust evaluation of the damping values at the identified frequencies. The effectiveness of the approach was assessed on an asphalt road structure.

1. Introduction

High-level vibrations may lead to structural integrity issues, compromising the performance, durability, and reliability of structures. For this reason, energy dissipation under dynamic loads is often valuable in many engineering structures, where dampers are commonly employed to reduce vibrational effects [1]. Examples include earthquake-prone buildings [2], bridges [3], wind turbines [4], rotating machinery [5], and more.

When considering road structures, reducing vibrations may have beneficial effects, including mitigation of pavement material degradation, reduction of traffic noise, better control of vehicle-structure interaction, improved vehicle ride comfort, and minimized transmission of traffic-induced vibrations to adjacent buildings. The influence of the asphalt pavement type on the damping and resonance properties of light-weight bridges was shown in [6], while rubberized asphalt mixtures were suggested as damping layers in roads [7] to reduce detrimental effects on humans and buildings. Assessing the dissipative properties of road pavement structures may be thus of the utmost importance not only to reduce vibration and noise due to vehicle-road and road-bridge interactions but also to compare the overall performance of different pavement types.

The performance of road pavements has been extensively investigated in the last decades. Deflection methods were proposed to assess the mechanical moduli of road surface and subgrade. To this purpose, Benkelman Beam Deflectometers, lightweight deflectometers [8], and the Falling Weight Deflectometers (FWD) [9, 10] were mainly used. Recording vibrations from sensors bonded to or embedded within the pavement was also exploited to derive performance parameters of road structures under different climate and traffic conditions [11, 12, 13]. Crack propagation and fatigue damage were monitored by means of in situ measures [14] taken through piezoelectric sensors, which are in fact typically used to detect damage in structural components [15].

To account for traffic-induced vibrational effects, the dynamic properties of road systems, like natural frequencies and modal damping ratios, should however be evaluated. The real and imaginary parts of the dynamic complex modulus E^*, which define the elastic and viscous behavior of the material, are often evaluated for linear viscoelastic pavement materials like asphalt mixtures [16], Guidelines for laboratory tests for measuring E^* values on small unconfined cylindrical specimens are provided by the AASHTO TP 62-07 standard [17]. Based on these tests, the damping properties of different asphalt mixtures were estimated in [16] to quantify the effect on tire/road noise characteristics. Laboratory resonant column tests were also performed to determine the damping properties of rubber-modified asphalt and subgrade soils [18]. Numerical and analytical methods have been proposed for an empirical estimation of the Rayleigh damping matrix from the mass and stiffness matrices when two of the natural frequencies of the system are known [19]. These methods can also be used to assess the influence of subgrade damping on the pavement deflection during simulated FWD tests [20].

In-situ experimental campaigns are better suited than laboratory testing for characterizing the actual dissipative behavior of a road system, as they inherently account for real constraint conditions, pavement material properties, subgrade influences, and interaction phenomena. On the other hand, when non-linear and heterogeneous materials like road pavements and subgrades are involved, assessing modal damping ratios may not be a trivial task and classical methods may prove ineffective, also due to the presence of many heavily coupled modes. An approach to estimate the dissipative properties of pavement and subgrade through in-place tests conducted with a FWD device was proposed in [21]. In this approach the damping ratios were obtained by evaluating the decrease in the peak displacements measured at various distances from the impact given by the FWD. The applied procedure fails however to account for the well-known dependence of damping ratios on the vibration frequencies [22].

The present paper presents an alternative experimental approach to estimate the modal damping ratios of a road system based on impacting the road pavement and recording the signals from accelerometers attached to the pavement. The analysis of the recorded accelerations is carried out in both the frequency and time domains using two different tools: the Stabilization Diagram and the Energy Decay Method. Stabilization Diagrams provide the natural frequencies of the system and an initial estimation of the associated damping ratios. The Energy Decay Method is adopted to directly assess energy dissipation within frequency ranges that include the identified resonance frequencies. The combination of the two tools is expected to lead to a robust and reliable evaluation of the system's damping ratios.

2. In-situ testing set-up

The capability of the proposed approach was assessed on an asphalt road structure consisting of three layers: a subgrade, a granular base layer of 900 mm thickness, and a pavement asphalt layer, the latter composed of binder (80 mm thickness), intermediate (60 mm thickness) and a wearing surface (40 mm thickness).

The road structure was excited utilizing an impact hammer consisting of a steel mass of 9 kg instrumented with an *HBM CFT+ 50kN* piezoelectric force transducer. A rubber pad was interposed between the hammer and the road pavement to improve the uniformity of the load distribution over the irregular road surface and to control the range of excitation frequencies. A pad thickness of 3 mm was selected after a preliminary series of tests to ensure sufficient energy coverage for frequencies in the range of 0-300 Hz, which are typically associated to traffic loads. Impact forces with peak values of about 45 kN were applied to excite the road structure.

The response of the system was measured with a *Bruel & Kjaer 4370* piezoceramic accelerometer attached to the surface of the road using a thin butyl adhesive tape. The signals of the force and acceleration transducers were fed into *Bruel & Kjaer 2635* charge amplifiers to generate signal voltages, which were finally acquired by a National Instrument *NI9234* PC-controlled data acquisition unit at a sampling rate of 50 kSamples/s.

The inherent heterogeneous microstructure of the material, comprising various constituents (such as aggregates, fillers, and binders) of different shapes and size scales and characterized by the inevitable presence of irregularly distributed voids, cracks, and defects, is expected to lead to a substantial variability in the local dissipative properties of the road pavement. To reduce the effect of local microstructural features and assess the average damping properties over a sufficiently large section of the road structure, a minimum distance of 5 m was chosen between the excitation and the response measurement sites. In addition, to explore how variations in the specific locations of excitation and response acquisition affect the measured damping characteristics, the tests were conducted using three different pairs of input and output locations (testing configurations P1, P2 and P3 in figure 1).

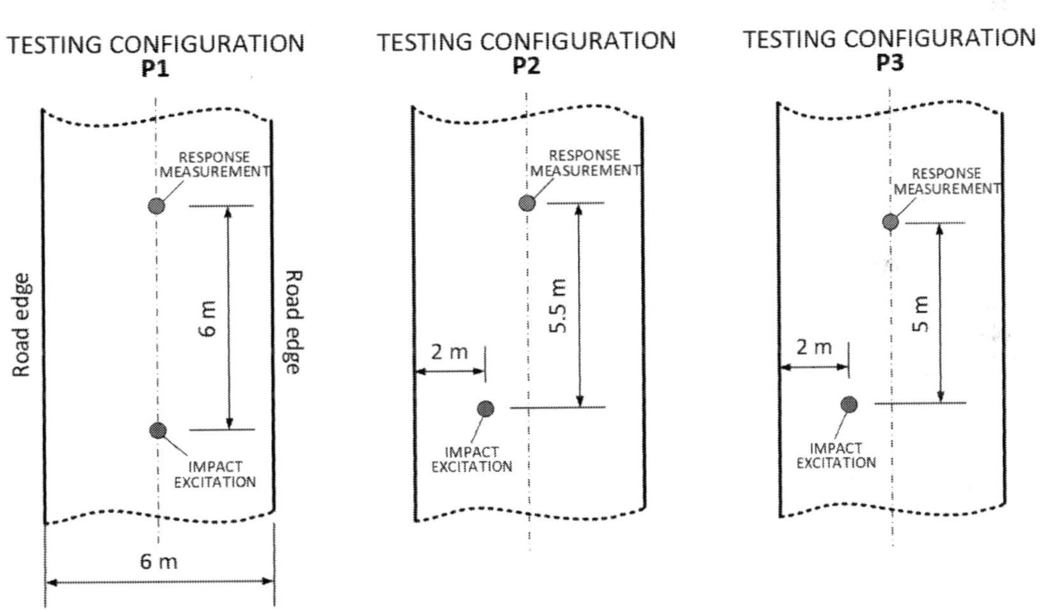

Figure 1. Locations of excitation and measurement points for the three testing configurations.

Figure 2 shows an example of the impact force and the acceleration response signals acquired for the testing configuration P1. The impact force signal has a duration of about 6 ms (figure 2) and a frequency content with a gradually decreasing trend up to approximately 600 Hz (figure 3). As expected, because of the heterogeneity and the wave-dispersive characteristic of the road pavement material, the response signal has a rather complex temporal pattern, characterized by a rapid decay involving multiple damped oscillations with different frequencies. The presence of a multitude of very closely spaced natural frequencies is clear in the frequency spectrum of the measured response, as shown in figure 4. From figure 2, it is immediately evident that the absence of a dominant oscillation with a roughly constant frequency makes it impossible to directly use the basic time-domain approaches (e.g., the Logarithmic Decrement Method) for estimating the damping properties of the system. At the same time, the absence

IOP Conf. Series: Materials Science and Engineering 1306 (2024) 012001 doi:10.1088/1757-899X/1306/1/012001

of well-separated natural frequencies in the response spectrum (see figure 4) hinders the clear identification of contributions from individual modes, thereby impeding the accurate evaluation of damping ratios through conventional frequency-based techniques, such as the Half Power or Circle Fitting methods.

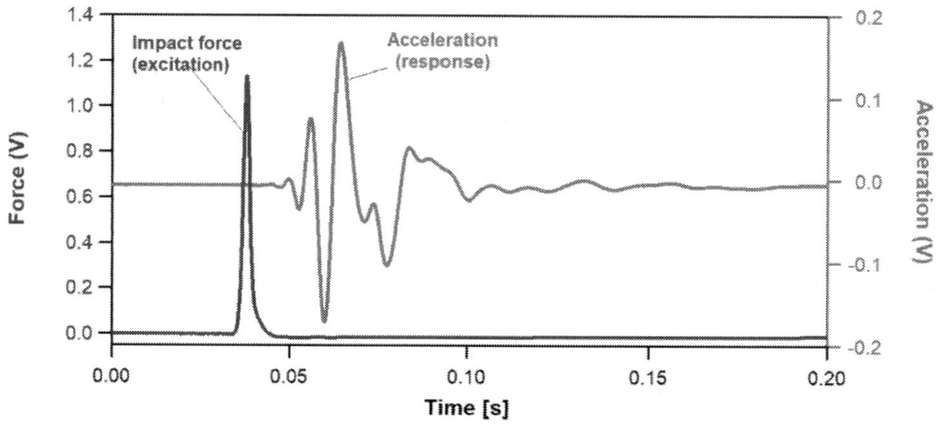

Figure 2. Histories of impact force excitation and acceleration response (testing configuration P1).

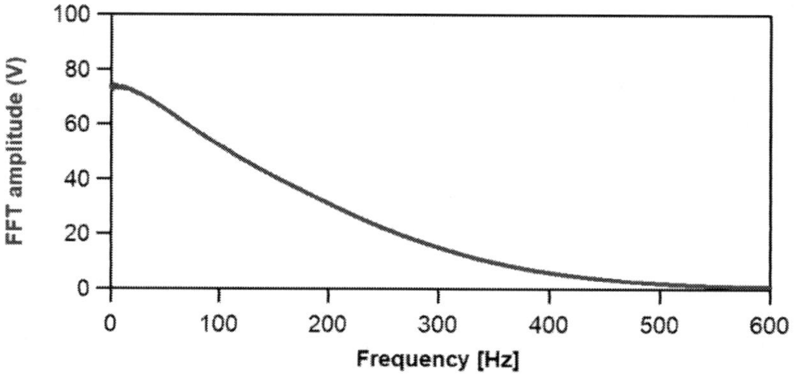

Figure 3. Frequency spectrum of the impact force signal (testing configuration P1).

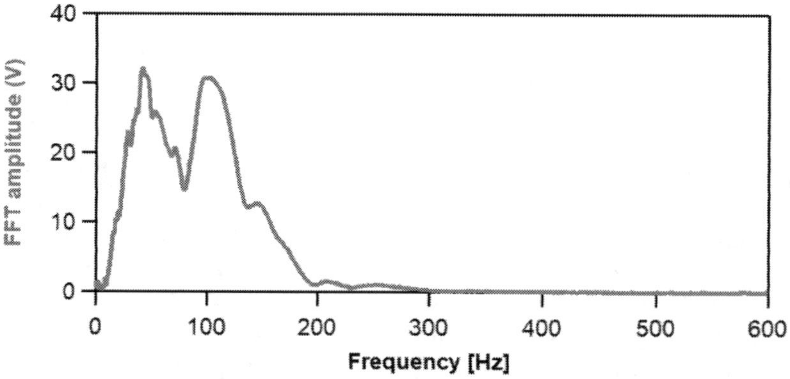

Figure 4. Frequency spectrum of the acceleration response signal (testing configuration P1)

To address these difficulties, a combination of two different techniques is proposed for experimentally determining the damping characteristics of the road pavement. A Stabilization Diagram tool is first applied to identify the natural frequencies of the system and generate a preliminary estimation of the associated damping ratios. An Energy Decay approach, based on the direct assessment of energy dissipation around the identified resonance frequencies, is then employed for a more robust and reliable evaluation of the damping ratios.

3. Results and discussion

Stabilization diagrams provide a concise representation of a system's modal parameters [22], which are extracted through iterative fitting procedures where a modal identification algorithm is repeatedly run by progressively increasing the order of the model (i.e., the number of modes). The main purpose of a stabilization diagram is assessing the accuracy and reliability of a solution, by helping to distinguish between physical modes (which represent the real dynamic response of the system) and spurious modes (generated by noise, overestimation of the model order, or numerical approximations).

The experience gained in many real applications [23] shows that physical modes are stable, i.e., are consistently very close to the same frequency, whereas spurious modes tend to exhibit a wide scatter over a frequency range. In a typical stabilization diagram the information obtained by the identification algorithm over the different model orders is presented with the frequency on the horizontal axis and the model order on the vertical axis. In these diagrams, therefore, physical modes appear as almost straight lines as opposed to spurious modes which tend to scatter over wider frequency ranges.

Figure 5 shows the stabilization diagrams obtained by applying the Direct Modal Parameter Estimation (DMPE) algorithm [24] to the frequency response function (FRF) of the road pavement acquired in the three examined input-output configurations. The criteria adopted to identify physical (i.e., stable) modes require that the difference between the solutions of two consecutive iterations is lower than 1% for the frequency and 5% for the damping ratio [23, 25]. Stable modes are represented as blue markers in the diagrams of figure 5. The data summarized in these charts reveal that the natural modes of the system are distributed in the frequency range 20 Hz –180 Hz. Within this relatively narrow frequency band, the applied algorithm identified several stable poles. In particular, the straight blue lines occurring at 18 Hz, 30 Hz, 46 Hz, 73 Hz, 98 Hz, and 117 Hz identify some of the system's natural frequencies.

It is worth noting that the stabilization diagrams obtained in the three testing configurations do not exhibit significant differences. In particular, the modes identified as stable in both frequency and damping appear to occur at almost the same frequency values, with just slight differences between each testing configuration. The similar responses acquired for the different testing configurations suggests that the size of the examined road portion is large enough to be assumed as representative of the bulk behavior of the road structure.

Stabilization diagrams are conventionally used to distinguish between physical modes and spurious modes, but they do not provide quantitative indications on the damping ratios of the extracted modes. The data condensed in the stabilization diagrams of figure 5 were further post-processed to determine the model order at which the difference between the experimental FRF ($FRF_{ex,i}$) and its estimation based on the extracted modes ($FRF_{num,i}$) is the lower. To assess the optimal model order, the following error function was defined:

$$\varepsilon = \frac{\sum_i \left(\left|FRF_{ex,i}\right| - \left|FRF_{num,i}\right|\right)^2}{\sum_i \left(\left|FRF_{ex,i}\right|\right)^2}$$

The graphs of figure 6 show the experimental FRFs and their estimation corresponding to the minimum error ε (models with a maximum order of 200 were considered) for the three different input-output configurations.

IOP Conf. Series: Materials Science and Engineering 1306 (2024) 012001 doi:10.1088/1757-899X/1306/1/012001

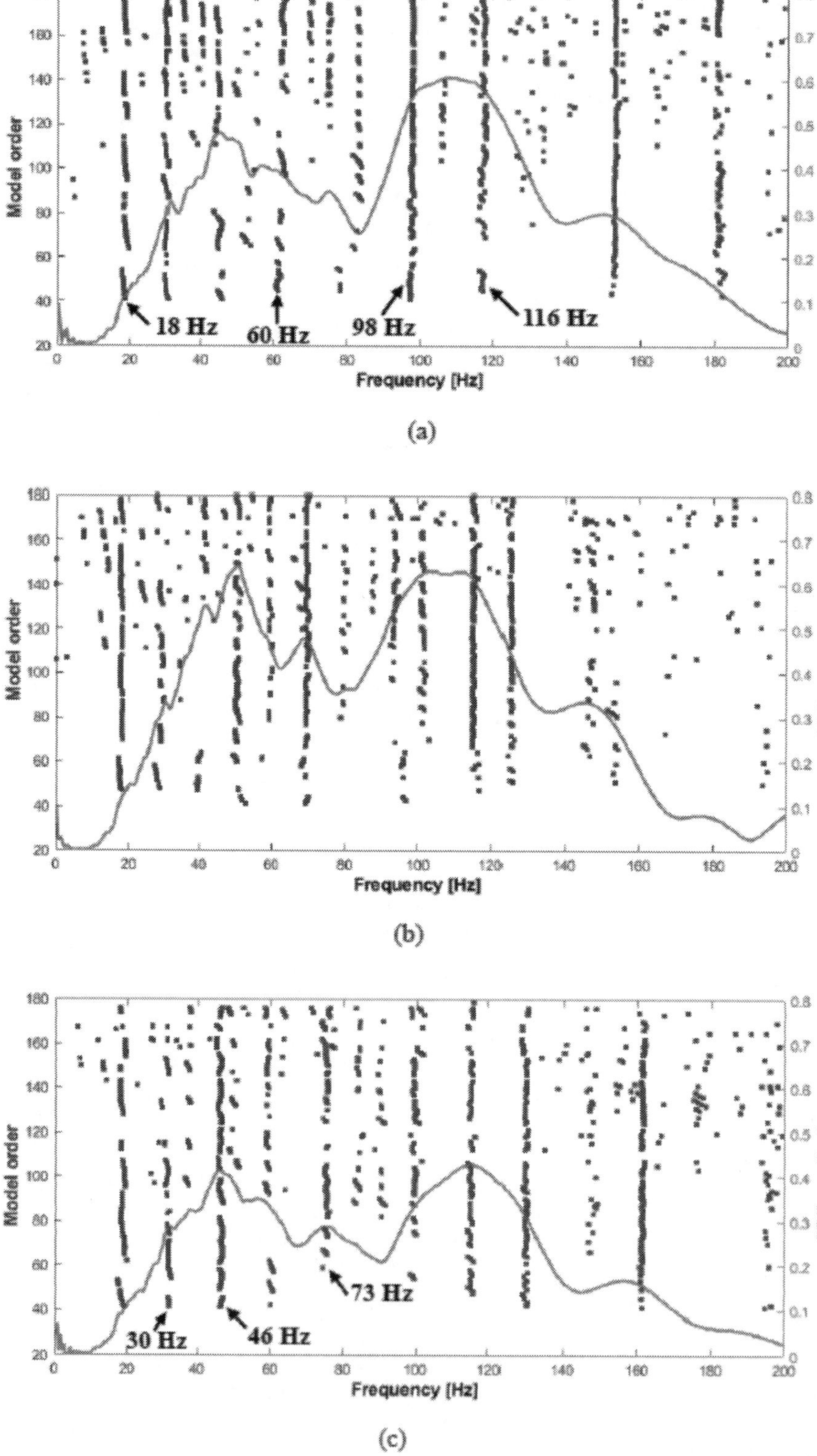

Figure 5. Stabilization diagrams for testing configuration P1 (a), P2 (b), and P3 (c).

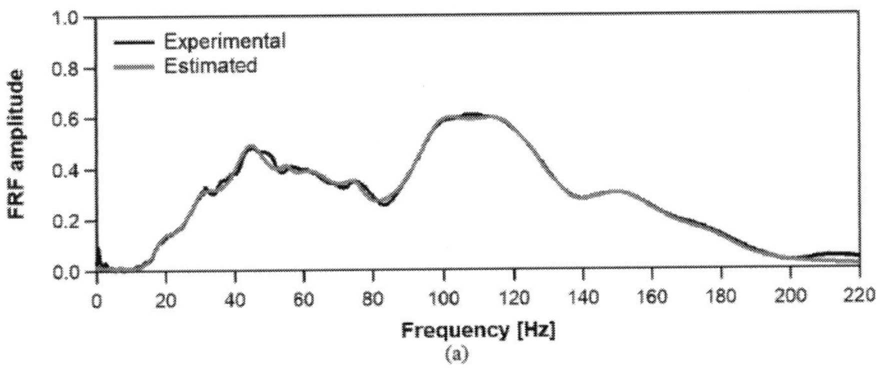

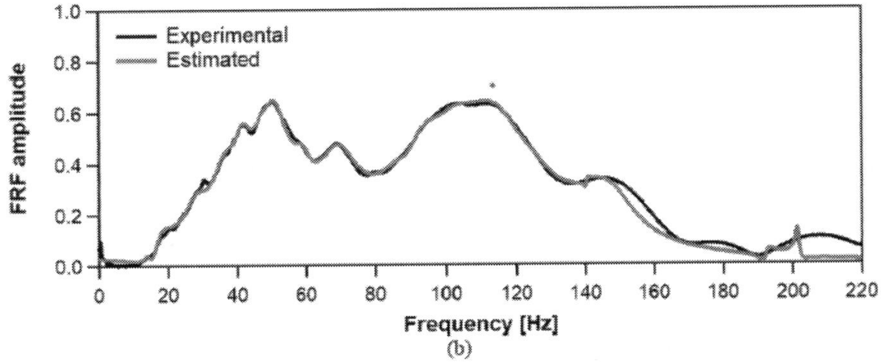

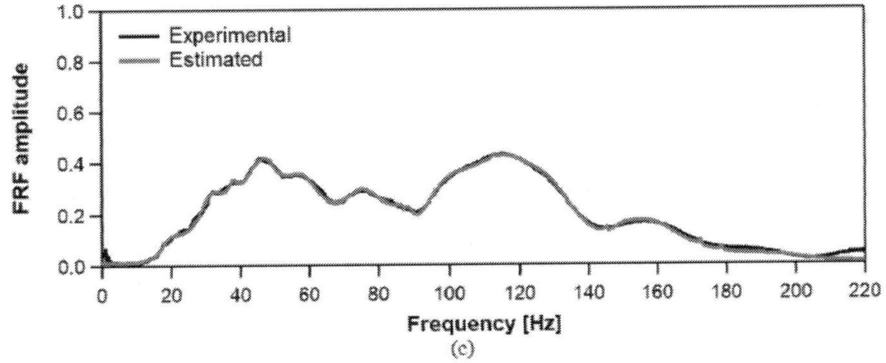

Figure 6. Comparison of experimental and estimated FRFs for testing configuration P1 (a), P2 (b), and P3 (c).

The damping ratios obtained at the optimal model order are reported as a function of the associated resonance frequency in figure 7, for the seven identified stable frequencies. The obtained results show that the damping ratio tends to decrease with increasing frequency. At the lower natural frequencies (i.e., 18 Hz and 30 Hz), the damping ratios range from 13% to 17%, but appear to stabilize around 9% for the 99 Hz and 116 Hz resonances. It can be noticed that for a few natural frequencies the estimated damping ratios exhibit significant dependence on the testing configuration (see for example the values of the damping ratios at the 46 Hz, 60 Hz and 73 Hz frequencies). The high variability of the estimation of the damping characteristics provided by this method for the different input-output locations may be related to both the multi-mode curve-fitting process over a wide frequency range and the presence of many heavily coupled modes.

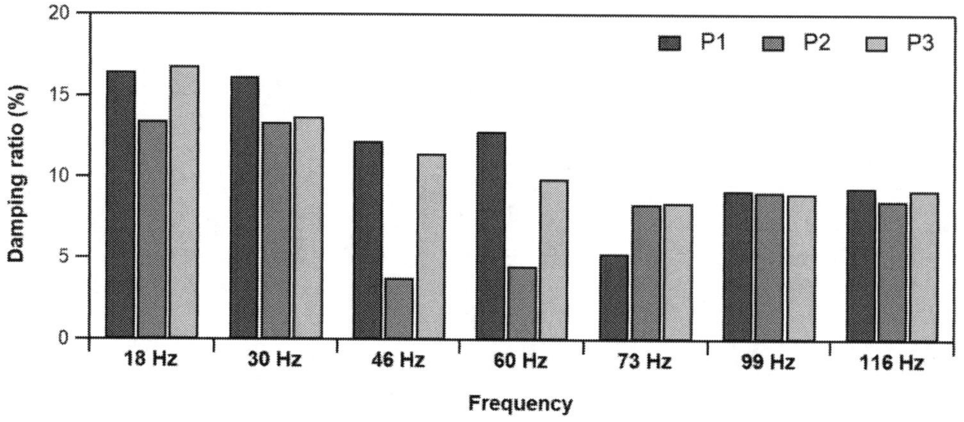

Figure 7. Damping ratios estimated through the Stabilization Diagram tool for the testing configurations P1, P2 and P3.

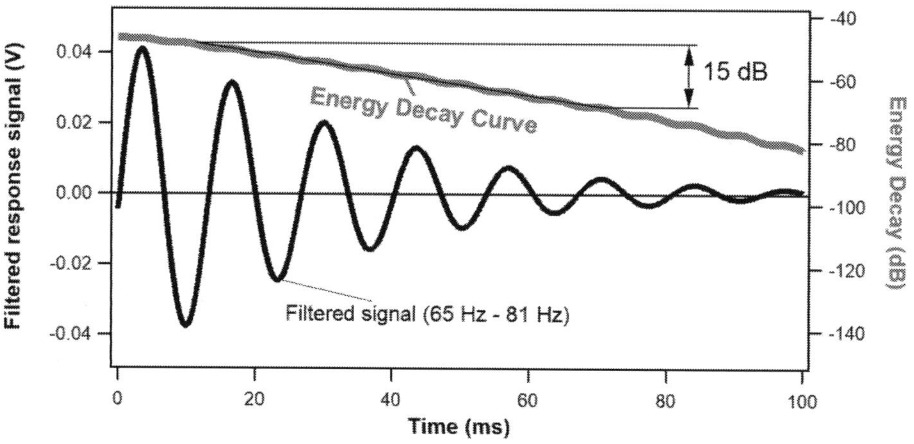

Figure 8. Evaluation of damping ratio using the Energy Decay method

To address this issue, the damping characteristics of the system were subsequently assessed using an Energy Decay approach, based on the evaluation of the decay rate of the time response to an impulse excitation [26, 27]. The analysis was carried out over different frequency bands to account for the dependence of the dissipative properties of the structure on the vibration frequency. The decay rate was evaluated by calculating the slope of a straight line fitted to the energy decay curve of the signal filtered over the investigated frequency band.

The energy decay curve, $E(t)$, was obtained using the backward integration scheme [28] as:

$$E(t) = \int_t^{\infty} h^2(\tau)d\tau$$

where $h(\tau)$ represents the response of the system after pass-band filtering in the frequency range of interest. A band-pass 5th order Butterworth filter with a 1/3 octave bandwidth centered on the desired frequency f_0 was used for filtering the response signal. The energy decay curve was then subjected to linear regression over the range between -2 dB and -17 dB from the maximum level (figure 8). Similar ranges were used in other studies to evaluate the loss ratios of vibrating structures [29, 30]. It is worth noting that all evaluations carried out on the data acquired in the tests consistently yielded linear regressions with correlation coefficients R^2 higher than 0.98. The slope k of the regression line was then used to calculate the reverberation time (T_{60}) and the damping ratio (ζ) according to the expressions [31]:

$$T_{60} = \frac{60}{k}\ ;\ \zeta = \frac{1.1}{T_{60} \cdot f_0}$$

The damping ratios calculated using the energy decay approach for the three testing configurations are presented in figure 9 as a function of the same frequencies examined with the stabilization diagram tool. It is immediately evident that the damping ratios show significantly less variability across the three examined input-output locations when compared to those obtained using the stabilization diagram. The two procedures provide nonetheless globally comparable results when considering the averages of the damping ratios over the three input-output configurations (figure 10). It should be noted that while the energy decay approach appears to be a more reliable and robust approach, it requires prior knowledge of the resonance frequencies of the inspected structure. Because of the inherent heterogeneity and structural complexity of the road structure, this knowledge cannot be easily achieved without the application of multi-mode fitting algorithms, thus suggesting a complementary use of the two approaches for the characterization of the dissipative properties of a road pavement.

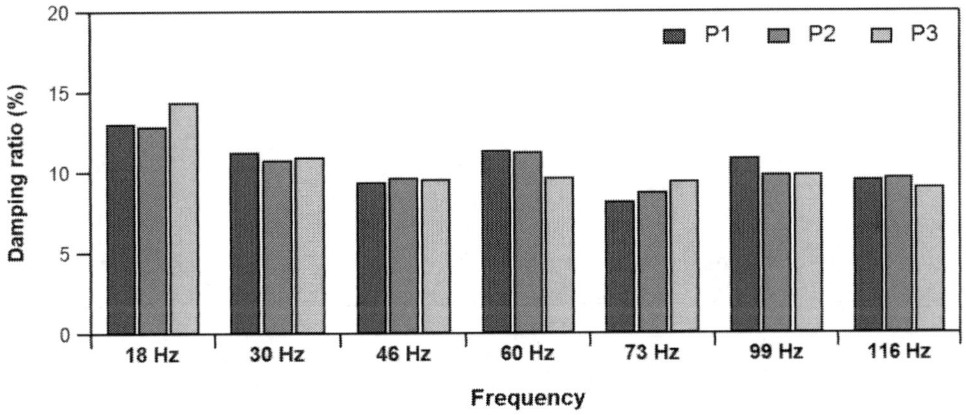

Figure 9. Damping ratios estimated through the Energy Decay approach for the testing configurations P1, P2 and P3.

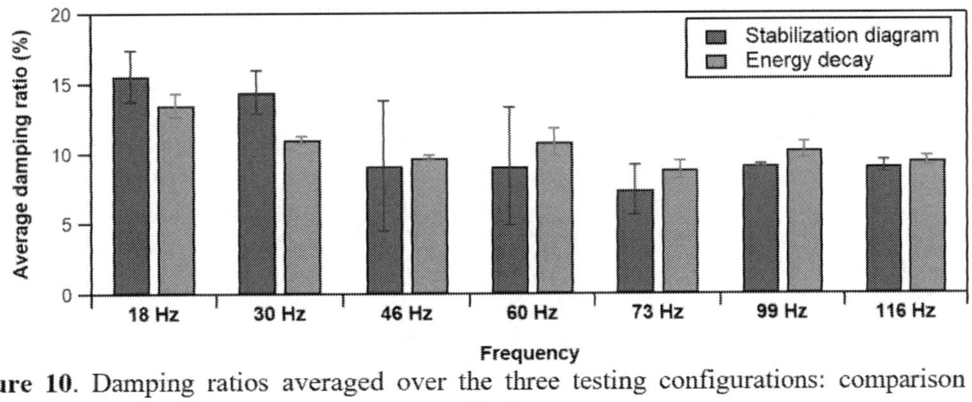

Figure 10. Damping ratios averaged over the three testing configurations: comparison between Stabilization Diagram and Energy Decay approaches.

IOP Conf. Series: Materials Science and Engineering 1306 (2024) 012001 doi:10.1088/1757-899X/1306/1/012001

4. Conclusions

The paper presents an experimental approach to characterize the dissipative properties of a road pavement, relying on signals recorded from accelerometers embedded in the pavement when subjected to instrumented impacts. The main findings of the study can be summarized as follows.

- The impulsive response of the examined road pavement rapidly decays with multiple and heavily coupled vibration modes, preventing the application of classical methods to evaluate damping properties.
- An approach is proposed based on the combined use of stabilization diagrams for the identification of natural frequencies and energy decay measurements for the evaluation of the damping properties of the system within frequency bands around the natural frequencies.
- The construction of the stabilization diagram allows for the identification of natural frequencies by examining the stability of the simulated modes as the model order increases. However, the damping ratios obtained through the stabilization diagram procedure exhibit significant variations across different excitation or measurement locations.
- An energy decay technique is thus subsequently applied to assess the energy dissipation of the system within frequency bands around the natural frequencies identified by the stabilization diagrams. The damping ratios calculated through the energy approach are in reasonable agreement with the average damping ratios obtained over the different testing configurations using the stabilization diagram. The damping ratios obtained using the energy decay method show however a much lower scatter, thus indicating greater reliability and robustness compared to the stabilization diagram tool.
- The potential of the approach will be explored in future work for the assessment of the dissipative properties of different types of engineering structures characterized by complex multi-mode dynamic responses.

References

[1] Occhiuzzi A 2009 Additional viscous dampers for civil structures: Analysis of design methods based on effective evaluation of modal damping ratios *Eng. Struct.* **31**(5) 1093-101
[2] Ghiani C, Linul E, Porcu M C, Marsavina L, Movahedi N and Aymerich F 2018 Metal foam-filled tubes as plastic dissipaters in earthquake-resistant steel buildings *IOP Conf. Series: Materials Science and Engineering* **416**(1) 012051
[3] Wang J F, Lin C C and Chen B L 2003 Vibration suppression for high-speed railway bridges using tuned mass dampers *Int. J. Solid Struct.* **40**(2) 465-91
[4] Colwell S and Basu B 2009 Tuned liquid column dampers in offshore wind turbines for structural control *Eng. Struct.* **31**(2) 358-68
[5] Zhao Q, Yuan J, Jiang H, Yao H and Wen B 2021 Vibration control of a rotor system by shear thickening fluid dampers *J. Sound Vib.* **494** 115883
[6] Schubert S, Gsell D, Steiger R and Feltrin G 2010 Influence of asphalt pavement on damping ratio and resonance frequencies of timber bridges *Eng. Struct.* **32**(10) 3122-29
[7] Huang J, Leandri P, Cucinello G and Losa M 2022 Mix design and laboratory characterisation of rubberised mixture used as damping layer in pavements *Int. J. Pavement Eng.* **23**(8) 2746-60
[8] Guzzarlapudi S D, Adigopula V K and Kumar R 2016 Comparative studies of lightweight deflectometer and Benkelman beam deflectometer in low volume roads *Journal of Traffic and Transportation Engineering (English Edition)* **3**(5) 438-47
[9] Kuo C M, Lin C C, Huang C H and Lai Y C 2016 Issues in simulating falling weight deflectometer test on concrete pavements *KSCE J. Civ. Eng.* **20** 702-8
[10] Rossi M, Wisén R, Vignoli G and Coni M 2022 Assessment of Distributed Acoustic Sensing (DAS) performance of geotechnical applications *Eng. Geol.* **306** 1-14
[11] Čygas D, Laurinavičius A, Paliukaitė M, Motiejūnas A, Žiliūtė L and Vaitkus A 2015 Monitoring

the mechanical and structural behavior of the pavement structure using electronic sensors *Comput.-Aided Civ. Infrastruct. Eng.* **30** 317-28

[12] Bahrani N, Blanc J, Hornych P and Menant F 2020 Alternate method of pavement assessment using geophones and accelerometers for measuring the pavement response *Infrastructures* **5**(3) 25

[13] Coni M, Mistretta F, Stochino F, Rimbi J, Sassu M and Puppio M L 2021 Fast Falling Weight Deflectometer method for condition assessment of RC bridges *Appl. Sci.* **11** 1-18

[14] Alavi A H, Hasni H, Lajnef N and Chatti K 2016 Continuous health monitoring of pavement systems using smart sensing technology *Constr. Build. Mater.* **114** 710-36

[15] Loi G, Marongiu G, Porcu M C and Aymerich F 2023 Vibro-Acoustic Modulation with broadband pump excitation for efficient impact damage detection in composite materials *IOP Conf. Series: Materials Science and Engineering* **1275**(1) 012008

[16] Biligiri K P 2013 Effect of pavement materials' damping properties on tyre/road noise characteristics *Constr. Build. Mater.* **49** 223-32

[17] AASHTO 2006 Designation: TP 62-07, Determining dynamic modulus of hot-mix asphalt concrete mixtures. AASHTO Provisional Standards

[18] Zhong X G, Zeng X and Rose J G 2002 Shear modulus and damping ratio of rubber-modified asphalt mixes and unsaturated subgrade soils. *J. Mater. Civ. Eng.* **14**(6) 496-502

[19] Huang J, Li X, Zhang J, Sun,Y and Ren J 2022 Determining the Rayleigh damping parameters of flexible pavements for finite element modeling *J. Vib.Control* **28**(21-22) 3181-94

[20] Kuo C M and Tsai T Y 2014 Significance of subgrade damping on structural evaluation of pavements *Road Mater. Pavement Des.* **15**(2) 455-64

[21] Uglova E and Tiraturyan A 2017 Calculation of the damping factors of the flexible pavement structure courses according to the in-place testing data *Procedia Eng.* **187** 742-8

[22] Avitabile P 2018 *Modal testing – A practitioner's guide* (Chichester: John Wiley & Sons Ltd) p 544

[23] Van der Auweraer H and Peeters B 2004 Discriminating physical poles from mathematical poles in high order systems: use and automation of the stabilization diagram *Proceedings of the 21st IMTC 2004 - Instrumentation and Measurement Technology Conference* (IEEE Cat. No. 04CH37510) 2193-8

[24] Chomette B and Mamou-Mani 2017 Modal control based on direct modal parameters estimation *J Vib. Control* **24**(12) 2389-99

[25] Wu C, Liu H, Qin X and Wang J 2017 Stabilization diagrams to distinguish physical modes and spurious modes for structural parameter identification *J. Vib. Eng.* **19**(4) 2777-94

[26] Cremer L, Heckl M and Petersson B A T. *Structure-Borne Sound - Structural Vibrations and Sound Radiation at Audio Frequencies* (Berlin: Springer Verlag) p 607

[27] Bloss B and Rao M D 2005 Estimation of frequency-averaged loss factors by the power injection and the impulse response decay methods *J. Acoust. Soc. Am.* **117**(1) 240-9

[28] Schroeder M R 1965 New method of measuring reverberation time *J. Acoust. Soc. Am.* **37**(6) 1187-8

[29] Cabell R, Schiller N, Allen A and Moeller M 2009 Loss factor estimation using the impulse response decay method on a stiffened structure *Proceedings Inter-noise 2009 - 38th International Congress and Exposition on Noise Control Engineering*

[30] Wu L, Agren A and Sundbäck U 1997 A study of the initial decay rate of two-dimensional vibrating structures in relation to estimates of loss factor *J. Sound Vib.* **206**(5) 663-84

[31] Gade S and Herlufsen H 1994 Digital filter techniques vs FFT techniques for damping measurements *Brüel & Kjær Technical Review* **1** 1-44.

IOP Conf. Series: Materials Science and Engineering 1306 (2024) 012002 doi:10.1088/1757-899X/1306/1/012002

Structural behaviour of single-deck floating roof tanks subject to seismic loading

Luca Chiappelloni[1]; Valerio Belardi[1]; Francesco Serraino[1]; Simone Trupiano[1]; Luca Gaetani[2]; Francesco Vivio[1]

[1]Department of Enterprise Engineering, University of Rome Tor Vergata, Via del Politecnico, 1, Rome, 00133, Italy.
[2]Eleo2 Engineering S.r.l., Via Benedetto Stay, 69, 00143 Rome, Italy.

E-mail: `Luca.Chiappelloni@uniroma2.it`

Abstract. During a seismic event, fluid storage tanks are subject to sloshing phenomena; their effect on the structural components of the tanks is one of the main aspects to be considered during the design. The movement in successive modal forms of the roof-fluid system determined by the earthquake can cause severe damage to both the containment shell and the floating roof. The fluid-structure interaction imposes extremely complex out-of-plane deformation fields, mainly in the presence of floating roofs with lower flexible resistance. Such deformations can cause severe structural damage and subsequent catastrophic events such as those experienced in the past in Japan and Turkey. In fact, recently recorded earthquakes in Japan and Turkey, characterized by oscillation periods between 3 and 8 sec (long-period ground motion), compatible with the natural first and second slosh frequencies of the fluid-structure system, led to conspicuous problems of instability failure with the consequent sinking of the floating roof. These events dictated a new starting point for the research into the seismic behaviour of roof-fluid-tank systems.

1. Introduction

The sloshing is one of the main issue when design liquid storage tanks [1],[2]. In the case of fixed roof tanks, the sloshing of free surface during an earthquake induces an extensive hydrodynamic force on the walls and roof of the tank [3]. On the other hand, the liquid storage tanks are generally covered with a FR (floating roof) to suppress oil vapor leakage from the free surface, the seismic loads could induce a large displacement of the FR due to the liquid sloshing. This FR motion possibly leads to severe structural damage and catastrophic fire accidents as experienced in Japan and Turkey during past earthquakes.

During the Tokachi-oki earthquake in 2003, several single-deck floating roofs (SDFR) located at Tomakomai were seriously damaged [4], the exposure of fuel oils to the atmosphere caused a large number of fires in the area. Other catastrophic events that brought greater attention to this phenomenon were found during the 1999 Kocaeli earthquake and the 1964 Niigata earthquake [5] in which the sinking of the roof led to the ignition of the hydrocarbons contained inside the tank.

After these catastrophes, the problem of sloshing has been studied in much more detail. The main theories have been formulated in Japan in the last two decades.

IOP Conf. Series: Materials Science and Engineering 1306 (2024) 012002 doi:10.1088/1757-899X/1306/1/012002

To obtain an analytical solution for the calculation of stress states in floating roofs during a seismic event, a linear solution based on the velocity potential theory, assuming the fluid is incompressible and inviscid, was established in 1970s [6]. Starting from this, Matsui presented the most complete solution for determining the displacement of a FR in a cylindrical storage tank subjected to seismic excitations. He proposed the explicit solutions for both single and double-deck type FRs, the double-deck FR was approximated by a uniform isotropic plate, and an SDFR was replaced by an inner deck with small bending stiffness and an outer pontoon with relatively large stiffness [7].

The developed analytical solutions do not consider the nonlinear effects of the free surface motion. To consider this effect, Yamauchi and Kamei used numerical methods to evaluate the nonlinearity of deformation during the sloshing phenomenon [8].

Starting from Matsui's analytical solution and considering nonlinearity effects, Yoshida et al. used axisymmetric finite element analysis to consider the hydrodynamic coupling of the fluid and a FR [9].

In this work an analytical procedure for stress evaluation in a specific floating roof single deck (SDFR) is presented. This method studies the relationship between vertical deck plate deformation and radial shrinkage. A specific application case will be described. Moreover, a mixed analytical/numerical procedure for the evaluation of the stress state in floating roofs of the Honeycomb type (HC) is also presented. First, the homogeneous mechanical characteristics of the HC panels are defined; in parallel, the out-of-plane displacement field is determined. The definition of the method is finalized with the analysis of a specific application case.

It was decided to consider only these two types of floating roofs (FR) for their very similar structural behavior. Both, in fact, due to their construction geometry impose extremely low stiffnesses. This leads to the simplifying assumption that the roof will not decisively influence the dynamic response of the system [8].

One of the most critical earthquakes for the sloshing phenomenon is the Tokachi-oki earthquake recorded in 2003 [8] Figure 1. Therefore, this specific case study is analyzed for the determination of the stress state in SD-type roofs.

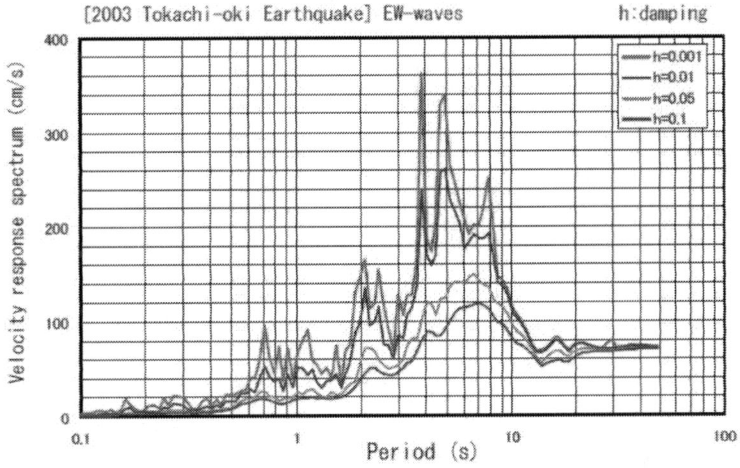

Figure 1: Records of the 2003 Takachi-oki earthquake at tomakomai [8].

For the HC roof, a smaller 10 Kl tank was chosen as a case study. The earthquake considered for the analysis is the catastrophic earthquake Izmit of Mw 7.6 [10]. Its accelerogram is shown in Figure 2.

IOP Conf. Series: Materials Science and Engineering 1306 (2024) 012002 doi:10.1088/1757-899X/1306/1/012002

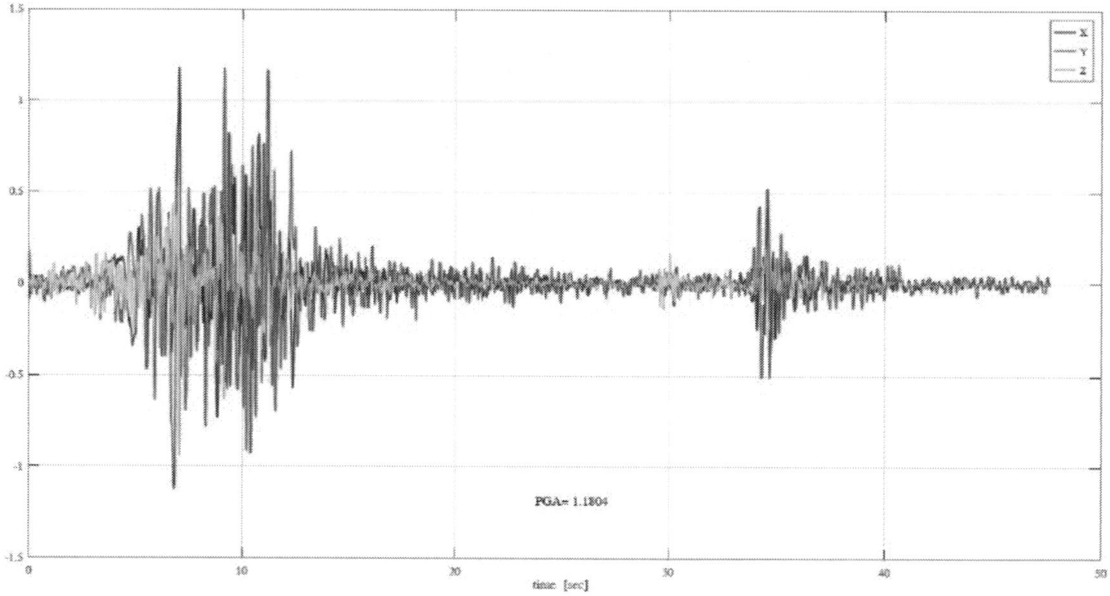

Figure 2: Record Acceleration [g] of the 1999 Izmit earthquake in Turkey (Mw 7.6).

2. Single-Deck roofs

Single-deck roofs (SD) have been evaluated via a fully analytical procedure in MATLAB. The main hypothesis, which is grounded in the proposed procedure, is to consider the structural stiffness of the roof small enough to have no influence on the natural frequencies and characteristic vibrational modes of the tank-fluid-roof system (Matsui [7]). The data available are spatial geometry of the roof, height of the free surface, density of the contained liquid and characteristic geometric parameters of the external pontoon.

Counting for the first two natural frequencies as the main cause of the global stress state of the system, a first computation of the angular frequencies of these two modes has been conducted:

$$\Omega = \sqrt{\frac{g}{R}\epsilon_i tanh(\epsilon_i \frac{g}{R})} \tag{1}$$

Under the assumption of non-viscous, incompressible, and irrotational fluid, the displacement of the free surface is entirely defined by a velocity potential function. The potential function ϕ which satisfies the Laplace equation, the wall condition, and the condition at the tank bottom, is:

$$\phi = \{x_g(t)r + \sum_{i=1}^{\infty} \dot{x}_i(t) \frac{2}{\epsilon_i^2 - 1} \frac{cosh[\epsilon_i(z/R]}{cosh[\epsilon_i(H/R]} \frac{J_i[\epsilon_i(z/R]}{J_i[\epsilon_i]} R\}cos(\theta) \tag{2}$$

where $x_i(t)$ is the associated displacement of the i-th mode, obtained by solving the differential equation of the SDOF forced system. Once the wave profile for a given accelerogram was defined, the effects of nonlinearity were considered following the theory described by Goudarzi in [5]. These effects, particularly present in conditions of high wave widths, decisively change the sloshing profile for the first mode (Figure 3).

To evaluate the stress state in the outer pontoon it has been hypothesized that this can be assimilated to a curved beam on an elastic foundation [8]. The maximum circumferential inflectional moment of the pontoon in $\theta = 0$ is:

$$M_\theta = 0.36\beta_1 \frac{EI_{theta}}{R_m} \frac{\eta^1_{max}}{R_m} \quad (3)$$

Where β_n is the ratio between the stiffness of the pontoon and the elastic constant of the hypothetical foundation system.

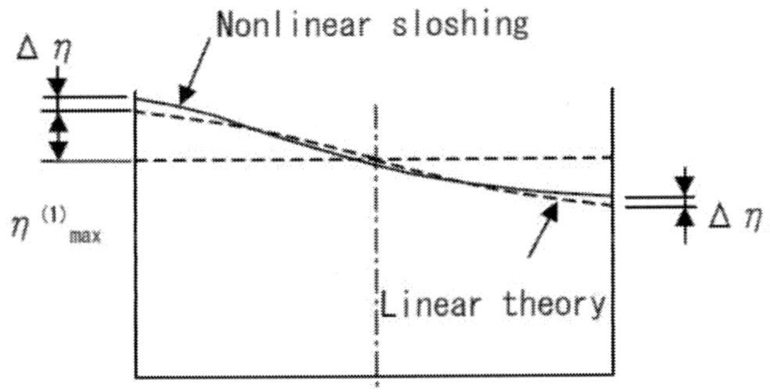

Figure 3: Schematic shape of linear and nonlinear free surface profiles.

As a result of the second sloshing mode, the internal deck plate of the floating roof deforms considerably out of plan. The membrane force in the radial direction causes a radial contraction and a geometric ovalization in the pontoon [11]. The pontoon considered as a beam on elastic foundation shows a deformation that can be seen as a superposition of both the ovalization and the radial contraction. In analogy with the approach proposed in [8], superimposing these two effects we can compute the bending moment and the normal stress associated with the second mode:

$$M_x = 6.25\beta_2 \frac{EI_x}{R_m} \left(\frac{\eta^2_{max}}{R_m}\right)^2 \quad (4)$$

$$N_\theta = 2.08\beta_2 EA \left(\frac{\eta^2_{max}}{R_m}\right)^2 \quad (5)$$

Figure (4.a) shows these forces acting on the cross section of the pontoon [12]. Once the forces are known, the stresses are acquired (the maximum tension is obtained through the SRSS method).

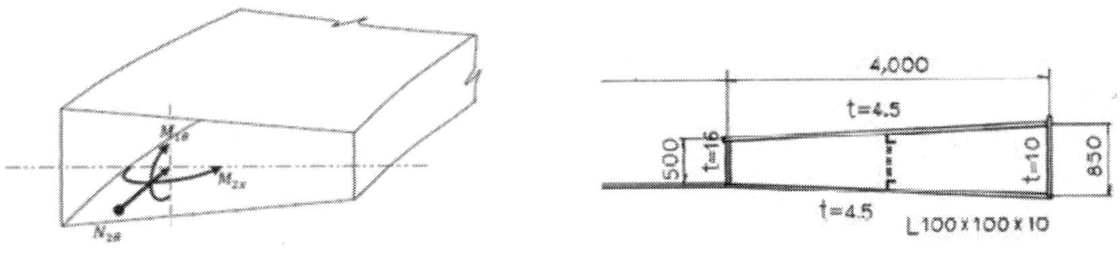

(a) (b)

Figure 4: (a) Seismic forces on the pontoon cross-section of a SDFR; (b) Geometry of the SDFRs [12].

IOP Conf. Series: Materials Science and Engineering 1306 (2024) 012002 doi:10.1088/1757-899X/1306/1/012002

2.1. *Case Study (SD)*

The chosen application case is a tank with a diameter size of $78m$ and a capacity of $100kl$, the fluid height is $15m$ with a density of $840\frac{kg}{m^3}$. The floating roof pontoon has the geometric characteristics shown in Figure (4.b) [12].

To determine the inertial properties of the closed section, a mirror model was created in a software CAD, and the results obtained are shown in Table 1.

Table 1: Inertia properties of the pontoon.

Geometry properties	Value
Area $[m^2]$	5.23e-2
$I_x[m^4]$	1.28e-1
$I_y[m^4]$	4.80e-3

Having defined the geometric properties and characterised the inertial properties of the pontoon, the differetial equation was solved for the two modes individually. In Figures 5 and 6 it is possible to evaluate the time course of the wave height for the two modes at the point $r = R$ and $\theta = 0$.

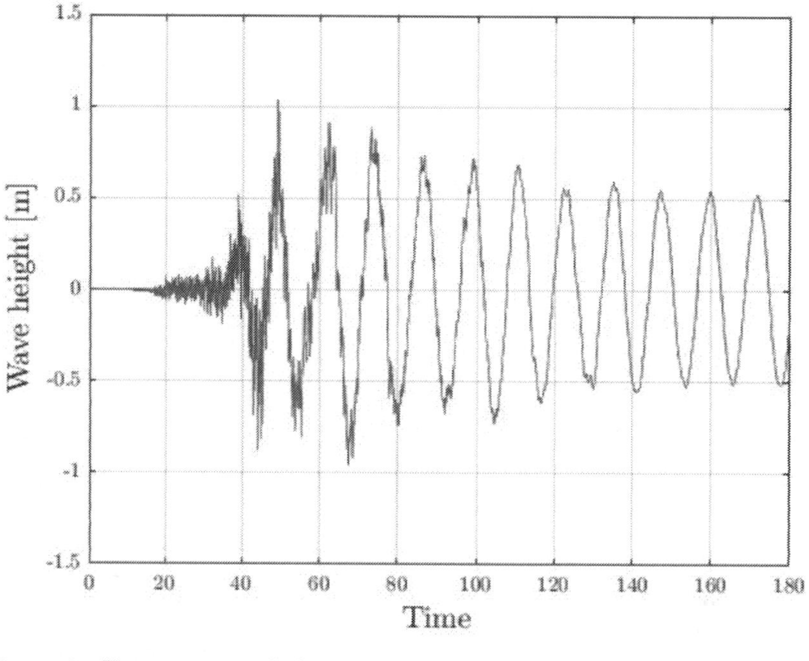

Figure 5: Time course of the wave for first slosh mode in r=R e $\theta = 0$.

Table 2 shows the results obtained in terms of maximum displacement. It should be remembered that the maximum displacements for the individual modes occur at different time; considering them as occurring simultaneously leads to conservative results in terms of overall stresses.

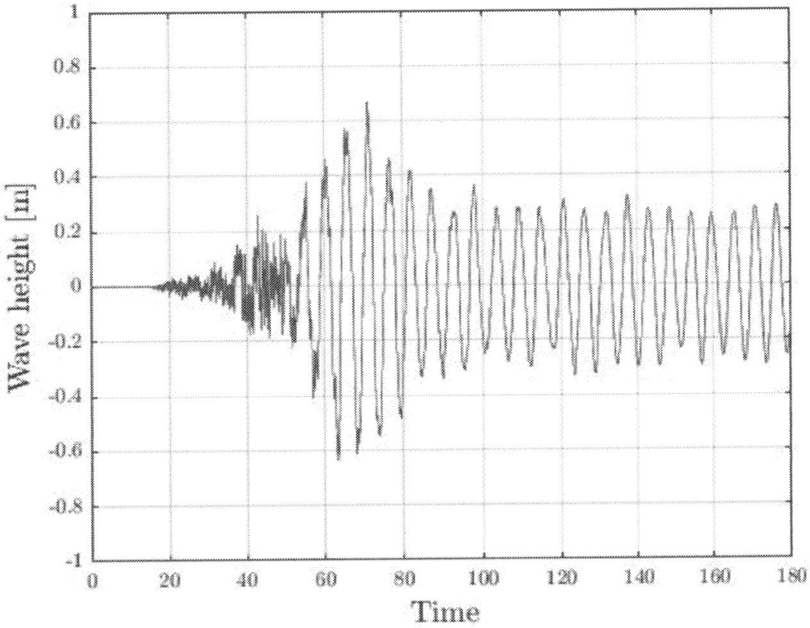

Figure 6: Time course of the wave for second slosh mode in $r = R$ e $\theta = 0$.

Table 2: Maximum slosh height for the first two modes.

	W 1st mode [m]	W 2nd Mode[m]
Analytical procedure	1.09	0.66

Given the maximum displacements, for the two main modes of fluid oscillation, following the previously reported relationships, the seismic forces acting on the cross-section of the pontoon were defined. Table 3 shows the results for the case study analysed.

Table 3: Seismic forces in the cross-section of the pontoon.

Seismic Force	
$M_{1,\theta}[Nm]$	4.55 e4
$N_{2,\theta}[N]$	1.43 e6
$M_{1,\theta}[Nm]$	7.27 e6

After calculating these forces, the seismic stresses were calculated. The stresses from the first and second slosh modes were ultimately combined through the SRSS (square root of the sum of the roots).

Table 4: Seismic stress for individual slosh mode and stress Total.

Seismic Stress [MPa]	
$\sigma M_{1,\theta}$	4.0
$\sigma N_{2,\theta}$	139.1
$\sigma M_{2\theta}$	14.7
$\sigma_{1,Tot}$	4.0
$\sigma_{2,Tot}$	139.9
σTOT	140.0

For a comparison with FEM a refined 3D model was defined (Figure 7) on which a vertical displacement field consistent with the solution of the differential problem was imposed.

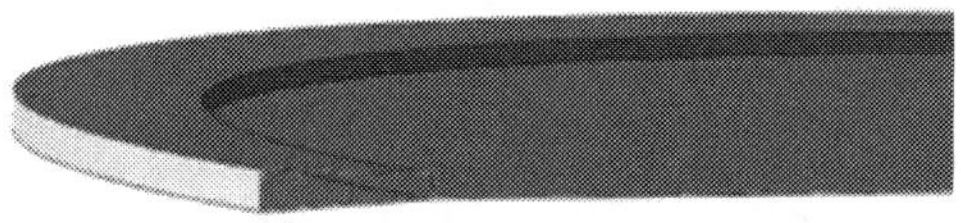

Figure 7: Refined 3D FE model of SDFR.

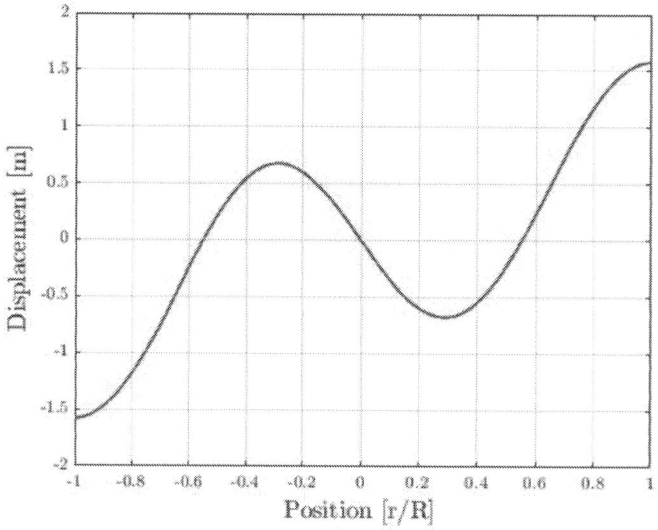

Figure 8: Vertical displacement field differential problem solution.

The vertical displacement field (Figure 8) led to the stress state shown in Figure 9.

As can be seen from Table 5, the stress values obtained for this SD floating roof, in this specific seismic excitation state, are perfectly consistent with the FEM simulation.

IOP Conf. Series: Materials Science and Engineering 1306 (2024) 012002 doi:10.1088/1757-899X/1306/1/012002

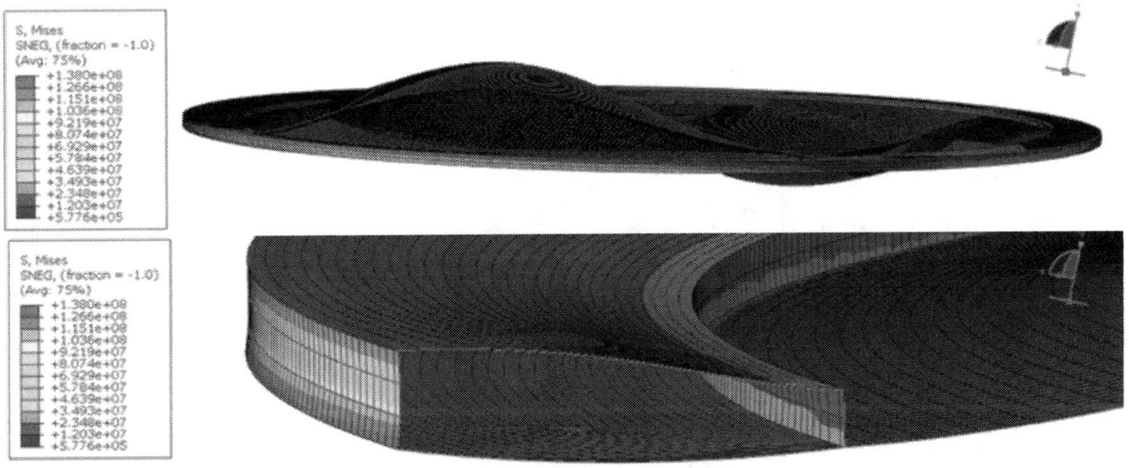

Figure 9: Stress state in the SDFR for Takachi-oki earthquake input accelerogram.

Table 5: Comparison of seismic stress obtained from model and FEM.

Method	Von Mises TOT [MPa]
Matlab	139.9
FEM	138.0

What happens if the tank has a smaller storage capacity?

To answer this question and understand how size influences the effects of the first and second slosh modes, a second test case with the same geometrical and physical properties but with a diameter of 1/2 has been analyzed. Table 6 compares the results obtained: a decrease in diameter causes an increase in the effects related to the first mode and a decrease in those related to the second.

Table 6: Results for tanks with the same input conditions but different diameters.

Diameter [m]	W 1st Mode [m]	W 2nd Mode [m]	$\sigma1[MPa]$	$\sigma2[MPa]$	Von Mises TOT [MPa]
78.2	1.09	0.66	4.0	139.9	140.0
39.1	2.54	0.24	24.3	18.4	30.5

3. HoneyComb roofs

For HoneyComb (HC) roofs simulation with the above technique is not possible due to their complex geometry. In fact, on the entire surface of the roof, there are C-shaped reinforcing partitions and Honeycomb panels.

This configuration did not allow for a fully analytical analysis to be performed. For this reason, it was necessary to develop a specific geometrical model, with FEM. First following the theory presented in [13], known as the geometric and physical properties of the HC panel, the mechanical properties in the three directions of an equivalent orthotropic material were computed. Therefore, it was possible to move from a complex geometry in the micro-scale to a macro-scale material with equivalent properties [14].

Having determined the characteristic properties of the homogenized material and the operative limits, the displacement field to be imposed on the roof has been defined. The hypothesis behind the procedure (as in the case of SD) is to consider the structural stiffness of the roof small enough not to clearly influence the natural frequencies and vibration modes characteristic of the tank-fluid-roof system. The presented method follow Matsui's approach [7].

$$\eta = -\frac{1}{g}(\frac{\partial\phi}{\delta t})_{z=H} = -\frac{1}{g}\{\ddot{x}_g(t)r + \sum_{i=1}^{\infty}\ddot{x}_i(t)\frac{2}{\epsilon_i^2 - 1}\frac{J_i[\epsilon_i(z/R]}{J_i[\epsilon_i]}R\}cos(\theta) \tag{6}$$

Once the characteristic properties of the homogenized material and the displacement field generated by the seismic event have been defined, A FE Model has been defined.

3.1. Case Study (HC)

The application case is a tank with a diameter of $24m$ and a capacity of $10kl$, the fluid height is $13.00m$ with a density of $740kg/m^3$. The roof is characterized by $40mm$ high aluminum HC panels, the cell has a regular hexagonal shape with a characteristic size of $15mm$ and $0.5mm$ thick walls. It can be seen in Figure (10.a) how the panels are connected to C-shaped aluminum connection elements.

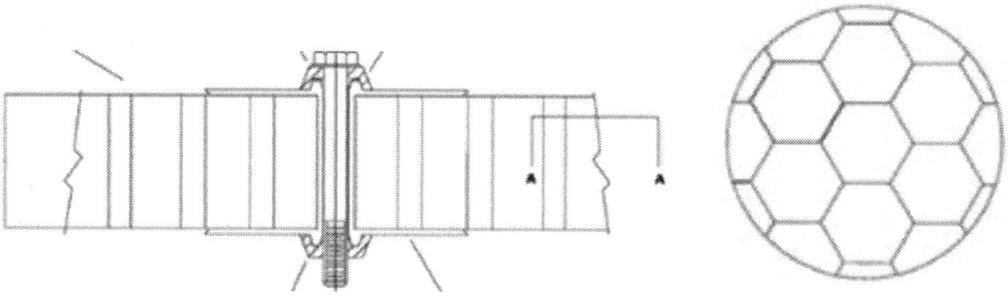
(a) (b)

Figure 10: HCFR: (a) Connection between HC panel and C-shape; (b) Honeycomb section

With the relations presented in [13] the elastic characteristics and the plastic instability limits of the homogenized material were determined. The obtained values are reported in Table 7 e 8.

Table 7: Elastic characteristics of the Honeycomb core.

$\rho[kg/m^3]$	E1 = E2 [MPa]	E3 [MPa]	ν12	ν13=ν23	G12 [MPa]	G13 [MPa]=G23 [MPa]
71.5	1.9	183.0	1	0	0.4	352.5

Table 8: Plastic characteristics of the Honeycomb core.

σ critical plastic stress [Pa]	Equivalent plastic strain for densification
49700	~ 60% [13]

In Figure (11.a) it is possible to see the C-shaped reinforcement (white) and the HC panels (green).

Defined the geometry and the elasto-plastic properties of the roof, it was loaded with the displacement field (Figure (11.b)) previously defined through the solution of the SDOF (Matsui [7]).

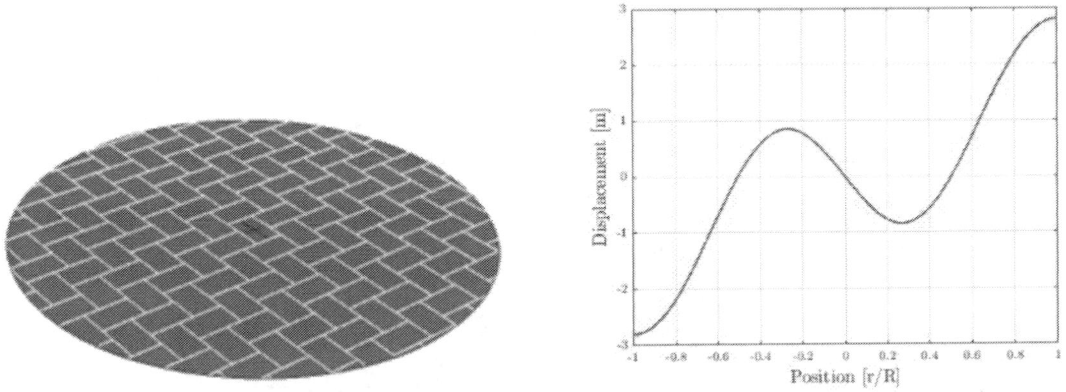

(a) (b)

Figure 11: HCFR: (a) FE model of Honeycomb roof; (b) Displacement field.

Once the vertical displacement field is set as a boundary condition in Figure (11.b), a static analysis has been conducted.

The results obtained can be viewed in Figures 12-16 and summarized in Tables 9 and 10.

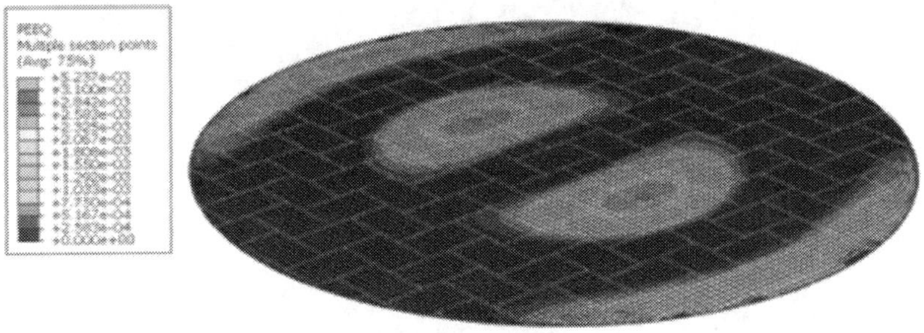

Figure 12: Max equivalent plastic deformation for HCFR - IZMIT earthquake.

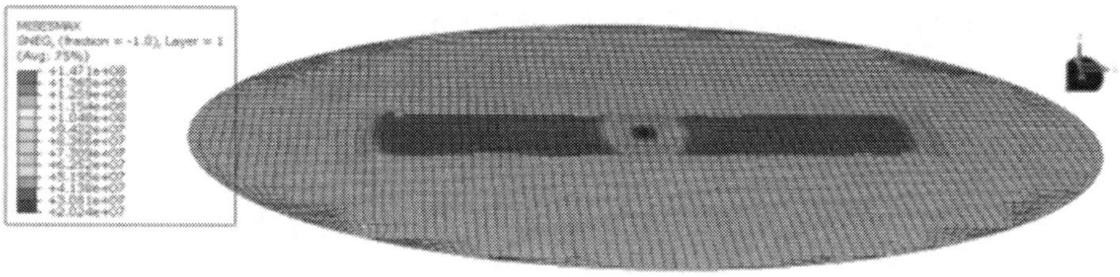

Figure 13: Von Mises stress [Pa] for HC panel - IZMIT earthquake.

Figure 14: Von Mises stress [Pa] for beam elements - IZMIT earthquake.

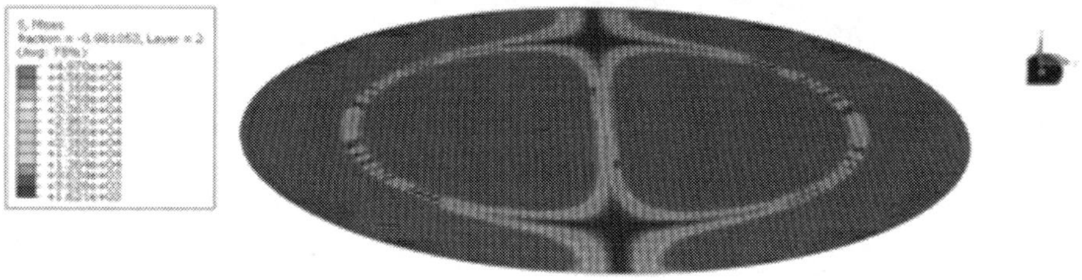

Figure 15: Von Mises stress for HC core plane - IZMIT earthquake.

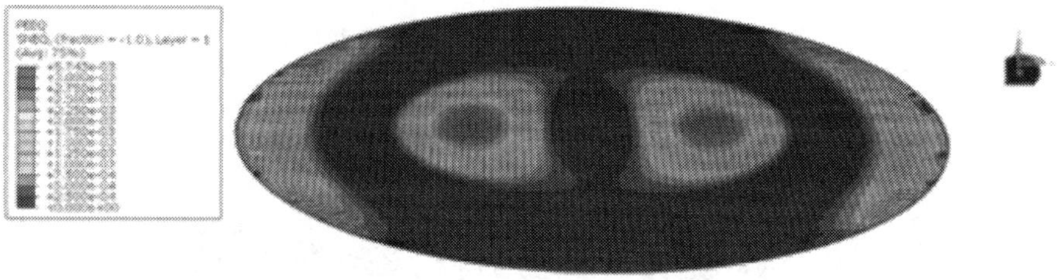

Figure 16: Max equivalent plastic deformation for HC core plane - IZMIT earthquake.

Table 9: Max equivalent plastic deformation for beam and ply.

	BEAM	PLY
PEEQ	0.34%	0.35%

IOP Conf. Series: Materials Science and Engineering 1306 (2024) 012002 doi:10.1088/1757-899X/1306/1/012002

Table 10: Max Von Mises stress for beam and ply.

	BEAM	PLY
σ MISES MAX [MPa]	174.4	147.1

4. Conclusion

In this study, a seismic design procedure for evaluating the seismic stress in a Single Deck Floating Roof (SDFR) and in a Honeycomb Deck Floating Roof (HCFR) is proposed.

For the SD, the relationship between the vertical out-of-plane deformation and elliptical deformation of the pontoon is analyzed. A modus operandi is provided for the design of SDFR, evaluating the effects of the first and second sloshing mode on stresses. The following conclusions can be made.

(i) The stress caused by the net radial compression of the pontoon is much higher than the force caused by its oval deformation. Furthermore, the compression stress caused by the contraction of deck plate due to the second sloshing mode dominates in pontoon plates.

(ii) The second sloshing mode effects are more significant for tanks with storage capacity like that of the case study. Decreasing the storage capacity, the effects related to the first mode will become more important.

For the HC, the procedure studied presents an analytical-numerical development, the punctual characterization of the tension and deformation state required the use of a complete model. The following conclusions can be deduced from the results obtained.

(i) The connecting C-shapes and HC panels deform continuously, showing significant levels of plastic deformation. The zones of maximum plastic deformation coincide with the horizontal tangent zones of the vertical displacement field.

(ii) The central core of the Honeycomb panel plasticizes (Figure 15) due to the formation of plastic hinges at the junction points of the base cell septa. Not reaching the PEEQ (Figure 16) value at which the densification process is activated leads to the conclusion that despite the activation of the plastic process, the roof will not suffer macroscopic damage. If the densification phenomenon were to be activated, the macroscopic spatial configuration of the roof would be lost.

The proposed methods can be used for the future design of SDFRs and HCFRs. They are also applicable for the selection of existing SDFRs and HCFRs for which repair is required as well as to determine the maximum filling level of the tanks.

References

[1] Goudarzi, M. A., Sabbagh-Yazdi, S. R. (2008). Evaluating 3D earthquake effects on sloshing wave height of liquid storage tanks using finite element method. Journal of Seismology and Earthquake Engineering, 10(3), 123-136.

[2] GOUDARZI, M. A., SABAGH, YAZDI SR. Numerical investigation on accuracy of mass spring models for cylindrical tanks under seismic excitation. 2009.

[3] Goudarzi, M. A., Sabbagh-Yazdi, S. R., Marx, W. (2010). Seismic analysis of hydrodynamic sloshing force on storage tank roofs. Earthquake Spectra, 26(1), 131-152.

[4] Hatayama, K. (2005, December). Long-Period Strong Ground Motions Having Fired Large Oil Storage Tanks During the 2003 Tokachi-Oki, Japan, Earthquake. In AGU Fall Meeting Abstracts (Vol. 2005, pp. S21B-0212).

[5] Goudarzi, M. A. (2015). New design method to evaluate the seismic stress of single deck floating roof for storage tanks. Earthquake Spectra, 31(1), 421-439.

[6] Goudarzi, M. A. (2013). Seismic behavior of a single deck floating roof due to second sloshing mode. Journal of Pressure Vessel Technology, 135(1), 011801.

[7] Matsui, T. (2007). Sloshing in a cylindrical liquid storage tank with floating roof under earthquake excitation: Analytical solution. Transactions of Architectural Institute of Japan, Journal of Structural and Construction Engineering, 594, 167-173.

[8] Yamauchi, Y., Kamei, A., Zama, S., Uchida, Y. (2006, January). Seismic design of floating roof of oil storage tanks under liquid sloshing. In ASME Pressure Vessels and Piping Conference (Vol. 47551, pp. 1407-1415).

[9] Yoshida, S., Sekine, K., Mitsuta, T. (2010). Axisymmetric finite element analysis for sloshing response of floating roofs in cylindrical storage tanks. Journal of Environment and Engineering, 5(1), 27-38.

[10] Barka, A. (1999). The 17 august 1999 Izmit earthquake. Science, 285(5435), 1858-1859.

[11] Sakai, F., Nishimura, M., Ogawa, H. (1984). Sloshing behavior of floating-roof oil storage tanks. Computers Structures, 19(1-2), 183-192.

[12] Sakai, F., Inoue, R., Hayashi, S. (2006, January). Fluid-elastic analysis and design of sloshing in floating-roof tanks subjected to earthquake motions. In ASME Pressure Vessels and Piping Conference (Vol. 47551, pp. 1437-1446).

[13] Gibson, L. J., Ashby, M. F., Harley, B. A. (2010). Cellular materials in nature and medicine. Cambridge University Press.

[14] Chen, S., McGregor, O. P. L., Endruweit, A., Harper, L. T., Warrior, N. A. (2020). Simulation of the forming process for curved composite sandwich panels. International Journal of Material Forming, 13, 967-980.

The oxymoron of damage assessment in dynamics by static approach

F.Cianetti[1], M. Palmieri[1*], G. Zucca[2], C. Braccesi[1]

[1]University of Perugia, Department of Engineering, Via G. Duranti 93, 06125 Perugia, Italy
[2] Italian Air Force, Aeronautical and Space Test Division, Via Pratica di Mare, 00040 Pomezia, Italy

E-mail: massimiliano.palmieri@unipg.it

Abstract. The evaluation of fatigue damage of structural components within dynamic systems subjected to random loads is typically addressed using multibody models wherein one or more components are modeled as flexible using the modal approach. While incorporating flexible elements allows for consideration of their influence on the overall dynamic behavior of the system, certain components are intentionally designed to function as rigid bodies. Consequently, in this case the incorporation of flexible elements within multibody models merely leads to complex and time-consuming analysis. Hence, it would be more pragmatic to realize rigid body multibody models, with components characterized solely by their inertial properties, and subsequently extract the dynamic forces applied to the components to be verified. The assessment of stress can then be carried out by exploiting the principle of superposition of effects in the time domain. In this context, the objective of the present study is to develop this methodology in the frequency domain. This approach not only capitalizes on the simplicity of rigid multibody models but also harnesses the computational capabilities of the frequency-domain method for evaluating fatigue damage in components subjected to random loading conditions. This work, therefore, provides a rapid, effective, and robust method for verifying and designing rigid mechanical components integrated into dynamic systems subjected to random loads.

1. Introduction

The frequency domain approach for the fatigue behavior assessment of systems and components subjected to random stationary-Guassian [1] loads in frequency domain is generally reduced to the evaluation of fatigue damage (spectral methods [2]) based on the Power Spectral Density (PSD) matrix of a specific point of the component that is typically modeled by the Finite Element Method (FEM) [3]. In the analysis of structures subjected to random loads, dynamic analysis plays a fundamental role.

By analyzing this step and incorporating it into a design process in which the component has not yet assumed a definition that allows for a faithful Finite Element Modeling, the authors wanted to highlight how, with knowledge of the inputs expressed in the time domain and/or the PSD matrix, it is possible to quickly obtain the necessary information for calculating stress cumulatives or rather internal forces without undertaking a dedicated dynamic analysis aimed to obtain stress PSDs.

Under the assumption that the component to design and/or verify does not participate to the dynamic of the system (thus it is characterized by a stiffness that does not modify the natural frequencies of the system or does not interact with the frequency range of the excitation), the

authors have demonstrated analytically and then, with a simple test case, how it is possible to obtain, starting from results expressed in the form of PSD matrices, the spectral moments [2] of individual stress characteristics or the entire stress tensor or its equivalent uniaxial synthesis for each node or element of the finite element model, using the so-called static approach [4].
In the case of a rigid behavior of the component, one could consider adopting a multibody model and dynamic simulation [5, 6, 7] of the system to assess fatigue behavior without the need to introduce the component as flexible bodies into the multibody model. Thus, the only required information needed by the simulation would be related to the set of forces acting on the body, expressed either as time histories or as PSD matrices. These forces can be used to reconstruct the real internal forces of the real state of stress. Moreover, assuming a linear behavior of the system, through linearization and extraction of state-space matrices of the system, having as inputs the forces to the system and outputs the forces acting on individual components it is possible to obtain the state of stress of the component without the need of transient dynamic simulation. In this way, the dynamic simulation would no longer be a serial step in the design process but would run in parallel. Indeed, under the assumption of a rigid behavior of the component, if the inertial properties of the component remain approximately unchanged, the input forces to the component also unvary, and the entire detailed design phase of the single component no longer requires dynamic analyses but only progressively more complex FEM analyses.
Therefore, this work represents a methodological proposal in the design of components integrated into dynamic models subjected to dynamic loads, seemingly in contrast with the dynamic nature of the problem. In fact, it suggests the use of multibody dynamic simulation for the system, a static approach to reconstruct the component's loading conditions, and frequency-based spectral methods to obtain load spectra and, consequently, the fatigue damage. This approach identifies a fast and error-free methodology if the assumption of the dynamic "rigidity" of the components is met.

2. The static approach

The assessment of the stress state of a component through a static approach in the frequency domain follows the classical modal approach [8]. It starts with a multibody model (MBS) in which the component to be designed/analyzed is inserted as a rigid body and thus defined only by its inertia properties. From this model, the state space matrices must be extracted [9, 10], where the inputs consists of the forces applied to the MBS system and the output are the forces applied to the component to be designed/verified. Thus, once the input PSD matrix of dimension $[r \times r \times 1]$ defined as $\mathbf{G_x}(\omega)$ is known and thanks to the frequency response function matrix $\mathbf{H}(\omega)$ of dimension $[r \times u]$ (where u are the input to the system) between the forces acting on the component and the forces acting to the system derivable from the state space representation [11], it is possible to calculate the PSD matrix of forces acting to the component $\mathbf{G_l}(\omega)$ as:

$$\mathbf{G_l}(\omega) = \mathbf{H}(\omega) \cdot \mathbf{G_x}(\omega) \cdot \mathbf{H}(\omega)'^t \tag{1}$$

Once the PSD matrix of forces acting to the component $G_l(\omega)$ is known, it is possible to easily determine the spectral moments of the PSDs as:

$$\mathbf{\Theta_n} = \int_0^\infty f^n \cdot \mathbf{G_l}(\omega) d\omega \tag{2}$$

This calculus is performed only to the forces acting on the component and since it must not be done on each element/node of the component, the computational effort is strongly reduced. However, the spectral moments just calculated are based on the forces PSD matrix and in their actual definition are not viable for the evaluation of the stress state on the component. To this aim, a finite element (FE) model is required (it can be a simplified model made by only beam/shell element or more complex made by solid element) from which obtain the needed information finalized to the evaluation of the spectral moments of real stress state or of internal forces. These information, defined as static shape of stress or of internal forces, can thus be obtained doing as many static analysis, applying a unitary load, as many constraints on the component to design. However, it's important to emphasize that it's necessary to use the inertia relief approach [12, 13], which allows not applying constraints to the system because equilibrium is achieved by applying an acceleration field to the center of gravity of the body. The correct evaluation of the absolute position of the "rigid body" may not be guaranteed, but the assessment of stress, deformation, or internal forces is assured. The output of these analysis are the static shape of stress $\boldsymbol{\phi}^{\sigma}$ or of internal forces $\boldsymbol{\phi}^{IF}$ with a unitary load and allows determining the real state of stress as:

$$\mathbf{m_n} = \{\boldsymbol{\phi}^{\sigma} \cdot \boldsymbol{\Theta}_\mathbf{n} \cdot \boldsymbol{\phi}^{\sigma t}\} \tag{3}$$

where t indicates the transpose.
What obtained is further extendible to multiaxial state of stress. In such case it is sufficient to adopt the approach proposed by Pitoiset [14] to reduce a multiaxial state of stress into an equivalent one:

$$m_{n,eqv} = trace\left(\boldsymbol{Q}_{\sigma} \cdot \{\boldsymbol{\phi}^{\sigma} \cdot \boldsymbol{\Theta}_{\boldsymbol{n}} \cdot \boldsymbol{\phi}^{\sigma t}\}\right) \tag{4}$$

where:

$$\mathbf{Q}_{\sigma} = \begin{bmatrix} 1 & -1/2 & -1/2 & 0 & 0 & 0 \\ -1/2 & 1 & -1/2 & 0 & 0 & 0 \\ -1/2 & -1/2 & 1 & 0 & 0 & 0 \\ 0 & 0 & 0 & 3 & 0 & 0 \\ 0 & 0 & 0 & 0 & 3 & 0 \\ 0 & 0 & 0 & 0 & 0 & 3 \end{bmatrix} \tag{5}$$

Other methods for multiaxial synthesis can be adopted at the same manner [15, 16, 17]. Once the spectral moments of internal forces or stress are known, it is possible to use any frequency domain approach (such as narrow-band, Dirlik, etc.) [2, 18, 19, 20] for assessing load cumulatives and damage.

3. Test case

To validate the proposed static approach, until now only theoretical valid, a real model has been used as test case. In particular, the aerial platform shown in figure 1 is considered.

Two multibody models of the aerial platform were created, one with the drive chassis modeled as a rigid body and the other with the drive chassis modeled as a flexible body. The multibody models are shown in figure 2. As visible, the model includes all the real parts of the structure except for the drive chassis, which is represented by an ellipse in one case and by a finite element model imported into the multibody code in the other case. For the flexible body, 40 modes of vibration were considered, and the first 5 natural frequencies are: 242.6 Hz, 441.3 Hz, 533 Hz, 792 Hz, and 848.6 Hz. In both variants of the MBS model, the drive chassis has consistent inertial properties with the following values defined in the reference system shown in figure 2:

- Mass= 795 Kg
- $I_{xx} = 9.4 \cdot 10^7 Kg \cdot mm^2$
- $I_{yy} = 5.8 \cdot 10^8 Kg \cdot mm^2$
- $I_{zz} = 6.3 \cdot 10^8 Kg \cdot mm^2$
- $I_{xy} = 1.4 \cdot 10^4 Kg \cdot mm^2$
- $I_{zx} = 40 \cdot 10^6 Kg \cdot mm^2$
- $I_{yz} = -3021 Kg \cdot mm^2$

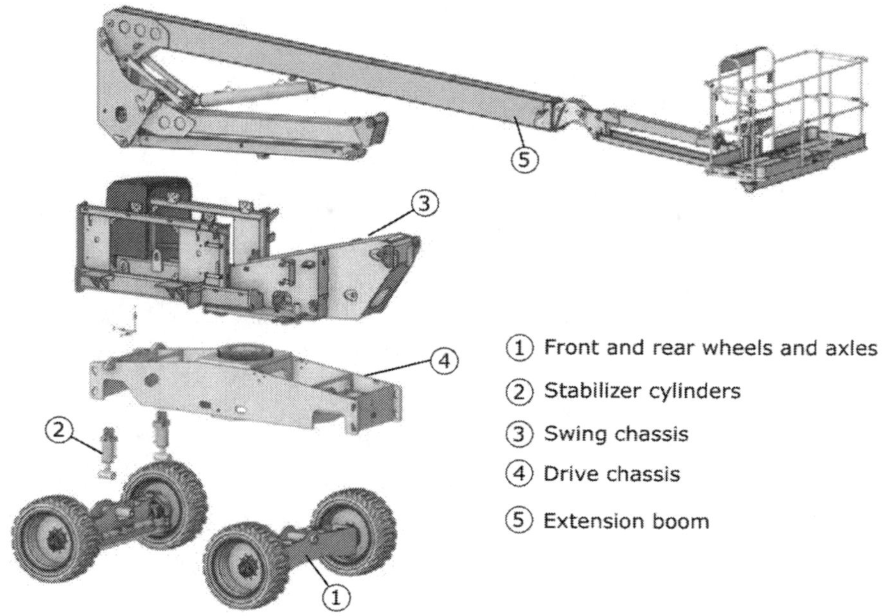

Figure 1. Aerial platform used for the validation of the proposed approach

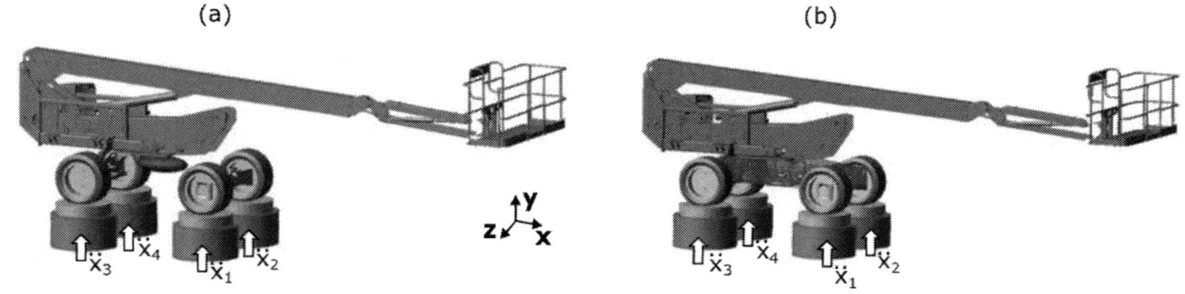

Figure 2. Multibody models realized for the validation of the proposed approach. (a) MBS model with rigid frame, (b) Multibody model with flexible frame

As shown from figure 1, the drive chassis is constrained to the rest of the components in 5 points for a total of 22 reaction forces/moments.

To excite the model as it works in real-life condition (random excitation introduced by the road asperity), a four-post test-rig has been created and the MBS models are places over [21]. The model is thus excited by the four-post test-rig with four uncorrelated acceleration random excitation, defined by a PSD matrix with terms only on the main diagonal (figure 2). The input PSD has a trapezoidal shape, defined in a frequency between 5-10-25 Hz and as amplitude 30-30-0 $(m/s^2)^2$ respectively.

The so-realized models were linearized in the equilibrium position and from the model, the state space matrices were obtained. For the rigid model the output of the state-space representation are the input forces to the drive chassis (22 in total) while for the flexible model the output of the state-space are the modal coordinates (40 in total) with which the flexible body was created. The input of the state space form are instead, for both models, the forces acting on the entire system (4 in total). During the linearization, a modal analysis of the system was then performed.

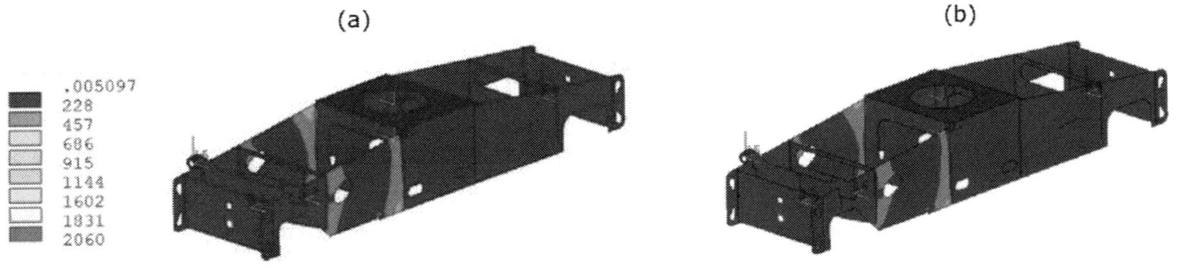

Figure 3. Comparison between 0-order spectral moment of the equivalent stress PSD: (a) Proposed approach, (b) Classical modal approach

Table 1 shows a comparison between the obtained natural frequencies of both system.

Table 1. Comparison between the natural frequencies of the two MBS models.

Mode nr.	Rigid	Flex	Unit
1	5.69	5.69	Hz
2	12.4	12.5	Hz
3	14.9	14.8	Hz
4	117.2	106.8	Hz

As visible from table 1, the first four vibrating modes obtained by the two models are identical, while differences occur from the fourth vibrating modes. This is due to the presence of the flexible body in one model. By the way, this aspect can be omitted since, as previously introduced, the maximum frequency of the excitation PSD is equal to 25 Hz and thus the fourth vibrating modes is outside the frequency range of the excitation. This allows attesting the two models are dynamically equivalent [4].
According to this, it is possible to adopt the proposed static approach for the evaluation of the spectral moments and thus of the load spectra and of the fatigue damage of the analyzed rigid component to be compared with those derivable with the classic approach in frequency domain. As previously introduce, in order to evaluate the stress state by the static approach, it is necessary to have the static shape of the stress. These are obtained for each element of the model by performing 22 static analysis in FE environment with the inertia relief approach, each of them performed by imposing a unitary load/moment in each point of the model where it is constrained to the rest of the components. For what concern the classical approach in frequency domain, the same FE model has been used to perform a modal analysis considering the first forty vibrating modes extracting the modal shapes of stress for each element of the model. By adopting the results of the multibody model and of the FE analysis it was possible to evaluate the state of stress with both methods (the proposed and the reference one). To compare the accuracy of the proposed static approach, figure 3 shows a comparison between the zero-order spectral moment m_0 of the equivalent stress PSD.

As shown in figure 3, there are not visible differences between the results obtained with both methods. The results are almost identical even if small differences (lower than 2 MPa) exist. The difference is due to the flexibility of the drive chassis that, despite is practically indifferent, is real. However, a differences lower than 2 MPa is acceptable, especially in a preliminary design phase. Similar results are obtained also for the higher-order spectral moments. According to the obtained results, it is possible to attest that the proposed model is a valid tool for the design of rigid component, especially in the preliminary design phase, where a detailed FE model is not viable and only the inertia properties of the components are known. With this approach, the design phase is much faster since there is no need to upgrade the dynamic model with flexible components step by step more detailed.

4. Conclusion

In this work, a design/analysis method for rigid components (i.e., components whose flexibility does not significantly affect the system's dynamics) inserted into complex dynamic systems subjected to dynamic loads, specifically random, has been proposed. The proposed approach leverages the advantages of rigid body multibody modeling and the frequency domain to obtain spectral moments of internal forces or stress, which are useful for calculating load spectra or fatigue damage. The proposed approach has been validated using a real test case, comparing the results with those derived from the classical frequency domain approach. The obtained results demonstrate excellent agreement, indicating this method as a valuable tool, especially in the design phase but also during verification.

References

[1] Palmieri M, Česnik M, Slavič J, Cianetti F and Boltežar M 2017 Non-Gaussianity and non-stationarity in vibration fatigue *International Journal of Fatigue* **97** 9-19
[2] Slavič J, Mrsnik M, Česnik M, Javh J, and Boltežar M 2013 Vibration fatigue by spectral methods: from structural dynamics to fatigue damage *Elsevier*
[3] Braccesi C, Cianetti F and Tomassini L 2016 An innovative modal approach for frequency domain stress recovery and fatigue damage evaluation *International Journal of Fatigue* **127** 382-396
[4] Preumont L 2013 Twelve Lectures on Structural Dynamics *Springer Dordrecht*
[5] Peeters J, Vandepitte D, Sas P 2006 Structural analysis of a wind turbine and its drive train using the flexible multibody simulation technique *Proceedings of ISMA 2006*
[6] Cianetti F, Cetrini A, Becchetti M, Castellani F, Braccesi C 2018 Dynamic modeling of wind turbines. Experimental tuning of a multibody model *Procedia structural integrity* **8** 56-66
[7] Braccesi C and Cianetti, F 2004 Development of selection methodologies and procedures of the modal set for the generation of flexible body models for multi-body simulation *Part K: Journal of Multi-body Dynamics* **218** 1 19-30
[8] Braccesi C, Cianetti F, and Tomassini L 2017 Fast evaluation of stress state spectral moments *International Journal of Fatigue* **127** 4-9
[9] Valasek M, Sika Z, Vaculin O 2007 Multibody formalism for real-time application using natural coordinates and modified state space *Multibody System Dynamics* **17** 209-227
[10] Schwertassek R and Roberson RE 1984 A state-space dynamical representation for multibody mechanical systems part I: Systems with tree configuration *Acta Mechanica* **50** 141-161
[11] Ogata K 2001 Modern Control Engineering *Prentice Hall PTR*
[12] Vdovin DS, Chichekin IV, Levenkov YY and Shabolin ML 2019 Automation of wheeled vehicles load bearing frames finite-element models loading procedure by using inertia relief method and vehicle multi-body dynamics model *IOP Conference Series: Materials Science and Engineering* **534** 012024
[13] Farrahi GH 2020 A comparative study on the fatigue life of the vehicle body spot welds using different numerical techniques: Inertia relief and Modal dynamic analyses *Fracture and Structural Integrity* **14** 52
[14] Pitoiset X and Preumont A 2000 Spectral methods for multiaxial random fatigue analysis of metallic structures *International Journal of Fatigue* **22** 7 541-550
[15] Braccesi C, Morettini G, Cianetti F, Palmieri M 2018 Development of a new simple energy method for life prediction in multiaxial fatigue *International Journal of Fatigue* **112** 1-8
[16] Benasciutti D, Sherratt F, Cristofori A 2016 Recent developments in frequency domain multi-axial fatigue analysis *International Journal of Fatigue* **91** 2 397-413
[17] Braccesi C, Cianetti F, Lori G, Pioli D 2015 Random multiaxial fatigue: A comparative analysis among selected frequency and time domain fatigue evaluation methods *International Journal of Fatigue* **74** 107-118

IOP Conf. Series: Materials Science and Engineering 1306 (2024) 012003 doi:10.1088/1757-899X/1306/1/012003

[18] Dirlik T and Benasciutti D 2021 Dirlik and Tovo-Benasciutti spectral methods in vibration fatigue: a review with a historical perspective *Metals* **11** 1333

[19] Zorman A, Slavič J and Boltežar M 2023 Vibration fatigue by spectral methods—A review with open-source support *Mechanical Systems and Signal Processing* **190** 110149

[20] Larsen C E and Irvine T 2015 A review of spectral methods for variable amplitude fatigue prediction and new results *Procedia Engineering* **101** 1

[21] Braccesi C, Cianetti F, and Landi L 2005 A methodology for active control of multibody test-rig for virtual simulation of vehicles through acceleration inputs *Proceedings of IDETC/CIE* 2359-2366

Innovative solutions to make travel more comfortable

E V Arcieri[1], S Baragetti[1] and L Monticelli

[1]Department of Management, Information and Production Engineering, University of Bergamo, Viale Marconi 5, Dalmine 24044, Italy

emanuelevincenzo.arcieri@unibg.it

Abstract. Traveling in various vehicles and means of transport (ambulance, airplane, ...) can be characterized by sudden accelerations, decelerations and vibrations/perturbations. All these loads are transferred to the passenger, making the journey uncomfortable. The study for the improvement of comfort in such situations was the subject of a collaboration between the authors of the present document. The designed solutions are reported in a patent application published with the number WO2023111937A1.

1. Introduction

Passenger comfort is an important aspect in designing vehicles and means of transport. It is essential, for instance, to improve the comfort of patients and medical staff present in an ambulance. The primary approach to ensure a comfortable travel is to minimize the transmission of loads to the people present in the compartment since the transportation can include sudden accelerations, decelerations, vibrations and changes of direction due to the emergency situation. Similar considerations can be made for airplane transportation, where perturbations affect passenger comfort. It is for this reason that people continuously search for solutions that improve passenger comfort by decreasing the transfer of such loads during transportation.

A method for reducing the vibrations transferred to the people in an ambulance is described in CN109124903A [1]. According to this invention, four rods are positioned crosswise, with one end connected to one of the vertices in the lower region of the cabin and the other end to a pressurized element. The system created by the rods and the pressure produced by cylinders mounted on the bottom region of the vehicle enable the damping of any vertical movements of the cabin.

Document KR100855613B1 [2] presents a system to improve the comfort for a patient on a stretcher only during the phases of loading and unloading from the ambulance. The rear height of the vehicle is adjusted with an air spring, which is placed between the frame and a leaf spring attached to the bottom portion of the vehicle chassis.

A system for reducing the pitching and rolling of a cab mounted on a vehicle chassis is described in JP2001301651A [3]. The solution involves the presence of a connecting element inclined with respect to the horizontal direction, which connects the front part of the cabin to the vehicle frame. An air spring is also interposed between the cabin and the inclined element. The rear suspension system comprises two inclined air springs positioned between the cab and the chassis.

JPS6218374A [4] refers to a truck cab suspension system with a structure which adjusts the stiffness of an elastic element interposed between the cab and the vehicle chassis based on the external factors which can reduce the travel comfort. There are a cab support element and an arch element which connects the two side members of the vehicle chassis in the rear area of the cab. The front suspension system involves an element connected by hinges to two brackets fixed respectively to the vehicle frame and to the cab support element. A compressed coil spring and shock absorber are located between the cab and the element connected by hinges. The rear suspension system consists of a shock absorber and a coil spring compressed between a bracket attached to the arch element and another component. Both shock absorbers are composed of an electromagnetic actuator that adjusts the damping. Each actuator is controlled by a unit which manages the operating parameters based on external data evaluated by special sensors.

Document EP2886377A1 [5] presents a solution for maintaining the cab of a truck or tractor horizontal when parked on uneven areas. The front axle supports the frame with springs. Each spring is associated with a shock absorber which absorbs the shocks suffered by the chassis when the vehicle is in motion. There is also a front stabilizer bar that reduces rotation of the vehicle upper body during movement. The cabin is suspended by vertically arranged coil springs, while the lateral rotation of the cabin is controlled by a stabilizer only during movement.

US2014319876A1 [6] shows a suspension system capable of damping the shocks suffered by the driver cab of a vehicle. The solution described comprises a spring-shock absorber unit and a hydraulic system consisting of two double-acting cylinders positioned respectively on the right and left side of the cab. Each cylinder has one chamber at the top and one chamber at the bottom, for a total of two chambers per cylinder. The two cylinders are then cross-connected with respect to the described chambers. To limit the transverse oscillations of the vehicle, the solution also includes two transverse shock absorbers, connected to two cross-bridge structures.

A suspension system for the driver cab of a vehicle and, in particular, of an agricultural machine is presented in document WO2018203124A1 [7]. One of the solutions presented has a counter-frame on which the cabin is mounted. The suspension system consists of a front unit fixed to a bracket mounted on the counter-frame and a rear unit fixed to the vehicle chassis. There are two transverse elements and two longitudinal elements which keep the cabin in position avoiding lateral movements and provide torsional resistance to the structure.

Document MXPA02004380A [8] reports a suspension for a truck driver cab that reduces the twisting forces transmitted to the frame. Two air springs are positioned below the rear sill of the truck and two inclined shock absorbers are placed between the air springs.

Improving the passenger comfort was the objective of a collaboration among the authors of the present document. The designed solutions are reported in a patent application published with the number WO2023111937A1 [9]. The proposed solutions consist of an alternative suspension system and a vehicle comprising such a suspension system. The suspension system can be used on different types of vehicles.

2. Finite element analyses

The Abaqus finite element code [10] was used to conduct preliminary simulations for the conception of different solutions to improve passenger comfort in vehicles and means of transport. The case study analyzed is the improvement of the comfort in an ambulance compartment. An implicit integration scheme was used instead of the explicit one, commonly adopted for impact simulation [11-14], since this is not the objective of the study.

The model created for the analysis of the dynamic behavior of the ambulance compartment is shown in Figure 1. 2D rigid elements were used to model the ambulance compartment (Figure 1a), whose inertial properties were assigned to the centroid. Concentrated masses were used to model the presence of the equipment and human beings in the compartment, of the cabin and of the vehicle frame according to Figure 1b and Table 1.

A multi-point constraint type beam (C1 in Figure 1c) was used to constrain the movement of the primary suspension (suspension of the vehicle) to the frame centroid, the centroid of the cabin and the secondary suspension (suspension to improve comfort). Another multi-point constraint type beam (C2 in Figure 1c) was used to constrain the movement of the compartment centroid and the equipment inside to the movement of the secondary suspension. The tires of the vehicle were modelled using four

IOP Conf. Series: Materials Science and Engineering 1306 (2024) 012004 doi:10.1088/1757-899X/1306/1/012004

connector elements (A_1A_2, B_1B_2, C_1C_2 and D_1D_2 in Figure 1d) with an axial stiffness of 250000 N/m per single element. Each shock absorber of the primary suspension was modelled with a connector element (A_2A_3, B_2B_3, C_2C_3 and D_2D_3 in Figure 1d) having an axial stiffness of 50000 N/m and an axial damping coefficient of 5123.5 Ns/m. Each shock absorber of the secondary suspension (other elements in Figure 1d) was modelled with a connector element having an axial stiffness of 160000 N/m and an axial damping coefficient of 4000 Ns/m.

In the first simulation step, a gravity load in the vertical direction (y) was applied to the whole model for 10 s in order to reach the static equilibrium condition of the system due to its own weight. Subsequently, a further gravity load was applied in the three directions x, y and z for 50 ms. In the final simulation step, a gravity load in the vertical direction was applied for 10 s in order to reach again the static equilibrium condition of the system after its perturbation. In all the simulation steps the nodes A_1, B_1, C_1 and D_1 were fixed while only the vertical movement of the nodes A_2, B_2, C_2, D_2, A_3, B_3, C_3 and D_3 was allowed.

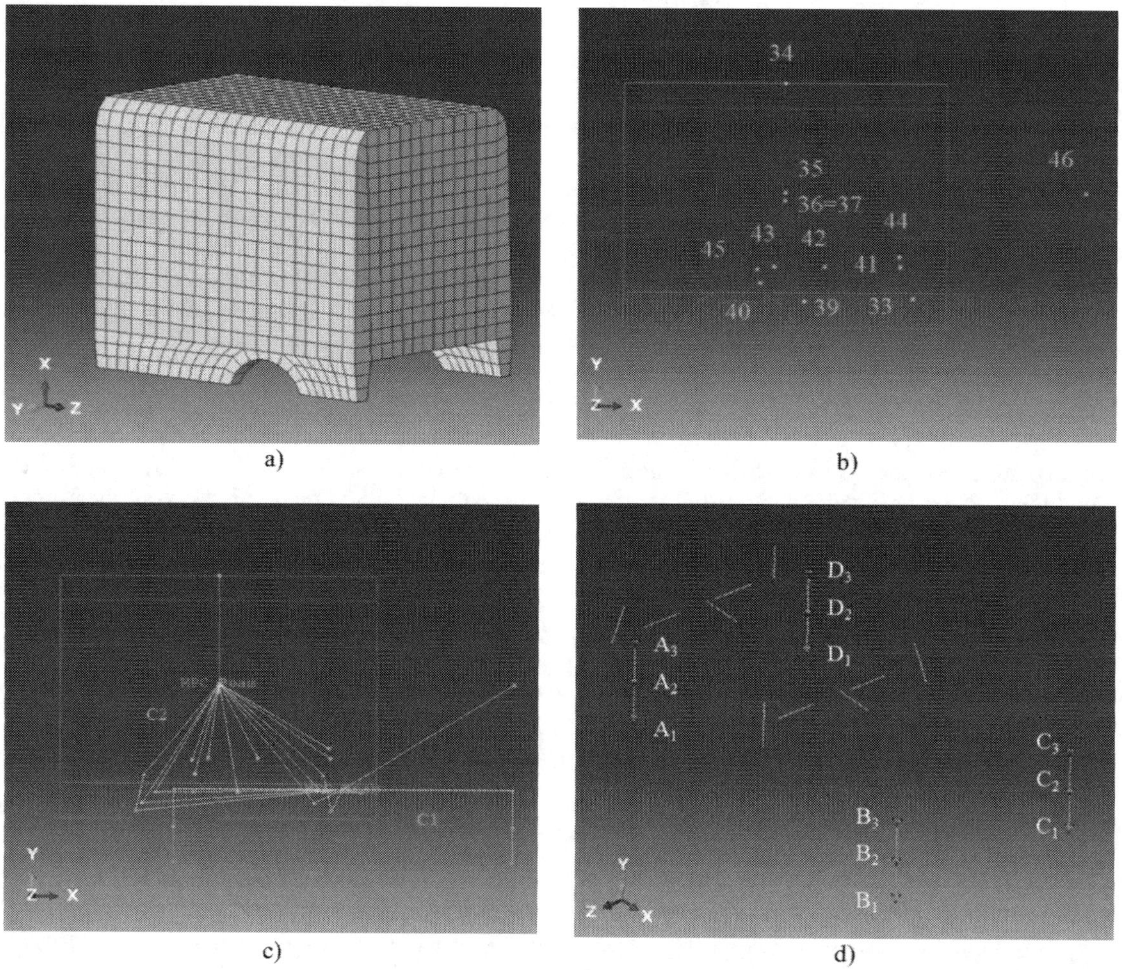

Figure 1. Finite element model: a) compartment; b) concentrated masses; c) rigid connections; d) connectors for modelling tires, primary and secondary suspensions.

Table 1. Concentrated masses.

Point	Description	Mass (kg)	Notes
40	Stretcher support	89	
41	Seat 1 (+ 1 person)	140	
42	Seat 2 (+ 1 person)	140	
43	Seat 3 (+ 1 person)	140	
44	Support	5	
36,37	Furniture + cables + ventilation system	170 (85x2)	Distributed in the two side walls Positioned in the center of gravity of the walls
34	Air conditioning system	40	Positioned in the center of gravity of the ceiling
45	Stretcher (+ 1 person and equipment)	215	
39	Secondary suspension	250	
35	Compartment	411	Placed in the centroid Moments of inertia with respect to the x, y and z axes: 2110 kgm^2
46	Cabin	1000	
33	Frame	900	Moments of inertia with respect to the x, y and z axes: 2850 kgm^2

In this document the analysis focuses on accelerations and displacements, Table 2 and Figure 2. Accelerations are related to travel comfort [15,16]. Displacements were measured to verify that the compartment and the other vehicle components do not move much and do not collide with each other or with objects eventually present on the road. The minimum and maximum values of the acceleration along the travel direction (x), the vertical direction (y), the transversal direction (z) and in magnitude are shown in Table 2. These values do not consider the first simulation step for reaching the static equilibrium condition, which is not related to travel comfort, and refer only to the ambulance compartment. The minimum values in the x and z directions are equal in magnitude to the increment in gravity load applied in the three directions in the simulation step which provides the perturbation load. The maximum acceleration obtained in the compartment is 17.28 m/s^2 in magnitude (Figure 2a), equal to slightly less than double the acceleration of gravity. The minimum and maximum displacements of the whole model along the x, y and z directions and in magnitude are shown in Table 2. The maximum displacement in magnitude is 348 mm (Figure 2b). However, these values of displacements take into account the achievement of the static equilibrium condition, obtained at the end of the first simulation step. The maximum actual displacements resulting from the implemented perturbation, i.e. the relative displacements, were approximately estimated as the maximum difference between the displacements obtained after the beginning of the second simulation step and the displacements at the end of the first simulation step for reaching the equilibrium condition, as shown in Figure 2c for the displacement in the y direction. Combining the maximum relative displacements in the three directions x, y and z, the maximum relative displacement in magnitude is certainly less than 160 mm. The maximum displacement values obtained, both absolute and relative, are small compared to the dimensions of the ambulance and therefore can be considered acceptable.

IOP Conf. Series: Materials Science and Engineering 1306 (2024) 012004 doi:10.1088/1757-899X/1306/1/012004

Table 2. Results of the finite element analysis.

Direction	Min. acc. (m/s^2)	Max. acc. (m/s^2)	Min. displ. (mm)	Max. displ. (mm)	Max. rel. displ. (mm)
x	-9.81	5.76	-33	17	<50
y	-12.73	7.60	-348	0	<150
z	-9.81	8.65	-32	25	<40
Magn.		17.28		348	<160

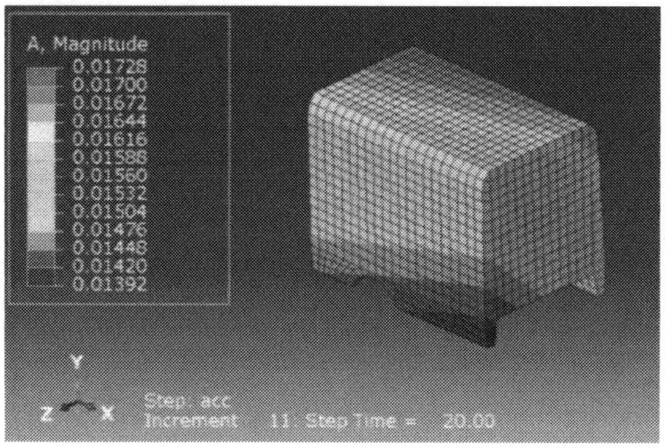

a)

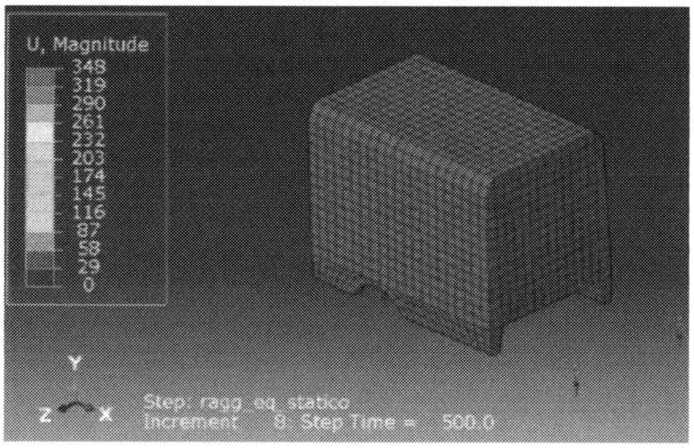

b)

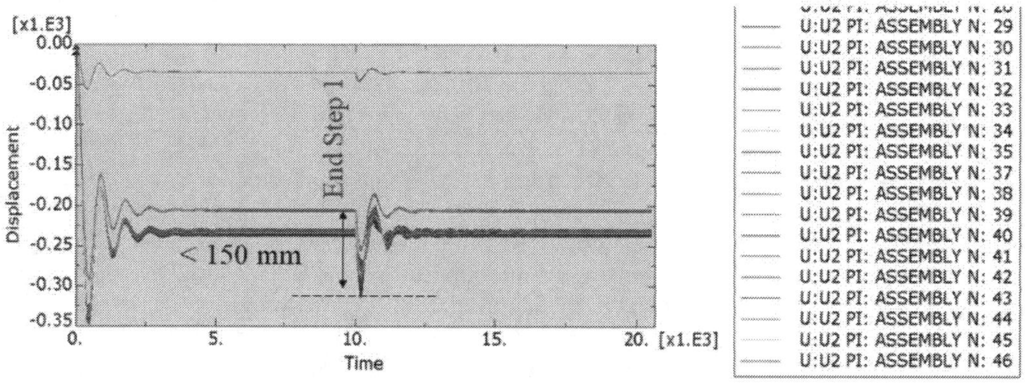

c)

Figure 2. Finite element results: a) acceleration (magnitude) field when the maximum value is achieved (mm/ms^2); b) displacement (magnitude) field when the maximum value is achieved (mm); c) estimate of the relative displacement in the y direction (time in ms, displacement in mm).

3. Patent application WO2023111937A1

The different solutions identified to improve passenger comfort are reported in the patent application published with the number WO2023111937A1 [9], Figure 3.

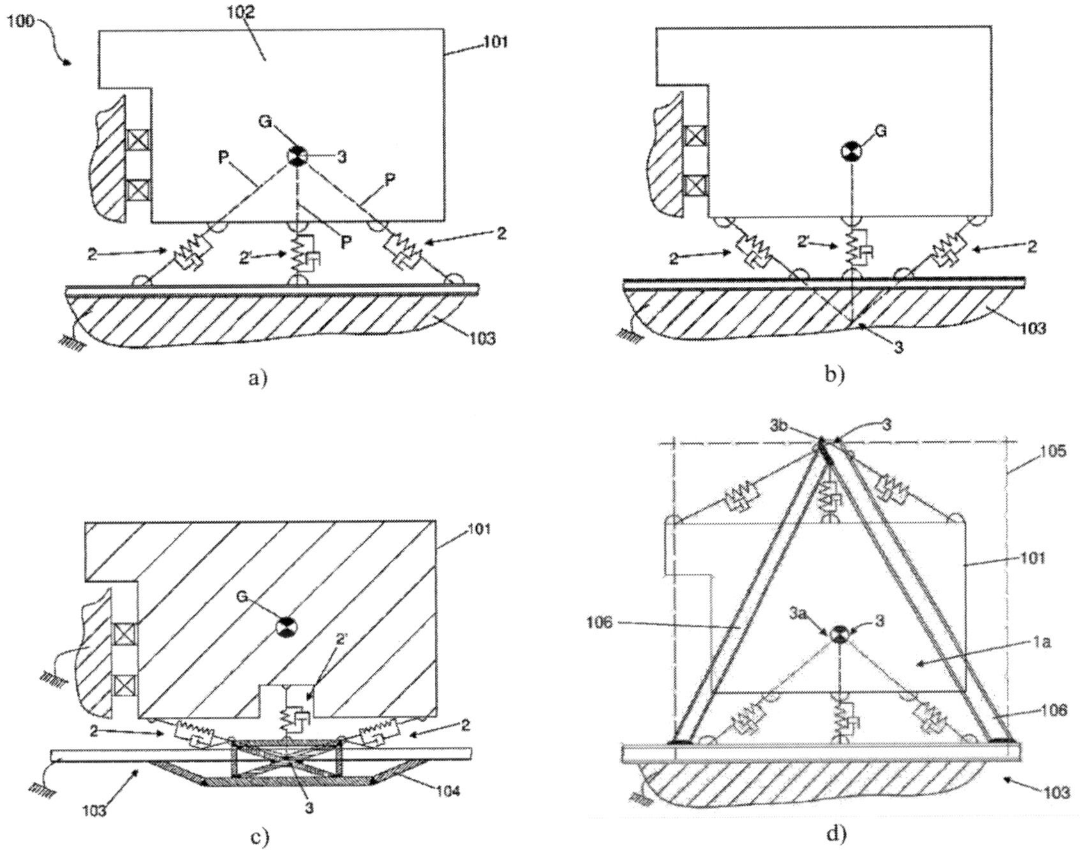

Figure 3. Possible solutions to improve passenger comfort: a) straight pyramid; b) overturned pyramid; c) with counter-frame; d) suspended compartment [9].

Figure 3a shows a suspension setup that could be mounted in a vehicle 100, between the compartment 101 and the vehicle frame 103. The presented configuration contains several suspensions, indicated with number 2 in the figure, composed of at least one elastic element and/or a damping element. The suspensions are arranged forming a straight pyramidal configuration. The system may also include one or more supplementary suspensions, located transversely to the pyramidal configuration.

The solution presented in Figure 3b is similar to that presented in Figure 3a but the suspension system is arranged according to an overturned pyramidal configuration.

Figure 3c shows a configuration in which the vehicle frame 103 incorporates a counter-frame 104, capable of optimizing the distribution of loads in the vehicle frame. The counter-frame acts as an extension of the suspension system for the compartment 101. It is suitably sized and has bending and torsional stiffness. This influences the dynamic and shock-absorbing behavior of the vehicle. As a result, in this particular solution, there is a sequence of suspension elements: a vehicle suspension system, a suspension system associated with the existence of the counter-frame and one or more suspension systems arranged according to one of the pyramidal configurations of Figures 3a and b.

According to the solution presented in Figure 3d, the vehicle incorporates a second body 105, which contains the compartment 101. The casing 105 is fixed to the vehicle frame 103, or to the counter-frame. An additional suspension system, similar to the one described above, is present in the vehicle, between the casings 105 and 101. The two suspension systems are positioned on opposite sides of the compartment 101, one above and the other below. The purpose of this design solution is to have a suspended compartment. One or more inclined beams 106 may be present and converge at the common vertex of the suspension system located above the compartment to increase the stiffness of the casing 105. In Figure 3d, the suspension systems are arranged in a straight pyramid configuration, but variations with one or both of the suspension systems assuming the overturned pyramid arrangement are admissible. The solution presented in Figure 3d can also include the presence of a counter-frame, with defined torsional-bending stiffness.

4. Conclusions

Passengers of vehicles and means of transport, such as ambulance and airplane, may be subjected to sudden accelerations, decelerations and vibrations/perturbations which make travel uncomfortable. Possible solutions designed to improve passenger comfort are reported in a patent application published with the number WO2023111937A1 and can be described according to four schemes: (i) straight pyramid, (ii) overturned pyramid, (iii) vehicle with a counter-frame, (iv) suspended compartment. Future studies could involve theoretical and experimental approaches as well as further numerical analyses in order to deepen the behavior of the conceived systems as done in [17-19] for other mechanical problems.

References

[1] Ran J 2019 Patent CN109124903A
[2] Kang S H 2008 Patent KR100855613B1
[3] Hiramatsu M and Momiyama F 2001 Patent JP2001301651A
[4] Ozawa T, Noguchi M, Okawa T and Kutsukake Y 1987 Patent JPS6218374A
[5] Vreede R J 2015 Patent EP2886377A1
[6] Schauer M, Eismann J, Helm E, Quaing M and Bublies H 2014 Patent US2014319876A1
[7] Soldan D J and Jenning J M 2018 Patent WO2018203124A1
[8] Taylor M A 2004 Patent MXPA02004380A
[9] Baragetti S, Monticelli L and Arcieri E V 2023 Patent WO2023111937A1
[10] ABAQUS, ABAQUS Documentation, Dassault Systèmes, Providence, RI, USA, 2017
[11] Arcieri E V, Baragetti S, Fustinoni M, Lanzini S and Papalia R 2018 *Procedia Struct. Integr.* **8** 212
[12] Arcieri E V, Baragetti S and Božić Ž 2021 *Procedia Struct. Integr.* **31** 22
[13] Baragetti S and Arcieri E V 2019 *Procedia Struct. Integr.* **24** 91
[14] Baragetti S and Arcieri E V 2020 *Eng. Fail. Anal.* **113** 104564
[15] UNI, UNI ISO 2631-1:2014
[16] BS, BS 6841:1987
[17] Baragetti S 2006 *Meccanica* **41** 443
[18] Baragetti S, Guagliano M and Vergani L 2000 *Int. J. Mater. Prod. Technol.* **15** 91
[19] Baragetti S and Tordini F 2007 *Int. J. Fatigue* **29** 1832

IOP Conf. Series: Materials Science and Engineering 1306 (2024) 012005 doi:10.1088/1757-899X/1306/1/012005

True stress-strain identification accounting for anisotropy of sheet metals

Giuseppe Mirone[a*], Raffaele Barbagallo[a], Giuseppe Bua[a,b], Daniele De Caro[c], Matteo Ferrea[c], Michele M. Tedesco[c]

[a] Department of Civil Engineering and Architecture – University of Catania – Viale A. Doria 6, 95125 Catania – Italy

[b] Department of Electrical, Electronic and Computer Engineering – University of Catania – Via Santa Sofia 64, 95123 Catania – Italy

[c] C.R.F. S.C.p.A, Metals, Corso Settembrini 40, 10135 Turin, Italy

*gmirone@dii.unict.it

Abstract. Sheet metals for the automotive industry are subjected to continuous research efforts aiming at ever increasing mechanical performance. A remarkable feature of modern high strength sheet metals is their anisotropy, intrinsic in the technologic process of their production. When the effect of anisotropy on the mechanical response of a material cannot be neglected, specimens along different directions are usually tested, possibly under different stress states, to assess the flow curves and the deforming ratios for each direction and each loading mode. Such data are then used to calibrate many possible plastic anisotropy models available in the literature. In this work, the experimental procedures for determining the stress-strain curve and the anisotropic straining ratio are studied in detail, referring to representative tensile tests along the rolling direction of two anisotropic sheet metals, respectively PHS-1800 steel and 6181 aluminium alloy. Both alloys are ductile and exhibit remarkably long post-necking phases in tension, revealing that, in such cases, the standard procedures for the experimental derivation of the hardening curves and of the anisotropic strain ratios are limited to the very early phases of the material life and miss to cover the major part of the strain range up to failure. Different alternative procedures for the derivation of experimental data and for their postprocessing are considered and compared to each other, identifying a set of guidelines for achieving a good engineering accuracy up to failure in deriving both the stress-strain curves and the anisotropic strains ratios. The above analyses are made on the results of tensile tests at static, intermediate and high strain rate, confirming the generality of the identified procedure.

1. Introduction

Advanced high-strength sheet metals are often characterized by pronounced anisotropy resulting from the technological processes to which they are subjected, making their characterization more challenging. In the literature, numerous testing methodologies [1] and advanced material models [2] have been proposed. However, regardless the approach, the fundamental step is to obtain the material's flow curve through a classical tensile test on a dog-bone specimen.

Many researchers employ indirect approaches to obtain the material's flow curve, relying on the iterative reduction of target quantities representing the differences between simulation results and experimental data [3–8]. One of the proposed inverse approaches is the Virtual Fields Method (VFM), which involves calibrating a predetermined material model to comply with the principle of virtual work [9–11]. Macek et al. [12] proposed an hybrid indirect/direct approach, considering experimental data from classical tensile tests and a novel heterogeneous strain field test.

With highly anisotropic complex materials, indirect approaches could give not-unique solutions to the characterization problem due to the multitude of parameters to be calibrated. Consequently, researchers have also devoted significant efforts to direct approaches. They typically involve evaluating the true stress-strain curve of the material from a tensile test, based on the total axial load acting on the specimen and the corresponding minimum cross-section area. Indeed, such curve is identical to the flow curve before necking and, afterwards, it must be corrected in order to take into account the growing triaxiality. Bridgman [13] proposed the most well-known correction based on the curvature of the deforming specimen, while Mirone [14] proposed a material-independent corrective function applicable to both flat and cylindrical specimens.

The pre-necking true curve can be simply obtained using the conventional length-based approach, which is still commonly employed today [15]. However, deriving the true curve after necking onset becomes challenging for anisotropic materials. Indeed, in isotropic or at least transversely isotropic materials, the specimen maintains its initial geometric proportions until fracture. Thus, by measuring the width or diameter of the specimen, the current cross-sectional area can be determined [16,17]. However, this is not the case for anisotropic materials, making it more difficult to evaluate the actual cross-sectional area [16,18]. Consequently, complex experimental setups such as 3D Digital Image Correlation (DIC) have been proposed to directly reconstruct the full 3D geometry of the specimen [19,20].

In this work, a relatively simple method is proposed to obtain the true curve of anisotropic materials from tensile tests on dog-bone specimens throughout their entire straining life until fracture. The novel procedure is based on optically analyzing the evolution of the thickness relative to the width before necking onset. Then, the obtained trend is assumed to apply in the post-necking phase as well, allowing for the derivation of the effective cross-section resistant area of the specimen and, consequently, the true stress and strain until fracture. The robustness of the proposed procedure is validated through optical analysis of the geometric proportions of the fractured specimens. Moreover, the proposed methodology is compared with other classical approaches, demonstrating its superior accuracy.

2. Materials and methods

2.1. Experimental campaigns

The experimental campaigns considered in this work have been carried out on tensile specimens of two materials, the high strength steel PHS1800 and the aluminium alloy AL6181. All the specimens had the geometry shown in Figure 1, with thicknesses of 1.5 mm for the PHS1800 ones and 2.0 mm for the AL6181 ones. For both materials, the tensile tests have been carried out at the quasistatic rate of $3\text{x}10^{-4}$ s^{-1} with an electromechanical testing machine, at the intermediate strain rate of 10 s^{-1} with a hydraulic testing machine, and at the high strain rate of 250 s^{-1} with the direct-tension Hopkinson bar developed by the authors [21]. The load histories have been acquired with the load cells in the static and intermediate strain rate tests and with the three waves procedure in the high strain rate tests. They have been coupled with the images of the deforming specimens obtained with a standard high-resolution camera (static rate) or with a high-speed camera (intermediate and high strain rates) in order to employ the DIC technique and to obtain optical measurements of the gauge length L and of the width w of the specimens.

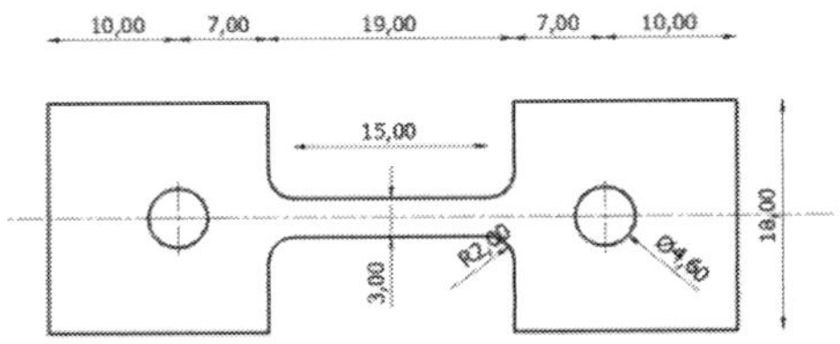

Figure 1. Specimens' geometry.

2.2. Stress and strain evaluation in flat specimens

Several characterization approaches have been considered and compared, aiming at the evaluation of the true curve of the anisotropic metals at hand. By definition, the true stress σ_{True} and strain ε_{True} are obtained by eqs. (1) and (2), in which A is the resistant cross-section area of the specimen and F is the total axial force applied to the specimen. The subscript "0" refer to the initial value of every variable.

$$\varepsilon_{True} = Ln\left(\frac{A_0}{A}\right) \tag{1}$$

$$\sigma_{True} = \frac{F}{\mathrm{A}} \tag{2}$$

Given the difficulty in evaluating the actual resistant cross-section area, different estimations of such variables can be done experimentally with diverse approaches. Given the gauge length of the specimen L, the classical *length-based* one makes use of the expressions shown in eqs. (3) and (4) to obtain the corresponding true stress σ_{True-L} and strain ε_{True-L}. It is well known that these expressions, based on the volume conservation, are rigorously valid only until the necking onset and then diverge from the actual values quite consistently for ductile materials.

$$\varepsilon_{True-L} = Ln\left(\frac{L}{L_0}\right) \tag{3}$$

$$\sigma_{True-L} = \frac{F}{A_0} \cdot \frac{L_0}{L} \tag{4}$$

If the material is isotropic, then it is possible to consider that the proportions of the resistant cross-section of the specimen remain constant during the test. Then, measuring the width w of the specimen, it is possible to derive the corresponding thickness t. With this hypothesis, the *width-based* true stress σ_{True-w} and strain ε_{True-w} can be calculated as shown in eqs. (5) and (6).

$$\varepsilon_{True-w} = 2 \cdot Ln\left(\frac{w_0}{w}\right) \tag{5}$$

$$\sigma_{True-w} = \frac{F}{\frac{t_0}{w_0} \cdot w^2} \tag{6}$$

Depending on the material and on the specimen shape at hand, the local peak strain measured with the DIC in the outer surface of the necking section can be higher or lower than the actual mean strain in that section [16]. However, considering them approximately equal, it is possible to obtain the *DIC Peak-based* true stress $\sigma_{True-DIC}$ and strain $\varepsilon_{True-DIC}$ as shown in eqs. (7) and (8).

$$\varepsilon_{True-DIC} = \varepsilon_{DIC-Peak} \tag{7}$$

$$\sigma_{True-DIC} = \frac{F}{\mathrm{A}_0} \cdot \mathrm{e}^{-\varepsilon_{True-DIC}} \tag{8}$$

The deviation of $\varepsilon_{True-DIC}$ from the effective true strain directly impacts the accuracy of eqs. (7) - (8), and the magnitude of such deviations cannot be estimated a priori.

If the material is anisotropic, the proportions of the resistant cross-section do not remain constant. Then, the *width-based* approach is not valid during the entire test, while the *length-based* one is valid only before the necking inception (volume conservation still applies). Then, there are no options for the correct evaluation of the post-necking part of the true curve. Therefore, a new procedure is proposed here. The main new concept is to evaluate the anisotropic behaviour of the material in the pre-necking phase and to consider that it applies also to the post-necking one. In particular, thanks to the volume conservation, the trend of the evolving thickness can be evaluated in the pre-neck phase as shown in eq. (9). Then, such trend is prolonged all over the post-necking range up to failure. This hypothesis will be also checked analysing the broken samples and evaluating if the final proportions of the cross-sections are compatible with it.

$$\frac{t}{t_0} = \left(\frac{L_0 \cdot w_0}{L \cdot w}\right)_{pre-neck} \tag{9}$$

3. Results and discussion

3.1. Strain histories

The strain histories obtained with the different approaches are shown in Figure 2 for two representative static tests of the two materials (similar trends are found for all tests).

For the PHS1800 (left part of Figure 2), the necking strain is roughly 0.05. It is possible to see that in the pre-necking stage all three estimations are essentially overlapped. The equality between the *DIC Peak-based* and the *L-based approaches* is trivial in this stage. On the other hand, the fact that also the *w-based* approach gives the same trend means that anisotropy has negligible effect on the equality $t_0/w_0 = t/w$. In the post-necking stage, obviously the *L-based* approach loses accuracy while the other two give very similar estimations. Then, the pre-necking strain estimation can be done with all three approaches, while, for the reasonings explained earlier, no reliable prediction is present for the post-necking range.

The AL6181 (right part of Figure 2) presents a necking strain of 0.13. In this case, the *w-based* pre-necking estimation is very different from the other two evidencing a consistent anisotropic behaviour. Therefore, the pre-necking strain evaluation must be done with the *L-based* approach. Again, no reliable post-necking estimation is present.

Given that both metals exhibit a long post-necking ductile behaviour, the above uncertainties apply for more than half of their straining life.

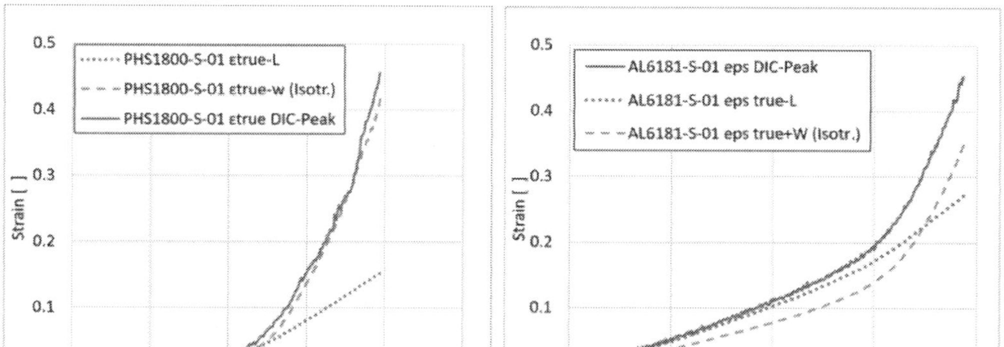

Figure 2. *L-based*, *w-based* and *DIC-Peak* strain histories for representative tests on PHS1800 steel (left) and AL6181 alloy (right) at static rates.

IOP Conf. Series: Materials Science and Engineering 1306 (2024) 012005 doi:10.1088/1757-899X/1306/1/012005

3.2. Evaluation of the post-necking straining life of anisotropic materials

The trend of the thickness reduction is obtained for every test with eq. (9). As already mentioned, such approach is valid only before the necking onset. However, through the optical analysis of the broken specimens (Figure 3), it is possible to evaluate the final proportions of the resistant cross-sections.

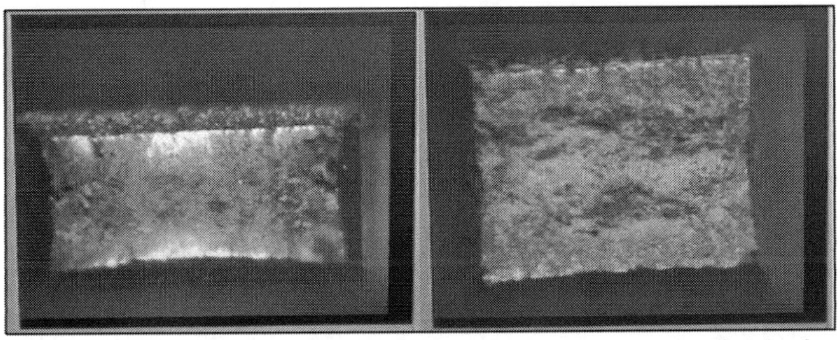

Figure 3. Thickness and width at failure from two representative broken specimens of PHS1800 (left) and AL6181 (right)

Considering three representative tests for each material, in Figure 4 are presented (PHS1800 left and AL6181 right), in terms of t/t_0 vs w/w_0, the results obtained with eq. (9) (scattered curves for the pre-necking phase and solid curves for the post-necking phase), together with circular points representing the specimens' geometries at failure. The initial condition of every test is the point (1;1). The pre-necking points of all tests of each material are found to be approximable with a power law according to eq. (10), with exponent n equal to 1.35 for PHS1800 and 1.8 for AL6181 (green dotted lines in Figure 4).

$$\frac{t}{t_0} = \left(\frac{w}{w_0}\right)^n \tag{10}$$

As expected, the post-necking data represented by the solid lines quit to be approximable by those power laws, since they are not representative of the actual behaviour of the material. The black dashed lines represent a hypothetical isotropic behaviour descripted by a power low according to eq. (10) with the exponent n equal to 1. It is easy to demonstrate that the exponent n is equal to the inverse of the classic Lankford coefficient $R_{yz} = \varepsilon_y/\varepsilon_z$.

From Figure 4, it is very clear that the power laws approximating the pre-neck data, comply really well also with the fracture circular points for both materials. Then, such trends are considered to be valid also in the post-necking range. Therefore, with this hypothesis, it is possible to evaluate the actual resistant cross-section of the specimens during the entire test and, in turn, the actual true strain and stress values.

It is important to underline that, if the Lankford coefficient is directly derived in the pre-necking phase from the strains instead as the inverse of the exponent n, more uncertainties arise, especially for very early necking materials. This concept is clearly shown in Figure 5, in which the Lankford coefficients calculated directly from the strains obtained with optical measurements (green scattered lines) and with local DIC (yellow scattered lines) are compared to the inverse of the obtained exponents (blue dotted lines). Indeed, at the beginning of the plastic straining, the classical Lankford coefficient is an intrinsically undetermined ratio $\varepsilon_y/\varepsilon_z = 0/0$. Therefore, in the case of the very early necking PHS1800 (left column of Figure 5), it is almost impossible to obtain useful data without considering the sections at fracture. For the AL6181 (right column of Figure 5) it is possible to obtain useful information in the pre-necking phase but it is possible to see that the DIC and length-based evaluations quite differ from each other due to measuring uncertainties. Then, the proposed power-law based approach demonstrates to be more robust than the classical strain-based one.

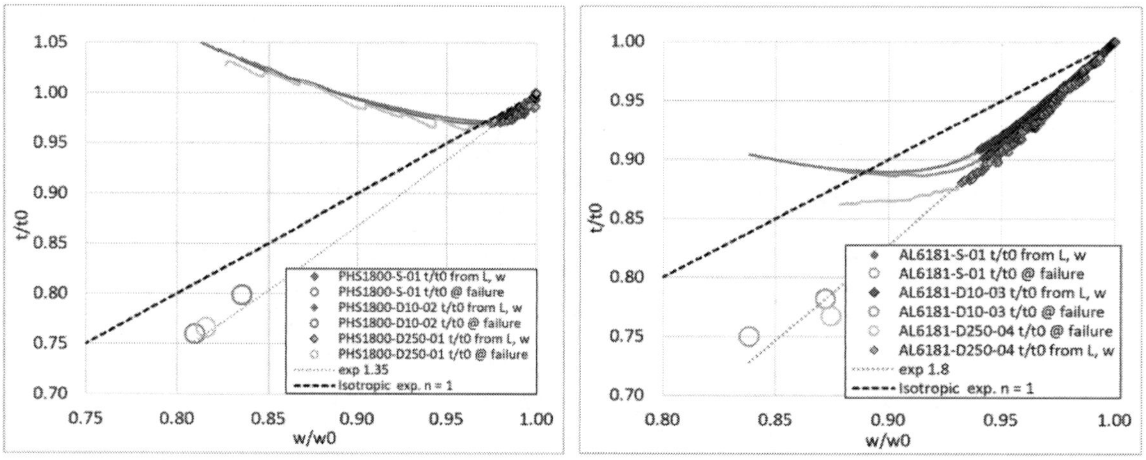

Figure 4. Thickness Reduction vs. width reduction for representative tests on PHS1800 (left) and AL6181 (right).

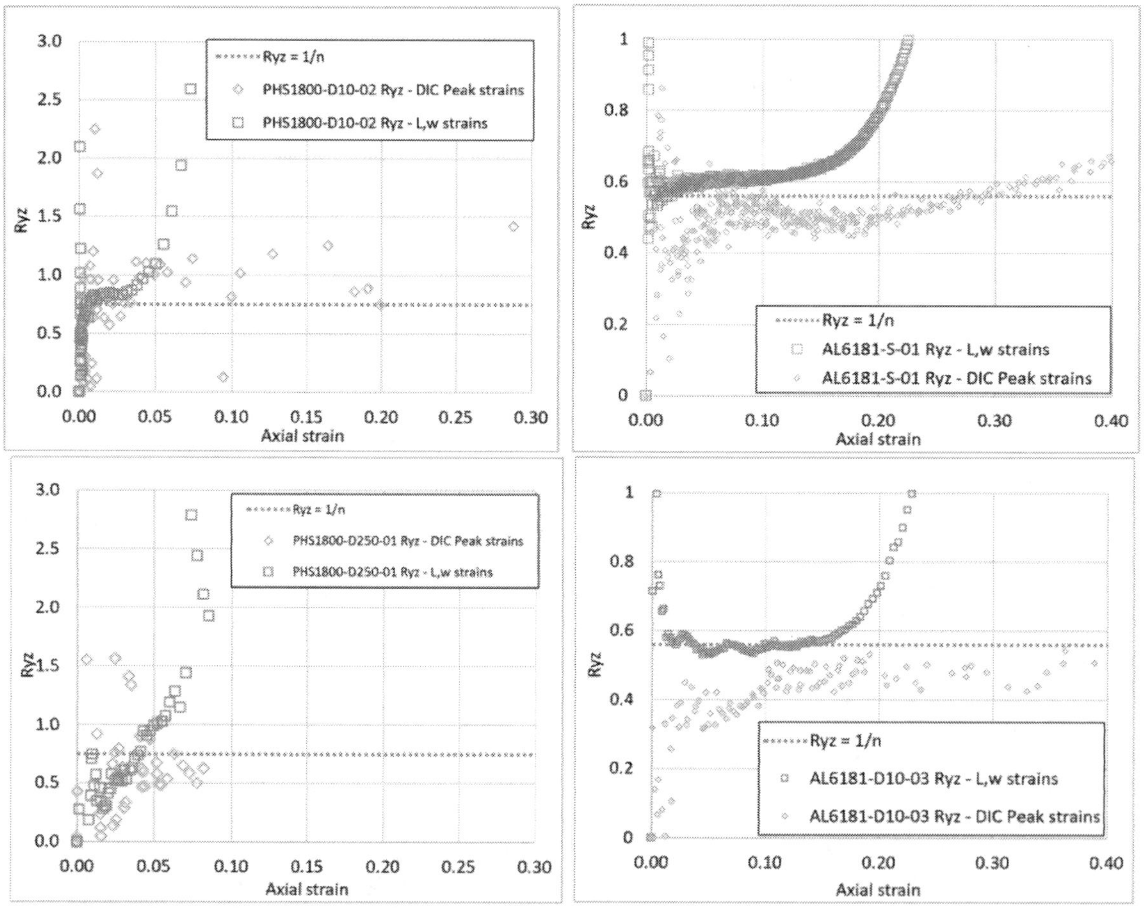

Figure 5. Lankford coefficients from DIC strains, from L-w strains and from power laws of the t-vs.-w reduction

3.3. *Final true stress-strain curves*

As already pointed out, considering that the trend obtained with eq. (10) in the pre-necking phase of every test always applies until fracture, it is possible to evaluate the actual resistant cross-section of the specimen during the entire test and, in turn, the actual true strain and stress values as in eqs. (11) and (12). It is worth noting that with $n = 1$, eqs. (11) and (12) degenerate to eqs. (5) and (6) which are valid for isotropic materials.

$$\varepsilon_{True} = Ln\left(\frac{A_0}{A}\right) = Ln\left(\frac{t_0}{t}\cdot\frac{w_0}{w}\right) = Ln\left(\left(\frac{w_0}{w}\right)^{n+1}\right) = (n+1)\cdot Ln\left(\frac{w_0}{w}\right) \tag{11}$$

$$\sigma_{True} = \frac{F}{A} = \frac{F}{\left(\frac{w}{w_0}\right)^n \cdot t_0 \cdot w} = \frac{F \cdot {w_0}^n}{w^{(n+1)} \cdot t_0} \tag{12}$$

The obtained curves with the different approaches, called *L-based*, *w-based*, *DIC-peak*, and *Expo*, are shown, together with the fracture points, in Figure 6 for representative tests at different strain rates for both materials. Analysing such figure, it is possible to highlight the differences between the approaches both in the pre-necking and in the post-necking phase.

In the pre-necking phase, the *L-based, DIC-peak*, and *Expo* approaches give obviously the same curve, while the *w-based* differs from the formers depending on the magnitude of anisotropy (more evident for the AL6181).

In the post-necking phase, evidently the *L-based* stops to be representative of the actual behaviour of the material, being the corresponding true curve much shorter and lower than the others, not adequately representing the local behaviour of the material. On the other hand, the *w-based* and *Expo* approaches give different estimates of the true curve depending on the magnitude of the anisotropy: for the PHS1800 there is a noticeable difference but for the AL6181 the difference is very significant, with the fracture strain increased of over 40%. Moreover, the *Expo* curves nicely comply also with the fracture points, confirming once again the reliability of the approach. The *DIC-peak* curves, depending on the single test, are more or less scattered due to the speckle degradation during the tests but are quite similar to the *Expo* curves. As already pointed out earlier and better explained in [16], the accuracy of this approach cannot be predicted since it depends on the relationship between the evaluated DIC peak strain and the actual strain distribution in the resistant cross-section. Given the results, for these materials combined with the corresponding geometries of the specimens, there is a quite good agreement between such two strains. Then, the *Expo* approach is the only one effectively reliable in evaluating the post-necking part of the curve, significantly improving the characterization procedure.

It is useful to recall that the proposed approach regards only the evaluation of the true curve of the material, which still must be corrected to obtain the actual corresponding flow curve taking into account the triaxiality [13,14]. Moreover, additional tests would be necessary to assess the full anisotropic behaviour of the materials in different directions and stress states in order to correctly calibrate an appropriate anisotropic model. However, the proposed approach gives a fundamental basepoint in order to correctly evaluate the tensile behaviour of a material.

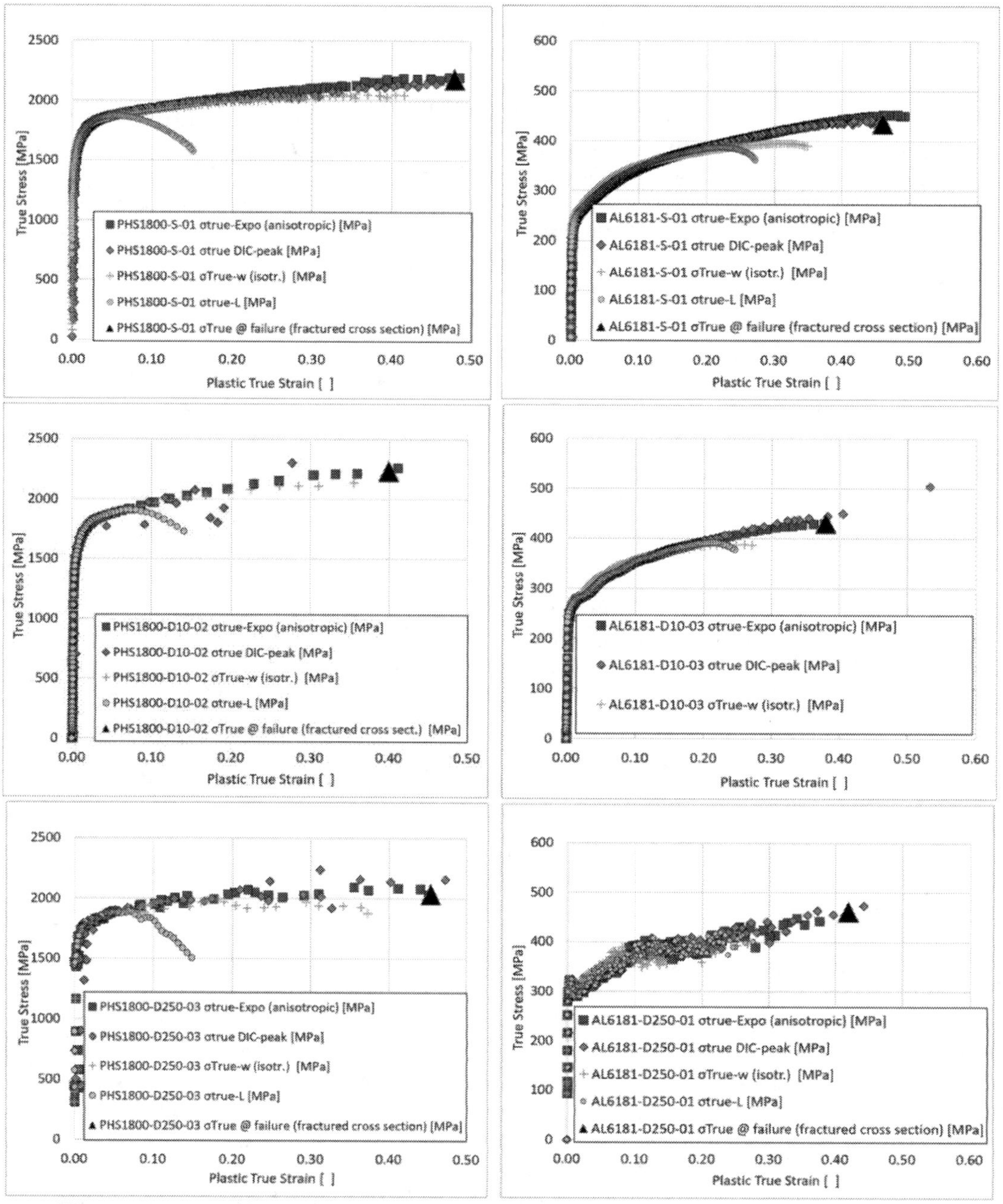

Figure 6. Comparison of experimental true curves according to different approaches.

Conclusions

In this work, a novel procedure for the evaluation of the true curve of an anisotropic material from a tensile test on a dog-bone specimen has been proposed. The novel procedure has been compared to other possible approaches, highlighting the differences, and demonstrating the better performance of the proposed one. The different characterization approaches have been applied to tensile tests at different strain rates on dog-bone specimens made of PHS1800 steel and AL6191 aluminium alloy.

The novel characterization approach is based on the evaluation of the anisotropy of the material at hand thanks to optical evaluation of the geometry of the specimen during the test. In particular, the evolution of the normalized thickness t/t_0 related to that of the normalized width w/w_0, is derived from optical measurements of elongation and width before the necking onset, when it is rigorously valid. Then, the obtained trend is supposed to apply also in the post-necking phase, deriving consequently the effective cross-section resistant area of the specimen and, in turn, the true stress and strain until fracture. The adopted hypothesis has been checked analysing optically the shape of the minimum section of the fractured halves of the specimens, confirming that their t/w proportions comply with the pre-necking trend of anisotropy. It was demonstrated that the proposed methodology based on the evolution of the normalized thickness t/t_0 related to that of the normalized width w/w_0, is much more effective and robust than the classical Lankford coefficient-based one, especially for low necking strain materials in which the latter gives almost no useful information about the material anisotropy.

Finally, the true curves obtained from the different tests with the proposed methodology are compared with the curves obtained with the other classical approaches, demonstrating that it is the only one delivering accurate results in the post-necking range, which for the steel and Al alloy at hand, respectively covers nearly 90% and 75% of their whole straining life.

Acknowledgements
This work has been partially funded by European Union (NextGeneration EU), through the MUR-PNRR project SAMOTHRACE (ECS00000022)

References

[1] Rossi M, Lattanzi A, Morichelli L, Martins J M P, Thuillier S, Andrade-Campos A and Coppieters S 2022 Testing methodologies for the calibration of advanced plasticity models for sheet metals: A review *Strain*

[2] Jia X, Hao K, Luo Z and Fan Z 2022 Plastic Deformation Behavior of Metal Materials: A Review of Constitutive Models *Metals* **12** 2077

[3] Joun M, Eom J G and Lee M C 2008 A new method for acquiring true stress–strain curves over a large range of strains using a tensile test and finite element method *Mechanics of Materials* **40** 586–93

[4] Cortis G, Nalli F, Sasso M, Cortese L and Mancini E 2022 Effects of Temperature and Strain Rate on the Ductility of an API X65 Grade Steel *Applied Sciences* **12** 2444

[5] Abedini A, Narayanan A and Butcher C 2022 An investigation into the characterization of the hardening response of sheet metals using tensile and shear tests with surface strain measurement *Forces in Mechanics* **7** 100090

[6] Habraken A M, Aksen T A, Alves J L, Amaral R L, Betaieb E, Chandola N, Corallo L, Cruz D J, Duchêne L, Engel B, Esener E, Firat M, Frohn-Sörensen P, Galán-López J, Ghiabakloo H, Kestens L A I, Lian J, Lingam R, Liu W, Ma J, Menezes L F, Nguyen-Minh T, Miranda S S, Neto D M, Pereira A F G, Prates P A, Reuter J, Revil-Baudard B, Rojas-Ulloa C, Sener B, Shen F, Van Bael A, Verleysen P, Barlat F, Cazacu O, Kuwabara T, Lopes A, Oliveira M C, Santos A D and Vincze G 2022 Analysis of ESAFORM 2021 cup drawing benchmark of an Al alloy, critical factors for accuracy and efficiency of FE simulations *Int J Mater Form* **15** 61

[7] Wang L and Tong W 2015 Identification of post-necking strain hardening behavior of thin sheet metals from image-based surface strain data in uniaxial tension tests *International Journal of Solids and Structures* **75–76** 12–31

[8] Wu H, Zhuang X, Zhang W and Zhao Z 2022 Anisotropic ductile fracture: experiments, modeling, and numerical simulations *Journal of Materials Research and Technology* **20** 833–56

[9] Rossi M, Pierron F and Štamborská M 2016 Application of the virtual fields method to large strain anisotropic plasticity *International Journal of Solids and Structures* **97–98** 322–35

[10] Lattanzi A, Barlat F, Pierron F, Marek A and Rossi M 2020 Inverse identification strategies for the characterization of transformation-based anisotropic plasticity models with the non-linear VFM *International Journal of Mechanical Sciences* **173** 105422

[11] Rossi M, Lattanzi A, Barlat F and Kim J-H 2022 Inverse identification of large strain plasticity using the hydraulic bulge-test and full-field measurements *International Journal of Solids and Structures* **242** 111532
[12] Maček A, Starman B, Mole N and Halilovič M 2020 Calibration of Advanced Yield Criteria Using Uniaxial and Heterogeneous Tensile Test Data *Metals* **10** 542
[13] Bridgman P W 1964 Studies in Large Plastic Flow and Fracture: With Special Emphasis on the Effects of Hydrostatic Pressure *Harvard University Press*
[14] Mirone G 2004 Approximate Model of the Necking Behaviour and Application to the Void Growth Prediction *International Journal of Damage Mechanics* **13** 241–61
[15] Corallo L, Revil-Baudard B, Cazacu O and Verleysen P 2023 Role of anisotropy on strain localization at high-strain rates in Ti6Al4V *European Journal of Mechanics - A/Solids* **97** 104856
[16] Mirone G, Barbagallo R, Tedesco M M, De Caro D and Ferrea M 2022 Extended Stress–Strain Characterization of Automotive Steels at Dynamic Rates *Metals* **12** 960
[17] Mirone G, Verleysen P and Barbagallo R 2019 Tensile testing of metals: Relationship between macroscopic engineering data and hardening variables at the semi-local scale *International Journal of Mechanical Sciences* **150** 154–67
[18] Mirone G and Corallo D 2013 Stress–strain and ductile fracture characterization of an X100 anisotropic steel: Experiments and modelling *Engineering Fracture Mechanics* **102** 118–45
[19] Larour P, Freudenthaler J and Weissböck T Reduction of cross section area at fracture in tensile *Journal of Physics* 9
[20] Gu G H, Kim Y, Kwon H, Ahn S Y, Seo M H, Ahn K and Kim H S 2022 A New Proposal for a Method to Measure Orthogonal R-Value Using a Single Tensile Test with Three-Dimensional Digital Image Correlation *Exp Mech* **62** 999–1006
[21] Mirone G, Barbagallo R and Cadoni E 2017 Tensile Test of a HSLA Steel at High Strain Rates with Two Different SHTB Facilities *Procedia Engineering* **197** 89–98

Static and dynamic tensile behaviour of polycarbonate connected to the deforming specimen shape

Giuseppe Mirone[a*], Raffaele Barbagallo[a], Giuseppe Bua[a,b]

[a] Department of Civil Engineering and Architecture – University of Catania – Viale A. Doria 6, 95125 Catania – Italy

[b] Department of Electrical, Electronic and Computer Engineering – University of Catania – Via Santa Sofia 64, 95123 Catania – Italy

*gmirone@dii.unict.it

Abstract. Like other structural thermoplastic polymers, polycarbonate (PC) exhibits a typical two-stage plastic deformation mode, including a first phase of strain localization and a second phase of strain propagation. At the same time, the stress-strain curves at the local scale undergo two distinct loading phases connected by a softening-instability decreasing phase. This mechanical response is not yet fully recognized and the reproduction of the peculiar phenomena in numerical simulations is not yet commonly standardized. In this work, the relationship between the straining/loading macroscopic response and the material flow curve at the local scale is investigated in detail, aiming at the two-way relationships enabling to infer the latter curve from the former data. Experimental tests at static and dynamic rates are carried out to verify how the special features of the PC are affected by the strain rate and to estimate its rate sensitivity.

1. Introduction

Polycarbonate (PC) is a thermoplastic polymer used in many engineering applications for its remarkable properties as lightness, impact resistance, and transparency. The mechanical response of PC is highly dependent on the average molecular weight of the polymeric chains [1], temperature, strain rate, and physical ageing [2–5].

Typical true stress-strain curves of PC exhibit, after the first elastic phase, a peak and a more or less pronounced drop. Then, after a valley point, they present a new smooth increase lasting up to failure and characterized by an upward concavity. Therefore, direct procedures commonly used to transform the true curve into the corresponding material flow curve for metals, like the most known Bridgman's correction [6] or other polynomial functions [7,8], cannot be used for glassy polymers like the PC, because the shape evolution of the necked zone is completely different from that of metals.

Several constitutive models for glassy polymers have been proposed in the literature in the last decades. Among the first researchers who tried to characterize this kind of materials, G'Sell et al. [9] conducted a static tensile test at room temperature on a High-Density Polyethylene (HDPE) round specimen. They analysed the conditions of necking initiation and stabilization and provided a model for the material flow curve depending on the strain and the strain rate. Moreover, they focused on the stress state at the minimum cross-section, highlighting the effect of triaxiality on necking propagation. Boyce et al. [10] elaborated an analytical model for Polymethyl methacrylate (PMMA) based on the energetic and entropic modelling of the polymeric macromolecules interactions during plastic deformations. Muhammad & Jar [11] proposed a model in which the stress-strain relationship was expressed using four equations for the elastic–plastic deformation and one for the creep deformation. The material constitutive parameters were calibrated by inverse procedures based on Finite Element (FE) simulations, minimizing the error between experimental data and the output from FE analysis.

Regarding the dynamic behaviour of this kind of materials, Boyce et al. [12] tested PC at high strain rates, highlighting the relationship between the presence of a second Considère point in the true curve (condition for which the slope of the true curve is equal to the corresponding true stress [13]) and necking propagation. An analogous observation is reported in the work of Crist et al. [14] for Polyethylene (PE). Tzibula et al. [15] performed quasi-static tensile tests and dynamic compression tests using a Kolsky-Bar apparatus on PC round specimens. They characterized the material behaviour using a simple four-point piecewise flow curve and reverse engineering method.

Despite the several studies conducted on the tensile response of glassy polymers, many key aspects of their deformation mechanism are not completely understood and the characterization procedure to adopt with these materials, including PC, is not consolidated. Then, the aim of this work is to understand the special phenomena occurring in the deforming process of PC flat tensile specimen and to identify their possible impact on the stress-strain response, at quasi-static and dynamic strain rates.

The analysis of the evolving shape of the specimen is used to analyse in detail the deforming process and to outline the mechanism causing the deviation between the true stress and the flow stress.

2. Materials and methods

2.1. Experimental campaign

The experimental campaign considered in this work has been carried out on PC specimens characterized by the geometry shown in Figure 1. The tensile tests have been run at quasistatic rate ($3x10^{-4}$ s^{-1}) with an electromechanical testing machine and at the dynamic strain rate of 7 s^{-1} with a hydraulic testing machine. Making use of a standard high-definition camera and a fast camera, all tests have been recorded and the images coupled to the corresponding load and displacement data from the testing machines. For every test, the images have been analysed in order to obtain the evolution of the geometry of the specimen (in particular, its gauge length L and its minimum width w).

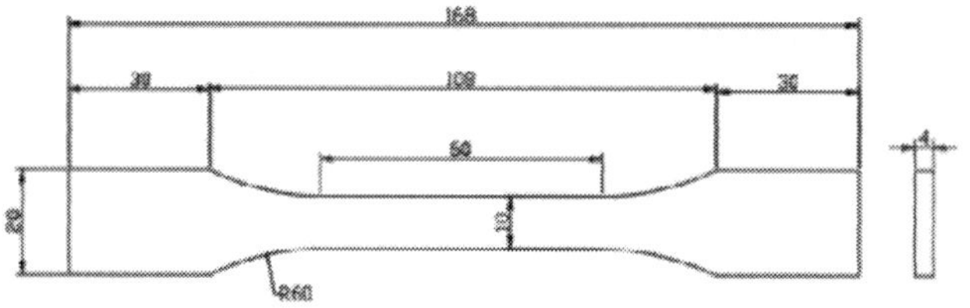

Figure 1. Specimens' geometry.

2.2. Stress and strain evaluation

Two characterization approaches have been implemented to evaluate the true curve of the material from each test. In the first adopted approach, the *length-based* strain ε_{True-L} and stress σ_{True-L} are obtained as in eqs. (1) and (2), where A is the resistant cross-section area of the specimen, F is the total axial load, and L is the considered gauge length of the specimen.

$$\varepsilon_{True-L} = Ln\left(\frac{L}{L_0}\right) \tag{1}$$

$$\sigma_{True-L} = \frac{F}{A_0} \cdot \frac{L_0}{L} \tag{2}$$

If the material is isotropic, as PC is considered here, given that the proportions of the resistant cross-section remain constant during the test, it is possible to obtain its thickness t from the corresponding width w.

$$t = \frac{t_0}{w_0} w \tag{3}$$

$$A = t \cdot w = \frac{t_0}{w_0} w^2 \tag{4}$$

Then, the *area-based* true stress and strain can be calculated as shown in eqs. (5) and (6); for obvious reasons, they will be also addressed as *width-based* true strain ε_{True-w} and stress σ_{True-w}.

$$\varepsilon_{True-w} = Ln\left(\frac{A_0}{A}\right) = 2 \cdot Ln\left(\frac{w_0}{w}\right) \tag{5}$$

$$\sigma_{True-w} = \frac{F}{A} = \frac{F}{\frac{t_0}{w_0} \cdot w^2} \tag{6}$$

As will be shown in the next sections, the analysis of the experimental tests considering two characterization approaches will give important information on the peculiar behaviour of PC.

3. Results and discussion

3.1. Static tests

Considering the static tests, Figure 2 shows the obtained *length-based* (a) and *width-based* (b) true curves. In this figure, it is possible to see that the post-peak phase of both curves can be accurately described by an exponential function. Indeed, after the softening phase, all static tests are characterized by a nearly constant load phase. Then, in this phase it must be true the well-known instability condition $\partial\sigma_{True}/\partial\varepsilon_{True} = \sigma_{True}$. Such mathematical condition, in case of common structural materials (e.g., metals) only applies to the single instant of necking onset, whereas for the PC and for similar polymers it applies over a finite time range extending over more than 50% of the duration of a tensile test. Then, in the latter case the above condition yields a differential equation which, in turn, delivers the true stress as an exponential function of the true strain, $\sigma_{True} = k \cdot e^{\varepsilon_{True}}$, with the parameter approximately equal to the first-yield strength, $k = \sigma_{yield} = 47.5\ MPa$.

IOP Conf. Series: Materials Science and Engineering 1306 (2024) 012006 doi:10.1088/1757-899X/1306/1/012006

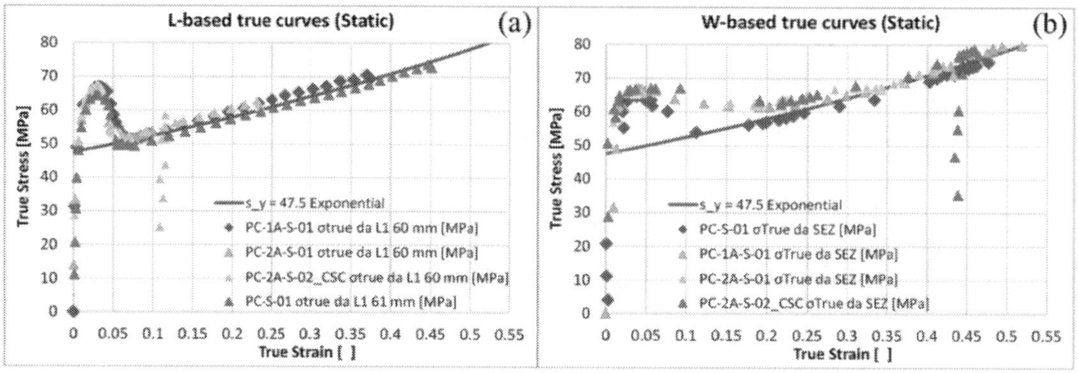

Figure 2. Static tests: length-based (a) and width-based (b) true curves.

The most peculiar phase of the PC load-time curve is the softening one, which follows the onset of the diffused and localized necking. Then, it is useful to descript it in detail, as shown in Figure 3 and Figure 4, identifying the following characteristic points:

- *N* = Necking onset
- *LDS* = Load Drop Start
- *LDE* = Load Drop End (minimum distorted cross-section due to the necking band)
- *CS* = Cross-section Straight (the cross-section returns roughly straight).
- *TE* = Necking zone elongates and becomes roughly square.

In all tests, after TE until failure, an almost straight portion of the specimen with narrower width is present in the most deformed part, with an approximately uniaxial stress state.

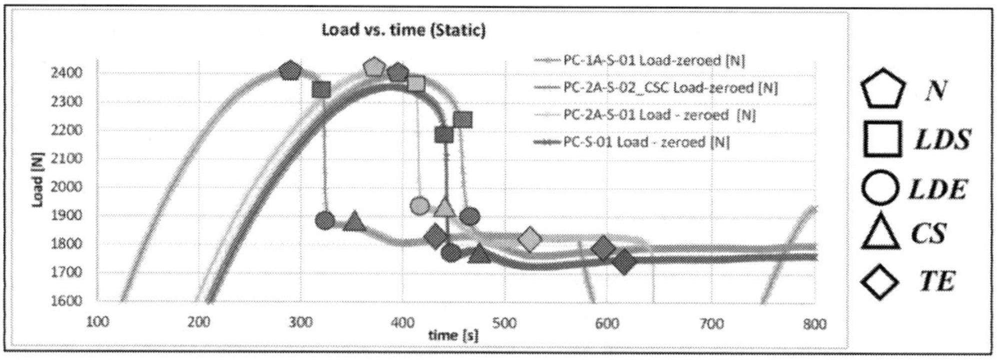

Figure 3. Static tests: Characteristic points in the load-time graph.

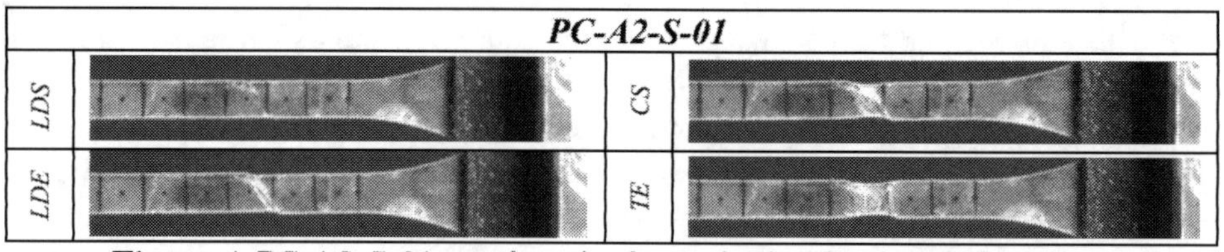

Figure 4. PC-A2-S-01 specimen's shapes in the characteristic points.

The same characteristic points are also depicted in the *w-based* vs *L-based* true strains graph in Figure 5, together with the corresponding tests' phases. In all tests, the machine speed has been maintained constant. Therefore, the *L-based* strain rate $\dot{\varepsilon}_{True-L}$ remains roughly constant. On the other hand, the local *w-based* strain rate $\dot{\varepsilon}_{True-w}$ changes depending on the overall strain localization/propagation behaviour in the specimen. Then, four distinct phases shown in Figure 5 can be identified:

1. Uniform straining until *N* in which ε_{True-L} and ε_{True-w} grow at the same speed.
2. From *N* to *LDE*, the ε_{True-w} grows quicker and quicker in respect to ε_{True-L} due to the diffused and localized necking.
3. From *LDE* to CS, the ε_{True-w} stops accelerating and grows at a nearly constant speed, higher then the ε_{True-L} one.
4. From *CS* to failure, the ε_{True-w} starts decelerating until it becomes almost constant during full strain propagation after *TE*.

Looking at the specimen's shapes in Figure 4 and considering the graph in Figure 5, it is clear that during phases 2 and 3 the whole concept of deriving the true curve from load and width measurements is a remarkable approximation since the deformation mode recalls a mix of tension and shear and the minimum cross section is not even rectangular.

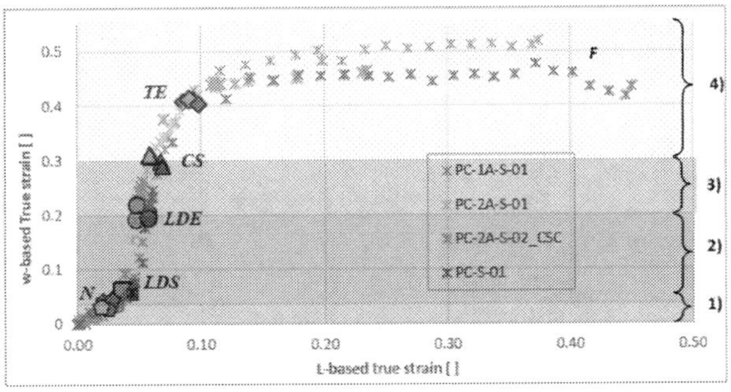

Figure 5. Static tests: *W-based* vs *L-based* true strains.

3.2. Dynamic tests

Analogously to the static case, also in dynamic tests both the *length-based* and *width-based* true curves shown in Figure 6 can be accurately approximated by an exponential function in the post-peak phase. In this case, the yield stress is roughly $\sigma_y = k = 60\ MPa$. There is a consistent scattering in the failure strain between the tests due to premature fracture of some specimens. However, it is clear that the higher strain rate limits the strain propagation in respect to the static case.
In fact, while the maximum value of local *width-based* fracture strain is almost equal in static and dynamic tests (0.55 in both cases), instead the maximum value of global *length-based* fracture strain is quite lower in dynamic than it is at static rate (0.22 vs. 0.45).

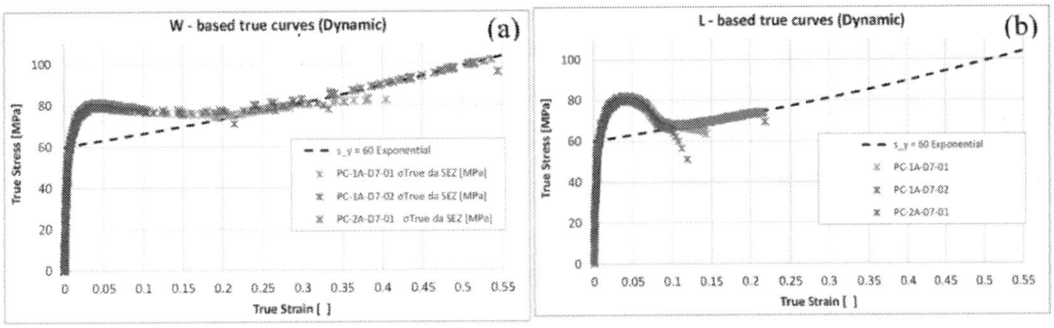

Figure 6. Dynamic tests: Length-based (a) and width-based (b) true curves.

The *W-based* vs *L-based* true strains graph with the characteristic points of Figure 7 shows again that the higher strain rate limits the strain propagation, with the slope of the curve after *CS* and *TE* that barely decreases. Comparing Figure 5 with Figure 7, it is evident that the strains at stages *N*, *LDS*, *LDE* and *CS* are nearly equal to those from static tests, instead the stage *LDS* is rather delayed.

IOP Conf. Series: Materials Science and Engineering 1306 (2024) 012006 doi:10.1088/1757-899X/1306/1/012006

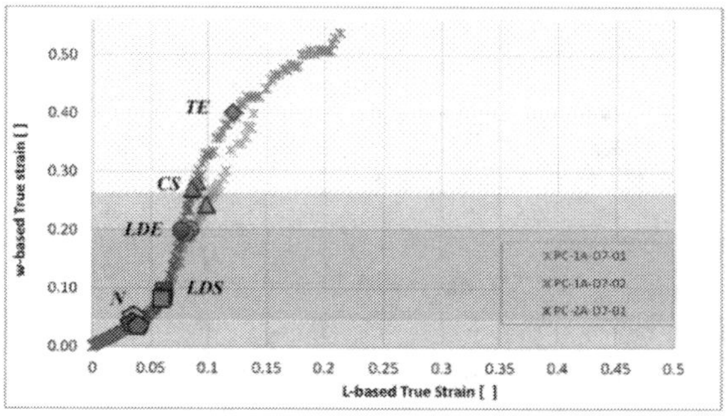

Figure 7. Static tests: *W-based* vs *L-based* true strains.

3.3. *Dynamic amplification*

The *w-based* true curves from static and dynamic tests are shown in Figure 8 together with the corresponding best-fit functio

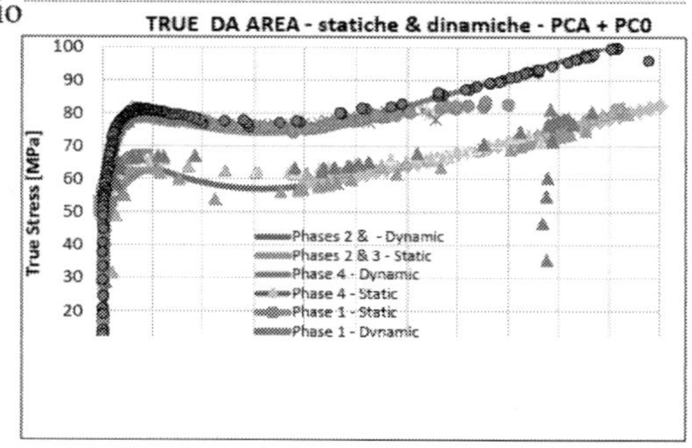

Figure 8. Static and dynamic tests: Width-based true curves and corresponding best-fit functions.

The ratio between the true curves best-fit functions is shown in Figure 9 together with the strain rate histories from the dynamic tests. The ratio between the true curves is an approximation of the ratio between the flow curves. In particular, it is exact before the necking onset, becomes less accurate when the specimen is distorted, and rebecomes more and more accurate when the specimen straights again. Then, considering also that a similar triaxiality correction should apply in both the static and dynamic cases, the true curves ratio is a good approximation of the flow curves one and, therefore, is a good description of the strain rate sensitivity of the material at hand.

The dynamic amplification shown in Figure 9a grows to a value of roughly 1.25 at strain 0.05 (necking onset) and then remains almost constant, despite the strain rate grows consistently after such deformation stage (up to ten times its value at necking onset). Then, a 25% increase demonstrates the high strain rate sensitivity of PC. Moreover, the fact that the dynamic amplification stops evolving at the necking onset despite the strain rate grows ten times afterwards demonstrates that the locking of the strain rate effect, already found by the authors for metals in previous works [16–18], also applies to the PC at hand.

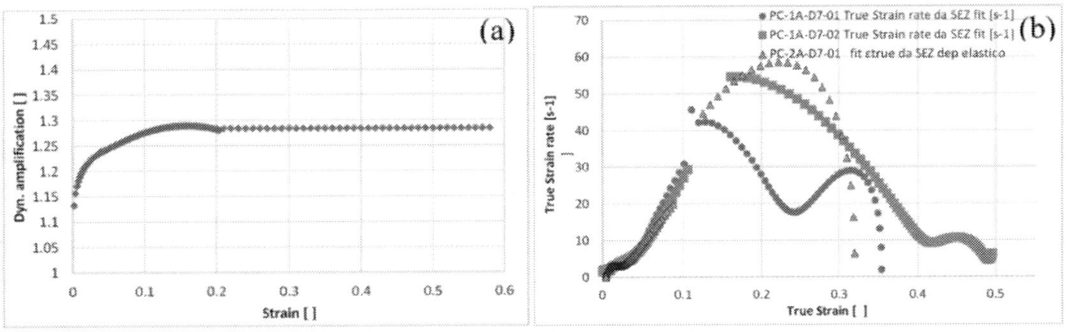

Figure 9. Dynamic amplification (a) and strain rate histories from the dynamic tests (b).

The performed analysis allows to obtain the following useful information about the correction to be made to the experimental true curves, in order to obtain an estimate of the corresponding flow curves:

- Phase 1: No correction is needed before the necking onset since the stress distribution is uniaxial.
- Phase 2: The correction grows very quickly while the specimen becomes not straight nor rectangular until the maximum correction value is reached at point *LDE*.
- Phase 3: The localized necking starts to relax and stretch axially and the cross-section rebecomes more and more rectangular. Then, the triaxiality correction decreases accordingly.
- Phase 4: The strain developed in the necking region starts to propagate, making the stress distribution in the most deformed part of the specimen more and more uniaxial. Then, the triaxiality correction decreases and tends to ideally vanish at indefinitely high plastic strains.

The qualitative picture of how such correction should evolve at different strain levels is reported in Figure 10, where the magnitude of the deviation is normalized to its maximum value, occurring at the stage *LDE*.

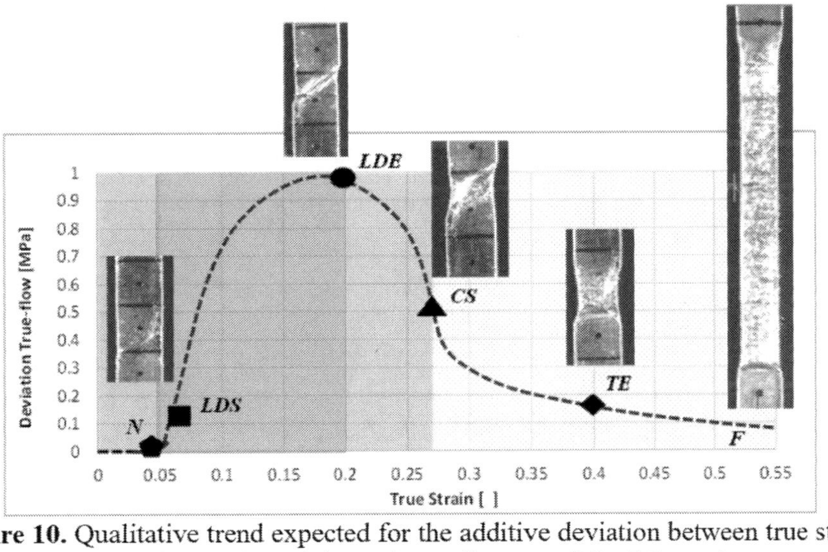

Figure 10. Qualitative trend expected for the additive deviation between true stress and flow stress during static and dynamic tensile tests of flat PC specimens.

Assuming a piecewise approximation of the above trend by a quadratic function with peak at LDE (phases 2, 3) joined to a negative-power function asymptotically vanishing at high strains (phase 4), it would require five unknown parameters. Imposing continuity at stage CS and null deviation at stage N, only three parameters remain, while four conditions can be approximated (deviation values at stages LDE, CS and TE), making the identification process suitable for reverse engineering approaches.

Conclusions

The mechanical response of PC has been investigated by tensile tests at static and intermediate dynamic rates. The load constancy lasting over large part of the duration of the test is demonstrated to imply that the corresponding segment of the true curve evolves as an exponential law. Four distinct test phases are identified in static and dynamic tests, each phase being delimited by key stages where the deforming specimen assumes peculiar shapes. The different deforming modes of the four phases explained and described the very special straining process of polymers like PC: after the uniform straining and the successive strain localization, the special hardening response of PC makes the strain of the necked zone to propagate along the specimen axis. The analysis of the evolving shape of the specimen also allowed to derive a qualitative trend of the deviation between flow stress and true stress, useful for iterative reverse identification. An estimate of the dynamic stress amplification is made as the ratio of dynamic to static true stresses, delivering a stress increase around 25% at the nominal strain rate of 7 s^{-1} which reflects a remarkable rate sensitivity of the PC material at hand. Moreover, it was highlighted the presence of the necking-induced locking of the strain rate effect, already found by the authors for other materials in previous works.

Acknowledgements

This work has been partially funded by European Union (NextGeneration EU), through the MUR-PNRR project SAMOTHRACE (ECS00000022)

References

[1] Song P, Trived A R and Siviour C R 2021 Mechanical response of two different molecular weight polycarbonates at varying rates and temperatures ed F Gálvez Díaz-Rubio and D A Cendón Franco *EPJ Web Conf.* **250** 06013

[2] Wang H, Zhang Y, Zhou H and Huang Z 2017 Constitutive modelling of polycarbonate at low, moderate and high strain rates *IOP Conf. Ser.: Mater. Sci. Eng.* **187** 012023

[3] Wang H, Zhou H, Huang Z, Zhang Y and Zhao X 2017 Constitutive modeling of polycarbonate over a wide range of strain rates and temperatures *Mech Time-Depend Mater* **21** 97–117

[4] Song P, Trivedi A R and Siviour C R 2023 Mechanical response of four polycarbonates at a wide range of strain rates and temperatures *Polymer Testing* **121** 107986

[5] Ohara A and Kodama H 2019 Correlation between enthalpy relaxation and mechanical response on physical aging of polycarbonate in relation to the effect of molecular weight on ductile-brittle transition *Polymer* **181** 121720

[6] Bridgman P W 1964 Studies in Large Plastic Flow and Fracture: With Special Emphasis on the Effects of Hydrostatic Pressure *Harvard University Press*

[7] Mirone G 2004 Approximate Model of the Necking Behaviour and Application to the Void Growth Prediction *International Journal of Damage Mechanics* **13** 241–61

[8] Choung J M and Cho S R 2008 Study on true stress correction from tensile tests *J Mech Sci Technol* **22** 1039–51

[9] G'sell C, Aly-Helal N A and Jonas J J 1983 Effect of stress triaxiality on neck propagation during the tensile stretching of solid polymers *J Mater Sci* **18** 1731–42

[10] Boyce M C, Parks D M and Argon A S 1988 Large inelastic deformation of glassy polymers. part I: rate dependent constitutive model *Mechanics of Materials* **7** 15–33

[11] Muhammad S and Jar P-Y B 2013 Determining stress–strain relationship for necking in polymers based on macro deformation behavior *Finite Elements in Analysis and Design* **70–71** 36–43

[12] Sarva S and Boyce M 2007 Mechanics of polycarbonate during high-rate tension *JOMMS* **2** 1853–80

[13] Considère, Armand Annales des Ponts et Chaussées *Annales des Ponts et Chaussées* **9** 574

[14] Crist B and Metaxas C 2004 Neck propagation in polyethylene *J. Polym. Sci. B Polym. Phys.* **42** 2081–91

[15] Tzibula S, Lovinger Z and Rittel D 2018 Dynamic tension of ductile polymers: Experimentation and modelling *Mechanics of Materials* **123** 30–42

[16] Mirone G, Barbagallo R and Giudice F 2019 Locking of the strain rate effect in Hopkinson bar testing of a mild steel *International Journal of Impact Engineering* **130** 97–112

[17] Mirone G, Barbagallo R, Giudice F and Di Bella S 2020 Analysis and modelling of tensile and torsional behaviour at different strain rates of Ti6Al4V alloy additive manufactured by electron beam melting (EBM) *Materials Science and Engineering: A* **793** 139916

[18] Mirone G, Barbagallo R, Corallo D and Di Bella S 2016 Static and dynamic response of titanium alloy produced by electron beam melting *Procedia Structural Integrity* **2** 2355–66

IOP Conf. Series: Materials Science and Engineering 1306 (2024) 012007 doi:10.1088/1757-899X/1306/1/012007

Modeling of composite combat helmet subjected to blunt impacts

Edison Shehu, Loris Ferrari, Álvaro González Jiménez, Marco Giglio, Andrea Manes

Politecnico di Milano, Dipartimento di Meccanica, Via La Masa n.1, 20156, Milano, Italy

E-mail: edison.shehu@polimi.it

Abstract. Combat helmets are a complex system that is crucial for users in different adverse scenarios. Historically, the primary focus in the development of these items has been on compliance with high-velocity impact requirements, such as bullets or blast loads. In recent years, the investigation of blunt impact scenarios has gained momentum. Consequently, standards have been updated; however, they still lack a consolidated methodology for assessing helmet safety. A virtual testing approach is designed to replicate experimental tests. The main advantage of this model is its ability to assess loading conditions of interest and evaluate existing engineering solutions against new standards requirements.

1. Introduction

The aim of the research in the field of military helmets is to conjugate the reduction of weight together with the improvement of protective performances and the increase of comfort. The helmets adopted during the two World Wars were made of steel and particularly heavy to be carried. By the end of World War II, in order to face the growing lethality of modern weapons, the possibility of increasing the thickness and, consequentially, the weight of the existing steel helmets was discarded. The armies were pushed to look for more innovative solutions [1].The turning point in this research process was the introduction of the aramid fiber known by the commercial name of Kevlar® during the 1960s. Thanks to its mechanical properties, like high tensile strength and modulus, Kevlar® demonstrated to be suitable for bulletproof applications and was selected by the U.S. Government as the material for new generation helmets. With respect to ballistic impacts, blunt impacts are commonly associated with non-penetration, closed-head injuries, larger areas of impact and lower velocities and accelerations involved [2]. The severity of a blunt impact is strictly correlated to the levels of acceleration to which the head is subjected after the contact. For a long time, the relevance of blunt impact consequences was underestimated. Blast and ballistic threats were always taken in higher consideration by the military authorities [3]. An indication of this tendency is that ballistic regulations such as the STANAG 2920 [4] or the NIJ 0106.01 [5] set specific prescriptions for helmets subjected to high-impact velocities for the different kinds of ammunition. On the other hand, the standards for blunt impacts refer entirely to the civilian regulations used for motorbike helmets such as the ECE 22.06 [6] and the FMVSS 218 [7]. With reference to blunt impacts, the Advanced Combat Helmet (ACH) model was taken as a case study to investigate the impact absorption effectiveness of the composite shell combined with several padding systems. Indeed, the introduction of foams inside the helmet is one of the main novelty of the ACH. Many alternatives are feasible to obtain foams with good absorption properties for a high range of impact velocities [8]. Moss and King [9] compared several foam materials adopted in football helmets and in ACH. They obtained stress-strain curves at different strain rates through compression tests. They noticed also that the airtight bags in which the pads are sealed have a non-negligible effect on the response to an impact. For this reason, they suggested to consider an air material model to account for the additional damping effect. Terpsma et al. [10, 11] conducted hemispherical and oblique impact tests using a biofidelic model of the human head to investigate the brain tissue response. They found that the distribution of strains which is more critical is for an oblique impact scenario because of the rotational components of the acceleration. Franklyn and Laing [12] developed some quantitative analysis related to the vertical impacts on the internal roof of military vehicles with the aim of developing a new standard for minimum head clearance. With the aim of understanding the effects of rotational accelerations on the human brain, Begonia et al. [13] performed military helmet drop tests using two different dummy neck models. They used the collected acceleration data to calculate the different brain strains using the SIMon Finite Element brain model. Fitek et Meyer [14] tested different polymeric foams and configurations for the ACH padding system developing prototypes with the aim of investigating possible improvements for offering low-velocity impact protection in ambient and hot conditions. Kayhart et al. [15] organized a study on the feasibility of adopting an external blunt impact attenuator. The concept involves creating a detachable shell of rigid foam to be applied on the external surface of the ACH during the operations in which there is a higher probability of incurring in blunt impacts. Their tests showed how these external shells can effectively reduce peak accelerations. The design of a helmet capable of withstanding various types of impact loads is an active research area. No consolidated solution exists because different feasible approaches prioritize specific aspects while neglecting other requirements. The aim of this work is to replicate a high-fidelity numerical model of the experimental test required to certify helmets. Standard specifications evolve every few years by enforcing higher performance at each new release. Therefore, the ability to obtain results with a virtual test approach in advance is crucial to enhance the engineering design process.

This work is organized as follows. The Section *Methodology* is divided into two parts. Subsection *Testing method* introduces the framework used to evaluate traumatic brain injury (TBI) according to standards. While subsection *Model of the system* provides a description of the finite element model that is used to perform numerical blunt impact tests. The numerical model is then used to perform a numerical validation with data taken from literature in Section *Numerical validation*. Results for further numerical impacts are described in Section *Results*. Conclusions are drawn out from this work in the Section *Conclusions*.

2. Methodology

2.1. Testing method

The North Atlantic Treaty Organization (NATO) updated its regulation regarding the blunt impact evaluation on combat helmets in 2019 releasing the new AEP-2902 [16], to ensure good non-ballistic protection performances to the users of the Member States. The National Authorities are advised to perform blunt tests, adopting the methods and acceptance criteria most appropriate according to their national standard assessment [16]. The testing procedures prescribed by the AEP-2902 lean on several standards used to assess motorcycle helmet safety such as the European standard ECE 22.05 [17] and the American standard FMVSS 218 [7]. The absence of prescriptions to assess rotational kinematics for blunt impacts is considered one of the limits of the current AEP-2902. Several studies claim the relevance of rotational acceleration components in the damaging mechanisms of brain tissues, but the AEP-2902 lacks any indication. When this normative was released in 2019, there was not present an internationally accepted threshold and method for assessing rotational blunt impact injury risk for different helmet types. Such absence is thought to be corrected in a further release of AEP, based on ECE 22.06. The ability of a helmet to absorb an impact is measured by monitoring the acceleration of a headform, equipped with the helmet, as it is dropped in guided free fall onto a stationary steel anvil at a designated impact velocity [6]. Several metric indexes are available in the literature that are able to quantify a certain kind of TBI. Different criteria are able to describe different aspects of the head which is subjected to an impact. The guidelines followed for developing the numerical simulations present in this work are provided by the *G method* prescribed by the NATO AEP-2902, which refers to the European regulation only in terms of headform, anvils, test methodology and measurement tools. Among the most common criteria used to evaluate injuries, there are Peak Linear Acceleration (PLA) and Head Injury Criterion (HIC). The simplest parameter used for assessing head injuries is the PLA, which is the maximum linear acceleration value recorded by the measurement instruments placed at the center of mass, as reported in Equation 1. A relevant limitation of this criterion is that it does not consider the impact duration, which plays a role in the mechanisms of head injuries [18].

$$PLA = \max_{t} a(t) \tag{1}$$

According to the ECE R22.06 [6] standard, a PLA over 275g indicates serious injuries. The most popular and widely used head injury predictor index is the HIC, which is spread worldwide as a criterion used for designing protective measures to prevent severe brain injury. HIC is calculated considering the acceleration of the center of gravity according to Equation 2.

$$HIC = \max_{t_1,t_2} \left\{ (t_2 - t_1) \left[\frac{1}{t_2 - t_1} \int_{t_1}^{t_2} a(t)dt \right]^{2.5} \right\} \tag{2}$$

The time span (t_2 - t_1) is called the acceleration window and it is conventionally set to 15 ms or to 36 ms, depending on the impact condition. In the FMVSS 208 [7], a HIC of 1000 is prescribed as an injury threshold.

2.2. Model of the system

2.2.1. Helmet The reproduction of the geometry of ACH was pursued in the absence of access to the official drawings by a reverse engineering approach. The photogrammetry methodology was selected to replicate the ACH geometrical feature. It is a powerful technique that enables the determination of point positions, shape, and dimension proportions of an object by processing information extracted from a series of photographs. To replicate the geometry, proportions, and curvature of the ACH helmet through the photogrammetry process, an ABS replica of the ACH was procured. Typically, ACH helmets are commercially available in a plain color, this aesthetic feature could pose challenges for the photogrammetry process. In the presence of an exceedingly uniform surface, photogrammetry software algorithms may encounter difficulties in extracting points from images, potentially leading to inaccuracies in point cloud generation. The initial step involved applying an extremely thin layer of acrylic paint to the outer shell of the helmet. This was done to maximize surface opacification and mitigate undesirable light reflections that could disrupt the image processing workflow. Secondly, on the newly opaque surface, irregular patterns and points were created randomly. The results of this pre-processing phase are visible in Figure 1a and Figure 1b. The following phase involved alignment and feature extraction, during which the images were systematically scanned and compared with others within the batch to identify matches through specialized algorithms. The output of this alignment step is the sparse point cloud, as depicted in Figure 2a. Subsequently, the acquired tie points played a pivotal role in the next workflow step. These tie points were necessary in

Figure 1a. Frontal view

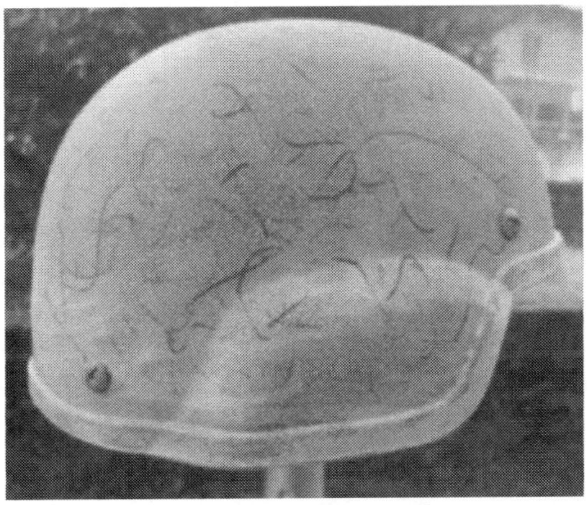

Figure 1b. Lateral view

calculating the camera positions and distortion parameters by solving systems of equations. The camera positions elucidated the precise point from which each image had been captured, while the distortion parameters elucidated the behavior of the camera during image acquisition. With both sets of information, an algorithm was employed to match pixels between pairs of images, culminating in the generation of the dense point cloud, illustrated in Figure 2b.

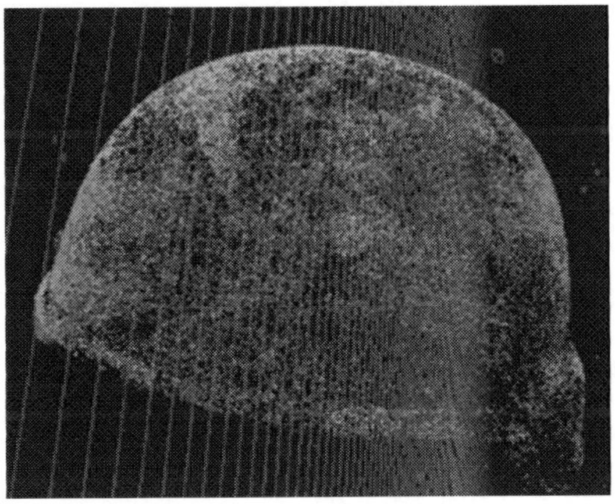

Figure 2a. Sparse point cloud

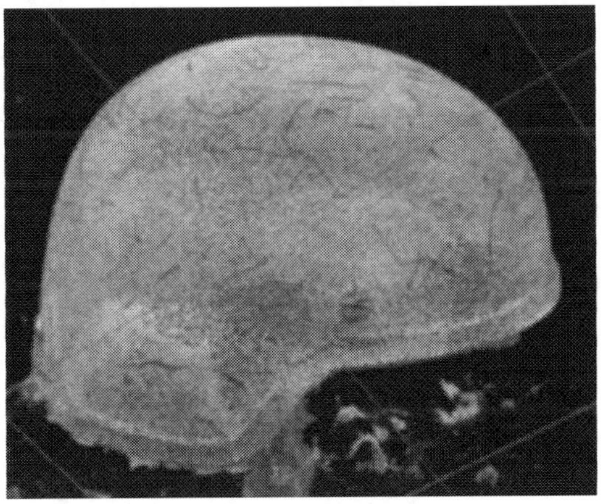

Figure 2b. Dense point cloud

This surface representation was subsequently exported and scaled to represent an ACH Size M helmet in accordance with the specifications. Picture of the final models of the helmet shell are represented in Figure 3a and Figure 3b.

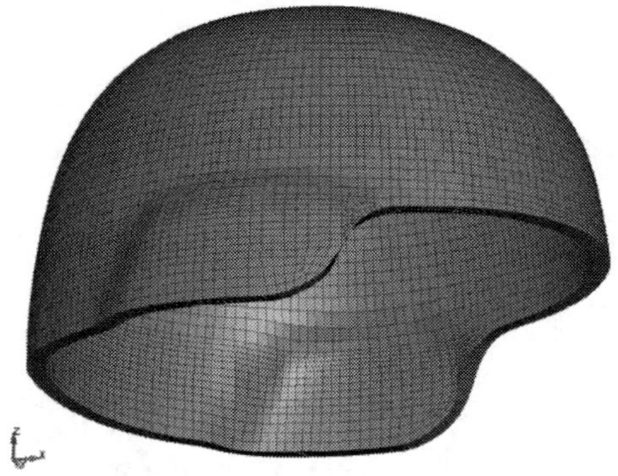

Figure 3a. Side view

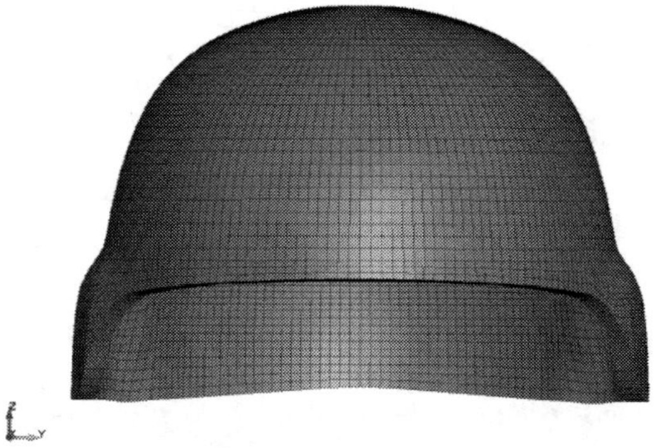

Figure 3b. Frontal view

The ACH models under investigation were fabricated using a composite construction method involving multiple layers of woven aramid fibers fabric embedded within a phenolic resin matrix. In order to represent the material behavior of the helmet within the LS-DYNA® FE environment, a total of twelve internal layers are defined. This approach differs from creating the entire helmet shell as a singular entity, as it provides the advantage of detecting potential delamination occurrences between the distinct material layers. Each individual layer comprises Kevlar® K129 fabric combined with a PVB-phenolic resin and possesses a thickness of 0.655 mm. This design ensures that the resultant ACH model, sized as M, attains the specified thickness of 7.86 mm [19]. The mechanical properties of the material are reported in Table 1.

Symbol	Variable Name	Value	Unit
ρ	Density	1230	kg/m^3
E_a	Elastic Modulus 1	22	GPa
E_b	Elastic Modulus 2	22	GPa
E_c	Elastic Modulus 3	9	GPa
v_{ba}	Poisson's ratio 21	0.25	–
v_{ca}	Poisson's ratio 31	0.33	–
v_{cb}	Poisson's ratio 32	0.33	–
G_{ab}	Shear modulus 12	0.77	GPa
G_{bc}	Shear modulus 23	2.715	GPa
G_{ca}	Shear modulus 31	2.715	GPa
S_{ba}	In-plane shear strength	77	MPa
S_{ca}	Transverse shear strength	1,086	MPa
S_{cb}	Transverse shear strength	1,086	MPa
X_c	Compressive strength 1	1200	MPa
Y_c	Compressive strength 2	1200	MPa
Z_c	Compressive strength 3	1200	MPa
X_t	Tensile strength 1	800	MPa
Y_t	Tensile strength 2	800	MPa
Z_t	Tensile strength 3	1200	MPa

Table 1. Properties of K129/PVB-phenolic [19]

In order to model the K129/PVB-phenolic material behavior of the helmet shell layers it was selected MAT_59 (MAT_COMPOSITE_FAILURE_MODEL) present in the LS-DYNA library [20]. MAT_059 is one of the possible constitutive failure models available for composite materials. More complex material laws, such as MAT_162, may be used in modeling composite materials.

Given that the focus of this study is the effects on the headform of blunt impacts, MAT_059 was evaluated as a valid compromise with a good level of accuracy and lower computational costs. No localized material failure is expected, indeed blunt impacts provoke a global response on the structures involved.

2.2.2. Padding system The ACH was the first model to adopt a foam-based padding system. The pads are contained inside a coverage, usually made of fabric, and they can be detached and substituted after degradation. Once inserted, they are held in place by a velcro-based system present on the internal shell of the helmet. Among the seven pads attached to the helmet shell, Figure 4a, one has a circular section while the others have trapezoidal profiles. The circular pad is commonly called *"Crown pad"*. It is positioned on the top of the head and it covers the largest internal area. Each pad is composed of two layers of foam with different densities and mechanical properties. The numerical model of a single side pad is represented in Figure 4b. The soft foam layer is needed to provide comfort to the user, while the harder layer is in charge of dissipating the high energies involved during an impact.

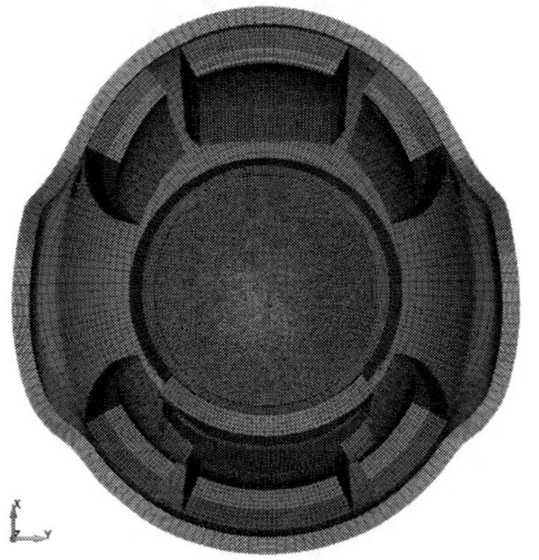

Figure 4a. Positioning of the padding system in ACH helmet

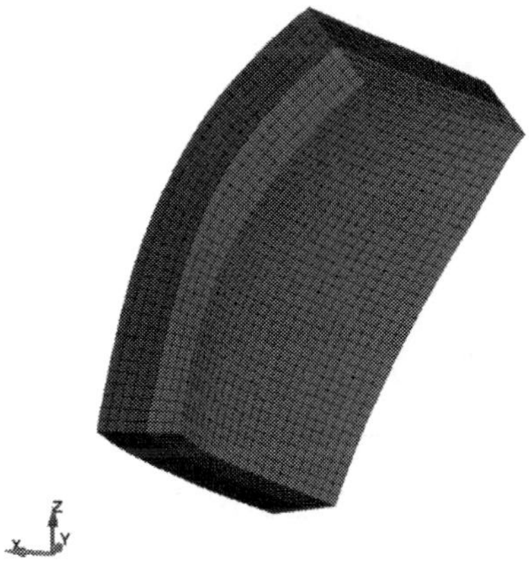

Figure 4b. Single pad composed of a soft (red) and hard pad (blue)

In this study, the foams are modeled by choosing MAT_57(MAT_LOW_DENSITY_FOAM) since this is the most common solution adopted in literature and more data are available. This material card requires the input of a stress-strain curve and the addition of other parameters for characterizing the damping and the unloading in order to model the behavior of highly compressible low-density foams [20]. The viscous coefficient (DAMP) is needed in order to model the damping effects of the material. The hysteretic unloading factor (HU) and the shape factor for unloading (SHAPE) describe the behavior of the unloading curve in terms of energy dissipation while the decay constant β is used to model creep in unloading [20]. In particular, data from [21] are used and reported in Figure 5a and Figure 5b.

IOP Conf. Series: Materials Science and Engineering 1306 (2024) 012007 doi:10.1088/1757-899X/1306/1/012007

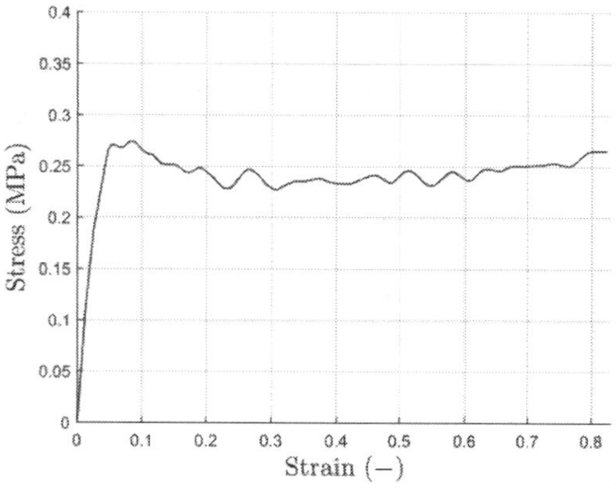

Figure 5a. Constitutive curve of a hard pad

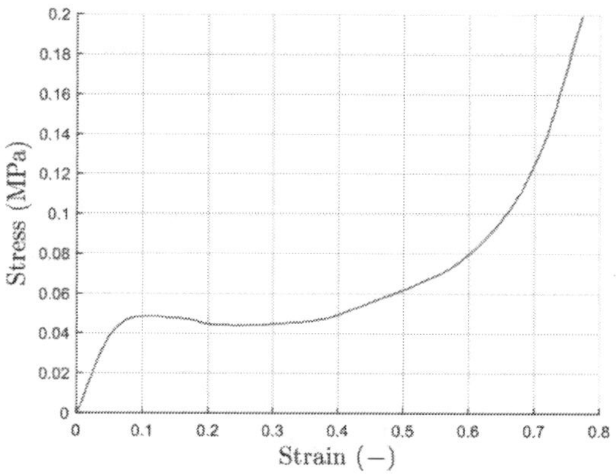

Figure 5b. Constitutive curve of a soft pad

According to [22], the usage of MAT_057 material card in the case in which the testing device used in the laboratory is not able to fully capture the foam behavior up to the desired strain level may be limiting. It is suggested that the curve can be extrapolated by using a 3rd order hyperbolic function with a good level of approximation. This extrapolation was performed in order to obtain a stress-strain curve with values up to strain equal to 1 for the soft pad.

Symbol	Variable Name	Value	Units
ρ	Density	63	kg / m^3
E	Young's Modulus in tension	8.4	MPa
HU	Hysteretic unloading factor	0.25	/
β	Creep decay constant in unloading	5	ms^{-1}
DAMP	Viscous coefficient for damping	0.5	/
SHAPE	Shape factor for unloading	5	/

Table 2. MAT_057 input parameters for the hard foam [21].

Symbol	Variable Name	Value	Units
ρ	Density	61	kg / m^3
E	Young's Modulus in tension	0.84	MPa
HU	Hysteretic unloading factor	0.25	/
β	Creep decay constant in unloading	5	ms^{-1}
DAMP	Viscous coefficient for damping	0.5	/
SHAPE	Shape factor for unloading	5	/

Table 3. MAT_057 input parameters for the soft foam [21].

One important remark to underline is that the stress-strain curves may be strain rate dependent, meaning that the curves show a certain variability according to the strain rate at which they are subjected. During quasi-static compression, the material shows low levels of stress, while, by increasing the strain rate, the foam materials demonstrate a much stiffer behavior [11]. The consequence is that using as input the stress-strain curve at an adequate strain rate for the phenomenon that is under investigation is of fundamental importance for obtaining reliable results.

2.2.3. Headform The surrogate of the human head was adopted in order to simulate the blunt impact testing procedures prescribed by the NATO AEP-2902. It was realized following the specific characteristics provided in the BS EN 960:2006 *"Headforms for use in the testing of protective helmets"* [23] as indicated by the European standard ECE 22.06 [6]. For each specific designation and headform circumference, the standard prescribes the spherical coordinates of the points which lie on the external surface of the headform with respect to the reference point R. After the insertion of the spherical coordinates for the headform J 575, it was possible to obtain the headform external surface as represented in Figure 6a.

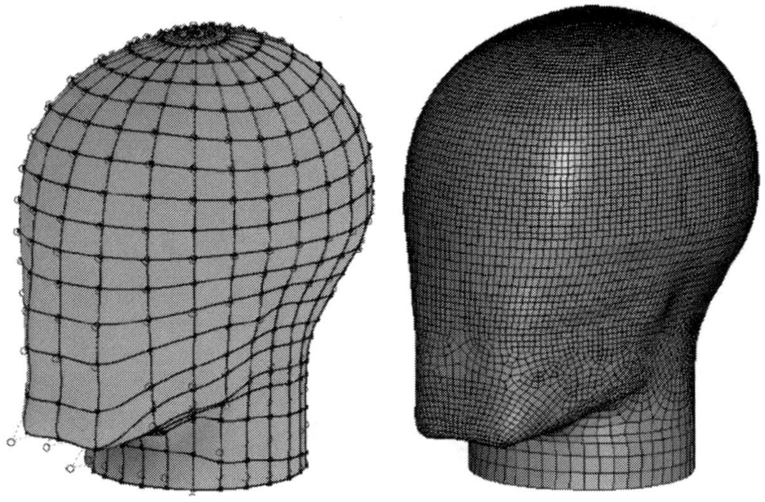

Figure 6. Headform J 575 outer surface

The headform is made of K1A magnesium, an alloy that is widely used for producing testing headforms. Since the standard prescribes that the headform should not deform during the impact testing, the material was modeled as rigid by adopting the MAT_20 (RIGID) material card. The Elastic modulus is 42000 MPa, the Poisson coefficient is 0.3 and the density is 1770 kg/m^3. In the center of gravity (CoG) of the headform, a virtual accelerometer is placed and all the quantities of interest are recorded. The values collected by these accelerometers, able to track linear and rotational accelerations, are the data used for calculating the injury metrics to be compared with the threshold values indicated by the standards. The choice of modelling the headform as a surface reduces the number of moving nodes present in the simulation, but alters the mass and the inertial properties of the headform that are calculated by the software algorithms. For this reason, the inertial properties have been defined with respect to the center of gravity (CoG) and the local reference system, by using the option PART_INERTIA present in LS-DYNA®. The inertial values reported in the PART_INERTIA card are taken from the ECE 22.06 [6] which indicates the mass and the moments of inertia with respect to the principal axis for the J 575 headform. In the ECE 22.06 regulation, it is reported that the maximum mass for the accelerometer positioned at the CoG of the headform should be 50 grams [6]. In order to account for this presence in the following simulations, to the mass value of 4.7 kg characterizing the headform, it was added a mass of 50 g.

IOP Conf. Series: Materials Science and Engineering 1306 (2024) 012007 doi:10.1088/1757-899X/1306/1/012007

Headform type	Circumference mm	Mass kg	I_{xx} cm^2	I_{yy} cm^2	I_{zz} cm^2
J	575	4.7	264.0	318.3	193.1

Table 4. Headform J inertial properties in principal directions [6]

2.2.4. Retention system The retention system plays a fundamental role in the right functioning of an ACH. It is also known as chinstrap and operating together with the padding system provides stability, comfort and balance [24]. The chinstrap used in this case study is made of a nylon webbing large 15.875 mm with a wicking fabric liner to be in contact with the skin [25]. The frontal laces and the neck laces are large 15.875 mm, 2 mm thick and were modeled using solid elements. The frontal laces adhere and slide on the headform surface when tensioned. Four circular-shaped components called strap plates represent the four attaching points where there is located the adhesive fixing the retention system on the helmet. They are constituted of shell elements and they are connected to the helmet through a CONTACT_TIED_SURFACE_TO_SURFACE since their position is fixed in time.

Figure 7. Retention system

Six beam elements are used to connect different parts together. They have been modeled using the MAT_71 (MAT_CABLE_DISCRETE_BEAM). This material card allows the definition of an initial load force F_0, set to 5 N, which can be used for simulating the action of tightening the straps, making the laces adherent to the headform. All these components have been modeled using the nylon material characteristics defined through the MAT_001 (MAT_ELASTIC) material card as reported by [3]. The Elastic modulus is 1000 MPa, the Poisson coefficient is 0.44 and the density is 1400 kg/m^3.

2.2.5. Anvil The anvils were modeled following the dimensions indications provided in the ECE 22.06 [6] and in the FMVSS 218 [7] standards. Since anvils should not deform during the testing procedure, all the anvils were modeled as rigid surfaces in order to reduce the number of elements present in the simulations. Steel is the material defined for all the anvils using the MAT_20 (MAT_RIGID) material card. The Elastic modulus is 210000 MPa, the Poisson coefficient is 0.3 and the density is 7890 kg/m^3. The two different types of anvils used in the study are represented in Figure 8a and Figure 8b.

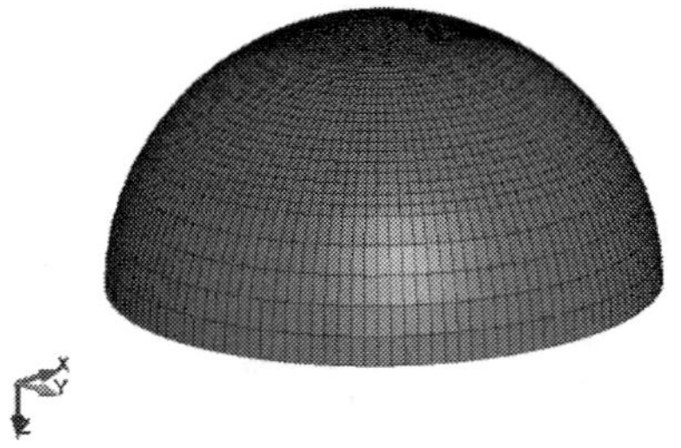

Figure 8a. Hemispherical anvil

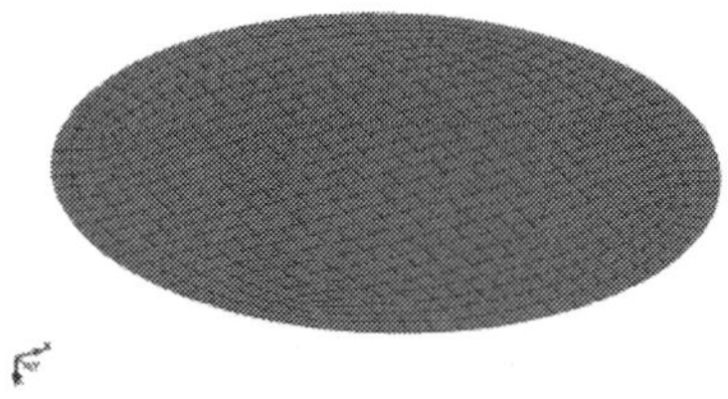

Figure 8b. Flat anvil

3. Numerical validation

In order to understand if the proposed model is coherent with the experimental results available in the literature, two impact validation tests have been settled. Both these blunt impacts were carried on the hemispherical anvil at the velocities of 3.05 m/s and 4.30 m/s. The initial setup is represented in Figure 9 where the anvil is fixed and the helmet drops at the specified velocity. The peak linear acceleration values obtained have been compared with some of the experimental test results at the crown site retrieved in the literature [9, 10, 14, 26, 27, 28, 29] as represented in Figure 10a and Figure 10b. The first column represents the values obtained from the current simulation. The numerical model gives values that are aligned with available data and are comparable with the scatter from different sources. These results are regarded as a satisfactory validation of the likelihood of the model when compared to experimental reality. Consequently, new impacts are conducted at points prescribed by the NATO AEP-2902 where a direct experimental reference is not present.

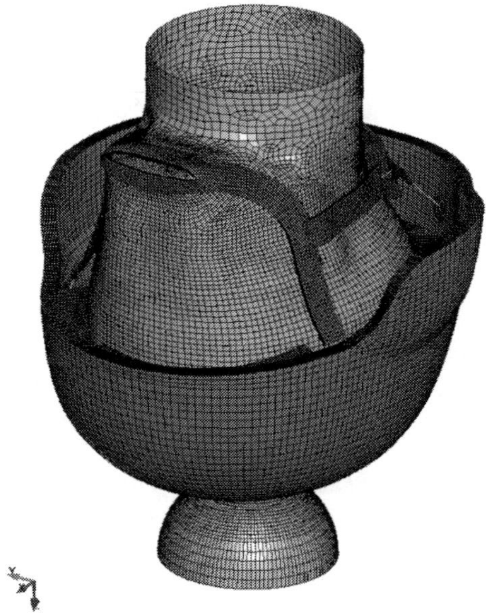

Figure 9. Crown impact on hemispherical anvil

IOP Conf. Series: Materials Science and Engineering 1306 (2024) 012007 doi:10.1088/1757-899X/1306/1/012007

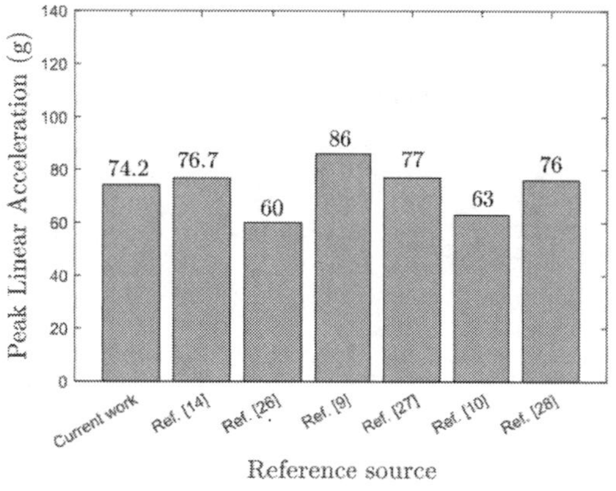

Figure 10a. PLA at 3.05 m/s

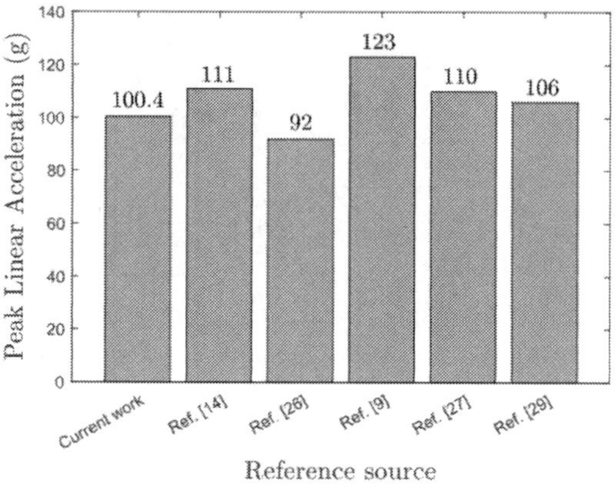

Figure 10b. PLA at 4.30 m/s

4. Results

The *Method G* of NATO AEP-2902 prescribe different site locations where to perform the impact. In particular, the crown, the front and the rear impact are simulated. The selected injury metrics are HIC and PLA, whose threshold values are set at 1000 and 275g, respectively. The velocity of impact is 4.40 m/s and a flat anvil is adopted. The three numerical cases are represented in Figure 11a, Figure 11b and Figure 11c at their initial condition before the contact with the anvil.

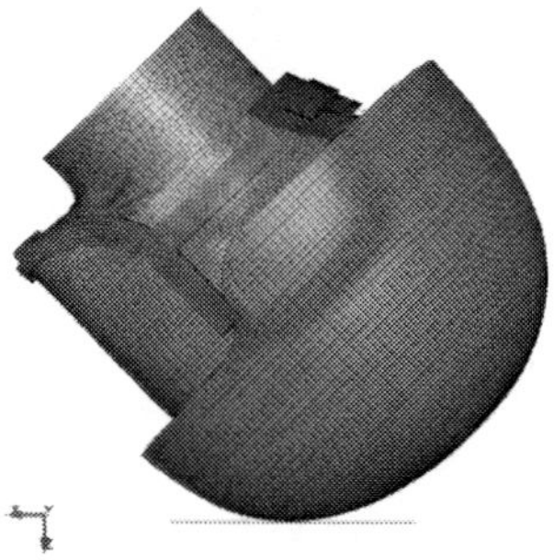

Figure 11a. Frontal impact

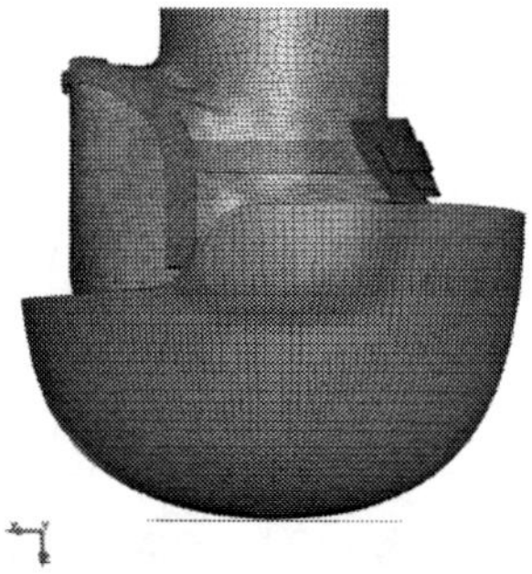

Figure 11b. Crown impact

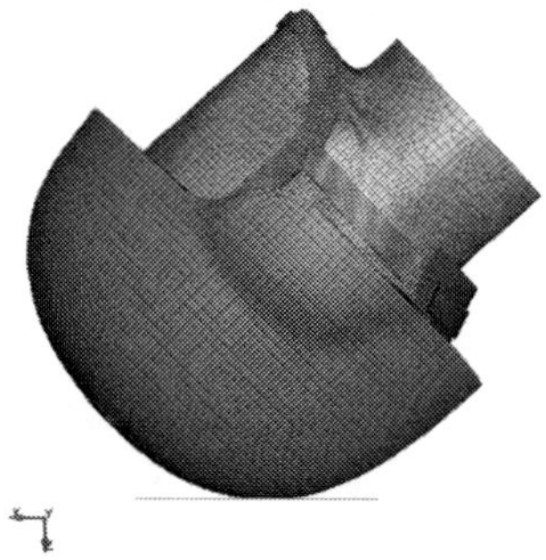

Figure 11c. Rear impact

All results reported in Figures 12a and 12b are compliant with the standards. Values of PLA and HIC are below the critical threshold for the model under investigation. The highest value of HIC is recorded by the impact performed in the rear site. Instead, the highest PLA is experienced during the crown impact. This result may be attributed to the fact that in crown impacts the helmet velocity direction is normal to the surface, thereby avoiding the decomposition of velocity and acceleration that could lead to potential rotations. On the other hand, rear and frontal impacts, which have a dominant component of accelerations that is not normal to the anvil when the impact occurs, may produce a significant rotational component, thus splitting the impact energy into translational and rotational contributions. This may lower the overall value of PLA, which measures only translational quantities.

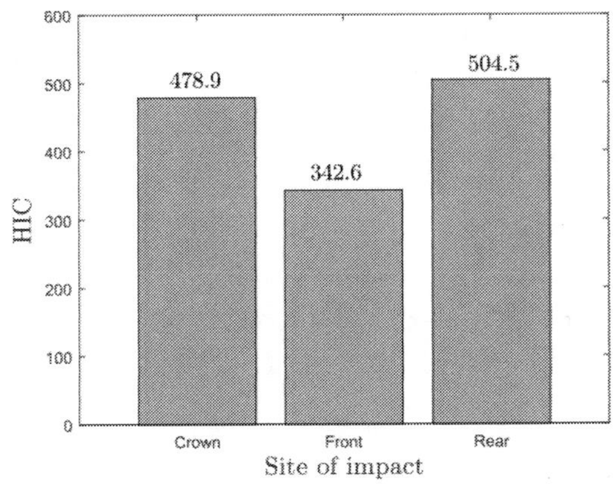

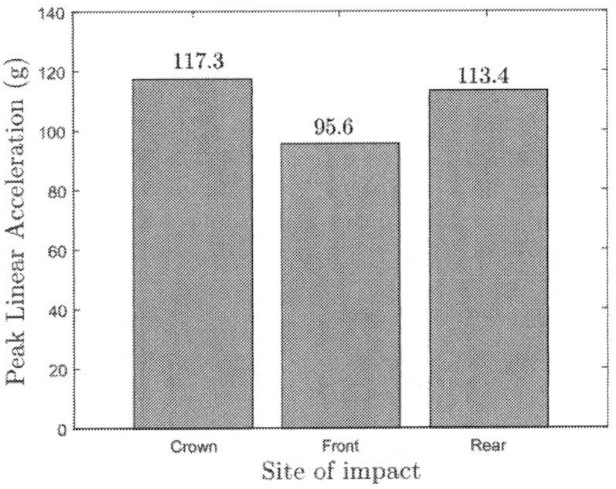

Figure 12a. HIC values for different impacts **Figure 12b.** PLA values for different impacts

5. Conclusions

The proposed work aims to define a virtual test approach to validate the results of the blunt impact on ACH. Firstly the geometry of the ACH shell was retrieved by using photogrammetry, and then a finite element model was built according to data retrieved from the literature. The accordance between the numerical and experimental values suggests that the model may be used to perform numerical tests before arranging the experimental equivalence. Such a model is able to provide acceleration histories that can be post-processed to derive injury metrics. The capability to analyze such a complex system with a reliable model can be exploited in the engineering design process to improve the performance of helmets or to anticipate possible future testing requirements before the release of future standard updates.

References

[1] Council N R 2014 *Review of Department of Defense Test Protocols for Combat Helmets* (Washington, DC: The National Academies Press) ISBN 978-0-309-29866-7
[2] Young L R, Rule G, Bocchieri R T, Walilko T J, Burns J M and Ling G S F 2015 When physics meets biology: Low and high-velocity penetration, blunt impact, and blast injuries to the brain vol 6 URL `https://doi.org/10.3389/fneur.2015.00089`
[3] Moure-Guardiola C, Rubio I, Antona-Makoshi J, Olmedo Á, Loya J A and Rodríguez-Millán M 2020 *Applied Sciences* URL `https://doi.org/10.3390/app10238470`
[4] 2016 "procedures for the evaluation and classification of personal armour - bullet and frafmentation threats" Standard NATO Standardization Office (NSO)
[5] 1981 Nij standard for ballistic helmets nij standard-0106.01 Standard National Institute of Justice
[6] 2021 Uniform provisions concerning the approval of protective helmets and their visors for drivers and passengers of motor cycles and mopeds Standard United Nations
[7] 2011 Nhtsa. fmvss 218 "laboratory test procedure for motorcycle helmets". Standard FMVSS
[8] Team wendy. innovate, 2023. URL `https://www.teamwendy.com/about/science-and-technology/innovate`
[9] Moss W C, King M J and Blackman E G 2012 Towards reducing impact-induced brain injury: lessons from a computational study of army and football helmet pads vol 17 pp 1173 – 1184 URL `https://doi.org/10.1080/10255842.2012.739162`
[10] Terpsma R J and Hovey C B 2020 Blunt impact brain injury using cellular injury criterion Tech. rep. URL `https://www.sandia.gov/app/uploads/sites/147/2022/01/Terpsma_2020_SAND2020_11444.pdf`
[11] Terpsma R J, Carlsen R W, Szalkowski R, Malave S, Fawzi A L, Franck C and Hovey C B 2021 Head impact modeling to support a rotational combat helmet drop test. URL `https://doi.org/10.1093/milmed/usab374`
[12] Franklyn M and Laing S 2016 *Traffic Injury Prevention* **17** 750 – 757 URL `https://doi.org/10.1080/15389588.2016.1146946`

[13] Begonia M, Rooks T F, Pintar F A and Yoganandan N 2019 *Military medicine* **184 Suppl 1** 237–244 URL `https://doi.org/10.1093/milmed/usy282`

[14] Fitek J and Meyer E 2013 Design of a helmet liner for improved low velocity impact protection URL `https://apps.dtic.mil/dtic/tr/fulltext/u2/a578032.pdf`

[15] Kayhart T J and Hewitt C 2022 Feasibility of an impact shell for combat helmets Tech. rep. URL `https://apps.dtic.mil/sti/pdfs/AD1167588.pdf`

[16] (NSO) N S O 2019 Nato standard aep-2902 non-ballistic test methods and evaluation criteria for combat helmets Tech. rep. NATO Standardization Office (NSO)

[17] 2000 Uniform provisions concerning the approval of protective helmets and their visors for drivers and passengers of motor cycles and mopeds Standard United Nations

[18] Fernandes F A O and de Sousa R A 2015 *Proceedings of the Institution of Mechanical Engineers, Part H: Journal of Engineering in Medicine* **229** 592 – 608 URL `https://doi.org/10.1177/0954411915592906`

[19] Li Y, Li X and Gao X L 2015 Modeling of advanced combat helmet under ballistic impact vol 82 p 111004 URL `https://doi.org/10.1115/1.4031095`

[20] (LST) L S T 2020 Ls-dyna keyword user's manual, volume ii, material models Tech. rep. LIVERMORE SOFTWARE TECHNOLOGY (LST)

[21] Zhang L, Makwana R and Sharma S 2013 Brain response to primary blast wave using validated finite element models of human head and advanced combat helmet vol 4 URL `https://doi.org/10.3389/fneur.2013.00088`

[22] Sambamoorthy B and Halder T 2001 Characterization and component level correlation of energy absorbing (ea) polyurethane foams (pu) using ls-dyna URL `https://www.dynalook.com/conferences/european-conf-2001/3.pdf`

[23] 2021 British standards. bs en 960:2006 "headforms for use in the testing of protective helmets" Tech. rep.

[24] of the Army D 2015 Technical manual operator manual for advanced combat helmet (ach) Tech. rep. Department of the Army

[25] Wendy T 2022 Team wendy brochure - standard chinstrap Tech. rep. Team Wendy

[26] Neice R and Plaisted T 2020 Evaluation of a combat helmet under combined translational and rotational impact loading Tech. rep. URL `https://apps.dtic.mil/sti/pdfs/AD1120852.pdf`

[27] Hopping J A, Bass C, Shridharani J, Perritt C, Haynes N and Masters K 2010 Methodologies for blunt trauma assessment in military helmets Tech. rep. URL `https://apps.dtic.mil/sti/tr/pdf/ADA585970.pdf`

[28] Vasquez K B, Logsdon K P, Dorman D B and Chancey V C 2011 Combat helmet-headform coupling characterized from blunt impact events URL `https://doi.org/10.1115/IMECE2011-64213`

[29] Petrone N, Candiotto G, Marzella E, Uriati F, Carraro G, Bäckström M and Koptyug A V 2019 *Journal of science and medicine in sport* URL `https://doi.org/10.1016/j.jsams.2019.05.015`

IOP Conf. Series: Materials Science and Engineering 1306 (2024) 012008 doi:10.1088/1757-899X/1306/1/012008

Taguchi analysis on finite element results to identify the most influential parameters on stresses after impact-damage

E V Arcieri[1] and S Baragetti[1]

[1]Department of Management, Information and Production Engineering, University of Bergamo, Viale Marconi 5, Dalmine 24044, Italy

emanuelevincenzo.arcieri@unibg.it

Abstract. Impact-damage can reduce the fatigue life of mechanical components as a consequence of premature crack growth due to the occurrence of high residual stresses. In this work, finite element analyses were conducted to evaluate the residual stresses in a 7075-T6 aluminium hourglass specimen for various impact cases. The Taguchi method was applied to the finite element results to identify which parameters most influence the residual stresses in the specimen. The outcomes showed that the impact speed is the most influential factor for axial stresses, followed by the material and the size of the impacting object. The factor that most affects the von Mises stresses is the object size followed by its material and the impact speed.

1. Introduction

Impact-damage is one of the possible causes of failure of mechanical components, especially in the aeronautical sector [1]. For this reason, this phenomenon represents a hot topic for the scientific community.

Ding et al. [2] identified the occurrence of microcracks and residual stresses as the principal causes of reduction in fatigue strength of components subjected to impact-damage. According to Peters and Ritchie [3], stress risers in the damaged area induce early failure of components hit by foreign bodies. Boyce et al. [4] performed numerical calculations to determine the residual stresses and damage induced by a body impacting a plate. The results showed that for low-speed impacts quasi-static numerical analyses ignoring material strain-rate sensitivity, wave, and inertia effects are adequate, but dynamic simulations are required for severe impact loading. According to Martinez and colleagues [5], the principal reasons for failure in components with impact-damage are post-impact stress risers and residual stresses. Nowell et al. [6] found that the depth of the impact-induced dent has a significant impact on fatigue strength. The consequences of the notch presence are nonetheless mitigated by a compressive stress condition seen near its root. In a numerical analysis of the damage resulting from a ball striking the edge of a thin plate, Chen [7] found that a static stress concentration factor smaller than 1 occurs at the bulge tip while the concentration factor was higher than 1 at the base of the impact-generated dent. In order to assess the impact of residual stresses, Ruschau et al. [8] conducted a stress relief annealing after impact-damage. The stress relief was unsuccessful in improving the fatigue resistance of specimens damaged at high speeds because of the resulting material loss.

In summary, residual stresses significantly influence the fatigue behavior of impact-damaged components, as also demonstrated in [9,10] for other mechanical situations. This study examines the residual stress distribution induced by the impact-damage in an hourglass specimen made of 7075-T6 aluminium alloy, which is a material commonly used for manufacturing aeronautical components. The

IOP Conf. Series: Materials Science and Engineering 1306 (2024) 012008 doi:10.1088/1757-899X/1306/1/012008

post-impact stresses are modelled with the Abaqus Explicit finite element code. Taguchi analysis was conducted on the finite element results in order to quickly and efficiently identify the parameters that most influence the residual stresses in the specimen.

2. Finite element model

The Abaqus Explicit finite element code [11] was used to simulate the impact of the ball on the hourglass specimen, as is commonly done for impact problems, which involve geometric and material nonlinearities [12-16].

The model for the impact simulation is shown in Figure 1 [17-19]. The 7075-T6 specimen was modeled using general purpose linear brick elements with reduced integration and an overall mesh size of 0.25 mm in the impact area. The 7075-T6 specimen was considered to have an isotropic elastic perfectly plastic material behavior, with the properties shown in the figure. Linear quadrilateral rigid elements were used to model the ball, with an overall mesh size of 0.25 mm. The ball inertial properties were assigned to its center, which was positioned on the same plane of the minimum cross-section of the hourglass specimen. The impact ball materials implemented in the finite element analyses are steel and ceramic, with the properties reported in the figure [20]. The tested diameters D are 5 and 7 mm.

A space of 0.1 mm was left between the specimen outer surface and the ball outer surface. All analyses were run with a friction coefficient of 0.6 [21] between the ball and the hourglass specimen. The surface nodes in the areas of the hourglass specimen marked in Figure 1 were fixed. Normal and inclined impact were analyzed, with impact angles α=0° or α=20° and β =0° or β=20°. The angle α is measured on the plane xz and β on the plane yz (Figure 1). The tested initial ball speeds v are 80 m/s or 120 m/s. The stress state used for the Taguchi analysis is that obtained after a simulated time of 7 ms from the beginning of the movement of the sphere, when the stresses can be considered constant in the specimen over time.

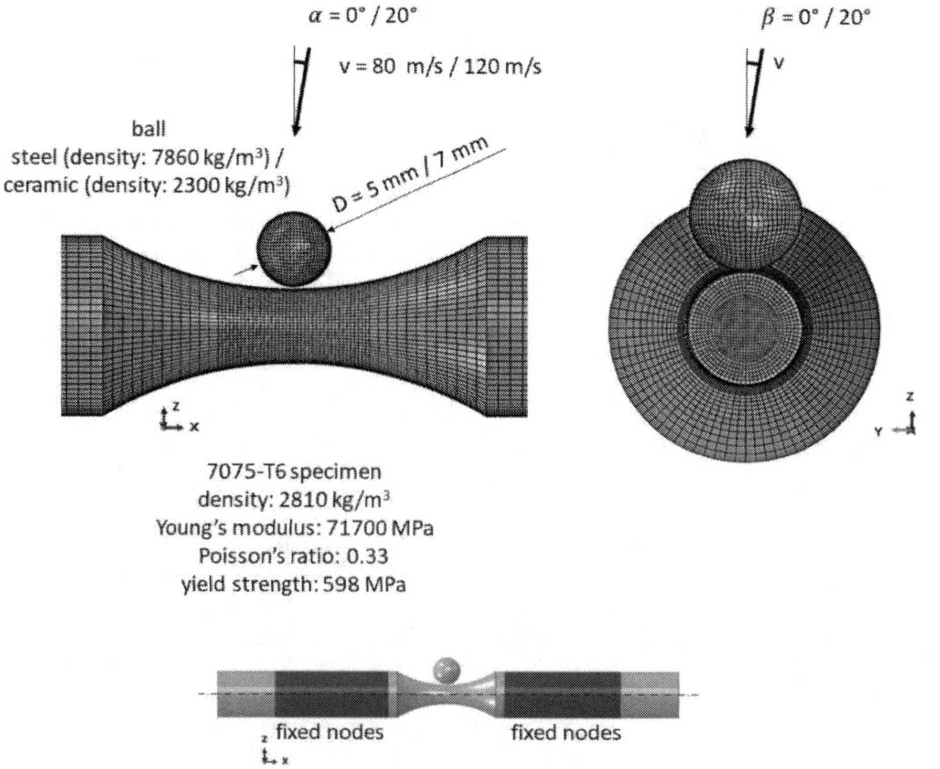

Figure 1. Finite element model [17-19].

3. Taguchi analysis on axial stresses

The most significant factors for impact-damage were identified thanks to the Taguchi analysis [22]. Fatigue cracks may propagate from the locations of high tensile stresses and therefore the Taguchi method was used to determine which factors most generate high residual stresses in the specimen. The Taguchi L8(2^7) array was used, which requires only eight runs to examine the influence of seven factors with two levels. Table 1 contains the array including the parameters and their corresponding levels [17,20]. The impact speed, v, the ball material, Material, the ball diameter, D, the impact angles, α and β, are the elements taken into consideration in the study. It was possible to check if all the parameters that significantly influence the residual stresses were considered by introducing the empty columns e1 and e2.

Table 1. Taguchi L8(2^7) array employed for the study [17,20].

Run	v (m/s)	Material	D (mm)	α (°)	β (°)	e1	e2
1	80	Steel	5	0	0	1	1
2	80	Steel	5	20	20	2	2
3	80	Ceramic	7	0	0	2	2
4	80	Ceramic	7	20	20	1	1
5	120	Steel	7	0	20	1	2
6	120	Steel	7	20	0	2	1
7	120	Ceramic	5	0	20	2	1
8	120	Ceramic	5	20	0	1	2

After impact, it was supposed that the specimen was exposed to axial or bending cyclic loading. Since both loading conditions produce a variable axial stress distribution, the axial residual stresses were analyzed [17].

Figure 2 shows the distributions of the axial stress, S11, obtained at the end of the eight simulations. The Taguchi analysis was conducted on the maximum axial stress in the specimen. Table 2 is the 'Response table' and was constructed with the results of Figure 2 according to the methodology reported in [22]. The column Y of the table reports the maximum axial residual stress in each run. Low tensile residual stresses are preferable from a fatigue point of view. For this reason, the analysis was conducted according to the modality 'smaller is better' and the mean square deviation MSD was calculated as $MSD=\Sigma Y^2/n$, with the number of simulations per run, n, equal to 1 due to the deterministic nature of the results. The subsequent steps in the Taguchi method consist in the calculation of $S/N=-10\log(MSD)$, of Y' and of S/N', where S/N is the signal-to-noise ratio, Y' is the average of the stresses obtained in the runs in which the factor x is at level y and S/N' is the average of the S/N values obtained when a considered factor x is at a considered level y. The quantity ΔS/N' represents the difference between the S/N' calculations for the two levels of the parameter taken into account. The significance of a parameter is higher than that of another if its ΔS/N' value is greater [22]. The impact velocity v (ΔS/N'=2.804) is the most influential factor, followed by the material (ΔS/N'=2.723) and the diameter D (ΔS/N'=2.630) of the ball. The values of ΔS/N' for the other factors are very small compared to those of v, Material and D. Therefore, the influence of the other parameters on the axial stresses in the specimen is very low. The analysis of the empty columns reveals that no significant factors were neglected.

The level of the important parameters which allows to minimize the stresses in the specimens corresponds to the highest S/N' [22]. Low impact speed, low material density and small size of the impacting objects induce low axial stresses.

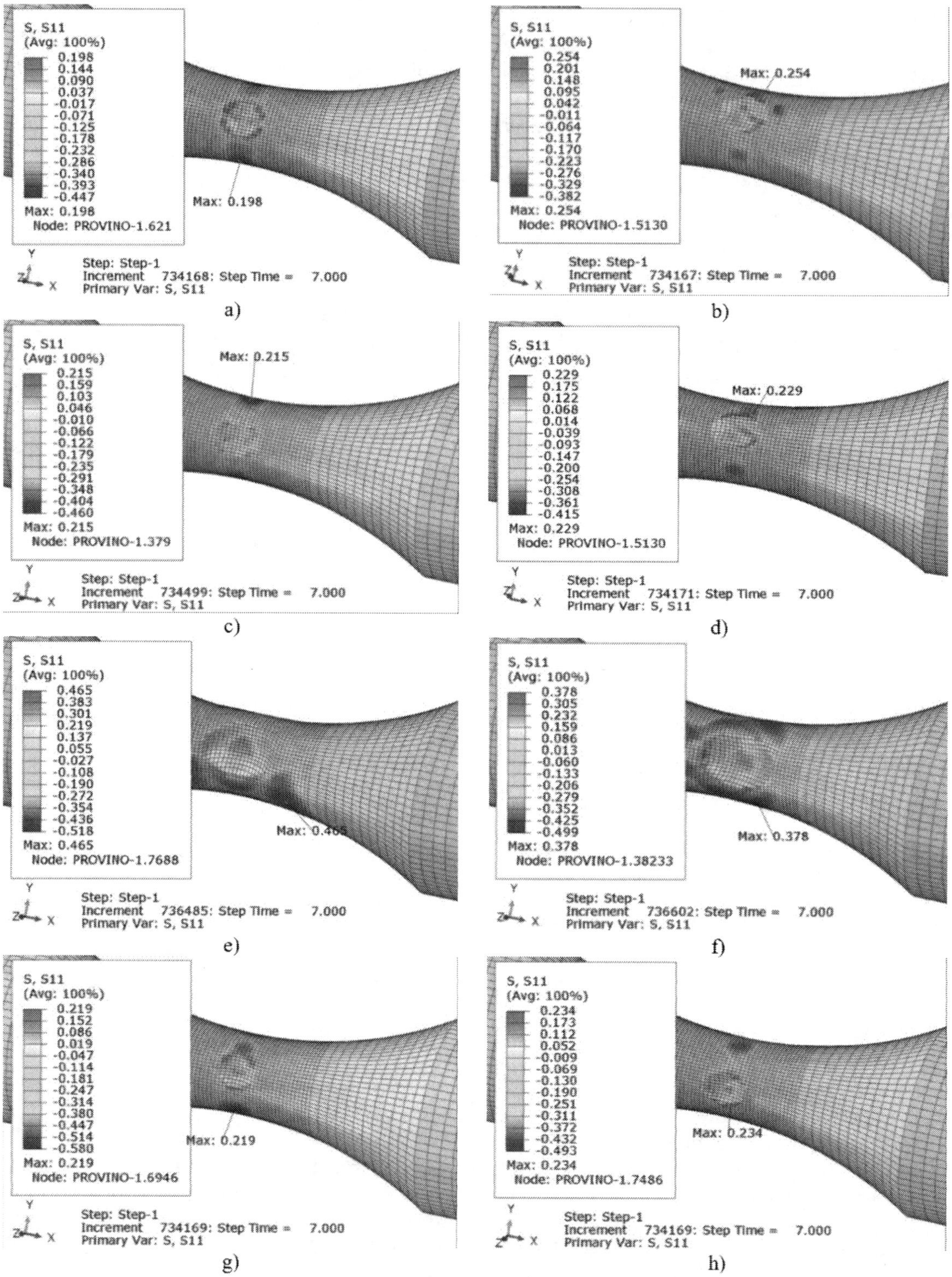

Figure 2. Axial stresses (GPa): a) Run1, b) Run 2, c) Run 3, d) Run 4, e) Run 5, f) Run 6, g) Run 7, h) Run 8 [17].

Table 2. Response table for maximum axial stress [17].

Run	Y	MSD	S/N	Factor	Level	Y'	S/N'	ΔS/N'
1	198	39204	-45.933	v (m/s)	80 m/s	224	-46.969	2.804
2	254	64516	-48.097		120 m/s	324	-49.773	
3	215	46225	-46.649	Material	steel	324	-49.732	2.723
4	229	52441	-47.197		ceramic	224	-47.010	
5	465	216225	-53.349	D (mm)	5	226	-47.056	2.630
6	378	142884	-51.550		7	322	-49.686	
7	219	47961	-46.809	α (°)	0	274	-48.185	0.372
8	234	54756	-47.384		20	274	-48.557	
				β (°)	0	256	-47.879	0.984
					20	292	-48.863	
				e1	1	282	-48.466	0.190
					2	267	-48.276	
				e2	1	256	-47.872	0.998
					2	292	-48.870	

4. Taguchi analysis on von Mises stresses

In order to evaluate the complexity of the stress field induced by the impact, the Taguchi method was also applied to identify the factors that mostly influence von Mises stresses in the specimen.

Figure 3 shows the von Mises stress distributions at the end of the eight runs. The Taguchi analysis was conducted on the maximum axial von Mises stress in the specimen. Table 3 is the 'Response table' for maximum von Mises stress. Column Y of the table reports the maximum von Mises stress after each simulated impact. The Taguchi analysis was again conducted according to the modality 'smaller is better'. Even if all the values of ΔS/N' are very small, it can be stated that the impact parameter which most influences the von Mises stresses in the specimen is the diameter D of the ball (ΔS/N'=0.799), followed by the material of the impacting object (ΔS/N'=0.730) and the impact speed v (ΔS/N'=0.572). The ΔS/N' values of the other factors are smaller and therefore their influence on the von Mises stresses is very low. The analysis of the empty columns points out that no significant factors were ignored.

Looking at the S/N' values, low impact speed, low material density and small size of the impacting objects induce low von Mises stresses.

5. Discussion

In both the analyses on axial and von Mises stress, it was found that the diameter and the material of the ball and the impact speed are the most influential parameters and that low values of these quantities induce low residual stresses. The density and size of the impacting object influence its mass, which multiplied by the impact speed gives its initial momentum. A high momentum means a high impact force and therefore high deformations and stresses in the impacted object. However, the order of parameter importance is different. Impact velocity is the most significant factor for axial stresses, followed by ball material and diameter. The impact parameter that most influences the von Mises stresses in the specimen is the diameter of the ball followed by its material and the impact speed.

IOP Conf. Series: Materials Science and Engineering 1306 (2024) 012008 doi:10.1088/1757-899X/1306/1/012008

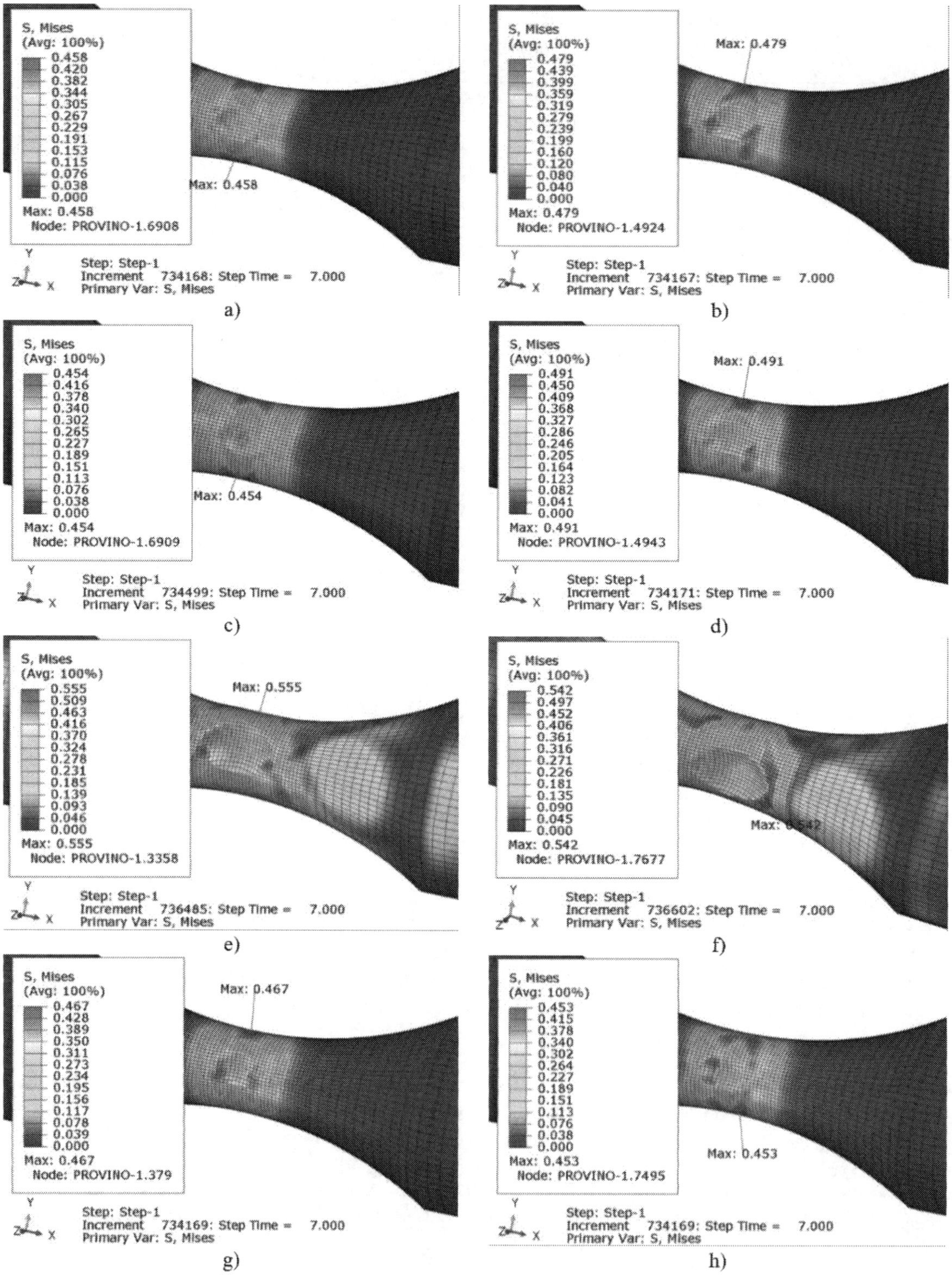

Figure 3. Von Mises stresses (GPa): a) Run1, b) Run 2, c) Run 3, d) Run 4, e) Run 5, f) Run 6, g) Run 7, h) Run 8.

Table 3. Response table for maximum von Mises stress.

Run	Y	MSD	S/N	Factor	Level	Y'	S/N'	ΔS/N'
1	458	209764	-53.217	v (m/s)	80 m/s	471	-53.447	0.572
2	479	229441	-53.607		120 m/s	504	-54.019	
3	454	206116	-53.141	Material	steel	509	-54.097	0.730
4	491	241081	-53.822		ceramic	466	-53.368	
5	555	308025	-54.886	D (mm)	5	464	-53.333	0.799
6	542	293764	-54.680		7	511	-54.132	
7	467	218089	-53.386	α (°)	0	484	-53.658	0.150
8	453	205209	-53.122		20	491	-53.808	
				β (°)	0	477	-53.540	0.385
					20	498	-53.925	
				e1	1	489	-53.762	0.058
					2	486	-53.704	
				e2	1	490	-53.776	0.087
					2	485	-53.689	

The maximum von Mises stress values obtained in the eight runs are approximately twice the maximum axial stress values. This means that the axial stress distribution could not fully describe the complex stress state in the specimen. This aspect would influence the fatigue behavior of the specimen.

The maximum stress values were analyzed without considering the position in which they occurred. Under fatigue loading, stress concentrations due to impact can induce high total stresses in other positions in the specimen and therefore the contribution of stress risers should be evaluated.

6. Conclusions

In this work, the Taguchi method was applied to the finite element results to identify the parameters that most influence the residual stresses in a 7075-T6 specimen subjected to impact-damage. The impact velocity is the factor that most influences axial stresses, followed by the material and the size of the impacting object. The factor that most influences the von Mises stresses in the specimen is the diameter of the object, followed by its material and the impact velocity. The comparison between the von Mises stress and the axial stresses obtained in the eight runs leads to the conclusion that axial stresses could not fully describe the complex stress state in the specimen after impact-damage. Future developments could include the creation of a theoretical model that allows to predict the shape of the damage after impact and the associated stress distribution as done in [23] for another mechanical problem.

References

[1] Nicholas T 2006 Foreign Object Damage *High Cycle Fatigue*, Elsevier Science Ltd (Oxford) pp x–xiv
[2] Ding J, Hall R F, Byrne J and Tong J 2007 *Int. J. Fatigue* **29** 1339
[3] Peters J O and Ritchie R O 2000 *Eng. Fract. Mech.* **67** 193
[4] Boyce B L, Chen X, Hutchinson J W and Ritchie R O 2001 *Mech. Mater.* **33** 44
[5] Martinez C M, Eylon D, Nicholas T, Thompson S R, Ruschau J J, Birkbeck J and Porter W J 2002 *Mat. Sci. Eng.* **A325** 465
[6] Nowell D, Duó P and Stewart I F 2003 *Int. J. Fatigue* **25** 963
[7] Chen X 2005 *Mech. Mater.* **37** 447

IOP Conf. Series: Materials Science and Engineering 1306 (2024) 012008 doi:10.1088/1757-899X/1306/1/012008

[8] Ruschau J, Thompson S R and Nicholas T 2003 *Int. J. Fatigue* **25** 955
[9] Baragetti S and Tordini F 2007 *Int. J. Fatigue* **29** 1832
[10] Mlikota M, Schmauder S, Dogahe K and Božić Ž 2021 *Procedia Struct. Integr.* **31** 3
[11] ABAQUS, ABAQUS Documentation, Dassault Systèmes, Providence, RI, USA, 2017
[12] Baragetti S, Guagliano M and Vergani L 2000 *Int. J. Mater. Prod. Technol.* **15** 91
[13] Arcieri E V, Baragetti S, Fustinoni M, Lanzini S and Papalia R 2018 *Procedia Struct. Integr.* **8** 212
[14] Baragetti S and Arcieri E V 2019 *Procedia Struct. Integr.* **24** 91
[15] Wang W and Shi J 2013 *Int. J. Impact Eng.* **60** 67
[16] Baragetti S and Arcieri E V 2020 *Eng. Fai.l Anal.* **113** 104564
[17] Arcieri E V, Baragetti S and Božić Ž 2021 *Procedia Struct. Integr.* **31** 22
[18] Arcieri E V, Baragetti S and Božić Ž 2022 *Eng. Fail. Anal.* **138** 106380
[19] Arcieri E V, Baragetti S, Božić Ž 2023 *Procedia Struct. Integr.* **46** 24
[20] Baragetti S 1997 *Int. J. Mat. Prod. Tech.* **12** 83
[21] Javadi M and Tajadri M 2013 *Tribol. Ind.* **35** 286
[22] Condra L W 1993 *Reliability improvement with Design of Experiments* Marcel Dekker Inc (New York)
[23] Baragetti S 2006 *Meccanica* **41** 443

IOP Conf. Series: Materials Science and Engineering 1306 (2024) 012009 doi:10.1088/1757-899X/1306/1/012009

Analytical computation of stress intensity factor for multi-physics problems

Francesca Pistorio, Davide Clerici

Department of Mechanical and Aerospace Engineering, Politecnico di Torino, Corso duca degli Abruzzi 24, 10129 Torino, Italy

E-mail: davide.clerici@polito.it

Abstract. This work presents a methodology for the analytical calculation of the stress intensity factor when the stress distribution on the crack surfaces is non-homogeneous. At first, a polynomial function is used to express the non-homogenous stress distribution. Subsequently, the principle of superposition of effects is applied, and the stress intensity factor is computed by multiplying each polynomial term by its respective geometric factor. Finite element fracture model is used to compute the geometric factor of the single polynomial grade. To explain the method, a spherical body is considered, with central and superficial cracks. Each geometric factor depends on a normalized geometrical parameter (the ratio between the crack length and sphere radius). The proposed methodology is applied to determine the stress intensity factor in the case of a crack driving force caused by diffusive fields, such as the concentration gradient in particles of electrodes active material in lithium-ion batteries. The methodology allows to speed up the fracture computation, then it is used to give electrode design guidelines to limit the fracture likeliness and mechanical degradation in lithium-ion batteries, as well as it is the basis for the development of algorithms assessing the capacity loss and the remaining useful life of lithium-ion batteries in real-time.

1. Introduction

Fracture is one of the most common failure modes in several engineering applications. Indeed, cracks propagation and fatigue phenomena occur well below the yield strength of the material and may lead to the failure of the component.

It is known that cracks initiate and propagate from geometrical discontinuities in structures under loading, such as micro-defects, edges, holes, and notches, acting as strain and stress raisers [1, 2]. However, the crack propagation size and the propagation rate depend on the geometry of the discontinuity and on the loading condition.

Different parameters can be used to assess the level of fracture, such as the energy release rate G, the J-integral, or the stress intensity factor (SIF) [3, 4]. Among these, the SIF is a well-known parameter assessing the singularity of the stress field near the crack region, according to linear elastic fracture mechanics (LEFM) [5]. Fracture criteria describing crack propagation and material failure are usually expressed as a function of SIF both for brittle [3] and fatigue fracture [6]. For this reason, the accurate evaluation of SIF is of great importance.

Several analytical expressions for the SIF are available in the literature [7, 8, 9, 10]. However, finite element method (FEM) is often preferred, despite the ease of analytical computation, because analytical solutions are only applicable to a limited number of standard cases.

In recent times, the fracture mechanics theory has been extended to fields beyond purely structural applications, such as the field of micro electro-mechanical systems [11, 12, 13] and lithium-ion batteries (LIBs) [14, 15, 16].

Nowadays, LIBs technology is experiencing rapid growth, driven by the decarbonization challenge and the spread of LIBs in new fields like automotive, and stationary energy storage [17, 18, 19, 20, 21, 22].
It is demonstrated that mechanical fracture is among the main causes of aging and performance decay in LIBs [23]. As a result, fracture in LIBs is receiving significant attention from the scientific community, especially with the aim of developing LIBs with long life cycles and high performance.

The basic working principle of lithium-ion batteries is the movement of electrons in the external circuit (current flow), balanced by the movement of lithium ions between electrodes of opposite polarity and the insertion in their structure, driven by electrochemical reactions. The electrodes are composite materials, among other components, the most important material is the active material where electrochemical reactions take place. The active material is a particulate matter, so when lithium ions get into the electrode, they intercalate into these particles, occupying the interstices in the crystalline structure of the active material. Since lithium ions are much smaller than active material particles ($\approx 10\,\mu m$ in radius) they diffuse within the active material particles. Then, lithium concentration distribution within the particles is inhomogeneous, being higher in the particle core with respect to the surface during lithium extraction, and the opposite during insertion.
The intercalation of lithium-ions in the crystalline structure of the electrodes causes the local deformation of the material, proportional to the concentration of lithium-ions. Then, strain mismatch arises in the particles due to the inhomogeneous swelling caused by the inhomogeneous distribution of lithium concentration. This differential strain results in the so-called diffusion-induced stress (DIS), leading to crack propagation in electrode microstructure ultimately [24, 25, 26, 27, 28, 29, 30]. Finally, crack propagation causes the capacity fade and impedance rise in LIBs because it hinders the passage of lithium -ions due to the isolation of some portions of active material, and it triggers undesired side reactions on newly created crack surfaces (solid electrolyte interface growth), consuming lithium ions [15, 16].

Fracture problem in LIBs is usually solved numerically with FEM in the literature because of two reasons: firstly, a coupling exists between the mechanical and electrochemical domains which makes the problem multi-physics, secondly, stress distribution on the crack surfaces is non-homogeneous and no analytical solutions are available for SIF computation. However, numerical computation with FEM is often time-consuming, then it is unsuitable for integration into a damage model for real-time scenarios.

Two methodologies are reported to compute SIF in case of an arbitrary stress distribution over the crack surface, the weight factor method [3, 15] and the Green's function method [31]. The weight function method is particularly difficult to implement. Green's function method is easier to handle, but it still requires solving integrals. For this reason, a more easy and analytical way to compute SIF is pursued, on the basis of geometric factors.

This work provides a general analytical methodology for calculating the SIF resulting from any arbitrary stress profile on the surfaces of the crack. This methodology is effective for accurately and rapidly assessing the fracture level in LIBs electrodes. First, a polynomial function is employed to approximate the arbitrary stress profile. Then, the SIF is calculated by applying the principle of superposition of effects, involving a summation of products between each term of the polynomial stress function and their corresponding geometric factors. Geometric factors for spheres with cracks at both the center and the surface are obtained from a FEM model. Subsequently, the SIF in graphite particles of LIBs electrodes is computed using these geometric factors. The analytical results are validated against the results of the multi-physics FEM fracture

IOP Conf. Series: Materials Science and Engineering 1306 (2024) 012009 doi:10.1088/1757-899X/1306/1/012009

model. Finally, the methodology is used to assess the SIF and the fracture likeliness as a function of different electrode design solutions, at different current rates.

The article is organized as follows. Section 2.1 provides the main Equations solving the mechanics of LIBs. The fundamentals of LEFM theory with special regard to multi-physics applications are provided in Section 2.2. Section 2.3 describes the analytical methodology for SIF computation due to arbitrary stress profile on the surfaces of the crack. The geometric factors for spheres with cracks located at the center and on the surface are provided in 3.1 and the results of the application of the analytical procedure to the case of graphite particles in LIBs are discussed in Section 3.2. Section 3.3 reports the influence of electrode design parameters on fracture in LIBs and Section 3.4 describes the correlation between mechanical and electrochemical (capacity loss) damage.

2. Material and methods

2.1. Mechanics of lithium ion batteries

Active material particles of LIB electrodes undergo mechanical stress and strain due to lithium diffusion during battery operation. The problem is multi-physics because it involves computing the chemical concentration of lithium ions to get the stress.

The main equations of the electrochemical-mechanical model and their boundary conditions are reported in Table 1. The following assumptions are considered: the active material particle is spherical, the material is linear elastic, homogeneous, and isotropic [32]. Then, equations are written as a function of the radial coordinate (r) thanks to the hypothesis of axisymmetry.

The chemical field is governed by the mass transfer equation (Equation 1), which describes the variation of lithium concentration c due to lithium diffusion, in analogy to the heat transfer equation.

Stress-strain equations, i.e. constitutive (Equation 3), congruence (Equation 4), and equilibrium (Equation 5) equations, govern the mechanical field caused by the inhomogeneous lithium concentration, in analogy with the thermal-mechanical problem.

The constitutive equation shows that the total strain has two contributions, namely the elastic strain contribution (first term on the right-hand side of Equation 3) and the chemical strain contribution depending on the concentration c (second term on the right-hand side of Equation 3). It is pointed out that chemical strain ($\frac{\Omega(c-c_R)}{3}$), is equivalent to a thermal strain, where $\frac{\Omega}{3}$ is replaced by the thermal expansion coefficient α and the concentration c is replaced by temperature T [26, 28, 33].

Once the diffusive problem (Equation 7) is solved and the lithium concentration c is got, the mechanical field is solved in turn. The equilibrium equation is written as a function of the displacement u using the constitutive (Equation 3) and congruence (Equation 4) equations. Then, the resulting equation is integrated twice and the displacement u (Equation 8) is got using the boundary conditions in Equation 6. The displacement u is replaced in the congruence equation (Equation 4) to compute radial and hoop strains, then, the constitutive equation (Equation 3) is used to obtain radial (Equation 9) and hoop stress (Equation 10).

Chemical equations		
Mass conservation	$\frac{\partial c}{\partial t} = \frac{D}{r^2}\frac{\partial}{\partial r}\left(r^2\frac{\partial c}{\partial r}\right)$	(1)
Boundary conditions	$\begin{cases} \left.\frac{\partial c}{\partial r}\right\vert_{r=0} = 0 & \text{for } t \geq 0 \\ \left.\frac{\partial c}{\partial r}\right\vert_{r=R} = J_{flux} & \text{for } t \geq 0 \end{cases}$	(2)
Mechanical equations		
Constitutive	$\varepsilon_r = \frac{1}{E}\left[\sigma_r - 2\nu\sigma_c\right] + \frac{\Omega}{3}(c - c_R) \quad \varepsilon_c = \frac{1}{E}\left[(1-\nu)\sigma_c - \nu\sigma_r\right] + \frac{\Omega}{3}(c - c_R)$	(3)
Congruence	$\varepsilon_r = \frac{du}{dr} \quad \varepsilon_c = \frac{u}{r}$	(4)
Equilibrium	$\frac{\partial \sigma_r}{\partial r} + \frac{2}{r}\left(\sigma_r - \sigma_c\right) = 0$	(5)
Boundary conditions	$u\Big\vert_{r=0} = 0 \quad \left.\frac{\partial \sigma_r}{\partial r}\right\vert_{r=R} = 0$	(6)
Solutions		
Concentration	$c(r,t) = c_0 + \frac{IR}{FD}\left[3\tau + \frac{1}{2}\left(\frac{r}{R}\right)^2 - \frac{3}{10} - 2\frac{R}{r}\sum_{n=1}^{\infty}\left(\frac{sin(\lambda_n r/R)}{\lambda_n^2 sin(\lambda_n)}e^{-\lambda_n^2\tau}\right)\right]$	(7)
Displacement	$u(r) = \frac{\Omega}{3(1-\nu)}\left[(1+\nu)\frac{1}{r^2}\int_0^r (c - c_R)r^2\,dr + 2(1-2\nu)\frac{r}{R^3}\int_0^R (c - c_R)r^2\,dr\right]$	(8)
Radial stress	$\sigma_r(r) = \frac{2\Omega}{3}\frac{E}{1-\nu}\left[\frac{1}{R^3}\int_0^R (c - c_R)r^2\,dr - \frac{1}{r^3}\int_0^r (c - c_R)r^2\,dr\right]$	(9)
Hoop stress	$\sigma_c(r) = \frac{\Omega}{3}\frac{E}{1-\nu}\left[\frac{2}{R^3}\int_0^R (c - c_R)r^2\,dr + \frac{1}{r^3}\int_0^r (c - c_R)r^2\,dr - (c - c_R)\right]$	(10)

Table 1. Electrochemical-mechanical model equations [26].

In previous authors' work [26] it was shown that lithium transport is affected by DIS through the gradient of hydrostatic stress, which depends on the lithium concentration in turn. In this case, lithium transport is governed by Equation 11, which fully couples mechanical and chemical fields.

$$\frac{\partial c}{\partial t} = \frac{D}{r^2}\frac{\partial}{\partial r}\left(r^2\frac{\partial c}{\partial r} - r^2\frac{\Omega(c - c_R)}{R_g T}\frac{\partial \sigma_h}{\partial r}\right) \tag{11}$$

Where D is the diffusion coefficient, c_R is the reference concentration, $\sigma_h = \frac{\sigma_r + 2\sigma_c}{3}$ is the hydrostatic stress, R_g is the gas constant, and T is the temperature.

Equation 11 can be rewritten in the same form as the heat transfer equation and the solutions reported in Equation 7-10 can be employed. Then, an equivalent diffusion coefficient depending on lithium concentration substitutes the physical diffusion coefficient D in Equation 1, according to Equation 12 [26].

$$D_{eqv}(c) = D\left[1 + k_m(c - c_R)\right] \tag{12}$$

Where $k_m = \frac{2\Omega^2 E}{9R_gT(1-\nu)}$ is the coefficient for the mechanical-chemical coupling, ν is the Poisson ratio, and E is the Young modulus.

2.2. Fracture mechanics theory for multi-physics problems

Pre-existent defects in electrode microstructure mainly propagate due to mode-I caused by tensile hoop stress, despite the numerous fracture mechanisms reported in the literature according to the specific active material [15, 32, 33].

According to LEFM theory, the stress field near the crack tip is infinite and can be expressed as a function of the SIF K, according to Equation 13 [5].

$$\sigma_{ij} = \frac{K}{\sqrt{2\pi r_c}} f_{ij}(\theta) \tag{13}$$

Where σ_{ij} is the stress tensor, r_c and θ are the polar coordinates with the origin at the crack tip, and $f_{ij}(\theta)$ is a dimensionless shape function.

SIF can be computed as a function of a dimensionless geometric factor Y (Equation 14), which considers both the geometry of the specimen and of the crack (size, location, and shape).

$$K = Y\sigma\sqrt{a} \tag{14}$$

Where a is the crack size and σ is the constant far-field stress, which is the stress unaffected by the crack presence.

Numerous works in the literature provide various geometric factors for different cracks and body geometries [7, 8, 34]. However, only solutions for simple cases have been developed, such as those involving constant or linear stress on the crack surfaces (caused by tension, bending, or their combination).

Tabulated geometric factors cannot be used when the stress distribution on crack surfaces is non-homogeneous, such as for stress resulting from diffusive fields (thermal or chemical). Then, SIF is usually computed using FEM from the J-integral developed by Rice [35] according to Equation 15, when the hypotheses of LEFM hold.

$$K = \begin{cases} \sqrt{JE} & \text{plane stress} \\ \sqrt{\frac{JE}{(1-\nu^2)}} & \text{plane strain} \end{cases} \tag{15}$$

The J-integral for a 2D fracture problem is the contour integral reported in Equation 16, considering an arbitrary counterclockwise path Γ around the crack tip (Figure 1). This contour integral is path-independent, meaning that its value is independent of the particular choice for the path Γ.

$$J = \int_\Gamma \left(W dx_2 - \mathbf{t} \cdot \frac{\partial \mathbf{u}}{\partial x_1} ds \right) \tag{16}$$

Where $W = \int_0^{\epsilon_{ij}} \sigma_{ij} d\epsilon_{ij}$ (Einstein notation) is the strain energy density, σ_{ij} and ε_{ij} are the components of the stress and strain tensors, $\mathbf{u}$ is the displacement vector, $\mathbf{t}$ is the traction vector

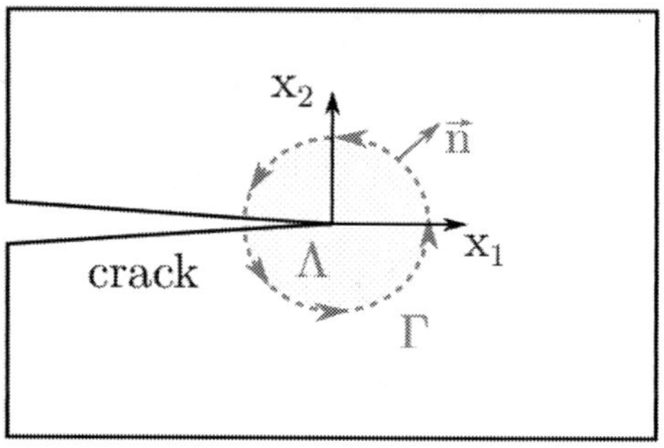

Figure 1. Scheme of the crack area, where Γ is an arbitrary counterclockwise path surrounding the crack tip, Λ is the area enclosed by the path Γ, x_1, x_2 are the coordinate directions and $\vec{n}$ is the versor normal to Γ.

acting on the path Γ and its components are $t_i = \sigma_{ij} n_j$, n_j is the versor normal to Γ, x_1 and x_2 are the coordinate directions, and ds is the element length along the path Γ.

The standard form J-integral does not keep the path-independence property in case of multi-physics fracture problems, such as thermal-mechanical [36, 37] or mechanical-diffusive fracture problems [38, 39, 40]. The modified version of path-independent J-integral for these multi-physics problems is reported in Equation 17.

$$J^{MF} = \int_{\Gamma} \left(W dx_2 - \mathbf{t} \cdot \frac{\partial \mathbf{u}}{\partial x_1} ds \right) + \int_{\Lambda} \left(\boldsymbol{\sigma}^P \cdot \frac{\partial \boldsymbol{\varepsilon}^{MF}}{\partial x_1} d\Lambda \right) \tag{17}$$

Where Λ is the area enclosed within the path Γ, $\boldsymbol{\sigma}^P = [\sigma_{11}, \sigma_{22}, \sigma_{33}]$ is the principal stresses vector, $\boldsymbol{\varepsilon}^{MF}$ is the strain vector caused by the multi-physics fields, i.e. the chemical ($\frac{\Omega(c-c_R)}{3}$) or thermal field ($\alpha(T - T_R)$).
The additional area integral (the second term on the right-hand side of Equation 17) is added to the standard Rice's expression of J-integral (the first term on the right-hand side of Equation 17) to ensure that the energy balance is satisfied and the J-integral is path-independent.

2.3. Analytical computation of SIF

The principle of superposition of effects can be exploited to analytically calculate the SIF as a function of geometric factors even when the stress distribution on crack surfaces is arbitrary.
First, a polynomial of grade n is used to approximate the arbitrary nominal stress distribution (the stress unaffected by the crack presence) at the location of the crack, as reported in Equation 18.

$$\sigma(x) = \sum_{i=0}^{n} \sigma_i x^i \tag{18}$$

Where σ_i are the polynomial coefficients, and x is the coordinate along the crack surface as reported in Figure 2).

Then, Equation 18 is combined with Equation 13, and the principle of superposition of effects is applied, resulting in the generic expression of SIF as reported in Equation 19.

$$K = \sum_{i=0}^{n} Y_i \sigma_i x^i \sqrt{a} \tag{19}$$

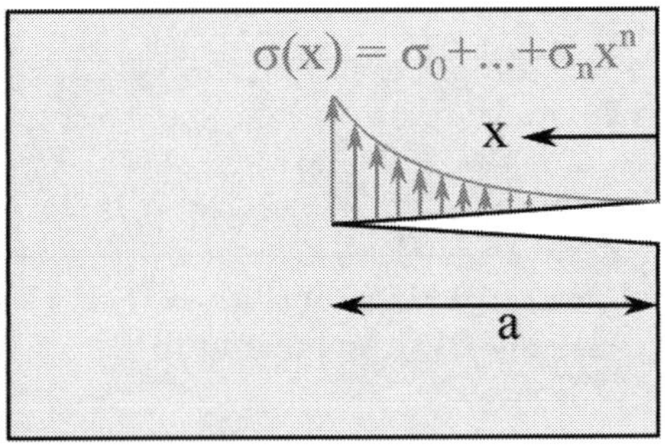

Figure 2. Crack surface with polynomial stress distribution, where x is the coordinate along the crack surface and a is the crack length.

Where Y_i is the geometric factor corresponding to the i-th term of the polynomial function approximating the stress for the specific crack configuration.

Referring to Figure 3, the procedure used in this work for the analytical computation of SIF in active material particles with central and superficial cracks, and for the validation of the results, is summarized below.

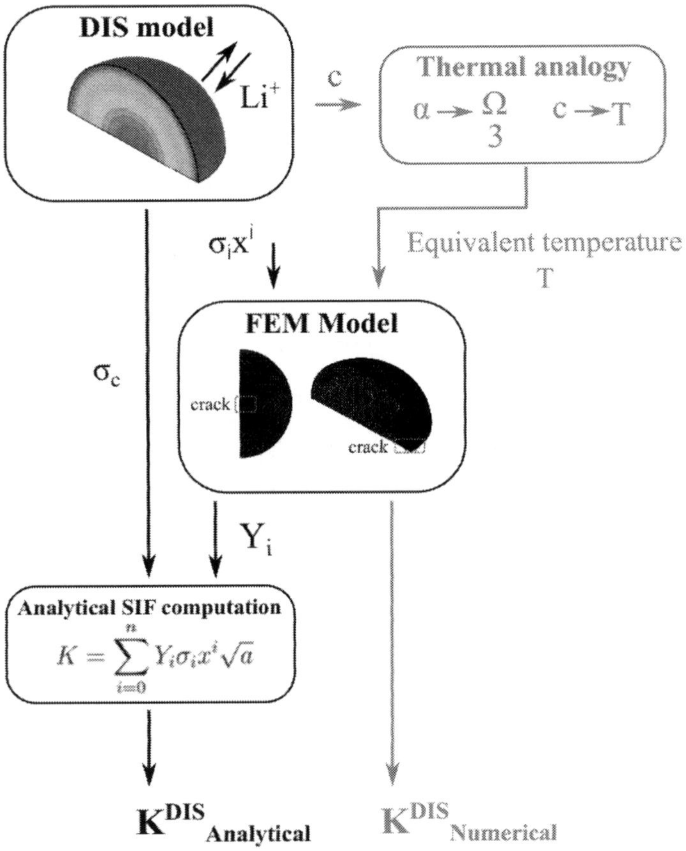

Figure 3. Methodology for the analytical computation of SIF, where the validation stage is highlighted in green.

IOP Conf. Series: Materials Science and Engineering 1306 (2024) 012009 doi:10.1088/1757-899X/1306/1/012009

- **Geometric factor computation.** The computation of each geometric factor Y_i for each i-th grade of the polynomial function approximating the stress profile is performed considering different ratios a/R, where R is the radius of the sphere. First, a FEM model of a sphere with cracks located at the center and the surface is built in Ansys Mechanical APDL. Then, a pressure load equal to the single grande of the polynomial stress function ($\sigma_i x^i$) is applied on the nodes of the crack surface. Then, the SIF (K_i^{FEM}) is computed from the J-integral (Equation 15) using the CINT command in Ansys. Finally, the geometric factor Y_i is computed according to Equation 20.

$$Y_i = \frac{K_i^{FEM}}{\sigma_i a^i \sqrt{a}} \qquad i \in \{0, 1, ..., n\} \tag{20}$$

  This procedure is carried out just one time, once geometric factors are computed, they can be applied to different stress distributions and crack/body sizes.
- **Analytical SIF computation in case of DIS.** The nominal hoop stress distribution over the crack region is computed according to the equations of the electrochemical-mechanical model reported in Table 1. Then, the hoop stress is fitted with a polynomial, and SIF is computed using Equation 19, using the geometric factors computed at the beginning.
- **Numerical SIF computation in case of DIS.** SIF is computed using a FEM model built in Ansys Mechanical APDL. Lithium concentration in the active material particle is computed according to the equations of the electrochemical-mechanical model reported in Table 1. Then, the thermal analogy [32, 33] is used to convert the concentration into an equivalent temperature. Finally, the equivalent temperature is mapped on structural nodes of the FEM model, and SIF is obtained from the J-integral through the CINT command (Equation 15).
- **Validation.** The results of the analytical computation of SIF are compared with the ones resulting from numerical computation with FEM.

A detailed discussion on the strategy used to build the FEM model and mesh is not reported here but a more comprehensive explanation is present in previous authors' works [33, 32].

3. Results and Discussion

This section provides the results of the computation of the geometric factors for a sphere with a crack located at the center and surface. Subsequently, the analytical computation of SIF due to the hoop stress caused by lithium diffusion in active material particles is performed. Finally, a comparison between the analytical and FEM model results is performed.

3.1. Geometric factor results

In this work, a polynomial stress distribution of grade 6 is considered, leading to seven geometric factors. The geometric factors computed are general and applicable to any polynomial stress distribution. In different scenarios, when the stress distribution can adequately be represented by a lower-grade polynomial, higher-order geometric factors can be neglected accordingly.

Figures 4 and 5 report the geometric factors for sphere with central and superficial cracks, respectively. The geometric factors exhibit quadratic dependency of the normalized crack length a/R. The analytical expressions for these geometric factors are provided in Tables 2 and 3.

3.2. Comparison between analytical and numerical SIF results in case of DIS

Analytical and numerical SIF results caused by lithium diffusion in active material particles of LIB electrodes are compared in this section.

IOP Conf. Series: Materials Science and Engineering 1306 (2024) 012009 doi:10.1088/1757-899X/1306/1/012009

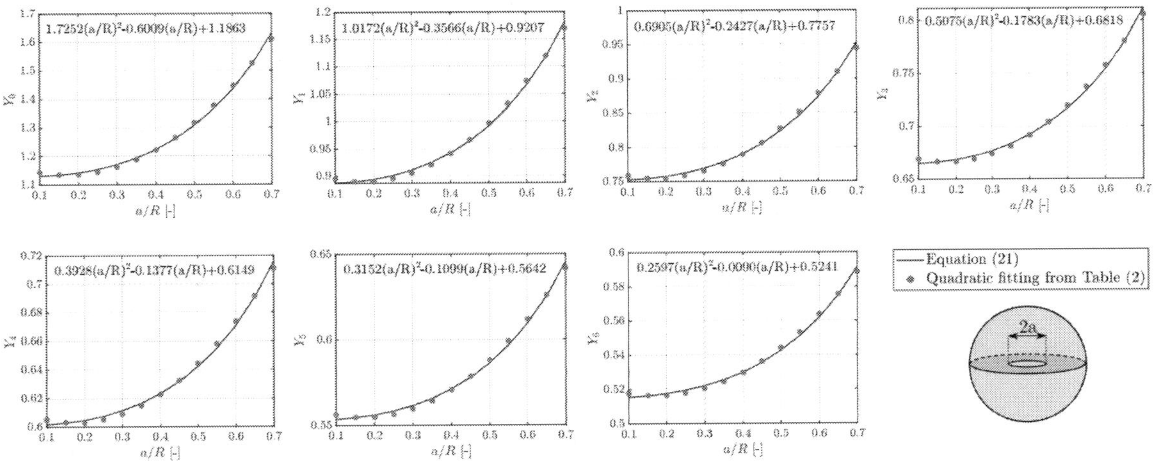

Figure 4. Geometric factors Y_i for sphere with central crack. The black curves correspond to the values computed with FEM and Equation 20, and the red dots correspond to the fitting using quadratic functions $(p(\frac{a}{R})^2 + q(\frac{a}{R}) + r)$.

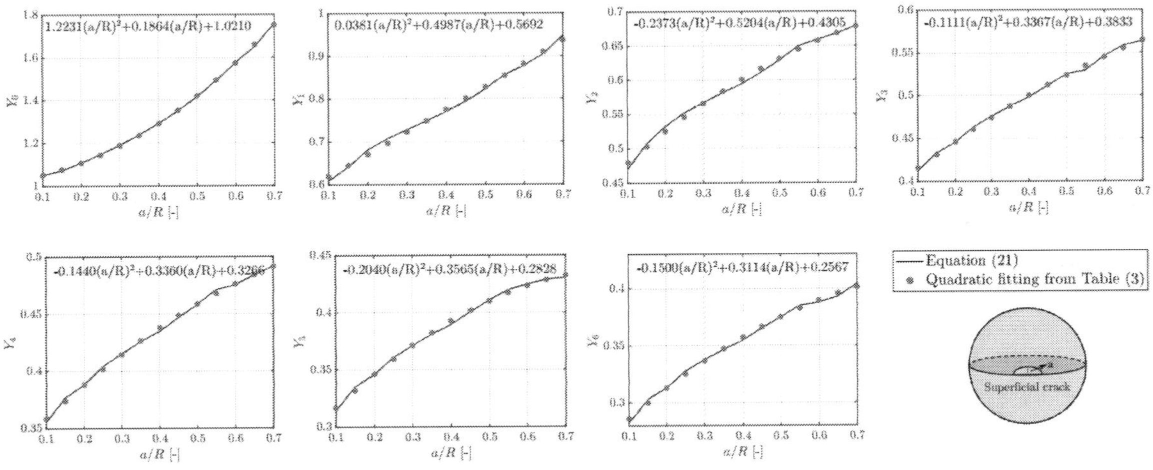

Figure 5. Geometric factors Y_i for sphere with superficial crack. The black curves correspond to the values computed with FEM and Equation 20, and the red dots correspond to the fitting with quadratic functions $(p(\frac{a}{R})^2 + q(\frac{a}{R}) + r)$.

Geometric factor	p	q	r
Y_0	1.7252	-0.6009	1.1863
Y_1	1.0172	-0.3566	0.9207
Y_2	0.6905	-0.2427	0.7757
Y_3	0.5075	-0.1783	0.6818
Y_4	0.3928	-0.1377	0.6149
Y_5	0.3152	-0.1099	0.5642
Y_6	0.2597	-0.0900	0.5241

Table 2. Coefficients of the quadratic function $p(\frac{a}{R})^2 + q(\frac{a}{R}) + r$ for geometric factors of sphere with central crack.

Geometric factor	p	q	r
Y_0	1.2231	0.1864	1.0210
Y_1	0.0381	0.4987	0.5692
Y_2	-0.2373	0.5204	0.4305
Y_3	-0.1111	0.3367	0.3833
Y_4	-0.1440	0.3360	0.3266
Y_5	-0.2040	0.3565	0.2828
Y_6	-0.1500	0.3114	0.2567

Table 3. Coefficients of the quadratic function $p(\frac{a}{R})^2 + q(\frac{a}{R}) + r$ for geometric factors of sphere with superficial crack.

The equations in Section 2.1 are used to analytically solve the electrochemical-mechanical problem, including the coupling mechanical-chemical coupling. The galvanostatic operation is considered, corresponding to a constant flux on the particle surface (Equation 2), which depends on the current I delivered by the battery. The case study is a graphite particle, considering the following physical parameters: Young modulus $E = 15\,\text{GPa}$, Poisson ratio $\nu = 0.3$, partial molar volume $\Omega = 4.2\times10^{-6}\,\text{m}^3/\text{mol}$, diffusion coefficient $D = 2\times10^{-14}\,\text{m}^2/\text{s}$, maximum concentration $c_{max} = 2.9155 \times 10^4\,\text{mol/m}^3$, temperature $T = 298\,\text{K}$.
Tensile hoop stress occurs at the particle centre during lithiation, vice-versa it occurs on the surface during delithiation, then the particle with central crack is simulated during lithiation, on the other hand, particle with superficial crack is simulated during delithiation.
Normalized current rate is considered, then the C-rate equal to 1 corresponds to the current enables complete filling (or emptying) of the active material particle within one hour. In addition, SOC is the percentage fraction of lithium concentration with respect to the maximum concentration ($SOC = \frac{c}{c_{max}}$).

Figures 6b,d compare SIF values computed analytically and numerically. The results obtained with the two methodologies are in good agreement, demonstrating the validity of the proposed analytical procedure for SIF computation due to a generic stress distribution over the crack.

SIF depends on hoop stress and crack size, according to Equation 19. The distribution of hoop stress within the particle at 50% SOC and different C-rates is shown in Figure 6a during lithium insertion and in Figure 6c during lithium extraction. As also demonstrated in previous authors' works [26, 28, 27], higher C-rates cause higher concentration gradients within active material particles, giving rise to higher stress. Similarly, SIF values increase with the increase of C-rate, as shown in Figure 6b,d. Then, higher C-rate is more detrimental because it increases fracture and accelerates LIB degradation, as confirmed by previous authors' work [32] and experimental measurements [41, 42].
On the other hand, as the crack length increases, the SIF increases accordingly, for the crack length below $a/R = 0.5$ for the particle with central crack and $a/R = 0.15$ Then, the SIF decreases as the crack length increases. This occurs because the hoop stress distribution varies along the radial coordinate. Indeed, the positive (tension) hoop stress decreases and becomes negative (compressive) when moving from the particle core to the surface during lithium insertion (Figure 6a), and from the surface to the core during extraction (Figure 6c). Then, the crack tip of longer cracks undergoes lower hoop stress because it is closer to the less tension/compressive region. Then, referring to Equation 19, the SIF increases with the increase in crack length as long as the increase in crack length balances the decrease in the hoop stress.

Several works in the literature compute the SIF analytically in order to avoid multi-physics FEM simulations with high computational cost, especially when the aim is to integrate the fracture results into a degradation model for the estimation of capacity fade during battery

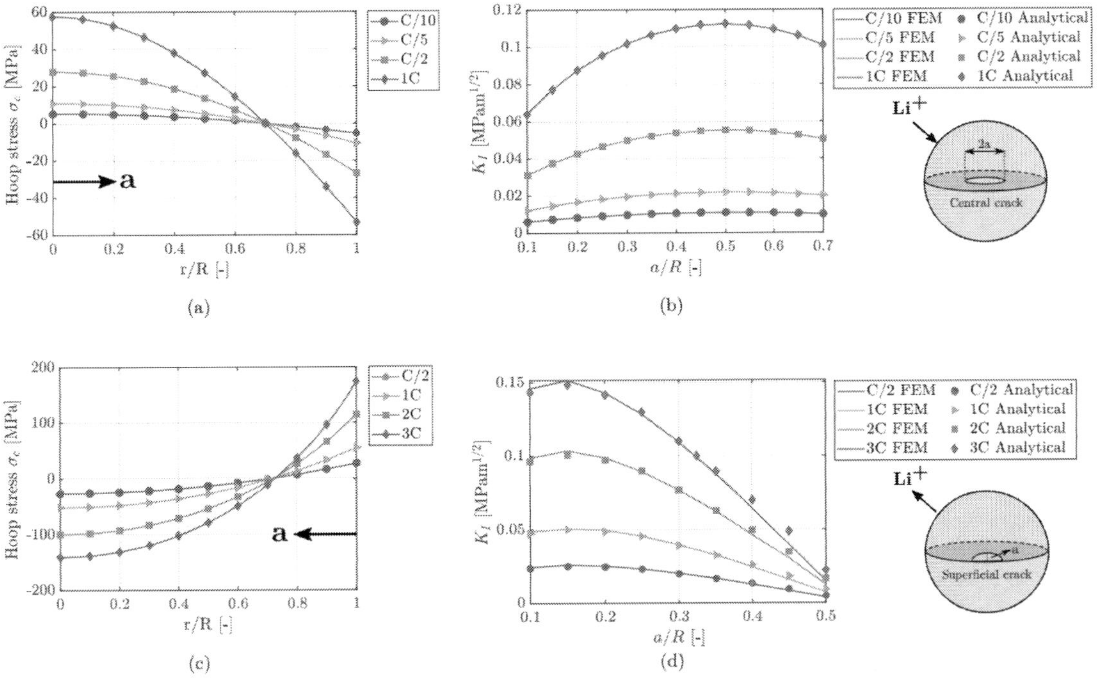

Figure 6. Results of DIS and fracture model. The graphite particle is simulated at different C-rates and results correspond to 50% SOC. Distribution of hoop stress (a) during lithiation and (c) during delithiation. SIF values for different a/R for particle (b) with a crack at the center during lithiation and (d) a crack at the surface during delithiation.

operation [43, 44, 45, 46, 47]. In these works, SIF is directly computed as $K = Y\sigma\sqrt{a}$, making two approximations: (a) constant far-field stress is used, neglecting that the stress distribution on crack surfaces is not uniform; (b) the geometric factor of the plate undergoing uniaxial stress is used ($Y_0 = 1.12\sqrt{\pi}$ [3]) instead of using the one relative to the spherical case.

Figure 7a provides an insight into the error regarding the first approximation.

The SIF for spherical particle with central crack computed with the analytical procedure proposed in this work is compared with the SIF computed as $K = Y_0\sigma\sqrt{a}$, where σ is equal to the stress in the uncracked particle at the crack tip location, and Y_0 is the geometric factor for constant stress distribution on the crack surfaces (Y_0 in Figure 4). The comparison demonstrates that the non-homogeneous stress distribution less affects the SIF for shorter cracks. On the other hand, SIF values obtained considering uniform stress distribution decrease faster as the crack length increases, eventually reaching zero when the stress at the crack tip becomes compressive. Figure 7b compares SIF values obtained with the analytical computation methodology proposed in this work and the one computed with the approximation of the plate's shape factor and constant superficial stress, which is common in the literature. A particle with a superficial crack is considered, and the far-field stress σ is the hoop stress on the surface of the particle without the crack. The results show a great difference between SIF values, which still increases as a/R increases. Indeed, the decrease in SIF as the crack length increases is not observed when only focusing on the superficial hoop stress, thereby neglecting the complete stress distribution along the crack surfaces. Additionally, the geometric factor of the plate significantly differs from that of the sphere.

The approximations usually made in the literature affect the accuracy of the SIF computation,

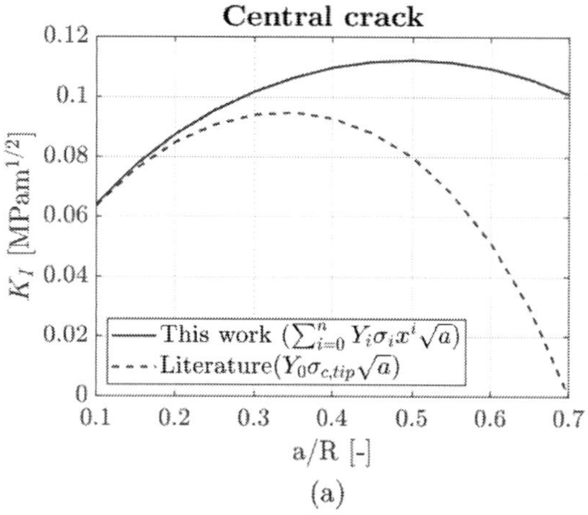

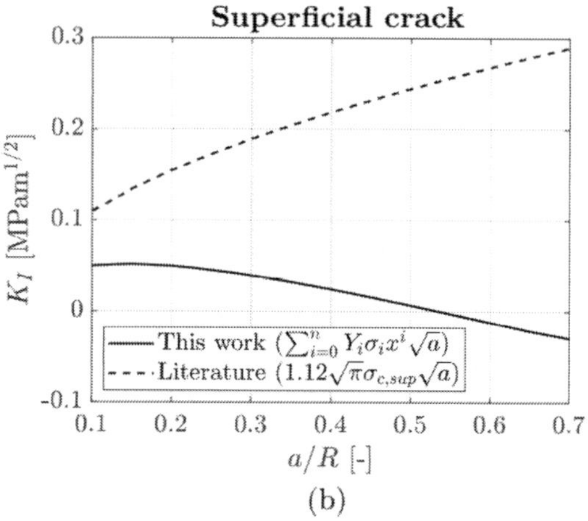

Figure 7. Comparison between SIF values resulting from literature and obtained with the procedure presented in this work. (a) Graphite particle with central crack during lithiation at 1C and 50% SOC, where $\sigma_{c,tip}$ is the hoop stress in the uncracked particle at the crack tip location. (b) Graphite particle with superficial crack during delithiation at 1C and 50% SOC, where $\sigma_{c,sup}$ is the hoop stress on the particle surface.

and then, the results of the degradation model assessing the decay of LIB performance during its usage. On the contrary, the analytical procedure developed in this work is as accurate as performing multi-physics FEM fracture model simulation. The advantage is that analytical computation is significantly faster and requires less computational resources as compared to the FEM model (fractions of a second for the analytical computation compared to minutes/hours for the FEM model) because it requires just a sum of products to compute the SIF. Then, the proposed analytical methodology for SIF computation is the optimal compromise to assess the impact of fracture on battery damage, balancing accuracy with computational cost.

3.3. Impact of fracture on the electrode design

The effect of the electrode design parameters on fracture is quantified in this section, such as the size of the active material particles and the thickness of the electrode layers. In this way, it is possible to give electrode design guidelines to limit fracture and mechanical degradation in LIBs.

Electrode design parameters affect the electrochemistry of the LIB, namely the current flowing into each elementary cell (sequence of current collectors, cathode, separator, anode), then the lithium flux on the surface of active material particles, as well as the resulting lithium concentration and stress distributions.

First, lithium flux on particle surfaces (J_{flux}) is computed by solving the electrochemical model of the entire LIB as described in previous authors' works [32], varying the electrode design parameters. Then, equations in Section 2.1 are solved using the computed lithium flux in the boundary condition reported in Equation 2.

The effect of the active material particle size on the fracture behavior is reported in Figure 8. The concentration gradient within smaller particles is lower than bigger particles, then the former are less affected by fracture because of the lower stress. However, electrodes with smaller particles result in LIB with lower tap density and energy density. In addition, the manufacturing cost for electrodes with smaller particles is higher. Then, a trade-off is necessary between lifespan (which increases as active material particle size is reduced), electrochemical performance, and cost.

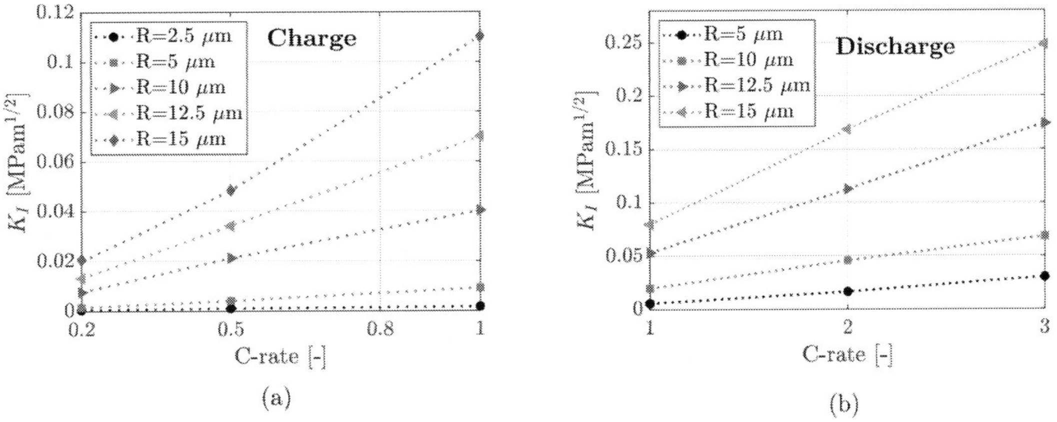

Figure 8. Influence of active material particle size on SIF during (a) charge (lithium insertion in graphite particle) and (b) discharge (lithium extraction from graphite particle).

Figure 9 reports the influence of electrode thickness on the SIF. Thicker electrodes allow to store greater amount of energy per unit of volume (or mass) because the number of elementary cells connected in parallel is lower, and then the amount of inactive components, i.e. separator and current collectors, is reduced. In this condition, considering the battery capacity and the current delivered by the battery as constant, a lower number of elementary cells results in a higher current flowing into each cell. This higher current results in higher lithium flux (J_{flux}) on active material particles surface, then steeper concentration gradient, higher stress, and SIF. On the other hand, thinner electrodes are generally used for high-power applications thanks to the lower internal resistance and greater current capabilities. Furthermore, thinner electrodes are preferred for higher C-rates than thicker electrodes because they are less affected by fracture, as shown in Figure 9.

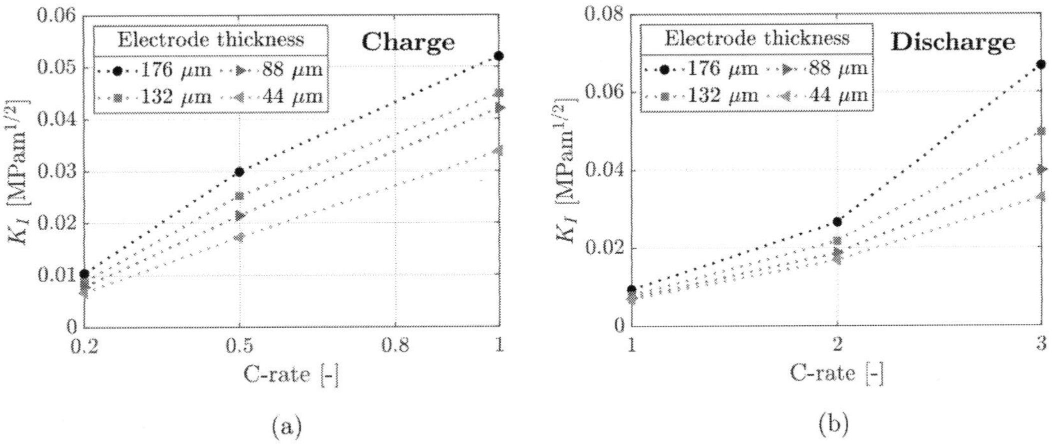

Figure 9. Influence of the electrode thickness on SIF during (a) charge (lithium insertion in graphite particle) and (b) discharge (lithium extraction from graphite particle).

3.4. Electrochemical-mechanical damage model

Fracture of electrode microstructure is among the main causes of LIB degradation [15].
As summarized in the scheme reported in Figure 10, fatigue phenomena and cracks growth occurring during battery operation lead to capacity fade because of two reasons: (a) isolation of active material, which becomes inactive and no longer able to host lithium ions; (b) creation of new surfaces which are exposed to the electrolyte, giving rise to side reactions such as the formation of solid electrolyte interphase (SEI) layer. These reactions consume lithium ions, which cannot be cycled further.

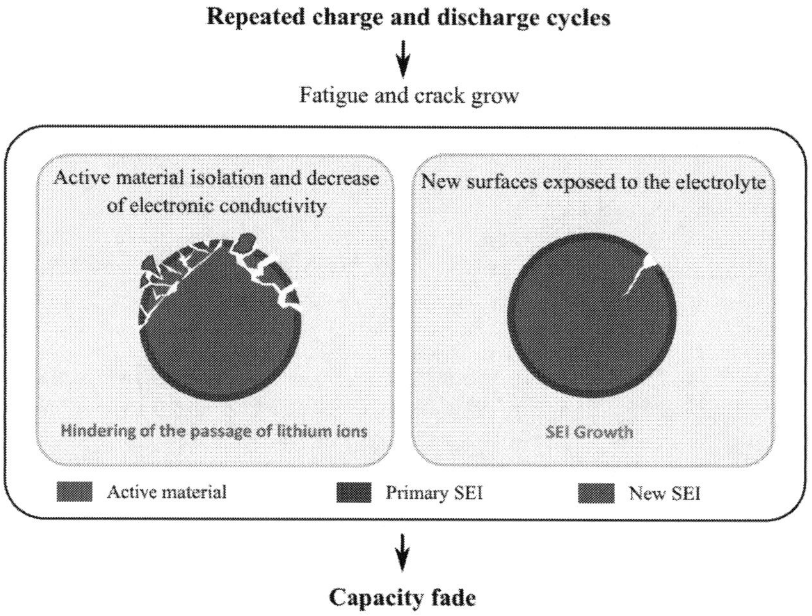

Figure 10. Sketch of the damaging mechanisms triggered by fracture during LIB operation.

The rate of capacity fade caused by the formation of the SEI layer on the cracks surfaces created during repeated charge/discharge cycles is computed according to Equation 21 [15].

$$\frac{dQ}{dN} = X\frac{dA_c}{dN} \tag{21}$$

Where Q is the capacity, N the number of cycles, A_c is the crack surface and X is a constant taking into account the amount of lithium ions consumed during the SEI formation reaction and the rate of the reaction. The increase of the crack surface with the number of cycles ($\frac{dA_c}{dN}$) is computed according to Equations 22a-b [15].

$$\frac{dA_c}{dN} = \frac{dA_c}{da}\frac{da}{dN} \tag{22a}$$

$$\frac{da}{dN} = C\Delta K^m \tag{22b}$$

Where a is the crack length, C and m are material coefficients and Equation 22b is the Paris' law, correlating the increase of crack length ($\frac{da}{dN}$) to the SIF range (ΔK).
Equations 21-22 show that the degradation in LIB is a multi-physics phenomenon because the mechanical damage (the increase in crack length in Equation 22) is correlated with the chemical damage (SEI formation and loss of lithium inventory in Equation 21).

Since the capacity loss depends on the SIF, the analytical computation of SIF developed in this work will significantly speed up the capacity loss computation. Then, this work is the basis for the development of the algorithms assessing the capacity loss and the remaining useful life of LIB during usage, which will be the focus of the future research of the authors.

4. Conclusions

A generic methodology for the analytical computation of SIF in case of arbitrary stress profile on crack surfaces is presented in this work. The methodology can be effectively applied to assess fracture resulting from multi-physics phenomena, such as fracture arising in active material of LIBs electrodes due to lithium diffusion. In this case, an analytical solution for SIF is not available in the literature because the stress distribution on the crack surface is non-homogeneous.

Firstly, the arbitrary stress profile is expressed using a polynomial function, then the principle of superposition of effects is applied, and SIF is computed as a summation of products between each term of the polynomial stress function and their corresponding geometric factors. A FEM model is built to compute the geometric factors for a sphere with a crack at the center and surface. The expression of the geometric factors is generic and can be used for any kind of stress profile and crack/radius geometrical sizes.

The developed methodology is applied for the analytical computation of SIF caused by diffusion induced stress arising in graphite particles during lithium insertion and extraction. The analytical results are compared to the numerical results, showing good agreement.

SIF due to DIS is usually computed in the literature using multi-physics FEM model, which is time-consuming (the simulation time can range between minutes to some hours), or analytically. In the latter case, the computation is inaccurate because the non-homogeneous stress distribution on crack surfaces is neglected, and the geometric factor of plates is often used, which is far from the case of spherical geometry. On the contrary, the proposed analytical computation of SIF is as accurate as performing FEM fracture model simulations, but it is less time-consuming because the SIF computation consists of just a sum of products, running in fractions of seconds.

The methodology is effectively applied to assess the influence of electrodes parameters on fracture, in order to give electrode design guidelines limiting fracture and then LIB degradation. It is observed that thinner electrodes with smaller active material particles are preferred from a fracture mechanical point of view because they result in lower SIF values.

Finally, the developed analytical methodology can be effectively used to describe fracture in active material particles with LIB cycling, being eventually integrated into a degradation model for the estimation of LIB performance decay in real-time.

Nomenclature

Symbols

σ_i	i-th coefficient of polynomial stress distribution (MPa)
θ	Polar coordinate in crack reference frame (−)
a	Crack length (m)
c	Concentration (mol/m^3)
c_R	Reference concentration (mol/m^3)
D	Diffusion coefficient (m^2/s)
D_{eqv}	Equivalent diffusion coefficient for coupled model (m^2/s)
E	Young modulus (MPa)
F	Faraday constant (As/mol)
f_{ij}	Dimensionless shape function
I	Current density (A/m^2)
J_{flux}	Flux on the particle surface (mol/m^{2}s)
K	Stress intensity factor (MPam$^{0.5}$)
k_m	Mechanical coupling parameter (m^3/mol)
n_i	Normal versor (−)
R	Active material particle radius (m)
r	Radial coordinate (−)
r_c	Polar coordinate in crack reference frame (−)
R_g	Gas constant (J/molK)
T	Temperature (K)

IOP Conf. Series: Materials Science and Engineering 1306 (2024) 012009 doi:10.1088/1757-899X/1306/1/012009

t	Time (s)
t_i	Traction vector (MPa)
T_R	Reference temperature (K)
u	Particle level displacement (m)
W	Strain energy density (J/m^3)
x_1	Coordinate direction along crack extension
x_2	Coordinate direction perpendicular to crack extension
Y	Geometric factor

Acronyms

DIS	Diffusion induced stress
FEM	Finite element model
LEFM	Linear elastic fracture mechanics
LIB	Lithium ions battery
MF	Multi-physics field
SIF	Stress intensity factor
SOC	State of charge (%)

Greek Symbols

α	Expansion coefficient (1/K)
ϵ^c	Diffusion strain (−)
ϵ^T	Thermal strain (−)
ϵ_c	Hoop strain (−)
ϵ_r	Radial strain (−)
Γ	Path
Λ	Area enclosed in the path Γ
ν	Poisson ratio (−)
Ω	Partial molar volume (m^3/mol)
σ_c	Hoop stress (MPa)
σ_h	Hydrostatic stress (MPa)
σ_r	Radial stress (MPa)

Recursive superscripts

max	Maximum

References

[1] Mingchao Liu, Yixiang Gan, Dorian AH Hanaor, Bin Liu, and Changqing Chen. An improved semi-analytical solution for stress at round-tip notches. *Engineering fracture mechanics*, 149:134–143, 2015.

[2] Aurelio Somà, Francesca Pistorio, and Muhammad Mubasher Saleem. Study of notched mems specimen: elasto-plastic modeling and experimental testing. *Journal of Micromechanics and Microengineering*, 32(2):025006, 2022.

[3] T L Anderson. *Fracture Mechanics: Fumdamental and Applications.* 2017.

[4] David Broek. *Elementary engineering fracture mechanics.* Springer Science & Business Media, 2012.

[5] G. R. Irwin. Analysis of Stresses and Strains Near the End of a Crack Traversing a Plate. *Journal of Applied Mechanics*, 24(3):361–364, 06 2021.

[6] P. Paris and F. Erdogan. A Critical Analysis of Crack Propagation Laws. *Journal of Basic Engineering*, 85(4):528–533, 12 1963.

[7] Y. Murakami and L. M. Keer. Stress intensity factors handbook, vol. 3. *Journal of Applied Mechanics*, 60(4):1063–1063, 12 1993.

[8] Hiroshi Tada, Paul C. Paris, and George R. Irwin. *The Stress Analysis of Cracks Handbook, Third Edition.* ASME Press, 01 2000.

[9] George C Sih. Handbook of stress-intensity factors: Stress-intensity factor solutions and formulas for reference. *Bethlehem, Pa., Lehigh University, 1973. 815 p*, 1973.

[10] S Al Laham and Structural Integrity Branch. *Stress intensity factor and limit load handbook*, volume 3. British Energy Generation Limited Gloucester, UK, 1998.

[11] Allyson L Hartzell, Mark G Da Silva, and Herbert R Shea. *MEMS reliability.* Springer Science & Business Media, 2010.

[12] Alberto Corigliano, ALDO Ghisi, Giacomo Langfelder, A Longoni, Federico Zaraga, and A Merassi. A microsystem for the fracture characterization of polysilicon at the micro-scale. *European Journal of Mechanics-A/Solids*, 30(2):127–136, 2011.

[13] Francesca Pistorio and Aurelio Soma'. Fatigue fracture mechanics in gold-based mems notched specimens: experimental and numerical study. *Journal of Micromechanics and Microengineering*, 2023.

[14] Ping Li, Yibo Zhao, Yongxing Shen, and Shou-Hang Bo. Fracture behavior in battery materials. *Journal of Physics: Energy*, 2(2):022002, 2020.

[15] Francesca Pistorio, Davide Clerici, Francesco Mocera, and Aurelio Somà. Review on the experimental characterization of fracture in active material for lithium-ion batteries. *Energies*, 15(23):9168, 2022.

[16] Francesca Pistorio, Davide Clerici, Francesco Mocera, and Aurelio Somà. Review on the numerical modeling of fracture in active materials for lithium ion batteries. *Journal of Power Sources*, 566:232875, 2023.

[17] Michel Armand, Peter Axmann, Dominic Bresser, Mark Copley, Kristina Edström, Christian Ekberg, Dominique Guyomard, Bernard Lestriez, Petr Novák, Martina Petranikova, et al. Lithium-ion batteries–current state of the art and anticipated developments. *Journal of Power Sources*, 479:228708, 2020.

[18] Francesco Mocera, Aurelio Somà, Salvatore Martelli, and Valerio Martini. Trends and future perspective of electrification in agricultural tractor-implement applications. *Energies*, 16(18):6601, 2023.

[19] Francesco Mocera, Valerio Martini, and Aurelio Somà. Comparative analysis of hybrid electric architectures for specialized agricultural tractors. *Energies*, 15(5):1944, 2022.

[20] Francesco Mocera and Aurelio Somà. Analysis of a parallel hybrid electric tractor for agricultural applications. *Energies*, 13(12):3055, 2020.

[21] Francesco Mocera, Elena Vergori, and Aurelio Somà. Battery performance analysis for working vehicle applications. *IEEE Transactions on Industry Applications*, 56(1):644–653, 2019.

[22] Aurelio Somà. Trends and hybridization factor for heavy-duty working vehicles. *Hybrid Electric Vehicles*, pages 3–32, 2017.

[23] Jorn M Reniers, Grietus Mulder, and David A Howey. Review and performance comparison of mechanical-chemical degradation models for lithium-ion batteries. *Journal of The Electrochemical Society*, 166(14):A3189–A3200, 2019.

[24] Rutooj Deshpande, Yang-Tse Cheng, and Mark W Verbrugge. Modeling diffusion-induced stress in nanowire electrode structures. *Journal of Power Sources*, 195(15):5081–5088, 2010.

[25] Ying Zhao, Peter Stein, Yang Bai, Mamun Al-Siraj, Yangyiwei Yang, and Bai-Xiang Xu. A review on modeling of electro-chemo-mechanics in lithium-ion batteries. *Journal of Power Sources*, 413:259–283, 2019.

[26] Davide Clerici, Francesco Mocera, and Aurelio Somà. Analytical solution for coupled diffusion induced stress model for lithium-ion battery. *Energies*, 13(7):1717, 2020.

[27] Davide Clerici, Francesco Mocera, and Aurelio Somà. Shape influence of active material micro-structure on diffusion and contact stress in lithium-ion batteries. *Energies*, 14(1):134, 2020.

[28] Davide Clerici and Francesco Mocera. Micro-scale modeling of lithium-ion battery. In *IOP Conference Series: Materials Science and Engineering*, volume 1038, page 012007. IOP Publishing, 2021.

[29] Davide Clerici, Francesco Mocera, and Aurelio Somà. Experimental characterization of lithium-ion cell strain using laser sensors. *Energies*, 14(19):6281, 2021.

[30] Davide Clerici, Francesco Mocera, and Aurelio Soma. Electrochemical–mechanical multi-scale model and validation with thickness change measurements in prismatic lithium-ion batteries. *Journal of Power Sources*, 542:231735, 2022.

[31] David John Cartwright and DP Rooke. Green's functions in fracture mechanics. In *Fracture Mechanics*, pages 91–123. Elsevier, 1979.

[32] Francesca Pistorio, Davide Clerici, Francesco Mocera, and Aurelio Somà. Coupled electrochemical–mechanical model for fracture analysis in active materials of lithium ion batteries. *Journal of Power Sources*, 580:233378, 2023.

[33] D Clerici, F Mocera, and F Pistorio. Analysis of fracture behaviour in active materials for lithium ion batteries. In *IOP Conference Series: Materials Science and Engineering*, volume 1214, page 012018. IOP Publishing, 2022.

[34] David Percy Rooke and David John Cartwright. Compendium of stress intensity factors. *Procurement Executive, Ministry of Defence. H. M. S. O. 1976, 330 p(Book).*, 1976.

[35] J. R. Rice. A path independent integral and the approximate analysis of strain concentration by notches and cracks. *Journal of Applied Mechanics*, 35(2):379–386, 06 1968.

[36] S. Aoki, K. Kishimoto, and M. Sakata. Elastic-plastic analysis of crack in thermally-loaded structures. *Engineering Fracture Mechanics*, 16(3):405–413, 1982.

[37] K. Kishimoto, S. Aoki, and M. Sakata. On the path independent integral- J. *Engineering Fracture Mechanics*, 13(4):841–850, 1980.

[38] Y. F. Gao and M. Zhou. Coupled mechano-diffusional driving forces for fracture in electrode materials. *Journal of Power Sources*, 230:176–193, 2013.

[39] Hamed Haftbaradaran and Jianmin Qu. A path-independent integral for fracture of solids under combined electrochemical and mechanical loadings. *Journal of the Mechanics and Physics of Solids*, 71(1):1–14, 2014.

[40] Min Zhang, Jianmin Qu, and James R. Rice. Path independent integrals in equilibrium electro-chemo-elasticity. *Journal of the Mechanics and Physics of Solids*, 107:525–541, 2017.

[41] Di Chen, Dominik Kramer, and Reiner Mönig. Chemomechanical fatigue of limn1. 95al0. 05o4 electrodes for lithium-ion batteries. *Electrochimica Acta*, 259:939–948, 2018.

[42] Na Lin, Zhe Jia, Zhihui Wang, Hui Zhao, Guo Ai, Xiangyun Song, Ying Bai, Vincent Battaglia, Chengdong Sun, Juan Qiao, et al. Understanding the crack formation of graphite particles in cycled commercial lithium-ion batteries by focused ion beam-scanning electron microscopy. *Journal of Power Sources*, 365:235–239, 2017.

[43] Rutooj Deshpande, Mark Verbrugge, Yang-Tse Cheng, John Wang, and Ping Liu. Battery cycle life

prediction with coupled chemical degradation and fatigue mechanics. *Journal of the Electrochemical Society*, 159(10):A1730, 2012.

[44] Junya Shao, Junfu Li, Weizhe Yuan, Changsong Dai, Zhenbo Wang, Ming Zhao, and Michael Pecht. A novel method of discharge capacity prediction based on simplified electrochemical model-aging mechanism for lithium-ion batteries. *Journal of Energy Storage*, 61:106788, 2023.

[45] Justin Purewal, John Wang, Jason Graetz, Souren Soukiazian, Harshad Tataria, and Mark W Verbrugge. Degradation of lithium ion batteries employing graphite negatives and nickel–cobalt–manganese oxide+ spinel manganese oxide positives: Part 2, chemical–mechanical degradation model. *Journal of power sources*, 272:1154–1161, 2014.

[46] Gyuyeong Hwang, Niranjan Sitapure, Jiyoung Moon, Hyeonggeon Lee, Sungwon Hwang, and Joseph Sang-Il Kwon. Model predictive control of lithium-ion batteries: Development of optimal charging profile for reduced intracycle capacity fade using an enhanced single particle model (spm) with first-principled chemical/mechanical degradation mechanisms. *Chemical Engineering Journal*, 435:134768, 2022.

[47] Kai Zhang, Junwu Zhou, Tian Tian, Yue Kai, Yong Li, Bailin Zheng, and Fuqian Yang. Cycling-induced damage of silicon-based lithium-ion batteries: Modeling and experimental validation. *International Journal of Fatigue*, 172:107660, 2023.

Feasibility of the tower crane jib made of composite material

L Solazzi

University of Brescia, Department of Mechanical and Industrial Engineering, via Branze 38, 25123 Brescia, Italy

E-mail: luigi.solazzi@unibs.it

Abstract. The main purpose of this research is to design a jib of a tower crane using composite material and to compare the effects/performance on the whole machine due to this lightweight process. The research starts by sizing the jib with steel material. Many different load conditions were involved and different criteria were assumed; those are: stress safety factor, stiffness, dynamic performance (modal) and buckling phenomenon which is the most important parameter. Then other non-standard load conditions were applied to the tower crane: moving load and jib (rotation and elevating) and adopting wind variable in time. The results of these load conditions were assumed to design a new jib made of pultruded composite material. The next step involves the design of additional elements of the crane: ropes, counter boom, counterweight, electric motor, etc. The last step concerns the economic feasibility of the new solution. The results show that the weight of the jib made by composite material is about 25% of the one made by steel and the economic payout for making the composite arm can be equalized in about a year and a half of machine operation, making the proposed solution very worthwhile.

1. Introduction

Nowadays energy efficiency is one of the most important issue, at which many efforts are oriented. In the mechanical field, and in particular in the machine design sector, this topic can be studied in almost two aspects. The first one is by increasing the motor efficiency and the second one is by lightweighting the components. For example, this problem and its approach to study can be seen in the automotive sectors [1,2,3]. The materials choice for lightweighting the structures can be done by adopting some criteria, for example through the Asby diagram [4,5,6]. These maps correlate the mechanical properties of different materials: strength, Young modulus, density, and so on, to help the designers choose the best solution.

The lightweight of the structural component is already affronted in different fields, such as the excavator [7], pressure vessel [8,9,10], hydraulic cylinder [11], etc.

The present research is part of this issue i.e. the structural lightweihting of the machines and components correlated to the lifting and transport industry. In particular, the present research aims to evaluate the feasibility study inherent the implementation of a boom in composite material to a very big tower crane to assess the implications of this application on the entire machine and to estimate, including an economic point of view, the convenience of this innovative application.

2. Description of the tower crane

The tower crane adopted for this research is a classical tower crane with the jib that can be raised. In the horizontal position, the angle concerning the horizontal plane is 14,5° while in the vertical position is 85°. The length of the jib is 46,5 m and the tower height is 29 m. The maximum height below the hook is 75,5 m. These dimensions were acquired by a similar tower crane built by Liebherr company. The working load limit adopted is 11000 N. Figure 1 shows the crane and the jib.

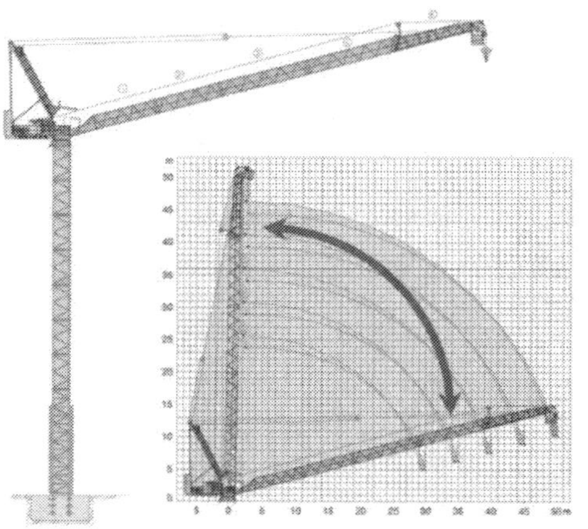

Figure 1. Tower crane size and the two different positions of the jib: horizontal and vertical.

3. Jib design in steel material

Three are the main criteria adopted to design the jib. The first one considers the stiffness of the jib, i.e. the displacement. This value can be imposed by adopting the ratio between the length of the jib and the displacement greater than 200. The second criterion considers the safety factor with respect to the yield of the materials. The value adopted is 3,5 to include the dynamic effect due to the moving load and the fatigue phenomena (that must be verified after the jib design). The last criterion imposed regards the safety factor concerning the buckling phenomena evaluated through the classical Euler equations. [12,13]. By a very simplified schematization, it is possible to estimate the action in the jib (in two different configurations i.e. 14,5° and 85°) and then the action on each element (truss) that composes the reticular structure. These forces can be estimated through two different methods. The first one is the method of joints, and the second one is the sections method or Ritter method. The joints method is based on the equilibrium of each node that compose the structure while the Ritter method considers the equilibrium of the portion of the section. The material adopted is the classical structural steel S355 JR (UNI EN 10025) the main mechanical properties are: ultimate strength σR= 510 MPa, yield strength σY =355 MPa, Young modulus E=210000 MPa and the density equal to ρ=7810 kg/m^3.

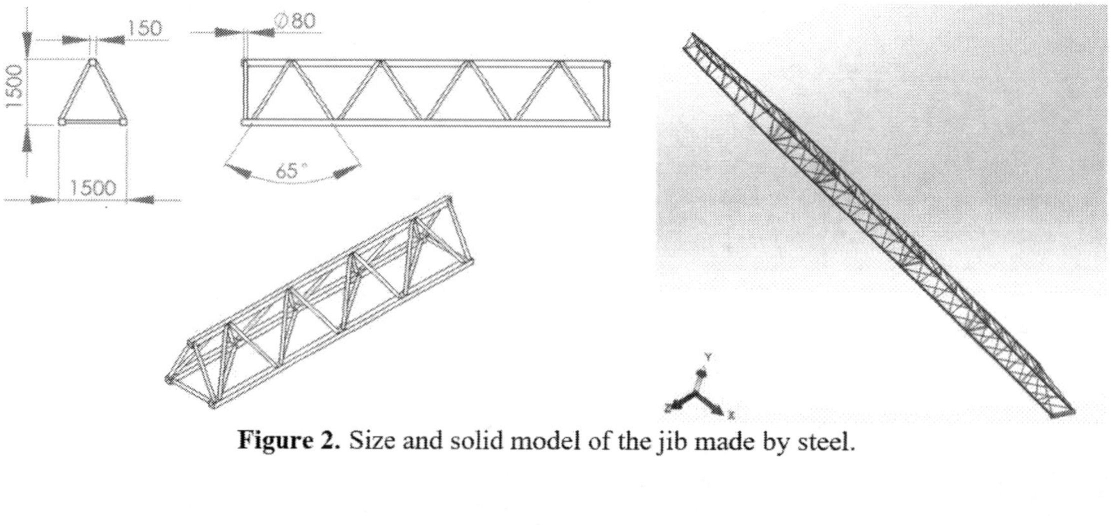

Figure 2. Size and solid model of the jib made by steel.

After the preliminary design, the subsequent phase concerns making the solid model and carrying out the fem analyses, which confirm the hypothesis and boundary assumed to size the jib. It is very important to underline that the geometry of the jib is not double symmetric and the length of free bending defined in Euler's equation is not the same in two planes. In the lateral plane, the free length is equal to the jib length (the schematization is hinge and the roller); while in the front plane, the free length in the Euler's equation is a double of the jib length because the schematization is fixed (ad base) and free at jib end.

Table 1 shows the critical buckling load and the safety factor. Table 2 reports the first natural frequencies of the jib and the mass participation factors. The structure is characterized by low values of natural frequencies (as in the other crane types) and high values of the participation factor within the first frequencies.

Table 1. Critical buckling load and safety factor.

	Numerical value [kN]	Safety factor
Frontal plane	512,9	3,9
Lateral plane	3140,3	24,1

Table 2. Natural frequencies and participation factor for the jib.

N°	Freq. [Hz]	Participation factor [x]	Participation factor [y]	Participation factor [z]
1	0,592	9,3E-07	6,3E-05	63,57
2	2,220	5,678	72,92	0,044
3	2,293	0,0117	0,1106	16,34

3.1. New load conditions.

Over the previous load conditions, many others were involved in the design process to deeper characterise the mechanical behaviour of the jib. These actions are moving arm (rotation and rise) and wind adopting changing in time.

To carry out these analyses (time-varying) one of the most important parameters is the damping value. For this structure type (reticular /lattice) the damping values change from 0,5 to 12 % of critical damping [14,15]. In this study, the value adopted is 3%. The numerical analyses were carried out through Solid Works ® software and the methodology adopted is the superpositions modal method; the damping value applied is the Rayleigh, which is characterized by two values (α and β), reported in the equation (1) These values depend on the natural frequencies and the values adopted are: : α=0,49 e β=6.6e-4 the values are the same for 14,5° and 85° configurations.

$$\alpha = \frac{\omega_1 * \omega_2 * \xi * (\omega_2 - \omega_1)}{(\omega_2{}^2 - \omega_1{}^2)} \ ; \beta = \frac{2 * \xi * (\omega_2 - \omega_1)}{(\omega_2{}^2 - \omega_1{}^2)} \quad (1)$$

3.1.1 Moving jib.

The moving of the jib can be done by many different motion laws. The parameters adopted are acquired from the brochure of the tower crane and concern the maximum velocity and the power installed on the machine; with these parameters, the classical three segments law of motion was applied to the model. Figure 3 shows the law of motion while figure 4 shows the displacement at the end of the jib. The trend is a classical motion of the damping system. The displacement is very low; when the jib is in vertical configuration i.e. 85° the displacement is minimum because the jib's barycentre is near the vertical axis of rotation. The trend in the vertical motion (lift or lower) is very similar even if the displacement magnitude is strictly correlated to the rope stiffness for moving the jib.

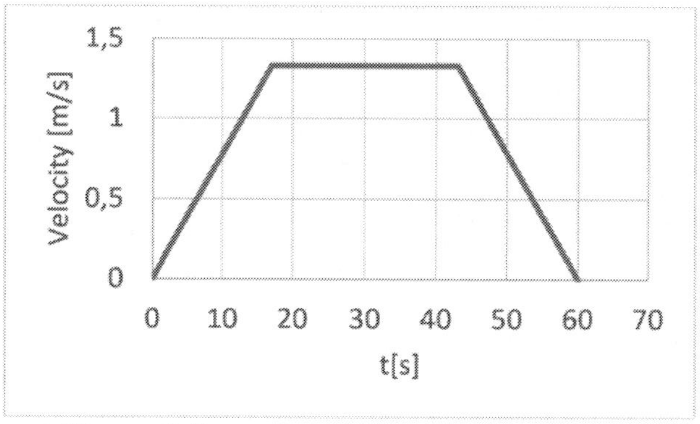

Figure 3. Law of motions applied to the jib.

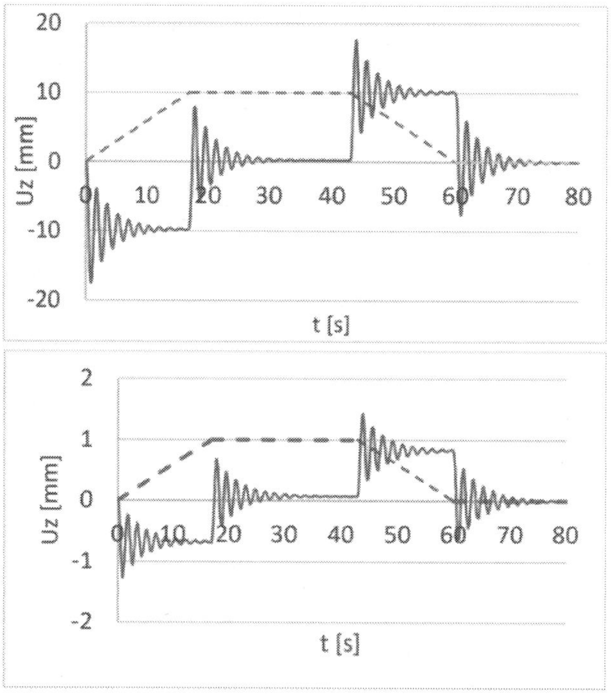

Figure 4. Displacement at the end of the jib in two different geometrical configurations (upper: 14,5°, down: 85°).

3.1.2 Wind actions.
It is known that the wind actions induce forces on the component; these forces are correlated to many different parameters, and one of the most important is the wind velocity. Equation 2 reports these formulae.

$$F_D = \frac{\rho * [V(x,y,z,t)]^2}{2} * C_D * A \,;\; F_L = \frac{\rho * [V(x,y,z,t)]^2}{2} * C_L * A \quad (2)$$

Where:
ρ is the air density;
C_D; C_L are the aerodynamic coefficients (Drag and Lift);
A is the relative area.

The constants CD and d CL are the aerodynamic coefficients and can be estimated by bibliography, numerical analysis (CFD) or experimental analyses (Wind tunnel) in this case the values adopted are: C_D=1,65 and C_L=0,3.
The wind model adopted for this research is stochastic stationary i.e. the wind velocity is a sum of two terms. The first one corresponds to the constant velocity correlated only to the height (Z) of the point in the exam, while the second term is variable in time and correlates to the point position (equation 3).

$$V(x, y, z, t) = V_m(z) + v(x, y, t) \tag{3}$$

The standards for crane design and lifting equipment (EN 13001 series) consider only the constant contribution which is correlated to the height and the environment (near the sea, hill, mountain, close to the buildings, etc.).
The variable component in the wind velocity can be estimated through the power spectral density function. In the literature, there are many different formulations (also in Eurocode). These formulations are characterized by the magnitude of the peek values and its frequency value. [16,17,18,19].
In this research, the model adopted is the Kimal [17,18].
Through a specific program developed by Mathcad software, it is possible to determine the wind speed in each zone into which the arm has been divided, and then determine the forces (by the equation 2) to be included in the fem model. Figure 5 shows the trend velocity in the wo zone near the base (29 m) ant at the end of the jib (75,5 m).

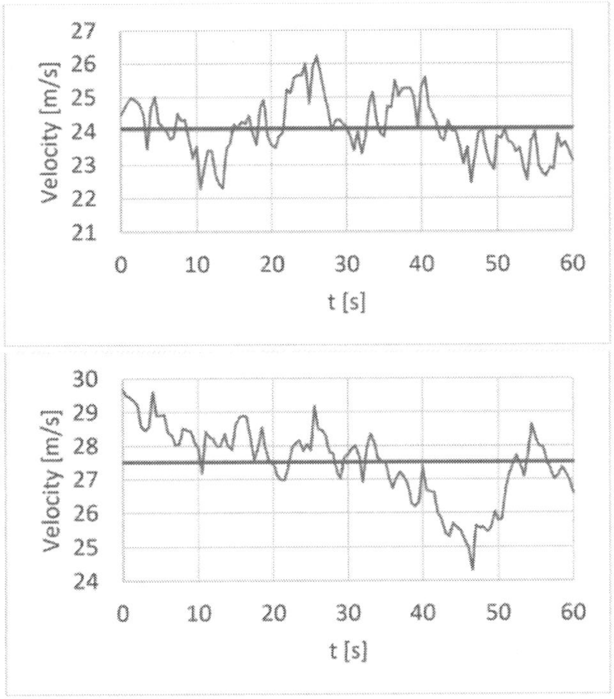

Figure 5. Wind velocity trend in two different jib zones (upper at 29 m; down at 75 m).

Figure 6 shows the displacement trend at the end of the crane. It is very important to point out some issues. Assuming a wind velocity variable in time induces the displacement (and the stresses) variable in time. It is possible, therefore, to estimate the dynamic effect made by the wind and observe that the wind induces fatigue phenomena in the jib structure. It is very important to underline that the dynamic effect generated by assuming a calculation scheme with time-varying speed can cause overturning or buckling phenomena which are the main technical reasons why the crane collapses during an extreme event, like, for example hurricane.

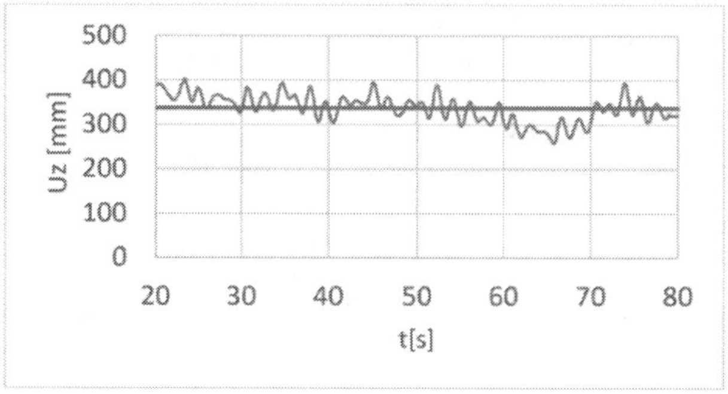

Figure 6. Displacement at the end of the jib due to the wind actions.

4. Jib design in composite material

The material used to design the jib with composite material is a commercial product, it is the pultruded composite material with carbon fiber (65% in volume) and epoxy resin (35% in volume). The main mechanical properties are ultimate strength σR≈ 1000 MPa, Young modulus E=135000 MPa and a density equal to ρ=1900 kg/m3. The principal criteria assumed to design the jib with this material are as follows. First of all, the new jib must have a similar mechanical behaviour to the one made by steel i.e. the same performance with respect to the stiffness and buckling parameters. To do this, the product of the Young modulus with the inertia must be very similar for the two solutions, obviously in the two planes, the lateral and frontal plane (equation 4).

$$E_{Steel} * J_{Steel_{x,y}} \leq E_{Composite} * J_{Composite} \tag{4}$$

Other boundary regards the transportability of the solution and the use of already commercially available geometries. With these assumptions, figure 7 shows the final section of the jib. In particular the section changes from (bxh) 1500x1500 mm to 1800x1800 mm, current from 150x150 with thickness 5 mm to 100x100 thickness 8 mm and diagonal/upright from Øe=80 mm and Øi=74 mm to Øe=45 mm and Øi=35 mm for composite solution.

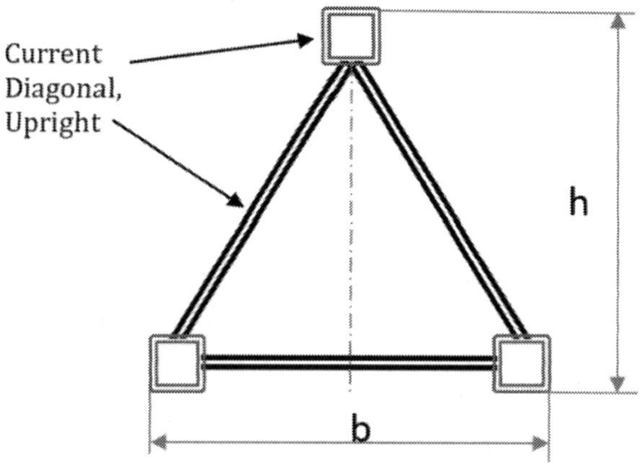

Figure 7. Geometry sections of the jib.

The numerical results are reported in the tables 3 and 4. It can be seen from the reported results that the two solutions are absolutely comparable. Figure 8 shows the jib displacement due to the wind variable in time; for the solution made of composite material the magnitude is a bit larger but totally acceptable.

Table 3. Displacement [mm].

	Steel	Composite	Ratio [%]
14,5°	91,3	28,6	68,6%
85°	5,7	1,4	75,5%

Table 4. Critical buckling load [kN] and safety factor.

	Steel	Composite	Safety factor (Composite)
Frontal plane	512,9	481,7	3,7
Lateral plane	3140,3	2566,4	19,6

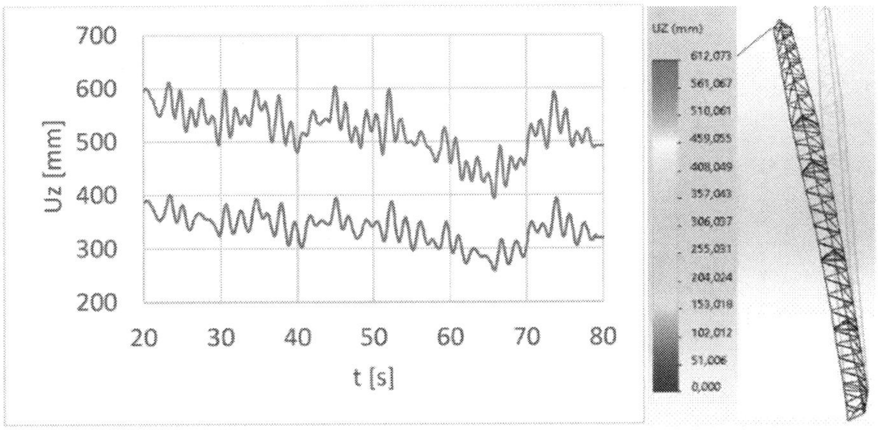

Figure 8. Displacement to the wind actions (red = steel; blue= composite).

5. Build the jib in composite material

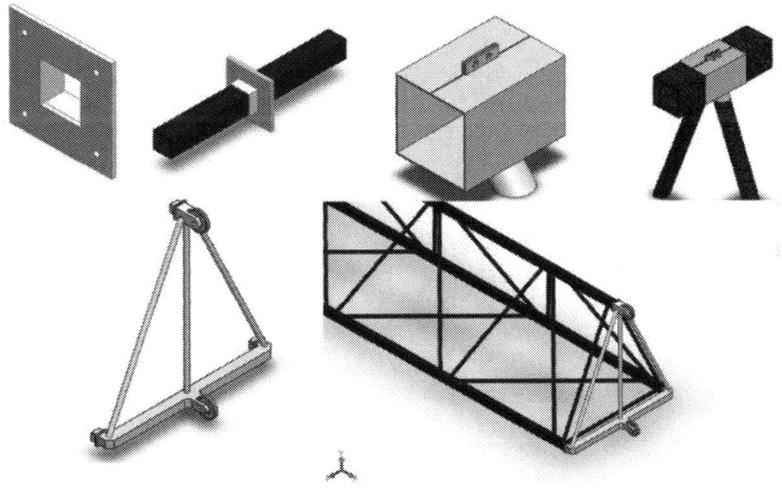

Figure 9. Solutions to join the elements of the jib.

The lattice structure was composted by commercial pultruded composite material (currents, diagonals and uprights). These elements must be assembled. The solutions proposed are reported in figure 9. The current must be joined by a plate in aluminium alloy bolted together and fixed to the pultruded composite material by glue. The same methodology was adopted to join the diagonals and uprights with the current. At the end of the jib was positioned a triangular structure (made of aluminium alloy) to increase the stiffness of the jib and fix the pulleys to move the jib and the payload. At the base of the jib a similar structure also made of aluminium alloy to make a cylindrical hinge to allow rotation of the jib in the lateral plane.

6. Results

The whole weight of the jib made of composite material is 12,17 kN, while in the steel solution the value is 49,37 kN, i.e. the weight in composite is about 25% of the one made by steel. With this performance it is possible to study and lightweight others tower elements. Figure 10 reports the main crane elements (black, in steel solution, while in pink the new solution). Due to the heavy lightweight of the jib it is possible to reduce the pawer to move the jib (obviously with the same performance) both for rotation and lift movements. The total power to move the jib changes from 185 kW to 92 kW. The nominal values change from 60,3 kW to 30,3 for lifting and 98 kw to 32 kW for rotation; the power installed is correlated do the power of commercially available motors.

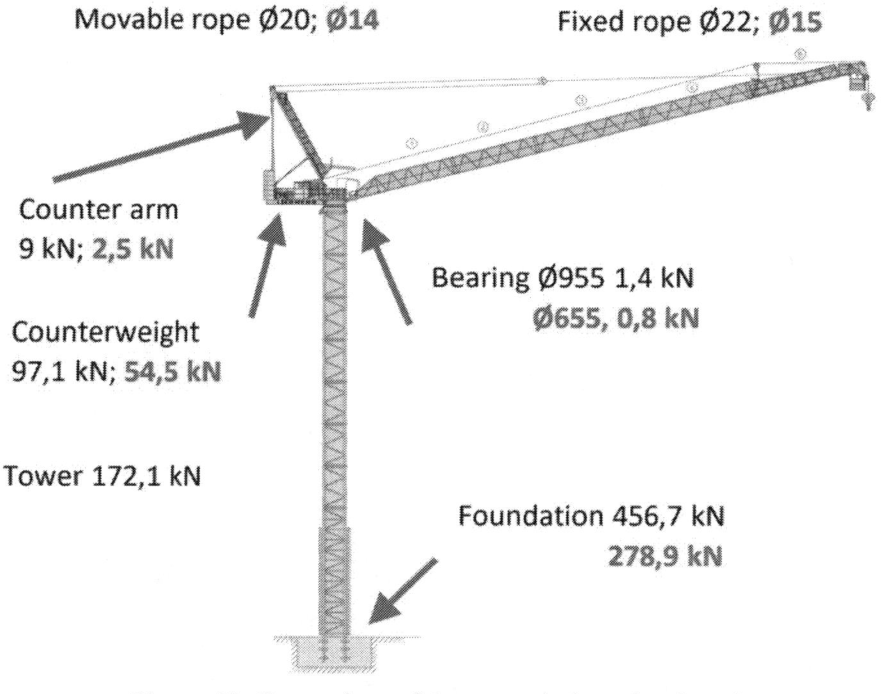

Figure 10. Comparison of the two solutions developed.

7. Economic feasibility

Following the previous results this chapter reports a preliminary analysis in term of the economic point of view. Table 5 reposts the estimate cost to build the composite solution and the operating cost.
This value was evaluated considering 5h per day and 255 days per year the electricity price and a classic load cycle: lifting, rotating and downloading the jib.

IOP Conf. Series: Materials Science and Engineering 1306 (2024) 012010 doi:10.1088/1757-899X/1306/1/012010

Table 5. Cost estimate for the purchase and implementation of the arm [k€].

	Made Jib	Lifting motor	Rotation motor	Operating
Steel	8	9,9	11,5	35
Composite	40	5	7,7	17,5

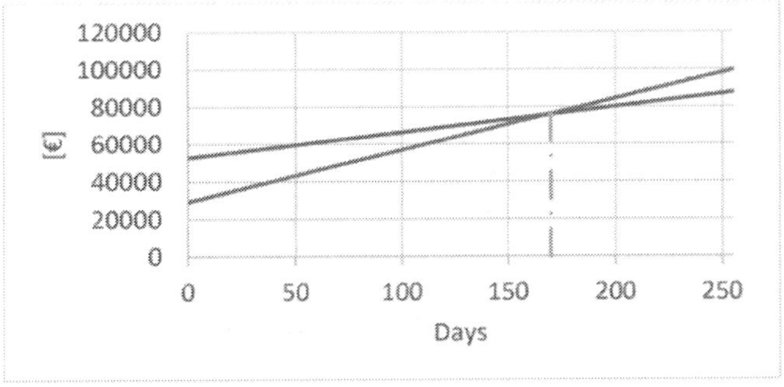

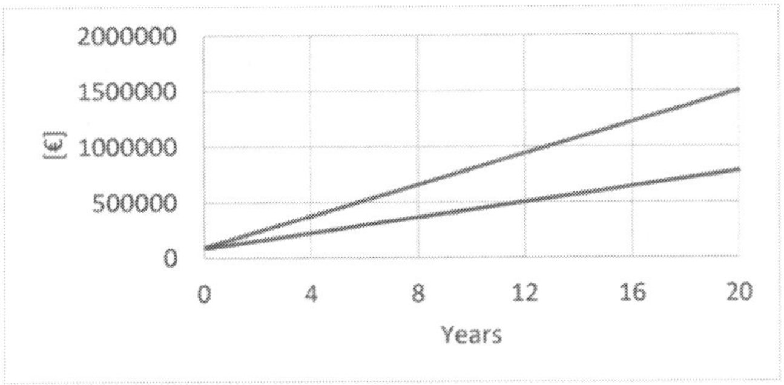

Figure 11. Upper: determination of the equilibrium point; down economic advantage estimate after 20 years. Red for composite solution while blue for steel solution.

8. Conclusions

This research reports the feasibility study of the implementation of an innovative jib made of composite material instead of classical steel. The design was carried out considering different loading conditions in addition to those defined by the standards.

The research shows the preliminary solution to build the component in the composite material and the research was extended to the implications of the lightweight on the other mechanical components of the tower crane.

The lightweight of the jib (the final weight is about 25% of the one made of steel) induces the weight reduction of many crane parts, particularly for the counterweight.

Another important issue is the reduction of the installed power on the crane. Using a new jib, it is possible to adopt half power to move it with the same performance. This aspect cuts the operating cost. The jib made of composite material is more expensive than the one made of steel (5 times). Involving also the operating cost (5 h per 255 days per year) the composite material solution shows a high economic advantage considering 20 years as the useful life of this machine type. Based on the excellent results achieved, research is being developed in at least two aspects. The first one is to evaluate the environmental effects on the structural behaviour of the composite material (temperature, humidity, etc.); on the market there are protective films and their effects must be investigated. The second one is to estimate more precisely the real operations performed with this crane type (work cycle, assembly, transportation, maintenance, etc.) for which the lightening of the machine increases the economic benefits for the solution developed by adopting the composite material.

References

[1] Michael A F, Didac F B and Jordi S 2016 *Materials and Sustainable Development* (Butterworth-Heinemann)
[2] Omar F, Jimi T and Mohini S 2017 *Lightweight and Sustainable Materials for Automotive Applications* (CRC Press)
[3] Mike A and Kara J 2014 *Materials and Design, The Art and Science of Material Selection in Product Design* (Butterworth-Heinemann)
[4] Ashby M F and Jones D R H 2018 *Engineering Materials 1 An Introduction to Properties, Applications, and Design, 5th Edition* (Butterworth-Heinemann)
[5] Ashby M F and Jones D R H 2012 *Engineering Materials 2 An Introduction to Microstructures and Processing, 4th Edition* (Butterworth-Heinemann)
[6] Md Abdul M and Mohd Sapuan S 2013 *Materials Selection and Design* (Singapore: Springer)
[7] Solazzi L, Assi A and Ceresoli F 2017 New design concept for an excavator arms by using composite material, *Appl. Compos. Mater.* **25** (3) p 601 – 617
[8] Solazzi L and Vaccari M 2022 Reliability design of a pressure vessel made of composite materials, *Compos. Struct.* **279**, 114726
[9] Lin J, Zheng C, Dai Y, Wang Z and Lu J 2023 Prediction of composite pressure vessel dome contour and strength analysis based on a new fiber thickness calculation method, *Compos. Struct.* **306**,116590
[10] Coskun T and Sahin O S 2023 Comparison of mechanical properties of the Type 1 and Type 2 composite hydraulic cylinder designs: A numerical study, *Poly. Composite.* **44**, 5, p 2941 – 2955
[11] Solazzi L and Buffoli A 2021 Fatigue design of hydraulic cylinder made of composite material. *Compos. Struct.* **277**, 114647
[12] Reza M E 2018 *Buckling and Postbuckling of Beams, Plates, and Shells* (Springer International Publishing)
[13] Falzon B G and Aliabadi M H 2008 *Computational and Experimental Methods in Structures Buckling and Postbuckling Structures Experimental, Analytical and Numerical Studies* (Imperial College Press)
[14] Taillon J Y, Légeron F and Prud'Homme S 2012 Variation of damping and stiffness of lattice towers with load level *J. Constr. Steel Res.* **71**, p 111 - 118
[15] Ostendorp M 1997 Damping characteristics of a steel lattice structure *Structures Congress – Proceedings* **1** p 358 – 362
[16] Ambrosini R D, Riera J D and Danesi F 2022 Analysis of structure subject to random wind loading by simulation in the frequencies domain, *Probabilist. Eng. Mech.* **17** 233-239
[17] Simiu E and Dong Hun 2019 *Wind effects on structures* (John Wiley & Sons)
[18] Holmes J D 2015 *Wind loading of structures*, (Taylor & Francis Group)
Davenport A G and Riera J D 1998 *Wind effects on building and structures*, (CRC Press)

On the Development of an Open-Source Toolbox for the Structural Analysis of Composite Materials

E Mancini[a,b], **L Mussolin**[a], **G Ambrosi**[b], **F Cianetti**[a] **and C Braccesi**[a]

[a] University of Perugia Department of Engineering Via G. Duranti 93, Perugia 06124, Italy
[b] National Institute of Nuclear Physics (INFN), Section of Perugia, Perugia 06100, Italy

E-mail: edoardo.mancini2@studenti.unipg.it

Abstract. Commonly used in structural applications since the '90s, composite materials represent a valid alternative to metallic solutions. An intense research activity accompanied the spread of this technology to the point that, nowadays, a large variety of modeling techniques and tools are at the designers' disposal. Although the literature on this topic is vast, detailed, and thorough, the need for new, faster, and more efficient design tools and methodologies remains. In this context, this work covers the development of an Open-Source computational laminate analysis toolbox, its structure, and its validation. The objective of the research was the creation of a simple and accessible mean for the calculation of laminates' stresses in complex structural components with a reduced computational cost. To this end, the authors implemented the basics of micro, meso, and micromechanics in a Python-based toolbox and then performed a validation campaign against the results of an FE analysis. The work resulted in an input-generation/output-analysis instrument to be used by designers alongside the FEA.

1 Introduction

Widely used in structural and non-structural applications, composite materials represent a valid alternative to the standard metallic solution. Indeed, the high stiffness-to-mass ratio (when compared to metallic materials), and, most importantly, the possibility to design the directional material properties allowed the spread of the technology in almost all engineering fields. Despite these benefits, the solution comes with new challenges and issues. Although most of the above-cited problems relate to the manufacturing process, the possibility of designing the material mechanical properties is simultaneously an advantage and a drawback. Certainly, if on one hand the resulting component's mechanical properties are optimized for specific loading conditions, on the other the design process has to include the new and non-trivial material design phase, thus leading to increased workloads and the project intricacy.

The authors have grown acquainted with the problem's complexity in the last two years through the collaboration with the Italian National Institute of Nuclear Physics (INFN) concerning the design of composite-based structures for space-borne particle physics experiments[1],[2],[3]. The former project presented the group with the composite design challenge and the associated issues. And on the experience gained during the design process, the toolbox was conceived and developed.

To pursue accessibility and outreach, the development took place in the Python[4],[5],[6] environment.

IOP Conf. Series: Materials Science and Engineering 1306 (2024) 012011 doi:10.1088/1757-899X/1306/1/012011

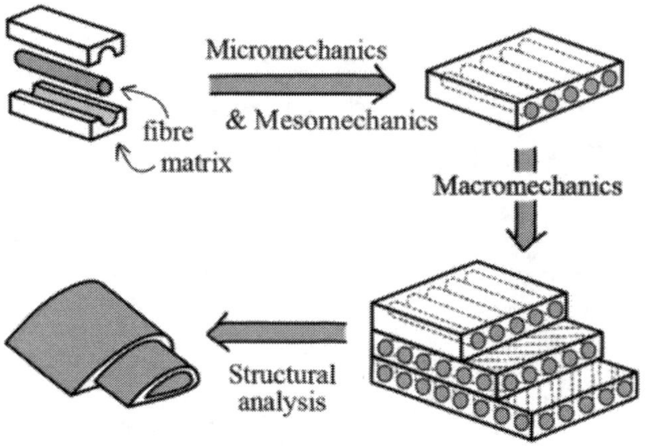

Figure 1: Micro, meso, and macro scale approach.

2 The toolbox

Being the project mentioned in section 1 one of the first research group's approaches to the composite topic, the authors started from the basics[7],[8],[9],[10],[11],[12],[13] and put together a simple but effective toolbox. Two different parts compose the toolbox: the *lamina* class implementing the micro-mechanics and the *laminate* class implementing the meso and macro mechanics. Although either toolset is independent of the other, the outputs from the first can be inputs for the second, thus allowing the definition of a laminate whose lamina properties are not provided by the user but inherited from another Python class.

In literature, the study of composite materials splits into three scales: the micro-scale, the mesoscale, and the macro-scale. The first concerns the material's constituents and deals with them as merged but distinct entities, the second works with homogeneous composite materials laminae (making no distinction between the components), and the third deals with the materials and the structures as a whole without differentiating between constituents and layers. Concerning figure 1, consider a laminated wind turbine blade: analyzing the object micro-mechanically would mean studying the interaction between the fiber and the matrix of the constituents' laminae; instead, the meso-mechanics would concern the stresses and deformations of the single laminae in their reference system; closing with macro-mechanics which deals with the overall laminate with flexural and tensional stiffness matrix associating the punctual deformation with the stress resultants. Due to the close relation between the laminate deformation and the stress in the lamine (often regarded as "ply-by-ply stress state") meso and macro mechanics are entangled, whereas micro-mechanics can be treated as a separate topic, hence the split in the toolbox.

As previously stated, micro-mechanics starts from the material's constituents, their properties, and their relative proportions to extrapolate equivalent composite properties. As a whole, the code requires inputs on the fiber and matrix to compute the lamina's mechanical properties. The laminate tools start from the ply attributes (which can be either external or derived with the former tools) and the stack-up definition to compute properties as the stiffness matrices (A, B, D, and H) or the laminate equivalent orthotropic material constants. In addition, the code allows the computation of the ply-by-ply stress state under a given set of actions (N, M, V). Lastly, the code can detect First Ply Failure and compute safety factors according to the

Tsai-Wu, max stress, or max strain criteria.

3 The code structure

Before moving on to the toolbox applications and validation let us spend some words on the rationale behind the coding. The code follows the Python Enhancement Programming (PEP) 8[14] format and exploits various Python packages like Numpy[15],[16] and SciPy[17],[18]. The code uses class properties to store the majority of the computed outputs, including the list of dictionaries containing the plies characteristic. Additionally, the exploitment of methods (either internal or user accessible) allows the performance of the requested operations. The usage of class methods is interesting from a code flexibility point of view: more of those can be added to the class, thus introducing new functionalities or new formulations can be added to perform the same calculation (e.g. the computation of lamina properties can be performed with different criteria, with this structure it is easy to add new theories to the existing methods and switch between those). The code is currently available on GitHub and PyPi at the following addresses:

- **GitHub:** *https://github.com/edo-147/PyComps*
- **PyPi:** *https://pypi.org/project/PyComps/*

4 Possible applications

The application of such general tools can be multiple. Given the impossibility of covering every usage, the present section deals with the authors' application of the code.

The most obvious and common usage for the code is the calculation of equivalent properties at all scales. As an example, meso-scale properties (ply properties) are a required input for meso-scale FE composite analysis, in the same way as equivalent laminate mechanical properties are for macro-scale ones.

In more sophisticated applications, the toolbox carries on a part of the analysis. Starting from the FEA stress resultants in a specific point as input to a Python routine to perform a preliminary laminate ranking. Specifically, refer to the workflow schematized in figure 2: the design of composite components often exploits the black metal approach where the design starts with a preliminary simulation of isotropic components from which it is possible to identify critical areas and load paths, and which precedes the stack-up ranking and choice, and the final simulation. Computationally wise, the most critical step of the process is the stack-up ranking, a design phase where several different laminae configurations have to be weighed against each other to define the best material solution. If the designer performs this trade-off using only the FE method the computational cost could be high. Instead, the ranking can be performed in Python considering only critical areas and thus evaluating different stack-ups on specific points. Being the stress-resultants the only input needed to perform this analysis, the output from the first simulation(isotropic material) is sufficient to complete the raking. Additionally, if more accurate results are sought the user could perform parametric analysis using an equivalent material formulation and getting the FE input from the Python code, thus switching from a mesoscopic to a macroscopic (less accurate but cheaper) FE model.

5 Validation

Although there was a relevant effort in making the toolbox as general as possible, the authors mainly employed it alongside the FEA. Given the former, it is easy to understand the choice of taking the commercial tool as the validation reference.

As previously stated, the code is divided into two parts, and so it is the validation. In detail, the authors set up two different models in the commercial FE software ANSYS®: one to validate the micromechanical aspects and developed in the Material Design environment, and another for the meso and micromechanical aspects implemented in the Mechanical and ACP environment.

IOP Conf. Series: Materials Science and Engineering 1306 (2024) 012011 doi:10.1088/1757-899X/1306/1/012011

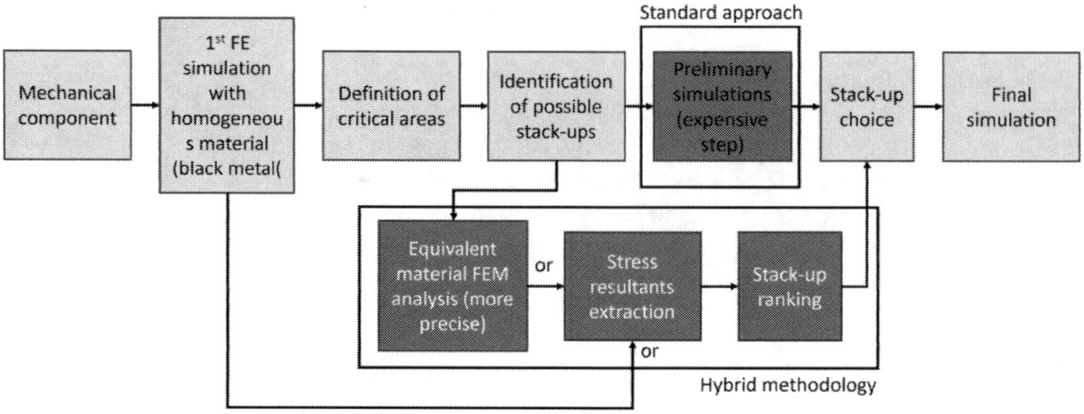

Figure 2: Work-flow of a possible tool application.

5.1 *Micromechanics validation*

Again, ANSYS® Material Designer module is the reference for the micromechanics validation. The commercial toolbox's purpose is the calculation of the mechanical properties of a general (in terms of geometry and materials) Representative Volume Element (RVE). Typical Material Designer applications are sandwich cells, trabecular structures, and composite laminae. The preliminary validation of PyComps considered a unidirectional long-fiber Carbon/Epoxy composite RVE (figure 3). The RVE calculated equivalent properties were then compared with the one from a PyComps analysis with similar inputs. Results, presented in table 1, show good agreement.

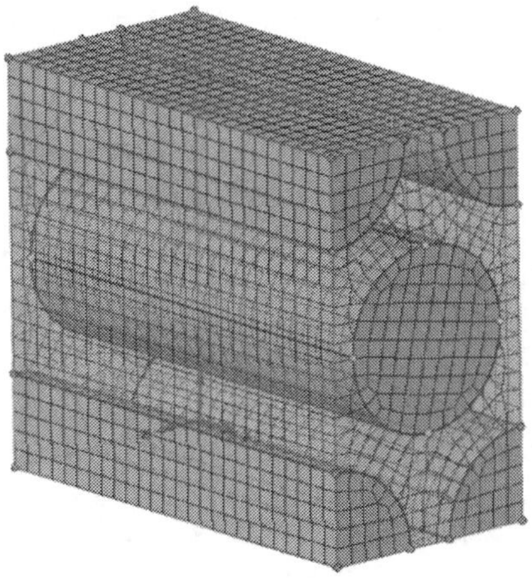

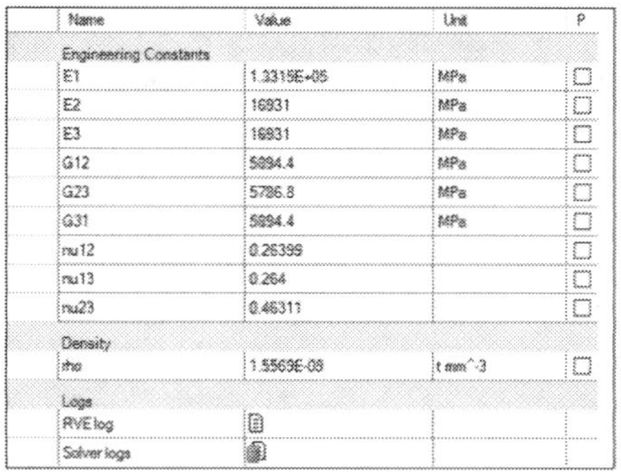

Name	Value	Unit	P
Engineering Constants			
E1	1.3315E+05	MPa	☐
E2	16931	MPa	☐
E3	16931	MPa	☐
G12	5894.4	MPa	☐
G23	5786.8	MPa	☐
G31	5894.4	MPa	☐
nu12	0.26399		☐
nu13	0.264		☐
nu23	0.46311		☐
Density			
rho	1.5569E-08	t mm^-3	☐
Logs			
RVE log			
Solver logs			

Figure 3: Reference model for the micromechanics validation.

5.2 *Meso/macromechanics validation*

Similarly, the PyComps laminate's features were validated. In this case, the reference is an FE model of a composite beam, modeled as a surface (shell elements) and with an imposed lamination (operation performed within the ACP environment). The beam, shown in figure 4,

Table 1: Micromechanics. RVE equivalent properties calculation. Commercial SW vs. toolbox.

Property	Reference	PyComps	Error % PyComps
E_1 - $[GPa]$	133.15	133.152	0.00200
E_2 - $[GPa]$	16.931	17.3883	2.70100
E_3 - $[GPa]$	16.931	17.3883	2.70100
ν_{12}	0.264	0.26352	0.47780
ν_{23}	0.4361	0.45219	3.45770
ν_{13}	0.264	0.26352	0.47780
G_{12} - $[GPa]$	5.8944	5.92256	0.18130
G_{23} - $[GPa]$	5.7868	5.98689	3.69140
G_{13} - $[GPa]$	5.8944	5.92256	0.18130
ρ - $[kg/m^3]$	1556.9	1556.92	0.00193

is clamped on one extremity and loaded with an oblique load on the other. Additionally, the FE computes equivalent composite properties (those of a homogeneous material behaving as the laminated) and provides the stress resultants in a point of the structure. The computation

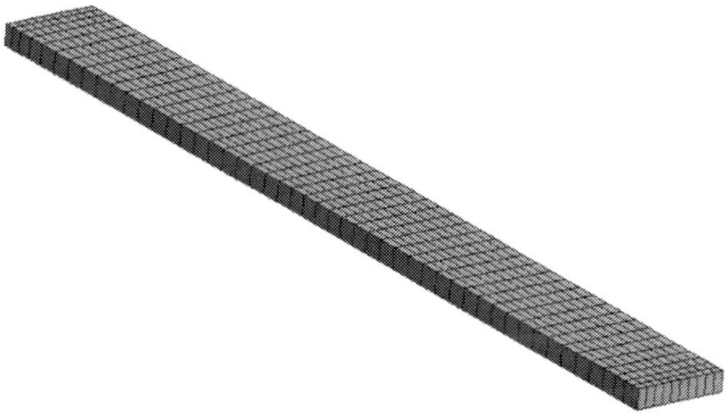

Figure 4: Laminate validation model.

of equivalent properties is performed in PyComps as well, and the delta between the results is presented in table 2. Finally, using the extracted stress resultants on a given FE element, it was possible to compare the PyComps' and the FE's ply-by-ply stress state. For the sake of clarity, the authors propose both a visual result representation (figure 5) and a tabular one (table 3).

6 Conclusions

The presented document briefly describes the work done by the machine design group of the University of Perugia on an Open Source composite structural analysis toolbox. The code follows the PEP-8 and the Object-Oriented Programming approach. A validated version of the code is available on two of the most common Python open-source sharing platforms: GitHub[19] and PyPi[20]. The toolbox generates lamina and laminate objects from two different classes and uses class attributes to store mechanical and physical properties. Although the possible applications

IOP Conf. Series: Materials Science and Engineering 1306 (2024) 012011 doi:10.1088/1757-899X/1306/1/012011

Table 2: Equivalent composite lamina properties.

Modulus	Reference	Python	Error %
G_{12}	9.66 x 10^9	9.63 x 10^9	0.30%
E_1	4.18 x 10^{10}	4.17 x 10^{10}	0.03%
E_2	4.18 x 10^{10}	4.17 x 10^{10}	0.03%
G_{12_flex}	1.35 x 10^{10}	1.34 x 10^{10}	0.04%
E_{1_flex}	3.57 x 10^{10}	3.58 x 10^{10}	−0.34%
E_{2_flex}	3.57 x 10^{10}	3.58 x 10^{10}	−0.34%
G_{13_flex}	2.30 x 10^9	2.25 x 10^9	2.301%
G_{23_flex}	2.30 x 10^9	2.25 x 10^9	2.301%

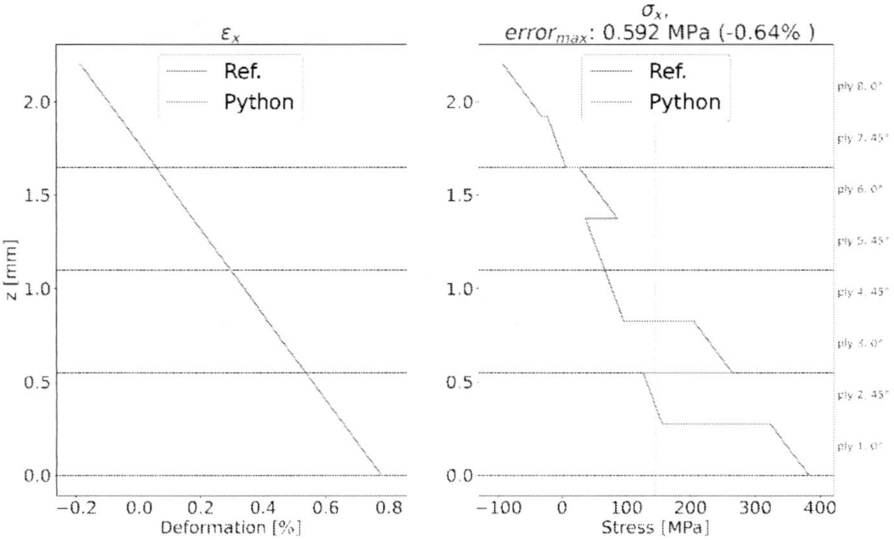

Figure 5: Ply-by-ply stress analysis.

Table 3: Error committed by the toolbox on the stress calculation in different plies.

Ply no. and FEA element	err. σ_x	err. σ_y	err. σ_{xy}
ply no. 0	0.41%	0.33%	0.52%
ply no. 1	0.40%	0.55%	0.52%
ply no. 2	0.33%	0.31%	0.27%
ply no. 3	0.00%	0.00%	0.00%
ply no. 4	0.43%	0.43%	0.44%
ply no. 5	0.33%	0.31%	0.27%
ply no. 6	0.40%	0.55%	0.52%
ply no. 7	0.41%	0.33%	0.52%

are various, the authors briefly dwelt on the toolbox-FEA interactions, input provision, and co-simulation. Finally, the present document describes the preliminary validation of said toolbox, performed against commercial FE software. The validation results show good agreement and justify the code release on free-access platforms.

References

[1] Zhang S N , Adriani O, Albergo S, Ambrosi G et al. 2017 *Proceeding of Science - 35th International Cosmic Rays Conference* **301**
[2] Zhang S N , Adriani O, Albergo S, Ambrosi G et al. 2014 *Proc. of SPIE - The International Society for Optical Engineering* **9114**
[3] Mancini E et al. 2023 *IOP Conference Series: Materials Science and Engineering,* **1257** 1 – 10
[4] Lutz M 2009 *Learning Python* (O'Reilly)
[5] Amos D, Bader D, Jablonski J and Heisler F 2021 *Python Basics: A Practical Introduction to Python 3* (Real Python)
[6] Hunt J 2019 *Advanced Guide to Python 3 Programming* (Springer)
[7] Johnes R M 2014 *Mechanics of composite materials* (CRC Press)
[8] Barbero E J 2017 *Introduction to Composite Materials Design* (CRC Press)
[9] Barbero E J 2008 *Finite Element Analysis of Composite Materials* (CRC press)
[10] Tsai S W 1986 *Composite Design* (Think Composites)
[11] Tsai S W 1980 *Introduction to Composite Materials* (Technomic)
[12] Kattan V 2008 *Mechanics of Composite Materials with MATLAB* (CRC press)
[13] Schramm U, Kitis L, Kang W and Pilkey W D 1994 *Finite Elements in Analysis and Design* **16** 141–162
[14] Van Rossum G, Warsaw B and Coghlan N Python enhancement proposal 08 - pep08 `https://peps.python.org/pep-0008/` url visited on 2023-07-06
[15] Charles R et al. 2020 *Nature Methods* **585** 357 – 362
[16] NumPy Developers Numpy python package documentation `https://numpy.org/doc/` url visited on 2023-07-06
[17] Virtanen, Pauli et al. 2020 *Nature Methods* **17** 261 – 272
[18] The SciPy community Scipy python package documentation `https://docs.scipy.org/doc/scipy/` url visited on 2023-07-06
[19] Mancini E `https://github.com/edo-147/PyComps` url visited on 2023-09-18
[20] Mancini E `https://pypi.org/project/PyComps/` url visited on 2023-09-18

Thermal-structural Modeling of power electronic package: effects of deposition geometry and dry spot on the stress distributions

Giuseppe Mirone[1], Raffaele Barbagallo[1], Giuseppe Bua[1,2], Alessandro Sitta[3], Michele Calabretta[3], and Gaetano Sequenzia[2]

[1] *Department of Civil Engineering and Architecture – University of Catania – Viale A. Doria 6, 95125 Catania – Italy*
[2] *Department of Electrical, Electronic and Computer Engineering – University of Catania – Via Santa Sofia 64, 95123 Catania – Italy*
[3] *Automotive and Discrete Group, R&D department, STMicroelectronics, Catania, 95121, Italy*

Corresponding author: *Giuseppe Mirone, gmirone@dii.unict.it*

Abstract. A Power Electronics package is a heterogeneous system made of semiconductor devices (dies), metallic-ceramic substrate, baseplate and encapsulating material.

The electronic devices can be MOSFETs (Metal Oxide Semiconductors Field Effect Transistors), diodes, IGBTs (Insulated Gate Bipolar Transistors) and passive components such as capacitors, resistors and sensors integrated on support circuit; the complete set of semiconductor devices installed on a package form the so-called "power modulus".

These devices play an important role in the transmission and conversion of energy in electric and hybrid vehicles. Due to the elevated currents and, consequently, the elevated temperatures at which these systems work, and due to the mismatch of the thermal-mechanical properties of the materials from which they are made, stresses and strains develop within the devices.

Such stresses can be locally concentrated and amplified, depending on the deposition geometry of the various layers constituting the semiconductor device. Edge termination structures are essential to decrease the electric field at die's edges, and, including brittle compounds in their composition, like Si_3N_4 or SiO, they are quite sensitive to mechanical stress.

Another reason that may cause the stresses concentrations is the presence of dry spots. Dry spots are areas of various size, located at the interface between die or leadframe and the encapsulating resin, where there is a total or partial lack of adhesion.

The aims of this work are, mainly, two. At first, four linear finite element analyses have been performed in order to evaluate the stresses concentrations at the corners of edge termination structure.

The first case concerns sharp corners and the other cases concern differently filleted corners. In the second part of this work, it has been analysed the dry spot effect on stress distributions. Due to the small size of the defect with respect to the whole package, a Global-Local FEM approach has been used, creating a local subdomain where the mesh is denser than the global domain and a running a further separate numerical analysis which takes its boundary conditions from the global analysis, so delivering reliable small-scale stress and strain gradients. This approach permits to achieve accurate analyses focused on zones of interest of the global domain, limiting the increase of the computational cost of the modelling.

Nomenclature and acronyms:

EMC	Epoxy Molding Compound
GaN	Gallium Nitride
GLFEM	Global Local Finite Element Method
IGBT	Insulated Gate Bipolar Transitor
MOSFET	Metal Oxide Field Effect Transistor
PbSnAg	Lead-Tin-Silver Alloy
Si	Silicon
Si_3N_4	Silicon Nitride
SiC	Silicon Carbide
SiO_2	Silicon Dioxide
TEOS	Tetraethyl Orthosilicate
WBG	Wideband gap

1. Introduction

Power Electronics package is a heterogeneous system that comprises several components, including semiconductor devices (dies), passive components like resistors or capacitance, copper frame, metallic-ceramic substrate, baseplate, and encapsulating material. The dies are the active components of an electronic device, such as a transistor or a diode. They are usually very small, typically a few millimeters in size, and are mounted onto a substrate using brazing, sintering, wire bonding or flip-chip technology [1,2]. Dies, and eventual passive components such as capacitors, resistors and sensors, can be integrated on support circuit; the complete set of semiconductor devices installed on a package form the so-called "power modulus" [3].

Transistors are mainly MOSFETs (Metal Oxide Semiconductors Field Effect Transistors) or IGBTs (Insulated Gate Bipolar Transistors), they are made by silicon (Si) and wideband gap (WBG) compounds like silicon carbide (SiC) and gallium nitride (GaN).

Nowadays, SiC-based and GaN-based power MOSFETs play an important role on the transmission and conversion of the energy. They are gaining more and more attention thanks to their excellent properties in terms of electric resistance and capacitance, withstanding high temperature and voltage [4–6]. WBG compounds are enabling factors to boost performance and efficiency of current and next converters' technology, which plays a key role in industry and automotive frameworks. For example, converters and, more in general, Power Electronics are deployed in several functions of electric and hybrid vehicles, such as on-board charger, DC-DC converter, and traction inverters [7]. The spreading of such technologies pushes the improvement of power semiconductors devices and packages reliability, required to fit the string requirements from the market.

A key aspect is the design and reliability of passivation structures of dies. Passivation scheme is made by one or more thin films, which are deposited on the surface of the device. Its function is to protect the surface of the device from environmental factors such as moisture, dust, and other contaminants. Passivation can also improve the electrical properties of the device by reducing surface recombination and leakage currents. They also support the function of edge termination to reduce of electric field close to die edges, avoiding undesired avalanche breakdown [8].

Some compounds that ensure passivation are silicon nitride (Si_3N_4) or silicon dioxide (SiO_2). They behave like brittle materials, being quite sensitive to mechanical stress.

On the contrary, other elements of the package like the copper baseplate and the encapsulating resin may behave in a rather ductile or viscoplastic way, with lower stress sensitivity and with the possibility of accommodating larger, possibly permanent strains.
The above elements of a package may also exhibit significantly different coefficients of thermal expansion.
The above mismatch between thermal and mechanical properties implies that semiconductor packages are stressed and deformed whenever temperature variations occur within the package [9], which can either be induced by the whole packaging processes , such as the molding compound injection or wire boding process [10], or by the temperature swings typical of their operating conditions in common applications [11]. Stress gradients at weaker interface could lead the degradation of the bonding between two layers, causing delamination or damage induced by thermal-fatigue. All these phenomena may affect the reliability of the whole system. The experimental assessment of thermal stresses and failure modes in power electronics packages is very difficult to perform due to the complex configuration of the integrated devices and their microscopic dimensions, which require precise instruments for destructive and non-destructive testing [12,13]. In recent years, Finite Element and Multi-physics simulations are used from industry and researchers as power tools for the prediction and the evaluation of stresses and damage in these systems, combining structural, thermal and electromagnetic effects [14–17] .
The stresses generated by the temperature changes in non-homogeneous packages can be locally concentrated and amplified, depending on the deposition geometry of the various layers constituting the semiconductor device.

Another reason that can induce local stress magnification is the presence of dry spots. Dry spots are areas of various dimensions, located at the interface between die or leadframe and the encapsulating resin, where a total or partial lack of adhesion occurs. Due to the small size of the defect with respect to the whole package, a Global-Local FEM approach has been used, creating a local subdomain where the mesh is denser than the global domain and a running a further separate numerical analysis which takes its boundary conditions from the global analysis, so delivering reliable small-scale stress and strain gradients [18–20].

2. Materials and Models

2.1. Sharp Edge Model

The aim of this work is an advanced Finite Element modelling of a power electronics package subjected to a quasi-static heating from -65 °C to 150 °C. Under this condition, it is reasonable to assume that thermal equilibrium is ensured instant by instants, for each bodies belonging to the model. Then, thermal transients are neglected and uniform temperature distribution is imposed on all points.

Looking the whole package from the bottom to the top, the layers constituting it are following: the copper layer having the function to form the electrical circuitry for the semiconductor devices (dies) installed on it, the solder layer (typically PbSnAg alloy) which ensures the connection between copper and dies, the Silicon Carbide (SiC), Tetraethyl Orthosilicate (TEOS), and Silicon Nitride (Si_3N_4) layers.

In order to protect the materials sensitive to thermal oxidative phenomena and guarantee the electrical insulation of the electronic components, a thermosetting resin filled with silica particles (Epoxy Molding Compound or EMC) encapsulates the entire system described before.

For the modelling of the thermal-mechanical behavior of the materials, the assumption of linear-elastic-isotropic response was adopted. The thermal-mechanical properties of each materials are reported in the following Table 1.

Table 1: Thermal-mechanical properties of the materials used in the package

	Young Modulus [GPa]	Poisson ratio [-]	Coefficient of Thermal Expansion (CTE) [ppm/K]
Copper	125	0.3	17.5
Solder	5.6	0.4	29
SiC	410	0.28	4
TEOS	59	0.25	1
Si_3N_4	300	0.29	3.4
Resin	24	0.34	12

The entire modeled package has the following dimensions: 14.5 mm x 10 mm x 5 mm. Taking as reference the isometric view of the model (see Figure 1) , the extended surfaces of each layers are parallel to the x-y plane of the fixed cartesian reference system, while the thicknesses are parallel to the z direction. The in-plane dimensions and the thicknesses of the various layers are reported in the following Table 2 .

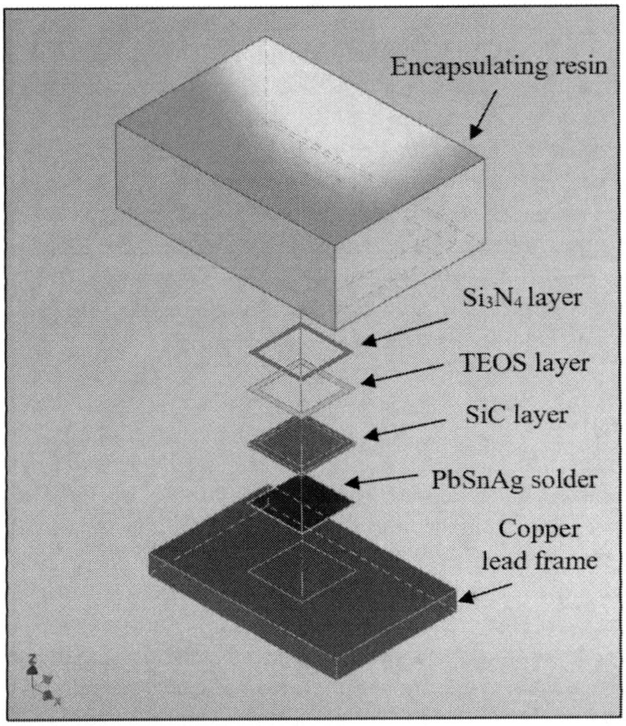

Figure 1: Exploded isometric view of the layers constituting the electronic package

Table 2: In-plane dimensions and thicknesses for each layers

	In-plane dimensions [mm]	Thickness [μm]
Copper	12.5 x 8	1270
Solder	3.6 x 3.6	45
SiC	3.6 x 3.6	180
TEOS	3.6 x 3.6	1.6
Si_3N_4	3.6 x 3.6	1
Resin	14.5 x 10	5000

Furthermore, TEOS and Si_3N_4 layers have a square inner cavity of side 2.85 mm.

It is very important to notice, from Table 2, that the proportions between the in-plane dimensions and thicknesses, for TEOS and Si_3N_4 layers, are about 2000:1 and 3600:1 respectively. Then, in the Finite Element modelling of the whole package, the mesh of the "thick" layers is realized with hexaedric 8-nodes elements, while "thin" layers are discretized with planar quadrilateral 4-nodes elements with thin shell formulation. The commercial software used for the simulations performed for this work is MSC MARC Mentat ®.

The package is supposed to be free to expand and warpage in all directions, as should happen if it was placed and heated on a bench into a climatic chamber. Therefore, two orthogonal roller constrains in x-y plane and a spherical hinge constrain have been imposed on three vertex nodes of the package's base.

For each interface within the package, the bonding between two layers is supposed to be infinite rigid. Glued contact has been implemented both for the contact between solid element/solid element interfaces and planar element/solid element interfaces. Rigid links type n to n have been adopted for the contact between planar/planar elements since they do not support the glued interaction both on top and bottom face.

All the analyses for this work have been performed under the hypothesis of "large displacements", then geometric non-linearities have been taken into account.

2.2. Filleted edge Models

The main objective of this work is the assessment of the effect of the deposition geometry of TEOS and Si_3N_4 layers on stresses distributions around the edge termination structures. The design of the edge termination structures play a crucial role in the mitigation of the electrical field intensity at the die's edge, increasing the energy efficiency of the electronic devices. Also stresses can be locally concentrated and amplified depending on the geometry of these regions. Then, four Finite Element simulations have been performed in order to compare the stresses distributions along paths extracted at representative interfaces between two adjacent layers, for sharp edge geometry and filleted edge geometry.

Looking at one corner of the TEOS (or Si_3N_4) frame (Figure 2), three values of the frame external boundary's fillet radius Rext have been chosen: 0.1 mm, 0.25 mm, and 0.32 mm, while the frame internal boundary's fillet radius Rint are respectively: 0.05 mm, 0.125 mm, and 0.16 mm.

The procedure for the modelling described before is suitable for each filleted model.

IOP Conf. Series: Materials Science and Engineering 1306 (2024) 012012 doi:10.1088/1757-899X/1306/1/012012

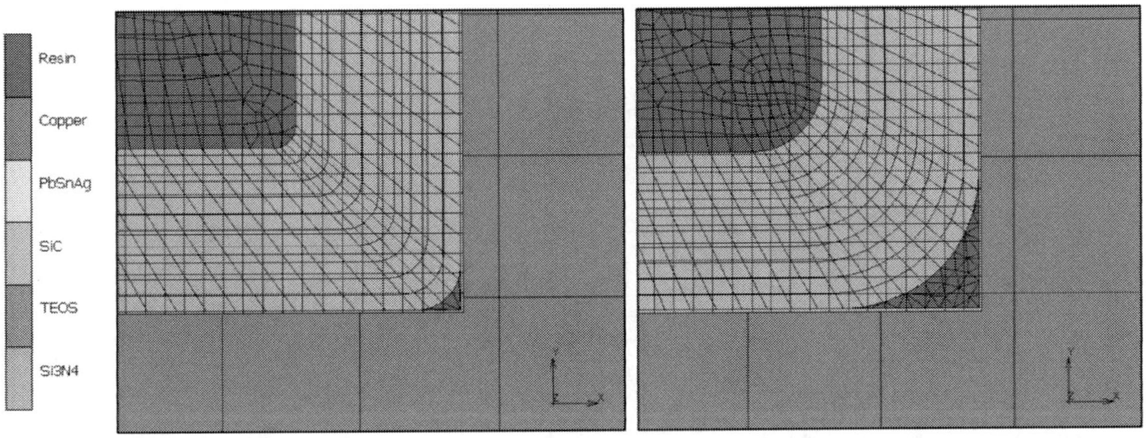

Figure 2: Particular views of two filleted corners: Rext=0.1 mm (on the left), and Rext=0.32 mm (on the right)

2.3. Global-Local approach for the effect of dry spots on stresses distributions

The second objective of this work is the modelling of the package when an artificial defect, in terms of de-bonded area, has been introduced at the Si_3N_4/Resin interface reproducing a so-called "dry spot".

During the process in which the thermosetting resin is injected into a mold containing the metallic-ceramic layers, partial or total lack of adhesion may occur between this one and die or copper leadframe. The de-bonded areas have small dimensions (an average about of 100 μm x 100 μm) with respect to the in-plane dimension of Si_3N_4 frame, so the modelling of these critical interfacial regions requires more finer mesh than other parts of the package. Then, the Global-Local Finite Element Analysis has been used in order to investigate more accurately the stresses distributions near the discontinuity induced by the dry spot.

The Global-Local approach consists on the modelling of two physical domains: the so-called "global" one which is the whole model discretized with a coarse mesh, and a "local" one representing a spatial sub-domain belonging to the global domain. From the analysis on global model, the displacement field is determined for each points with a certain degree of accuracy, depending on spatial and temporal discretization. Then, the global displacements and corresponding nodal forces are assigned to the nodes belonging to the local domain's boundary.

In this work, a dry spot with size of 0.128 mm x 0.332 mm has been modelled in the middle region of the Si_3N_4 frame, sufficiently far from the two corners, neglecting other edge effects.
The representative volume element chosen for the analysis has the following dimensions: 0.5 mm x 0.17 mm x 0.0426 mm (see Figure 3). A portion of resin lying on the two lateral sides of the TEOS and Si_3N_4 frames has been included in the local model in order to capture the in-plane transition from a layer to the encapsulating resin. For the modelling of the dry spot, the elements belonging to the Si_3N_4 contact body at the interface with resin elements have been excluded from the glued interaction.

3. Results and discussion

3.1. Stresses distributions on sharp edge model

Given the geometry, constrains and thermal loads described in section 2, the modelled system has two symmetry planes as shown in **Errore. L'origine riferimento non è stata trovata.**. Due to these symmetries, Equivalent Von Mises stress distributions along each corner's diagonal path are perfectly coincident.

IOP Conf. Series: Materials Science and Engineering 1306 (2024) 012012 doi:10.1088/1757-899X/1306/1/012012

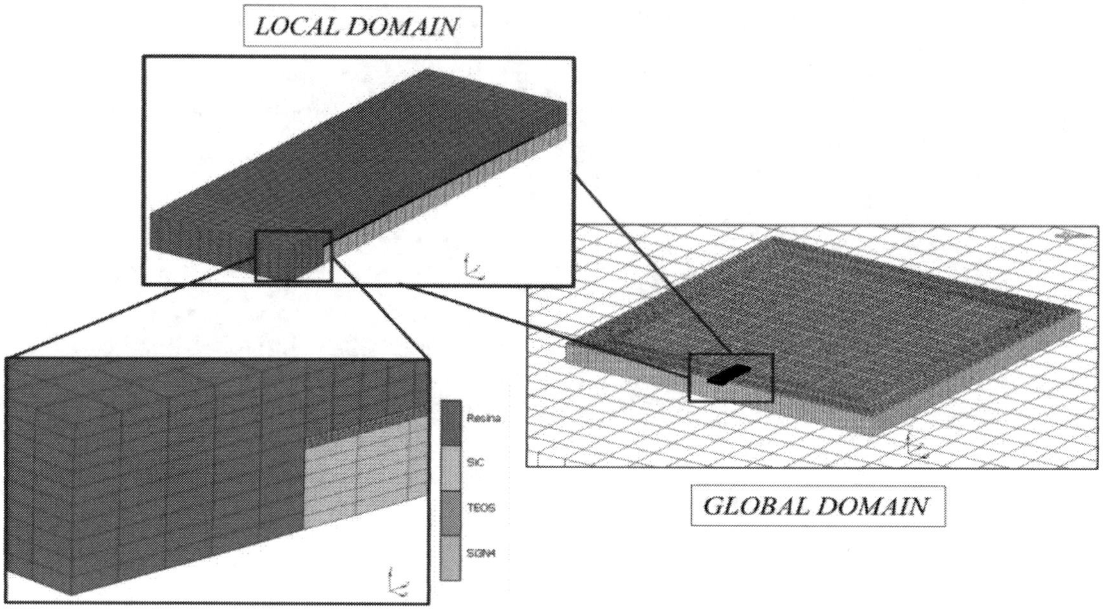

Figure 3: Two axonometric views of the global (on the right) and local (on the left) physical domains

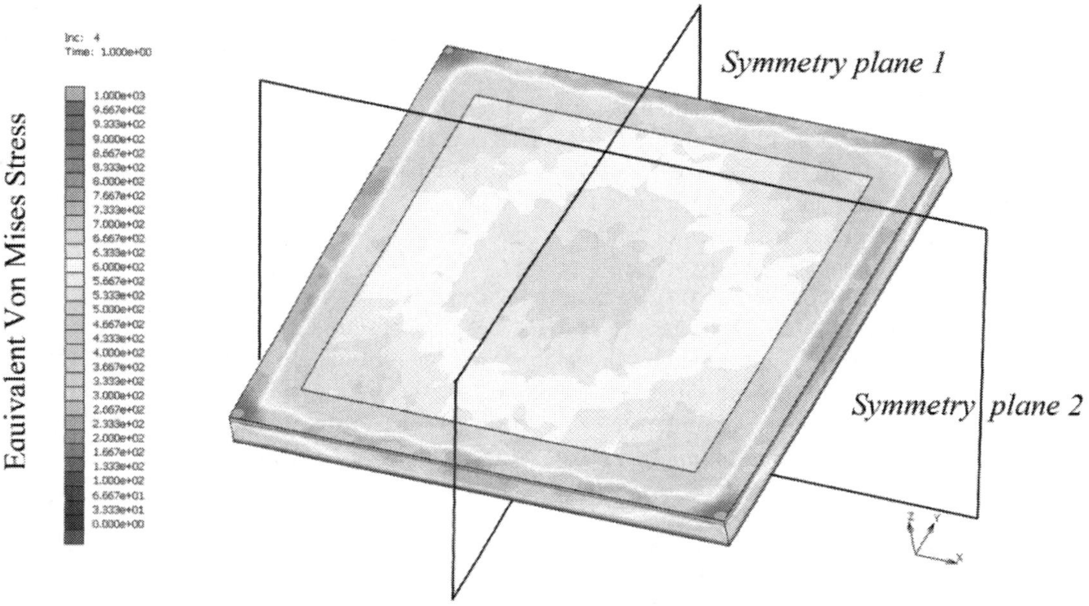

Figure 4: Equivalent Von Mises stress distribution on die at T_{max}=150 °C and symmetry planes

Then, we focused only on one corner of the package. Taking as reference the Figure 5, the in-plane path plots have been extracted starting from point P1 (Arc length = 0 mm) to point P2 (Arc length = 0.53 mm) of the representative corner's diagonal, at four different levels: SiC/TEOS interface, TEOS middle plane, Si_3N_4 middle plane, and Si_3N_4/Resin interface.

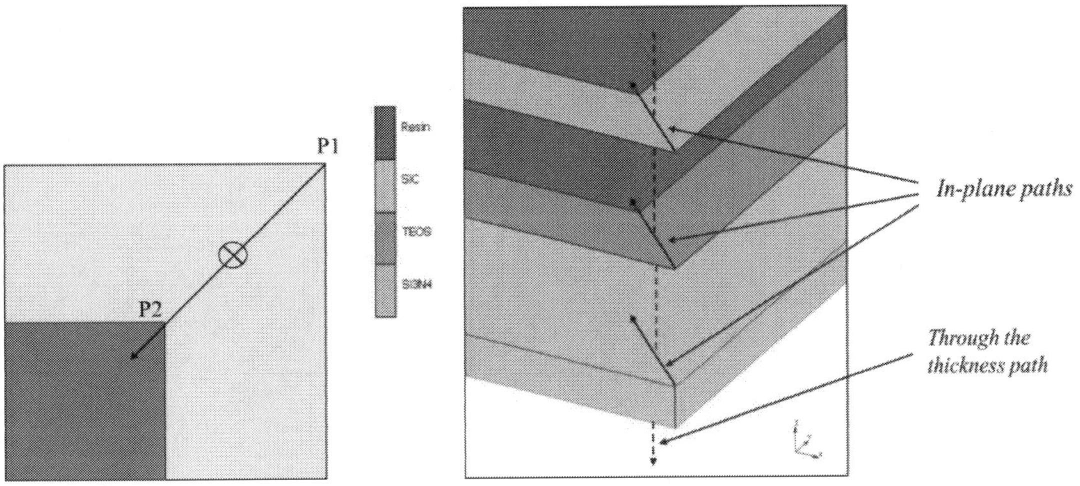

Figure 5: Schematic in-plane (left) and exploded axonometric (right) views of a representative corner of the package and path plot's extremes

For the layers discretized with shell elements, information about interlaminar shear stress are not available at the interfaces with other materials. Therefore, the components of the stress tensor chosen for the following comparisons are: normal stress along x direction, normal stress along y direction, and tangential stress on x-y plane. All these components are extracted at maximum temperature (T=150 °C).

For the "sharp edge" package, stresses distributions are shown in Figure 6. From these diagrams, it is possible to appreciate the effect of stresses concentrations induced by sharp edge configuration particularly on SiC/TEOS interface and Si_3N_4 middle plane. Stress components increase along the path extracted at the four levels defined before, from a minimum value reached "far" from sharp edge to a maximum one near the edge termination, excepted for normal components of stress at SiC/TEOS interface which present the maximum on a point 0.1 mm distant from the sharp termination.

3.2. Effect of filleted corners on stress distributions

For the assessment of the effect of layer's deposition geometry on in-plane stresses distributions, diagonal paths on Si_3N_4 and TEOS middle planes are chosen (results in Figure 7). Moreover, taking as a reference Figure 5, additional path along a straight line passing through the middle point of the corner's diagonal and orthogonal to the x-y plane is proposed, in order to observe stresses gradients through the thickness. This last path starts from a point 5 μm above Si_3N_4/Resin interface (inside encapsulating resin) and finishes 11.6 μm below SiC/TEOS interface (inside SiC layer). The obtained results are shown in Figure 8, where colored rectangles have been added in the diagrams in order to remark the changing of the materials along the arc length: red for resin, green for Si_3N_4, blue for TEOS, and orange for SiC.

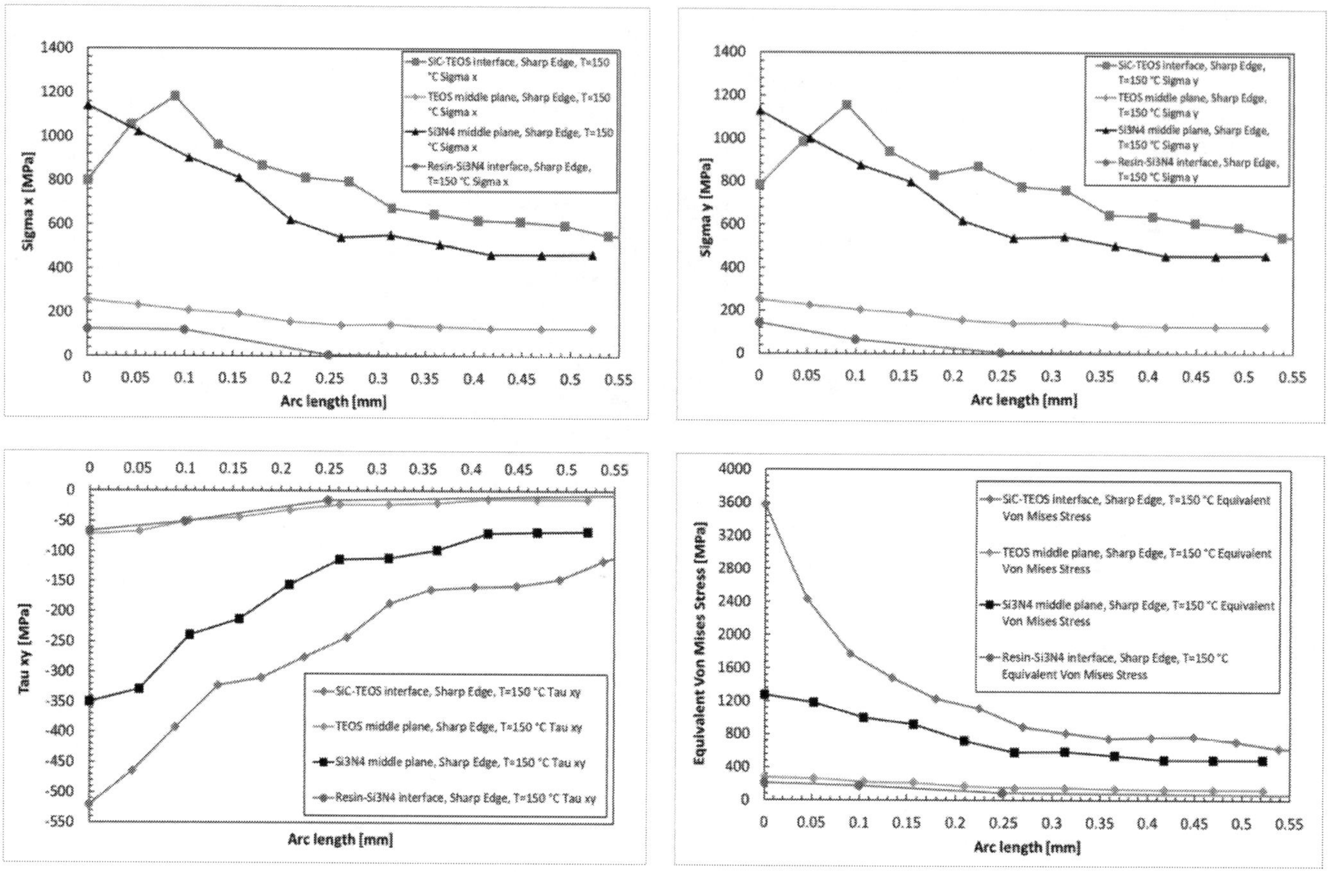

Figure 6: Stress distributions extracted along sharp edge corner's diagonal, at four levels and T_{max}=150 °C

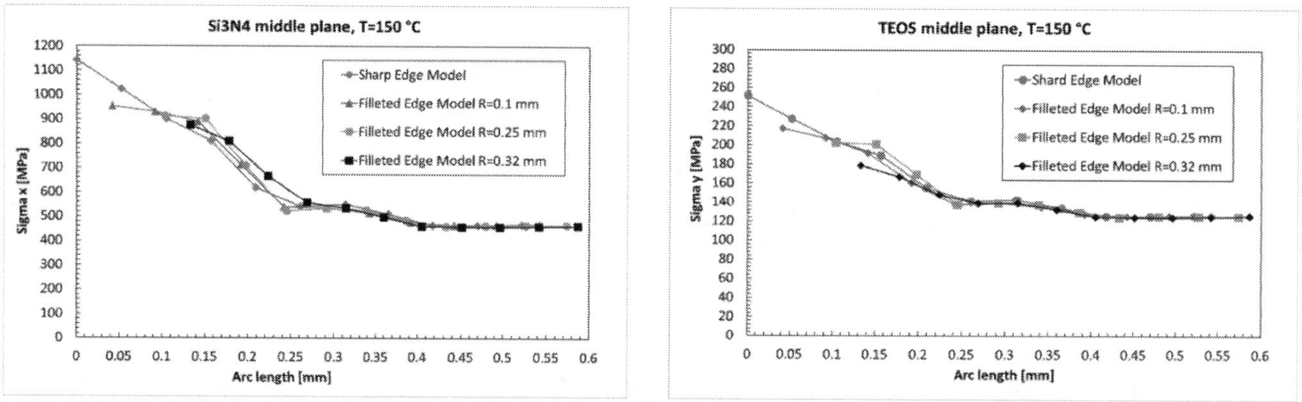

Figure 7: Comparisons between in-plane stresses distributions along corner's diagonal in sharp edge and filleted edge configurations

Looking at the stresses distributions in Figure 7, the filleted and sharp configurations lie on the same curve. Only the maximum values of stresses differ depending on the external fillet radius.

Having this useful information, by a single simulation on sharp edge model is possible to predict the maximum stresses on filleted layers model.

If path along the thickness pass through the middle point of the corner's diagonal, any effect of fillet radius is observed on stresses components (see Figure 7 and Figure 8) . However, these diagrams show the amplitude of the stresses discontinuities due to the transition from one layer to the adjacent one. Taking as reference the component sigma y, stress gradient is 550 MPa at Resin/Si_3N_4 interface, 400 MPa passing from Si_3N_4 middle plane to TEOS middle plane, and about 450 MPa at SiC/TEOS interface.

For the component tau xy, stress gradients assume these following values: 90 MPa at Resin/Si_3N_4 interface, 75 MPa along the transition from Si_3N_4 middle plane to TEOS middle plane, and about 110 MPa at SiC/TEOS interface.

These huge discontinuities are caused by the high mismatch between the thermal-mechanical properties of the materials constituting the model. It is important to remark that performed analyses are linear and elastic and these predicted stresses could be overestimate the real ones since plasticity or damage effects have not been taken into account.

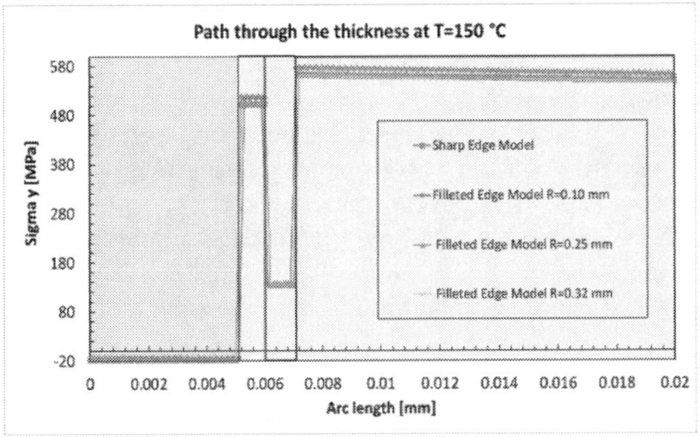

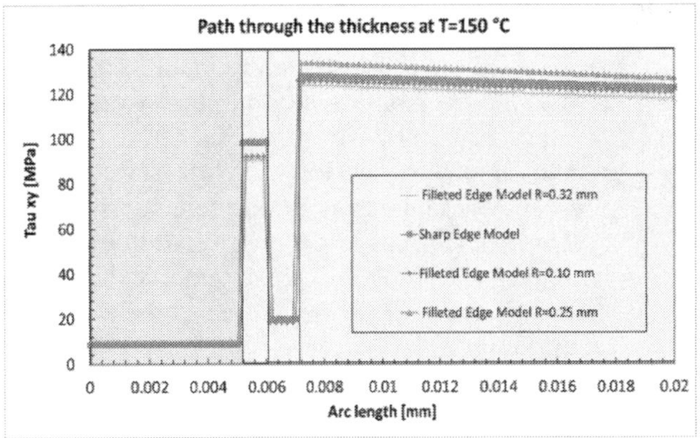

Figure 8: Components sigma y and tau xy through the layers' thickness: comparisons between sharp edge and filleted edge corner

3.3. Global-Local approach: Effect of dry spot on stress distribution

Taking as reference the local model shown in Figure 3, three orthogonal path's directions passing through the centre of the dry spot have been chosen for the evaluation of this de-bonded area on stresses amplifications. The components of stress tensor used for the comparisons are the same of the previous case (subsection 3.1 and 3.2), excepted for the shear stress tau yz which substitutes tau xy since this last one assumed values closed to zero. The obtained results are shown in Figure 9, where the arc length interval between the two dashed lines indicates the region where dry spot is present.

The effect of this defect is negligible if we observe the component of stress sigma x. Indeed, the mean value of the difference between stress in presence and in absence of dry spot (delta sigma reported in secondary axis) is about 15 MPa compared with the mean value of sigma x of 600 MPa (difference of 2.5 %). Local peaks of stress of 25 MPa and 80 MPa are located, respectively, at arc length 0.024 mm and 0.147 mm, where there is the in-plane transition from the "glued zone" to the de-bonded one.
A more relevant stress amplification effect is observed on the tau yz stress distribution. Within the dry spot region, far from the boundary edges, the interlaminar shear passes from a mean value of -50 MPa to zero MPa. Also in this case, the peak stresses of -350 MPa and -400 MPa occur at the glued-dry spot transition.

As expectable, the presence of the fillets and the change of their radii do not induce any change in the predicted through-the-thickness distributions of stresses, as the whole path is taken at the midpoint of the 300 microns-wide track of TEOS and Si_3N_4 (see Figure 3).

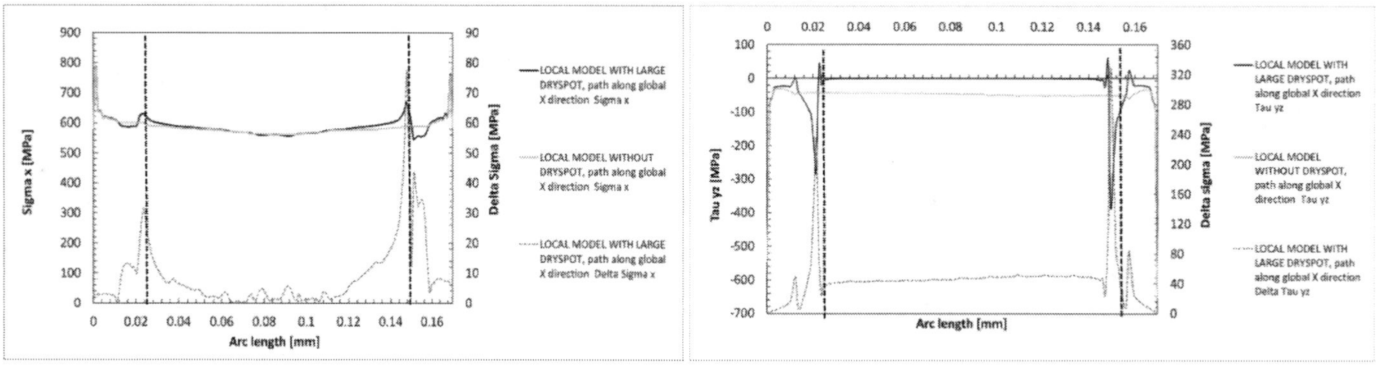

Figure 9: Effect of dry spot on in-plane stresses distributions

4. Conclusions

The present work deals with the prediction of the stress distributions induced within a power electronics package by a reference temperature change, focusing at the interfaces of resin/silicon nitride, TEOS/silicon carbide and within the layers of TEOS and silicon nitride.

The sensitivity of the stress distributions to the deposition layout of the layers has been assessed by numerical analyses performed with different corner's geometric configurations, namely a sharp-edges reference configuration one and three derived configurations with different radii fillets.

The whole package geometry initially modelled in the sharp edge configuration exhibited a double symmetry in the plan view together with its thermomechanical loading system, which successively allowed to model a quarter/corner of the entire system to evaluate the mitigation effect of the fillet radii on the stress maps.

For the sharp edge model, the x, y and xy components of the stress tensor were analysed along four diagonal paths at different heights within the model, in order to assess the reference effect of stress concentrations at the corners of the deposition layout which is likely the critical location where mechanical damage may occur.

The same stresses predicted by models with filleted edge deposition layouts shown a relevant role of the filled radius in the reduction of the maximum stresses.

More in detail, the overall stresses distributions are nearly unaffected by the radii which, instead, tend to truncate such distributions at their outer ends where the increasing spatial trends deliver the higher values: the larger is the fillet radius, the longer is the segment of increasing stress distribution trimmed-out with respect to the sharp-edge configuration. This finding potentially allows to decide the fillet radius to be realized depending on the maximum acceptable stress level, simply based on the stress distributions of the sharp-edge configuration, with no need for further time-consuming finite elements analyses.

The distributions through the thickness of x, y and xy stresses have also been shown in order to quantify stresses gradients and discontinuities along the transition zones between adjacent layers.

The second part of this work concerned the modelling of a de-bonded area located at Resin/Si_3N_4 interface which is called "dry spot". Since the dimension of this defect are too much small compared with the whole model, a global-local approach has been implemented for this task, in order to discretize a portion (local domain) of the entire system (global domain) with a finer mesh, performing a dedicated finite elements analysis with without further increasing the computational cost of the analysis.

Two orthogonal paths, lying in the x-y plane and passing through the centre of the dry spot have been taken to compare the stress distributions with and without this defect.

The dry spot exhibited an irrelevant influence on the normal components of stress: sigma x and sigma y, while it nullifies interlaminar shear stress tau yz on the inner region of the dry spot, far from its boundary edges. Peaks of shear stress occur in the proximity of the in-plane transition zone between glued interface and de-bonded elements.

Acknowledgments
This work has been partially funded by European Union (NextGeneration EU), through the MUR-PNRR project SAMOTHRACE (ECS00000022).

References

[1] Caccuri V, Milhet X, Gadaud P, Bertheau D and Gerland M 2014 Mechanical Properties of Sintered Ag as a New Material for Die Bonding: Influence of the Density *Journal of Elec Materi* **43** 4510–4

[2] Huang Y, Luo Y, Xiao F and Liu B 2019 Failure Mechanism of Die-Attach Solder Joints in IGBT Modules Under Pulse High-Current Power Cycling *IEEE J. Emerg. Sel. Topics Power Electron.* **7** 99–107

[3] Yang Y, Dorn-Gomba L, Rodriguez R, Mak C and Emadi A 2020 Automotive Power Module Packaging: Current Status and Future Trends *IEEE Access* **8** 160126–44

[4] Coffa S, Saggio M and Patti A 2015 SiC- and GaN-based power devices: Technologies, products and applications *2015 IEEE International Electron Devices Meeting (IEDM)* 2015 IEEE International Electron Devices Meeting (IEDM) (Washington, DC, USA: IEEE) p 16.8.1-16.8.5

[5] Mauromicale G, Sitta A, Calabretta M, Oliveri S M and Sequenzia G 2021 Integrated Electromagnetic-Thermal Approach to Simulate a GaN-Based Monolithic Half-Bridge for Automotive DC-DC Converter *Applied Sciences* **11** 8302

[6] Huazhong University of Science and Technology and Chen C 2017 A Review of SiC Power Module Packaging: Layout, Material System and Integration *CPSS TPEA* **2** 170–86

[7] Morya A K, Gardner M C, Anvari B, Liu L, Yepes A G, Doval-Gandoy J and Toliyat H A 2019 Wide Bandgap Devices in AC Electric Drives: Opportunities and Challenges *IEEE Trans. Transp. Electrific.* **5** 3–20

[8] Baliga B J 2008 *Fundamentals of power semiconductor devices* (New York, NY: Springer)

[9] Timoshenko S 1925 Analysis of Bi-Metal Thermostats *J. Opt. Soc. Am.* **11** 233

[10] Dornic N, Khatir Z, Tran S H, Ibrahim A, Lallemand R, Ousten J-P, Ewanchuk J and Mollov S V 2019 Stress-Based Model for Lifetime Estimation of Bond Wire Contacts Using Power Cycling Tests and Finite-Element Modeling *IEEE J. Emerg. Sel. Topics Power Electron.* **7** 1659–67

[11] Mirone G, Sitta A, D'Arrigo G and Calabretta M 2019 Material Characterization and Warpage Modeling for Power Devices Active Metal Brazed Substrates *IEEE Trans. Device Mater. Relib.* **19** 537–42

[12] Ju Y, Saka M and Abé H 2001 Detection of delamination in IC packages using the phase of microwaves *NDT & E International* **34** 49–56

[13] Ong R and Cheong K Y 2016 Non-destructive electrical test detection on copper wire micro-crack weld defect in semiconductor device *2016 IEEE 37th International Electronics Manufacturing Technology (IEMT) & 18th Electronics Materials and Packaging (EMAP) Conference* 2016 IEEE 37th International Electronics Manufacturing Technology (IEMT) & 18th Electronics Materials and Packaging (EMAP) Conference (Georgetown, Penang, Malaysia: IEEE) pp 1–6

[14] Sunil J, Aslam S, Perinpanyagam S and Wileman A 2022 Finite Element Analysis of Si and SiC Power Modules for Accurate Reliability Assessment *2022 IEEE Aerospace Conference (AERO)* 2022 IEEE Aerospace Conference (AERO) (Big Sky, MT, USA: IEEE) pp 1–10

[15] Ye G, Fan X and Zhang G 2022 Practical aspects of thermomechanical modeling in electronics packaging: A case study with a SiC power package *Microelectronics Reliability* **132** 114514

[16] Zhu N, Van Wyk J D and Liang Z X 2004 Thermal-mechanical stress analysis in embedded power modules *2004 IEEE 35th Annual Power Electronics Specialists Conference (IEEE Cat. No.04CH37551)* 2004 IEEE 35th Annual Power Electronics Specialists Conference (Aachen, Germany: IEEE) pp 4503–8

[17] Ye H, Lin M and Basaran C 2002 Failure modes and FEM analysis of power electronic packaging *Finite Elements in Analysis and Design* **38** 601–12

[18] Mirone G, Barbagallo R, Bua G and Rosa G L 2023 Finite Element Simulation and Sensitivity Analysis of the Cohesive Parameters for Delamination Modeling in Power Electronics Packages

[19] Conversion A, Ubando A and Gonzaga J 2022 Global-to-local finite element model of shear stress analysis on Fan-out wafer-level package *2022 IEEE 14th International Conference on Humanoid, Nanotechnology, Information Technology, Communication and Control, Environment, and Management (HNICEM)* 2022 IEEE 14th International Conference on Humanoid, Nanotechnology, Information Technology, Communication and Control, Environment, and Management (HNICEM) (Boracay Island, Philippines: IEEE) pp 1–5

[20] Brinkfeldt K, Edwards M, Ottosson J, Neumaier K, Zschieschang O, Otto A, Kaulfersch E and Andersson D 2015 Thermo-mechanical simulations of SiC power modules with single and double sided cooling *2015 16th International Conference on Thermal, Mechanical and Multi-Physics Simulation and Experiments in Microelectronics and Microsystems* 2015 16th International Conference on Thermal, Mechanical and Multi-Physics Simulation and Experiments in Microelectronics and Microsystems (EuroSimE) (Budapest, Hungary: IEEE) pp 1–7

IOP Conf. Series: Materials Science and Engineering 1306 (2024) 012013 doi:10.1088/1757-899X/1306/1/012013

Multimethodology based on Design-to-Value (DtV), integrated with simulation techniques and prioritization of teamwork for the optimization of a pneumatic rack & pinion actuator

G Maiocchi[1,2,3], **L Vigliano**[2], **G Nicoletto**[1]

[1] Department of Engineering, University of Parma, Via delle Scienze, 181/1, 43124, Parma (PR), Italy
[2] Flowserve S.r.l., Mezzago Campus, Via dele Industrie 23, 20883, Mezzago (MB), Italy
[3] Author to whom any correspondence should be addressed. E-mail: giovanni.maiocchi@unipr.it, gmaiocchi@flowserve.com

Abstract. The investigations here presented focus on the redesign and innovation of a pneumatic rack and pinion actuator for valve actuation, as a case study for investigating the potential of a multimethodology based on the Design to Value (DtoV) process, coupled with design techniques utilizing FEA simulations and giving high priority to teamwork. The final objective of this case study is to show how it is possible to optimize the design, increasing weight efficiency while maintaining performance, and to simplify it, with a reduction of components construction complexity, according to the growing demand for a lean production. The principle that guided all the activities was valorizing the power of teamwork, focusing the team on Safety and Reliability. In an initial phase, all the instruments foreseen by "Design-to-Value" process have been applied, obtaining a classification of the contents of the product constituting its sources of value. Subsequent outputs are proposals for efficient construction solutions, driving a second phase, dedicated to the re-design of the actuator. The peculiarity of this project has been to combine the "Design by Formulas" techniques with advanced FEA simulations ("Design by Analysis"), aiming to stress, deformation, and topology optimization. A two-step experimental validation is used, based on a preliminary "mockup" prototype followed by a complete detailed prototype, for confirming the results of the calculations and simulations, by directly performing a series of in-depth tests. Preliminary obtained results show that the approach based on the described multimethodology, makes it possible to optimize the design of the actuator, maintaining safety, reliability, and performance. In the case studied, the weight reduction is expected to be 8% and economic efficiency increase is expected to be near 20%.

1. Objectives of the activity

In this paper are reported the results of the investigations and of the activities performed during a project of optimization of a rack & pinion actuator, as a case study of application of a multimethodology based on Design to Value (DtoV) process, coupled with design techniques exploiting the potential of FEA. During all the activities the prioritization of teamwork has been a fundamental approach, as show in figure 1.

The rack & pinion actuator is a mature product which however has constant demand from the valve automation sector. First objective of the optimization is to align its features with those that have been identified as the most important sources of value perceived for this product by customers and operators in the sector. Second one is to ensure that the design complies with the most stringent requirements of

IOP Conf. Series: Materials Science and Engineering 1306 (2024) 012013 doi:10.1088/1757-899X/1306/1/012013

the latest versions of the international reference standards (actuators, pressure vessels). Concomitant objectives driving the design choices are to simplify the design, following the growing demand for lean production, increase lightness, maximize economic efficiency. The main final purpose is to design a safety and reliable product.

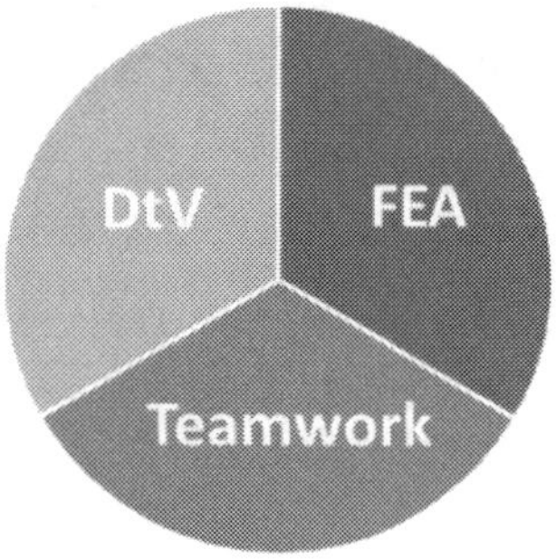

Figure 1. Fundamental components of the studied methodology applied for Actuator Optimization.

2. Product description

The studied product is the "rack and pinion" pneumatic actuator, designed in 1984, shown in Figure2.

An aluminum alloy hollow center body contains two pneumatic pistons, mounting linear gears (the racks), engaging a circular gear located on a vertical shaft (the pinion). This mechanism converts the linear motion of the pistons into the rotary motion of the pinion. The actuator is suitably powered by a pneumatic control system so that to obtain clockwise and anti-clockwise rotations and operate valves with 90° stroke, such as ball or butterfly valves. In the external chambers between pistons and lateral closing flanges, compression springs can be located, making a "spring return" function available. In this configuration the force for performing one of the two strokes is generated by the potential energy accumulated by compressing the springs under pressure action during the previous stroke. In "double acting" function both strokes are operated by pressure.

Figure 2. old design version of the studied rack and pinion actuator.

3. Design-to-Value: introduction to the process and initial activities

Design-to-Value (DtV) is an effective process, developed in the field of "value engineering", for obtaining product innovation, involving several activities aimed to maximize the value of the new design [1]. DtV approach focuses on the value attributed to the product by the customers, in function of available features. Its maximization guides the decisions for the design optimizations [2]. Among main elements of value, DtV takes into consideration safety, reliability, ease of use and maintenance, weight optimization and economic efficiency. Through DtV an attempt is made to understand which characteristics of the product are the most interesting for customers and industry experts, and to translate them into design choices. DtV allows modeling the characteristics of a product, based on what satisfies the real needs, ensuring that the product is perceived as having a high value [3].

During the application of DtV process, some fundamental steps can be identified, based on cooperation and exchange of inter-functional knowledge among the people involved in the project:

1 – *Creation of an Inter-functional Team* - A team is set up, made up of representatives of all company functions interacting with the product: design, production, assembly, quality assurance, sales, marketing, after market and customer care.

2 - *Understanding the value for the customer* - It is fundamental during this step to acquire and analyze information about needs, problems, and suggestions from the customers, by means of investigations, including quantitative and qualitative surveys.

3 – *Generating improvement ideas* – this activity is based on periodic inter-functional meetings and workshops, actively involving the entire team. Brainstorming activities, based on teamwork, allow the team to generate ideas about new features and innovative solutions, that could improve the design, always giving the highest priority to product safety and reliability. In the studied re-design, concerning a product that has been on the market for many years, opportunities have been investigated also concerning materials and production processes not available at the time of initial design. Ideas in the following areas have been identified:

a. improvement of the topology of components,
b. ideas for reducing the number of components,
c. ideas for simplifying the functioning of product functions (for example, angular strokes adjustment system),
d. ideas for simplifying the assembling activities.

In the context of DtV, improvement indices can be introduced, for monitoring performances of the new product. Here a mass reduction index ***Iw*** and a cost efficiency index ***Ic*** have been introduced, showing the percentage improvements, with equal safety and reliability, with respect to the old model, taken as a baseline to evaluate the impact of new design improvements.

Improvements of social and environmental value also are among the objectives of this application of DtV. The re-design is in fact an occasion to move towards solutions having a low environmental footprint. Investigations about the major or minor energy-intensive production processes have been carried out. In addition, components simplification with reduction of machining time allows for a reduction in the total energy required for production. Finally, it has been possible to estimate the total CO_2 reduction linked to actuator manufacturing by means of a software dedicated to these evaluations.

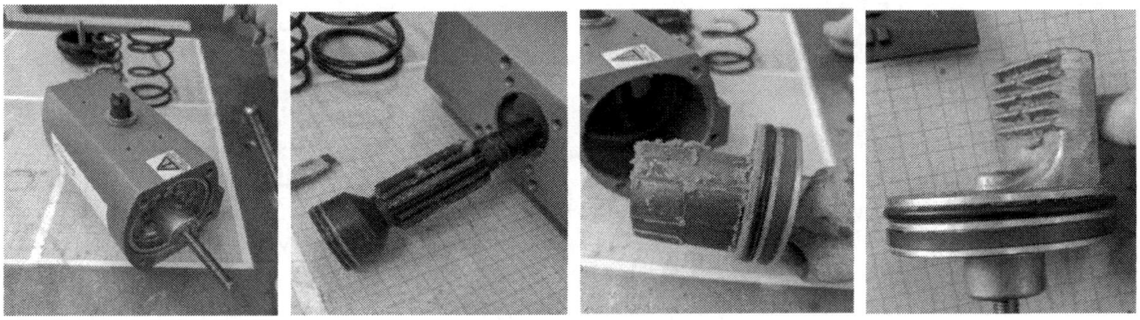

Figure 3. Moments of the teardown Analysis, during in-depth study of each component.

4. "Teardown Analysis"

In the context of DtV process a very effective technique consists in examining a sample of the product having old design, by completely disassembling it and focusing on the analysis of its components during brainstorming meetings [1], as shown in figure 3. Thanks to Complete disassembly it is possible for the

IOP Conf. Series: Materials Science and Engineering 1306 (2024) 012013 doi:10.1088/1757-899X/1306/1/012013

team to have physically available on a workbench every single component, so that it can be easy to in-depth analyze it.

Furthermore, during teardown it is possible to study the actions to be performed during the procedures for assembling/disassembling, in search for design changes making them faster and easier. Finally, entire functions of the product are analyzed, for example, the angular adjustment of the strokes, in search of more efficient construction solutions. Ideas born during the teardown are then added to the bucket of ideas identified during previous DtV steps.

5. "Design Change Proposals"

The proposals emerged during surveys, teamwork meetings and "teardown analysis" constitute a large pool, from which to draw the best ideas to be included in the final plan of the design improvement initiatives [4]. For each idea, the team highlighted the pros and cons, and ideas receiving a final positive opinion have been formalized in final "Design Change Proposals" (DCP), implemented in the next project phase, dedicated to detailed design. The main agreed DCPs are the following ones:

1. Re-design of the stroke adjustment system, by implementing a solution based on a cam (Figure 4 and Figure 5).
2. Optimization of the geometry of the casted lateral closing flanges.
3. Optimization of the geometry of the extruded central body.
4. Optimization of the geometry of the casted pistons.
5. Re-design of piston guide pads.
6. Re-design of pinion bushings.
7. Simplification of the system for fastening the pinion to the center body.
8. Introduction of a new nameplate using QR code technology.

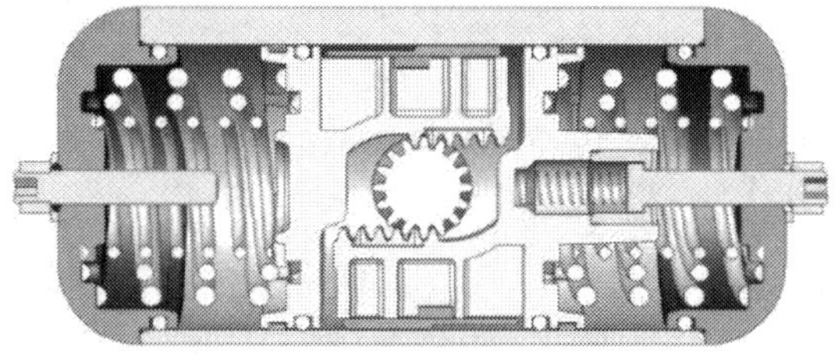

Figure 4. Old Angular Strokes Adjustment system based on lateral screws and complex machining.

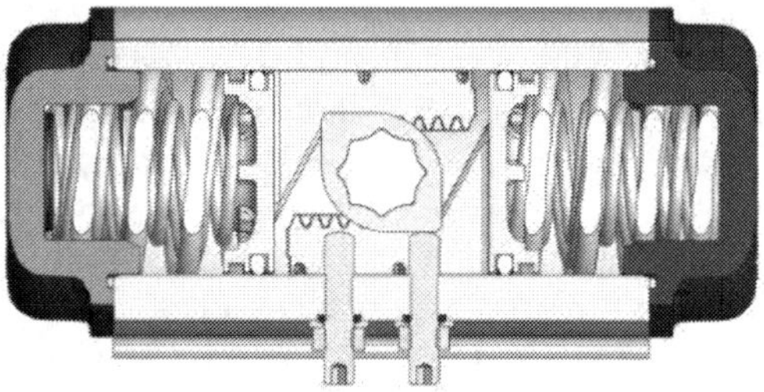

Figure 5. New system based on a cam system and simplified components.

6. Detailed design phase

The path followed for executing the design plan is based on two successive and complementary phases. The first phase has been named "design by Formulas", borrowing this term from a standard for pressure vessels, EN 13445-3 "*Unfired pressure vessels - part 3: design*". It includes verifications performed by issuing electronic spreadsheets with structural verifications. These analyses are used to make the first choices regarding the structural characteristics of each component.

The second phase focuses on the use of simulation techniques, based on Finite Element Method (FEM). During this phase different design concepts are simulated, so that it is possible to arrive to experimental validation phase with design solutions that proved to be most promising, after careful optimization activities, based on "topology optimization" concept [5].

7. "Design by Formulas"

During this phase the verification of the safety and reliability of the components is performed by means of formulas for structural verifications, included in electronic spreadsheets. These formulas constitute the basic tool for choosing the minimum dimensions and thicknesses guaranteeing the components mechanical strength under the mechanical and pressure solicitations.

Starting shapes could then be refined and optimized thanks to simulation techniques, in the following design phase. For this case study the spreadsheet containing structural calculations was elaborated by using the formulas present in the manual "*Roark's Formulas for stress and strain*" [6]. During verifications the components have been simplified, by schematizing them with simple geometric shapes and by dividing them into sections, for which suitable formulas are present.

In addition, being powered by pneumatic pressure, a rack and pinion actuator also represents a pressure vessel, even if its main function is delivering a torque. The safety of the design in presence of pressure loads is an absolute priority, and main industry standards require that actuator pressure containing parts are designed in compliance with recognized international standard for pressure vessels (EN13445-3, ASME BPVC Sec. VIII, …).

A spreadsheet complying with formulas of EN13445-3 standard has been then issued. Pressure containing parts are central body, lateral closing flanges and flange fixing screws. Pistons are housed inside the central body, and therefore do not contain the pressure towards the external environment. Pressure containing parts can simultaneously be subjected to loads of a different nature, in addition to pressure load, for which no formulas are available in the standards (for example, concentrated loads or earthquake solicitations). For these situations, main standards provide for the possibility to use FEA, defined as "design by analysis". In studied actuator this situation occurs for central body and lateral closing flanges (Figure 6), and the performed verifications are described in next paragraph.

At the end of "design by formulas", a prototype, consisting of a detailed 3D model where all the parts have been preliminary verified, is ready to enter in the next phase.

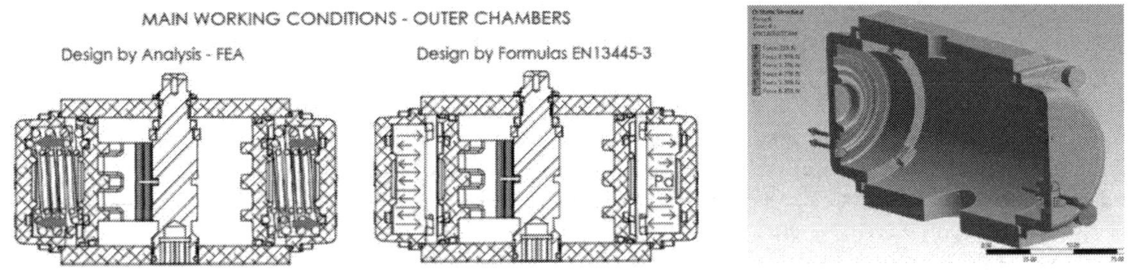

Figure 6. Lateral closing flanges are subjected to pressure or spring loads. According to EN13445-3, for spring loads verifications by means of FEA are possible.

8. "Design by Analysis"

FEA simulations are a very powerful tool, here used for two concomitant purposes. On one side, to carry out a further in-depth verification of the design choices, on the other one, to achieve an accurate "design optimization".

In a first step, simulations allow to verify that stress levels in the components are aligned with those found by means of formulas [7]. In addition, they are used for verifying the conformity to pressure vessels international standards, in this case to requirements of "*Annex B - Design by Analysis - Direct Route*" of EN13445-3. In a second extremely important step, FEA are used to perform an optimization of the shape and dimensions of each component, a set of characteristics often defined as "topology" [8]. It is possible to model the component geometry, reinforcing the areas where a high strength is needed and lightening other ones where reinforcing material is not necessary [9, 10]. Two examples are reported below, for describing more clearly how this approach was used.

1) Optimization of a component - piston

An integrated approach between structural calculations and FEA was extensively used for one of most important actuator components, the piston. Structural checks by formulas are very difficult for this component, having a complex shape, connected to other components through constraints in different points, and is possible only by schematizing it in a simplified way. Simplifications introduced in the calculations do not allow for a precise evaluation of the distributions of deformations and stresses in the different areas, and consequently it is not possible to precisely evaluate if, with hypothesized thicknesses, the material is used in the best way. Alternatively, accepting to have lower safety factors in calculations and then lower thicknesses, could lead to increase the probability to revise the design during the experimental validation. To complement the calculation-based method, it is therefore possible to use FEA. For the piston it was possible:

- to simulate the real constraints and real contacts existing with surrounding components;
- to identify areas subject to high stress concentrations and verify the possibility to accept small local plasticizations under certain conditions;
- to study the detailed maps and trends of deformation and stress in the component.

A detailed representation of the distribution of these quantities makes it possible:

A) to strengthen weak areas, if present,

B) to identify, vice versa, less stressed areas, and, if possible, remove or redistribute material.

This shape improvement process is also called "topology optimization" and is a very effective tool for increasing the weight efficiency, while maintaining the functionality and respecting the constraints given by connections with other components [11].

The deformation levels present in the old piston were used as a "baseline" to define levels of acceptability, causing not excessive stress and then wear on the sliding ring guiding the piston inside the center body. The old piston did not present problems of this type and therefore the hypothesis was to maintain in the new prototype aligned values of deformations. In a first phase then simulations were carried out on old piston, to find reference values to compare with the new ones. The model for the simulations of the new prototype was initially a subassembly, where interactions with other components through different types of contacts ("bonded", "frictionless", "frictional") are possible, so as to have a detailed representation of the stresses in the region. In a second phase, FEA on models of the single piston, with constraints calibrated by means of the results of the complex model, allowed to speed up the analyzes, with the possibility to perform many simulations, each one associated to small geometry improvements. When the component is subject to design pressure (10,5 barg), stress values with a minimum safety factor of 1.5 with respect to yield value are considered as acceptable, with exception of small surface areas considered as "peak stress". The expected average stress, when the supply pressure is about half of the design pressure, can be considered at values below of the fatigue limit.

Several iterations have been carried out, as shown in figure 7, gradually optimizing the use of the material, and checking to remain within acceptable stress and deformation limits [12, 13]. At the end of the optimization process it has been possible to achieve a total weight reduction equal to 6%.

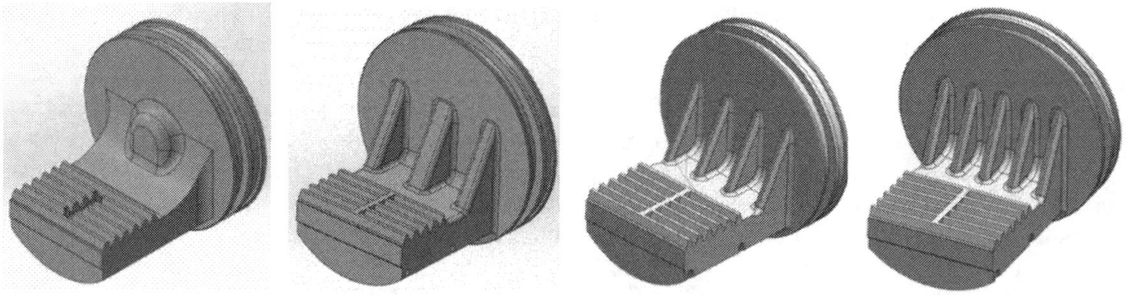

Figure 7. Steps of the topology optimization analysis of the piston.

1) Verifications of center body and angular strokes regulation system

Thanks to FEA simulations it has been possible to verify the central body accurately and in accordance with EN134445-3 requirements. This component is in fact subjected to an additional concentrated load with respect to pressure, for which "Design by Analysis" is suggested. The component contains in fact the threaded holes for the grub screws of the angular strokes adjustment system. This system has been totally re-designed, with a construction solution based on a cam. The cam is connected to the pinion, allowing to stop its angular travel when it goes into contact with two grub screws, located frontally in the center body. Regulating the cam travel it is then possible to regulate actuator strokes.

A FEA model of the complete actuator assembly was prepared, where the contacts and reactions among all the components were simulated. By applying a pressure to the pistons, it has been possible to evaluate the entities and directions of the forces exchanged among the components, in particular, at the contact between cam and grub screw, as shown in figures 8 and 10. A special function was also included, allowing to simulate the presence of a real thread between grub screw and center body, even if, from a 3D point of view, components are modeled as smooth cylindrical surfaces.

Using this approach, the new cam system has been thoroughly verified and the geometry of the center body has been verified following topology optimization principles [14], guaranteeing at same time compliance with EN13445-3 requirements.

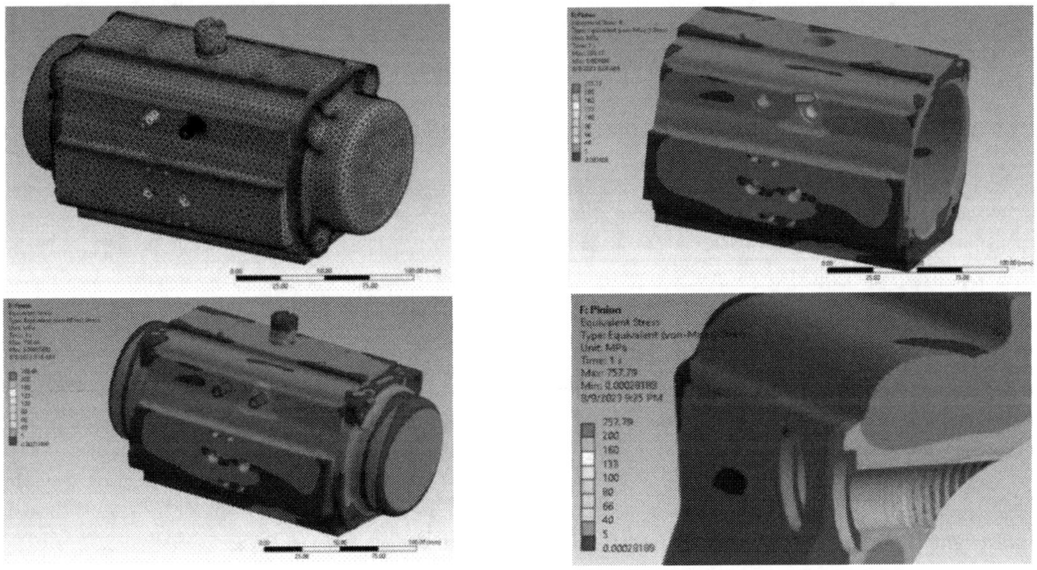

Figure 8. FEA simulations used for topology optimization and verification of compliance to EN13445-3 of the center body.

In summary, accurate FEAs on critical components are a big help to confirm and optimize the design choices, correcting situations that could present criticalities (for example by updating fillet radii and other areas subjected too high peak stresses). Thanks to them, during the subsequent phase of the project dedicated to experimental tests on prototypes, it is expected to encounter a lower number of situations requiring modifications to the design, and in that case to deal with minor changes.

9. Experimental validation of the new design

Once the detailed re-design of the product was completed, a validation plan has been drawn up, containing all the tests necessary to discover and fix all possible critical points before proceeding with mass production [15]. For this case study, a special validation procedure, based on two subsequent steps, has been adopted:

1- a first phase is focused on the creation of a partial prototype called "mockup";

IOP Conf. Series: Materials Science and Engineering 1306 (2024) 012013 doi:10.1088/1757-899X/1306/1/012013

2- a second phase is focused on a sample identical to the actuator that will go into production.

Prototypes are essential for multiple purposes:

I. to validate the safety of the pressure containing parts, by performing the tests required by the standards,

II. to confirm the good reliability of the product, ensuring that it can perform the emergency functions, when required, with a very low probability of failure. The actuator must in fact be certified as "SIL 3 Capable" (SIL = Safety Integrity Level), according to EN61508 "*Functional safety of electrical/electronic/programmable electronic safety related systems*".

III. to validate good fatigue life under pressure and mechanical loads for all the components. The actuator must in fact be suitable for special applications, requiring performing up to 500,000 cycles during its operational life [16].

IV. to check assembling activities, and to confirm the chosen dimensional and geometric tolerances.

9.1 Validation based on a "Mockup" prototype

A partial prototype here called "Mockup" has been assembled with the special aim to validate some functions of the actuator. Checks are focused on the new strokes adjustment system and on the new pinion bushings. In fact, dimensions of other important components, the casted pistons and the extruded central body, depend on them. The production processes of these components require to produce stamps, making very difficult to perform changes in component shapes once produced. Then the the chosen approach was to create a preliminary model, as shown in figure 9, composed by:

1) Components in the new version for pinion, pinion guide bushings, stroke adjustment grub screws.
2) The new designed cam.
3) A central body machined from solid. The component features detailed machining for the areas of interest for the tests while for other sections it does not present all the details present in final version.
4) For all other parts not directly affected by the tests, components with old design taken from old model.

During the assembly phases of the prototype, a problem had to be faced, regarding the assembly tolerances of the cam, that was never previously produced. Excessive backslash was noticed, requiring a design modification for indicating tighter dimensional tolerances.

Checks were performed about the correct assembling and functioning of the pinion bushings, completely re-designed for simplifying the assembly operations. Also, the ease of regulation of the new strokes adjusting system was tested. Once these preliminary checks were completed, tests under load allowed to verify the structural resistance of the new designed parts in static and dynamic conditions.

Figure 9. Moments of the validation by means of a "mockup" prototype. From left: (a) center body obtained by machining a solid block, (b) assembling of the prototype, (c) execution of cyclic test with angular deformations measurement.

The "mockup" was tested by performing a cyclic test, powering it with high pressure and high flow rates, to obtain high torques and high stroke execution speeds, causing impacts between the cam and the grub screws. The possible cumulative damage caused by the repetition of the impacts can in this way be assessed. After completing 900 cycles, the actuator was disassembled, and components examined. The absence of permanent deformations in grub screws was verified, and pinion and cam were checked for ensuring absence of cracks. Finally, the conditions of the pinion bushings were assessed, for excluding premature deterioration.

The observations on the components after the tests confirmed some behaviors foreseen by FEA with good approximation. For example, the formation of an acceptable small denting on the cam surface, in correspondence to the contact position with grub screw, was noticed, as shown in figure 10. This result is very important and confirms the useful predictions obtained with FEA.

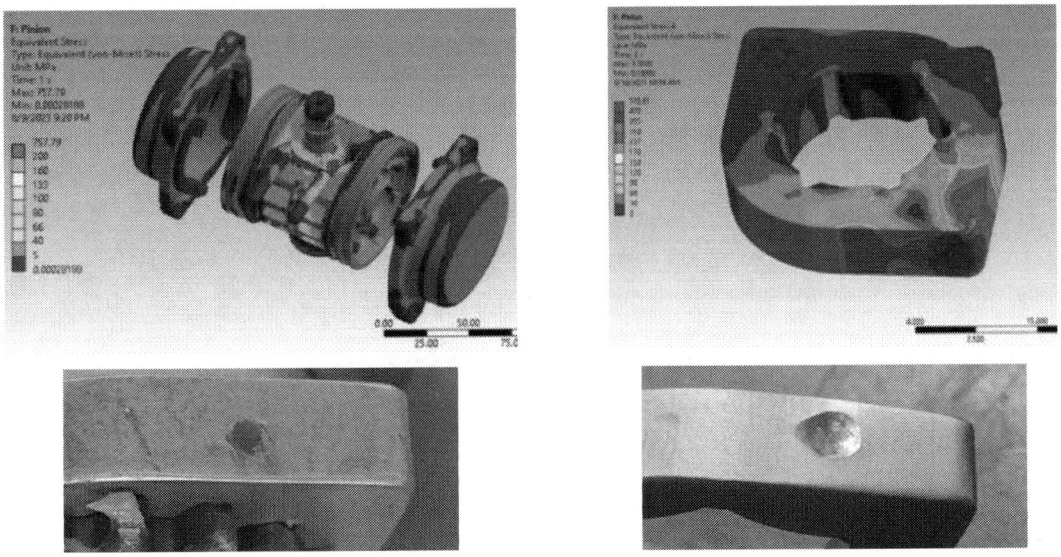

Figure 10. Images showing Consistency between the results obtained with FEA and results of experimental tests on the cam - grub screw connection.

The tests on "mockup" prototype confirmed the validity of the completely new design solutions introduced after DtV analyzes, drastically reducing the probability of having to make changes to them in the final prototype.

At the time of writing this text, the tests on the "mockup" have been successfully completed and the procurement of the final parts made by casting and extrusion processes is ongoing. Considering that the experimental results examined in the "mockup" model are aligned with calculations and FEA predictions, it is expected that in final prototype also a certain alignment of the results will be found, allowing to avoid major modifications of big impact.

9.2 Validation of a complete prototype

The second part of the validation will be based on a prototype identical to the final model intended to production. All possible problems regarding the components not already checked with "mockup" could be discovered, and final changes could be carried out before the start of production. Tests of fundamental importance for the safety and reliability of the product will be performed, as briefly described below:

I) Quality, dimensional, tolerances and assembling checks.

II) Pressure tests – Tests listed below are essential to preserve the health of the operators, placing safe products on the market. These tests are the first ones to be carried out, both during prototype validation and once production started.

A) Leakage test at low pressure, for checking the correct functioning of sealing systems,

B) Structural pressure tests, at pressure equal to 1.5 times the design pressure:

$$1{,}5 \times Pd = 1{,}5 \times 10.5\ barg = 15{,}75\ barg$$

C) "Burst pressure test", performed only during validation, for checking the behavior of the actuator before being subject to structural collapse, in case of improper use or failures of the pneumatic supply system. Actuator is gradually powered up to a pressure equal to 2.5 times the design pressure:

$$2{,}5 \times Pd = 2{,}5 \times 10.5\ barg = 26{,}25\ barg$$

Components are expected to plasticize and have permanent deformations. Some O-rings may be extruded, and the actuator begins to leak. Such a behavior without explosions can be assessed as positive.

III) Verification of actuator performances – Torques delivered by the actuator at different angular positions are measured by means of a calibrated dynamometric brake. Torques aligned with the expected ones will confirm that the new components involved in torque transmission do not present unwanted deformations, increasing friction, and then causing a reduction in transmissible torque.

IV) Cyclic endurance Test – Test finalized to verify the reliability and the good fatigue life of actuator components, by analyzing their behavior when subjected to a very high number of cycles [16]. In accordance with reference standard EN15714-3:2022 "*Industrial valves - Actuators - Part 3: Pneumatic part-turn actuators for industrial valves - Basic requirements*", the minimum number of cycles to be performed without failures for actuators in this torque range is equal to 500,000 cycles.

V) OK to launch the first production batch

10. Conclusions

The results of the activities carried out in the context of this case study show that a multimethodology based on the Design-to-Value process, integrated with innovative design techniques based on FEA simulations, and valorization/maximization of teamwork, make it possible to obtain a significant improvement in the re-design of a product, as shown in figure 11, maintaining constant safety, reliability, quality and performances.

Figure 11. Comparison between the actuator before and after the value-driven design optimization activites.

The case study up to now demonstrated that is possible to obtain significant product improvements about the following aspects:

1 – compliance of the design with most up-to-date reference standards. Optimized actuator is compliant with requirements of product standards (EN15714-3:2022, ISO12490:2011 "*Petroleum and natural gas industries — Mechanical integrity and sizing of actuators and mounting kits for pipeline valves*", API

6DX:2020 "*Standard for Actuators and Mounting Kits for Valves*"), pressure vessel standards (EN13445-3) and of main reference customers specifications (for example, S-707 "*Supplementary Specification to ISO 12490 for Actuators for On-off Valves*" by IOGP "*International Association of Oil&Gas Producers*"). Finally, the reliability of the new product is suitable for SIL certification obtainment (EN61508).

2 – by integrating formulas-based structural verifications with FEA, it is possible to obtain the following important product improvements:

A) Re-design of product functions, like new strokes adjustment system, featuring a simpler design, made of few light-weight components and user-friendly;

B) improvement of components topology. At the time of the first design, the tools available for the design were reliable structural calculations. It is now possible thanks to the simulations to perform a step forward, redistributing more easily and in detailed way the material where it is more necessary.

The examined actuator underwent a profound transformation, many components were simplified and where possible lightened. The results in terms of increase of efficiencies are shown in the chart of figure 12, where it is possible to notice that the weight of the actuator has been reduced of the following mass improvement index ***Iw***.

$$Iw = (new\ mass - old\ mass) / old\ mass = (6050\ gr. - 6575\ gr) / 6575\ gr. = -8\ \% \qquad (1)$$

3 – Improvement of environmental efficiency. Reduction in weight and simplification of components machining allow for a lower consumption of raw materials and for a lower quantity of energy required to production. These reductions in turn lead to lower carbon dioxide emissions into the environment, making the product more sustainable.

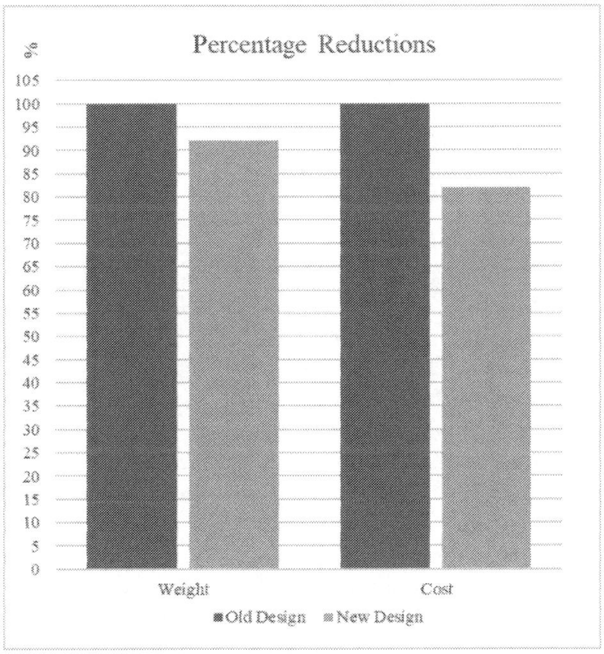

Figure 12. Increase of product efficiency in terms of weight and cost reduction.

4 - Last but not least, increase of economic efficiency, maintain high standards of safety and reliability. Weight efficiency brings to an immediate benefit in terms of costs. Furthermore, components simplification with a reduction in machining time, led to a consequent further reduction of costs.

IOP Conf. Series: Materials Science and Engineering 1306 (2024) 012013 doi:10.1088/1757-899X/1306/1/012013

Finally, studies about simplification of assembling activities led to increase productivity. The expected achievable cost reduction for the present case study can be estimated with the following cost efficiency index **Ic**.

$$\boldsymbol{Ic} = (new\ cost - old\ cost) / old\ cost = -18\% \qquad (2)$$

The next steps of the investigations will be the application of FEA tools dedicated to Explicit Dynamics and Fatigue, for augmenting the details of predictions achievable with simulations, and entering in validation phase of other actuator sizes with a still more verified design. The current limitations to the integration in the design of components obtained by addictive manufacturing will also be investigated.

"Secret ingredient" fundamental for the good result of all the activities has been the collaboration. Teamwork and know-how exchange aimed at mutual enrichment are essential components for increasing the commitment of the team during all the project phases.

References

[1] Henrich J, Kothari A and Makarova E 2012 Design to Value: a smart asset for smart products *McKinsey&Company*

[2] Sharma A. and Belokar R 2021 Achieving Success through Value Engineering: A Case Study. *Proc. of the World Congress on Engineering and Computer Science 2012* Vol **2**

[3] Ben Hamida S 2017 Innovate by Designing for Value - Towards a Design-to-Value Methodology in Early Design Stages. *PhD Thesis Université Paris-Saclay,* HAL Id: tel-01811739

[4] Sharma A. and Harshit S 2011 A case study analysis through the implementation of value engineering. *International J. of Engineering Science and Technology* Vol **3**

[5] Gülbahçe E, Sezgen H and Çakan A. 2019 Topology Design and Modal Analysis of a Bracket via FEA. *Applied Engineering Letters: J. of Engineering and Applied Sciences.* Vol **4** p 102-5

[6] Roark R, Young W, Budynas R and Sadegh A 2020 *Roark's Formulas for Stress and Strain - 9th edition* (McGraw-Hill Companies)

[7] Agrawal A and Shahazad A 2023 Investigation of stresses at a critical location for shape optimization of connecting rod using finite element analysis. *Songklanakarin J. Sci. Technol.* Vol **45 (2)** p 264-9

[8] Shanmuga S, Dharanidharan M, Vishwa D and Kumar AP 2021 Design, analysis and topology optimization of connecting rod. *Materials Today: Proceedings.* Vol **46**

[9] Dämmer G, Gablenz S, Neumann R and Major Z 2023 Design, Topology Optimization, and Additive Manufacturing of a Pneumatically Actuated Lightweight Robot. *Actuators* Vol **12**

[10] Cheirmakani B, Balamurugan S, Balamurugan P and Beneston L *2018* Weight reduction of motorcycle frame by topology optimization. *J. of Achievements in Materials and Manufacturing Engineering* Vol **2** p 67-77

[11] Abdelghani E, Nassraoui M and Bouksour O *2023* Optimisation topologique par la méthode SIMP d'un système multicorps. *Incertitudes et fiabilité des systèmes multiphysiques* Vol **6**

[12] Yadu k, Athul K and Thency G *2020* Optimization of engine truss mounting bracket. *Int. J. of Engineering Applied Sciences and Technology* Vol **5** Issue 2 p 258-263

[13] Dheeraj G and Anadi M 2012 Topology Optimization of Sheet Metal Brackets Using ANSYS. *MIT Int. J. of Mechanical Engineering* Vol **2** No 2 p 120-6

[14] Tyflopoulos E, Hofset T, Olsen A and Steinert M 2021 Simulation-based design: a case study in combining optimization methodologies for angle-ply composite laminates. *Procedia CIRP* Vol **100** p 607-12

[15] Weiping H, Mutlu R, Li W and Alici G 2021 A Structural Optimisation Method for a Soft Pneumatic Actuator. *Robotics* Vol **7(2):24**

[16] Bergonzoni M, Campagnolo A, Pellinghelli D, Riboli M, Spagnoli A and Meneghetti G 2018 Experimental tests and fatigue strength assessment of a scotch yoke valve actuator. *Procedia Engineering* Vol **213** p 58-68

IOP Conf. Series: Materials Science and Engineering 1306 (2024) 012014 doi:10.1088/1757-899X/1306/1/012014

FMEA, monitoring, retrofit and redesign of insulated rail joints

A Bracciali, G Megna

Department of Industrial Engineering, University of Florence, via di Santa Marta 3, 50139 Florence, Italy

andrea.bracciali@unifi.it

Abstract. Insulated rail joints (IRJ) are a fundamental component of railway signalling that ensure train protection using "track circuits". A track circuit is a section of line delimited by rail joints electrically insulated, powered at one end with an electric power supply to establish a different voltage that is checked at the other end with a so-called "track circuit relay". When the section is short-circuited by an axle the relay drops, and the section is defined as "occupied". IRJ are subjected to continuous and dangerous failures as they represent a singularity in the rail continuity. Continuous battering produces the classical train impact noise while damage in the rail accumulates until the IRJ fails with extremely unpleasant consequences. The paper discusses the main failure modes of an IRJ starting from a historical perspective, showing how an appropriate failure analysis would have prevented a catastrophic accident that recently occurred in Italy. Several countermeasures are shown and a completely new IRJ potentially completely solving the issue is introduced.

1. Introduction

On January 25th, 2018, a railway accident [1] happened on the outskirts of Milan involving a train coming from Treviglio near the village of Pioltello, causing the death of 3 passengers and major damages to vehicles and infrastructure. The push-pull train consisted of a driving trailer, four passenger cars and a pushing locomotive. It was running at 140 km/h when the first bogie of the third passenger car derailed suddenly over a broken insulated rail joint (IRJ).

Investigations readily discovered that the two rail ends that constitute the joint where were separated by approximately 25 mm and that the downstream rail was heavily battered by the passing wheels. An around 23 cm piece of rail eventually detached, ending under one of the train wheels, producing a "jump" and the consequent derailment. The fragment showed signs of the classical "defect 135 – star crack from fishbolt hole" as described by the UIC 712 code, that is probably the most common critical and symptomatic defect that regularly affects bolted joints.

2. Brief IRJ description

An insulated railway joint guarantees both the rail mechanical continuity to support the loads of the trains passing on the line and the electrical insulation for separating trains and protecting them from collisions with the "track circuit" system. As it will be shown in this paper, these requirements conflicted at least in the Pioltello accident. Insulated rail joints are made by several different parts made of different materials. The following description applies to prefabricated IRJ, the most common type, that are inserted in the track by welding the IRJ to existing rails.

Figure 1. The joint broken in Pioltello on 25.01.2018. The train travels right to left. It can be seen that downstream portion of the IRJ is completely destroyed. The flown away portion of the railhead got stuck under a wheel causing the derailment of the train.

Two rail bars of length varying in a wide range (approx. from 1.5 to 12 m) are cut, deburred and their web is drilled with 2, 3 or 4 holes (called "fishbolt holes") to result in an IRJ with 4, 6 or 8 holes. The rails are then connected by using two steel profiles (called "fishplates") fastened to the rail by using either conventional bolts (called "fishbolts") or hydraulically pulled irreversible locking bolts (like rivets). Insulating sheets made of fiberglass guarantee the electrical insulation between the fishplates and the rails as well as the insulation between the fishbolts and the fishbolt holes. An "end post" around 5 mm thick between the rail ends completes the insulation.

An IRJ is assembled by bonding all the components with a high-strength structural epoxy glue to provide the requested mechanical strength as friction would not result to be sufficient to prevent longitudinal displacements under large axial forces arising from the thermal behavior of an ordinary CWR (Continuous Welded Rail) track. After an appropriate curing time (2 to 4 weeks) the assembled IRJ is delivered to the application point.

3. Design and criticalities in conventional IRJ

3.1. Historical background

The extensive use of IRJ dates to the massive introduction in early XXth century of track circuits to guarantee train protection. At that time the use of relatively short, jointed rails was common (typically 18 m long) as the development of currently used rail welding processes (aluminothermic and flash butt) was still in progress. The IRJ were assembled with a specific kit directly on track without gluing them.

With the adoption of CWR (Continuous Welded Rail), the IRJ strength was increased by bonding all the elements. As this process revealed to be critical on the track, IRJ were started to be assembled in a plant and the "prefabricated joint" was delivered to the desired location and welded to the existing track with special attention to keep the desired stress state of the rail if temperature is not close to the SFT (stress-free temperature) of the rail.

The drawing of IRJ may date back to the 60s. Small modifications may have been introduced in materials and or in the sequence of operations but basically the design remained the same. The drawing in figure 2 shows the original prefabricated joint that was introduced in Italy in 1968. The current drawing is very similar, the only difference being the replacement of C960 fishbolt with Huck bolts.

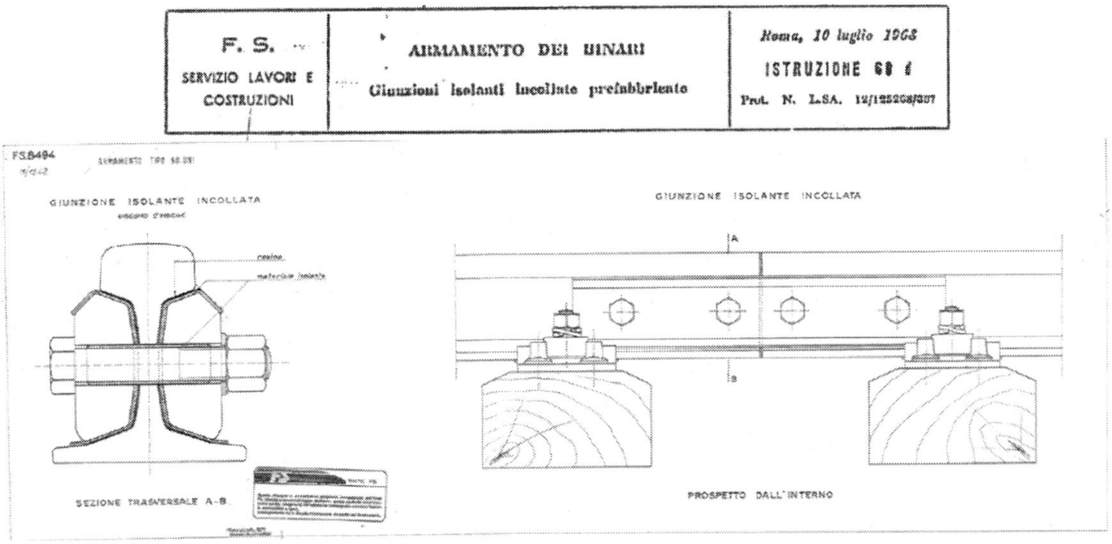

Figure 2. Reproduction of the drawing of the standard prefabricated IRJ in Italy since 1968.

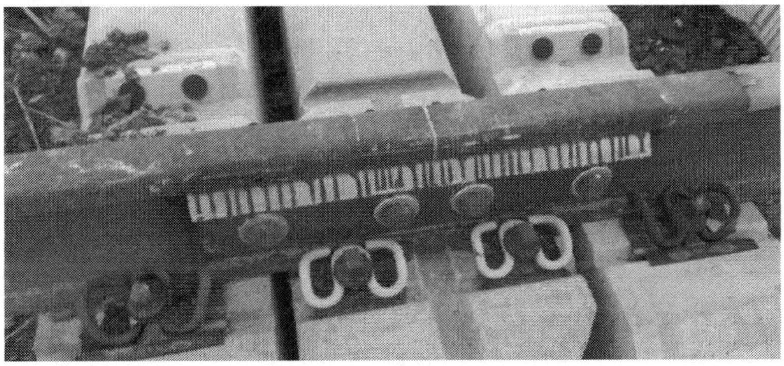

Figure 3. The IRJ currently used in Italy pictured before installation. A specifically widened sleepers was developed to (theoretically) better support the IRJ. Hollow sleepers to pass cables can be seen. Sleeper spacing is adjusted during installation but results to be non-standard with tamping difficulties.

3.2. FMEA of IRJ and criticality analysis

Failure Mode and Effects Analysis (FMEA) is the process of reviewing as many components, assemblies, and subsystems as possible to identify potential failure modes in a system and their causes and effects [2]. While procedures for conducting FMECA (Failure Mode, Effects and Criticality Analysis) were described in US Armed Forces Military Procedures document MIL-P-1629 in 1949, no such analyses were common practice in the railway environment until recent times. IRJ were not certainly analyzed in a formally planned FMEA framework. No trace of such analyses was found in the literature and, what's more, no design calculations or test results were found about conventional, old IRJ.

Although a full FMEA according to nowadays approach was performed, a more conventional analysis of failure modes and their consequences can be conducted by using engineering judgment and the experience arising from more that one century of use of IRJ throughout the world.

Unfortunately, IRJ failure is relatively common and descends basically from the interaction between the (loaded) wheels and a rail with important stiffness variations along its length, generating a "dip joint" irregularity. The shocks given by wheels passing over the discontinuity (gap) between the rails result in rail steel flowing, high stresses in the fishbolt holes, high stresses in the fishplates [3-5].

Mechanical FMEA of IRJ leads to the following failure modes:

- debonding of the joint under the combined action of high pulling forces due to thermal actions of the CWR in cold season and the aforementioned shocks;
- failure of the rail cross section with 45°-oriented cracks originating from fishbolt holes (so-called "star cracks"), due to both the stress rising given by holes (worsened by bad drilling and/or absence of hole deburring) and the aforementioned shocks;
- failure of the fishplates for high stresses when IRJ support (ballast) has lost its bearing capability (worsened by the fact that IRJ can hardly be tamped because of the presence of cables on the track), enhanced by the aforementioned shocks;
- failure of the fishbolts when the axial displacements of a debonded IRJ is such that the design clearance between the fishbolts is recovered and the fishbolt is sheared instead of simply pulled. Crack propagation is speeded up by the aforementioned shocks.

It should be noted that while some of these failures are consequential to the joint debonding (e.g. bolts shearing), some others are independent (e.g. star cracks and fishplates failures). A common root cause for all the failures can be easily identified in the shocks naturally occurring in the IRJ because of the unavoidable endpost gap.

Criticality analysis needs little explanation considering the consequences of the disaster described above. All literature considers the debonded joint (the only easily degraded condition observable with the naked eye) to be replaced in the shortest possible time with temporary speed restrictions until the replacement,

Electrical FMEA of IRJ leads to the following failure modes:

- loss of rail insulation due to fiberglass crushing in the fishbolt holes, possible only if the IRJ is debonded and already seriously malfunctioning;
- loss of insulation on the rail head due to "lipping" of rail ends because of repeated plastic deformation of steel that brings the rails in contact;
- loss of insulation of fishplates insulation (rare).

It must be underlined that loss of insulation has a negative impact on service regularity (the track circuits resulting "occupied" stopping train circulation) but leads to a safe condition as the "electrically broken" IRJ prevents any derailment. It can be said that from this point of view the IRJ is "self-protecting", and that the failure mode is not critical as it leads to the typical *fail-safe* condition.

It is worth mentioning a safety-related condition non directly regarding the IRJ itself. Conventional track circuits have the intrinsic capability to detected broken rails as the track relay drops if the track circuit is opened. With axle counting block or radio block systems there is no need for IRJ anymore, but the continuous check of the overall integrity (continuity) of the track is lost.

3.3. Considerations about certification process of conventional IRJ

As rails are welded today and CWR approach is extensively used throughout the world, joints remain as a signaling-related component only. The safety analysis of any modification (new technology or improvement) to an existing railway system must follow a so-called *Common Safety Method (CSM)*, that in Europe is formalized by the Regulation 402 of EU.

The application of such method starts from a preliminary analysis of the modification proposed; if it is deemed to relevant, a formal risk assessment process must be strictly followed under the supervision of third-party competent assessor.

Signaling components belong to control-command systems. The are described in a Technical Specification for Interoperability of the European Union (TSI-CCS). Basically, CCS components are classified according to the SIL (*Safety Integrity Level*) of the application, that can be shortly defined as the level of reliability that the CCS system must guarantee during service.

As for the FMEA described above, none of these formalities were present decades ago, and all IRJ were internally approved by national railways.

4. Root cause of the Pioltello accident

It was said previously that the train derailment was originated by the interaction of a trail wheel with a slice of rail section that detached from the downstream portion of the joint.

Starting from this evidence, a possible reverse sequence of the events is likely the following:

- star cracks appeared starting from the downstream hole close to endpost, and cracks progressed until the rail slice detached;
- as observed, the rails displaced by more than 20 mm, so debonding was a clear contributor to the final failure;
- during the movement the fishbolts certainly crushed the small fiberglass tubes insulating them from the fishplates and the rail web.

This last condition conflicts with the expected (and declared above) *fail-safe* condition guaranteed by an IRJ, i.e. the loss of insulation (and the consequential interruption of train circulation) of the joint. What happened may be explained by figure 4.

Figure 4a shows the conventional joint as assembled in a plant, tested and correctly installed on a railway line. Track circuits power supply and insulation element geometry are such that rails facing in the IRJ are at different electrical voltage (red and yellow). The fishplates are neutral, as they are insulated from both the rails and the fishbolts.

Figure 4b describes what happened in Pioltello. The left (yellow) rail displaced by a large distance (more than 20 mm) destroying both fishbolts/rails and fishbolts/fishplates tubes. As a result, both fishplates "became yellow", therefore getting the same electrical potential as the left rail. This condition stayed for a long time until the whole joint was destroyed without losing its insulation and therefore without shortening the track circuit. **This is the root cause of the Pioltello accident: the missed track circuit occupation given by an IRJ that was mechanically destroyed**.

A short redesign scenario was then approached that readily showed that there are several possible insulation arrangements that fulfil the requested condition "mechanical failure → electrical failure → interruption of service → no risks". Figure 4c shows a possible layout in which:

a) each rail in electrically connected to one of the fishplates and
b) fishbolts are insulated only on one side by an insulated washer (no fiberglass tubes)

It can be easily proven that any movement of any component immediately leads to short-circuiting the rails, causing the immediate stop of the traffic along the life. This is a *fail-safe* condition reached by a simplified FMEA.

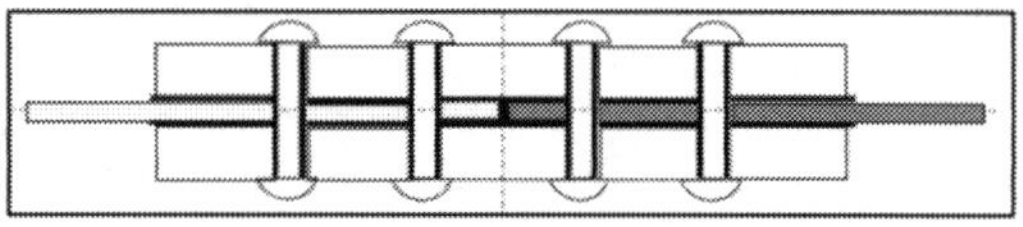

Figure 4a. Horizontal cross section of an IRJ in standard conditions.

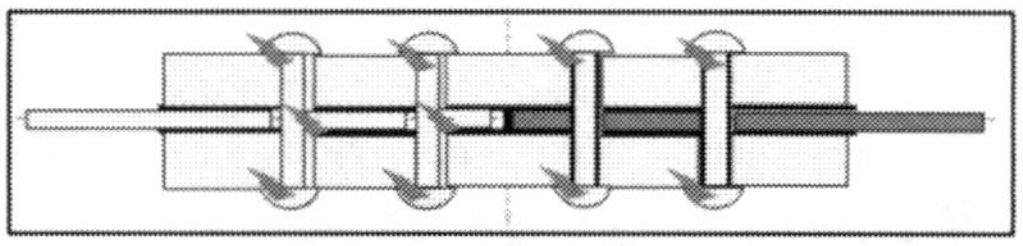

Figure 4b. Horizontal cross section of an IRJ when the left rail moved to the left bringing the fishbolts in contact with the rail web and the fishplates.

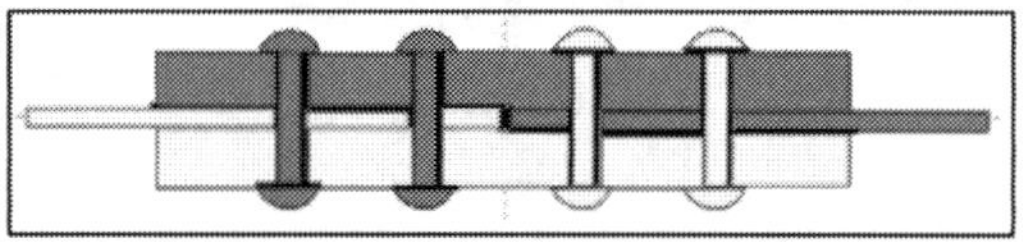

Figure 4c. Horizontal cross section of a modified IRJ in standard condition. Any movement leads to short circuiting the rails.

5. In-service check and improvements of existing joints

5.1. Measurement of the endpost gap

Several commercial devices were developed in Italy before and after the Pioltello accident [6,7]. They are basically designed on the principle of checking any increase in the endpost gap design (5 mm) by using either mechanical, electromechanical, or optical sensors (figure 5).

Figure 5. Left: electromechanical IRJ found detached along a railway line (photo A. Bracciali). Right: optical IRJ sensor (from [6]).

One of these designs was extensively used showing some limitations:

- the sensor is applied in the IRJ manufacturing plant but tends to be detached during handling of the joint during the trackworks;
- On-Track Machines (OTMs) with rubber tyres used during trackworks interfere with the bonded sensor, further detaching it;
- no connection to the signalling system was provided until today, making the check nearly useless (one type of device has just a mechanical device that is redundant as IRJ opening can easily be checked visually).

It must be stressed that this kind of device is unique of Italy and, to the best knowledge of the authors, is not used elsewhere. None of these devices leads automatically to an increased safety, as the relevant data (endpost gap) must be measured, transmitted, processed and the relevant actions must be taken to ensure that the adequate level of safety is achieved, and all these steps must comply with the aforementioned regulations on signaling systems.

5.2. Addition of an intrinsically safe switch

A different approach to intrinsic safety was proposed by the authors taking inspiration from the past practice of short-circuiting the rails when many yards were electrified. As shown in figure 6, a rather common practice was to weld conductive cables to the field side of railheads to ensure the return of current to electric substations. Similarly to this design, a pair of stainless steel elements forming a switch was developed. The use of conductive glue was forecasted to avoid welding that is not accepted anymore in main line applications.

Beyond the greatest advantage of simplicity of manufacturing and installation and a ridiculous cost, the main advantage of this device is that it automatically stops traffic when the IRJ debonds, ensuring the highest possible safety. A formal FMEA analysis was not conducted because of the simplicity of the device and because the proposal was not even considered by the railway industry.

Figure 6. Left: current return conductors in an electrified yard with old rails resting on wooden sleepers (photo A. Bracciali). Safety switch produced with rapid prototyping technique bonded to field side of railheads (centre) and to rail bottom (right).

6. Introduction of a new IRJ to remove all failures

According to the FMEA conducted on existing IRJ, the common mode of failure with direct and potentially immediate effects on safety is debonding. If debonding doesn't appear other failures may appear (e.g., nucleation of cracks in fishbolt holes) but with possibly lower occurrence and effects. Other failures, such as rail lipping or burring, can be removed restoring the joint operation without safety implications.

Debonding certainly happens because of tensile forces in cold seasons although qualification tests require that IRJ must withstand a tensile force of 1500 kN or even 2000 kN. Even though this value may be unrealistically too high, IRJ debond *because of impacts*. The only way to avoid any joint failure in principle is *removing the shocks*.

This evidence is at the basis of the development of a (patented) completely new IRJ obtained by coupling two shallow depth switch rails properly machined with a lower thick plate (joint cover) and proper insulating components. While the joint is described thoroughly on the website www.absolutelybetterjoint.com and on the paper [8] to which the reader is referred, the concept is quite simple: using an shallow cut (3°) of thick web rails to increase the "handover" length up to 400 mm completely avoiding shocks. While this type of joints was attempted in the past by using thick web rails for expansion joints, the innovation introduced by the lower plate enormously increases bending strength while the geometry is fully compatible with existing tracks with standard sleepers (figure 7)

IOP Conf. Series: Materials Science and Engineering 1306 (2024) 012014 doi:10.1088/1757-899X/1306/1/012014

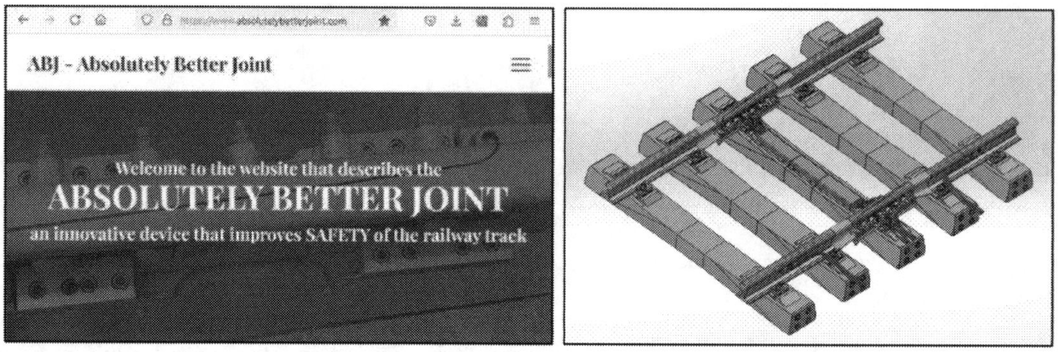

Figure 7. Left: home page of the *Absolutely Better Joint* (ABJ) website. Right: General view of a track jointed with ABJ (note cabling that allows continuous tamping and the unchanged sleeper type and spacing)

7. Conclusions and further developments

Insulated rail joints are still a source of headaches for infrastructure managers. Despite the use of several technique to reduce the failure rate, such as the increase of the number of fishbolts, the increase in the length of the fishplates and the use of *cold bolt expansion* technique to reduce star cracks from fishbolt holes, IRJ remain an Achille's heel in a heavily loaded railway track, without mentioning the impact noise always present also in IRJ in mint conditions.

FMEA analysis showed that one the most critical accident arising from a broken IRJ was due to the lack of consideration of the criticality of a failure mode. Some devices to monitor existing joints were described, together with a simple retrofit safety switch that could avoid in principle accidents of the kind.

But the most important innovation is the introduction of a new joint (called ABJ) that generates no shocks at all. As shocks are the root cause of all damages, this new joint promises to be very effective in removing forever the issues related to conventional IRJ use. Currently a first batch of ABJ joints is in production to make official homologation tests. Line tests are expected to start by 1Q of 2024.

References

[1] https://en.wikipedia.org/wiki/Pioltello_train_derailment (accessed 20/09/2023)
[2] Stamatis D H 2003 *Failure Mode and Effect Analysis* (Milwaukee, Wisconsin; ASQ Quality Press).
[3] Mandal NK, Bracciali A, Megna G. Evaluation of railhead material damage of insulated rail joints due to cyclic wheel loadings. *Wear*. 2023 Sep;528–529:204966.
[4] Soylemez E and Ciloglu K. Influence of Track Variables and Product Design on Insulated Rail Joints. Transportation Research Record. 2016 Jan;2545(1):1–10.
[5] Mayers A. The effect of heavy haul train speed on insulated rail joint bar strains. Australian Journal of Structural Engineering. 2017 Jul 3;18(3):148–59.
[6] https://www.generalecostruzioniferroviarie.com/en/gcf-news/news-2022/296-safer-with-cogi (accessed 20/09/2023)
[7] https://www.mariniimpianti.it/en/cgm-mechanical-joint-controller/ (accessed 20/09/2023)
[8] Megna G, Bracciali A and Mandal NK. Design, wheel-rail interaction and testing of an innovative reinforced smooth transition insulated rail joint. *Wear*. 2023 Oct;530–531:205038.

Optimization and validation of a finite element methodology for thermo-structural analysis of polyhuretane wheels for roller coaster application

Massimiliano Palmieri[a], Alberto Ferri[b], Filippo Cianetti[a], Claudio Braccesi[a]

[a] University of Perugia, Department of Engineering, Via G. Duranti 93, 06125 Perugia (Italy)
[b] Antonio Zamperla S.p.a., Via Monte Grappa 15, 36077 Altavilla Vicentina (Italy)

E-mail: massimiliano.palmieri@unipg.it

Abstract. Polymeric materials are widely used in several engineering sectors. Among these, a particularly critical sector for this type of material is that of roller coasters. The wheels are indeed made with an aluminum hub and a compact polyurethane coating, which, being in contact with the track, is subject to high speeds dynamic loads. Due to the viscoelastic behavior typical of these materials, such loads induce overheating of the coating and therefore a rapid degradation of the wheel. This results in machine downtime and, consequently, significant waste of time and money. In this context, the authors have developed a methodology for finite element thermo-structural analysis capable of quickly evaluating the temperature reached during work cycles and proving very useful in selecting the type of wheels to use. In this work, this methodology was firstly computationally developed and then validated by comparing the analysis results with data obtained from experimental tests conducted by the manifacturer. The comparison demonstrated the effectiveness of the proposed method, highlighting, however, a constant error in terms of maximum temperature reached attributable to a non-exact material characterization.

1. Introduction

Roller coasters are a engineering sector that is continuously evolving, particularly in terms of track length, speed (exceeding 100 km/h), and forces. Roller coasters are essentially trains composed of carriages (typically 10) that move along a track consisting of two or three steel tubes. The connection between the carriages and the track is achieved through a system known as the 'three-wheel locking system'. This system involves the use of at least 4 'main wheels' for each axle, which support the weight of the carriage and maintain the train's movement on the track, 4 'lateral wheels' that keep the train centered on the track to prevent derailment, and 4 'stop wheels' that prevent the train from detaching from the track during maneuvers. Overall, there are approximately 240 wheels for each roller coaster (considering an average of two axles per carriage for an average of 10 carriages). These wheels, although of different sizes depending on their function, are made with a steel or aluminum hub and a polyurethane coating. The hysteresis behavior typical of these materials [1-3], while useful for absorbing shocks and vibrations, dissipates energy and thus generates heat, which can lead to overheating of the material [4-5]. Such overheating is one of the main causes of damage to the polyurethane coating

IOP Conf. Series: Materials Science and Engineering 1306 (2024) 012015 doi:10.1088/1757-899X/1306/1/012015

(other types of damage are due to mechanical fatigue and photo-degradation) [6-7]. These damages can take various forms, such as surface wear, crumbling, cracking and/or fissuring of the coating, and the formation of bubbles.
Due to the high speeds and loads to which the polyurethane coating of the wheels is subjected, especially with the latest developments in roller coasters, these damage phenomena occur on average three times a year for standard applications and approximately every 30 days for particularly demanding applications. Considering that the average cost to replace the coating of a single wheel is approximately €500, the maintenance cost of such installations solely due to wheel damage is very high. For this reason, the purpose of this work is to develop and experimentally test a simulation methodology capable of assessing the temperature reached inside the wheel coating [8]. This way, the company, during the design phase of the installation, can choose the appropriate type of wheel to maximize its lifespan, significantly reducing maintenance costs.
The most immediate approach to simulate this phenomenon would be to perform a coupled transient thermo-structural analysis using finite elements [9]. However, this type of simulation, due to the large deformations (caused by the viscoelastic behavior of polyurethane), wheel-track contact, and dynamic phenomena, results in very long computation times and is therefore impractical [10]. For this reason, the authors have developed an analysis methodology that, although based on the decoupling of thermal and structural aspects, has been shown to assess the temperature distribution in the wheel during actual operating conditions with relatively short computation times [8].
This methodology is based on the assumption that the material's behavior remains unchanged with varying temperatures and is divided into two phases. The first phase involves conducting structural transient analyses to evaluate the energy dissipated by the material and the heat per unit volume. The results of the structural analyses are then used as input for a transient thermal analysis (phase two) performed on a finite element model identical to the one used in the structural analyses, but aimed at obtaining the temperature during the machine's operating cycles. The assumption of decoupling structural phenomena from thermal ones, although it may seem limiting, is not actually so when considering the use of this method in the design phase, aimed at selecting the most suitable wheel to maximize its lifespan. However, in this study, it was decided to test the method through a numerical-experimental comparison of two different types of wheels, tested using a twin-disc machine capable of replicating the real load and speed time-history. The numerical-experimental comparison has shown that the proposed method yields results in line with experimental data. What is evident is an overestimation of the temperature, which can, however, be attributed to the material parameters used, which were derived from literature data of a similar material. Nonetheless, the results obtained are encouraging and allow the company to choose the most appropriate wheel. However, it is necessary to conduct material characterization in order to define a material model that can reasonably describe the real behavior.

2. The proposed approach

The direct approach to assess wheel temperature involves a coupled finite element thermo-structural analysis. However, this method is hindered by the long computational time due to nonlinearity and transient nature of the analysis, frequency sampling (about 1000 Hz) required for accuracy and the need for fine meshing to assess the exact deformation. This makes such kind of simulation nearly impractical, especially when considering various wheel types and multiple track circuits.
To address these challenges, the authors devised a faster finite element technique for temperature estimation [8]. This method decouples thermal and structural analyses, assuming material behavior remains unchanged with temperature changes. The proposed method comprises three

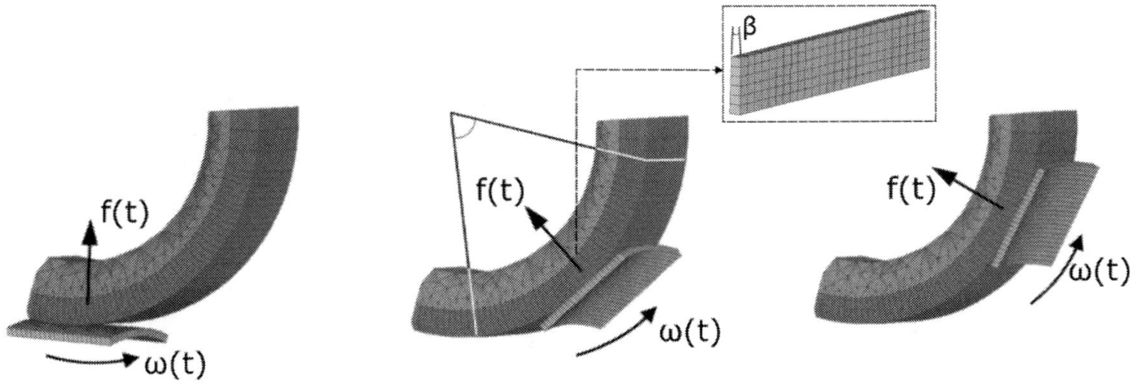

Figure 1. Speed and vertical load application in transient structural analysis

phases: first, extracting vertical load and carriage linear velocity time histories from a roller coaster multibody model. Instead of specific time histories (essential in coupled analyses), the latter are discretized into suitable ranges. Next, a finite element model of a wheel segment (approximately $\theta \approx 90$ degrees) is created, conducting transient structural analyses for various force and velocity combinations (see Figure 1).

As shown in Figure 1 the track is pressed against the wheel and then it is rotated around the wheel with an angular velocity proportional to the carriage's forward velocity. Introducing frictional contact between the wheel and track, the wheel experiences both rolling and sliding conditions. Structural simulations are used to determine the dissipated power, calculated as the product of the nodal forces vector and nodal velocities vector, for each element within a slice of elements located at the center of the considered segment (Figure 1). To significantly reduce computational effort and account for the effect of deformation in elements before or after those in the considered slice, it has been observed that, in the case of a regular mesh, each element positioned before or after the slice experiences the same time histories of force and velocity and, consequently, dissipated power but with a time phase shift of $\delta t = \beta/\omega$ (where β is the angular width of the element slice, and ω is the angular velocity of the wheel). Thus, it is possible to calculate the energy dissipated in the time interval Δt (the temporal span during which the radial row of considered elements influences energy dissipation) for each ring k of elements, as indicated in equation 1, also considering the influence of n_F sections upstream and downstream.

$$E_{k,\Delta t} = \sum_{j=1}^{n_F} \int_{\Delta t} |P(t)_{k,j}| \, dt \tag{1}$$

In equation 1, j represents the index for the n_F considered sections. To account for the influence of deformation in sections upstream and downstream of the slice of elements being considered, it's sufficient to shift forward and backward in time, with increments of δt, the power curves obtained and verify whether they still fall within the Δt interval. Once the dissipated energy for each ring of elements is known, it's possible to calculate the power dissipated in the time interval Δt, as described in equation 2.

$$P_k = \frac{E_{k,\Delta t}}{\Delta t} \tag{2}$$

By repeating this process for each pair of force and carriage velocity, it is possible to obtain power dissipation surfaces for each element of the slice, as a function of the normal force at the contact point and the carriage's forward velocity (Figure 3). These surfaces can be interpolated

IOP Conf. Series: Materials Science and Engineering 1306 (2024) 012015 doi:10.1088/1757-899X/1306/1/012015

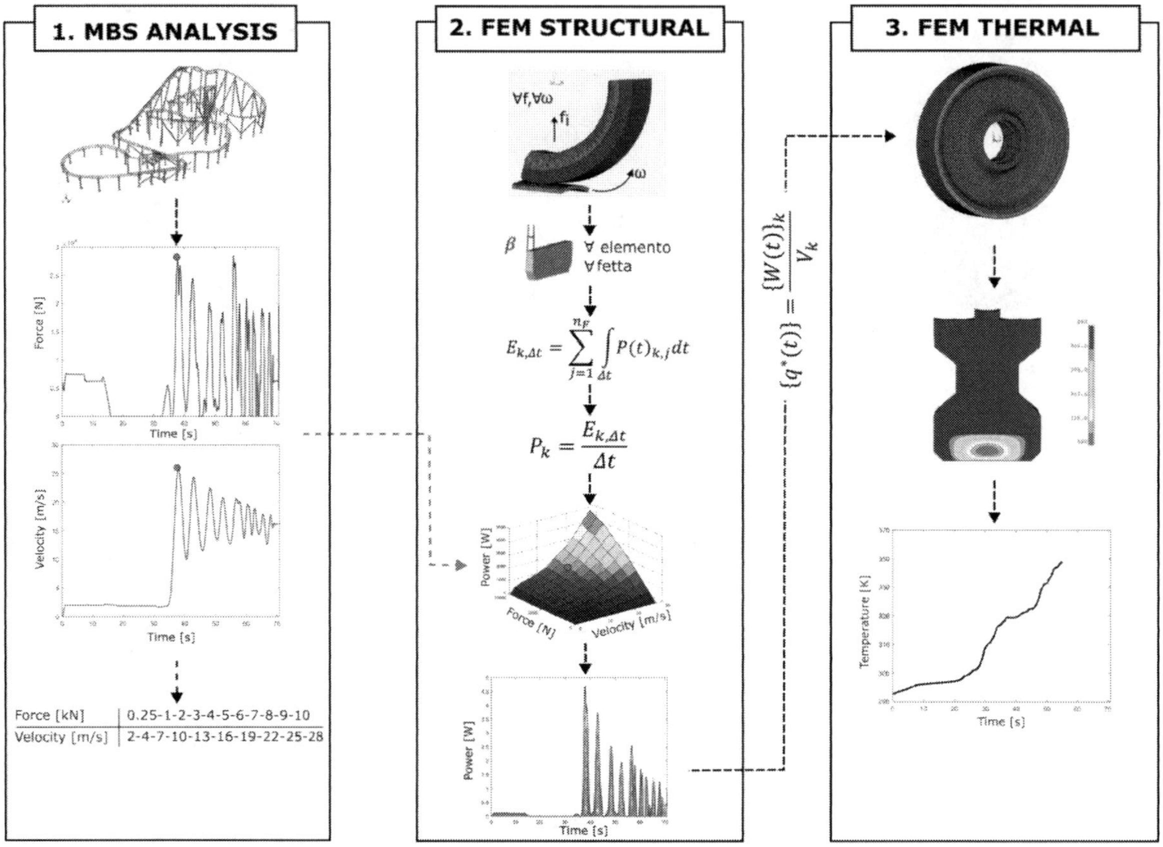

Figure 2. Flowchart of the procedure used for the thermo-structural simulation of polyurethane wheels.

instant by instant to obtain a time-history of dissipated power $W(t)$. This result can be used to perform a transient thermal analysis [11-12]. In fact, at the core of the thermal analysis, the equation to be solved for each element of the model is the one shown in equation 3.

$$\rho C \left(\frac{\partial T(t)}{\partial t} + \mathbf{V}^T \mathbf{L} T(t) \right) + \mathbf{L}^T \mathbf{q}(t) = \mathbf{q}^*(t) \tag{3}$$

where ρ is the density, C is the specific heat, T is the temperature, t is the time vector, $\{L\}$ is the is the derivative operator, $\{V\}$ is the relative velocity vector for heat exchange mass transport, $\{q(t)\}$ is the thermal flux vector between the wall and the fluid and $\{q^*(t)\}$ represents the internally generated heat per unit volume. This last quantity represents the input to be provided for a transient thermal analysis and can be calculated as shown in equation 4.

$$\{q^*(t)\} = \frac{\{W(t)\}_k}{V_k} \tag{4}$$

where $W(t)$ is the time history of dissipated power, calculated by interpolating the surface, element by element, with the instantaneous values of the normal force at the contact point and the forward velocity, and V_k is the volume of the row of elements associated with the i$-th$ element of the considered slice. Figure 2 depicts a flowchart of the procedure used.

The illustrated method allows for the evaluation of the temperature reached within the coating for various types of tracks (provided that the forces and velocities fall within the initially

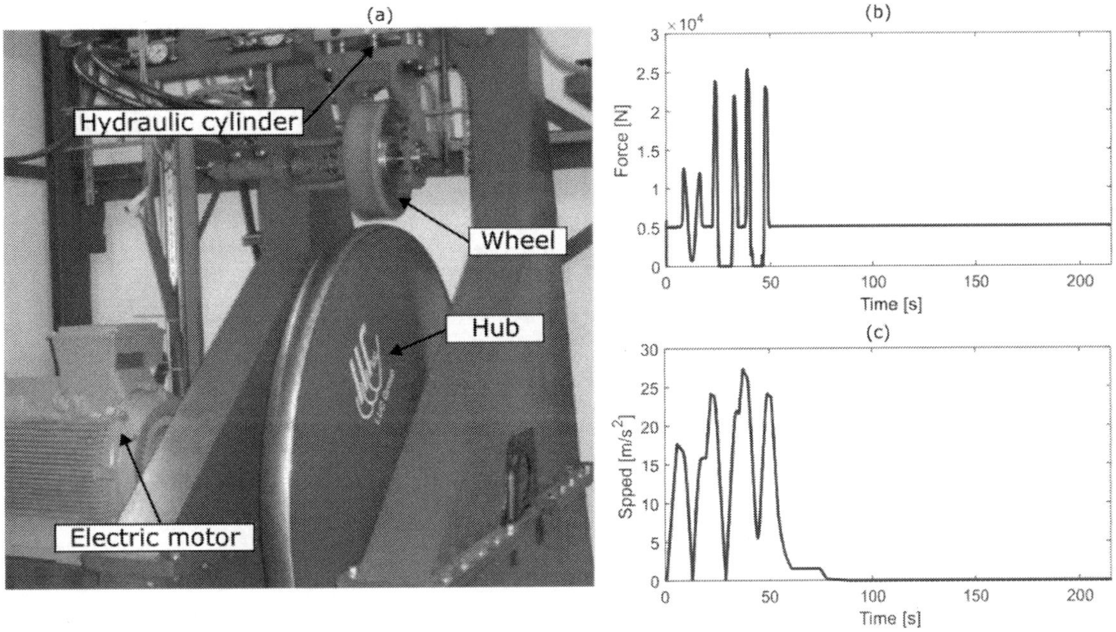

Figure 3. Experimental setup. (a) Testing Machine, (b) Applied vertical load, (c) Applied velocity

considered range) with significantly reduced computation times compared to a coupled transient thermo-structural analysis. It is essential to note that the assumption that the material behavior remains unchanged with temperature variation must hold. To further speed-up the computation time, given the simple geometry of the wheel, symmetry has been exploited to greatly reduce the number of degrees of freedom [13]. Symmetry has been utilized both in structural and thermal analysis.
In this manner, the computation time required to conduct a complete analysis is approximately 4 days, making it compatible with the company's design schedules. Another crucial aspect to highlight is that the power dissipation surfaces can be reused to assess the temperature reached by the same wheel on different types of tracks without the need to repeat the entire characterization process aimed at obtaining the surfaces of dissipated power through transient structural analyses.

3. Validation of the proposed method with experimental data

The method presented in the previous section has been employed solely in the design phase until now, *i.e.*, for selecting the type of wheel capable of achieving lower temperatures. However, this methodology has never been compared with experimental data. For this reason, in this study, the proposed method was tested by comparing the simulation results with experimental data conducted by the wheels manufactures.

3.1. Experimental setup

To carry out the experimental tests required for the evaluation of the proposed method, the experimental setup shown in Figure 3a was utilized.

As shown in Figure 3a, the wheel is pressed against a flywheel using a hydraulic cylinder. This cylinder replicates a force load history at the contact point provided by the roller coaster manufacturer, while the flywheel is driven by an electric motor that imposes an angular velocity

proportional to the forward speed of the roller coaster. The time histories are obtained from a multibody simulation and those used during the tests are depicted in figure 3b and c.
The force and velocity time histories provided by the company were replicated multiple times, thus simulating multiple track circuits. Two types of wheels were tested: one with an outer diameter of 516 mm and the other with an outer diameter of 570 mm. The coating width is 75 mm and 85 mm respectively, while the coating thickness is the same for both wheels at 15 mm. Each wheel was tested at different load percentages: 70%, 85%, and 100% of the load provided by the company, while the velocity remained unchanged. The tests were conducted at room temperature (22 °C), and the temperature was measured on the outer surface of the wheel after a sufficient number of cycles to ensure temperature stabilization.

3.2. Numerical - experimental comparison

In order to validate the proposed approach, the experimental tests were simulated using finite element analysis (FEA) following the method introduced in the previous section. For this purpose, FEA models were created for the two wheels experimentally tested. To correctly model the viscoelastic behavior of the polyurethane coating in finite element analysis [14-16], the generalized Maxwell model was used. The elasticity was modelled using the Mooney-Rivlin hyperelastic material model [17-19] with two parameters, while the viscosity was modelled using the Prony series with 5 coefficients [20-21]. The parameters used for the viscoelastic material model are presented in Table 1 and Table 2.

Table 1. Adopted parameters for elastic potential of Mooney-Rivlin

Parameter	Value	Unit
C_{10}	2.1428571×10^7	$[Pa]$
C_{01}	0	$[Pa]$
d	9.33×10^{-9}	$[Pa^{-1}]$

Table 2. Adopted parameters for the Prony series

$\alpha_i^G[-]$	$\tau_i^G[s]$
0.532	2.6×10^{-5}
0.034	0.007
0.072	0.009
0.016	0.0062
0.260	2.6×10^{-5}

The coefficients of the Mooney-Rivlin hyperelastic material model and the coefficients of the Prony series shown in Table 1 and in table 2 were obtained from quasi-static compression tests or DMA tests conducted by the authors in the past on a similar material. The hub of the wheel was modeled using aluminum with the following properties: Young's modulus $E = 2 \times 10^{11}$ Pa, Poisson's ratio $\nu = 0.34$, and density $\rho = 2700$ kg/m^3. For the track, steel with the following

properties was considered: Young's modulus $E = 7.2 \times 10^{10}$ Pa, Poisson's ratio $\nu = 0.3$, and density $\rho = 7850$ kg/m^3. To determine the thermal power, the wheels were numerically mapped considering the load and velocity time history used in the experimental tests shown in Figure 3. A friction coeffient equal to 0.1 was used between the track and the wheel. The ranges considered for obtaining the thermal power surfaces as a function of vertical force at the contact point and carriage advancement velocity ranged from 0 N to 30000 N with a step of 2000 N for force and from 0 m/s^2 to 30 m/s^2 with a step of 5 m/s^2 for acceleration. In this way, it was necessary to conduct 112 transient structural analyses to obtain the information required for thermal power calculation. By interpolating these surfaces instant by instant with the values of vertical force at the contact point and the roller coaster forward velocity, following the procedure shown in Figure 2, it was possible to obtain the time histories of dissipated power for each element and thus the heat internally generated per unit volume $\{q^*(t)\}$ to be applied in a transient thermal analysis.
Exploiting the symmetry it was possible to simulate multiple laps. For this the time histories of dissipated power were replicated to simulate 7 consecutive laps of the track. The thermal analysis were conducted considering the entire wheel model, consisting of both the aluminum hub and the coating. However, the track was not considered, and thus, neither the thermal conduction between the covering and the track. Nonetheless, convective heat transfer with the external air was taken into account, assigning a convection coefficient of 60 W/(m^2 K) at 293 K. The material parameters used in the thermal analysis are listed in Table 3.

Table 3. Material parameters for transient thermal analysis

Parameter	Value	Unit
	Aluminium	
Density	2700	Kg/m^3
Thermal conductivity	240	W/(m·K)
Specific heat	880	J/(Kg·K)
	Polyurethane	
Density	1000	Kg/m^3
Thermal conductivity	0.2	W/(m·K)
Specific heat	1100	J/(Kg·K)

The conducted thermal analyses allowed evaluating the temperatures reached by the wheel. Figure 4 shows a comparison between the temperature maps for the two models at the instant when the maximum temperature is reached (T = 1342 s).

As shown in Figure 4, the maximum temperature is reached at the core of the coating. Indeed, due to its low thermal conductivity, polyurethane retains heat inside, and there is no heat dissipation either to the outside or to the wheel hub. However, what is particularly interesting is the temperature evolution over time. Indeed, after a certain number of track circuits, the temperature stabilizes at a constant value and no longer increases. This phenomenon has also been experimentally observed in the results obtained by the wheel manufacturer.
The results obtained from these simulations (shown in Figure 4) were then compared with the experimental results obtained by the company. To this aim, Table 4 shows a comparison between the numerical and experimental results for all test conditions for both wheels.

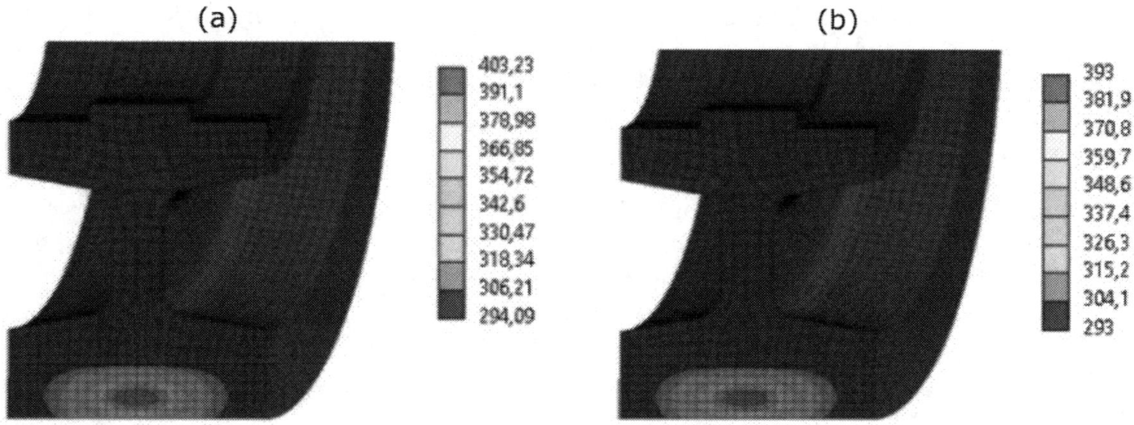

Figure 4. Temperature maps at instant T = 1342 s for 100% load simulation: (a) 516 mm wheel and (b) 570 mm wheel.

Table 4. Numerical-experimental comparison of the temperatures reached on the external surface of the wheels

Load %	Experiment [k]	Experiment mean [k]	Simulation [k]	Ratio
		516 mm wheel		
70	328.7			
70	330.37	330.59	334.22	0.94
70	333.15			
85	333.15			
85	335.92	336.07	342.47	0.91
85	338.71			
100	341.48	341.56	351.60	0.88
100	342.03			
		570 mm wheel		
70	327.03	326.48	330.3	0.93
70	325.37			
85	332.03	331.06	338.70	0.88
100	334.26	335.65	347.40	0.85
100	336.48			

The comparison of Table 4 shows that the proposed allows obtaining temperature values very close to the experimental data, although there is an overestimation of the reached temperatures. However, this difference could be attributed to the parameters used in the material model that being obtained from similar materials (from tests conducted by the authors or from bibliographic data) may not accurately describe the actual behavior of the wheel coating.
Table 4 also reports the ratio (normalized to 293K) between the numerical values and those obtained by linearly interpolating the experimental results for each load percentage. As evident, the slope of the line interpolating the data is not equal in both tests. This is because when

no load is applied (0% of the imposed load), the obtained temperature must tend to ambient temperature (293 K). However, as previously introduced, this may be caused by the material model and a correct characterization of the actual material will be the topic of future research.

4. Conclusion

This work has introduced and experimentally validated a method for the thermo-structural analysis of wheels with polyurethane coatings. The proposed method involves decoupling the structural and thermal simulation, assuming that the material's behavior remains constant with temperature variations. This calculation methodology allows for transient analyses to evaluate the temperatures reached in the wheels during operation, within timeframes compatible with the company's design phase. This result would not be achievable with a coupled thermo-structural analysis due to the various nonlinearities present in the model (contact, material, large deformations). The proposed method was validated by comparing the results obtained from simulations with experimental data provided by the wheels manufactures. Tests were conducted on two different types of wheels with three different load percentages. The numerical results show a good correspondence with experimental data, although there are still differences in terms of the maximum temperature reached. However, this can be attributed to the parameters used to describe the viscoelastic behavior of polyurethane. These parameters, obtained from tests on a similar material, represent a significant source of error, and it is therefore necessary to characterize the material used in order to define a material model capable of representing its real behavior.

References

[1] Nielsen L E, Landen R F 1993 Mechanical properties of polymers and composites – 2nd edition, *CRC Press.*
[2] Gogos C G, Tadmor Z 2006 Principles of Polymer Processing *Wiley–Interscience.*
[3] Ebewele R O 2000 Polymer Science and Technology *CRC Press.*
[4] De Cazenove J, Rade D A, De Lima A M G, Araújo C A 2012 A numerical and experimental investigation on self-heating effects in viscoelastic dampers, *Mechanical Systems and Signal Processing* **27** 433-445.
[5] Bakirani M, Hepburn C 1992 Internal Heat Generation and Fatigue Life Behaviour of Polyurethane Elastomer Based on Trans 1,4 – Cyclohexane Diisocinate, *Iranian Journal of Polymer Science & Technology* **1** 21-24.
[6] Xie F, Zhang T, Bryant P, Kurusingal V, Colwell J M, Laycock B 2019 Degradation and stabilization of polyurethane elastomers, *Progress in Polymer Science* **90** 211-268.
[7] Mead J L 1996 Mechanical Degradation of a Polyurethane Elastomer, Ph.D. Thesis at the Massachusetts Institute of Technology, Boston, USA.
[8] Braccesi C, Cianetti F, Ferri A 2019 Numerical evaluation of internal heat generation of roller coaster polyurethane wheels, *Procedia Structural Integrity* **24** 612-624.
[9] Caiying H, Zuyao L, Hongxia W, Lei W, Feng L, Hongfeng R 2009 Thermo–Mechanical Simulation and Optimization Analysis for Warpage-Induced PBGA Solder Joint Failures *Proc. of SMTA International Conference.*
[10] Gopalakrishna H s , Lai M L 1998 Finite element heat transfer analysis of viscoelastic damper for wind applications, *Journal of Wind Engineering and Industrial Aerodynamics* **77** 283-295.
[11] Ansys inc. 2018 ANSYS Mechanical APDL Thermal Analysis Guide, release 18.2.
[12] Ansys inc. 2018 ANSYS Mechanical APDL Theory Reference, release 18.2.
[13] Ansys inc.2018 ANSYS Mechanical APDL Advance Analysis Techniques Guide, release 18.2.
[14] Dal H, Kaliske M 2009 Bergström–Boyce Model for Nonlinear Finite Rubber Viscoelasticity: Theoretical Aspects and Algorithmic Treatment for the FE Method, *Computational Mechanics* **44** 809-823.
[15] Briody C, Duigan B, Jerrams S, Tierman J 2012 The Implementation of a Visco–hyperelastic Numerical Material Model for Simulating the Behaviour of Polymer Foam Materials, *Computational Materials Science* **64** 47-51.
[16] Bottoni M, Mazzotti C, Savoia M 2008 A finite element model for linear viscoelastic behavior of pultruded thin-walled beams under general loadings, *International Journal of Solids and Structures* **45** 770-793.
[17] Ali A, Hosseini M, Sahari B 2010 A review of constitutive models for rubber-like materials, *American Journal of Engineering and Applied Sciences* **3** 232-239.
[18] Dolwichai P, Limtragool J, Inban S, Piyasin S 2006 Hyperelastic Material Models for Finite Element Analysis with Commercial Rubber, *Technology and Innovation for Sustainable Development Conference* 769-774.

IOP Conf. Series: Materials Science and Engineering 1306 (2024) 012015 doi:10.1088/1757-899X/1306/1/012015

[19] Khajehsaeid H, Arghavani K, Naghdabadi R 2012 A hyperelastic constitutive model for rubber-like materials, *European Journal of Mechanics. A, Solids* **38** 144-151.
[20] Ghoreishy M H R 2012 Determination of the parameters of the Prony series in hyper-viscoelastic material models using the finite element method,*Materials & Design* **35** 791-797.
[21] Marvalova A 2007 Viscoelastic properties of filled rubber. Experimental observation and material modelling, *Engineering Mechanics,* **14** 81-89.
[22] López-Campos A J, Segade A, Casarejos E, Fernández J R 2019 Behavior characterization of viscoelastic materials for the finite element method calculation applying Prony series, *Computational and Mathematical Methods* **1**.

IOP Conf. Series: Materials Science and Engineering 1306 (2024) 012016 doi:10.1088/1757-899X/1306/1/012016

Preliminary study of the multi–physics design for the support structure of LEM–X space mission

G. Lombardi, Y. Evangelista and M. E. Biancolini

INAF/IAPS-Roma, Via del Fosso del Cavaliere 100, I-00133, Rome, Italy and University of Rome "Tor Vergata", Dept. Enterprise Engineering, Via Politecnico 1, I-00133, Rome, Italy

E-mail: giovanni.lombardi@inaf.it

Abstract. In this modern space exploration era, the lunar site offers new opportunities such as the deployment of an X–ray observatory capable of observing half of the sky simultaneously, but also may allowing access to the entire solid angle by exploiting the rotation of the moon around its own axis.

The Lunar Electromagnetic Monitor in X–rays (LEM–X) is an All Sky Monitor for the X–ray band (2–50 keV) based on the concept of coded aperture camera. The basic element is the camera of the Wide Field Monitor and the concept is to realize a "dome" composed of N identical modules that observe different directions, thus covering an overall simultaneous field of view of 2π in the sky. In this paper, we describe the preliminary study of a support structure for that camera, the design solutions adopted for the most important thermo–mechanical drivers, which have been elaborated and used for the demonstration of compliance to the system requirements with the environment. In particular, we reported some different thermal and mechanical scenario of this preliminary structure and the optimization of critical components with innovative and accurate fem analysis.

1. Introduction

Figure 1: logo of LEM–X space mission

We live in an era where, through the detection of gravitational waves, the discovery of their first electromagnetic counterparts, and with the full maturation of neutrino astronomy, astrophysics has entered a new phase; the "classic photon" astronomy must therefore integrate into a broader framework of "multi–messenger" astronomy. Multi–messenger astronomy is, in fact, a "time–domain astronomy", in which the classic pointing of know celestial objects with small–field instruments is accompanied and supported by instruments capable of observing or monitoring the entire sky. In this new scientific perspective, it is imperative for X–ray astronomy to contribute through the use of new instruments that can overcome the limitations of traditional space observatories. Global angular coverage of the sky is fundamental and extremely difficult to achieve with instruments on board satellites due to their necessarily limited resources.

The lunar site offers new opportunities to overcome these limitations, allowing for the deployment of an X–ray observatory capable of observing half of the sky simultaneously, but also allowing access to the entire solid angle by exploiting the rotation of the moon around its own axis.

The Lunar Electromagnetic Monitor in X–rays (LEM–X) is an All Sky Monitor for the X–ray band (2–50 keV) based on the concept of coded aperture camera. The basic element is the camera of the Wide Field Monitor, currently under study for the WFM experiment on the sino-European eXTP mission. The concept is to realize a "dome" composed of N identical modules that observe different directions, thus covering an overall simultaneous field of view of 2π.[1] Each module is autonomous, compact (about 35x35x35 cm^3), light (~10 kg), and with low power consumption (15 W). Moreover, thanks to its modular nature, LEM–X can be built gradually and can eventually reach dimensions, weight, and complexity impossible to achieve on a free–flyer.

The lunar site, as part of a wider infrastructure, can prove ideal for an all sky monitoring experiment. The stability of the site allows for full exploitation of the instrument's capabilities, obtaining high–resolution images of transient events and detected sources. If positioned appropriately (e.g. along the lunar equator), LEM–X will have simultaneous access to half of the celestial sphere and periodically to the entire sky. The site's accessibility will allow for maintenance and updates over time, making LEM–X a long-term observatory infrastructure, similar to a ground-based telescope[6][9].

Figure 2: LEM–X space mission concept

2. Workflow and Software

The engineering challenge of this work consists in designing structures suitable for supporting and orienting of the WFM cameras for the LEM–X space mission, which will have to work in extreme environmental conditions, ensuring compliance with the mechanical stability and pointing requirements[4] . The goal is to create a preliminary design that can be a baseline for the study of permanent structures on the Moon dedicated to specific types of scientific experiments. In particular, different types of materials and shapes, more or less complex, will be taken into consideration, which allow optimization of the structure and supports of the WFM chambers. The software used for this project are the following:

- **SolidWorks 2022**
  SolidWorks[13] is a three−dimensional parametric design and design software, produced and marketed by Dassault Systèmes. Useful for the design of mechanical devices, even complex ones, it involves the creation of 2D and 3D drawings of solids and surfaces, through a parametric geometric system. Parametric design allows to set up numerous types of relationships (parallelism, concentricity, perpendicularity, collinearity, equality, coaxiality, symmetry and many others) [11].
- **ANSYS SpaceClaim**
  ANSYS Space Claim is a parametric solid modeling software available in the ANSYS package that operates mainly according to the direct modeling logic, thanks to which it is possible to define the construction and modification operations of the desired geometry, without the need to define interdependent features.
- **ANSYS Meshing**
  ANSYS Meshing is the software responsible for generating the calculation grid. This process is automated in its simplest application although it is possible to add control settings to act directly on those parameters that define the mesh in terms of morphology, topology and dimensions of the constituent cells.
- **ANSYS Workbench**
  The Workbench platform is the tool designed to put in communication the various modeling and numerical analysis software that fall within the ANSYS landscape, facilitating the management of the project that is articulated there through various work steps. Its key feature is to allow the transfer and sharing of input and output data not only between the different tools used in pre and post processing, but also between multiple simulation models (which allows for the easy implementation of sophisticated multi−physical simulations) [8].

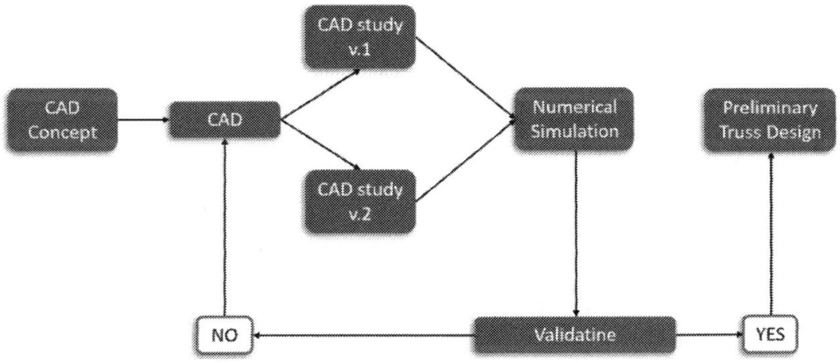

Figure 3: Workflow adopted

The CAD (Computer Aided Design) models of the truss was generated using Solidworks 2022 software, but the shapes were reconstructed and simplified using the standard features of the ANSYS SpaceClaim software. Furthermore, taking advantage of the advanced features of this software, a procedure of symmetry and conservation of structural constraints has been implemented even if the model is lightened. Therefore the study was carried out in the following steps:

- conceptual study of a complete structure that can support three different rows of WFM, each arranged at 30 degrees with respect to the ground or to the next row of chambers.

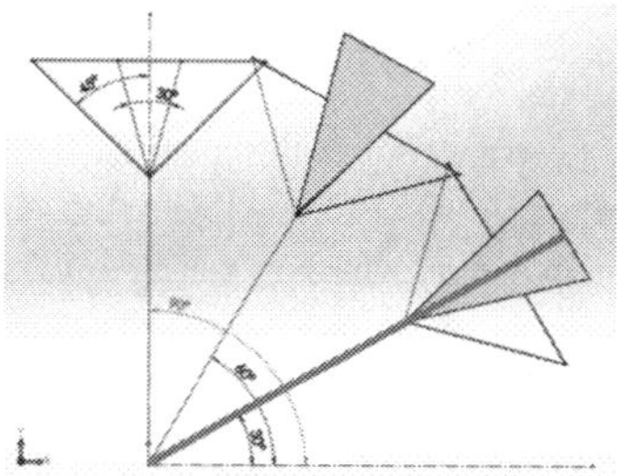

Figure 4: Camera angle

- preliminary study of a structure dedicated to a single row of WFM arranged at 60 degrees with respect to the ground.
- obtaining the simplified and defeatured models but compliant in terms of mass, volume and critical part of the structural truss on ANSYS SpaceClaim, meshing and import the simplified FEM model in the numerical simulation software. Subdivision of the models by appropriate axes of symmetry.
- numerical simulation, validation and control of the space mission requirement.

3. The Moon

The Moon is the only natural satellite of the Earth, located at an average distance from the Earth of approximately 384.600 km, it is the closest celestial body and easily visible even with the naked eye. It has an average diameter of about 3450 km and an orbital period of about 28 days. Man's goal of one day living on the Moon passes through the study of its physical-environmental characteristics. Below are listed the major differences found with Earth:

- absence of atmosphere;
- gravity lower than Earth;
- high thermal gradient;
- presence of regolith (moon dust);
- absence of a magnetic field.

Almost all the properties indicated above are linked to the mass of the Moon, which is approximately 1/80 of the Earth's mass. The reduced gravity value on the Moon, about 1/6 compared to that on Earth, had as its main consequence the absence of an atmosphere in the common sense of the term. We can in fact consider that there is a vacuum around the Moon although there is a very small layer formed by gas atoms released by the rocks following the impacts of the ions brought by the solar wind.
The composition of this layer mainly includes sodium and potassium, while the quantities of oxygen, nitrogen and carbon are unclear, but they are normally contained in the regolith, the rock that forms the lunar surface. The absence of an atmosphere, however, has a further great consequence: the rays coming from the Sun, which reach the surface, are not shielded. Therefore, sudden changes in temperature, even hundreds of degrees Celsius, are frequent within the same day.
When designing structures destined for the Moon, the resistance to the stressful temperature excursion suffered must therefore be analysed.
The lunar temperature is extremely variable both from a temporal point of view (day/night and seasonal alternation) and from a spatial point of view (geographical coordinates). Thanks to NASA studies, tables are available in the literature that describe the temperature variation as a function of position and the thermal gradient of this as a function of distance from the

ground. The lunar regolith must be taken into consideration when it is proposed to use electronic instrumentation, as in this case, since this would be exposed to contamination and corrosion, therefore, in the design a sufficiently high height with respect to the ground of the payloads is considered[2].
For the current mission, two possible candidate locations to host the structure have been identified:

- The "Shackleton Crater", near the south pole.
- The "Sea of Tranquility", at the latitude of the equator.

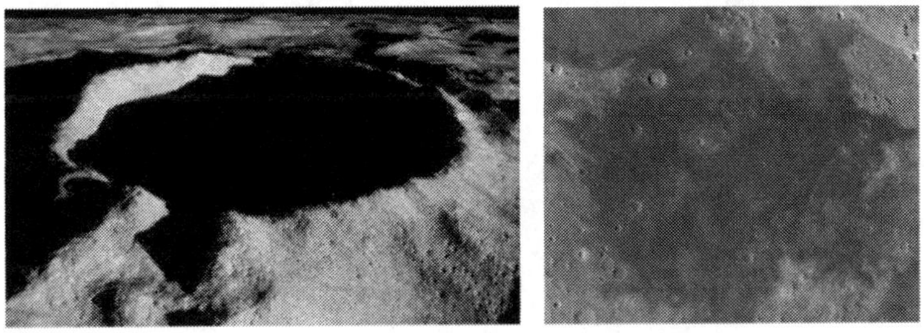

Figure 5: Shackleton Crater on the left and Sea of Tranquillity on the right

Given the rotation period of the Moon, the thermo−physical properties of the regolith and the absence of an atmosphere, the lunar surface is subject to more severe thermal conditions than any point on Earth, both in terms of values achieved and daily variation. Starting from the knowledge of the albedo of the lunar surface and the radiation of the Sun on it, tables have been developed by NASA which indicate the temperatures as geographical coordinates vary.

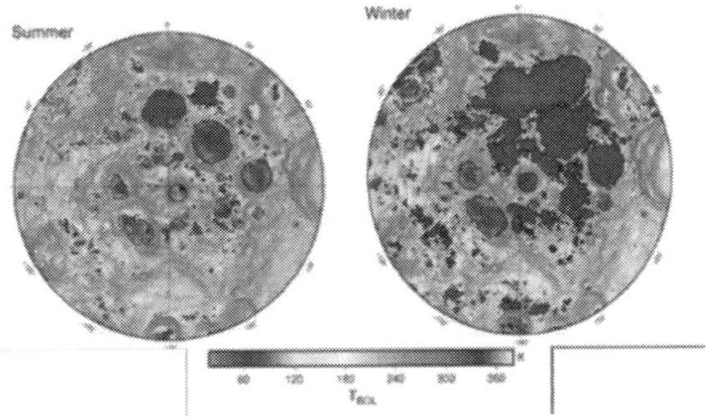

Figure 6: Temperature gradient on Moon surface

4. Structure proposal n.1

4.1. Mechanical design

In the first proposal, the starting idea is a compact structure, with a hemispherical shape, which allows the detection of X-rays coming from half the celestial sphere thanks to the correct arrangement of the WFM modules [3]. The design of the structure is divided into two phases:

- in a first part, the support structures for the WFM coded chambers are designed (made in CAD and subsequently a finite element analysis is carried out on them with the thermal loads and static loads to which the structure will be subjected).

- in the second part, once the design of the WFM truss modules has been completed, we move on to the design of the frame, the structure that connects all the various modules.

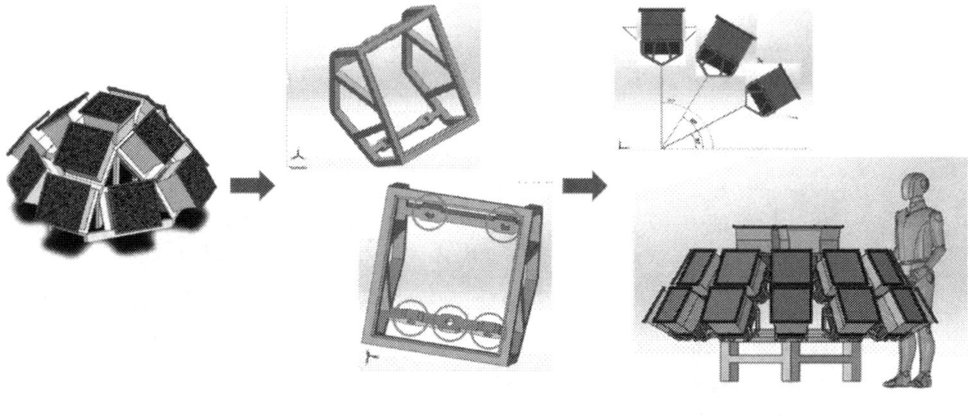

Figure 7: LEM−X from concept to the first geometry proposal

From an experimental point of view, each camera has a FoV (Field of View) of 90°×90°, of which only 30°×30° is fully illuminated, i.e. it is able to provide complete and low-noise coding of the image. To obtain optimal coverage of the vault, the rooms must therefore be spaced 30° apart. Due to environmental conditions, the presence of regolith and the irregularity of the Earth's surface, the cameras will have to be positioned in such a way as not to observe below 15° from the lunar surface, as below that measurement the data would be irrelevant.

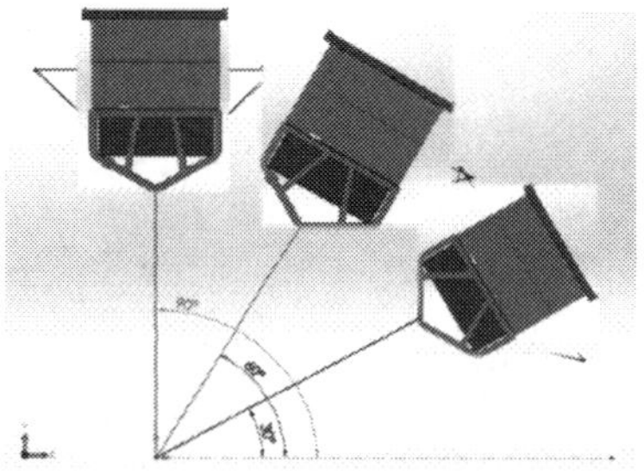

Figure 8: WFM experiment disposal

The requirements imposed for the construction of the structure are many. Since the layout of the chambers is subjected to the morphological conditions of the lunar surface, the first chamber will be placed raised from the ground to avoid incorrect or absent evaluations. The most stringent requirement concerns the deformation that the structure can undergo. Due to thermo−elastic deformations, the structure may not guarantee correct aiming and its stability for each chamber. For the stability of the mission pointing, it is therefore required that each camera (and therefore the structure that will support it) has a misalignment with respect to the support structure which must not exceed 1 arcmin.

4.2. Numerical Simulation

The use of finite element analysis or FEA allows complex problems to be solved precisely and quickly thanks to the geometric discretization of a solid body into smaller elements of regular

shape.
The FEM simulations were performed with four materials available on the Ansys Granta library, chosen from those typically used in aerospace missions:

alloy	α [K^(-1)]	density [Kg/m^3]	E [GPa]	TUS [MPa]
Ti6Al4V	8,79E-06	4429	111	918
Al 7075 T6	2,30E-05	2770	71	310
Al 6061 T6	2,28E-05	2713	69	306
Steel	1,7E-05	7750	193	586

Table 1: Material for space application

In the analyzed problem, the modules undergo the thermal gradients that characterize the lunar surface where the entire structure is located. They are also subjected to their own weight, which depends on the lunar gravity. Weight can be calculated as the product of the acceleration of gravity and the mass of the body under examination. The most important load is the thermal one which, due to the extreme temperatures reached, will be decisive in the design phase for the choice of geometry and material[7].
To determine the temperatures to be inserted as load conditions, use the manual provided by NASA:

Temperature	T [K]	T [°C]
Max	395,6	122,4
Min	95,2	-177,9

Table 2: temperature range for Sea of Tranquillity

Temperature	T [K]	T [°C]
Max	136	-137
Min	42	-232

Table 3: temperature range for Shackleton Crater

In the following work, however, it was decided to evaluate only one case, namely the Shackleton Crater, whose temperatures are more severe in absolute terms, thus providing a sizing that is safe compared to the real operating conditions. Once all the listed parameters were defined, a FEM type analysis was conducted with the primary objective of identifying the value of the misalignment of the structure, which must fall within the requirement of 1 arcmin. Once the simulations have been carried out, starting from the value of the absolute maximum deformation on the face (therefore sizing with respect to the absolute worst case) it was verified whether this falls within the requirement of 1 arcmin.

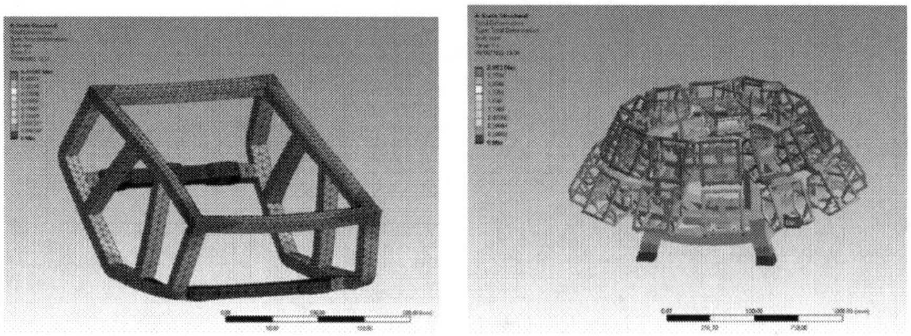

Figure 9: Numerical simulation in the proposal n.1

4.3. Results

In only one case, that of the titanium alloy, is it possible to remain within the deformation requirements. Aluminum alloys are much lighter, but due to the higher thermal expansion coefficient, they deform almost twice as much. Steel has an intermediate deformation between the two cases, but a density approximately double that of the others.

Alloy	Dis. XY [arcmin]	Dis. XZ [arcmin]	Weight [kg]
Ti6Al4V	0,74	0,97	0,55
Al 7075 T6	1,9	2,5	0,4
Al 6061 T6	1,9	2,5	0,3
Steel	1,1	1,3	0,9

Table 4: misalignment angle calculation

Analyzing, however, the state of tension with the use of the von Mises equivalent sigma, it is possible to observe how, with the exception of the areas close to the constraint, on almost the entire structure (and above all, for all the rooms), the value of the sigma is close to zero. In the areas close to the constraint, however, values of around 200 MPa are reached. Regarding the misalignment angle for the 30° modules, the requirement is respected but the same does not apply to the 60° and 90° modules, where the requirement is not respected in any of the cases. The results show that this solution is not suitable and that, only for the modules placed at 30° with parts in titanium alloy, have satisfactory values. Therefore, this study was discarded.

5. Structure proposal n.2

5.1. Mechanical design

In the second case study it was decided to "dismantle" the structure and focus on the strip of chambers placed at 60°. Preliminary prototypes were created based on previous lunar lander models and the best one was chosen, with the following characteristics:

- Tubular elements for greater rigidity
- Thin–walled beams for less weight
- Support feet at a suitable distance from the regolith

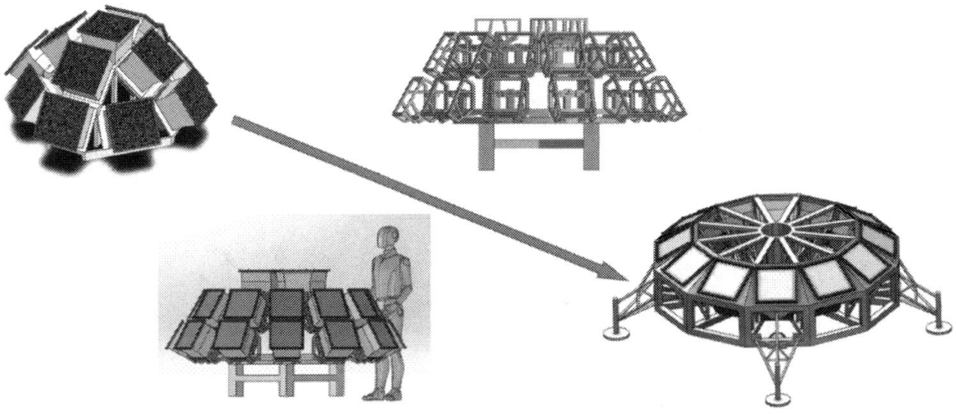

Figure 10: LEM−X from concept to the second geometry proposal

In the realization of the second proposal of the project, we based ourselves on the creation of a structure that is able to fulfill the specific requests. The concept to be created must be a support structure for 12 WFM cameras, inclined with respect to the horizontal by 60°, capable of guaranteeing, with mounted cameras, a maximum misalignment of 1 arcmin.
Furthermore, the individual components of the structure must present modes of vibration far from the frequency of 100 Hz, this being the main frequency that the launcher imparts during the launch phase, in this way the structure finds itself operating at frequencies far from those for which the resonance phenomenon would occur. In addition to these purely technical specifications, another important request lies in the containment of the overall weight of the structure, this object must not be excessively massive since it will have to be transported via a space carrier and consequently a greater weight turns into much greater costs. Three different concepts were developed in the construction of the structure: In all three structures

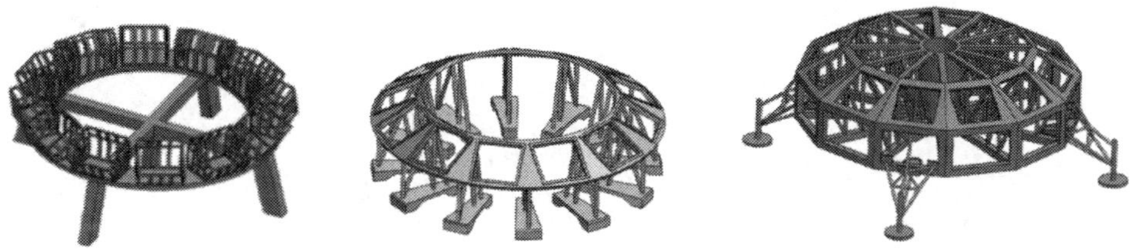

Figure 11: Concept design for the second proposal

the underlying idea was to exploit the symmetry that the problem possesses. Given these structures, we proceeded with the choice of the most valid layout typology and we chose to operate on the third structure designed.

IOP Conf. Series: Materials Science and Engineering 1306 (2024) 012016 doi:10.1088/1757-899X/1306/1/012016

5.2. Numerical Simulation

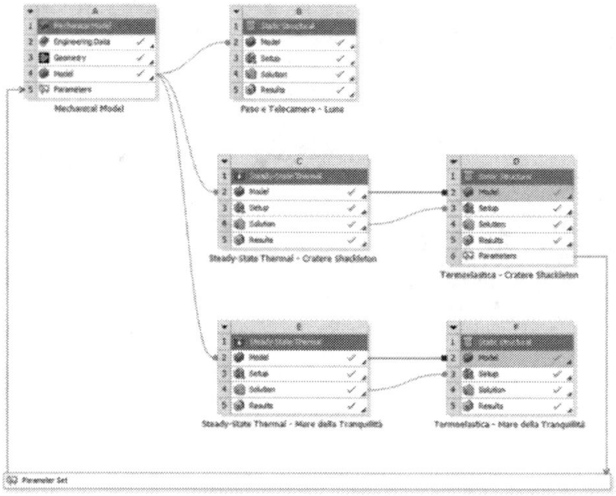

Figure 12: Numerical simulation workflow

The structure created, including supports, based on the chosen material configurations has a mass equal to:

Truss material	Support material	Weight [Kg]
Titanium	Titanium	360,5
Steel	Steel	638,9
Titanium	Steel	393,2
Steel	Titanium	606,3

Table 5: material for space application

Before proceeding with the finite element analyses, it was necessary to prepare the model, we proceeded following these general rules:

- Use of the symmetry of the structure.
- Simplification of the three–dimensional structure (solids) to two–dimensional (shell) or one–dimensional (beam) models where possible.
- Elimination of localized design complications.

The structure is a composition of four identical parts which in turn present a further plane of symmetry. Therefore, it was decided to exploit an "eighth" symmetry for some FEM analyzes (static and thermo–elastic) and a "quarter" symmetry for the modal analysis [5].

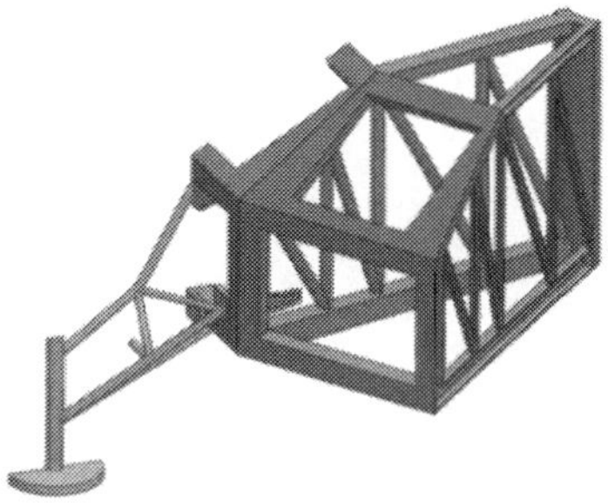

Figure 13: Eighth symmetry for static structural simulation

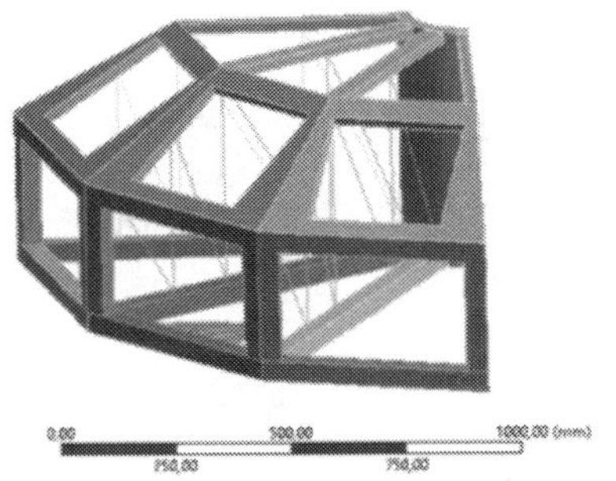

Figure 14: Quarter symmetry for modal analysis

Regarding the mesh, the structure has been divided into 4 areas on which to apply the mesh methods and sizing. The four areas of interest are the following: The validation of the model,

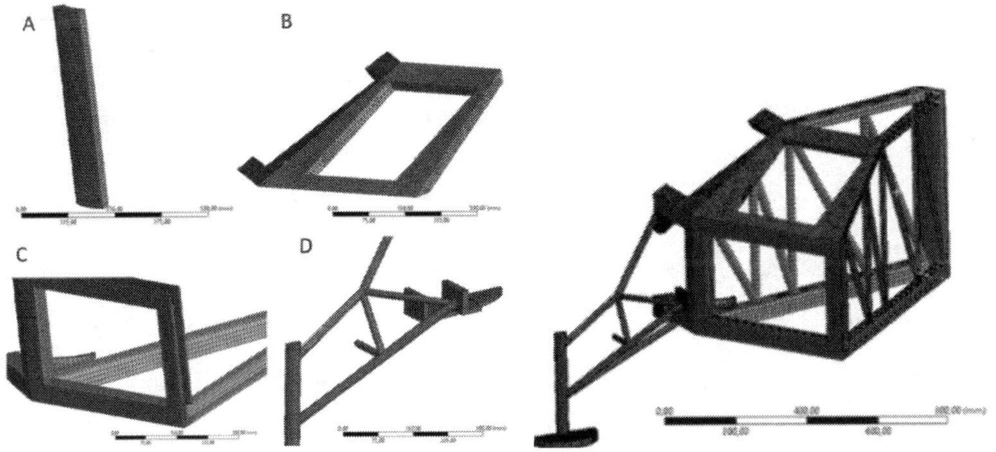

Figure 15: Local and global mesh

both in titanium alloy and in steel, is carried out through the following numerical analyses:

- Static analysis (structure weight and WFM)
- Thermoelastic analysis (temperatures for both places considered)

- Modal analysis (launch loads of SLS vector)

The requirements are:

- Maximum chambers misalignment of 1 arc minute.
- Maximum reference weight for the support structure: 800 kg
- First natural frequencies of the structure above 100 Hz

5.3. Results

The stresses resulting from the static analysis are small, in fact the maximum tension is 5,24 MPa compared to a yield stress of 845,7 MPa which divided by the safety coefficient for yielding of 1,5 (factor of safety) provides an admissible value of 563,8 MPa, therefore well below this threshold.
Furthermore, these maximum tensions occur in decidedly limited areas. On the basis of all these observations it is possible to highlight that the contribution of the own weight of the structure and the chambers in terms of displacements and tensions has a certainly marginal impact compared to what can be expected due to the thermal field.

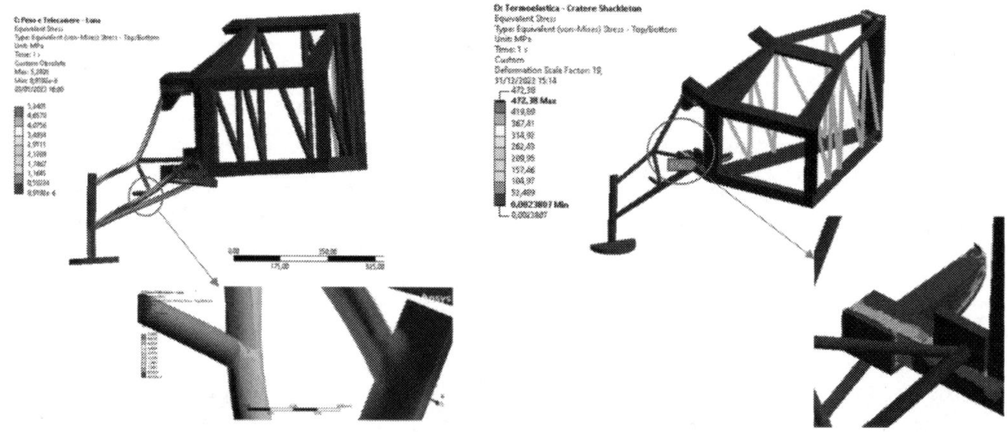

Figure 16: Static structural analysis

The best solution seems to be a structure made entirely of titanium alloy. Furthermore, it appears that the Sea of Tranquillity returns safety values for both yielding and rupture, while the case of Shackleton Crater presents more critical values.

	FOSY [MPa]	FOSU [MPa]
Steel	250	460
Ti6Al4V	846	918

IOP Conf. Series: Materials Science and Engineering 1306 (2024) 012016 doi:10.1088/1757-899X/1306/1/012016

	Shackleton Crater [MPa]	Sea of Tranquillity [MPa]
Support: Titanium Truss: Titanium	478	330
Support: Steel Truss: Steel	1529	833
Support: Titanium Truss: Steel	1416	785
Support: Steel Truss: Titanium	793	446

Table 6: static structural results

In order to validate the structure against the stresses due to launch via the SLS (Space Launch System) space carrier, it is necessary to conduct modal analyses. In this case the structure is considered packaged separately between the support foot and a quarter of the structure, given the dimensions of the dome.

	Configuration	
Mode	free-free [Hz]	bonded [Hz]
I	0	163
II	0	169
III	0	183
IV	0	186
V	0	202
VI	0	217
VII	131	224
VIII	138	227
IX	170	231
X	176	259

Table 7: modal analysis results

As can be seen from the results, the rigid motions in this case are three, as having placed a bilateral support constraint which inhibits the rotations along x and z and the translation in the y direction. Furthermore, the first proper mode of vibration is close to 160 Hz.
The embedded configuration represents a good approximation of the extremes of the most probable frequency spectrum. From the analysis of the results it can be seen that the behavior of the structure at the launch impulse is considered positive. The most restrictive requirement for the mission is undoubtedly that relating to maintaining orientation for the WFM. Even just tiny relative movements of the support area of the camera cause it to change its angle of inclination excessively. The calculation of the misalignment is carried out by considering the normal movements (X axes of the reference systems) with respect to the support surface of the four vertices of the hole for placing the single chamber.
In particular, it was necessary to recover these values for two different positions: the one located immediately above a support foot and the adjacent one.
The detected misalignment values are excellent in the Shackleton Crater and represent a further point in favor for future structural optimization. Also, due to the different thermal gradients the two positions have extremely different misalignment values.

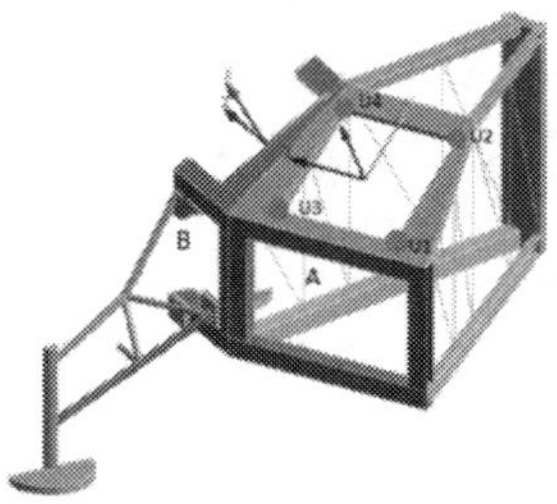

$$\theta_y = \frac{u_2 + u_4 - u_1 - u_2}{2 * b} \qquad \theta_z = \frac{u_1 + u_2 - u_3 - u_4}{2 * c} \quad \Rightarrow \quad \theta_{dis} = \sqrt{\theta_y^2 + \theta_z^2} * \frac{180}{\pi} * 60$$

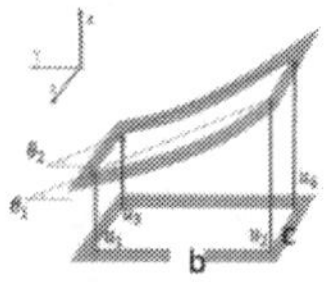

Camere A					
	U1 [mm]	U2 [mm]	U3 [mm]	U4 [mm]	θ_{dis} [arcmin]
Piedi: Titanio - Struttura: Titanio					
Cratere Shackleton	-1,872878	-1,8658	-1,879012	-1,875165	0,110
Mare della Tranquillità	-0,70582	-0,39435	-0,708952	-0,398975	3,597

Camere B					
	U1 [mm]	U2 [mm]	U3 [mm]	U4 [mm]	θ_{dis} [arcmin]
Piedi: Titanio - Struttura: Titanio					
Cratere Shackleton	-1,929277	-1,88052	-1,929277	-1,88052	0,564
Mare della Tranquillità	-0,732037	-0,40136	-0,732037	-0,40136	3,828

Figure 17: Thermo–elastic analysis

In the case of Shackleton Crater the gradient is less accentuated than in the Sea of Tranquillity and this leads to much larger relative displacements, which are the determining factor in the calculation of the misalignment rotation.

6. Summary and Outlook

The goal of being able to place a scientific experiment on the lunar surface is a dream for every scientist of this sector. The challenges are many and the use of the latest technologies and multi–physics finite element simulation will facilitate the optimization and the design work of LEM-X.

Figure 18: Astronaut with a WFM camera

The studies carried out on the structure are preliminary but the results obtained are satisfactory. In particular, the configuration that best copes with stresses and misalignments is the one made entirely of titanium alloy, although, as can be seen from the results obtained in the simulation case, in the Sea of Tranquility, the misalignment requirements are not respected while they are those relating to the maximum tensions acting on the structure are respected.
In the case of the Sea of Tranquility, the misalignment requirements are not respected while those relating to the maximum stresses acting on the structure are respected, instead in the Shackleton crater, the misalignment requirements are respected but, for a while, those on the tension are not maintained. In view of these results, the structure therefore presents a better behavior in the Shackleton crater.
Future works will be:

- optimization of geometry and realization of sub–modeling for stress concentration region;

- realization and study of the assembly areas and analysis of the threaded connections;
- introduce composite materials specifically designed for this application;
- create a realistic simulation environment and test the structure.

Among the possible sub–models to be created there are those relating only to the supports, to the interface areas between supports and structure and finally to the support faces of the WFM, in order to better visualize the areas that most influence the misalignment and where the maximum tensions.

Among the subsequent studies to be carried out is the refinement of the geometry of the structure, i.e. providing for the introduction of connecting radii in appropriate areas to reduce the concentration of tensions. An optimization of the thicknesses of the thin wall beams would have the dual function of increasing the resistance characteristics and respecting the misalignment of the structure where required and decreasing the overall weight by removing material in the less stressed areas. Related to this aspect is the fragile ductile transition of the materials in use, where extremely low temperatures are reached.

7. References

[1] ShuangNan, A Santangelo, M Feroci, Y Xu et al., The enhanced X–ray Timing and Polarimetry mission—eXTP. (feb. 2019).
[2] Hamill, B.D., Schunk, R. G., Erickson, L. R., Human Landing System – Lunar Thermal Analysis Guidebook, NASA, (2021)
[3] Master's thesis in Mechanical Engineering, University of Rome Tor Vergata, "Study on the design and optimization of on-board mechanics for the proposed LEM-X space mission", R. Fiacco, (2022)
[4] M Hernanz, S Brandt, M. Feroci, P. Orleansky et al., The wide field monitor onborad the eXTP mission, Astrophysics - Instrumentation and methods for astrophysics - High Energy Phenomena (2018)
[5] M. E. Biancolini, Fast radial basis functions for Engineers applications, Springers (2017)
[6] Y. Evangelista, M. Feroci, F. Ceraudo (IAPS), R. Campana (OAS), F. Fiore (OATs), R. Della Ceca (OAB), "A Lunar X-ray All Sky Monitor for Time Domain and Multi-messenger Astrophysics".
[7] Y. Li, Z.Z. Wang, J.S. Jiang, "Simulations on the influence of lunar surface temperature profiles on CE-1 lunar microwave sounder brightness temperature", 27/07/2010.
[8] O. C. Zienkiewics, R. L. Taylor and J. Z. Zhu, The finite element method: Its basis and fundamentals, Butterworth-Heinmann, 6th (2005)
[9] 18. G. Lombardi, A. Baldelli, R. Fiacco, E. Lombardi, M. E. Biancolini, R. Piazzolla, "The preliminary thermo-mechanical design, simulation and optimization, of LAD onboard the eXTP space mission, AIAS 2022.
[10] https://ecss.nl/
[11] Dassault Systemes, DS Solidworks 2022, www.solidworks.it

IOP Conf. Series: Materials Science and Engineering 1306 (2024) 012017 doi:10.1088/1757-899X/1306/1/012017

Biomechanical evaluation of a custom-made mandibular plate

R Mineo[1, 2], G La Rosa[1], T Cocomero[1], R Barbagallo[1], S Battaglia[3] and S Crimi[3]

[1]Department of Civil Engineering and Architecture, University of Catania, via S. Sofia 64, 95123 Catania, Italy
[2]MT Ortho s.r.l. Via Fossa Lupo, 95025 Aci Sant'Antonio (CT), Italy
[3]Maxillofacial Surgery Unit, Department of CHIRMED, Policlinico Rodolico-San Marco University Hospital, University of Catania, 95124 Catania, Italy

guido.larosa@unict.it

Abstract. In the field of maxillofacial surgery, mandibular reconstruction with plates is one of the operations that is performed when resection of part of the jaw is unavoidable in patients suffering from pathologies such as, for example, neoplasms and osteonecrosis. Through the use of standard prostheses, which are commercially available in various models, mandibular reconstruction is possible. However, in the presence of particularly serious and complex clinical pictures, in which it is necessary to resect large parts of the mandibular bone, such as in some oncological cases, the use of serial prostheses is not always resolving. In these cases, a customized implant, designed specifically for the patient, is preferable.
Aim of this work is the mechanical evaluation of a custom-made mandibular plate with an innovative bone-implant interface geometry through a finite element analysis. Three physiological loading conditions were analysed: RUC (right unilateral canine) bite, RGF (right group functional) bite and maximum opening. Finally, the maximum pull-out force acting on the fixation screws was evaluated through an experimental fatigue test.

1. Introduction

Thanks to improvements in technology and the advent of additive technologies to support maxillofacial surgery, it is possible to achieve better aesthetic and functional results for patients requiring major demolition surgery [1, 2]. The current gold standard for restoring extensive mandibular defects after surgical ablation is the insertion of a reconstructive titanium plate, however, a failure rate of such an implant related to its mechanical mobilization is still reported [3-5]. One of the main risk factors after reconstruction of mandibular bone defects using reconstructive plates is mechanical failure. The latter may be due to loosening of the fixation screws, fracture of the plate and aseptic loosening of the implant.

The muscular forces associated with the simple opening or closing of the mandible or in some cases the occlusion forces of incisors and molars in the healthy mandible result in shear and pull-out forces on the screw that can lead over time to loosening of the screw or in the most extreme cases to mechanical failure of the plate. An improved bone-prosthesis interface could reduce shear and pull-out forces acting on the screw during normal physiological postoperative movements and thus reduce the likelihood of loosening and implant failure.

The primary stability of a mandibular plate with innovative bone-implant interface geometry was evaluated in comparison to standard plate geometry by means of a four-point bending test simulating a real-life loading condition.

IOP Conf. Series: Materials Science and Engineering 1306 (2024) 012017 doi:10.1088/1757-899X/1306/1/012017

Furthermore, by means of a finite element analysis, a biomechanical evaluation of a custom-made mandibular prosthesis with a new bone-implant interface was carried out in collaboration with surgeons of the maxillofacial surgery department.

2. Description of the investigation

The traditional mandibular plates designed and manufactured in titanium alloy with 3D EBM technology, present the geometric characteristics shown as Plate A in Figure 1. Plate A has a constant thickness along the entire length, representing the state of the art currently used in maxillofacial surgery. The modified plate (Plate B) has similar geometric characteristics but different interface with the bone tissue. Plate B shows an innovative retention structure at the interface with the osteotomy plane and a fixation screw passing through it.

Plate A			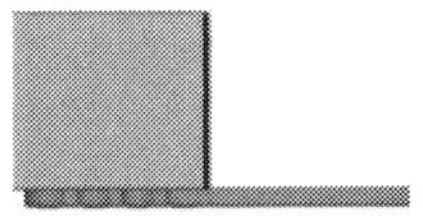
Plate B			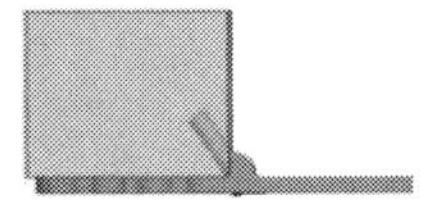

Figure 1. Frontal (left), lateral (centre) view and interface (right) for traditional and modified plates.

The case study considered concerned the mandibular reconstruction of a 60-year-old male patient suffering from bisphosphonate osteonecrosis. The patient's anatomical data obtained by CT examination were used to reconstruct the 3D model of the mandible using Mimics software (Figure 2).

The cutting planes were indicated by the prescribing physician and then used to virtually resect the jaw by dividing it into three parts: a central area to be removed and the two healthy lateral stumps.

The device shown in Figure 3 takes the form of a thin plate that follows a profile as close as possible to the lower profile of the original corrected jaw and present an innovative interface as seen previously in the plate B of experimental test. The implant is fixed with screws.

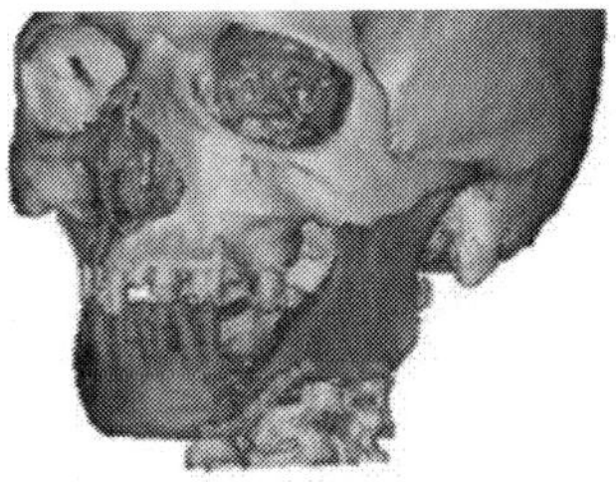

Figure 2. 3D model of the patient mandible.

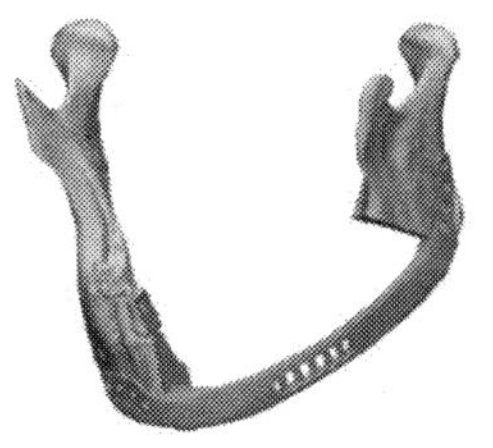

Figure 3. Custom-made plate design.

The plate has a height of 10 mm, a thickness of 3 mm throughout the main body and 2 mm at the attachment points. There are also two retention structures in the interface areas with the bone resection surfaces. The plate also has a grid structure, a lateral window to facilitate the reattachment of muscles and other soft tissues that, during surgery, are disconnected from the portion of the jaw to be removed. The fixation elements are bicortical (self-tapping) screws, specific for applications in which the thread must be gripped in both corticals of the bone. The mandibular prosthesis was manufactured from titanium alloy (Ti6Al4V ELI) using the EBM (Electron Beam Melting) process.

IOP Conf. Series: Materials Science and Engineering 1306 (2024) 012017 doi:10.1088/1757-899X/1306/1/012017

3. Finite element analysis

In this study, the biomechanical behaviour of the implant was analyzed with the aim of verifying the resistance of the prosthesis to physiological loads and the distribution of deformations that these induce in the bone tissue near the fixation elements, in order to predict whether screw loosening and bone resorption phenomena may occur, both of which must absolutely be avoided for a successful implantation.

The software used for the structural analyses was MSC Marc Mentat 2019 (MSC Software Corporation, Santa Ana, CA), a non-linear finite element analysis software developed to simulate the behavior of complex materials and interaction in the presence of large deformations.

Finite element analyses of the prosthesis-bone system were conducted by subjecting a model of the same to physiological loads. The following loading conditions were analyzed:

1. bite RUC (right unilateral canine);
2. bite RGF (right group functional);
3. maximum opening.

The first two are unilateral static bite configurations. Only unilateral bites were analysed as the mandible, after resection and implantation, has teeth only on the right hemiarch. In particular, the incisors, canine and right premolars were present. The condition of maximum opening is that of the mandible in fully open position, considering a rotation angle of 30 deg in the sagittal plane.

The constraints and loads applied to the system vary as the condition changes:

- in the RUC bite, the condyles and the occlusion surface, in this case the right canine only, were considered fixed,
- in the RGF bite, the occlusion surface is constituted by the whole canine and right premolars, therefore the surfaces of these three teeth and the condyles have been fixed,
- in the case of maximum opening, only the condyles are considered fixed.

The values and directions of the forces exerted by the muscles under the three conditions have been the subject of numerous studies, and can be found in the literature [6-10].

It was decided to carry out, for each loading condition, an analysis of the overall prosthesis-bone system considering, as a first approximation, the screws as infinitely rigid cylinders transmitting the load from the bone tissue to the prosthesis, and vice versa, as joints. Subsequently, in order to analyse in detail the stress-strain state present in the screw-bone interface zone with the real threaded profile, analyses were carried out on two sub-models: one comprising the fixation zone in the left abutment of the mandible, and another for the right one, considering real threaded profile of the screws [11].

The materials involved in the simulation are the cortical and trabecular bone tissues of the mandible, the titanium alloy Ti6Al4V ELI of the plate [12], and the titanium alloy of the Stryker self-tapping screws. In the analysis, the behaviour of all four materials was assumed to be linear elastic and isotropic, with the values of the respective elasticity constants, Young's Modulus (E) and Poisson's coefficient (ν), shown in Table 1

Table 1. Mechanical properties of the materials

Material	E [MPa]	ν
Cortical bone	13700	0.3
Spongy bone	1600	0.3
Ti6Al4V ELI	120000	0.3
Titanium alloy of Stryker screws	113800	0.342

The 3D models of all components (prosthesis, right cortical bone, right spongy bone, left cortical bone and left spongy bone) were imported in STEP format into Marc Mentat 2019 software (MSC software). Meshes of the five solids were created with tetrahedral elements. The mesh of each component was

IOP Conf. Series: Materials Science and Engineering 1306 (2024) 012017 doi:10.1088/1757-899X/1306/1/012017

defined as a meshed deformable contact body, so that the contacts between the bodies could subsequently be imposed.
To simulate the welded screws, Nodal Tie Type 100 (all degrees of freedom) links were set up, connecting nodes on the inner surface of each hole to a master node, positioned in the centre of the hole. This was performed for both the attachment holes on the prosthesis and the holes drilled in the bone. Subsequently, the master nodes of each pair corresponding to a screw were connected via a link of the same type (Nodal Tie Type 100). Through these, the force is transmitted between the prosthesis and the bone and represents the force acting on the screws. Two types of contact were defined, a glued type (G) to impose a perfect seal between the cortical and spongy tissues, and a touching type (T) with a friction coefficient of 0.3 to simulate contact at the plate-bone interface. Stepway forces were applied, to simulate a quasi-static loading
The mechanical behavior of the prosthesis was analyzed by observing how it responds to physiological loads in terms of stress distribution. It was observed that the Equivalent Von Mises Stress in the prosthesis is far below the yield stress (930 MPa) of the Ti6Al4V-ELI alloy, and also below the fatigue limit (600 MPa). This result guarantees the excellent mechanical resistance of the plate to stress. The prosthesis, therefore, should neither fracture nor plastic deformation under physiological loading conditions.
In order to analyze the stress and deformation state in the screw-bone interface zones, two sub-models were produced: one for the left fixation zone and another for the right (Figure 4).

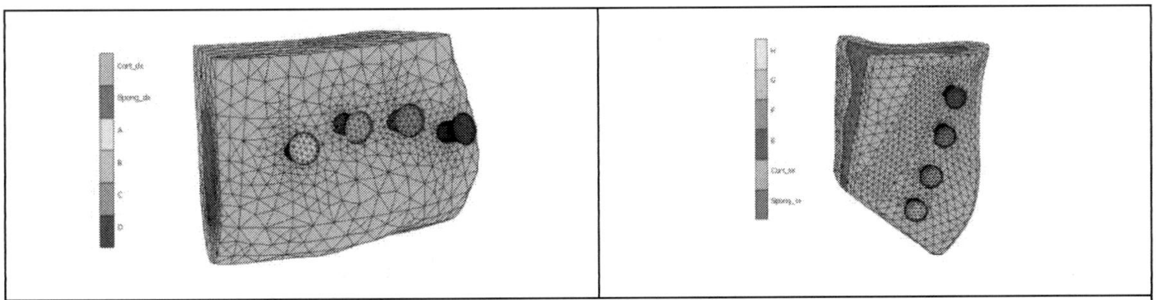

Figure 4. Submodels of the right and left fastening zones.

Forces were applied to the nodes belonging to the head of the screw, obtained from the analysis of the global model, connecting them to a master node to which the overall force is applied.

Simulations conducted on the sub-models of the right and left fixation zones were carried out to analyse in detail the deformations revealed in the bone surrounding the screws. In order to check whether the values of these deformations fall within the bone remodelling and modelling zone hypothesised by Frost [9], i.e. whether they are between 200 με and 4000 με, the values of the Equivalent Elastic Strain of the bone, obtained under the three loading conditions, were verified.

The strain values obtained in the bone in the area surrounding the fixation screws fall within the zone of bone remodelling and modelling assumed by Frost, confirming that there is no bone resorption in the vicinity of the fixation screws (Figure 5).

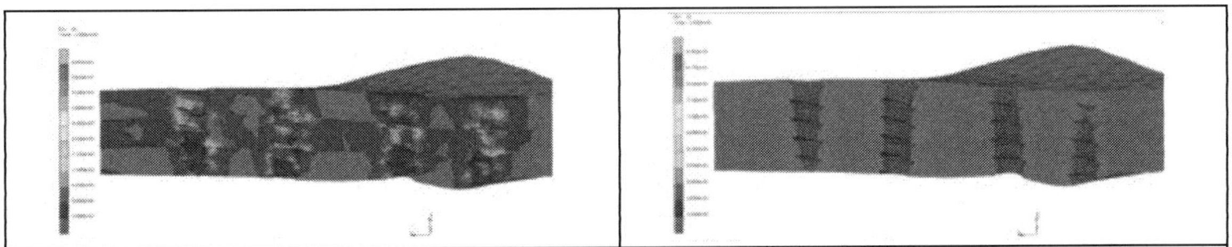

Figure 5. Equivalent Elastic Strain on the bone between 200 με and 4000 με (left) and over 4000 με (right).

IOP Conf. Series: Materials Science and Engineering 1306 (2024) 012017 doi:10.1088/1757-899X/1306/1/012017

4. Experimental tests of dynamic pull-outs

The FEM analysis made it possible to assess that the greatest force occurs on the screws positioned at the extreme points of the plate; the modulus value of this force was projected onto the direction of the axis of the screw, thus obtaining the pull-out force, which specifically for the case in question has a value of approximately 120N. The aim of the test campaign was to assess whether this force acting dynamically on the screw leads to its loosening over time and thus to the failure of the implant.

For this purpose, an experimental set-up was carried out that comprised:

- Grade 40 polyurethane foam block
- Screw gripping system (Figure 6) specially designed and manufactured for the purpose. This system consisted of two components. The first component consisted of a plate with a through hole in the centre, suitably shaped to accommodate the screw head and two small teeth that allowed it to engage with the second component, which instead ensured the connection with the test machine.
- Foam locking system
- Fixing screw with diameter 2mm and length 12mm.

The resulting system was then placed inside the test machine, as shown in Figure 6.

Figure 6. Upper and lower supports (left) and experimental pull-out test setup (right).

The test machine used was the Instron 8501, a servo-hydraulic machine on which a 100 kN load cell was mounted. The characteristics of the machine did not allow the execution of the dynamic test for such small loads in relation to the full scale of the load cell. To overcome this limitation, it was decided to interpose a spring with known elastic characteristics to the test system. Specifically, a spring with an elastic constant of 31.8N/mm was used. The test was then carried out in displacement control, applying a displacement of ±3.77mm, a value for which, multiplied by the spring's elastic constant, yields a load value of approximately 120N.

The dynamic test was performed by applying 50,000 cycles at a frequency of 5Hz to the system. This number was derived from two different evaluations

- the average number of chewing cycles performed by an individual over 5 years is 1 million [14].
- the time required for the implant to osseointegrate is approximately 3-5 weeks in a healthy individual [10] ,but can be up to 12 weeks in patients with diseases such as diabetes [15].

Therefore, an individual before the implant osseointegrates performs on average about 50,000 chewing cycles in the worst case, which is therefore the reference value for assessing the pull-out strength of the screw.

Tests were performed on three different samples and in all three cases tested the system showed no signs of failure, again confirming the performance of the implant from a functional point of view.

5. Conclusions

This study proposes a new mandibular plate with an improved bone-implant interface due to the presence of a retentive structure and a fixing screw.

The finite element analysis evaluated a case study of custom-made reconstruction of a mandibular plate with bone-tissue interface as plate B developed in collaboration with surgeons of the maxillofacial surgery department.

It showed that the resistance of the plate and the fixing screws, in the post-surgery loading conditions are lower than the yield stresses, as well as the deformation values obtained in the bone in the area surrounding the fixation screws fall within the area of bone remodeling and modeling.

Finally, the pull-out forces to which the screws are subjected are able to resist dynamic load conditions until the complete osseointegration of the implant.

Acknowledgments

The study was carried out as a PhD research activity performed in collaboration with MT Ortho company. The authors would like to thank MT Ortho for the support in carrying out the study.

References

[1] Di Bella S. and R. Mineo 2022 *3D Printing in Bone Surgery*, Springer Nature Switzerland AG C. Zoccali et al. (eds.)
[2] Battaglia et al. 2021 *Appl. Sci.*, 11, 11069
[3] Raphael L. et al. 2004, *J Oral and Maxillofacial Surgery,* vol. 62, pp. 421-426
[4] Shibahara T., et al. 2002, *J Oral and Maxillofacial Surgery,* vol. 60, pp. 182-185
[5] Lieger O., B. et al. 2015, *J Craniomaxillofacial Surgery,* vol. 43(8), pp. 1589-94
[6] Korioth W. P. and A. G. Hannam, *J Orofacial Pain,* vol. 8, pp. 178-1S9, 1994
[7] Nelson G. J. 1986, *Three Dimensional Computer Modeling of Human Mandibular Biomechanics*, The University of British Columbia
[8] Caraveo V. et al. 2021, *J Biosciences and Medicines,* vol. 9, pp. 158-178
[9] Peck C., et al. 2000, *Archives of Oral Biology,* vol. 45, pp. 963–982
[10] Van Eijden, T. et al. 1997, *The Anatomical Record,* vol. 248, p. 464–474
[11] La Rosa G. et al. 2016 *Procedia Structural Integrity*, 2, pp. 1295–1302
[12] Barbagallo R., et al. 2022 *Materials*, vol. 15 (12), 4217
[13] La Rosa G, et al. 2022 *IOP Conf. Series: Materials Science and Engineering AIAS-2021* 1214
[14] Beutel B., N. et al. 2016, *J Biomed Mater Res B Appl Biomater.,* pp. 1091-7
[15] Xiao L. et al.2022, *Orthop Surg,* vol. 14, pp. 1217-1228

Analysing the mechanical behaviour of thermoformed dental clear aligners using acoustic emission technique

C Barile[1], C Cianci[1], G Pappalettera[1], V P Kannan[1], C Pappalettere[1], C Casavola[1], D Ciavarella[2]

[1] Politecnico di Bari, Bari, Italy
[2] Università degli Studi di Foggia, Foggia – Italy

E-mail: pk.vimalathithan@poliba.it

Abstract. Clear aligners have been widely preferred in recent years for aesthetically pleasant orthodontic treatments. However, their efficacy in treating dental malposition and malocclusion is yet to be studied extensively. Clear aligners are mostly made from thermoplastic materials that guarantee sufficient elasticity while lowering the plastic deformations during their use. This is to guarantee a stable level of forces acting on teeth to be repositioned. This work aims to get insight into the mechanical behaviour of these devices during their period of use, by comparing aligners produced from two different thermoplastic materials: polyethylene terephthalate-glycol modified (PET-G) and polyurethane (PU) supplied by Dooris and Ghost Aligners, respectively. The aligners were subjected to cyclic compression loading, to simulate the swallowing act throughout 15 days of use. Tests were conducted by surrounding the aligners with artificial saliva to simulate the intraoral environment. The Acoustic Emission (AE) technique was used to analyse the damage progression in the aligners during the loading. Furthermore, the AE results were compared with the energy absorbed and the stiffness changes in the aligners. Finally, the damage evolution in the aligners post-loading was validated using optical microscopy. The AE results revealed a good correlation with mechanical and optical microscopy data, thus contributing to the understanding of the mechanical behaviour of the clear aligners.

1. Introduction

In the last 30 years, the number of adults seeking orthodontic treatment has increased significantly. Nonetheless, most adults are reluctant to use conventional treatments with wires, bands or brackets to treat malocclusion [1]. Clear plastic aligners have emerged as orthodontic devices to overcome the adult's indecisiveness in undergoing orthodontic treatments. The applications of medical-grade thermoplastic materials in dentistry are well-known [2]. Clear dental aligners are made from thermoplastic materials and are used for a short period of time, typically between 7 to 14 days. They are designed to apply incremental forces on the malpositioned tooth or a group of teeth and realign them to the desired position. During the use of these aligners, the teeth are programmed to move in small increments, say 0.1 mm to 0.2 mm, before changing them. Since they are used for a short duration, the risk of periodontal diseases such as gingivitis and tooth decay can be avoided. On maintaining oral hygiene, clear dental aligners are preferred over traditional treatments with wires and brackets [3].

Despite their advantages, clear dental aligners are susceptible to deformation due to the swallowing forces, since they are commonly produced from thermoplastic materials. Moreover, prolonged exposure to the intra-oral environment may result in their deformation, but it depends on the hydrophilic tendencies of the type of thermoplastic material used. Apart from these, the manufacturing method influences the mechanical performance, dimensional stability, and force delivery characteristics of the clear aligners [4]. Therefore, it is imperative to study the mechanical performance of the clear dental aligners under loading. The study of the mechanical performance of the material for dental aligners is enriched in literature, however, the number of research works on the performance of dental aligners is very few. The goal of this research work is to study the mechanical performance of thermoformed clear dental aligners produced using two different thermoplastic materials PU and PET-G.

In the authors' previous works, the mechanical performance of thermoformed clear aligners was studied and the strain evolution in different dental positions during the cyclic compression load was analysed using the Digital Image Correlation (DIC) technique [5]. Although the analysis of mechanical performance was successful, the damage evolution stages during the use of the aligners should not be identified. Therefore, in this research work, a passive Non-destructive Evaluation (NDE) tool, the Acoustic Emission (AE) technique is used to study the mechanical behaviour and damage evolution stages in the clear dental aligners. The AE technique is based on acquiring the stress waves (or acoustic waves) generated from a material under loading. These stress waves are generated when the stored strain energy in the material is released rapidly due to irreversible deformations [6]. Analysing the time and frequency characteristics of these stress waves may reveal information about their source.

The objective of this study is to compare the mechanical performances of two clear dental aligners made from PET-G (supplied by Dooris) and PU (supplied by Ghost Aligners), using the AE technique.

2. Materials and Methods

2.1. *Preparation of Clear Dental Aligners*

First, the dental alignment of a patient is scanned using a commercially available intra-oral scanner and reconstructed in 3D using commercial software with an accuracy of 6.9 μm. Then, the solid resin casts of the upper and lower jaw are constructed from the scanned output using daylight hard resin. The resin casts are constructed using the Liquid Crystal HR2 3D printer. Medical-grade sheets of PU and PET-G of thickness 0.75 mm are thermoformed over the hard resin casts using Erkoform 3D vacuum machine to prepare the clear dental aligners. The sheets are first heated by a medium-wave infrared heater before applying the vacuum. The respective aligners are named as TPET-G and TPU.

2.2. *Cyclic Compression Tests*

The mechanical tests are designed to simulate the forces applied to the dental aligners during the swallowing act during the period of use of the aligners, which is roughly equal to 15 days or 22500 cycles. The force induced by the upper jaw during swallowing is 50 N [7]. Therefore, each cycle of the compression test is designed into 4 stages. In the first stage, the load is ramped from 0 to 50 N in 1 s and in the second, the 50 N load is kept for a dwell period of 1 s. In the third and fourth cycles, the load is ramped down from 50 N to 0 in 1 s and a dwell period of 1 s at 0 N, respectively.

The cyclic tests are carried out on an Instron 3344 single-column universal testing machine (1 kN load capacity). The lower and upper resin casts are firmly held onto the pneumatic vices of the testing machine and the clear aligner is mounted onto the upper cast. In order to simulate an intra-oral environment during the loading, a sponge impregnated with artificial saliva is kept

IOP Conf. Series: Materials Science and Engineering 1306 (2024) 012018 doi:10.1088/1757-899X/1306/1/012018

in contact with the aligner throughout the test. The upper and lower casts are held firmly at a contact pressure of 10 MPa before the commencement of the test.

2.3. Acoustic Emission Test Setup

The stress waves generated during the compression cyclic loading are recorded using a piezoelectric transducer, PICO sensor (Physical Acoustics). It is a lightweight miniature sensor, which has an operating frequency of up to 750 kHz and has two resonant frequencies: one at 250 kHz and the other at 550 kHz. The stress waves that cross the detection threshold of 26 dB are acquired at a sampling rate of 2 MHz. The signals are preamplified by 60 dB and passed through a flat band-pass filter of 100 kHz to 1 MHz before acquisition.

The frequency characteristics of the stress waves generated from the complex-shaped clear aligners are unknown and not reported in the literature [6]. Therefore, a preliminary test is carried out to study their frequency characteristics. A previously tested aligner, which has a small crack in the first third molar is mounted onto the test machine and the compression cyclic load was applied for 100 cycles. However, this time, the loads during the ramping are set as 100 N (instead of 50 N), in order to propagate the crack. The stress waves acquired from this analysis are recorded and their frequency characteristics are reported in the following section.

3. Results and Discussions

3.1. Preliminary Cyclic Test Results

As reported in Section 2.3, a preliminary test is carried out to understand the frequency characteristics of the stress waves generated from the propagating crack in the dental aligners. The distribution of the acoustic waves in terms of their peak frequency and peak amplitude is presented in Figure 1.

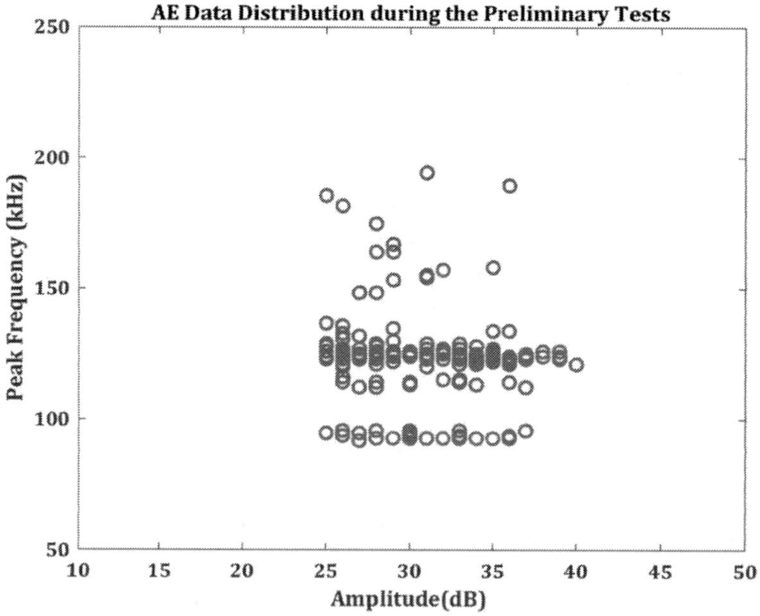

Figure 1. Distribution of Acoustic Waves as a function of their Peak Frequency and Peak Amplitude

The expected frequency components in the stress waves from the crack growth event can be observed in Figure 1. Most of the AE signals are expected to have a frequency around 125 kHz to 150 kHz, while generally no signals are expected to be above 200 kHz. Since the frequency characteristics of the signals are based on the stiffness of the material and the modulation frequency of the acquiring sensors and based on the previous results on similar thermoplastic materials, it can be assumed that the expected frequency of the AE signals will be in this frequency range [8].

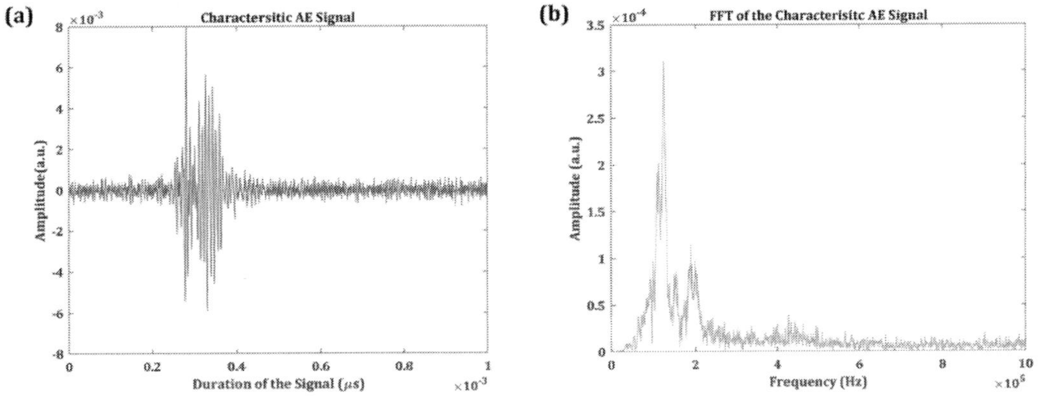

Figure 2. (a) Sample AE Signal in Time Domain and (b) Frequency Domain

A sample AE signal in its time domain and in its frequency domain (calculated using Fast Fourier Transform (FFT)) is presented in Figures 2a and 2b, respectively. Once again, the results show that most of the frequency components lie below the 200 kHz frequency band, while the central frequency is around 150 kHz.

3.2. Mechanical Performance of the Aligners

The mechanical performances of the aligners are analysed from the hysteresis loop of the load-displacement curve. The total area enclosed by the curve is used to calculate the energy absorbed by the aligner in each cycle and the first slope of the load-displacement curve is used to study the stiffness. The results are presented in Figures 3a and 3b, respectively.

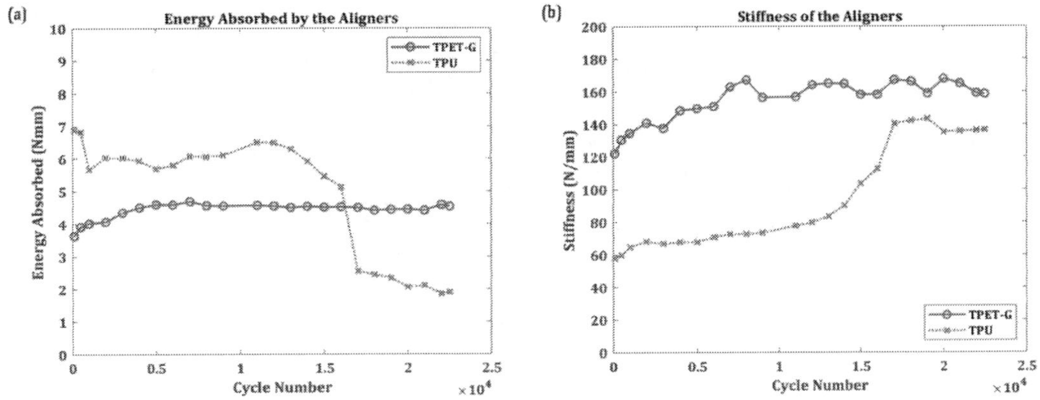

Figure 3. (a) Energy absorbed and (b) Stiffness variations in the aligners during the Compression Cycles

It can be observed in Figure 3a that the energy absorbed by TPET-G increases in the initial cycles (up to cycle 4000) and remains stable thereafter. On the other hand, the average energy absorbed by TPU is quite larger than TPET-G up to cycle 17000, and then it drops suddenly and decreases linearly until the end of the test. During the cyclic compression test on thermoplastic materials, the energy absorbed is proportional to the strain deformation in the material during the initial cycles. PU are thermoplastic materials characterized by large plastic deformation, which could be the reason why the average energy absorbed by TPU is greater than TPET-G. The large plastic deformation in thermoplastic material is followed by strain hardening, which results in an increase in stiffness. From Figure 3b, it can be observed that after 10000 cycles, the stiffness of TPU increases exponentially up to cycle 17000, where an energy drop is observed in Figure 3a.

However, the reason behind the sudden drop in the energy absorbed by TPU is quite unclear. In thermoplastic materials, the release of strain energy is associated with the formation and propagation of damage. Since the energy drop did not occur linearly but occurred in a rapid manner between cycles 17000 and 18000, it can be assumed that the damage that occurred was rapid. Based on the stiffness curve in Figure 3b, since the stiffness increases exponentially from cycle 10000, it can be assumed that the strain hardening commences to occur at this cycle and finally results in the propagation of damage at cycle 17000.

PET-G polymeric materials are often characterized by higher modulus and higher stiffness than PU [9, 10]. This is obvious from both Figures 3a and 3b that the energy absorbed by TPET-G is quite low compared to TPU, while its stiffness is higher. Moreover, both the energy and stiffness of this aligner remain more or less stable throughout the test. This clearly indicates that the performance of TPET-G is better than TPU.

3.3. Optical Microscopy Results

In order to validate the assumptions made in the previous section, both optical microscopy and AE results are used. First, the microscopic images of untested TPU aligners are presented in Figure 4. (Unfortunately, the microscopic images of untested TPET-G are unavailable at the moment).

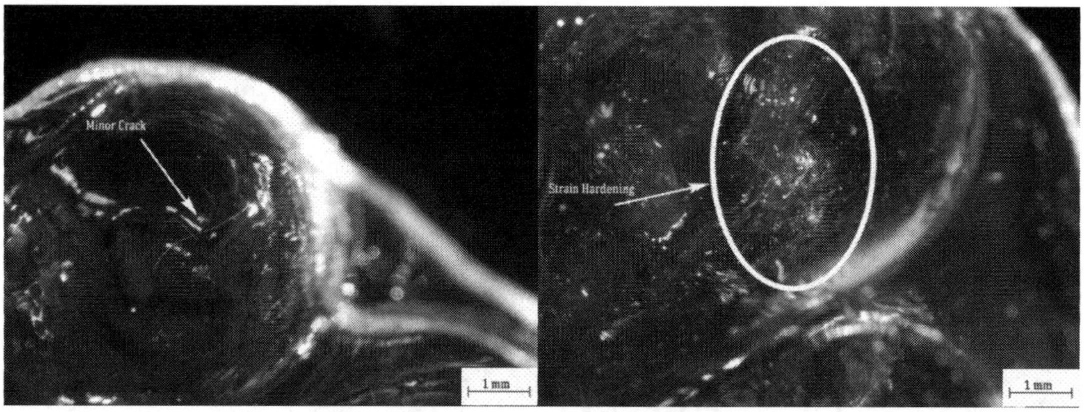

Figure 4. Microscopic Images of Untested TPU Aligner

The results show that there are some microcracks of length less than 1 mm and some strain-hardening regions that can be observed on the external surface of the aligner in Figure 4. These probably are due to the thermoforming process. It is safe to assume that these microcracks serve as the regions of stress concentration, which results in the damage progression in these aligners.

To verify this, the microscopic images of the tested TPU and TPET-G are presented in Figures 5 and 6, respectively.

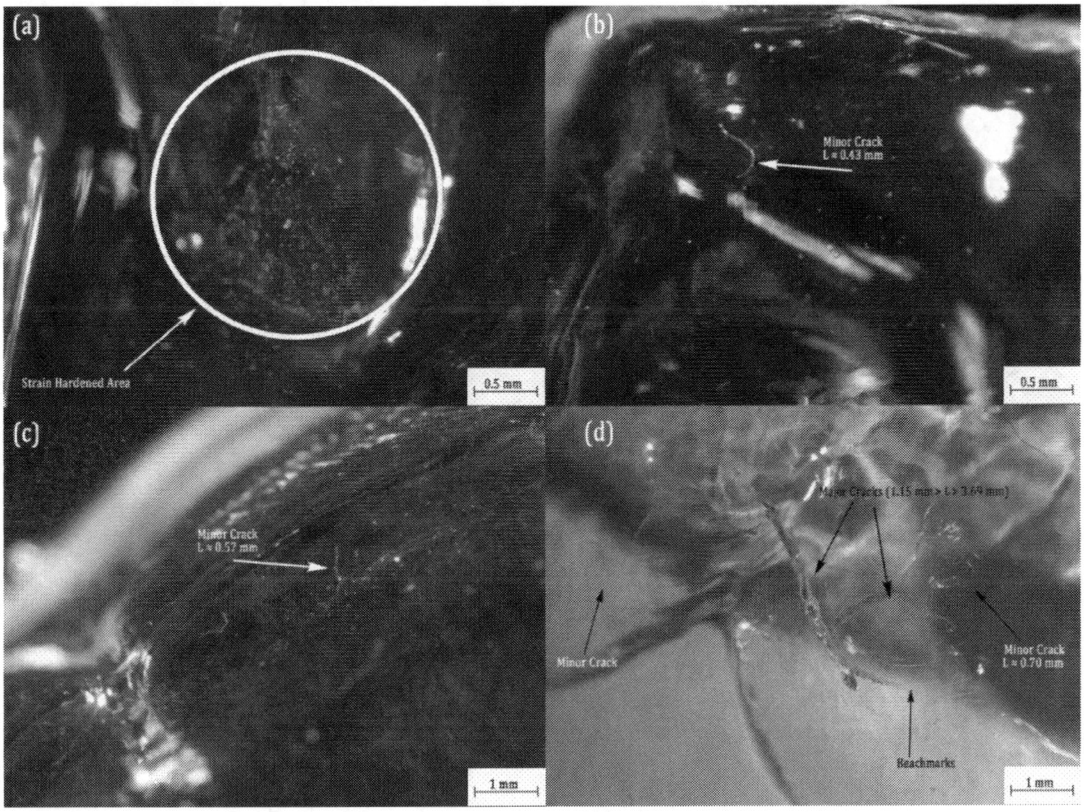

Figure 5. Microscopic Images of Tested TPU Aligner taken at different Tooth Positions

Similar to Figure 4 of the untested microscopic images, there are regions of strain hardening and microcracks of length less than 1 mm that can be observed in Figures 5a, 5b and 5c. However, in Figure 5d, several larger cracks with lengths greater than 1.15 mm and up to 3.69 mm can be observed. Moreover, several whitening regions of beachmarks indicating fatigue failure due to the cyclic load are also observed. This validates the assumption that the energy absorbed by TPU is released due to the formation of these cracks. Based on the stiffness curves and energy curves in Figure 3, it can be confirmed that this occurred at cycle 17000.

The microscopic images of the TPET-G aligner post-testing are presented in Figure 6. As expected from the mechanical results, only one major crack is observed in Figure 6a, whose length is significantly smaller than the one observed in TPU. Besides, there are no beachmarks around the failure, indicating that these are not due to the fatigue cycle. Most of the other microcracks and micro-failures in Figures 6b, 6c and 6d indicate the substantially good performance of TPET-G compared to TPU.

The microscopic images shed light on the regions of failure, the type of failure and the occurrence of failure when compared with the mechanical performance of the aligners. Despite identifying the occurrence of the final failure, an in-situ analysis is essential to understand the commencement of these failure modes. For this, the AE results are presented and discussed in the next section.

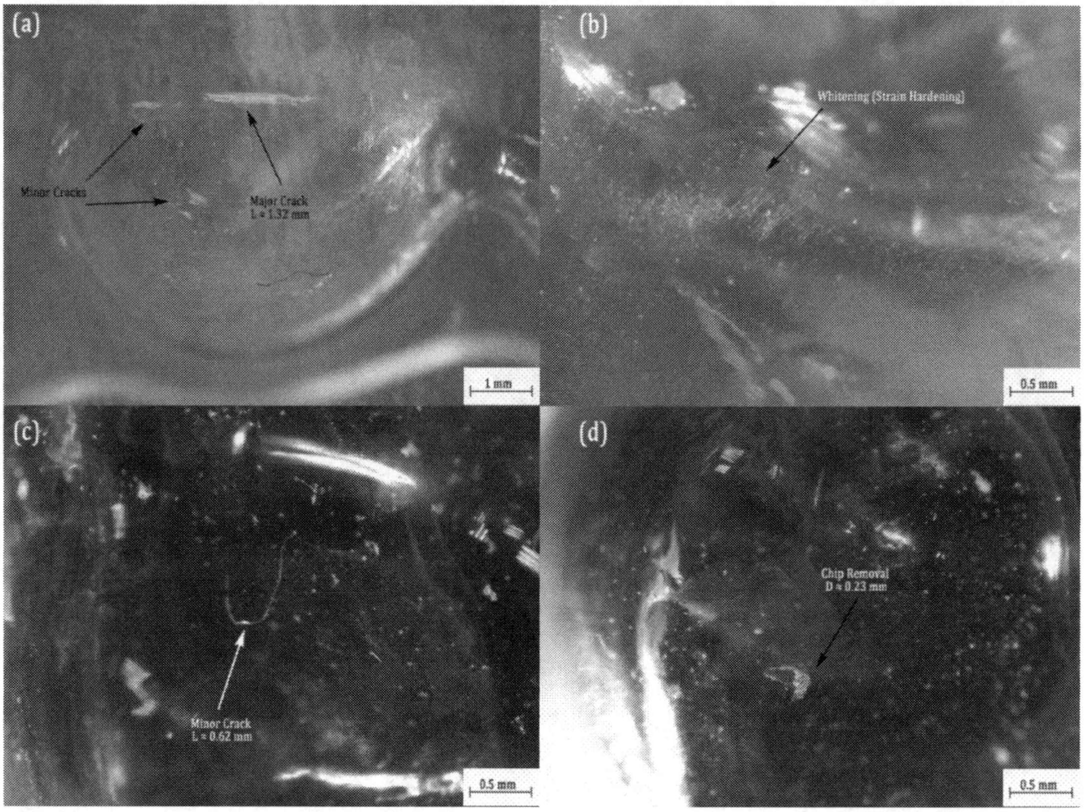

Figure 6. Microscopic Images of Tested TPET-G Aligner taken at different Tooth Positions

3.4. Acoustic Emission Results

The AE results are compared with the mechanical test results in this section. The time domain parameter of the AE signals, cumulative counts, which indicate the number of instances the amplitude of the stress waves crosses the detection threshold is compared with the energy absorbed by the aligner. The results for TPU and TPET-G are presented respectively in Figures 7a and 7b.

The first obvious difference between TPU and TPET-G can be found in the maximum number of cumulative counts recorded in each aligner. In TPU it is about 10500, while on TPET-G it is only 4750. This clearly shows that the number of stress waves generated in TPU is more than that of TPET-G. This validates the assumptions made in both the mechanical test results in Section 3.2 and the optical microscopy results in Section 3.2.

Secondly, it gives an indication of the onset of the damage in each of these aligners. For example, in Figure 7b, it can be seen that the cumulative counts remain close to 2 until cycle 7000 and it increases rapidly beyond that point and then it stabilises. This indicates that the crack observed in Figure 6a starts to propagate at this point and then it stabilises. Since there is no major change in the energy absorbed by this aligner during this phase, it can be concluded that this damage did not affect the mechanical performance of the TPET-G aligner.

Thirdly, from Figure 7a, the onset of different damages can be examined in the TPU aligner. The first damage onsets as early as cycle 2000 and it progresses until cycle 5000. Then there is a period of stabilisation up to cycle 9500. Beyond that cycle, the cumulative counts begin to increase gradually and it extends until the end of the test. The mechanical results indicate that the energy starts to drop from cycle 11000 but the AE results can precisely indicate the onset

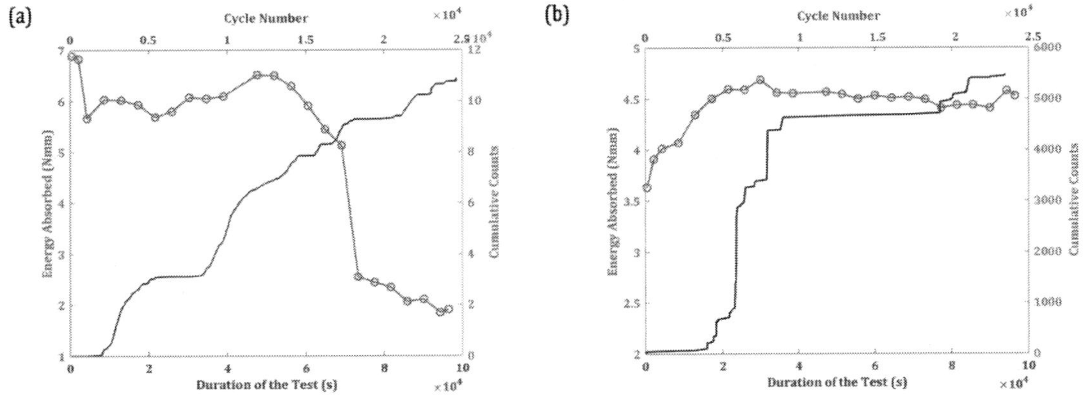

Figure 7. AE Results compared with Mechanical Results for (a) TPU and (b) TPET-G Clear Dental Aligners

of damage progression.

Finally, in order to verify that these AE signals are crack growth events, two signals from each aligner are randomly chosen and their frequency characteristics are studied using FFT. The results are presented in Figure 8.

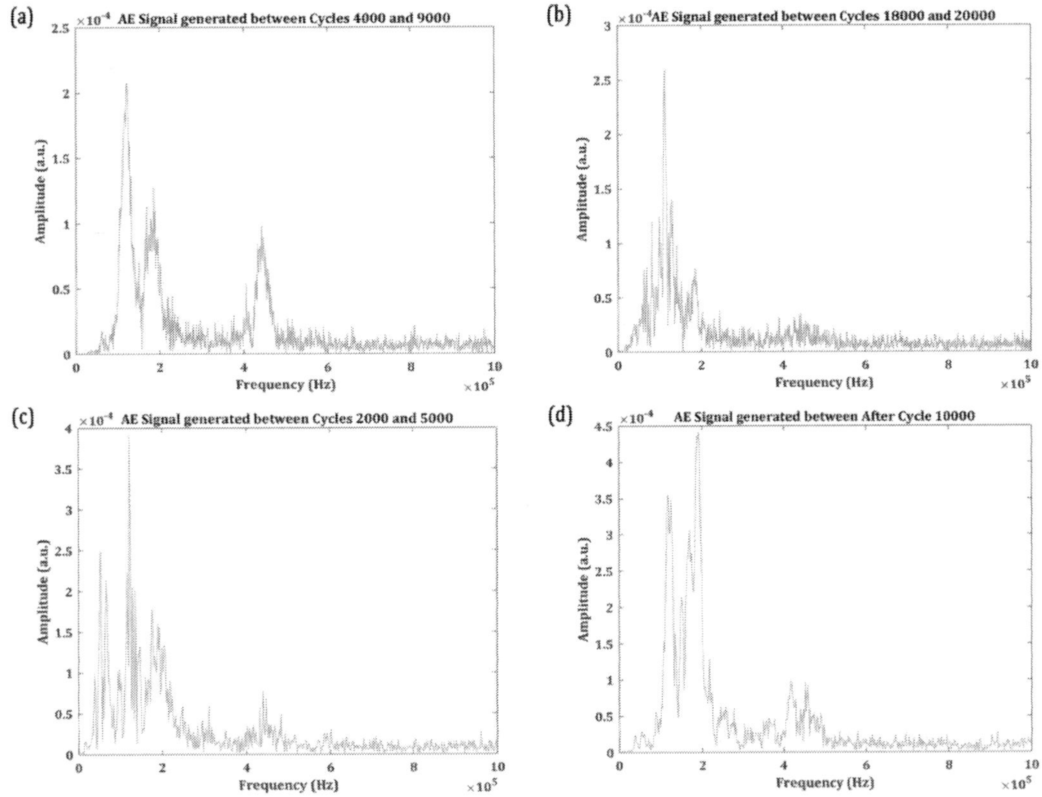

Figure 8. Frequency Characteristics of Randomly Selected AE Signals (a) and (b) from TPU; and (c) and (d) from TPET-G

IOP Conf. Series: Materials Science and Engineering 1306 (2024) 012018 doi:10.1088/1757-899X/1306/1/012018

The AE results clearly show that the major frequency components of the AE signals are centred between 100 kHz and 200 kHz, which was observed during the preliminary tests in Section 3.1. This confirms that these signals are generated from the crack growth events in the aligners.

4. Conclusion

The mechanical performances of two different thermoplastic clear dental aligners are evaluated and compared in this research work. The mechanical results show that the aligner thermoformed from PET-G has stable energy absorption and stiffness characteristics compared to that of PU. The AE technique is used to identify the damage evolution in these aligners under cyclic loading. The results are promising, particularly, the time-domain-based and frequency-based analyses. The time-domain-based shows the onset of the damage progression in the aligners, while the frequency-based analysis shows the type of damage. The results show that this technique has a greater potential in the evaluation of the mechanical performances of orthodontic devices.

5. References

[1] Buttke TM and Proffit W R 1999 *J. Am. Dent. Assoc.* **130(1)** 73
[2] Eliades T and Bourauel C 2005 *A. J. Orthod. Dentofac.* **127** 12
[3] Quinzi V, Orilisi G, Vitiello F, Notarstefano V, Marzo G and Orsini G 2023 *Sci. Total. Environ.* 161356
[4] Albilali AT, Baras BH and Aldosari MA 2023 *Polymers-Basel* **15** 1610
[5] Cianci C, Pappalettera G, Renna G, Casavola C, Laurenziello M, Battista G, Pappalettere C and Ciavarella D 2020 *Front. Mater.* **7** 104
[6] Barile C, Casavola C, Pappalettera G and Kannan VP 2020 *Eng. Fract. Mech.* **235** 107083
[7] Gibbs CH, Mahan PE, Lundeen HC, Brehnan K, Walsh EK and Holbrook WB 1981 *J. Prosthet. Dent.* **46** 443
[8] Barile C, Casavola C, Pappalettera G and Kannan VP 2022 *Appl. Acoust.* **185** 108425
[9] Qi HJ and Boyce MC 2005 *Mech. Mater.* **37** 817
[10] Scetta G, Ju J, Selles N, Heuillet P, Ciccotti M and Creton C 2021*J. Polym. Sci.* **59** 685

Assessment of different forecasting strategies for the breaking load resistance in notched specimens 3d-printed in PLA using additive manufacturing (FFF) technology

G. Morettini[1], M. Palmieri[1], F. Cianetti[1], C. Braccesi[1]

[1] Università degli Studi di Perugia - Dipartimento di Ingegneria, Via G. Duranti, 06124 Perugia

Corresponding author's e-mail address: giulia.morettini@unipg.it

Abstract. The introduction of fused deposition modelling in the industrial sector to produce functional components in thermoplastic materials, such as PLA, requires knowledge of the performance limits of such elements during the design phase. Predicting the breaking load resistance of notched specimens is essential to evaluate the mechanical performance of components manufactured using this technology. This study compares different methodologies based on the critical distance criterion (TCD) for predicting the static breaking strength of printed notched specimens made with Fused Filament Fabrication (FFF) technology using PLA material. Specimens with different topologies of V-notches were printed according to a standardized configuration. Various analysis techniques were applied to determine the breaking strength of each specimen. By comparing the results obtained with experimental data to assess their accuracy and reliability, it has been demonstrated that these methodologies, coupled with the concept of equivalent elastic material (EMC), can be useful for predicting static breaking load.

1. Introduction

The Nowadays, additive manufacturing is a widely used technology, even in the industrial sector, allowing for the easy production of highly complex geometries, as it is well-known. Fused deposition modeling (material extrusion) is undoubtedly the most utilized additive technology, especially for polymer materials [1]. In the industrial world, there is an increasing tendency to design structural components capable of withstanding significant static and dynamic loads. In various fields, such as aerospace, there have been considerations to replace aluminum in many significant applications [2-3]. This new trend is also due to the knowledge gained from scientific research in the mechanical characterization of these materials and the standardization of processes and technological controls that are gradually defining standard parameters for successful mechanical performance of finished components, both statically and dynamically [4-6].

In this context, there is still a lack of knowledge regarding the behavior of notched components produced through additive manufacturing. In other words, understanding the behavior of these components under specific manufacturing conditions, where notches do not represent material defects, is essential. As is well-known, the presence of notches has a direct effect on the material's strength. Therefore, this article focuses on the fracture behavior of PLA (polylactic acid) produced through additive manufacturing (3D printing), in the presence of V-notches with fillets. Different fillet radii

IOP Conf. Series: Materials Science and Engineering 1306 (2024) 012019 doi:10.1088/1757-899X/1306/1/012019

and various resisting specimen thicknesses will be analyzed, while using a specific parametric setting previously optimized by the authors [7].

The presence of a notch requires the use and/or development of specific approaches for predicting the fracture strength of these materials. While most of these methodologies were originally developed for brittle materials [8], the need to analyze the fracture strength of ductile materials arises from the fact that most structural engineering components are typically made from such materials. Numerous criteria have been presented in the literature, formulated through various approaches, including global criteria [9-10], energy-based criteria using the Strain Energy Density (SED) [11-14], cohesive zone methods [15-18], those based on fracture mechanics advance [19], and methods based on the Critical Distance [20-25]. The latter will be the focus of this study. The analysis of the effect of the notch on 3D printed PLA specimens will be based on the Theory of Critical Distances (TCD), a set of methodologies widely applied in recent years, especially for brittle or quasi-brittle materials, characterized by the use of a material length parameter. This methodology, first developed by Nember [26] and Peterson [27], has subsequently been elaborated in various forms by other authors such as Taylor [28, 29], who independently formulated the theory in slightly different ways for predicting fracture in brittle materials [30] and composites [31]. Only in recent years has attention shifted towards applying this methodology to polymeric plastic materials [32], leveraging specific calibration activities (experimental and/or numerical). The works of Cicero, Susmel, and Torabi [33, 24] are examples of this.This article, instead, aims to demonstrate that by leveraging the concept of Equivalent Elastic Material (EMC) introduced by Torabi [34], it is possible to derive all the necessary parameters for setting up the TCD methods in all their forms from the results of a simple tensile test of the material. Subsequently, a theoretical prediction of the method can be made using finite element analysis.

In conclusion, the results of two different TCD methodologies will be compared to identify the best prediction solution. Section 1 will present the methods at a theoretical level, along with the procedure for correction using the concept of Equivalent Elastic Material. Section 2 will gather experimentally obtained results, while Section 3 will present the main predictive results. Finally, in Section 4, the conclusions will be drawn.

2. Theoretical context

2.1. Theory of Critical Distance (TCD)Some text.

The origin of TCD can be traced back to the works of Neuber and Peterson [26, 27] in 1958, where it was used for predicting fractures, both static and fatigue, under the assumption of linear-elastic behavior of the material. Only in recent decades has this theory been extensively developed for the analysis of different types of materials, fracture processes, and loading conditions. In general, this theory states that the fracture of a notched component can be avoided as long as the effective stress σ_{eff} (calculated based on material-related parameters) remains below a value of intrinsic strength σ_0 (stress related to the type of applied load and the material itself), as stated in equation (1).

$$\sigma_{eff} < \sigma_0 \quad (1)$$

In other words, the entire linear-elastic stress field around the notch tip is represented by an appropriate effective stress σ_{eff}, while the material's strength is represented by a suitably chosen stress value related to the type of loading or material in use.

The effective stress can be calculated using different methodologies [29], all of which incorporate a length parameter known as the critical distance L. In static fracture analyses, this critical distance is calculated as expressed in equation (2):

$$L = \frac{1}{\pi}\left(\frac{K_C}{\sigma_0}\right)^2 \quad (2)$$

where $K_C$represents the fracture resistance of the material, and σ_0 (as mentioned above) is the intrinsic strength of the material. We can immediately state that only in particular situations where the behavior of the involved material is linearly elastic (both micro and macroscopically), σ_0 coincides with the ultimate tensile strength of the material σ_{UTS}, and K_C can be replaced by the material's fracture toughness K_{IC} [31]. In these rare cases, the application of the TCD method does not require any correction. However, in cases like the one analyzed by the authors, where the material, PLA, exhibits a clear plastic phase, a proper evaluation of these parameters is necessary for the method to function correctly. Regardless of the values of K_C and σ_0 identified as suitable for the analyzed situation, the TCD method can be formalized with different methodologies [20], the main ones of which, being rapid and efficient, are compatible with industrial requirements. In particular, among these methodologies, we find the Point Methodology (PM) and the Linear Method (LM). These will be analyzed and detailed below:

2.1.1. The Point Method (PM). This represents the simplest implementation of the TCD. This methodology suggests that the effective stress σ_{eff} is evaluated at a point located at a distance of $L/2$ along the bisector of the notch, as expressed in equation (3) and depicted in figure 1.

$$\sigma_{eff\_PM} = \sigma\left(\frac{L}{2}\right) \quad (3)$$

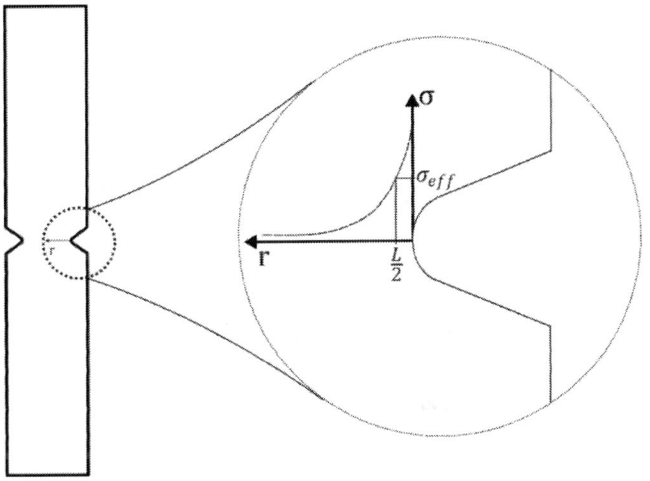

Figure 1. Example of Stress-distance curve from the notch tip and definition of the PM method.

2.1.2. Line Method (LM). This methodology suggests that the effective stress is evaluated as the average stress calculated over a length equal to 2L, along the bisector of the notch, as described by equation (4) based on figure 2.

$$\sigma_{eff\_LM} = \frac{1}{2L}\int_0^{2L} \sigma(\mathrm{r})\, dr \quad (4)$$

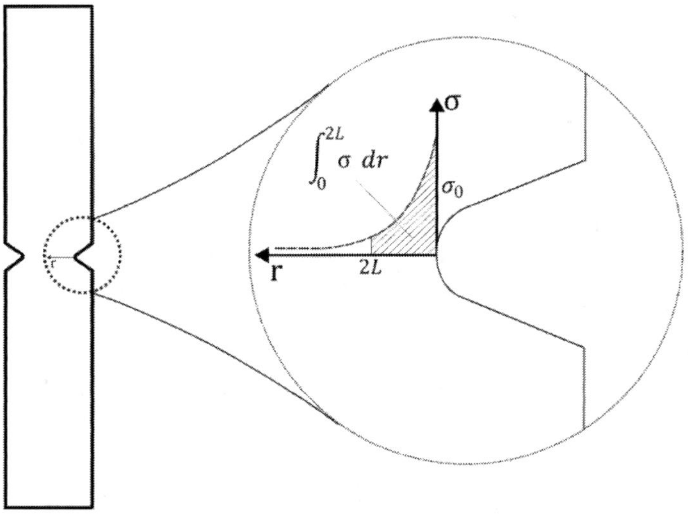

Figure 2. Stress-distance curve from the notch tip and definition of the LM method.

In most cases, the illustrated methodologies are developed with the support of finite element analysis, which allows replicating the loading condition and evaluating the stress distribution around the notch. The choice of which stress to consider will mainly depend on the type of loading in place. In our case, we will always refer to the maximum principal stress. However, it is not always straightforward, and in some cases, multi-axial phenomena may need to be taken into account by considering von Mises equivalent stress or Tresca stress.

As mentioned earlier, regardless of the chosen TCD method, accurate predictions rely on an appropriate selection of K_C and σ_0 parameters. For materials that do not exhibit linear-elastic behavior until fracture, σ_0 does not coincide with σ_{UTS}. In scientific literature, σ_0 is usually "calibrated" through a dedicated experimental program on specimens with different sizes and/or notch types. In this case, the critical distance L is defined as the value obtained from the intersection of two Stress-Distance curves (obtained through finite element modeling) of specimens with different notches. The calibration process represents a fundamental challenge in the application of TCD methodologies and is a clear obstacle to their widespread implementation in industrial practice [24].

For this reason, the authors have chosen to use an equivalent elastic material approach, which will be extensively described in the next paragraph, for determining σ_0 value. They will only later use experimental results to evaluate the effectiveness of the method used.

2.2. The Equivalent Material Concept (EMC)

As mentioned earlier, the TCD method is based on the strict assumption of having a linear-elastic behavior of the material. It is essential to respect this concept when applying the method. If we approximate a ductile material to have a brittle behavior, we would underestimate the energy required to cause its fracture. As shown in Figure 3a), the area under the real Stress-Strain curve of a ductile material is significantly larger than the line that approximates its behavior as that of a linear-elastic material. In other words, if we were to take σ_0 equal to the σ_{UTS} of the tensile test of a ductile material, we would lose a non-negligible energy contribution.

For this reason, the concept of equivalent material, introduced by Torabi [35], is used by the authors in this research activity to find a material with a fictitious linear-elastic behavior that stores the same energy contribution as the ductile material. In practical terms, this means evaluating the curve of linear-elastic fracture of the equivalent fictitious material, having the same elastic modulus and the area under the real curve resulting from a simple tensile test of the material under study, as shown in Figure 3b). From this curve, it is easy to obtain the value of σ_0, which corresponds to the ultimate tensile strength of the equivalent fictitious material σ_{UTS_EMC}.

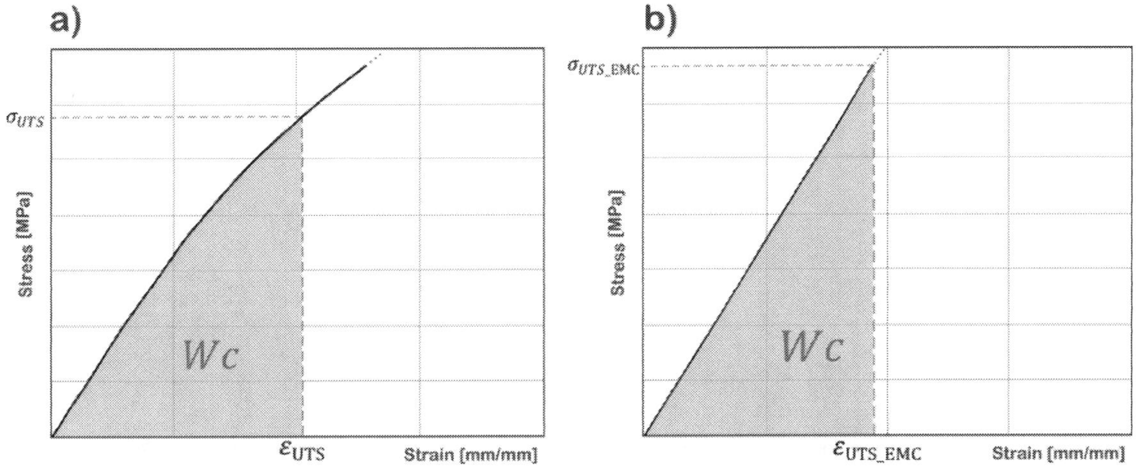

Figure 3. a) Real Stress-Strain curve of a ductile material in a tensile test. **b)** Linear-elastic fracture curve of the fictitious material with the same elastic modulus and the same area under the real curve.

Regarding the value of K_C, in this research activity, it has been replaced by the fracture toughness of the considered material, K_{IC} [36].

3. Fabrication of specimens and experimental tests

3.1. Specimens Fabrication

The fabrication of the specimens followed, whenever possible, the ASTM D638 standard [37], both for the smooth and notched sample production. All specimens were fabricated using an Ultimaker 3 Extended FDM printer with a 0.4 mm diameter nozzle, and green Ultimaker PLA filament with a diameter of 2.85 mm.

As known in the literature, the correct selection of process parameters is crucial for successful printing, especially to optimize the material's strength and load-carrying capacity. For this reason, all samples were printed using a single layer height of 0.2 mm, 100% infill created with a line pattern and a direction of +/- 45°, printed at a speed of 35 mm/s. The printing temperature was set at 215°C, while the build plate temperature was maintained at 60°C. For more details on the parameter settings, reference [4,38] can be consulted, where the same authors conduct research to optimize printing parameters. All specimens were printed in a horizontal position to optimize their mechanical properties.

Six dog bone samples were made to perform standard tensile tests on PLA, choosing Type I samples in accordance with the reference standard. Based on the geometries of the standard specimens, different notch geometries were created, as shown in Figure 4. Various notch sets were built by varying the notch thickness (H) and the fillet radius (ρ), while keeping the opening angle 2α=90° constant. A total of 9 different sets of samples, each containing 6 specimens, were fabricated, resulting in a total of 54 samples. The geometric parameters were varied as shown in the table in Figure 4.

IOP Conf. Series: Materials Science and Engineering 1306 (2024) 012019 doi:10.1088/1757-899X/1306/1/012019

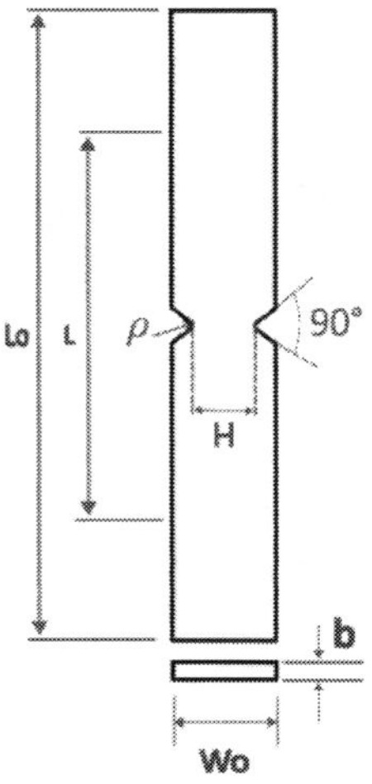

Set id	H10r05	H10r2	H10r4	H13r05	H13r2	H13r4	H15r05	H15r2	H15r4
H [mm]	10	10	10	13	13	13	15	15	15
ρ [mm]	0,5	2	4	0,5	2	4	0,5	2	4
L [mm]					60				
Wo [mm]					20				
Lo [mm]					100				
b [mm]					5				

Figura 4. Dog bone sample on the left with notched geometry and definition of characteristic geometric parameters for each set of samples.

For clarity, the latter will be referred to as **Hxry**, where x indicates the thickness of the sample and y the fillet radius. The different types of carvings will allow us to assess the influence of varying geometry on the strength of the specimens and the method's ability to predict their fracture.

For all samples, after removing any burrs following ASTM D638 standards, the angles and surfaces were smoothed using fine abrasive paper. Following this procedure, all surfaces were defect-free, and the samples were ready for testing procedures.

All tensile tests, both for standard specimens and notched ones, were performed in accordance with ASTM D638 [37], ASTM E132 [39], and ASTM D5045 [40] standards. Specifically, an Instron 3382 machine was used for force control during the tests. An Instron W-6280 extensometer was employed to determine deformation. Following the mentioned standards, the machine's speed of advancement was set to 2 mm/min, and data were acquired at a sampling frequency of 5 Hz.

3.2. Standard Tensile Test

From a standard tensile test, by monitoring the evolution of the specimen's area, it is possible to extrapolate the average results of the true Stress-Strain curve. Specifically, a value of the elastic modulus E equal to 3400 MPa and an ultimate tensile strength $\sigma_{UTS} = 62$ MPa at $\varepsilon = 0{,}023\ mm/mm$ elongation can be calculated. Consequently, it will also be possible to evaluate the value of the critical real energy density $Wc = 0{,}789\ mJ/\sqrt[3]{mm}$. The values of the Poisson's ratio $\nu = 0{,}36$ and the material's fracture toughness $K_{IC} = 126{,}491$ MPa $\sqrt[2]{mm}$ were obtained from the literature [38]. With this, all the necessary values for calibrating and implementing the TCD method are now available.

3.3. Test di trazione su campioni intagliati

Figure 5 shows the graphs of the average values, for each set, of the tensile tests on notched specimens. The figures are grouped by specimens with a constant H section and variable fillet radius.

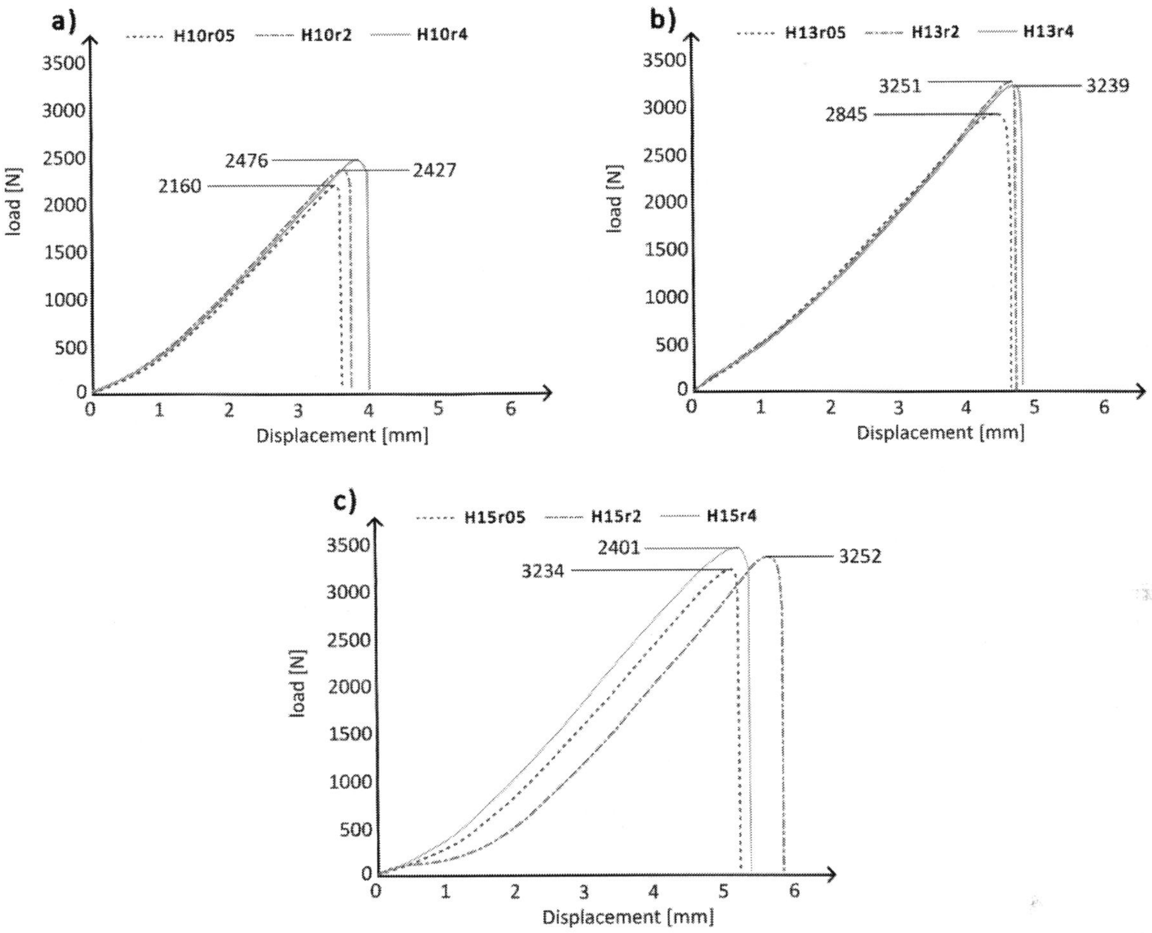

Figure 5. Graphs of the average values of the tensile tests on notch specimens grouped by specimens with a constant H section and variable fillet radius. **a)** section equal to 10 mm, **b)** section equal to 13 mm, **c)** section equal to 15 mm.

As expected, it can be observed that the presence of the presence of notch not only reduces the material's strength but also makes its behavior much more brittle. Figure 5 presents the average results in terms of the maximum force recorded at the material's fracture.

4. Calibration and application of the method using Finite Element Analysis (FEA)

4.1. Application of the concept of equivalent material and determination of the critical length

As extensively described in subsection 2.2, based on the data obtained from the standard tensile test on smooth dog bone specimens, it is now straightforward to derive the value of the ultimate tensile strength of the equivalent fictitious material, σ_{UTS_EMC}. As shown in Figure 4b) and from the knowledge of the Wc value, defined as the area under the real Stress-Strain curve, as shown in Figure 4a), it is possible to calculate the value of σ_{UTS_EMC} from equation (5), assuming the same slope of the elastic-linear curve (which is equivalent to the elastic modulus E).

$$\sigma_{UTS\_EMC} = \sqrt{2\,Wc\,E} \;=\; 73\ \text{Mpa} = \sigma_0 \tag{5}$$

Considering the known value of $K_c = K_{Ic}$ for the material, it is possible to derive, using equation (2), the critical length $L = 0{,}477\ mm$.

4.2. Application of the method using FEA: Workflow.

Figure 6 shows the workflow describing the procedure used to predict fracture force using the critical distance methods through the support to a finite element model (FEM). The process begins by accurately modeling the notched specimen geometries and their corresponding loading/boundary conditions within the finite element code. In this specific case, a shell representation of the specimens was used in Ansys Workbench software, assuming plane stress conditions, which are justified by the small thickness of the specimens and the typical stress distribution around the carving [24, 34]. The various sets of specimens were then loaded with a unit force applied to the gripping area in the machine (red area in figure 6 a)). The same specimen was fixed on the opposite side by restraining the displacement in the force direction (blue area).

Since TCD, different from other methodologies, is highly influenced by element size [41], an appropriate mesh pattern was applied around the carving to extract stress values at locations of interest for the various TDC methodologies using a simple linear-elastic analysis (example of the maximum stress results are shown in Figure 6b).

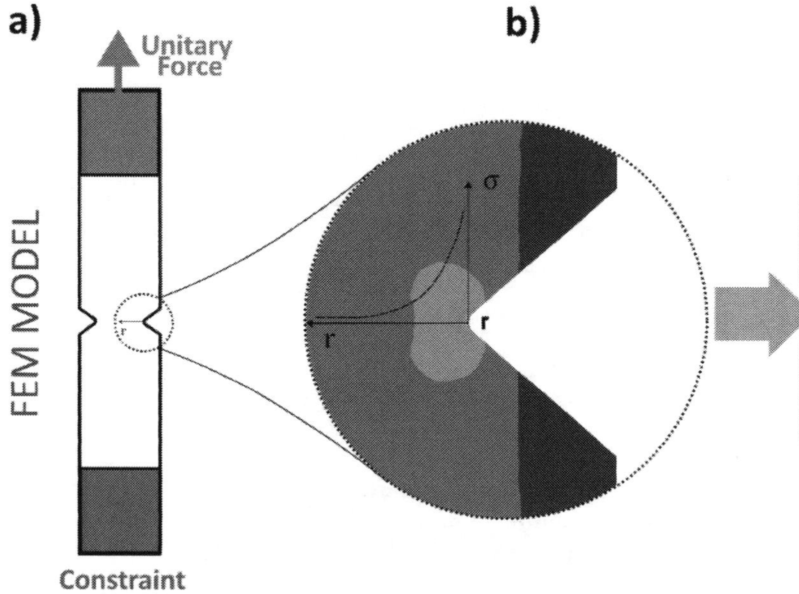

$$\sigma_{eff\_PM} = \sigma(r)\left(\frac{L}{2}\right)$$

$$\sigma_{eff\_LM} = \frac{1}{2L}\int_0^{2L} \sigma(r)\ dr$$

Figure 6. Workflow describing the procedure used to predict fracture using the critical distance methods. **a)** Finite element code schematization, application of unit load, and necessary constraints. **b)** Example of stress distribution around the notch. **c)** Equations (3) and (4) from which the values of the effective stress σ_eff can be derived for each of the proposed methods.

In this application, the maximum principal stress values were considered, as there were no multiaxial phenomena present, and the authors believe that this value represents the stress state of the specimen well. Thanks to the results of the numerical analysis, it is possible to extrapolate, for each experimental set, the trend of the maximum principal stress σ_{p_MAX} as a function of the distance r from the apex of the notch along its bisector. As visible in Figure 6c), from equations (3) and (4), it is then possible to obtain the values of the effective stress σ_{eff} for each of the proposed methods. Obtaining the values of the predicted maximum allowable force F_{prev} is now possible through the use of a simple ratio expressed by equation (6):

$$\frac{\sigma_{UTS\_EMC}}{\sigma_{eff}} = F_{prev} \tag{6}$$

5. Results and discussions

Table 1 presents the predictions obtained through the proposed TCD methodologies calibrated using the assumption of leveraging the concept of equivalent elastic material.

Table 1. Predictions obtained through the proposed TCD methodologies calibrated using EMC.

Id Set	F experimental [N]	PM		LM	
		F predicted [N]	E%	F predicted [N]	E%
H10r05	2160	2417	-11.9	2574	-19.1
H10r2	2427	2413	2.5	2712	-9.5
H10r4	2476	2578	-6.2	2881	-18.7
H13r05	2845	2930	-3.0	3138	-10.3
H13r2	3251	2947	9.3	3310	-1.8
H13r4	3239	3578	-10.5	3877	-19.7
H15r05	3234	3402	-5.2	3631	-12.3
H15r2	3252	3430	-5.4	3817	-17.4
H15r4	3401	3574	-5	4006	-17.8

From these results, it is possible to observe that: the predictions made using the PM method are highly accurate, with a maximum deviation from the experimental mean of -11.9% obtained for specimens with a cross-section of 10 mm and a notch fillet radius of 0.5 mm. The average estimation error for this method is 6.5%, and there is no apparent tendency to overestimate or underestimate the experimental data. The PM method's predictions are suitable for both large and small thicknesses and for wide and severe fillet radii. On the other hand, when using the LM method, the percentage error increases in almost all geometric notch conditions but remains below 20%. This method exhibits a clear tendency to overestimate the fracture load, consistent with other reported cases in the literature [24].

6. Conclusions

In this article, two simple methodologies (Point Method and Linear Method) derived from the Theory of Critical Distance (TCD) for predicting fracture load resistance were tested and compared with experimental results. PLA specimens with V-notch manufactured through additive manufacturing, having different cross-sections and notch fillet radii, were subjected to tensile tests to evaluate the accuracy and reliability of each method. A significant finding was that the use of the concept of equivalent elastic material (EMC) can be effectively beneficial for estimating the intrinsic strength σ_0, and consequently, the effective stress σ_{eff}. Thanks to the excellent results obtained from this research, it can be asserted that these methodologies, initially developed for application on brittle materials, can be applied to situations involving components with a considerable degree of plasticity before fracture when corrected using the EMC approach.

Under this assumption, the combination of the TCD methods and the concept of equivalent material leads to remarkably accurate predictions, especially when using the PM method, with deviations from correct predictions limited to a maximum of -11%. In conclusion, we can affirm that the combination of the TCD method and the equivalent material concept can be a valuable tool for designers. By employing simple linear finite element analyses, they can accurately predict the behavior of notched components with substantial plasticity before fracture, such as those made from

PLA used in this research activity. The future development of this work involves testing these methodologies on other ductile materials and/or different notch geometries to better understand their robustness.

References

[1] Corapi D.; Morettini G.; Pascoletti G.; Zitelli C.; Characterization of a polylactic acid (PLA) produced by Fused Deposition Modeling (FDM) technology. Procedia Structural Integrity, 24, 289–295 (2019). https://doi.org/10.1016/j.prostr.2020.02.026

[2] Rajput A.; Shukla S. K.; Thakur N.; Debnath A; Mangla B.; Advanced Polymeric Materials for Aerospace Applications. In Aerosp Poly Mat. https://doi.org/10.1002/9781119905264.ch5

[3] Landi L.; Stecconi A.; Morettini G.; Cianetti F.; Analytical procedure for the optimization of plastic gear tooth root. Mechanism and Machine Theory. Vol:166, (2021) https://doi.org/10.1016/j.mechmachtheory.2021.104496

[4] Palmieri M.; Zucca G.; Morettini G.; Landi L.; Cianetti F.; Vibration Fatigue of FDM 3D Printed Structures: The Use of Frequency Domain Approach. Materials. Vol:15, (2022) https://doi.org/10.3390/ma15030854

[5] Santonocito. D; Foti P.; Risitano G.; Berto F.; Fatigue damage assessment in AM polymers evaluating their energy release. Procedia Structural Integrity. Vol:34, (2021), https://doi.org/10.1016/j.prostr.2021.12.031

[6] Rossi, A.; Morettini, G.; Moretti, M.; Capponi, L; Filament Transport Control for Enhancing Mechanical Properties of Parts Realised by Fused Filament Fabrication. Materials. Vol:15. https://doi.org/10.3390/ma15103530

[7] Morettini G.; Palmieri M.; Capponi L.; Luca L.; Comprehensive characterization of mechanical and physical properties of PLA structures printed by FFF-3D-printing process in different directions. Prog Addit Manuf. Vol:7, (2022). https://doi.org/10.1007/s40964-022-00285-8

[8] Broberg K. B.; Critical review of some theories in fracture mechanics. Int J Fract 4, 11–19 (1968). https://doi.org/10.1007/BF00189139

[9] Niu L.S., Chehimi C., Pluvinage G.; Stress field near a large blunted V-notch and application of the concept of notch stress intensity factor to the fracture of very brittle materials. Eng Fract Mech. Vol 49 (1994).

[10] Pluvinage G.; Fatigue and fracture emanating from notch; the use of the notch stress intensity factor. Nucl Eng Des. Vol 185: 173-184. (1998)

[11] Kipp M.E.; Sih G.C.; The strain energy density failure criterion applied to notched elastic solids. Int J Solids Struct. Vol 11:153-173 (1975)

[12] Berto F.; Lazzarin P.; Recent developments in brittle and quasi-brittle failure assessment of engineering materials by means of local approaches. Mater Sci Eng R. 2014;75:1-48.

[13] Berto F.; A criterion based on the local strain energy density for the fracture assessment of cracked V notched components made of incompressible hyperelastic materials. Theor Appl Fract Mech. Vol 76:17-26. (2015)

[14] Foti P., Razavi N.; Marsavina L.; Berto F.; Volume free strain energy density method for applications to blunt V-notches, Procedia Structural Integrity Vol 28:734-742, (2020) https://doi.org/10.1016/j.prostr.2020.10.085

[15] Dugdale D.S.; Yielding of steel sheets containing slits. J Mech Phys Solids.Vol 8:100-108, (1960)

[16] Barenblatt G.I.; The formation of equilibrium cracks during brittle fracture. General ideas and hypothesis Axially symmetric cracks. J Appl Math Mech. 1959; Vol 23:622-636.

[17] Hilleborg A.; Modeer M.; Petersson P.E.; Analysis of crack formation and crack growth in concrete by means of fracture mechanics and finite elements. Cem Concr Res. 1976; Vol 6:777-782.

[18] Gómez F.J.; Elices M.; Valiente A.; Cracking in PMMA containing U-shaped notches. Fatigue

Fract Eng Mater Struct. 2000; Vol 23:795-803.
[19] Ritchie R.O.; Knott J.F.; Rice J.R.; On the relationship between critical tensile stress and fracture toughness in mild steel. J Mech Phys Solids. 1973; Vol 21:395-410.
[20] Taylor D.; The Theory of Critical Distances: A New Perspective in Fracture Mechanics. UK: Elsevier; 2007.
[21] Susmel L.; Taylor D.; On the use of the Theory of Critical Distances to predict static failures in ductile metallic materials containing different geometrical features. Eng Fract Mech. 2008; Vol 75:4410-4421.
[22] Susmel L.; Taylor D.; The theory of critical distances to predict static strength of notched brittle components subjected to mixed-mode loading. Eng Fract Mech. 2008; Vol 75:534-550.
[23] Taylor D.; Merlo M.; Pegley R.; Cavatorta M.P.; The effect of stress concentrations on the fracture strength of polymethylmethacrylate. Mat Sci Eng A. 2004; Vol 382:288-294.
[24] Cicero S.; Madrazo V.; Carrascal I.A.; Analysis of notch effect in PMMA using the theory of critical distances. Eng Fract Mech. 2012; Vol 86:56-72.
[25] Madrazo V.; Cicero S.; Carrascal I.A.; On the point method and the line method notch effect predictions in Al7075-T651. Eng Fract Mech. 2012; Vol 79:363-379.
[26] Neuber H.; Theory of notch stresses: principles for exact calculation of strength with reference to structural form and material. 2nd ed. Berlin: Springer Verlag; 1958. p. 292.
[27] Peterson R.E.; Notch-sensitivity. In: Sines G, Waisman JL, editors. Metal fatigue. New York: McGraw Hill. 1959; Vol 12:293–306.
[28] Taylor D.; The theory of critical distances, Engineering Fracture Mechanics. 2008; Vol 751:696-1705
[29] Taylor D.; Geometrical effects in fatigue: a unifying theoretical model, International Journal of Fatigue.1999; Vol 21:413–420
[30] Novozhilov V.V.; On a necessary and sufficient criterion for brittle strength. Prik Mat Mek 1969; Vol 33:201–10
[31] Whitney J.M.; Nuismer R.J.; Stress fracture criteria for laminated composites containing stress concentrations. J Compos Mater. 1974; Vol 8:253–65.
[32] Kinloch A.J.; Shaw S.J.; Hunston D.L.; Crack propagation in rubber-toughened epoxy. In: International conference on yield, deformation and fracture. London: Cambridge Plastics and Rubber Institute; 1982. Vol 29:1–6.
[33] Cicero S.; Mata M.; Jano L.C.; Estebanez A.A.; Arroyo B.; Analysis of notch effect in the fracture behaviour of additively manufactured PLA and graphene reinforced PLA, Theoretical and Applied Fracture Mechanics. 2021; Vol 114:103-102
[34] Torabi A.R.; Alaei M.; Application of the equivalent material concept to ductile failure prediction of blunt V-notches encountering moderate-scale yielding, Inte, rnational Journal of Damage Mechanics Vol 2:1-25; DOI: 10.1177/1056789515625451
[35] Torabi A.R.; Estimation of tensile load-bearing capacity of ductile metallic materials weakened by a V-notch: the equivalent material concept. Mat Sci Eng A. 2012; Vol 536:249-255.
[36] Taylor D.; Predicting the fracture strength of ceramic materials using the theory of critical distances. Engng Fract Mech 2004; Vol 71:2407–16.
[37] ASTM D638, Standard Test Method for Tensile Properties of Plastics, ASTM International, 2022, ASTM International. DOI: 10.1520/D0638-22
[38] Morettini G.; Razavi S.M.J.; Staffa A.; Palmieri M; Berto F.; Cianetti F.; Braccesi C.; On the combined use of averaged strain energy density criteria (ASED) and equivalent material concept (ECC) for the fracture load prediction of additively manufactured PLA v-notched specimens,Procedia Structural Integrity, 2023; Vol 47:296-309 https://doi.org/10.1016/j.prostr.2023.07.095.
[39] ASTM E132, Standard Test Method for Poisson's Ratio at Room Temperature, ASTM International, 2010, ASTM International DOI: 10.1520/E0132-04.
[40] ASTM D5045, Standard Test Methods for Plane-Strain Fracture Toughness and Strain Energy

Release Rate of Plastic Materials, ASTM International, 2022, ASTM International DOI: 10.1520/D5045-14R22

[41] Foti, P, Berto, F. Strain energy density evaluation with free coarse mesh model. Mat Design Process Comm. 2020; 2:e116. https://doi.org/10.1002/mdp2.116.

Thermal design considerations for a L-PBF built metal component: effects of Inter-Layer Cooling Time, Preheating Temperature and Gas Flow

N Baldi[1,2], A Giorgetti[3*], M Palladino[2], G Arcidiacono[1], P Citti[1]

[1] Università degli Studi Guglielmo Marconi - Dipartimento di Scienze Ingegneristiche, via Plinio 44, 00193 Rome, Italy
[2] Baker Hughes–Nuovo Pignone, 50127 Florence, Italy
[3] Università degli Studi Roma Tre - Dipartimento di Ingegneria Industriale Elettronica e Meccanica, via della Vasca Navale 79, 00146 Rome, Italy

E-mail:*alessandro.giorgetti@uniroma3.it

Abstract. The paper aims to investigate some important thermal effects that could affect the Additive Manufacturing (AM) process of Laser Powder Bed Fusion. This analysis starts with investigating the variation of the material substrate temperature due to a variation of the Interlayer Cooling-Time (ILCT); then, the paper analyzes the effect of Preheating temperature on the material microstructure of the first building layers. Finally, we assess the effect of variation in gas flow speed as a function of part position on the building platform. In addition, in this work, the previously mentioned thermal aspects are evaluated in detail under particular geometrical and printing conditions considered the most critical for the L-PBF process. All cases studied are performed on IN718 superalloy specimens. In particular, for ILCT investigation, 60 microns layered specimens are printed for Preheating temperature analysis 40 and 60 layered specimens and for gas flow speed evaluation 40 microns one. All the results are evaluated through a porosity and melt pool analysis. The results obtained in this work highlight a critical range for low ILCT, 2-6 seconds, for part integrity that could be affected by overheating effects. To avoid this criticality, inserting ghost parts during the printing or reducing the laser power value is suggested. Concerning the preheating temperature effect, the first 1.2 mm of printed layers are found to be critical and affected by melt pool instability. In this case, a sacrificial substrate used in the first layers could save the quality of a few layers height part. The gas flow analysis highlights how some areas of the building platform are affected by particular thermal conditions negatively influencing material printability. To minimize this issue as much as possible, modify the job layout to avoid printing parts in the critical zones.

1. Introduction

Additive Manufacturing (AM) facilitates the fabrication of components and structures in a layer-by-layer deposition and completely revolutionizes the manufacturing industry allowing near-net-shape parts with almost no material waste. Furthermore, AM compared to conventional subtractive manufacturing techniques significantly reduces the lead time for the manufactured parts not using expensive tooling or dies and enabling the fabrication of one-step parts [1-10].

Laser Powder Bed Fusion (L-PBF) is AM technique that sinks its roots in the selective Electron Beam Melting (EBM) process developed by Arcam AB, a Swedish company founded in 1997. During the 1990s, the L-PBF technique began to develop using a high-power laser instead of an electron beam to melt the powder, allowing higher precision in material melting and openings for the use of a wider range of materials. Early applications of L-PBF at this stage were mainly in the production of rapid prototypes and small runs of components.

During the early 2000s, the evolution of L-PBF technology saw significant improvements in hardware, materials, and software controls. These developments and improvements in L-PBF technique led to an improve in the quality of manufactured parts and the ability to print a wider range of metallic materials, including stainless steel, aluminum, titanium, and nickel alloys.

In recent years, the continuous research and development in L-PBF process allowed the widespread of this technique in various industrial sectors, making AM technologies the most versatile and promising ones [11]. In particular, several advantages can be pointed out compared to the other AM processes:

- L-PBF technique is suitable for many materials, especially for engineering metal alloys such as nickel-based ones [12];
- L-PBF can be conducted at a much higher speed, over 1 m/s, with a lower material cost;
- in L-PBF, lower heat induces lower distortions, allowing higher dimensional resolution and precision for the final part.[13]

All these advantages facilitate significant investments in developing industrial applications where the printed parts were increasingly larger and more complex and had to meet more stringent quality and functional requirements. Moreover, the ability to manufacture lightweight metal components while maintaining the mechanical and performance required allows the proliferation of L-PBF in the aerospace and aviation industry [13,14]. Furthermore, L-PBF enables the production of very complex lattice structures and customizing implants to each patient-specific anatomy, resulting in the development and implementation L-PBF parts in the biomedical industry. This technique is more and more widespread in the fabrication of turbomachinery components in Gas and Oil industry, such as Nozzle, Burners and Blades, allowing the printability of metal superalloy resistant to high temperatures that are hard and challenging to be machined and manufactured with subtractive techniques [12,15].

Despite all these advantages that facilitate the production of very geometrically complex parts in a wide range of materials, specific design rules for AM have to be followed for L-PBF parts to ensure the structural and dimensional integrity of this last one.

The Design for AM in L-PBF process has to consider different aspects, some related to the physics behind the process and others to machine architecture.

L-PBF is process thermally driven; controlling some parameters, such as the Inter-Layer Cooling Time (ILCT), plays an important role in the final part quality. The continuous variation of ILCT during the printing due to cross-sectional variation of components has to be investigated and considered during Job Layout design to avoid critical ILCT range [15].

Another important aspect to be not neglected while fabricating a few layers of height component is the effect of Preheating Temperature (PT) on the first layers of consolidated material. The PT is necessary to minimize the heat loss of the first printing layers due to the building platform's high thermal conductivity (BP) and to provide a proper adhesion between the first melted layers of material and BP itself [16,17].

Moreover, an important role is played by the Gas Flow system; the hypothetical gas flow speed variation as a function of position on BP could influence the printability of material. Thus, studying these parameters before creating the layout of a component on BP is necessary to avoid positioning parts of the zone affected by any criticality.

Other aspects related to machine components architecture, such as the recoating system and blade type, must be considered carefully. In the case of components characterized by several overhang surfaces, very frequently, the orientation of part has to be optimized in order to have the overhang surface in the direction of powder spreading. This is because of the difficulty of printability of the overhang surfaces due to the lack of layers of consolidated material below the overhang area to be melted. Considering these geometrical conditions, proper melting of the external area of the overhang part may be complicated and the adhesion between the previous layer and the next may not be complete due to a difficult heat transfer, being the powder thermal conductivity much lower than that of the consolidated material. Consequently, in this condition, an orientation of the overhang part in the opposite direction to that of powder spreading could promote an impact with the recoater, usually in stainless steel (hard), and the delamination of part of the layers. In addition, if the overhang angles reach values less than 40 degrees, the creation of support structures is recommended, which, in addition to the proper orientation, should promote proper heat transfer between the part and support, ensuring the structural and dimensional integrity of the component.

This work aims to evaluate and summarize the effect of some important parameters that influence the Design for AM considering the thermal point of view for nickel-based component manufactured through L-PBF technique. In particular the the following parameters are considered:

- ILCT effect on component integrity for components fabricated used Multi-lasers machine, in which the ILCT could reach very low and hypothetically critical values. All the results are evaluated through a porosity and melt pool analysis of specimens.
- Preheating Temperature effect on the first layers of material, setting a conventional preheating temperature. The results are evaluated in terms of a melt pool analysis of Single and Multi-tracks.
- Gas Flow effect as a function of specimen position on the building platform. The results are evaluated in terms of track stability and melt pool analysis.

2. Effect of ILCT

ILCT is calculated as the time between the exposure of one layer on the next, resulting different from the Inter-Layer Time (ILT) [18] which is calculated also considering the time required for powder recoating and spreading (variable depending on the machine used).

ILCT varies continuously as the cross-section area of the part varies, influencing the average input rate into the part and promoting the variation of substrate temperature during the printing. The difference in thermal history from layer to layer could affect the material density and promote defect formation, thereby deteriorating the material's mechanical properties.

From a thorough study of the state of the art, this parameter has not been extensively and adequately investigated for L-PBF process [19].

In the context of Multi-laser machines, this parameter could play a very important role, where the exposure times for each layer are considerably shorter than in a single laser machine as multiple lasers work simultaneously on the same building platform. An ILCT on the order of a few seconds could be reached, hypothetically promoting an increase in substrate temperature and a variation of the material's printability map. [20-23]

In particular, from previous literature studies [22,24,25], it has been observed that a variation in the melting regime of the powder due to an increase in substrate temperature promotes the transition of the melting regime from conductive mode to keyhole mode [15].

Thus, an in-depth study on the effect of low ILCT in Multi-laser machines context is of considerable importance when developing a set of process parameters robust to substrate temperature variation. Moreover, this parameter represents an important constraint during the thermal design of a component manufactured through L-PBF process, to be strongly taken into account to avoid unexpected temperature rises.

The effects of lower ILCT variation on the microstructure of IN718 superalloys processed by a Multi-laser machine have been evaluated through two main phases: first, a study on the effect of ILCT variation on material microstructure and second, based on the obtained results, evaluation on the effect of obstructing heat flux geometries on critical ILCT values. All the results are evaluated through porosity and melt pool analysis [13, 26-29]. All details regarding the machine, specimen's geometry, ILCT levels tested in each case study are fully described in [15]. The major findings of the two case studies are briefly shown below. In particular, during the investigation of the effect of ILCT variation, after the evaluation of the results of porosity and Melt Pool analysis, the ILCT range between 6-2 seconds turns out to be critical. In fact, high porosity percentages and tracks of keyhole porosity are measured and found in the specimens characterized by this range of ILCT. Moreover, the Melt Pool analysis remarks

IOP Conf. Series: Materials Science and Engineering 1306 (2024) 012020 doi:10.1088/1757-899X/1306/1/012020

how the Melt Pool formation, in these ILCT conditions, is completely governed by a Keyhole melting regime, resulting in a very deep and elongated Melt Pool shape (Figure 1).

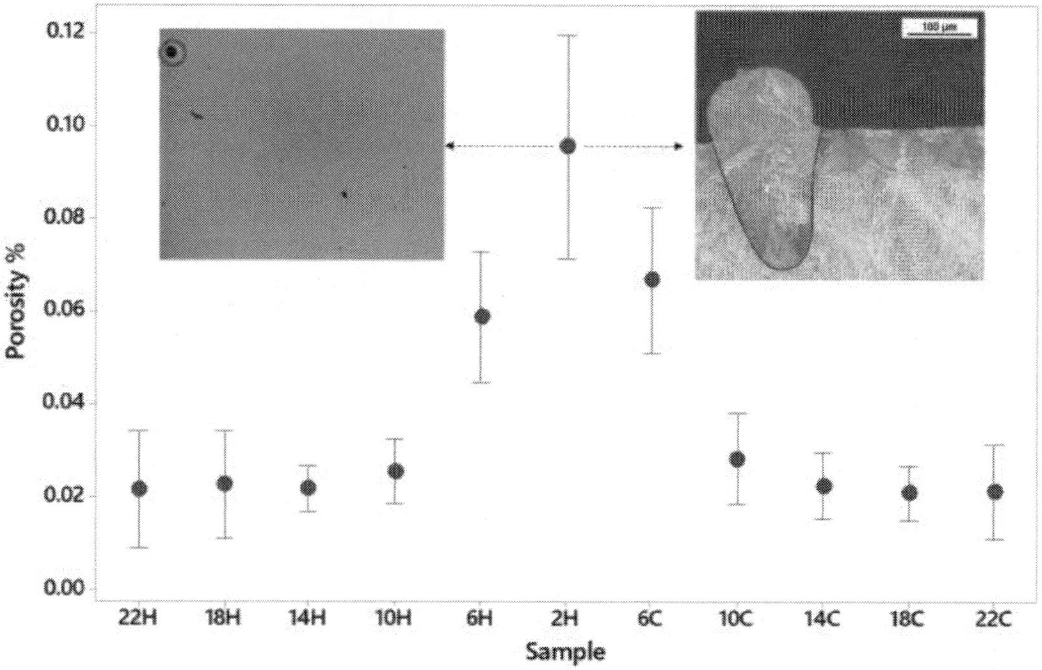

Figure 1 - Global overview of major findings in ILCT effect on material microstructure investigation.

The case study of obstructing heat flux geometries, tested for 2 seconds of ILCT, highlights how this peculiar, very common in industrial component design, increases the criticality detected in the previous condition (porosity percentage increased and defect formation promoted).

In conclusion, it is possible to state that low ILCTs, in the Multi-lasers context, are critical for material microstructure, promoting the expansion of the keyhole region in the material printability map, reducing the material density and facilitating defect formation [15].

Thus, an investigation of ILCT's effect on the final component printing plays an important role with a view to developing a robust thermal design.

3. Effect of Preheating Temperature

The possibility to manufacture micrometer-sized components with high resolution and quality through L-PBF process facilitates the proliferation of this technique for fabricating sensors for components in robotics, biomedical, automotive and aerospace industries. On the other hand, manufacturing components a few layers high could result critical for two main aspects. First, the beginning layers of consolidated material will not be as high as the desired layer thickness due to the difference in density between deposited powder and consolidated material; secondly, these layers could be affected by the effect of the preheating temperature of the building platform. Calculating the number of layers required to obtain a consolidated layer of material equal to the layer of powder deposited is easy to carry out depending on layer thickness value and powder and melted material density. In contrast, the estimation of the number of layers from which the material is no longer affected by any effect of the preheating temperature of the building platform is not yet well known and investigated. So, it is not clear whether this phenomenon depends on the layer thickness, the material investigated, or only on the distance from the building platform [17,30].

The preheating temperature turns out to be an important necessity for L-PBF process, commonly in a temperature range of 80 -250 °C, for different benefits:

- avoid heat loss in the first few layers due to the high thermal conductivity of the building platform, usually made of C40 Steel;
- minimize the high thermal gradients of L-PBF process, reducing residual stress formation in the final part [17];
- improve the adhesion between the first layers of consolidated material and the building platform itself.

It is interesting to evaluate the required number of layers to have a substrate of consolidated material that is no longer affected by the effect of the preheating temperature. This phenomenon is evaluated through two case studies using nickel-based superalloy Inconel 718 and investigating the effect on two layers thickness, respectively, of 40 and 60 microns. All the results are evaluated through the Melt Pool analysis [31-34] of several Single and Multi-Tracks [35-38]. More specifically, it will be investigated if the preheating temperature affects the Melt Pool shape and the Melting regime as a function of the substrate height of the tested specimen. All details regarding the machine, specimen's geometry, and substrate height levels tested in these experimental campaigns are fully described in [16]. The major findings for each case study are briefly shown below.

After the Melt Pool analysis for 40 microns case study, it is possible to highlight that the Melt Pool Depth of the specimen at a lower thickness (20 layers -> 0.8 mm) is considerably deeper than the other ones (Figure 2); the Melt Pool Depth is gradually reduced as the distance from building platform is increased. Moreover, by the evaluation of the Melt Pool Width/Depth ratio (Figure 3), used to calculate the melting regime that governs the Melt Pool formation [39], it is remarkable how the Melt Pool formation, in this configuration, is almost governed by Keyhole melting regime. These partial results on Single Tracks Melt Pool analysis indicate that the first 30 layers, equivalent to 1.2 mm of distance from the building platform, of consolidated material are affected by the thermal effect of Preheating Temperature. This outcome is also confirmed by the evaluation of the results of Multi-Tracks analysis.

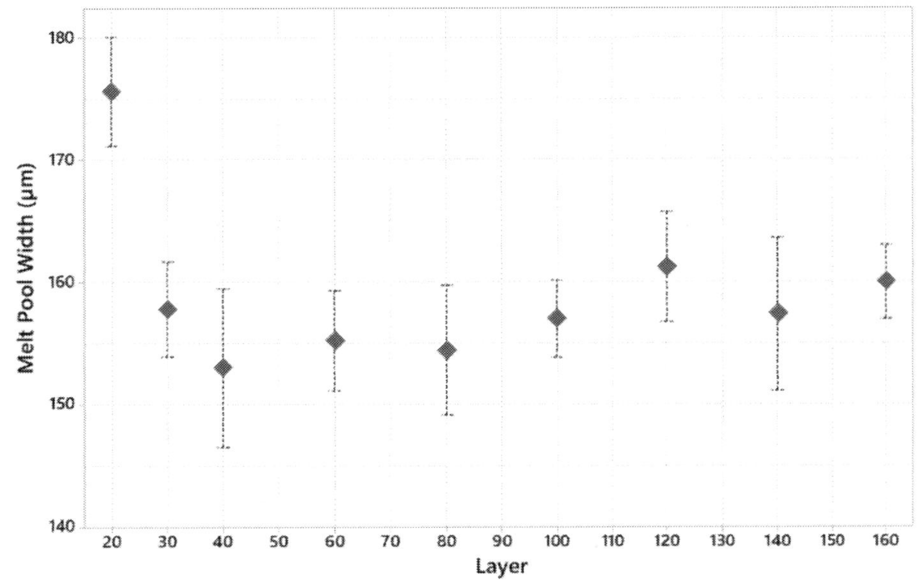

Figure 2 - Melt Pool Depth as number of layers plot.

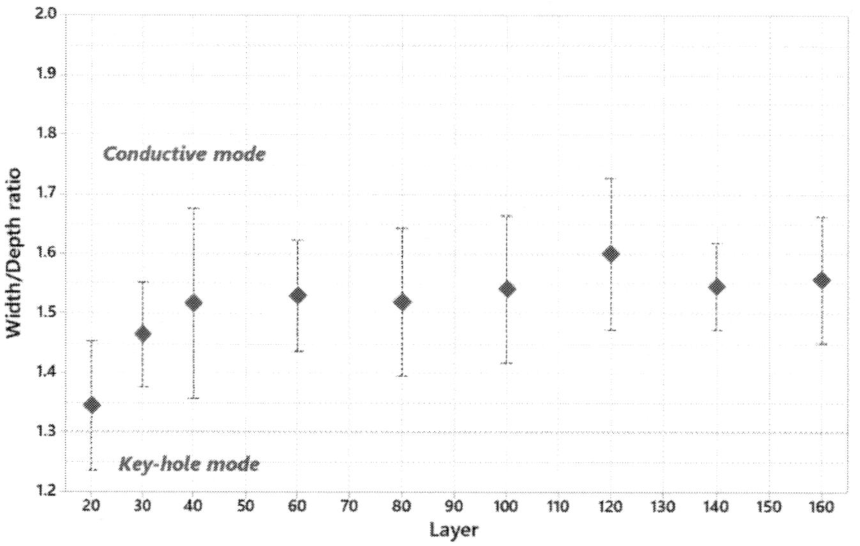

Figure 3 - Width/Depth ratio as a function of number of Layers.

The case study carried out on 60 layered specimens confirms the findings of 40 microns one, identifying the first 1.2 mm of consolidated material critical for the effect of Preheating Temperature. Thus, by virtue of these results, to avoid any thermal effect of Preheating Temperature on a few layers high components, it is suggested to use a sacrificial substrate (in this case, at least thick 1.2 mm) before starting the building of the part.

4. Effect of Gas Flow

The need to obtain an inert atmosphere within the building chamber in the L-PBF machines, reducing the pick-up of reactive gas by the liquid metal and avoiding oxidation and nitration required the use of a shielding gas flow during the part printing [40]. Moreover, the shielding gas flow is an effective tool in removing process by-products from the printing zone.

The melting and the vaporization of the metal powder promote the formation of these by-products, causing a welding plume; the possibility of interaction of these by-products with the laser beam depends on the orientation of scanning vectors relative to gas flow [41]. This phenomenon could reduce the absorption of laser energy, up to 10 percent [42], by the metal powder and could cause the formation of nano-scale particles from the rapid condensation of the welding plume. In addition, the gas flow facilitates the spatter removal from the printing zone, ejected liquid and solid particles whose formation is due to melt pool instabilities.

Previous literature studies [43], carried out using EOS Eosint M-280 machine, have highlighted how local gas flow speed promotes the re-deposition of by-products and the laser attenuation by the welding plume. In these gas flow conditions, the melt pool's stability in Single Tracks is compromised and these effects can lead to balling formations, resulting in incomplete melting of material. Other literature studies [44,45] show how improper removal of spatter from the powder bed can increase porosity and degradation of the material's mechanical properties.

Furthermore, in [41], it is emphasized that the defects formation due to spatter depends on the state of the spatter, molten or solidified, when impinging and on the powder bed conditions. As a result of the spatter falling onto the powder bed, the layer thickness locally increased, promoting the formation of Lack of Fusion defects in this area (increased layer thickness).

Therefore, having uniform gas flow ensures a powder bed free of by-products and spatter necessary to have microstructural integrity and defect-free parts throughout the building platform. The effect of

Gas Flow on melt pool stability and powder melting regime are evaluated through Single Tracks analysis.

4.1 Material and Methods

The specimens used in this experimental campaign are made in Inconel 718 alloy. The layer thickness used is 40 microns and the particle size of the powder is between 20 and 53 microns (obtained through a gas atomization process).

An EOS Eosint M-290 provided with one ytterbium fiber laser characterized by a beam wavelength of 1070 nm, a minimum spot size for the laser beam of 80 μm and a maximum laser power of 400 W was used for all tests. The building platform temperature was kept constant during all the printing at 80°C.

Argon gas flow was continuously pumped inside the building chamber to maintain oxygen content under 100 ppm; the gas flow speed was set to 2.5 m/s.

The gas flow system, which provided the EOS machine, including all its components, is summarized in Figure 4.

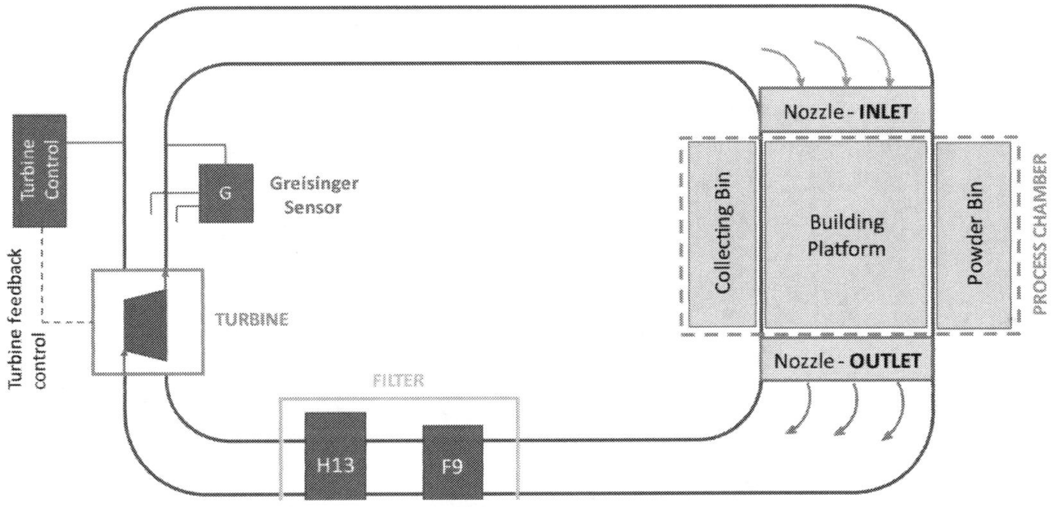

Figure 4 - EOS M290 Gas Flow System component overview.

All the specimens were exposed with the following process parameters, widely developed in previous literature studies [46,47]: Laser Power 280W, Scanning Speed 1000 mm/s and Hatch Distance 0.10 mm.

The specimens used in this experimental campaign were modeled as size 15 x 15 x 15 mm cubes to obtain a cross-section representative of the material microstructure. Moreover, ten Single Tracks were made on the top surface of each specimen to evaluate the melt pool stability and shape. After the printing, the specimens were detached from the building platform using a wire EDM machine (ECUT EU MS Genesi) and then cut along an axial plane, as shown in Figure 5.

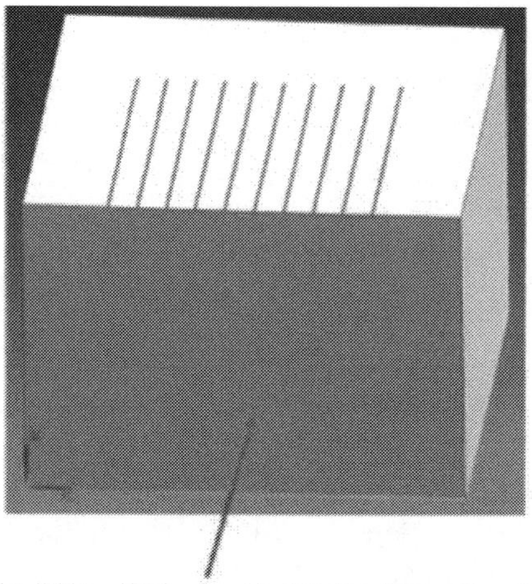

Figure 5 - Specimen geometry overview with the zone used for microstructural analysis.

The specimens were positioned on the building platform and labeled using Materialise Magics 24.1(Materialise NV, Leuven, Belgium). The build file was created using Eosprint Version 2.10.
In order to investigate as many areas of the building platform as possible, 49 specimens were printed and placed on the building platform, as shown in Figure 6 (step 12 mm, distance from the top and bottom edges 20mm, distance from lateral edges 30mm).

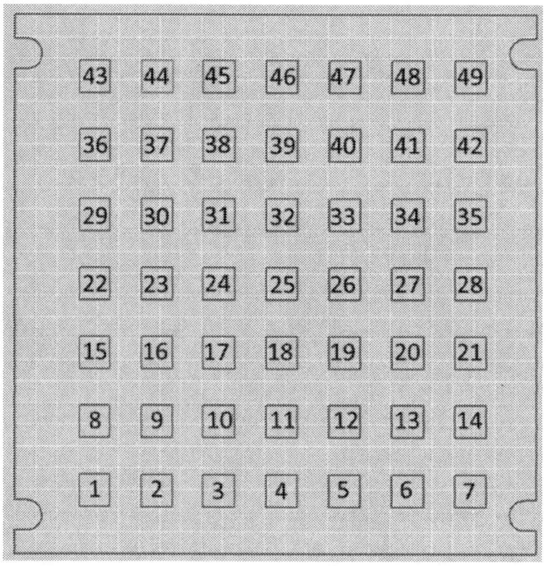

Figure 6 - Specimens' placement on the building platform.

The specimen was prepared following the procedure described in [12]. The Single Track's stability analysis was performed using a SEM Zeiss SUPRA 55 (Zeiss, Oberkochen, Germany) following the method widely described in [12]. The microstructural analyses, considering Porosity and Melt Pool, were performed using an Optical Microscope Leica Leitz DMRME (Leica Microsystems GmbH, Wetzlar, Germany). For further details about the methodology and tools used to carry out these analyses, refer to [12,15].

4.2 Results

4.2.1 Single Track's Stability analysis

In Figure 7 are tracked and shown the Single Track's stability level for each analyzed configuration.

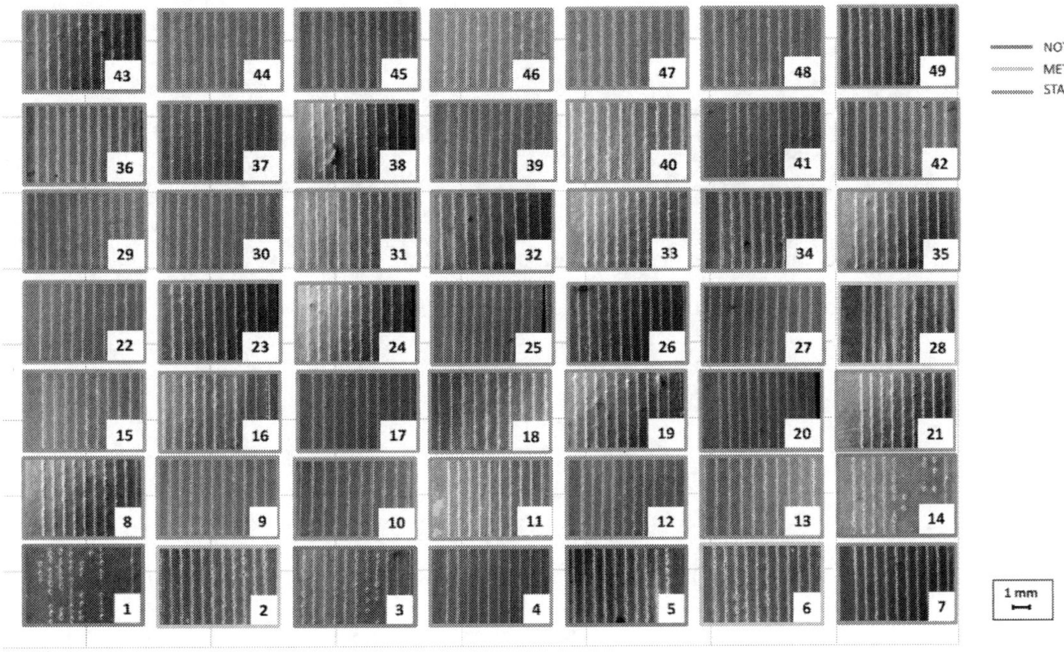

Figure 7 - Single Track's stability map.

4.2.2 Porosity analysis

For each stability level (not stable, metastable and stable) previously described three random configurations were analyzed in terms of porosity analysis, the results are tracked in Table 1.

Table 1 - Porosity tracking table.

Configurations	Stability level	Porosity (%)	
		Avg.	SD
1	Not stable	0.142	0.051
3	Not stable	0.098	0.037
5	Not stable	0.079	0.027
2	Metastable	0.046	0.017
6	Metastable	0.033	0.009
28	Metastable	0.032	0.016
4	Stable	0.013	0.008
7	Stable	0.017	0.009
8	Stable	0.086	0.034

IOP Conf. Series: Materials Science and Engineering 1306 (2024) 012020 doi:10.1088/1757-899X/1306/1/012020

Figure 8 shows the images corresponding to defects distribution and morphology for configurations 1, 2 and 4.

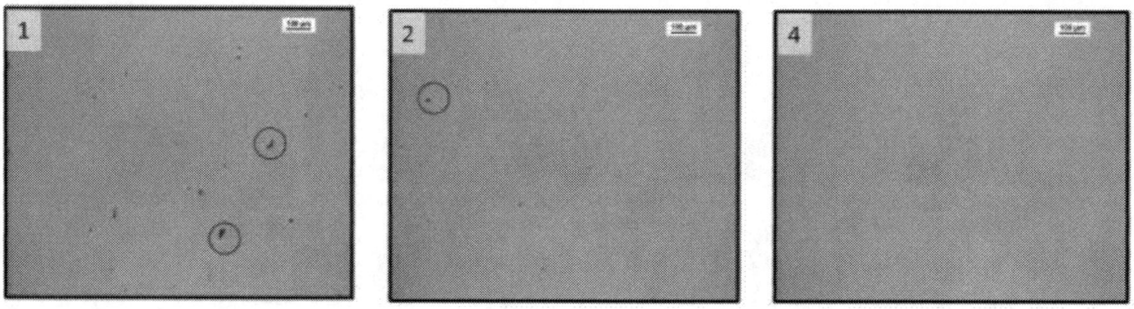

Figure 8 - Defects distribution and morphology, respectively, in configurations 1,2 and 4.

4.2.3 Melt Pool analysis

A melt Pool analysis was carried out on the same configuration previously analyzed in terms of porosity percentage measurement.

The results for each configuration in terms of Melt Pool Depth and Width are tracked in Table 2. Figure 9 shows the Melt Pool shape for configurations 1, 2 and 4.

Table 2 - Melt Pool analysis tracking table.

Configurations	Stability level	Melt Pool Depth (μm)		Melt Pool Width (μm)	
		Avg.	SD	Avg.	SD
1	Not stable	16.1	7.9	57.5	22.3
3	Not stable	20.3	8.3	53.5	10.1
5	Not stable	19.3	6.3	49.9	13.2
2	Metastable	63.6	5.1	126.6	6.2
6	Metastable	70.5	5.5	132.7	8.0
28	Metastable	80.6	4.8	137.8	5.9
4	Stable	118.6	4.5	157.6	5.5
7	Stable	100.7	4.3	149.2	5.8
8	Stable	104.9	5.6	154.8	6.1

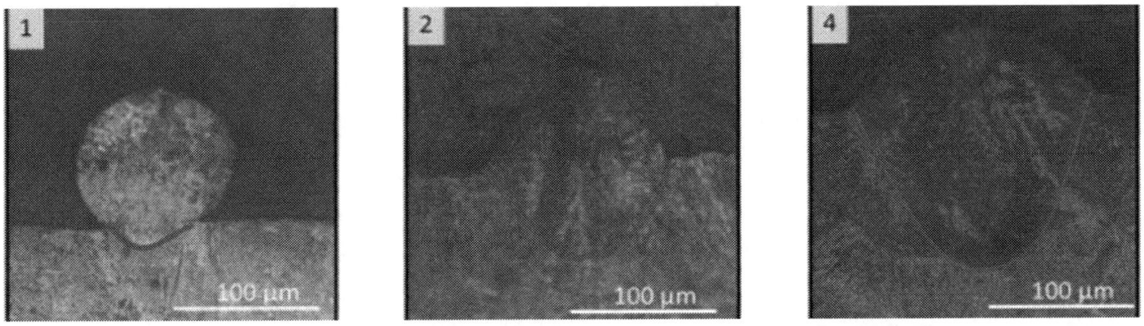

Figure 9 – Melt Pool morphology respectively in configurations 1,2 and 4.

5. Discussion

Single Track's stability results, shown in Figure 7, permit us to observe that most of the configurations affected by strong and widespread track instability phenomena are concentrated at the bottom of the building platform (Configuration 1, 2, 3, 5 and 6), particularly in the bottom-left area close to the suction Nozzle (except of configuration 14 far from this area). These results confirm the findings of previous literature studies [43] in which it was observed how that area of the building platform was affected by low gas flow speed. Thus, in that area, hypothetically, the removal of spatter and by-products from the shielding gas flow is more difficult and Melt Pool instability phenomena are promoted.

The porosity analysis results, shown in Table 1, highlight how the configuration affected by the track's instability is lower in material density, reaching the highest value of porosity percentage for Configuration 1 (0.142%). Moreover, by the evaluation of the morphology of porosity, shown in Figure 8, it is possible to state that in the specimens affected by the track's instability, there is a widespread Lack of Fusion porosity [48,49], commonly due to insufficient energy input to melt the powder properly. This finding is aligned with the study carried out by Schniedenharn et al. [41], in which an excessive deposit of spatter particles on the powder bed results locally in an increase in layer thickness. So, in this condition, to melt the powder properly, it is necessary to provide more energy input.

Also, after the Melt Pool analysis results, shown in Table 2, it is possible to state that in the configuration affected by the track's instability, the Melt Pool depth is lower than the layer thickness, so according to [39] in these conditions the melting regime of lack of fusion governates the Melt Pool formation promoting the defects formation in the material.

Based on all these results, it is possible to conclude that the gas flow influences the printability of material as a function of the location on the building platform.

Moreover, these results are of considerable importance to properly designing Job's layout, avoiding positioning the part/s in the critical area or possibly orienting the part/s so that support structures (sacrificial) are exposed in that zone.

6. Conclusions

Thermal design guidelines of a metal component, in particular made in Inconel 718 alloy manufactured through L-PBF process, can be summarized by evaluating the results presented in this paper.

- ILCT, in the low range of a few seconds (6 to 2 sec), plays an important role in the context of process parameters development for Multi-laser machines. Furthermore, the criticality of this parameter could be amplified when combined with special geometric conditions, such as obstructing heat flux geometries. Thus, ILCT has to be strongly considered and evaluated during the design phase of a component manufactured through L-PBF technique.
- The high resolution of L-PBF allows the fabrication of a few layers' high components, such as sensors. In this case, the evaluation of preheating temperature on material microstructure is very important to manufacture a fully dense part. Using a sacrificial substrate, at least 0.8 millimeters thick, can mitigate the effect of preheating temperature and ensure good printability for the final component.

- Another parameter that has to be included in the thermal design of a metal component in L-PBF is gas flow speed. The probable variation of this parameter as a function of part position on the building platform strongly affects the material printability, promoting powder melting regime changing and defects formation. Considering the results obtained so far, some no-printing zones are identified (bottom–left area) where it is suggested not to position the final part (only support structures are printable).

Further analysis should be carried out in future work on Gas Flow speed to quantify its variation as a function of location on the building platform using a hot wire anemometer and an optical tomography to estimate the possible thermal effect of gas flow on the powder bed as a function of the position on the building platform.

References

[1] Wang Z, Guan K, Gao M, Li X, Chen Xand Zeng X 2014 *J. Alloys Compd* **513** 518–23
[2] Das S 2003 *Adv. Eng. Mater.* **5** 701–11
[3] Osakada K, Shiomi M, 2006 *Int. J. Mach. Tools Manuf.* **46** 1188–93
[4] Kruth J P, Mercelis P, Van Vaerenbergh J, Froyen L and Rombouts M 2005 *Rapid Prototyp. J.* **11** 26–36
[5] Yadroitsev I, Gusarov A, Yadroitsava I and Smurov I 2010 *J. Mater.Process. Technol.* **210** 1624–31
[6] Gu D D, Meiners W, Wissenbach K and Poprawe R 2012 *Int. Mater. Rev.* 57 133–64
[7] Smith J, Xiong W, Yan W, Lin S, Cheng P, Kafka O, Wagner G J, Cao J and Liu W K 2016 *Computational Mechanics* **57** 583–610
[8] Jo A R, An J S, Kim S H, Park D Y, Moon Y H and Hwang S K 2023 *Met. Mater. Int.* 1–13
[9] Majeed A, Ahmed A, Lv J, Peng T and Muzamil M J. 2020 *Braz. Soc. Mech. Sci. Eng.* ***42*** 249
[10] Ciappi A, Giorgetti A, Ceccanti F and Canegallo G 2021 *Proc. Inst. Mech. Eng. Part C J. Mech. Eng. Sci.* **235** 1741–58
[11] Metelkova J, Kinds Y, Kempen K, de Formanoir C, Witvrouw A and Van Hooreweder B 2018 *Addit. Manuf.* **23** 161–9
[12] Giorgetti A, Baldi N, Palladino M, Ceccanti F, Arcidiacono G and Citti P 2023 A Method to Optimize Parameters Development in L-PBF Based on Single and Multitracks Analysis: A Case Study on Inconel 718 Alloy Metals **13**(2) 306
[13] Mahmoud D, Magolon M, Boer J, Elbestawi M A and Mohammadi M G 2021 *Appl. Sci.* **21** 11910
[14] Blakey-Milner B, Gradl P, Snedden G, Brooks M, Pitot J, Lopez E, Leary M, Berto F and Du Plessis 2021 *Mater. Des*. **209** 110008
[15] Baldi N, Giorgetti A, Palladino M, Giovannetti I, Arcidiacono G and Citti P. 2023 Study on the Effect of Inter-Layer Cooling Time on Porosity and Melt Pool in Inconel 718 Components Processed by Laser Powder Bed Fusion *Materials* **16**(11) 3920
[16] Baldi N, Giorgetti A, Palladino M, Giovannetti I, Arcidiacono G and Citti P. 2023 Study on the Effect of Preheating Temperature on Melt Pool Stability in Inconel 718 Components Processed by Laser Pow-der Bed Fusion *Metals* forthcoming
[17] Chen Q, Zhao Y, Strayer S, Zhao Y, Aoyagi K, Koizumi Y, Chiba A, Xiong W and To A C 2021 *Addit. Manuf.* **37** 101642
[18] Panwisawas C, Gong Y, Tang Y T, Reed R C and Shinjo J 2021 *Addit. Manuf.* **47** 102339
[19] Liu X, Wang K, Hu P, He X, Yan B and Zhao X. 2021 *Materials* **14** 991
[20] Khan H M, Karabulut Y, Kitay O, Kaynak Y and Jawahir I S 2021 *Mach. Sci. Technol.* **25** 118–76
[21] Guo Y, Jia L, Kong B, Wang N and Zhang H 2018 *Chin. J. Aeronaut.* **31** 860–66
[22] Shrestha S and Chou K 2018 *Procedia Manuf.* **26** 857–64
[23] Balbaa M, Mekhiel S, Elbestawi M and McIsaac J 2020 *Mater. Des.* **193** 108818
[24] Yadroitsava I, Els J, Booysen G and Yadroitsev I 2015 *S. Afr. J. Ind. Eng.* **26** 86–95

[25] Zheng H, Wang Y, Xie Y, Yang S, Hou R, Ge Y, Lang L, Gong S and Li H 2021 *Metals* **11** 937
[26] Li Y, Založnik M, Zollinger J, Dembinski L and Mathieu M 2021 *J. Mater. Process. Technol.* **289** 116930
[27] Chen Q, Zhao Y, Strayer S, Zhao Y, Aoyagi K, Koizumi Y, Chiba A, Xiong W and To A C 2021 *Addit. Manuf.* **37** 101642
[28] Yadroitsev I, Krakhmalev P and Yadroitsava I 2015 *Addit. Manuf.* **7** 45–56
[29] Tran H C and Lo Y L 2019 *Int. J. Adv. Manuf. Technol.* **105** 4443–60
[30] Panahi N, Åsberg M, Oikonomou C and Krakhmalev P 2022 *Procedia CIRP* **111** 166-70
[31] Singh S N, Chowdhury S, Nirsanametla Y, Deepati A K, Prakash C, Singh S, Wu L Y, Zheng, H Y and Pruncu C A 2021 *Materials* **14** 876
[32] Makona N W, Yadroitsava I, Moller H, Tlotleng M and Yadroitsev I 2016 *S. Afr. J. Ind. Eng.* **27** 210–18
[33] Ahmadi M, Tabary S B, Rahmatabadi D, Ebrahimi M S, Abrinia K and Hashemi R 2022 *J. Mater. Res. Technol.* **19** 1537–62
[34] Li S, Xiao H, Liu K, Xiao W, Li Y, Han X and Song, J M L 2017 *Mater. Des.* **119** 351–60
[35] Kan W H, Chiu L N S, Lim C V S, Zhu Y, Tian Y, Jiang D and Huang A 2022 *J. Mater. Sci.* **57** 9818–65
[36] Oliveira J P, Santos T G and Miranda R M 2020 Revisiting fundamental welding concepts to improve additive manufacturing: From theory to practice *Prog. Mater. Sci.* **107** 100590
[37] Mukherjee T, Zuback J S, De A and DebRoy T 2016 Printability of alloys for additive manufacturing *Sci. Rep.* **6** 19717
[38] Oliveira J P, LaLonde A D and Ma J 2020 Processing parameters in laser powder bed fusion metal additive manufacturing. *Mater. Des.* **193** 108762
[39] Johnson L, Mahmoudi M, Zhang B, Seede R, Huang X, Maier J T, Maier H J, Karaman I, Elwany A and Arróyave R 2019 Assessing printability maps in additive manufacturing of metal alloys. *Acta Mater.* **176** 199–210
[40] Philo A M, Sutcliffe C J, Sillars S, Sienz J, Brown S G R and Lavery N P 2017 A Study into the Effects of Gas Flow Inlet Design of the Renishaw AM250 Laser Powder Bed Fusion Machine Using Computational Modeling International *Solid Freeform Fabrication Symposium - An Additive Manufacturing Conference* University of Texas (Austin)
[41] Schniedenharn M, Wiedemann F and Schleifenbaum J H 2018 Visualization of the shielding gas flow in SLM machines by space-resolved thermal anemometry *Rapid rototyping Journal* **24**(8) 1296-1304
[42] Shcheglov P 2012 Study of vapour-plasma plume during high power fiber laser beam influence on metals. *Bundesanstalt für Materialforschung und-prüfung (BAM).*
[43] Ladewig A, Schlick G, Fisser M, Schulze V and Glatzel U 2016 Influence of the shielding gas flow on the removal of process by-products in the selective laser melting process *Additive Manufacturing* **10** 1-9
[44] Liu Y, Yang Y, Mai S, Wang D and Song C 2015 Investigation into spatter behavior during selective laser melting of AISI 316L stainless steel powder *Materials & Design* **87** 797-806
[45] Ferrar B, Mullen L, Jones E, Stamp R and Sutcliffe C J 2012 Gas flow effects on selective laser melting (SLM) manufacturing performance *Journal of Materials Processing Technology* **212**(2) 355-364
[46] Singh S N, Chowdhury S, Nirsanametla Y, Deepati A K, Prakash C, Singh S, Wu L Y, Zheng H Y and Pruncu C 2021 A Comparative Analysis of Laser Additive Manufacturing of High Layer Thickness Pure Ti and Inconel 718 Alloy Materials Using Finite Element Method *Materials* **14** 876
[47] King W A, Barth H D, Castillo V M, Gallegos G F, Gibbs J W, Hahn D E, Kamath C and Rubenchik A M 2014 Observation of keyhole-mode laser melting in laser powder-bed fusion additive manufacturing *J. Mater. Process. Technol.* **214** 2915–2925

IOP Conf. Series: Materials Science and Engineering 1306 (2024) 012020 doi:10.1088/1757-899X/1306/1/012020

[48] Ning J, Wang W, Zamorano B and Liang S Y 2019 Analytical modeling of lack-of-fusion porosity in metal additive manufacturing *Appl. Phys.* **125** 79
[49] Mukherjee T and DebRoy T 2018 Mitigation of lack of fusion defects in powder bed fusion additive manufacturing *J. Manuf. Process.* **36** 442–449

Integration of Piezoresistive Sensors into AM Structural Components: Evaluation of Sensor Properties and Its Impact on Component Mechanical Performance

A. Staffa[1], M. Palmieri[1], G. Morettini[1], F. Cianetti[1] and C. Braccesi[1]

[1] University of Perugia, Department of Engineering, Via G. Duranti 93, 06125 Perugia, Italy

E-mail: `agnese.staffa@studenti.unipg.it`

Abstract. Additive manufacturing (AM) technology has experienced significant growth in recent years, opening new prospects across various sectors, from biomedicine to aerospace. Among these, the Fused Filament Fabrication (FFF) additive manufacturing technology, which deposits material layer by layer, enables the integration of piezoelectric material during the printing process, thereby allowing for the creation of sensors within thermoplastic components. Such sensors have the potential to monitor and detect various internal physical parameters within the component at locations inaccessible to traditional sensors. In this field, scientific research is continuously advancing, primarily aimed at developing sensors with high performance.

In this context, the objective of this study is to assess the measurement properties, both static and dynamic, of sensors produced using this technology and to evaluate the influence on the static and dynamic behavior of the component caused by the incorporation of these sensors within the component itself.

1. Introduction

Additive Manufacturing technology, with its capability to produce objects of complex geometries, has opened new frontiers in design. One of these frontiers is the ability to integrate sensors during the component production process [1] [2]. By depositing material layer by layer [3], it allows for precise control over the manufacturing process and the creation of each individual layer. Moreover, it offers the advantage of incorporating different material layers at specific points within the structure, enhancing its functionality and potential applications.

This perspective finds applications in diverse fields [4], starting from the biomedical domain, where highly sensitive sensors have been studied for integration into wearable devices to monitor movement [5], all the way to environments like aviation [6] and aerospace [7]. It can also have relevance in everyday life, such as the ability to incorporate small Internet of Things (IoT) sensors into the Gravity Separation Spirals (GSS), enabling remote monitoring of operational conditions, prediction of failures, and utilization of collected data for production optimization [8].

As of today, the significant advancement in 3D printing technology has made it possible to have multiple extruders on the same machine, which can print different materials. This capability

IOP Conf. Series: Materials Science and Engineering 1306 (2024) 012021 doi:10.1088/1757-899X/1306/1/012021

allows for the simultaneous production of both the component and the integrated sensor in a single printing process. This not only enables designers to determine sensor placement in advance during the design phase but also facilitates their introduction into positions that are not normally accessible through external sensors.

The emerging frontier of embedded sensors for measuring parameters such as temperature and deformation, which can be used in algorithms for damage calculation and remaining component life, marks a turning point in structural monitoring [9] and early fault detection [10], especially in dynamic environments [11].

Structural monitoring is indeed a fundamental component in the realm of dynamic design, ensuring the safety and longevity of structures while reducing maintenance costs and risks to public safety. These measures are especially critical in civil or aerospace contexts where they are used for transporting people. However, they are also of paramount importance in situations where human lives are not directly at stake, such as in the aerospace industry. In such cases, mission failures can result in significant economic losses and safety concerns due to space debris.

The development of printed sensors directly during the printing process has become possible thanks to the advancement of new materials, such as Conductive PLA, that is a thermoplastic material primarily composed of polylactic acid (common PLA) filament with the introduction of conductive particles within it to make it electrically conductive. Commonly used conductive materials include graphite powder, carbon black (CB) powder, carbon fibers (CF), or carbon nanotubes (CNT) [12]. The conductive particles are evenly distributed within the filament during the production phase, and the material's conductivity properties can vary depending on the quantity and dispersion of these particles [13]. The resulting filament is, in essence, a *piezoresistive* material, meaning that its resistance changes proportionally to the deformation applied to the object.

In the literature, there are already studies on sensors created using this type of material that have led to the development of printed sensors for measuring deformations with the classical form of a strain gauge, assessing their functionality [14], the effects of certain printing parameters [15], and conducting studies on the impact of their integration within components and their reproducibility in a static environment [16].

Given the potential of this application, studies have also been conducted on sensors with temperature compensation [17] and on possible applications in high-temperature environments [18].

Indeed, the development of a printed sensor in Additive Manufacturing requires a comprehensive characterization of the material, both from a mechanical and electrical perspective [19]. This characterization is essential to ensure the functionality of the sensors.

The purpose of this work is to dynamically characterize a printed sensor of any shape using conductive PLA. The goal is to measure the deformation of a component subjected to dynamic loads, essentially creating a printed "strain gauge." This will be achieved by evaluating its linearity, repeatability, and assessing the range of usability before it is integrated into a structural component.

2. Dynamic Characterization

The environment in which the printed sensor operates is primarily subjected to dynamic loads and vibrations; therefore, the sensor must be capable of differential measurements. For these reasons, a *dynamic* characterization of the sensor has been conducted.

Dynamic characterization is very similar to static characterization, with the difference that the inputs vary over time. However, the quantities being defined are essentially the same, and they are used to assess the sensor's performance and make comparisons with other measuring instruments to determine the most suitable one.

The investigated characteristics are as follows:

- Linearity: the sensor's ability to provide responses that follow a linear trend with increasing inputs.
- Repeatability: it indicates the sensor's ability to provide the same result for the same input. Typically, repeatability is quantified by evaluating the maximum standard deviation of responses for the same input.
- Usable Range: it represents the frequency range within which the sensor can accurately measure. This is used to define the measurement linearity range and the upper and lower limits

To determine these characteristics, a rectangular test specimen measuring 150 *mm* x 25 *mm* x 5 *mm* was created, printed in Red PLA. Within this specimen, a rectangular sensor measuring 10 *mm* x 5 *mm* x 1 *mm*, was fabricated using conductive PLA, as depicted in the Figure 1.

The chosen printing parameters were optimized for PLA printing [8] [20] and for conductive PLA, the parameters recommended in the datasheet were used [21]. The printing directions selected were ± 45° for the red PLA and 90° for the conductive PLA, so that the deposited filament was parallel to the longer dimension of the test specimen.

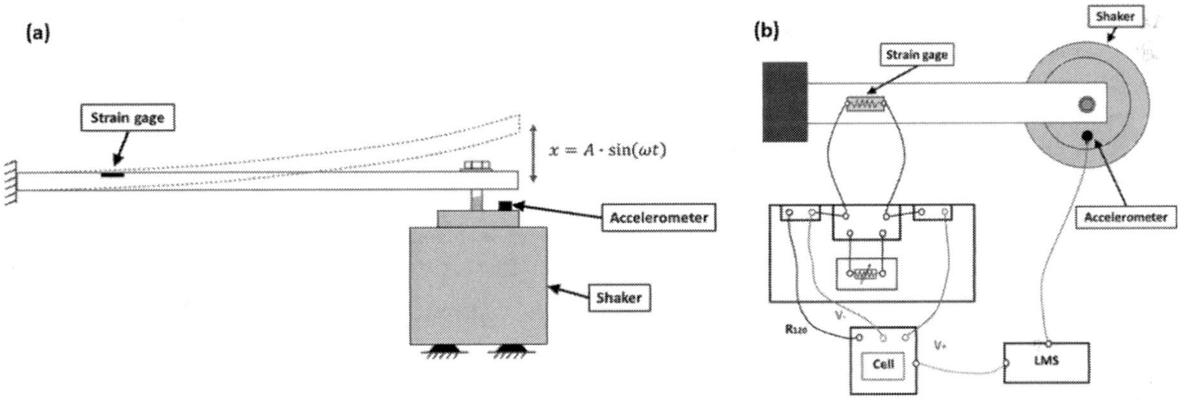

Figure 1. a) Experimental Setup for Dynamic Analysis Using a Shaker b) Block Diagram of the Data Acquisition System

To perform the dynamic characterization, an electrodynamic shaker was used for linearity and repeatability tests, and an instrumented hammer was used to analyze the frequency range within which the sensor can measure.

2.1. Dynamic analysis: Linearity and Repeatability tests

The dynamic tests were conducted using an electrodynamic shaker, which allows for the imposition of defined displacements, providing control over the test. One end of the test specimen was kept fixed, as depicted in the diagram in Figure 1a, while the other end was connected to the shaker.

Since the study pertains to dynamic measurements, the test specimen was excited using a sinusoidal signal with constant amplitude and frequency. Consequently, the test specimen underwent a dynamic flexural test.

In Figure 1b, the operating diagram of the test and the data acquisition system is shown. The shaker is controlled through a data acquisition system (LMS), which utilizes an accelerometer positioned at the base for closed-loop operation. Additionally, the same data acquisition system, not only manages the shaker, but also captures the signal output from the sensor. At both ends

IOP Conf. Series: Materials Science and Engineering 1306 (2024) 012021 doi:10.1088/1757-899X/1306/1/012021

of the sensor, tracks have been placed, from which two copper wires extend and are connected in parallel to a resistor, resulting in a total output resistance of 120 Ω from the sensor. Before the signal is sent to the strain gauge, it is routed through a Wheatstone bridge quarter-bridge system, which compensates for the signal and outputs the voltage value. This voltage value is recorded by the LMS data acquisition system.

The voltage value obtained is proportional to the deformation caused by the imposed displacement. However, the recorded signal is still in *Volt*, while the output from the accelerometer, which records the input signal, is in m/s^2. Therefore, a direct comparison in terms of amplitude is not possible, only in terms of phase. This is not a problem because, in this initial investigation, the goal is to determine the sensor's response time. Upon overlaying the output signals, it becomes evident that the two signals are in phase. This indicates that the sensor is capable of measuring the deformation variation at the moment it is applied

The procedure was repeated while keeping the frequency of the applied load constant and varying the amplitude value of the sine wave up to a maximum of 3 mm. This effectively covers a double range of displacement since the sine function is by definition an even function. The same procedure was also repeated for different frequency values, keeping them constant and varying the amplitude. The tests were conducted multiple times to assess the linearity and repeatability of the measurement.

The output signals from the sensor were analyzed in the time domain by averaging the maxima within the acquisition period. This process resulted in obtaining a voltage value for each different amplitude and frequency value.

The voltage values were plotted as a function of the imposed displacement for each analyzed frequency, and the results are shown in Figure 2. The graphs obtained for each frequency, as a function of various displacements, were linear and consistent across repetitions, indicating that the sensor exhibits linear behavior.

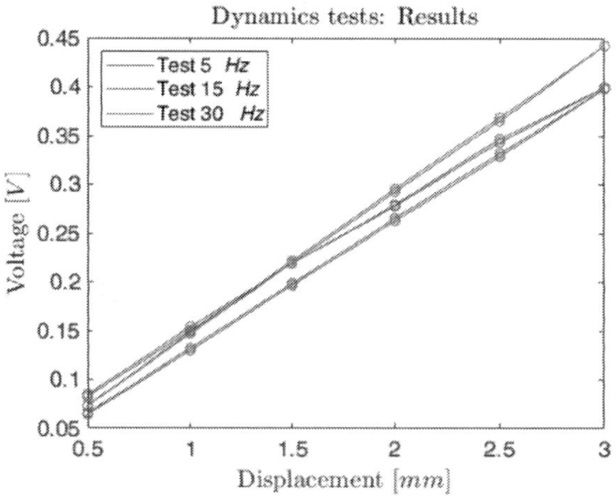

Figure 2. Results of Dynamic Analysis: Voltage Values Acquired as a Function of Displacement

In Equation 1, the law governing the strain gauges is presented.

$$e_0 = \frac{1}{4} \cdot \frac{\Delta R}{R} \cdot E = \frac{1}{4} \cdot K_s \cdot \varepsilon \cdot E \tag{1}$$

Where e_0 and E represent the output voltage value and the voltage applied to the strain gauge, respectively, ΔR is the change in resistance of the strain gauge, and R is the initial resistance value.

The voltage value e_0 is necessary to obtain information about the measurement being conducted, but a conversion factor is required to convert the voltage value from volts to strain in $\mu strain$. This conversion factor is represented by the gauge factor K_s.

The obtained curves were then related to the deformation caused by the imposed displacement value in the region where the sensor was printed. Deformation was calculated manually, taking into account all the factors that contribute to it.

Once the graph was converted into a Strain-Resistance curve, with displacement values converted to strain and voltage values using Equation 1 to obtain the resistance change $\Delta R/R$, a linear regression was performed on this data.

The slope of the line obtained represents the gauge factor K_s, and the values for the three curves are listed in Table 1.

Table 1. Values of the gage factor for three different frequencies

Frequency [*Hz*]	Gage Factor [*1/strain*]
5	2770.6
15	3118.7
30	2821.9
Mean	2903.7

2.2. Dynamic analysis: Frequency range
The analysis of the frequency range in which the sensor can operate was determined using an instrumented hammer, which allows for excitation of the structure up to 3000 *Hz*. The same test specimen used previously was employed for this test, with one end fixed and the other end left free, resembling a cantilever configuration.

To facilitate a comparison, an accelerometer was placed at the free end of the test specimen to monitor the natural frequencies of the structure. The data acquisition system remained the same as used previously, with the difference being the management of the input.

The tests were repeated 10 times, with an average of 5 hammer strikes for each test. The frequency response functions, shown in Figure 3, revealed good repeatability in the sensor's response, with the curves appearing nearly superimposed.

A direct comparison between the frequency response functions of the two instruments is not possible because they have different units of measurement, with the accelerometer using g/N and the strain gauge using *Volt/N*. However, to assess the quality of the measurement and the operating range, it is sufficient to analyze the peaks in the responses corresponding to the system's vibration modes

Table 2 contains the values of frequencies detected by the sensor and the accelerometer, along with the error between the two measurements.

As evident from the obtained results, the sensor can capture the natural frequencies of the system with an error of less than 2% compared to a calibrated accelerometer. Furthermore, these results demonstrate that the printed sensor is capable of detecting the system's response up to at least 2000 *Hz*.

3. Conclusion
In this study, the methodology for creating an integrated sensor using Additive Manufacturing for structural monitoring and stress analysis was verified. Through the dynamic characterization process, it was possible to assess the linearity and repeatability of the measurements. Using the

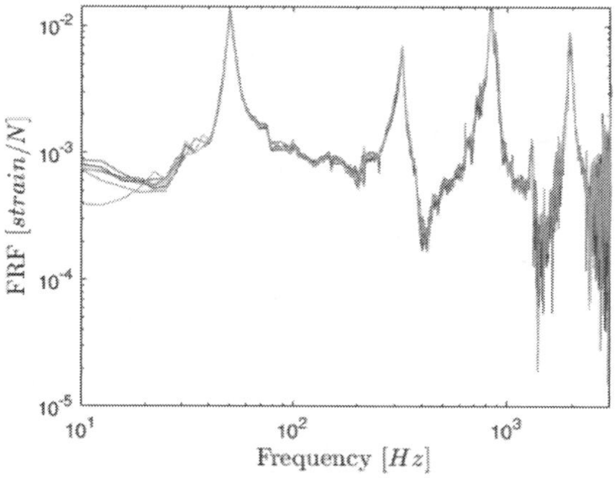

Figure 3. Experimental Frequency Response Functions for the Test with the Instrumented Hammer

Table 2. Comparison of Experimental Natural Frequencies and Error

Printed Sensor [Hz]	Accelerometer [Hz]	Error [%]
50	50	0
321.9	325	1.25
843.8	840.6	0.38
1931	1959	1.42

collected measurements, a regression line was calculated for the measurement points, allowing for the determination of the sensor's gauge factor. This gauge factor enables the conversion of measurements expressed in *Volt* to the actual deformation measurement. Furthermore, through a control system implemented with an accelerometer, the frequency range in which the printed sensor can make measurements was assessed and found to be at least up to 2000 *Hz*.

This work marks the beginning of a study on the creation of sensors integrated into the same 3D printing process as the component to be monitored. The next steps will involve investigating variations in printing parameters to optimize sensor performance and the miniaturization of the sensor's data acquisition system.

Acknowledgments

This work has been funded by the European Union - NextGenerationEU under the Italian Ministry of University and Research (MUR) National Innovation Ecosystem grant ECS00000041 - VITALITY. We acknowledge Università degli Studi di Perugia and MUR for support within the project Vitality.

References

[1] Mohammad Reza Khosravani and Tamara Reinicke. "3D-printed sensors: Current progress and future challenges". In: *Sensors and Actuators A: Physical* 305 (2020), p. 111916. ISSN: 0924-4247. DOI: https://doi.org/10.1016/j.sna.2020.111916. URL: https://www.sciencedirect.com/science/article/pii/S0924424720300868 (cit. on p. 1).

[2] S K Dhinesh and K L Senthil Kumar. "A Review on 3D Printed Sensors". In: *IOP Conference Series: Materials Science and Engineering* 764.1 (Feb. 2020), p. 012055. DOI: 10.1088/1757-899X/764/1/012055. URL: https://dx.doi.org/10.1088/1757-899X/764/1/012055 (cit. on p. 1).

[3] Rayappa Shrinivas Mahale et al. "Sensor-based additive manufacturing technologies". In: (2022) (cit. on p. 1).

[4] Yijie Jiang et al. "Recent advances in 3D printed sensors: materials, design, and manufacturing". In: *Advanced Materials Technologies* 8.2 (2023), p. 2200492 (cit. on p. 1).

[5] S. Shin, B. Ko, and H. So. "Structural effects of 3D printing resolution on the gauge factor of microcrack-based strain gauges for health care monitoring". In: *Microsyst Nanoeng* 8.12 (2022), pp. 891–921 (cit. on p. 1).

[6] Sudarsana Jena and Ankur Gupta. "Embedded sensors for health monitoring of an aircraft". In: *Sensors for Automotive and Aerospace Applications* (2019), pp. 77–91 (cit. on p. 1).

[7] Corey Shemelya et al. "Multi-functional 3D printed and embedded sensors for satellite qualification structures". In: *2015 IEEE SENSORS*. 2015, pp. 1–4. DOI: 10.1109/ICSENS.2015.7370541 (cit. on p. 1).

[8] Nuwan Munasinghe et al. "3-D Printed Strain Sensor for Structural Health Monitoring". In: *2019 IEEE International Conference on Cybernetics and Intelligent Systems (CIS) and IEEE Conference on Robotics, Automation and Mechatronics (RAM)*. 2019, pp. 275–280. DOI: 10.1109/CIS-RAM47153.2019.9095826 (cit. on pp. 1, 3).

[9] Xinhua Yao et al. "Evaluation of carbon fiber-embedded 3D printed structures for strengthening and structural-health monitoring". In: *Materials Design* 114 (2017), pp. 424–432. ISSN: 0264-1275. DOI: https://doi.org/10.1016/j.matdes.2016.10.078. URL: https://www.sciencedirect.com/science/article/pii/S0264127516313879 (cit. on p. 2).

[10] Pedro M. Ferreira et al. "Embedded Sensors for Structural Health Monitoring: Methodologies and Applications Review". In: *Sensors* 22.21 (2022). ISSN: 1424-8220. DOI: 10.3390/s22218320. URL: https://www.mdpi.com/1424-8220/22/21/8320 (cit. on p. 2).

[11] Agnese Staffa et al. "Development and Validation of a Low-Cost Device for Real-Time Detection of Fatigue Damage of Structures Subjected to Vibrations". In: *Sensors* 23.11 (2023). ISSN: 1424-8220. DOI: 10.3390/s23115143. URL: https://www.mdpi.com/1424-8220/23/11/5143 (cit. on p. 2).

[12] Juraj Beniak et al. "Resistance and Strength of Conductive PLA Processed by FDM Additive Manufacturing". In: *Polymers* 14.4 (2022). ISSN: 2073-4360. DOI: 10.3390/polym14040678. URL: https://www.mdpi.com/2073-4360/14/4/678 (cit. on p. 2).

[13] Riaan Stopforth. "Conductive polylactic acid filaments for 3D printed sensors: Experimental electrical and thermal characterization". In: *Scientific African* 14 (2021), e01040. ISSN: 2468-2276. DOI: https://doi.org/10.1016/j.sciaf.2021.e01040. URL: https://www.sciencedirect.com/science/article/pii/S2468227621003410 (cit. on p. 2).

[14] G Andria et al. "Design and characterization of innovative 3D printed embedded strain gauges". In: *2019 IEEE 5th International Workshop on Metrology for AeroSpace (MetroAeroSpace)*. IEEE. 2019, pp. 54–59 (cit. on p. 2).

[15] Gooding J. and Fields T. "3d printed strain gauge geometry and orientation for embedded sensing". In: *58th AIAA/ASCE/AHS/ASC Structures, Structural Dynamics, and Materials Conference*. 2017, p. 0350 (cit. on p. 2).

[16] Del Prete Z. Mileti I. Cortese L. and Palermo E. "Reproducibility and Embedding Effects on Static Performace of 3D Printed Strain Gauges". In: (2021), pp. 499–504. DOI: 10.1109/MetroInd4.0IoT51437.2021.9488430 (cit. on p. 2).

[17] Anna Maria Lucia Lanzolla et al. "Additive manufacturing for sensors: Piezoresistive strain gauge with temperature compensation". In: *Applied Sciences* 12.17 (2022), p. 8607 (cit. on p. 2).

[18] Md Taibur Rahman et al. "3D printed high performance strain sensors for high temperature applications". In: *Journal of Applied Physics* 123.2 (2018) (cit. on p. 2).

[19] Marco Maurizi et al. "Dynamic Measurements Using FDM 3D-Printed Embedded Strain Sensors". In: *Sensors* 19.12 (2019). ISSN: 1424-8220. DOI: 10.3390/s19122661. URL: https://www.mdpi.com/1424-8220/19/12/2661 (cit. on p. 2).

[20] Giulia Morettini et al. "Comprehensive characterization of mechanical and physical properties of PLA structures printed by FFF-3D-printing process in different directions". In: *Progress in Additive Manufacturing* 7 (Mar. 2022), pp. 1–12. DOI: 10.1007/s40964-022-00285-8 (cit. on p. 3).

[21] *Sito Web ProtoPasta*. URL: https://www.kickstarter.com/projects/1375236253/electrically-conductive-pla-3d-printer-filament (cit. on p. 3).

IOP Conf. Series: Materials Science and Engineering 1306 (2024) 012022 doi:10.1088/1757-899X/1306/1/012022

Structural and dynamic assessment of a friction buffer stop

G Megna, A Bracciali

Department of Industrial Engineering (DIEF), Università degli studi di Firenze, Via S. Marta 3, 50139, Firenze

gianluca.megna@unifi.it

Abstract. Under peculiar conditions railway braking system ca fail giving insufficient braking force during train ride or while standstill and buffer stops are used to avoid derailments at the end of railway tracks (dead tracks). Fixed stops are suitable only for quasi-static movements, while absorbing energy stops using hydraulic cylinders, friction shoes clamped on the rail head or a combination of both methods, are installed if crashes at higher speeds (about 25 km/h) are forecasted. Impact forces generated due to the interaction between train and buffer stops require a suitable design of the steel structure which is subjected to stress values depending on the braking performances of the buffer stop itself and a combined structural and dynamic analysis must be carried out to perform a proper design. Due to unavailability of detailed models and the uncertainties of the actual braking performances, the structure of currently available friction buffer stops is heavy and bulky, made of several steel profiles welded or bolted to each other. The paper describes an innovative and optimized design for friction buffer stops focusing on the dynamic analysis performed by Finite Element Method to assess the braking performances and structural strength of the buffer stop.

1. Introduction

As the braking system is designed with the aim to reduce passenger discomfort (low deceleration values) and to avoid wheel slip in almost all wheel-rail contact conditions, railway vehicles are characterized by high stopping distances. The braking force is evaluated considering a maximum available adhesion of f=0.15 [1] and anti-slip systems are installed on board to modulate the braking force in case of low adhesion conditions.

However, in certain situations, as for the *"leaves on the line"* case, which is a common issue in autumn especially in Great Britain, wheel slip cannot be avoided, furtherly increasing the stopping distances [2]. In other cases, trains can move from standstill position, especially if very long-time stops are planned, due to brake failure and insufficient immobilization given by handbrake or shoes place under the wheels.

Braking failures must be carefully evaluated and mitigated as they are directly related to safety involving collisions due to SPADs (Signals Passed At Danger), station overruns and derailments at dead end tracks and low adhesion conditions are still deeply studied [3-5]. To avoid derailments at the end of the tracks, an increasing number of buffer stops is installed on railway lines.

Buffer stops can be fixed, which are directly bolted to the rails and can withstands only quasi-static movements, or with absorbing energy systems if crashes at higher speed must be mitigated. Hydraulic cylinders, friction shoes clamped to the rails or a combination of both are the common methods used for this kind of buffer stops (see figure 1).

(a) (b) (c) (d)

Figure 1. Example of buffer stops with different configurations: a) fixed; (b) hydraulic; (c) friction; (d) hydraulic and friction.

The choice of the buffer stop depends on the design requirements in terms of vehicle mass m and speed v, resulting in the total kinetic energy to be dissipated according to equation (1).

$$E = \frac{1}{2} m v^2 \tag{1}$$

Depending on the combination of mass and speed the maximum energy to be dissipated is defined. For example, the Italian requirements for friction buffer stops [6] are described in table 1. In both cases the maximum deceleration rate is fixed at 2 m/s^2. Different specifications exist in other countries, especially for vehicle mass which can reach up to 3500 t [7], but in the following only the Italian standard is considered.

Table 1. Requirements of friction buffers stops according to Italian standard [6]

Load case	Mass [t]	Speed [km/h]	Energy [kJ]	Stopping distance [m]
1	650	15	5700	10
2	500	10	2000	5

About structural strength, buffer stops must be designed to survive for at least 40 years with one major impact (load case 1) and six medium impacts (load case 2) per year.

2. Design and limitations of existing friction buffer stops

Two main drawbacks are present in the current design of friction buffer stops:

- braking blocks (clamps) are made of several components and are clamped to rail head with several uncertainties about mounting position and preload conditions;
- reinforcing elements are needed to limit track lifting and rail stress in case of major impacts.

These issues increase the complexity of the buffer stop installation and the uncertainty of its dynamic performances, which are estimated by simple analytical models in which the braking forces exerted by the clamps are based on simplified assumptions.

A typical clamp is shown in figure 2(a). Friction forces are generated by the sliding of a cast iron block on the rail head while the preload is provided by two bent steel fishplates fastened by three M24 bolts. Two brass sliding chairs complete the assembly on the lower faces of the rail head. The mounting position is not unique as it suffers from uncertainties due to tolerances as shown in figure 2(b), and a proper calculation of the preload force and therefore of the friction force is not possible. Phosphoric P10 cast iron, i.e. the same material used on vehicles braking shoes [8] is used.

(a) (b)

Figure 2. Existing braking element (clamp) for friction buffer stop (a) and detail of its mounting position on the rail head (b).

The large amount of energy to be dissipated results in high impact forces and due to the height difference between the impact force and the braking blocks, about 1000 mm, an upward vertical force on the back of the buffer stop is generated. This force is such that the rails are raised, as the track in its common configuration, i.e. ballasted track, does not provide any resistance to upward movements. A prediction of the vertical displacement of the rail and therefore the generated stress is not easy to perform without the use of realistic models including the complete buffer stop and its interaction with the track.

Therefore, in the current design a reinforcing system of the track is required for major impacts, as shown in figure 3. The stiffening of the track consists in placing below the rail other steel beams, or another rail, connected to the main one by special connections. The buffer stop construction and installation costs are therefore noticeably increased. The simplified analytical models used to calculate the stress on the rails do not consider the dynamic effects of the impact and the actual contribution of the track components, such as the sleepers inertia, and the reaction forces ahead and back to the buffer stop. The reaction force ahead is given by the buffer stop itself, while the reaction force on the back is given by the vertical wheel load of the impacting vehicle. The design of buffer stops for load case 1 and for load case 2 is completely different, with resulting higher costs.

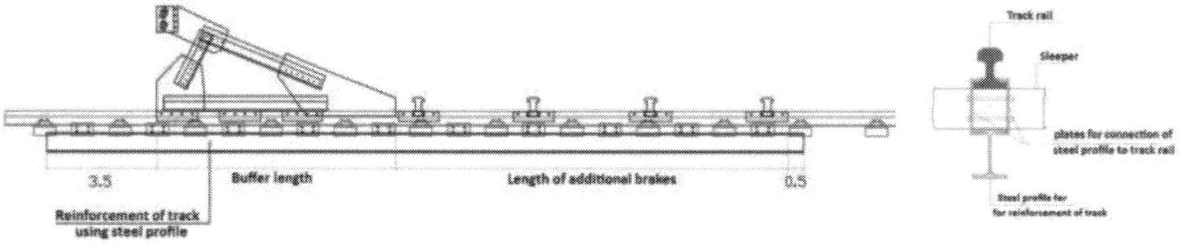

Figure 3. Example of track reinforcement [7]

The aim of this work is to design a new buffer stop with a simplified structure capable to satisfy all the requirements of the Italian standard [6] with a single configuration that does not require track reinforcements. For this scope a detailed model of the buffer stop and the track was developed including the actual friction behaviour of the clamps that were modified to reduce the number of components and to simplify the preload calculation.

3. Modelling and analysis of the new friction buffer stop

3.1. Description of the new buffer stop structure

The new buffer stop is shown in figure 4. Its innovative configuration consists in a main beam *a)* with the interfaces for the interaction with either the vehicle's buffers [9] or the central automatic coupling device [10]. The impact force is transferred from the buffer stop structure to the rail, where the clamps *c)* are fastened, by means of a strut made of a tubular steel beam *b)*.

The moment generated by the height difference is balanced by a vertical force passing through the rod *d)*, which is responsible for the vertical upward movement of the rail. This force is split in two identical components acting at 2 m distance to each other by a triangular structure *e)* made of two steel sheets clamped without preload to the rail head. All the components are connected by cylindrical joints, and the mechanism is closed by two rods *f)* connecting the front and the rear of the buffer stop. The structure is completed by two crossing steel ropes *g)* connecting the two sides of the buffer stop to cope with any possible asymmetric braking conditions.

The clamp is designed with only two cast iron components, connected by two M16 blots, with a rail-like shape to generate a pressure on the rail head surface when the bolts are preloaded. The forces acting on the clamp during the preloading phase are shown in figure 5(a) while the corresponding deformation is shown in figure 5(b). The preload is calculated considering a frictionless phase while the load of the bolts is applied, reaching the maximum value at the end the load application. The pressure values reached at the end of the preload application are shown in Figure 5(c). A friction coefficient of 0.2 is considered to reproduce the dynamic interaction between the cast iron clamp and the steel rail head, reaching a braking force of about 90 kN, which is nearly double than the one in the current configurations.

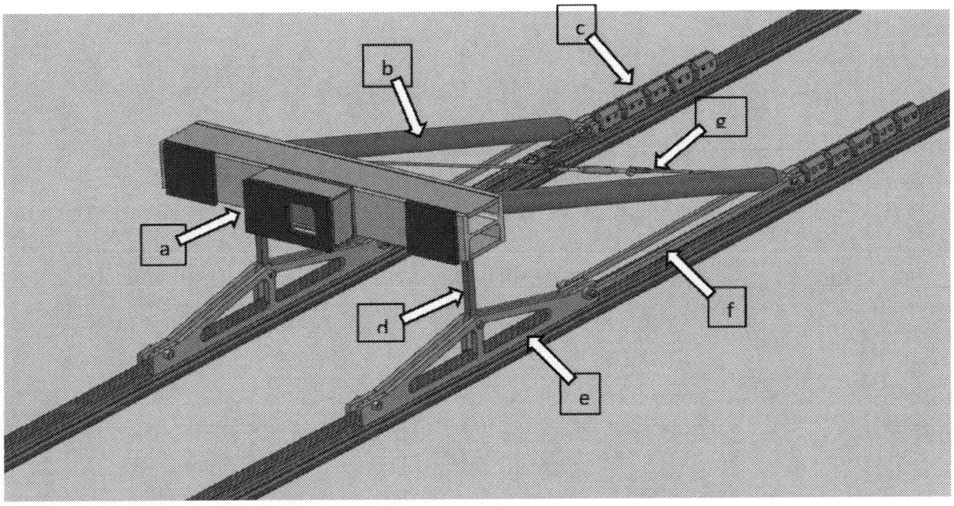

Figure 4. Description of the structure of the new friction buffer stop

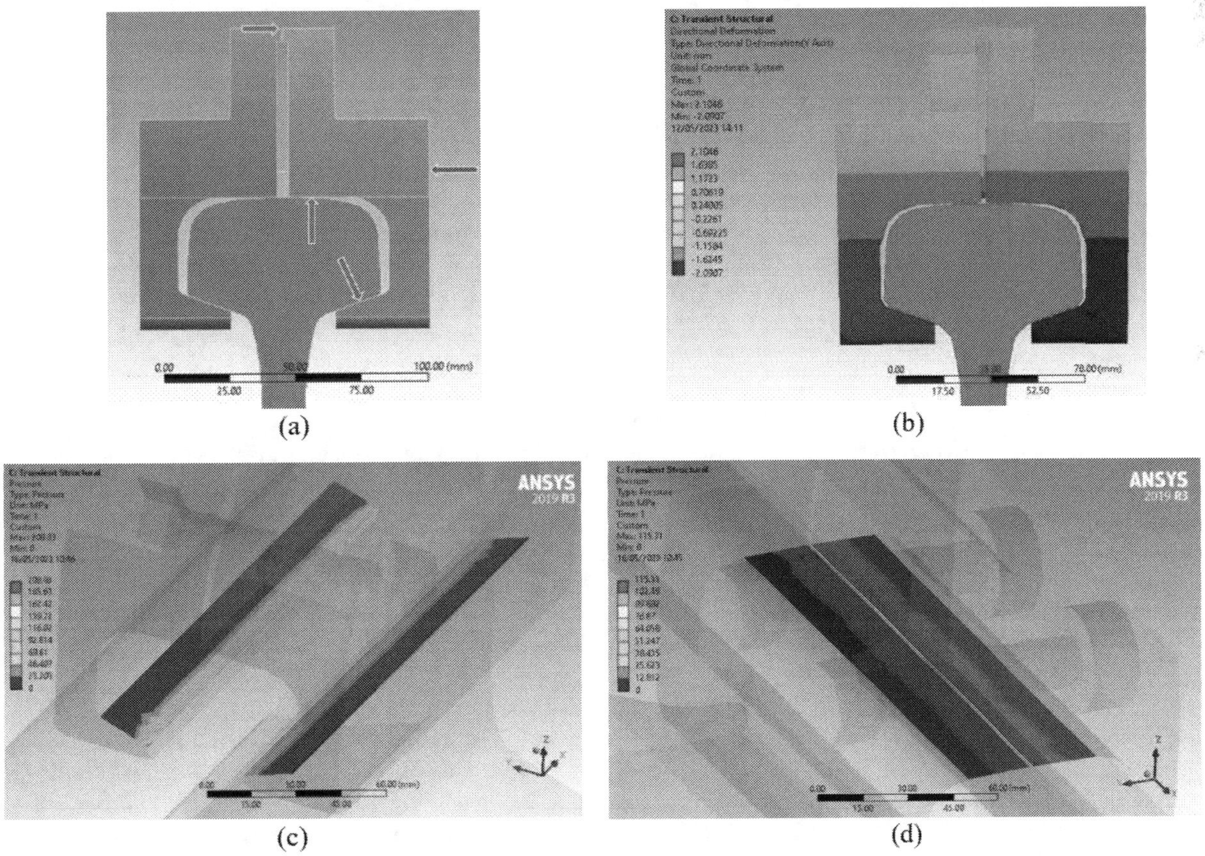

Figure 5. Forces acting on the clamp during the preloading phase (a). Displacements of the two cast iron blocks after the preload application to the bolts (b). Pressure map at the bottom faces of the rail head (c) Pressure map at the top of the rail head (d).

3.2. Dynamic model description

A strongly nonlinear dynamic model of the buffer stop was developed to simulate the complete interaction between the buffer stop and the track during the transient phases of the impact until a complete stop.

A general view of the model is shown in figure 6(a), which represent a 24 m long track section including 18 m of rail modelled with solid elements. When symmetric braking conditions are analysed only half model is considered to reduce the computational time. Sleepers are represented by point masses of 150 kg attached to the rail, while the vertical support of the rail is made of non-linear connections with k_z=0 for upward movements (free) and k_z=100 kN/mm for downward movements (reacted by the ballast).

Upward movements of the track are limited by inertial forces, by the elastic properties of the rail and by a downward vertical force due to the axleload of the first wheel of the vehicle hitting the buffer stop. The latter moves together with the buffer stop, locking their longitudinal connection at the impact to keep their distance constant (3 m). This kind of modelling allows to find the actual deformation of the track during the impact, evaluating the need of reinforcements if its strength is not sufficient. The rail is free to move longitudinally at the end of the track, while the lateral displacements and the longitudinal (about x axis) and vertical (about z axis) rotations are constrained at the sleeper positions (sleeper span = 600 mm).

The impact is simulated with two solids (one with zero mass simulating the vehicle buffer plate and one including the whole hitting mass) longitudinally connected by a flexible element representing the elastic behavior of the vehicle buffer when compressed. The non-linear stiffness of the vehicle buffer according to [9] was simplified with a bilinear curve for a maximum deformation of 100 mm, as shown in figure 6(b) and (c). The position of the masses, which is fixed during the whole simulation, is given by y=1750/2=875 mm and z=1050 mm. The contact between the first mass and the buffer stop beam was modelled as a frictionless contact with all the degrees of freedom free, except for the longitudinal displacement which is locked in the moment of the impact. Although, this simulates an inelastic collision between the impact masses and the buffer stop, the vehicle buffer mass is negligible compared to the hitting masses with no consequences on the total kinetic energy.

The distance between the masses and the buffer stop is 500 mm and the simulation is divided in two steps: in the first step the masses cover the 500 mm distance and at the same time the bolt preloads are applied to the clamps, while in the second step the impact occurs, and the braking force is applied by the frictional contacts between the clamps and the rail. The initial distance between clamps is 50 mm and they are pushed with frictionless contacts between their front and rear surfaces. The first clamp connected to the strut by a hinge is not preloaded, but it generates the vertical components of the reaction force and a frictional contact between the clamp and the rail head is also modelled contributing to the total braking force.

The vertical reaction force is balanced by the rod below the main beam which is connected to the triangular structure that splits the force by contact surfaces at the bottom faces of the rails. Friction is exerted also on these surfaces with a friction coefficient of 0.1 due to the absence of preload. The vertical rod is connected by hinges to the strut at its top and to the triangular structure at its bottom. The mechanism is completed by two rods connecting the front and the rear part of the buffer stop.

A complete model was developed to check the buffer stop behavior and the cross-bracing connections in case of asymmetric braking forces, as shown in figure 7 in which one clamp preload was removed on one side of the buffer stop.

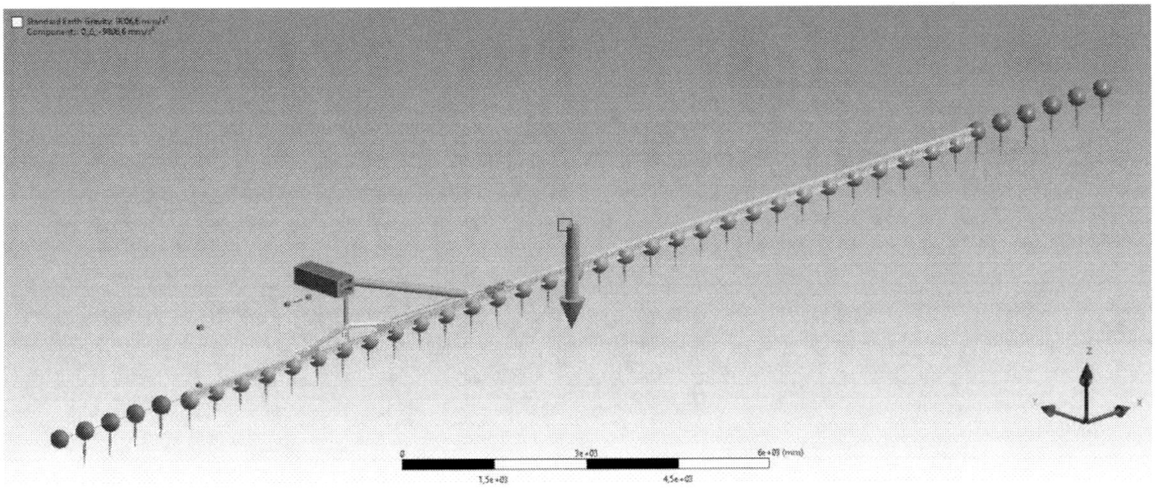

(a)

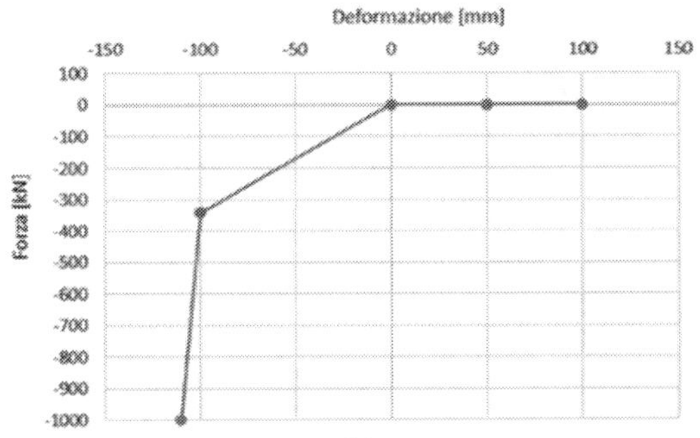

(b)

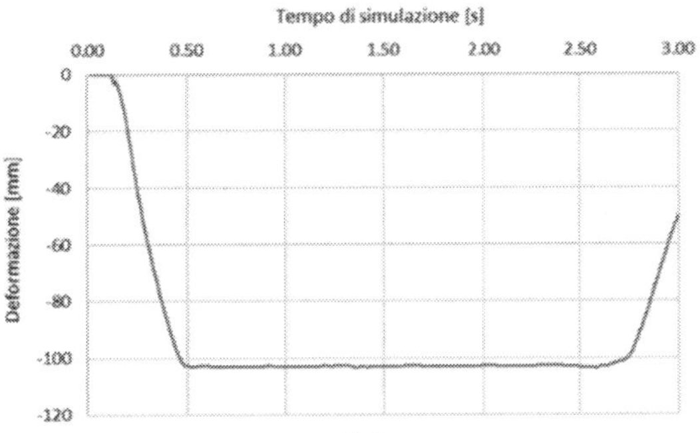

(c)

Figure 6. General view of the dynamic model (a). Force-displacement curve of the elastic element between the impact masses (b). Longitudinal deformation of the elastic element during a simulation.

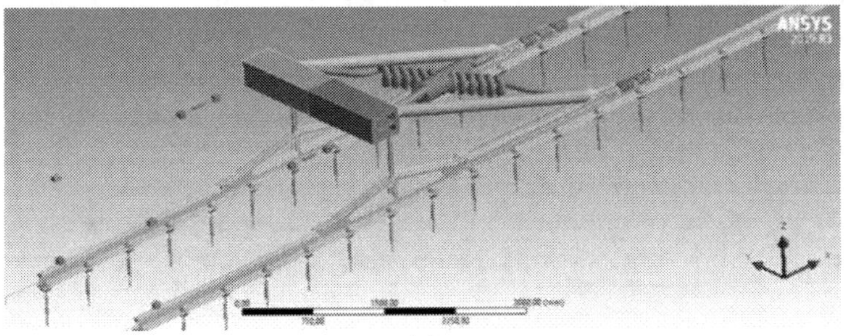

(a)

(b)

Figure 7. Complete buffer stop model including the crossing connections (a). Asymmetric braking conditions due missing preload on one side (b).

3.3. Results of the dynamic analyses

Several analyses were performed varying the impact mass and the speed to simulate the load cases according to table 1. The friction coefficient, the number of clamps and the wheel load were varied as well. Two vertical loads were considered in the case of empty (Q=50 kN) and laden vehicle (Q=100 kN). The input data and the main results for the simulated scenarios are shown in table 2.

Table 2. Input data and main results of the simulated scenarios

Input data					Main results		
Load case	# Left clamps	# Right clamps	Friction coefficient [-]	Wheel load [kN]	Stopping distance [m]	Acceleration [m/s²]	Rail lift [mm]
1	5	5	0.2	50	5.0	-1.7	160
1	5	5	0.2	100	5.0	-1.7	60
1	5	5	0.1	100	10.0	-0.8	21
1	3	3	0.2	100	9.0	-0.9	28
1	2	3	0.2	100	9.5	-0.9	28
2	2	2	0.2	100	4.5	-0.8	16

Results show that the requirements in terms of stopping distance and maximum deceleration are all satisfied. In the case of five clamps per rail, the resulting braking force is 550 kN per side of the buffer stop and the force that lifts the rail is 180 kN which is equally split at the two ends of triangular structures. The resulting force passing through the strut is 580 kN.

The maximum lift of the rail, i.e. 160 mm, is obtained in the unrealistic case in which load case 1 is combined with a stopping distance of 5 m only and the vertical wheel load is barely Q=50 kN. This deformation value is still compatible with the sleeper's height (common values are 180 and 220 mm), that remain in their position within the ballast, while the maximum stress on the rail is lower than the yield strength of the R260 rail steel (560 MPa). Some results from simulation are shown in figure 8. This confirms that the design is capable to keep both rail deformations and stresses low without the need to stiffen and reinforce the rails.

As maximum forces are exerted in the case with five clamps per rail, the stresses on the buffer stop components are evaluated only for that case and compared with the yield strength of the steel used (355 MPa). For the P10 cast iron used for the braking blocks, their Brinell hardness in the range of 197~255 HB corresponds to a tensile strength of 350 MPa according to 0. As cast iron does not shown relevant plastic deformation this value can be considered as the yield stress.

During the life span of the buffer stop (40 years), the structure will be subjected to a maximum number of 280 load cycles, which is lower than the commonly used minimum value (1000) that requires to consider fatigue. A static stress evaluation is therefore sufficient. Equivalent stresses calculated with the Von-Mises criterion for all the components of the buffer stop are shown in figure 9. A buckling analysis of the strut was performed to check potential stability issues due to the compressive force obtaining a safety factor of about 3.5.

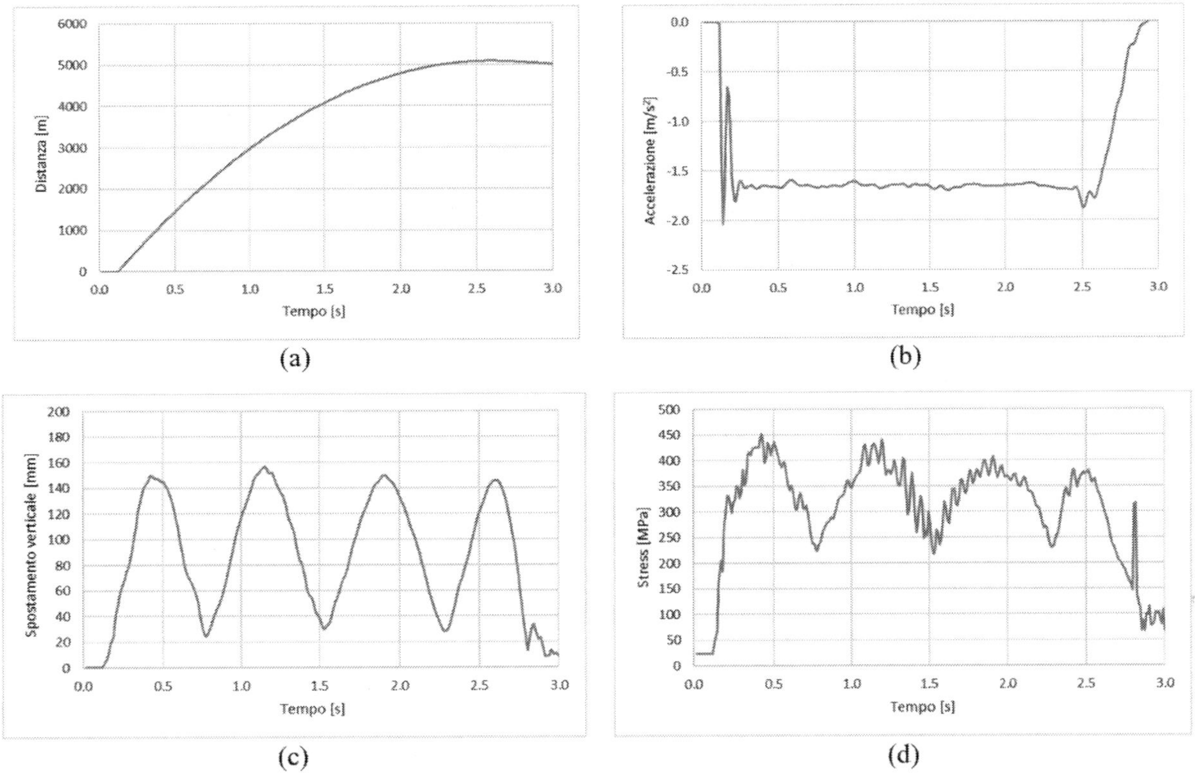

Figure 8. Results from the most unfavourable dynamic simulation: stopping distance (a), braking acceleration (b), rail lift (c) and rail stress (d).

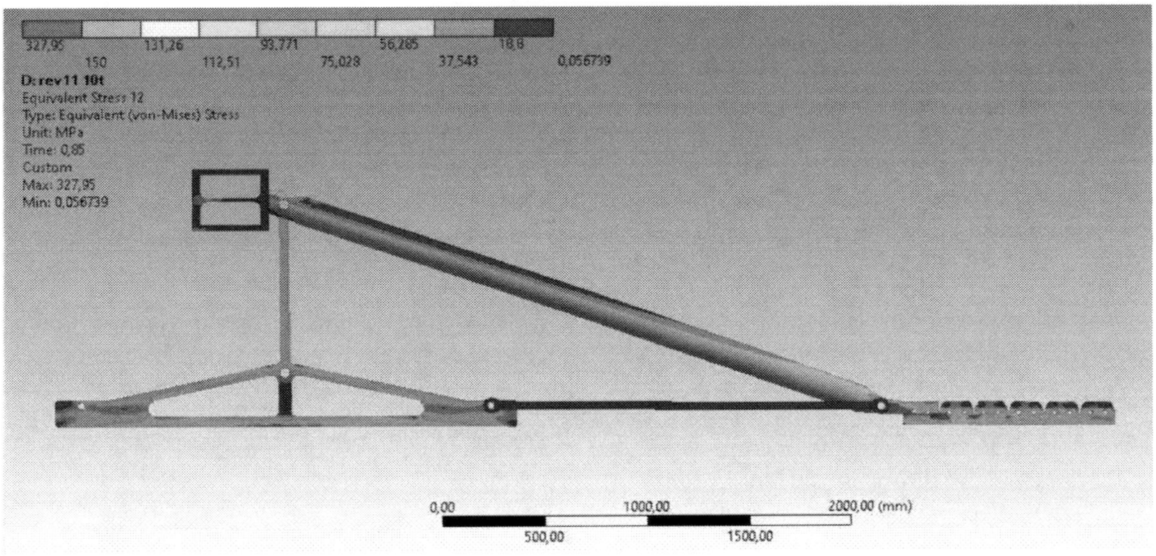

Figure 9. Von-Mises stresses as results of the dynamic simulation of the worst case.

IOP Conf. Series: Materials Science and Engineering 1306 (2024) 012022 doi:10.1088/1757-899X/1306/1/012022

4. Conclusions

Energy dissipation (friction) buffer stops are widely used at dead end tracks to mitigate the consequences of failed/insufficient braking of moving or theoretically standstill trains. Friction exerted at the interface between clamps and the rail head is used to dissipate the kinetic energy of the moving train hitting the buffer stop. Friction buffers stops are cheaper than hydraulic buffer stops and can absorb a large amount of energy depending on the number of clamps and their preload. Existing friction buffer stops are bulky and made of complex and expensive welded and/or bolted steel profiles. Installation costs are further increased by the need of reinforcing/stiffening the track to reduce the risk of rail failure. Moreover, the dynamic performances of buffer stops are not properly assessed considering the full interaction between the hitting mass, the buffer stop structure, the clamps and the track structure.

In this paper an innovative friction buffer stop was introduced, and the FEM model used to verify its dynamic behaviour was described. The results of the simulation showed that this simple and low-cost structure is suitable to withstand the impact forces resulting from the load cases of the Italian standard, which describes two different configurations. While existing designs need a special track reinforcing system in the case of major impacts, the complete model of the new buffer stop, of the track with the track and of the hitting vehicle shows that any damage is prevented. The simpler design of the clamps led to a more accurate estimation of the braking forces and therefore to a better verification of the dynamic performances of the buffer stop in terms of stopping distances and braking decelerations.

Further modelling improvements could be made considering the impact mass as a set of masses connected by elastic elements instead of a single mass. This could help to validate the behaviour of the buffer stop when the colliding mass is a real train. The peculiar non-linear behaviour of the connections between vehicles, i.e. screw and vehicle buffer, needs specific modelling strategies for an accurate assessment of the longitudinal dynamic of railway vehicles. This topic will be covered in a future paper.

References

[1] UIC 544-1. Brakes – Braking power. 2004

[2] Network Rail. Leaves on the line. https://www.networkrail.co.uk/running-the-railway/looking-after-the-railway/delays-explained/leaves/ (accessed 19.09.2023)

[3] White B, Watson M, Lewis R. A year-round analysis of railway station overruns due to low adhesion conditions. Proceedings of the Institution of Mechanical Engineers, Part F: Journal of Rail and Rapid Transit. 2023 Apr 1;237(4):458–69.

[4] Bosso N, Gugliotta A, Magelli M, Oresta IF, Zampieri N. Study of wheel-rail adhesion during braking maneuvers. Procedia Structural Integrity. 2019;24:680–91.

[5] Trummer G, Buckley-Johnstone LE, Voltr P, Meierhofer A, Lewis R, Six K. Wheel-rail creep force model for predicting water induced low adhesion phenomena. Tribology International. 2017 May 1;109:409–15.

[6] Rete Ferroviaria Italiana. Specifica Tecnica di Fornitura – Paraurti ad azione frenante. 2021.

[7] Israel Railways LTD. Technical specification for manufacture and supply of friction buffer-stops, fixed buffer stops and braking wheel stop sets. 2020.

[8] UIC 832. Technical specification for the supply of brake-shoes made from phosphoric iron for tractive and trailing stock. 2004

[9] EN 15551. Railway applications – Railway rolling stock – Buffers. 2017

[10] EN 16019. Railway applications – Automatic coupler – Performance requirments, specific interface geometry and test method. 2014.
EN 1561, Founding – Grey cast irons, 2011.

IOP Conf. Series: Materials Science and Engineering 1306 (2024) 012023 doi:10.1088/1757-899X/1306/1/012023

Vulnerability-driven multi-objective topological optimization of aircraft protections

Salvatore Annunziata, Luca Lomazzi, Marco Giglio, Andrea Manes

Politecnico di Milano, Dipartimento di Meccanica, Via La Masa n.1, Milano, Italy

E-mail: luca.lomazzi@polimi.it

Abstract. Enhancing the survivability of platforms is paramount for increasing their safety level and limiting the loss of lives. This is particularly important in the military field, where aircraft vulnerability has been studied for decades through specifically developed frameworks. So far, most of the contributions have focused on improving the accuracy of methods for assessing the vulnerability to ballistic projectiles, mainly by advancing our understanding of the penetration of the threat through the components of the platform. Although state-of-the-art solutions allow conducting accurate assessments, the information provided by such methods has not been fully exploited yet for automatically designing corrective solutions and improving the survivability of platforms. For this reason, this work proposes a vulnerability-driven multi-objective topological optimization algorithm for designing protective surfaces. A genetic algorithm is used to describe the Pareto front delimiting the space of solutions characterized by optimal vulnerability and weight. The number of clusters that constitute the protections proposed by the optimization algorithm is minimized by adopting a spatial filter that penalizes fragmented designs. The framework is demonstrated through a case study involving a portion of a realistic remotely piloted aircraft system impacted by ballistic projectiles.

1. Introduction

Aircraft survivability is defined as the capability of a platform to avoid or withstand a man-made hostile environment, while vulnerability identifies the inability of a platform to withstand enemy threats. These features have played a crucial role in the design of aviation platforms since the early 20th century, notably during World War I [1]. Initially, protective measures were restricted to flying at altitudes beyond the reach of threats and integrating basic safeguards, such as stove lids beneath pilot seats, aimed at reducing vulnerability and enhancing the platform's ability to withstand man-made threats [1]. Significant advancements in combat survivability emerged during the Cold War conflict in Southeast Asia during the '70s. In this context, the high intensity threat environment necessitated a more comprehensive approach to vulnerability assessment. This led to the development of platforms whose design was mainly based on vulnerability considerations, such as the AH-64 and the A-10 [2, 3]. The '80s marked a further increase in the importance of survivability and vulnerability, culminating in the establishment of the Department of Defense Survivability/Vulnerability Information Analysis Center (SURVIAC) in 1984, which provided technical support in non-nuclear survivability [4].
Enhancing the survivability features of civil and military aircraft is still paramount in the design process to ensure the safety of occupants and mission success. This emphasis on survivability assessments during the design phase is exemplified by initiatives like the Joint Strike Fighter

program (F-35 aircraft) [1]. Furthermore, the principles of survivability and vulnerability extend beyond the aeronautic field and are commonly applied to the design of any platform susceptible to external threats. In this context, designers focus on both susceptibility, i.e., the inability of a platform to not be identified by the enemy defense, and vulnerability assessments.

Vulnerability assessments are tools used for identifying critical components in platforms, and eventually taking corrective actions. The concept of vulnerabilty assessments was originally formalized by R. Ball in 1985 for combat aircraft [1]. Building upon Ball's work, ref.[5] delved deeper into the review and application of methods for reducing vulnerability in long-range Strike Fighter aircraft. A comprehensive methodology for conducting detailed single-hit vulnerability assessments during the conceptual design stage of generic aircraft was presented in ref.[6], based on Ball's earlier research. Vulnerability assessments typically adopt a statistical approach, demanding a substantial amount of data to evaluate the impact of selected threats from various angles. To manage this data, computer programs such as the FASTGEN and COVART software packages were developed [7, 8]. The FASTGEN software implemented a shotline-based procedure, which, when used as input for the COVART software, enabled the determination of critical indices in the vulnerability assessment framework, as extensively explained in ref.[9, 10]. It is worth noting that the shotline-based procedure employed in FASTGEN software has roots in the consolidated Ray-Tracing algorithms. The vulnerability concepts pioneered by Ball were further explored in ref.[11]. Here, a computable calculation model based on the shotline scanning method was introduced. This model had the capability to account for the threat's state of motion during the penetration process in a simplified manner. Subsequently, the authors refined the shotline method and introduced the evaluation of the platform's kill state to assess its vulnerability in a single-hit scenario, taking into consideration the potential overlap between components [11]. In ref.[12], a direct simulation method was introduced to assess the multiple-hit vulnerability of aircraft with overlapping components. This approach was based on the statistical framework of Monte Carlo simulations proposed in ref.[13]. Furthermore, ref.[14] discussed a detailed methodology for vulnerability assessment in multiple-hit scenarios involving fragmentation warheads, evaluating the characteristics of missile fragments, such as dispersion, hit locations, penetration into target components, and the overall vulnerability of the aircraft using an associated fault tree. More recently, in 2021, a work establishing the general framework within which a vulnerability assessment may be performed was published, with a specific focus on ballistic vulnerability [15]. However, to the authors' best knowledge, an optimization procedure to reduce the vulnerability of platforms based on vulnerability assessments has not been proposed yet.

In this context, this work aims at proposing a vulnerability-driven multi-objective topological optimization tool to design protective components, proving that vulnerability-driven optimization is possible. Specifically, a multi-objective Genetic Algorithm (GA) was combined with the framework proposed in ref.[15] to generate the Pareto front of the possible protection layouts. The software proposes solutions with minimum weight and minimum vulnerability of the target component. To obtain continuous optimized solutions and reduce the number of clusters within the proposed protections defined by the optimization algorithm, a spatial filter was introduced to penalize fragmented design.

The paper is organized as follows. In Section 2, the methodology is introduced, which includes a description of the vulnerability assessment based on ref.[15] and of the proposed optimization algorithm. Section 3 is dedicated to the case study, focusing on a section of a remotely piloted aircraft (RPAS) impacted by projectiles. Finally, Section 4 draws out the conclusions of this work and presents possible future developments.

2. Methodology

The workflow followed to generate the optimal layouts of protective surfaces for protecting critical components is shown in fig. 1 and summarized as follows:

- **Initial vulnerability assessment**: This is the first phase of the proposed algorithm, where the vulnerability assessment to projectiles is conducted. The results obtained are useful to identify the vulnerable parts, being able to initialize the protection surfaces that subsequently will be optimized. The phases that lead to the computation of the vulnerability of the platform are detailed in sec. 2.1.
- **Grid initialization**: In this phase, the software initializes the protection grid, which will serve as input for the subsequent optimization process. Building upon the shotline grid created and used in the previous phase to calculate the vulnerability of the platform's components, the software strategically positions protective elements to cover the components that need protection. Section 2.2 provides a comprehensive explanation of the generation of the protection grid.
- **Optimization**: The initialized grid serves as an input of the optimization process. Multi-objective genetic algorithms are used to generate protective solutions with optimized topology, minimizing weight and the vulnerability of the target component. With the aim of converging to continuous optimized topologies, a spatial filter is incorporated within the optimization loop to penalize fragmented solutions. This process, detailed in sec. 2.3, yields a solution space characterized by optimal vulnerability and weight, delimited by the Pareto front.

2.1. Initial vulnerability assessment

The first step of the software developed in ref. [15] to perform vulnerability assessment to projectiles consists in performing pre-processing and platform modelling phases. Here critical components are identified and then modelled reducing to simply geometric shapes in .STL format, which consists of triangles covering the surfaces of the parts. For each critical component diverse metadata are collected, such as: the system the component belongs to, eventual redundancy, the material is composed of and the vulnerability value $P_{k|h}$, defined as the conditional probability that the component is killed given a hit by a threat.
After this phases, the trajectory of threats must be described accurately, considering a sphere centered around the platform, allowing threats from any direction, or limiting it to a specific spherical sector for a defined range of directions. In order to do this it is convenient to define the trajectories in terms of azimuth and elevation angles. These angles are typically referenced to the platform model reference system, as depicted in Figure 2 (a) and (b) for the azimuth and elevation angles, respectively. Each unique combination of azimuth and elevation angles is known as an aspect. To analyze each specific aspect, the critical components' models are projected onto a plane perpendicular to the aspect's defined vector. A shot-lines grid is then superimposed on this projection, following a two-step process: (i) placing the grid edges to include the whole projection of the model, (ii) dividing the grid into equal square cells. Within each cell a shot-line is generated starting from their geometric centers, perpendicular to the grid plane. To determine the triangles intercepted by each shot-line, a ray tracing algorithm is implemented. This algorithm consists of two main steps: (i) computing the intersection point between the current shot-line and the plane containing the current triangle, (ii) evaluating if the computed intersection point lies within the boundaries of the current triangle. For each shot-line, this two-step ray tracing procedure is applied to all triangles on the surface of the 3D models of the critical components. Shot-lines that do not intersect any component are discarded, while those that intersect at least one triangle are stored (Figure 3). The code determines and records

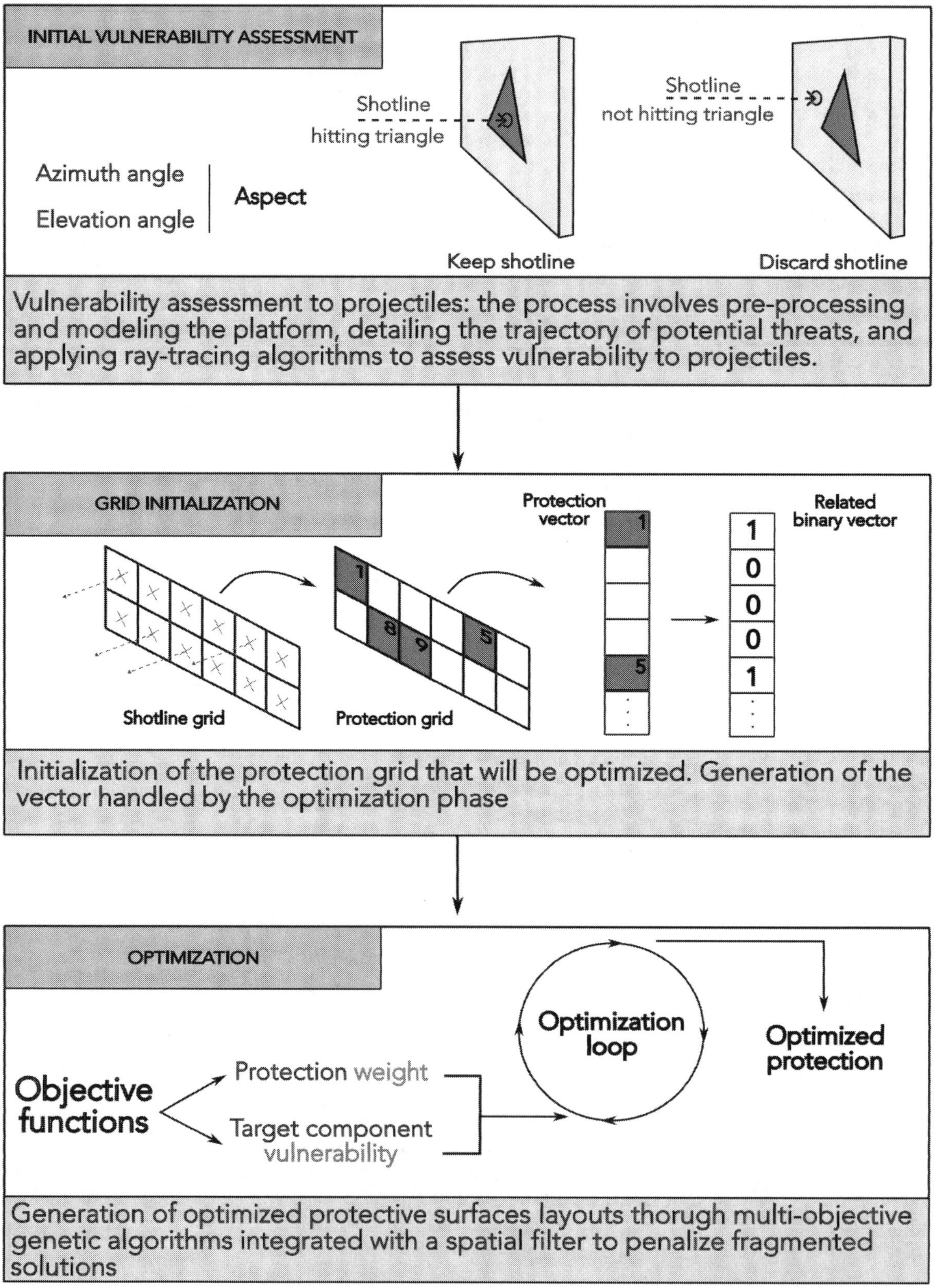

Figure 1. Algorithm block diagram

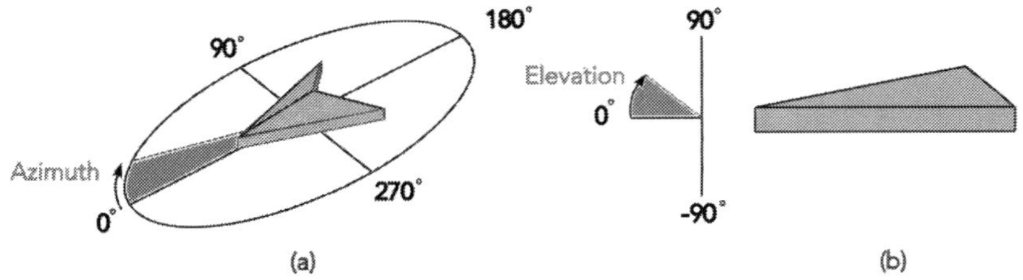

Figure 2. Aspect definition - azimuth angle (a) and elevation angle (b).

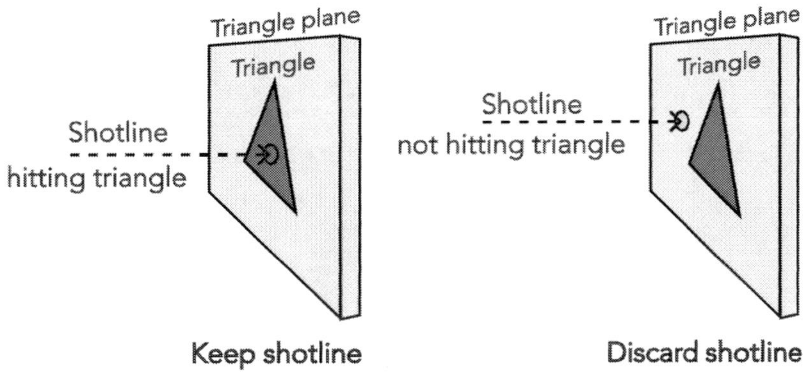

Figure 3. Ray tracing algorithm possible outcomes.

the complete list of intercepted components for each stored shot-line, saving this information in a structured variable. With all the data stored in the phases briefly explained before, now it's possible to perform the assessment of the vulnerability of components (that is of particular relevance for this specific study) and systems.
All the procedures previously illustrated are implemented in the in-house software presented in ref. [15] to perform the vulnerability assessment of a platform.
In this work these codes are exploited and modified to generate the protection structures that subsequently will be optimized. The new procedures will be described in the following paragraphs.

2.2. Grid initialization
The code, as showed in fig. 1, initialize a grid serving as the starting point for the shot-lines that will impact the target platform. Protective elements are then strategically positioned on this grid to cover the specific components requiring protection, according to the results obtained from the approach explained in sec. 2.1 (fig. 4). Once the grid designed to protect the selected component is generated, the implemented code will proceed to topologically optimize it, generating protective solutions with the least possible weight while ensuring optimal protection capabilities. Initializing the shot-lines grid, despite being recalculated during the vulnerability analysis, is necessary to determine the boundary conditions and the number of variables that the optimizer will handle. The protection is superimposed onto the initialized shot-lines grid, placing protective elements corresponding to the origins of projectile shot-lines impacting the target component. The code manage this initialized protection as a vector, with dimensions matching the number of elements in the shot grid, as shown in fig. 5. The elements within this vector, representing the protection, can assume values of either 0 or 1. Specifically, a value of 1 implies the placement of protective material at the corresponding grid element, while a value

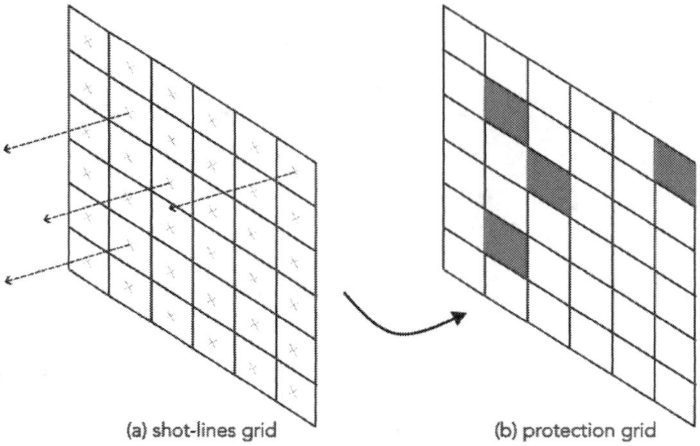

Figure 4. Shot-lines grid initialization and protection generation

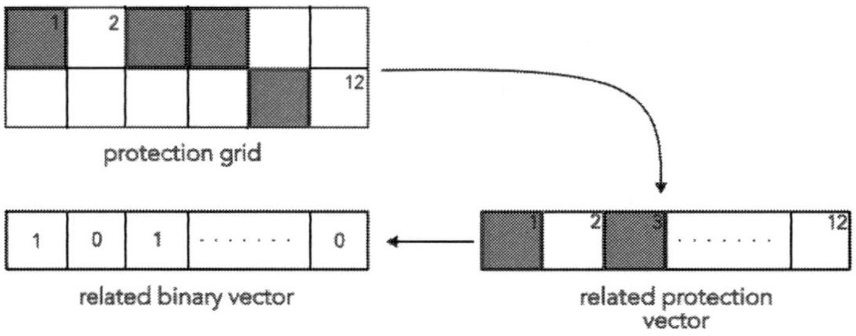

Figure 5. Binary protection vector generation

of 0 indicates its absence. It's worth noting that during this phase, a discontinuous protection with scattered elements is initialized to cover all the impacted areas of the target component.

2.3. Optimization

To optimize the protection and generate vectors with properly positioned unitary elements by minimizing the vulnerability of the target component and the weight of the protection, a multi-objective genetic algorithm is used. Two fitness functions are provided to achieve the optimal configuration: weight and vulnerability. To ensure the proper functioning of the genetic algorithm, it was necessary to express these two fitness functions based on the vector representing the protection, which is the object of optimization. The weight fitness function outputs the normalized weight of the protective plate by summing the number of unitary elements in the vector and dividing it by the total number of elements, as shown in eq. 1.

$$weight(x) = \frac{\sum_i^N x(i)}{N} \tag{1}$$

Where x is the protection binary vector and N is the number of elements of this vector, equal to the number of variables to be optimized.
Regarding the vulnerability of the target component, this is computed through the software described in sec. 2.1 and deeply explained in Ref. [15], modified so that this value also depends on the protection vector x. More specifically, in this version, the shot-lines grid is utilized to calculate the vulnerability using the given formulas but, when protective elements

are positioned on the grid, they effectively stop the corresponding shot-lines. Consequently, shot-lines originating from elements with protective material contribute to the vulnerability calculation as if they did not intersect any triangles of the .STL file, assuming that the protective measures have the capacity to completely stop the projectiles.
The following algorithm generates an optimized protection but may result in non-continuous solutions. To mitigate this possibility, a penalty factor is introduced as a filtering mechanism for the outcomes. This factor directly affects the fitness functions when the count of isolated elements exceeds one. Isolated elements refer to elements in the protection vector that have a value of 1 but are surrounded by elements with a value of 0. At each optimization iteration weight and vulnerability are computed in terms of the protection vector. Then the number of isolated elements are counted and the penalty factor is computed. If isolated elements are present, both fitness functions are multiplied by the penalty factor; otherwise, they remain unaltered. To ensure that the counting of isolated elements is performed every time the optimizer generates new solutions, this evaluation is integrated directly into the function generating the fitness functions. This means that for each newly generated solution, the penalty factor is applied based on the count of isolated elements, penalizing the corresponding fitness functions accordingly. By incorporating this counting mechanism into the optimization process, the algorithm consistently penalizes solutions with isolated elements and promotes the generation of more realistic outcomes. If the number of isolated structures is zero, the normal fitness functions are used.
In general, genetic algorithms are used to solve optimization problems based on natural selection, by repeatedly modifying a population of individual solutions. At each step, the genetic algorithm selects individuals from the current population to be parents and uses them to produce the children for the next generation. Over successive generations, the genetic algorithm is able to converge towards optimal solutions. This process yields a solution space characterized by optimal vulnerability and weight, delimited by the Pareto front.

3. Case study

The capabilities of the optimization framework were demonstrated through a case study considering the section of an RPAS shown in Figure 6. The demonstrator only included the outermost layer, i.e., the skin of the RPAS, and three representative critical components. The critical components were all modeled using thin-walled parallelepipeds with different dimensions and were numbered as shown in Figure 6. Although the proposed framework is able to automatically identify the most critical component to protect through an initial vulnerability assessment, the case study was conducted by skipping this first step. That is, C2 was selected as target component to protect, without conducting preliminary vulnerability analyses. This allowed saving computational resources, without limiting the validity of the case study. Two different aspects were considered: (i) 90° Azimuth and 0° Elevation, and (ii) 50° Azimuth and 30° Elevation.

3.1. Aspect 1: 90°-0°

Figure 7 shows the Pareto front of the results generated by the optimizer in the aspect 90° Azimuth and 0° Elevation. The plot represents the protection weight on the x-axis, and the vulnerability of the target component on the y-axis. The optimizer converged to the following four solutions:

- Solution 1: this solution represents the heaviest configuration among the four solutions. It offers the lowest vulnerability for the target component at the expense of increased weight.
- Solution 2: this layout is characterized by lower weight than solution 1, but higher vulnerability.

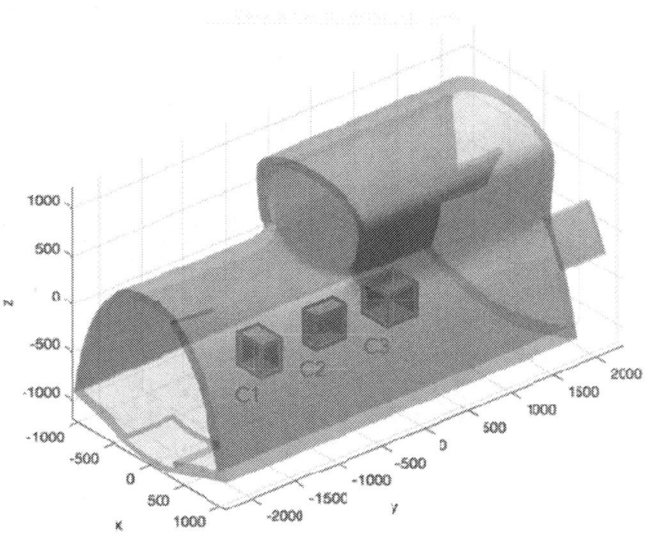

Figure 6. Demonstrator

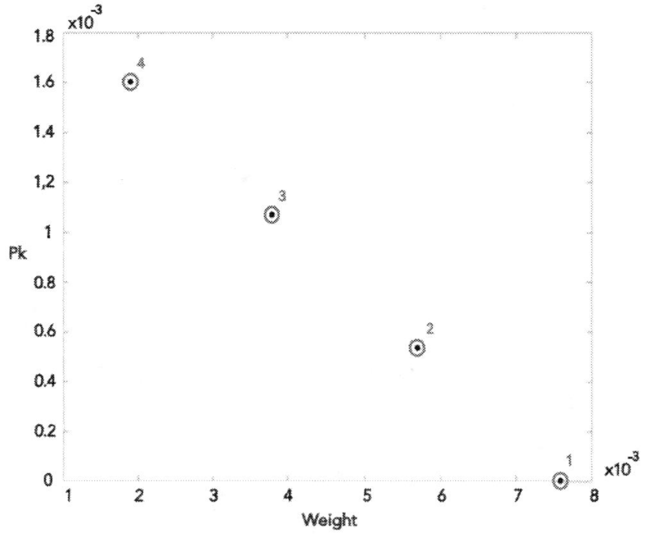

Figure 7. Pareto curve: Aspect 90° - 0°

- Solution 3: this layout is lighter than solution 2, but offers a reduced protection.
- Solution 4: this is the most lightweight configuration among the optimal ones, and only slightly reduces the component vulnerability with respect to the unprotected design.

These solutions represent different trade-offs between weight and vulnerability, allowing for flexibility in selecting the most suitable protection configuration based on specific requirements and constraints. The protection shown in Figure 8 represents Solution 1 of the Pareto front. Here, the shot-line grid is depicted as a grid of gray points, while the protection generated by the software is highlighted in red. It turns out that the layout is able to stop any projectile that may

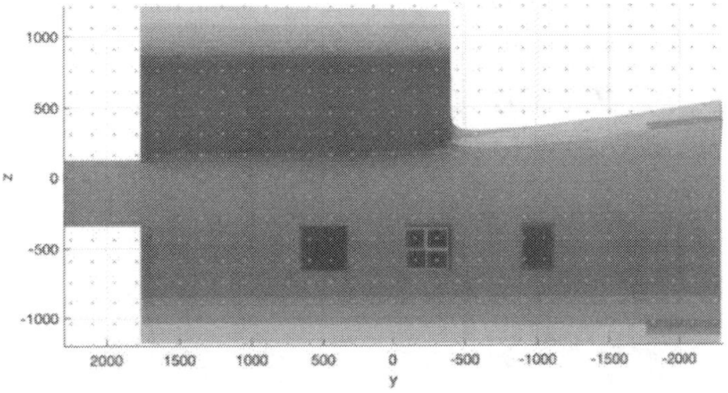

Figure 8. Protection solution 1

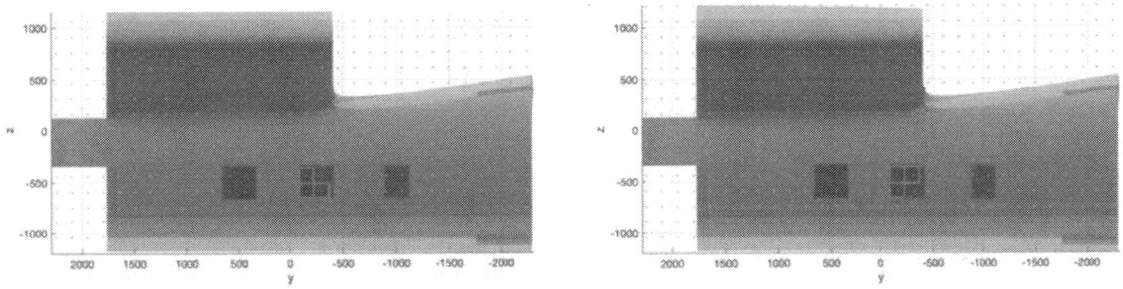

Figure 9. Protection solution 2

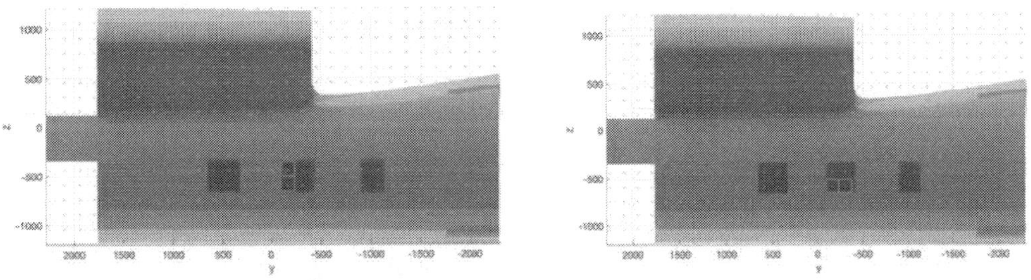

Figure 10. Protection solution 3

intersect component C2 in the aspect considered in the optimization algorithm, thus providing maximum protection. The layouts characterizing solution 2 are shown in Figure 9. Here, the optimizer generated two different topologies with the same weight and protection capability. In fact, the software generated a protection that blocks three out of the four shot-lines hitting the target component. Reducing the number of protective elements increased the vulnerability.
A similar reasoning can be applied to the protections proposed in solution 3 of the Pareto front, which are shown in Figure 10. Two solutions that differ only in terms of topology were generated, and therefore they were characterized by the same weight and vulnerability. In this case, the optimizer further reduced the number of protective elements by blocking only 2 out of the 4 shot-lines that intersect the target component. By doing so, the weight was significantly reduced, which is now halved compared to solution 1, and consequently the vulnerability of the target component was increased with respect to solutions 1 and 2.

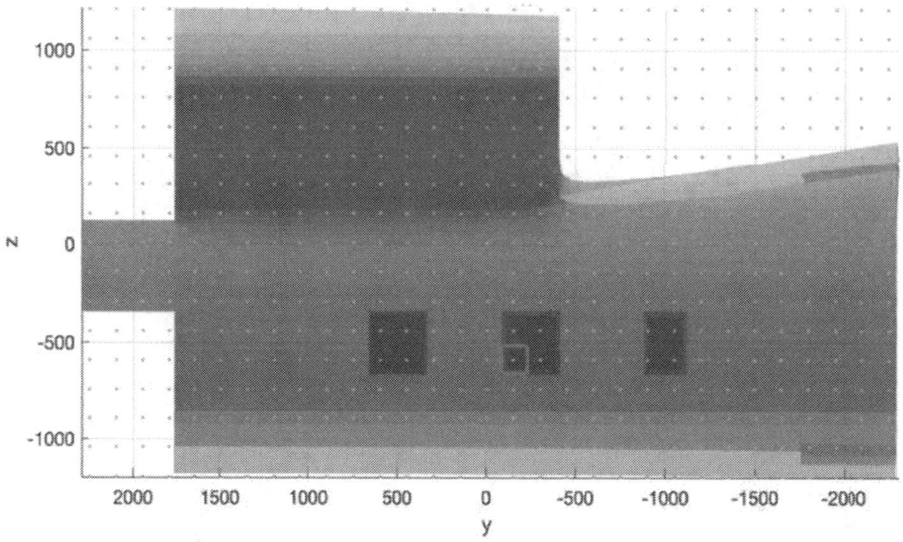

Figure 11. Protection solution 4

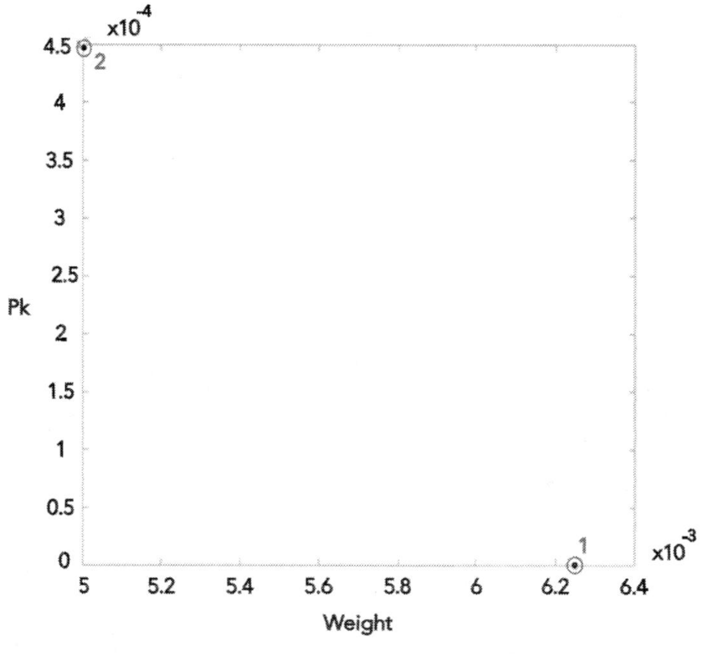

Figure 12. Pareto curve: aspect 50°-30°

Finally, the last generated solution is solution 4 of the Pareto front. In this case, the software produced a protective configuration that blocks only 1 out of the 4 shot-lines that intersect the target component, as shown in Figure 11. This results in a further reduction in weight compared to the previous solutions, at the expense of the vulnerability of the target component.

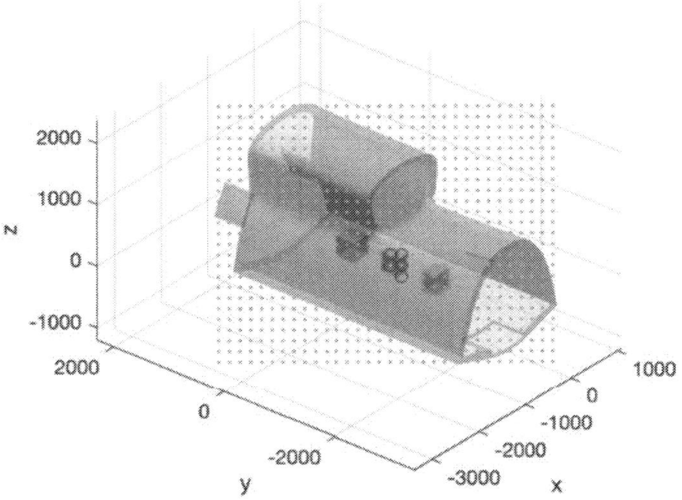

Figure 13. Protection solution 1

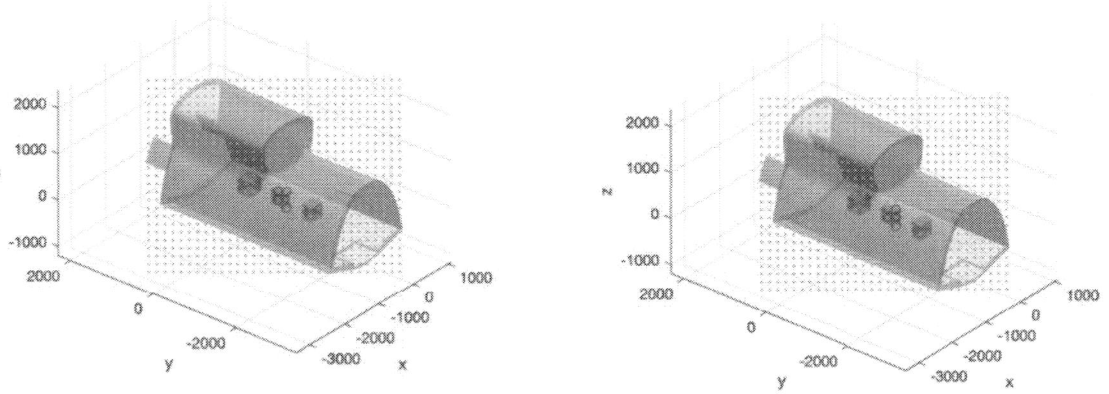

Figure 14. Protection solution 2

3.2. Aspect 2: 50°-30°

Figure 12 shows the Pareto front of the results generated by the optimizer in the aspect 50° Azimuth and 30° Elevation. Specifically, solution 1 zeroed the vulnerability of the target component, while solution 2 provided less weight increment, at the expense of a lower protection capability. The first protective solution is shown in Figure 13: it blocks all the shot-lines intersecting the target component. That is, the protection is formed by elements taken from the shot-line grid, aligned in accordance with the chosen aspect, and these elements correspond to the target component's projection onto the same oriented grid. The protections generated for solution 2 of the Pareto front are shown in Figure 14. Here, the optimizer generated lighter protections, resulting in an increased vulnerability of the target component. The solutions shown in Figure 14 differ from each other only in terms of the arrangement of protective elements.

4. Conclusions

The method proposed in this work aims to provide a general framework for generating optimized protective solutions that effectively mitigate platform vulnerability. The proposed methodology consists of performing the vulnerability assessment to projectiles, subsequently generating and optimizing the protective structure directly on the grid were the shot-lines originate from. The main conclusions drawn out from this work are as follows:

- To optimize protective structures, obtained through the vulnerability assessment, multi-objective genetic algorithms have been utilized. This represents a proof of concept that vulnerability-driven optimization is possible.
- To demonstrate the functionality of the implemented software, an application to a case study is proposed. Two different aspects were analyzed. The code was able to generate protective solutions that effectively blocked the shot-lines of the projectiles that would have hit the selected target component. Not only did it block all the shot-lines, resulting in a trivial solution, but it also provided alternative solutions by changing the topology of the protections, finding compromises between their weight and the vulnerability of the component, yielding a solution space delimited by the Pareto front. This demonstrates the capability of the proposed optimization framework to be applied in such a context.
- By implementing the filtering mechanism, the genetic algorithm converged to continuous and optimal solutions.
  As observed in case study section, the generated solution formed a continuous assembly of elements that intercepted the shotlines, thus further validating the efficacy of the implemented code in this context.

Different developments are still possible, multiple analyses could be conducted by generating different optimized protections for each aspect considered. Subsequently, these protections could be combined to create a protective structure around the target component, taking into account all the previously performed vulnerability assessments. Finally, the resulting protection could be tailored to the platform's outermost shape, thereby losing the direct connection to the grids from which the projectile shot-lines originate. Moreover, the code presented in this work could be applied to a more complex model compared to the one used in the presented case study. For example, by using a full-scale and complete platform as input file, with a significantly higher number of critical components modeled.

References

[1] Ball R E 1985 *The fundamentals of aircraft combat survivability analysis and design*
[2] Kopp C 2005 Are helicopters vulnerable? vol 59 pp 59–63 URL `https://www.ausairpower.net/PDF-A/TE-Helo-Mar-05-P.pdf`
[3] Arrighi J 1978 Survivability design features incorporated in the a-10a close air support weapon system *Proc. Of the third biennial aircraft survivability symposium* vol 1 (Naval Postgraduate School Monterey, California)
[4] Vice J 1986 Survivability/vulnerability information analysis center (surviac) - a tool for the aircraft survivability community vol 24 pp 511–515 URL `https://arc.aiaa.org/doi/abs/10.2514/3.45468?journalCode=ja`
[5] Gilman R J 1986 Survivability considerations during aircraft conceptual design. URL `https://apps.dtic.mil/sti/pdfs/ADA168555.pdf`
[6] Novak R 1986 A case study of a combat aircraft's single hit vulnerability URL `https://apps.dtic.mil/sti/pdfs/ADA175723.pdf`
[7] Aitken E D, Jones S L and Dean A W 1993 A guide to fastgen target geometric modeling URL `https://apps.dtic.mil/sti/pdfs/ADA273171.pdf`
[8] Joint Technical Coordination Group for Munitions. 1985 *Covart 3.0 - a simulation program for computation of vulnerable areas and repair time.*
[9] Pei Y and Cheng T 2014 Importance measure method for ranking the aircraft component vulnerability vol 51 pp 273–279 URL `https://arc.aiaa.org/doi/10.2514/1.C032334`

[10] Pei Y, Wang W and Li L 2016 Ranking the vulnerable components of aircraft by considering performance degradations vol 53 pp 1400–1410 URL `https://arc.aiaa.org/doi/full/10.2514/1.C033683`

[11] Yang P, Bifeng S and Qing H 2009 Shot line geometric description method for aircraft vulnerability calculation vol 22 pp 498–504 URL `https://www.sciencedirect.com/science/article/pii/S1000936108601326`

[12] Yang P, Bifeng S, Qing H and Baiyu O 2009 A direct simulation method for calculating multiple-hit vulnerability of aircraft with overlapping components vol 22 pp 612–619 URL `https://www.sciencedirect.com/science/article/pii/S1000936109603058`

[13] Pei Y and Song B 2006 Solving the combinatorial explosion problem when calculating the multiple-hit vulnerability of aircraft vol 43 pp 1190–1194 URL `https://arc.aiaa.org/doi/pdf/10.2514/1.18623`

[14] Konokman H E, Kayran A and Kaya M 2014 Analysis of aircraft survivability against fragmenting warhead threat URL `https://arc.aiaa.org/doi/pdf/10.2514/6.2014-0355`

[15] Lomazzi L, Cadini F, Giglio M and Manes A 2022 Vulnerability assessment to projectiles: Approach definition and application to helicopter platforms vol 18 pp 1523–1537 ISSN 2214-9147 URL `https://www.sciencedirect.com/science/article/pii/S2214914721001550`

Stress mitigation of a thermal engine head block using the bioinspired BGM-FEM method

C Groth[1], M Marra[1], S Porziani[1], M Schubert[2] and M E Biancolini[1]

[1] Department of Enterprise Engineering, University of Rome Tor Vergata, 00133 Rome, Italy
[2] Cummins Inc., Centenary House, Yarm Road, Darlington, DL14BS, UK

E-mail: corrado.groth@uniroma2.it

Abstract. Structural optimization plays a pivotal role in the design and development of engine heads, as it directly affects the performance, efficiency, and durability of internal combustion engines. In this paper, we propose a novel approach for the thermo-structural optimization of the internal surfaces of the engine heads, using the Biological Growth Method (BGM). The BGM is a bio-inspired technique that mimics the growth patterns observed in some biological organisms, able to achieve superior structural efficiency thus generating optimal designs. The effectiveness of the BGM methodology, coupled with mesh morphing techniques based on Radial Basis Functions (RBF), is in this work demonstrated for several districts, effectively reducing material usage and enhancing structural performance.

1. Introduction

The design and development of engine head blocks for internal combustion engines (ICE) demands meticulous attention to both the thermal and structural aspects. These critical components are one of the most complex parts of an ICE, subjected to extreme operating conditions, including high temperatures, pressure differentials, and mechanical stresses. The wide variety of loads acting on the engine head stems from various sources. For instance, mechanical loads can arise from the gas pressure within the cylinders or from the tightening of the screw bolts. On the other hand, thermal stresses primarily result from the thermal interaction between the coolant present in the water jacket and the high-temperature gases in the combustion chamber, as well as the exhaust line. For this reaseon, thermo-structural simulation plays a crucial role in the design and optimization of engine head blocks, particularly in light of the advancements in supercharging technology, the demand for lightweight design, and the stringent regulations governing engine emissions. The ever-increasing power density of engines necessitates innovative approaches to effectively manage the higher thermal and mechanical loads experienced by the engine cylinder heads. To address these challenges and enhance the design process, the integration of thermo-structural simulation techniques has become indispensable.

A vast number of works exist in literature regarding the study of the stress state in the ICE head block resulting from thermal and structural loads [1, 2]. In [3] and [4] the stress state was investigated taking into account also the residual stresses resulting from the casting process, while in [5] uncoupled Computational Fluid Dynamic (CFD) and Finite Element Method (FEM) simulations were carried in order to obtain the thermal distributions and to assess the strength of the cylinder head. The Fluid Structure Interaction (FSI) between the coolant inside the water jacket and the heated structural components can have an important impact on the structural reliablity ot the cilinder head [6]. Having a good understanding of the stress state of the engine head allows to faithfully reproduce the load-cycles [7] acting on the structural components. Several approaches at different level of detail exist in literature for the fatigue assessment of structural components [8, 9]. In [10] at first the S-N curve of the material was achieved both numerically and experimentally, then a coupled thermostructural simulation of an ICE head was carried to assess the fatigue life using ANSYS Workbench. A probabilistic approach based on neural networks is shown in [11].

Traditional design methods for engine head blocks often relied on empirical knowledge and conservative design margins. However, these approaches are no longer sufficient to meet the performance, efficiency, and environmental requirements of modern engines. To overcome these challenges, advanced simulation techniques that not only integrate thermal and structural analysis, but also optimisation methods, have become essential tools for engineers. While several approaches to structural optimisation exist, such as the one shown in [12] for the cylinder head, reliying on trial and error is not the most efficient approach especially when dealing with multiphisics shape optimizations such as the one under consideration. One possibility is to use a parameterisation technique, in which several shape variations are defined and blended in order to obtain multiple design points (DP). This approach, tipically identified by the use of a Design of Experiment (DoE) table, requires the generation and computation of many DPs. This task can be very time-consuming, having to remesh the domain for each new shape.

A viable alternative to rebuilding the model for each shape is the use of Radial Basis Functions (RBF) mesh morphing. Mesh morphing, as demonstrated in [13, 14] and [15], is a powerful technique that allows for the generation of different DPs by modifying the positions of mesh nodes, eliminating the need for remeshing and maintaining the original mesh topology. This technique has exhibited reliability across various engineering domains, from the stress state recovery of composite plates [16, 17] to fracture mechanics [18], FEM upscaling [19] and shape optimisation [20]. RBF-based mesh morphing can be also used, in view of its fast mesh update characteristics, for steady and unsteady FSI studies [21, 22, 23].

In this work a novel procedure for the thermo-structural optimization of engine head blocks is presented, addressing the need for lightweight designs while ensuring optimal performance under increased thermal and mechanical loads. The proposed approach utilizes the Biological Growth Method (BGM) [24] in combination with RBF mesh morphing, offering a unique solution for reducing material usage and enhancing the structural performance of engine head blocks.

Simulations were carried in the framework of the ANSYS® Mechanical™ FEA solver, using as RBF morpher the commercial tool RBF Morph.

The application of the BGM draws inspiration from biological growth patterns found in nature, which exhibit efficiency and resilience in structural formations. By mimicking the growth processes observed in biological organisms, the BGM provides an innovative approach to optimize the design of engine head blocks. The RBF mesh morphing technique complements the BGM by enabling seamless morphing and manipulation of the finite element mesh, facilitating the design iteration process in a fully automatic and evolutive fashion.

After a brief introduction to the RBF theory and the BGM, the proposed approach will be demonstrated and evaluated for several districts of a generic engine head. To close the paper, the effectiveness of the BGM and RBF mesh morphing will be further analyzed, showcasing the significant improvements achieved in reducing material usage while reducing the stress peaks of an industrial engine head block.

2. Theoretical Background

2.1. RBF

RBF are mathematical functions that have the capability to interpolate scalar information, which is known only at discrete points (referred to as source points), based on distance [26]. The quality and behavior of the interpolation are determined by both the chosen function and the type of basis function employed. These functions can be defined in an n-dimensional space and are dependent on the distance, which, in the case of morphing, can be represented by the Euclidean norm between two points in the space [25]. By defining the displacement at known points in space, the values at mesh nodes can be retrieved, resulting in a mesh deformation that preserves the original grid topology [27], [28].

The interpolation function consists of the basis function ϕ and a polynomial term h in the form:

$$s(\mathbf{x}) = \sum_{i=1}^{N} \gamma_i \phi(\|\mathbf{x} - \mathbf{x}_{k_i}\|) + h(\mathbf{x}) \tag{1}$$

In which N is the total number of source points. More information on the subject can be retrieved in several works from the field [14].

It must be noted that thanks to the analytical nature of RBF interpolation, it is possible to differentiate the kernel by using the Kansa method [30]. This feature allows to compute for instance the stress and strains from displacement fields [31, 32].

2.2. BGM

The Biological Growth Method (BGM) is a shape optimization technique specifically designed to act on the stress fields of structural components. This method draws inspiration from the growth patterns observed in biological structures such as tree trunks and animal bones, where new layers of biological material are added to the surface at a rate that corresponds to the level of stress. Building upon this concept, in [33] and [34] the material is added to surfaces experiencing high stresses and removed to those in which stresses are low. A more refined model was later developed in [35], to take into account the case of multiple stress peaks.

In this work, the implementation proposed in [24] and available in RBF Morph was adopted. In particular, the displacement S, normal to the surface, can be calculated for the i-th node as:

$$S_i = \frac{\sigma_i - \sigma_{th}}{\sigma_{max} - \sigma_{min}} d \tag{2}$$

where σ_i is stress value for the given node i, σ_{th} the threshold above which the material is added and below which the material removed and σ_{max} and σ_{min} the maximum and minimum values of stress of the whole domain. The value of the displacement for each node is tuned by multiplication with a parameter d, that can be chosen according to the maximum value of displacement wanted on the domain.

The displacement field obtained with equation (2) must be computed at each iteration, re-evaluating the maximum (σ_{max}) and minimum (σ_{min}) stresses for the set of nodes for which the BGM is computed. An advantage of employing a meshless morphing technique, such as RBFs, is that it enables the handling of complex three-dimensional (3D) problems without incurring in additional costs compared to simpler two-dimensional (2D) planar cases. This flexibility allows to take into account constraints such as the manufacturing ones, as shown in [36].

3. RBF-BGM setup for engine head optimisation

The Optimization procedure proposed in this work was implemented using ANSYS Mechanical as the Finite Element Analysis (FEA) solver. The RBF Morph ACT extension was used to update the numerical grid according to the BGM data and to perform the mesh morphing task [37]. RBF Morph allows for the application of two types of surface node offsets: a fixed surface offset and a driven surface offset. In both cases, the software can determine the surface normal at each node position and use it to impose a displacement along the normal, either inward or outward. With the fixed value surface offset, all nodes on the model surface will be translated by the same amount along the local surface normal. On the other hand, the driven value surface offset enables the definition of a movement along the surface normal direction for each surface node. The intensity of this movement is determined based on the BGM equation (2), which considers the stress/strain type, the threshold value σ_{th} and the parameter d.

By properly configuring the mesh morphing setup, it is possible to confine the deformation process to a specific portion of the domain, reducing the computational burden and acting only in the area interested by optimization. This approach is particularly useful for problems such as the cylinder head one, in which confined flows and complex geometries with internal ducts make difficult the selection of the surfaces to be sculpted and those to be kept fixed. This concept is shown in Figure 1, in which only the domain near the surfaces to be optimised is considered. The mesh elements interested by morphing are only those wrapped by the cylindrical domain as shown in the left image in Figure 1.

Being a meshless method, RBFs require the definition of a displacement field prescribed at points. To automatically build the complete problem the setup is split in two fields: one (*moving set*) prescribing displacements on the nodes designated to undergo movement, the other (*fixed set*) defining the nodes for which the displacement must be kept to zero. Being an evolutionary optimization, the first is computed at each step with a new evaluation while the latter is always maintained unchanged. The fixed set is for this reason extracted from the baseline mesh just once, on the baseline, when defining the morphing setup; this set is composed by several node selections joined together as shown in Figure 2 (b), where the mesh nodes intersecting the bottom (yellow), top (cyan) and lateral surfaces (blue) of the cylindrical domain are merged with the surfaces inside the domain that must be kept fixed (orange). The fixed set, resulting from this procedure, is shown in Figure 2 (c). In Ansys Mechanical this complex setup is automatically generated using tabular named selections, in which the rules employed to collect the nodes are defined by means of geometrical relationships.

In Figure 3 it is shown the complete setup in which the fixed set is enriched by the moving set (green points), composed by the nodes in the areas that must be sculpted. The resulting RBF problem is shown, on the right picture, superimposed to the domain. Displacements for the fixed set are computed using the driven surface offset modifier in the RBF Morph ACT extension. All the nodes inside the domain that are not part of the fixed or the moving sets are left free to deform, smoothly blending between the two.

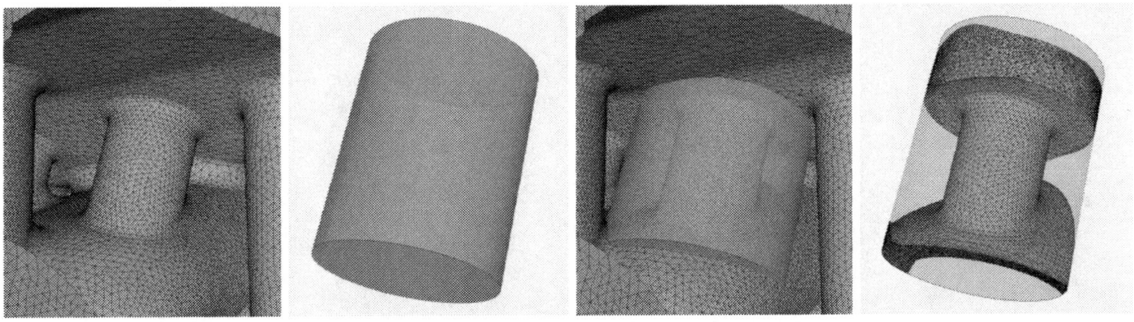

Figure 1. Domain interested by the morphing action and optimization

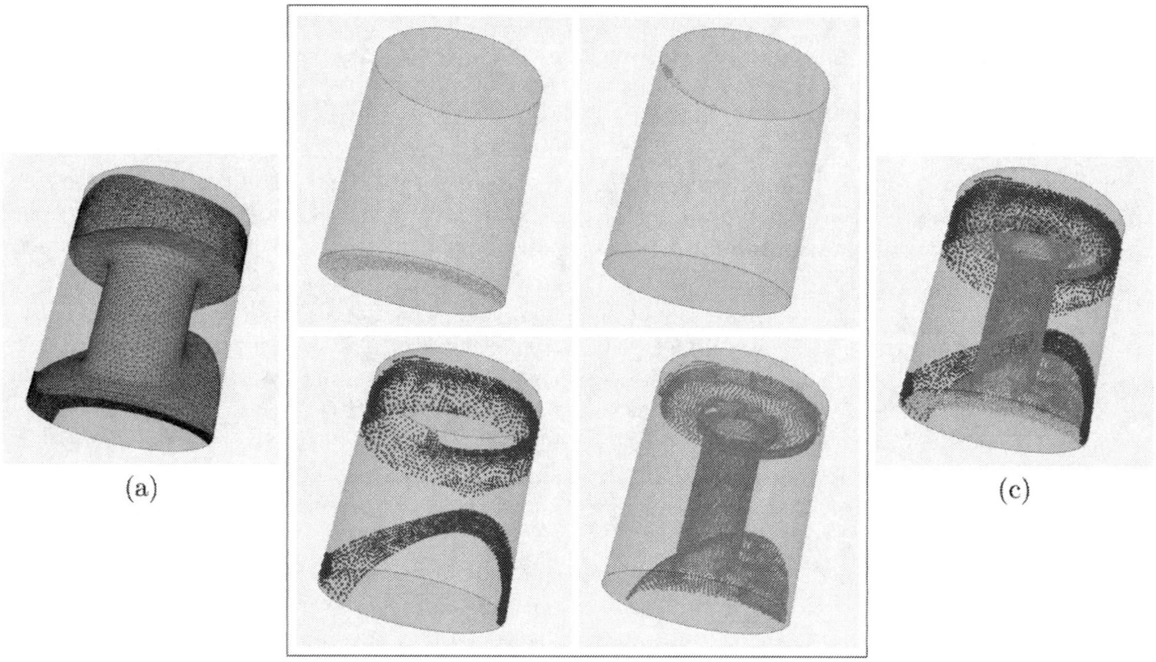

(a) (b) (c)

Figure 2. Generation of the fixed set on the morphing domain (a): the bottom, top and lateral surfaces of the cylinder domain are merged with the fixed surfaces (b) to obtain the complete set (c)

4. Generic engine head problem

In this section the proposed BGM approach will be demonstrated on a generic engine head model.

4.1. Baseline model

To demonstrate the method it was applied to a very generic model of engine head, whose baseline geometry is shown in Figure 4. To simplify the problem only the repetitive domain around a single cylinder was considered.

In Figure 4 on the right is also shown the computational mesh employed, obtained as the result of a mesh convergence study, and counting 2.9 Mill. nodes and 1.9 Mill. parabolic tetrahedral

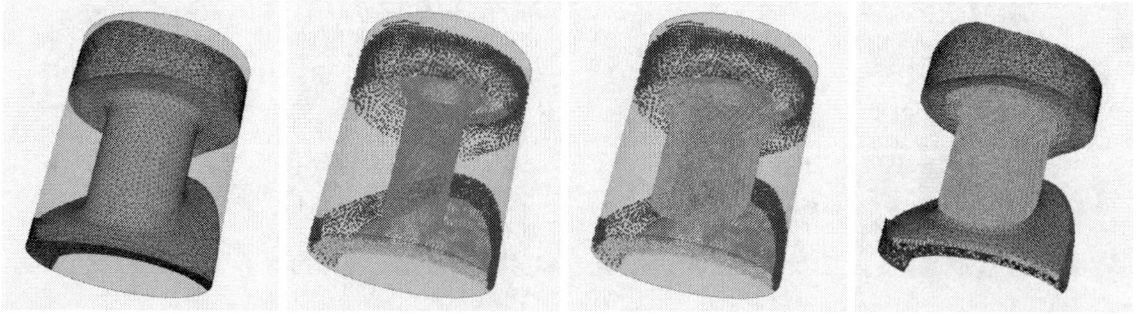

Figure 3. Complete morphing setup, obtained merging the moving and fixed sets

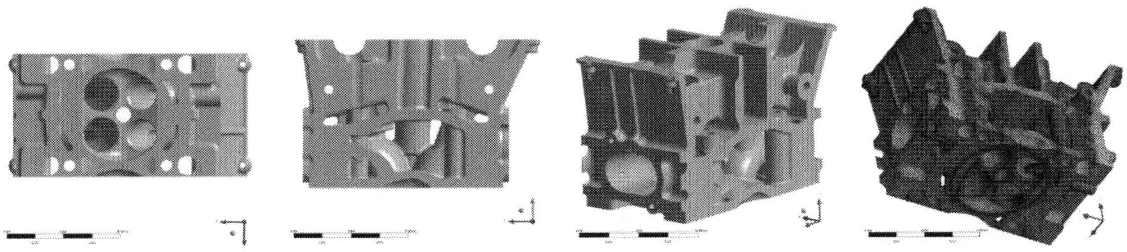

Figure 4. Baseline geometry for the generic engine head case

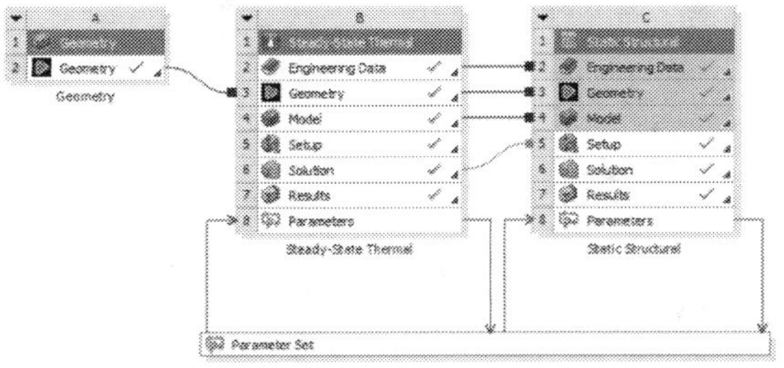

Figure 5. Workbench schematic for the thermostructural simulation

elements. As shown in Figure 5, to carry the thermostructural analysis in ANSYS Mechanical the same geometry and mesh was shared between a steady-state thermal and a static structural simulation. The temperature field obtained in the first was imported in the latter to compute the resulting stresses, taking into account also the structural constraints and loads.

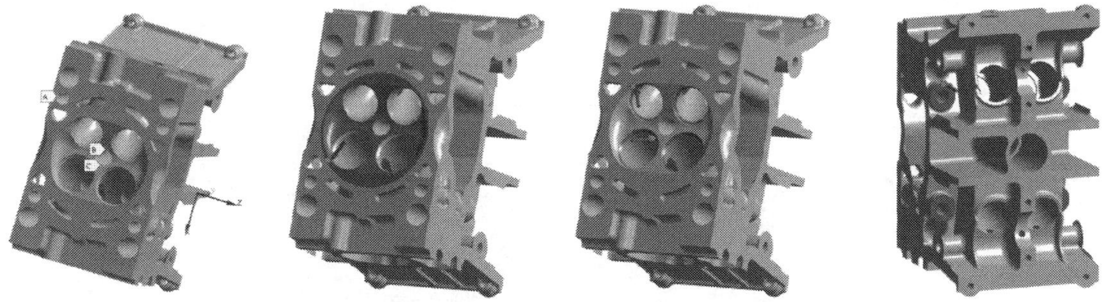

Figure 6. Structural boundary conditions

The displacement boundary conditions for the structural simulation, shown in Figure 6, were specified on two points near the injection hole and around the head studs holes. Specifically, and with reference to the first image in Figure 6, the edges A of the head stud holes where constrained along the Y direction, point B along Z and point C along both Z and X. Regarding the acting loads, pressures were defined on the combustion side of the cylinder head and on the valve seats, using different values for the exhaust and for the intake valves. Additionally, a pressure was defined on the bolts of the head struts as shown in the last picture of Figure 6.

Regarding the steady-state thermal simulation, convection boundary conditions were applied to the combustion side, dividing it in three concentric areas with different values of the convective film coefficient as shown in Figure 7. Convective boundary conditions were applied also to the exhaust and intake ducts and to the water jackets. The convective film coefficients depend on the fluid dynamic of the flows inside the ducts, that are strongly affected by an eventual shape change. In this optimization, however, this dependence was underestimated maintaining unchanged the convective heat transfer coefficients in all the optimization steps.

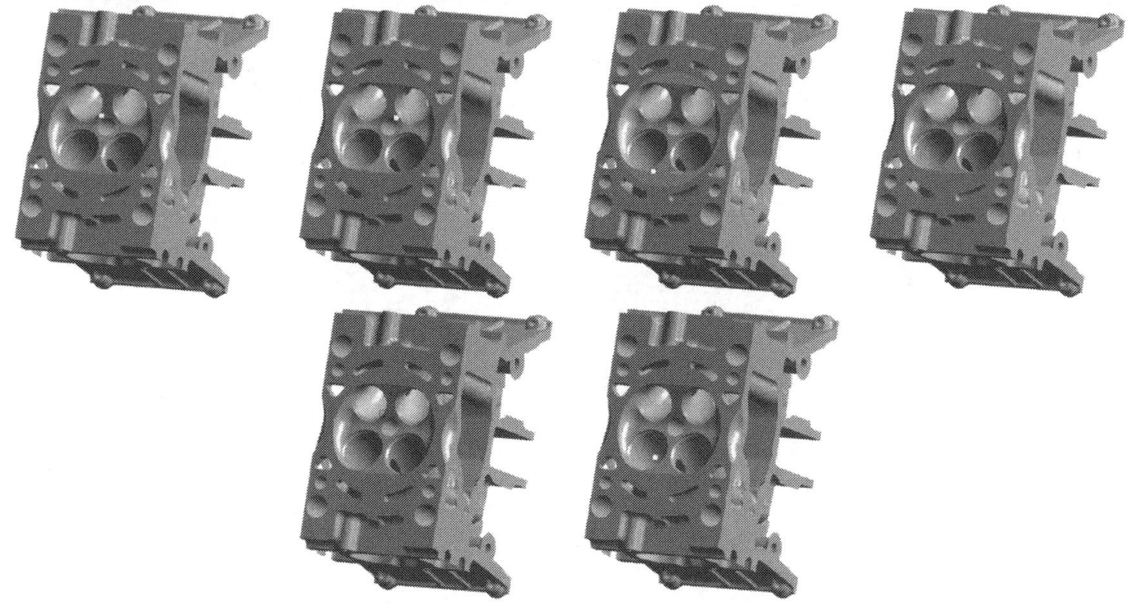

Figure 7. Thermal boundary conditions

An aluminium alloy (AlSi7MgCu0.5) was used for the whole model. Three areas of the domain were optimized, near the injector seat, the exhaust and the intake ducts. In Figure 8 the von Mises stress on the injector seat district is shown. The maximum value of stress is equal to 140 MPa.

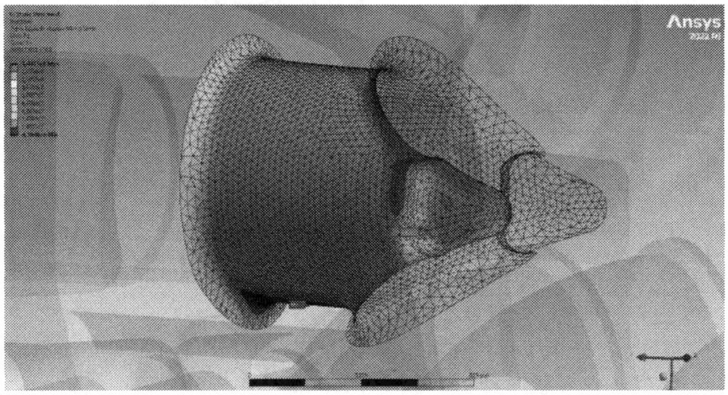

Figure 8. Von Mises stress on the injector seat

The contours for the von Mises on the exhaust ducts are shown in Figure 9, where it can be noted that the maximum stress value is equal to 344 MPa, far above the maximum Yielding stress for the material that is equal to 200 MPa.

Finally, in Figure 10, the contour for the von Mises stress is shown on the intake ducts, In this case the maximum stress is equal to 236 MPa.

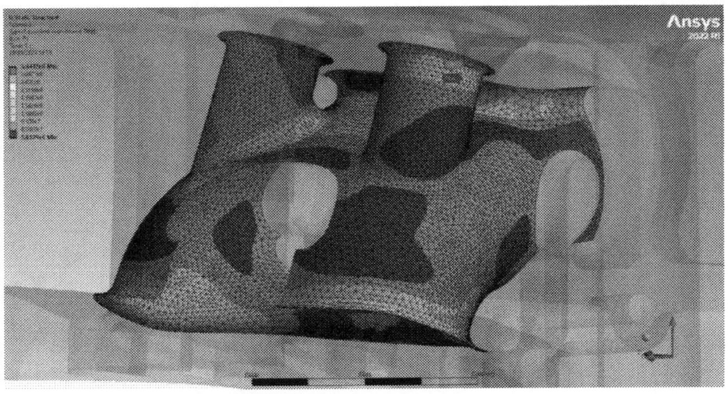

Figure 9. Von Mises stress on the exhaust ducts

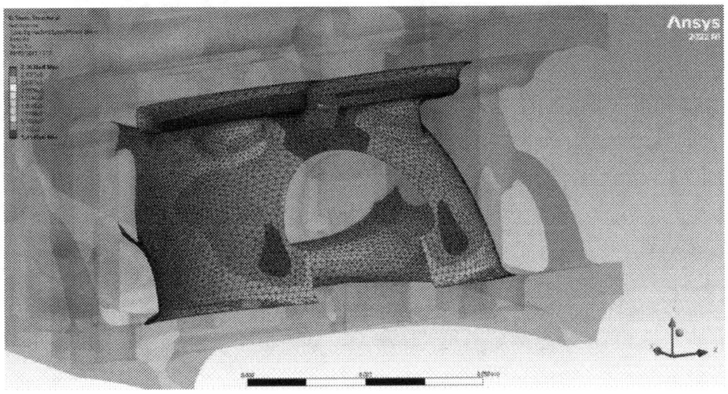

Figure 10. Von Mises stress on the intake ducts

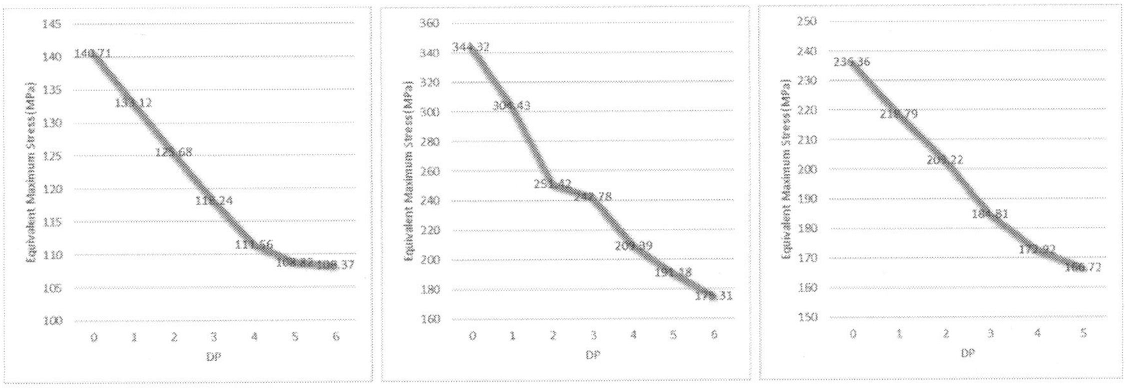

Figure 11. Von Mises stress reduction for the injector seat, the exhaust and the intake ducts

4.2. Optimization results

Following the application of the BGM method using RBF Morph a stress reduction was achieved, below the yielding stress, for all the three districts.

In Figure 11 the stress histories for the three cases shown previously are depicted. As it can be seen, 6 iterations were employed for the three cases but the intake duct one. The optimization results achieved are summarized also in Table 1, in which are shown also the parameters used

IOP Conf. Series: Materials Science and Engineering 1306 (2024) 012024 doi:10.1088/1757-899X/1306/1/012024

for the BGM method.

district	σ_{th}	d [mm]	σ_{max} baseline [MPa]	σ_{max} optimized [MPa]	reduction
Injection hole	0	0.5	140.71	108.37	25.5%
Exhaust ducts	58	1	344.32	175.31	49%
Intake ducts	0	0.5	236.36	166.72	29.5%

Table 1. Stress reduction

On the most critical case, the exhaust ducts, the von Mises stress was reduced from 344.32 MPa of the baseline to 175.31 MPa, well below the yielding value. Overall the maximum stress was reduced of 49%. The stress map for the optimized geometry is shown in Figure 12 on the left, using the same stress scale as in Figure 10.

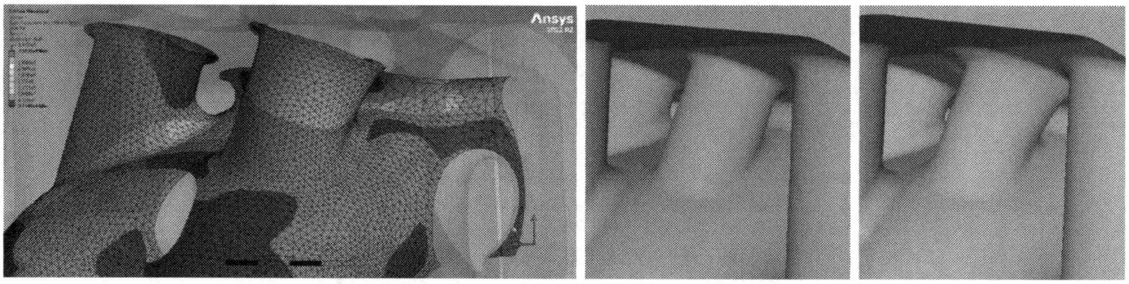

Figure 12. Von Mises stress for the optimized exhaust ducts and comparison between baseline and optimized geometries

In Figure 12 is also compared the baseline geometry (pink) with the optimized one (light green). it can be noted how the material was added in the areas with higher stress according to the BGM approach. By using, as shown in Table 1, a threshold stress value, material was added in presence of stresses higher than σ_{th} and removed elsewhere. This approach, that allows to reduce material usage, can be seen in the left images of Figure 12, where material removal can be spotted in the lower section of the exhaust duct.

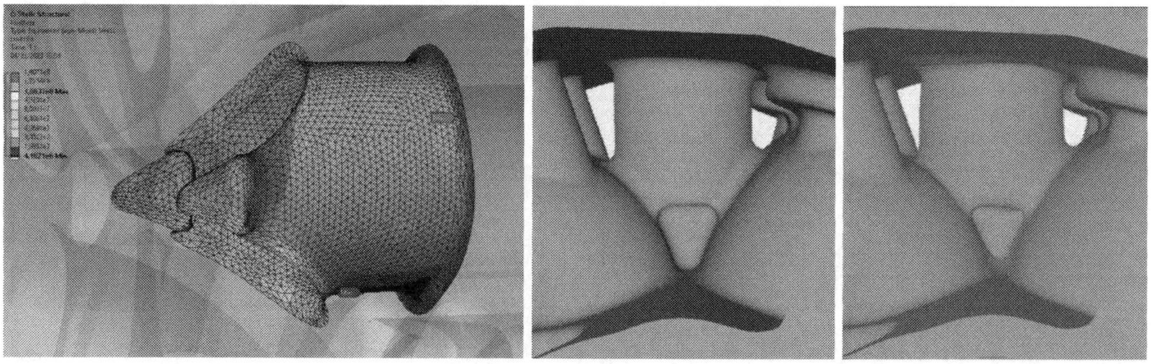

Figure 13. Von Mises stress for the optimized injection seat and comparison between baseline and optimized geometries

The Stress fields for the optimized geometry and the comparison between original and optimized geometries are shown also for the injection seat and intake ducts cases, respectively

IOP Conf. Series: Materials Science and Engineering 1306 (2024) 012024 doi:10.1088/1757-899X/1306/1/012024

in Figure 13 and Figure 14. Von Mises contour scales, for comparison pourposes, were left unchanged with respect to the baseline cases.

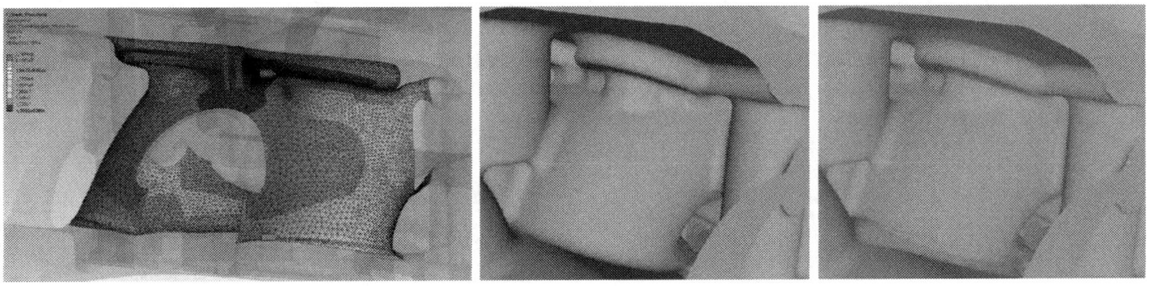

Figure 14. Von Mises stress for the optimized intake ducts and comparison between baseline and optimized geometries

5. Industrial application

FEA and Fatigue analysis was carried out on a new cast iron diesel engine cylinder head design. The analysis lead design process was highly iterative, whereby the analysis drove recommendations for design improvements, which upon implementation fed into a new analysis iteration until acceptance criteria for structural integrity was satisfactory. This workflow relied on close collaboration between the Structural Analyst, using Ansys Mechanical for the FEA, and the Design Engineer, using PTC Creo for CAD work, involving many manual and repetitive tasks leading to a turnaround time of approximately 2 weeks per iteration for a block-gasket-head assembly level model.

With this approach improvements in stress and fatigue FOS (factor of strength) were achieved everywhere except at 2 locations.

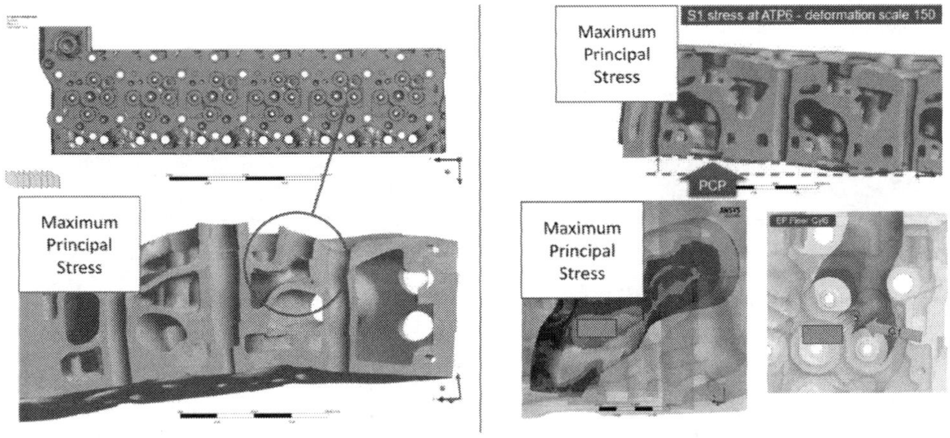

Figure 15. Maximum principal stresses for B1 (left) and G1 locations (right)

As shown in Figure 15, the B1 location is below the expansion plug opening in the upper water jacket, whereas G1 is located near the dividing wall where the exhaust port ducts merge. The stress at both locations are tensile in nature and highly sensitive to the cyclic combustion pressure loading, leading to unsatisfactory fatigue FOS. The lack of FOS improvement via the manual analysis and design iterations is a results of tight packaging constraints and high complexity of shape and load paths.

The described BGM based workflow was applied on a component level model via the RBF Morph ACT extension in Ansys Mechanical, using the Maximum Principal Stress as objective function for the shape optimization at both regions. A complete optimization was achieved in 30 design points, reducing the Maximum Principal Stress from 205MPa to 175MPa (14.6% reduction) on the B1 location and from 196MPa to 180MPa on the G1 location (8.1 % reduction). These stress reductions resulted in a FOS improvement of 13.4% and 9.2% respectively.

6. Conclusions

In this work was presented a methodology to perform automatic shape optimization via surface sculpting on a combustion engine cylinder head assembly model. The BGM based workflow was developed using Ansys Workbench and the RBF Morph ACT extensions, employing a mixed setup using fixed and moving sets. Proposed workflow was first demonstrated on a generic model of engine head, and proved to have several advantages such as: - Very fast turnaround time in comparison to the manual analysis/design iterations since the whole process is carried out automatically after initial set-up using the same model. Manual tasks such as CAD model changes and FE model updates are eliminated. - Improved clarity of analysis recommendations by exporting the final and optimal mesh for guidance with CAD modifications - The optimization method provides very good understanding of the limitations of the design before more major changes in topology or architecture may have to be considered. The method was finally applied on an industrial component level model via RBF Morph ACT extension in Ansys Mechanical with Maximum Principal Stress as the output parameter for the shape optimization. The resulting relative stress reduction on two hot spots, after 30 iterations, was 8.1% and 14.6% with a respective FOS improvement of 9.2% and 13.4%.

References

[1] Tichánek R, Španiel M and DiviŠ M 2005 Structural stress analysis of an engine cylinder head *Acta Polytechnica* **45** 3

[2] Shojaefard M H, Ghaffarpour M R, Noorpoor A R and Alizadehnia S 2006 Thermomechanical analysis of an engine cylinder head *Proceedings of the Institution of Mechanical Engineers, Part D: Journal of Automobile Engineering* **220(5)** 627-636

[3] Hartmann G and Sturm J 2002 Integrated Numerical Optimization of Highly Loaded Aluminum Cylinder Heads *SAE Technical Paper* **2002-01-3429**

[4] Mendes A S and de Azevedo Cardoso A 2007 Structural analysis of the aluminum cylinder head for a high-speed diesel engine *SAE Technical Paper* **2007-01-2562**

[5] Arnone L, Bella G, Vivio F and Vullo V 2011 Thermo-structural analysis of a new engine cylinder head *SAE Technical Paper* **2011-24-0165**

[6] Romanov V A, Lazarev E A and Khozeniuk N A 2015 The evaluation of the stress-strain state for the cylinder heads of high-powered diesel engines using the multiphysics ansys technology. *Procedia engineering* **129** 549-556

[7] Thomas J J, Verger L, Bignonnet A and Charkaluk E 2004 Thermomechanical design in the automotive industry *Fatigue & Fracture of Engineering Materials & Structures* **27(10)** 887-895

[8] Zhu S-P, Liu Q, Peng W and Zhang X-C 2018 Computational-experimental approaches for fatigue reliability assessment of turbine bladed disks *Int J Mech Sci* **502-17** 142-143

[9] Zhang Z and Qiu Z 2021 Fatigue reliability analysis for structures with hybrid uncertainties combining quadratic response surface and polynomial chaos expansion *Int J Fatigue* **144** 106071

[10] Vučetić N, Jovičić G, Krstić B, ivković M, Milovanović V, Kačmarčik J and Antunović R 2020 Research of an aircraft engine cylinder assembly integrity assessmentThermomechanical FEM analysis *Engineering Failure Analysis* **111** 104453

[11] Jing G, Li S, Xiao S, Ma T, Lyu Z, Sun S and Zhou H 2023 Research on fatigue reliability assessment of engine cylinder head based on neural network *International Journal of Fatigue* 107800

[12] Jing G, Lyu Z, Liu Y, Xiao S, Zhou H and Li S 2022 Reliability study for diesel engine cylinder head through fatigue failure analysis and structural optimization *Engineering Failure Analysis* **142** 106768.

[13] de Boer A, Van der Schoot M S and Bijl H 2007 Mesh deformation based on radial basis function interpolation *Computers & structures* **85** 784–795

[14] Biancolini M E 2018 Fast Radial Basis Functions for Engineering Applications, Springer, Berlin.
[15] Staten M L, Owen S J, Shontz S M, Salinger A G and Coffey T S 2011 A comparison of mesh morphing methods for 3D shape optimization *Proceedings of the 20th international meshing roundtable* Springer 293311
[16] Chiappa A, Groth C, Reali A and Biancolini M E 2020 A stress recovery procedure for laminated composite plates based on strong-form equilibrium enforced via the RBF Kansa method *Composite Structures* **244** 112292.
[17] Chiappa A, Groth C and Biancolini M E 2023 A two-scale RBF meshless method for the interface stress retrieval in simply bended and torqued long-fibres laminates *Composite Structures* **306** 116600.
[18] Chiappa A, Groth C, Brutti C, Salvini P and Biancolini M E 2019 Post-processing of 2D FEM Q1 models for fracture mechanics by radial basis functions and balance equations *Int. J. Mech* **13** 104-113
[19] Chiappa A, Groth C and Biancolini M E 2019 Improvement of 2D finite element analysis stress results by radial basis functions and balance equations *Int. J. Mech* **13** 9099
[20] Groth C, Chiappa A and Biancolini M E 2018 Shape optimization using structural adjoint and RBF mesh morphing *Procedia Structural Integrity* **8** 379-389
[21] Biancolini M E, Cella U, Groth C and Genta M 2016 Static aeroelastic analysis of an aircraft wind-tunnel model by means of modal RBF mesh updating *Journal of Aerospace Engineering* **29**(6), 04016061
[22] Groth C, Cella U, Costa E and Biancolini M E 2019 Fast high fidelity CFD/CSM fluid structure interaction using RBF mesh morphing and modal superposition method *Aircraft Engineering and Aerospace Technology* **91**(6), 893-904
[23] Groth C, Porziani S, Biancolini M E 2021 Radial basis functions vector fields interpolation for complex fluid structure interaction problems *Fluids* **6**(9) 314
[24] Porziani S, Groth C, Waldman W and Biancolini M E 2021 Automatic shape optimisation of structural parts driven by BGM and RBF mesh morphing *International Journal of Mechanical Sciences* **189**, 105976
[25] de Boer A, van der Schoot M S and Bijl H 2007 Mesh deformation based on radial basis function interpolation *Computers and Structures* **85** 784–795
[26] Buhmann M D 2000 Radial basis functions *Acta Numerica* **9**
[27] Beckert A and Wendland H 2001 Multivariate interpolation for fluid-structure-interaction problems using radial basis functions *Aerospace Science and Technology* **2** 125-134
[28] Biancolini M E 2012 Mesh Morphing and Smoothing by Means of Radial Basis Functions (RBF) *Handbook of Research on Computational Science and Engineering* 347-380
[29] Micchelli C A 1986 Interpolation of scattered data: Distance matrices and conditionally positive definite functions *Constructive Approximation* **1** 11-22
[30] Kansa E J 1990 Multiquadrics – A scattered data approximation scheme with applications to computational fluid-dynamics – surface approximations and partial derivative estimates. *Computers & Mathematics with applications* **19**(8-9) 127-145
[31] Groth C, Chiappa A, Porziani S, Salvini P, Biancolini M E 2022 An RBF Meshless Approach to Evaluate Strain Due to Large Displacements in Flexible Printed Circuit Boards *Micromachines* **13**(8) 1163
[32] Groth C, Chiappa A, Porziani S, Marotta E, Salvini P, Biancolini M E 2022 A Post-Processing Method Based on Radial Basis Functions for the Fast Retrieval of the Strain Field in Digital Image Correlation Methods *Materials* **15**(22) 7936
[33] Heywood R B 1969 Photoelasticity for designers *Pergamon Press* Oxford
[34] Mattheck C and Burkhardt S 1990 A new method of structural shape optimization based on biological growth *International Journal of Fatigue* **12**,185–190
[35] Waldman W and Heller M 2015 Shape optimization of holes in loaded plates by minimization of multiple stress peaks. Technical Report, *Defence Science and Technology Organisation* Fisherman Bend, Australia, Aerospace Div
[36] Porziani S, Groth C and Biancolini M E 2018 Automatic shape optimization of structural components with manufacturing constraints *Procedia Structural Integrity* **12** 416–428
[37] RBF Morph srl http://www.rbf-morph.com/act-module/ *Accessed: 2023-07-12*

IOP Conf. Series: Materials Science and Engineering 1306 (2024) 012025 doi:10.1088/1757-899X/1306/1/012025

Design And Optimization Of Aeronautical Components And Digital Twins Development

A Lopez[1], U Cella[1] , C Groth[1] and M E Biancolini[1]
[1] *Industrial Engineering Department, University of Rome Tor Vergata*

Abstract. In this work, the feasibility of using reduced models (ROMs) for optimizing a scoop air intake for aeronautical applications was evaluated. Since the air intake is exposed to aerodynamic loads, a two-way fluid-structure interaction workflow was used to characterize the component. The goal is to create an optimization dashboard that allows both scalar quantities (the parameters that are intended to be monitored during the design and optimization of the air intake) and field quantities to be evaluated in real time. In this way, the designer can have a full understanding of the physics of the problem and make more informed design choices. In addition, in this way it is possible to visualize results from different physics in a single dashboard, linking different components, interacting with models in real time. A mesh morphing technique based on Radial Basis Functions (RBFs) was used. The result was very interesting both from a structural point of view (mass reduction over 90% and maximum strain reduction of 36%) and from a fluid dynamic point of view (outlet pressure 86% higher and drag 32% lower) and the generated ROMs proved to be a very accurate (ROM relative error in the order of 7%) and flexible tool.

1. Introduction

The goal of this work is to create an optimization dashboard to visualize the digital twin of the real object. The digital twin is a digital copy of the physical object and can encapsulate a lot of information. Digital twins find many engineering applications in different fields. In the medical sector [1] [2] [3] [4], digital twins are used to make predictions of problematic behavior, to simulate surgical interventions, in the electronic field [5] they are used to reproduce the functioning of HPC and monitor its functioning to predict and anticipate possible malfunctions, avoiding service interruptions. In this project digital twins are used to create an advanced optimization platform, which allows you to visualize in real time both field quantities and scalar quantities coming from different physics, all integrated into the same dashboard, with the possibility of connecting the different submodels, different components of the same object.

In this study, the digital twin was created from CFD and FEM analysis. Thus, in the final dashboard, fluid-dynamic and structural field quantities can be visualized in real time as geometric and physical parameters change.

A Scoop air intake was considered as a testcase for two reasons:

- As an exposed, external air intake, it is subject to aerodynamic loads to be considered for sizing and structural optimization. Therefore, it is necessary to consider multi-physics analysis and study fluid-structure interaction. In fact, one of the goals of this work is to test the flexibility of ROMs and create a single environment for the different physics involved;
- Given the increasing electrification in aviation, cooling of eletronic systems is a central issue. Therefore, developing an updated workflow for air intake design covers a strategic role.

2. Methods

2.1. Mesh morphing and RBFs

In a shape optimization problem, a fundamental aspect is the definition of geometric parameters. A classic approach is to define a parametric CAD. Then at each Design Point the geometry is updated, a new mesh is created, and CFD or FEM analysis is performed. This approach has two main limitations related to the re-meshing phase: it is difficult to automate the re-meshing ensuring good mesh quality and very long computation times. Mesh morphing can be used to overcome these problems. The idea is to define parametric displacements and deform the pre-existing mesh. RBFs can be applied to interpolate the defined displacements on the mesh nodes. Source points are defined and a known displacement is applied. The displacements are interpolated on the mesh nodes using a proximity criterion: displacements proportional to the distance from the surrounding source points are applied on the mesh nodes.
From a mathematical point of view [6], it is necessary to solve the following linear system:

$$f^x(x) = \sum_{i=1}^{m} \gamma_i^x \phi(\|c_i - x\|) + \beta_1^x + \beta_2^x x_1 + \beta_3^x x_2 + \beta_4^x x_3 \quad (1)$$

$$f^y(x) = \sum_{i=1}^{m} \gamma_i^y \phi(\|c_i - x\|) + \beta_1^y + \beta_2^y x_1 + \beta_3^y x_2 + \beta_4^y x_3 \quad (2)$$

$$f^z(x) = \sum_{i=1}^{m} \gamma_i^z \phi(\|c_i - x\|) + \beta_1^z + \beta_2^z x_1 + \beta_3^z x_2 + \beta_4^z x_3 \quad (3)$$

Where γ are the weights and ϕ are the radial basis functions, the polynomial term has a stabilizing function. The system is solved by applying as boundary conditions the value of the known displacement at the source points and the orthogonality of the coefficients. There is a wide range of RBFs available. The interpolation behavior, while guaranteeing the value imposed on the center points, depends on the specific RBF chosen. The computational cost and approaches to RBF solutions also vary depending on the type of function.
In this study, RBFs were used to define shape parameters in the optimization of the air intake and to apply the displacement field of the FEM model in the CFD mesh. Source points can be defined anywhere, and the method is meshless, independent of the mesh. This aspect is very attractive for multi-physics problems, in which different meshes are generally used. In this way the same shape parameters can be used on different meshes. Finally, for ROM generation, mesh morphing is critical because it ensures that the mesh topology remains unchanged and it's necessary for ROM building.

2.2. Fluid-Structure-Interaction (FSI)

To study fluid-structure interaction [7] [8] [9] with high-fidelity analysis there are generally 3 approaches:

- One way
- Two ways
- Modal Superposition

The one way method is generally used when deformation effects are negligible in the aerodynamic behavior of the studied object. It is basically a mapping of the aerodynamic loads onto the FEM model.
The modal superposition method is less computationally onerous than the two-way method but also less accurate. Aerodynamic forces are projected in the direction of the main modes considered, and the resultant force is given by the linear superposition of the modal forces thus evaluated. Thus, a preliminary modal analysis on the FEM model is required. Next, an analysis of the forces exerted by the fluid on the structure is performed. To calculate the modal weights, a system of equations is solved for each natural mode. Each equation considers the balance of forces projected onto that mode. The modal weight for each mode represents the importance of that mode in the overall response of the

structure to the fluid-structure interaction. The ultimate goal of this method is to simplify the analysis of structural response to the presence of fluid loads by reducing the number of degrees of freedom to be considered, but maintaining good accuracy in the results.
The two-way method is the most computationally onerous but also the most accurate
From the CFD analysis, the aerodynamic loads are calculated and exported. The loads are imported into the FEM model and the displacements are estimated. The displacements are used to deform the CFD mesh and have a more accurate estimate of the loads. The workflow is iterated until forces and displacements converge. In this study, the focus is on the two-way method.

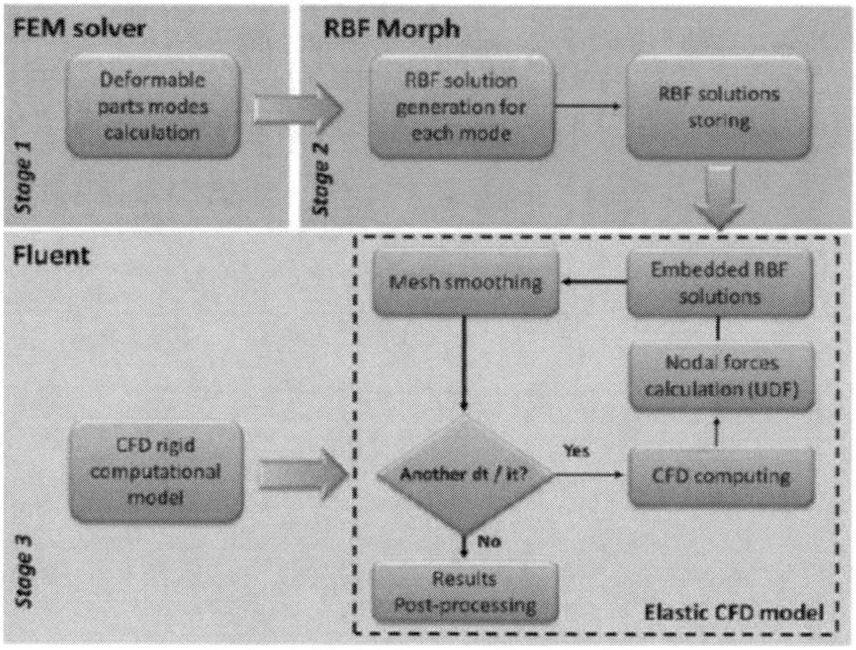

Figure 1. Modal Superposition Workflow

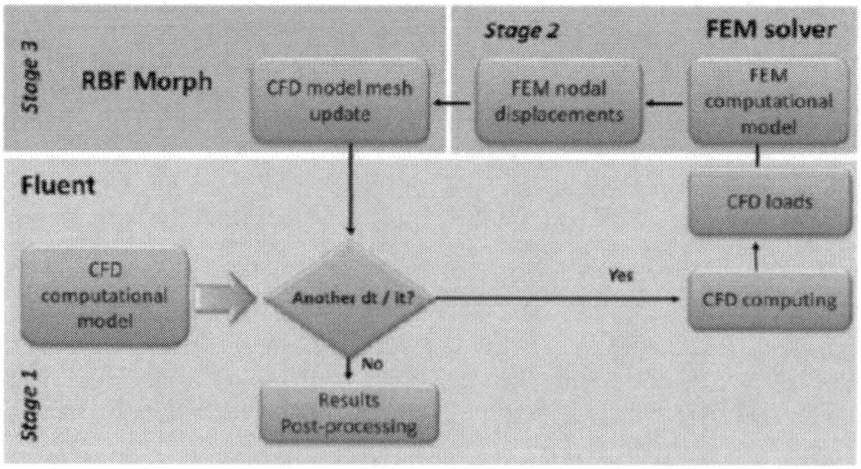

Figure 2. Two Ways Workflow

2.3. Optimization workflow: Response surface (RS)
The response surface method [10] [11] is a very robust and flexible method widely used in optimization problems. It lies within the zero-order optimization methods based on the Design of Experiment (DOE) and allows the identification of a global optimum. The workflow is shown in Figure 3.

IOP Conf. Series: Materials Science and Engineering 1306 (2024) 012025 doi:10.1088/1757-899X/1306/1/012025

Starting with the baseline geometry, the input and output parameters of the optimization are defined. Then a DOE is constructed, i.e., a table enclosing all DPs, each row is a DP, i.e., a combination of input parameters. There are several algorithms for optimally constructing a DOE. The goal is to have a distribution of points representative of the problem but as few points as possible. In fact, each DP is a CFD, or FEM analysis (or an FSI analysis), through which the values of the output parameters of interest associated with each row of the DOE are calculated.
Finally, it is necessary to interpolate the points obtained to identify an RS, i.e. a mathematical function that correlates input parameters and output parameters. In the literature there are many interpolation algorithms used, linear regression, RBFs, neural networks or more complex machine learning algorithms can be used. For optimization problems, linear regression with polynomial functions is very effective because it manages to capture the global trend of the global parameters and the surface is very smooth. This makes it easier to find the best direction.
Conversely, there are more precise methods that allow for a more accurate response, but often require a greater number of DPs and a search method for the optimum sensitive to local minima. Often the best strategy is to use polynomial regression to find the best zone, and then refine around this zone with more accurate methods.

Figure 3. RS Optimization Workflow

2.4. Digital twins development

Digital twins are digital copies of physical objects. They contain much more information than physical objects. They are built from data. In particular, in this study they were developed starting from high fidelity numerical simulations. Compared to response surfaces, which allow only scalar values to be predicted, to create a digital twin it is necessary to have a model that also allows field quantities, such as the pressure distribution or the stress and strain field, to be evaluated in real time. So the problem is very complex, in fact it is necessary to predict the value of the field quantity in n nodes of the mesh. For problems of industrial interest, the number of variables to be predicted becomes very high and the normal interpolation methods are poorly performing. This is why reduced order models (ROMs) were used. Single Value Decomposition (SVD) [12] was used to reduce the number of variables, identifying the main modes, whose linear superposition allows the field quantity to be reconstructed with less than a small error. Finally machine learning algorithms [13] [14] are used to correlate the input parameters to the weights of the principal modes.

2.4.1. POD. The use of numerical linear algebra plays a fundamental role in the field of scientific computing. In particular, throughout history, methods developed to enhance the efficiency of solving equations of the form Ax = b have been of crucial importance. Among these, matrix decompositions have served as the primary tools for optimizing computational calculations. However, among all decomposition techniques, one of the most critical is the Singular Value Decomposition (SVD). A is the snapshots matrix, it is a matrix of size m x n, where m represents the number of snapshots and n represents the number of grid points multiplied by the number of variables considered. Each snapshot represents a state of the problem, is the result of an FEM/CFD analysis, is the vector transposition of the monitored output magnitude. So if p is the number of nodes and the snapshots are referred to the velocity (3 components), then A will have dimensions m x 3p. Given a matrix A ∈ R m x n and given p = min(m, n), a SVD of A is a factorization of the form:

$$A = U \Sigma {}^{t}V \tag{4}$$

Where U = (u1 … um) ∈ R m x m and V = (v1 … vn) ∈ R n x n are orthogonal. Σ ∈ R m x n is (pseudo)diagonal with diagonal elements σ1 ≥ … ≥ σk ≥ 0. σ1, . . . , σk are the singular values of A. The first k modes are selected. The number of selected modes affects the accuracy of the model. Following matrix transformations A can be rewritten as:

$$A = \sum_1^k \alpha_i U_i \tag{5}$$

These formulae allow the original data A to be represented using a reduced number of modes and coefficients, enabling a more compact representation and a reduction in the dimensionality of the data. In fact, one goes from n variables to k, where k are the weights of the modes to be evaluated as the input parameters vary. Different methods can be used for fitting the data, in this study a neural network was trained.

2.4.2. Levenberg-Marquardt (LM) backpropagation. Artificial neural networks [15] [16] are computational models inspired by the functioning of the human brain. They consist of a set of artificial neurons connected to each other via synaptic weights. Each neuron processes the input, applies an activation function and passes the result on to the other connected neurons. The LM-based neural network consists of one or more layers of interconnected neurons. It has at least one input layer, one or more hidden layers and one output layer. The main goal of this network is to approximate an unknown function, i.e. to find a relationship between inputs and outputs. The LM algorithm is an advanced variant of the classical backpropagation algorithm.Initially, the network weights are assigned randomly or using a specific initialisation technique. Input is fed through the network, and intermediate results are calculated for each layer until output. The output of a neuron is calculated:

$$y = \sigma(\sum_{i=1}^{n} \omega_i x_i + b) \tag{6}$$

The error between the expected output and the desired output is calculated using an error function, such as mean square error (MSE).The error is propagated backwards through the network. The algorithm calculates the gradients of the error with respect to the weights of the neurons using differential calculus. The updates of the weights are calculated using the LM algorithm:

$$\omega_{k+1} = \omega_k - \frac{J^T J + \lambda I}{J^T e} \Delta\omega_k \tag{7}$$

This algorithm dynamically adapts the learning steps according to the geometry of the error curves, ensuring faster and more stable convergence than stochastic gradient optimisation. Training continues until a stopping criterion is met, e.g. when the error reaches a predefined threshold or stops decreasing. This algorithm is implemented using MATLAB's trainlm function.

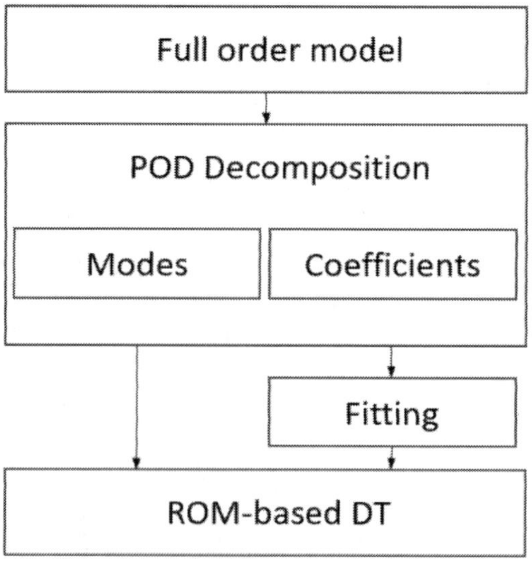

Figure 4. DT development workflow

3. Scoop air-intake design

The workflow in the figure 5 was used for the design of the FEM baseline. the reference standard was used for a preliminary design of the air intake. The resulting CAD model was used as a baseline for a CFD optimisation. Finally, the optimised geometry was studied with a FEM model to identify the correct number of composite plies. In this way, the FEM baseline was realised.

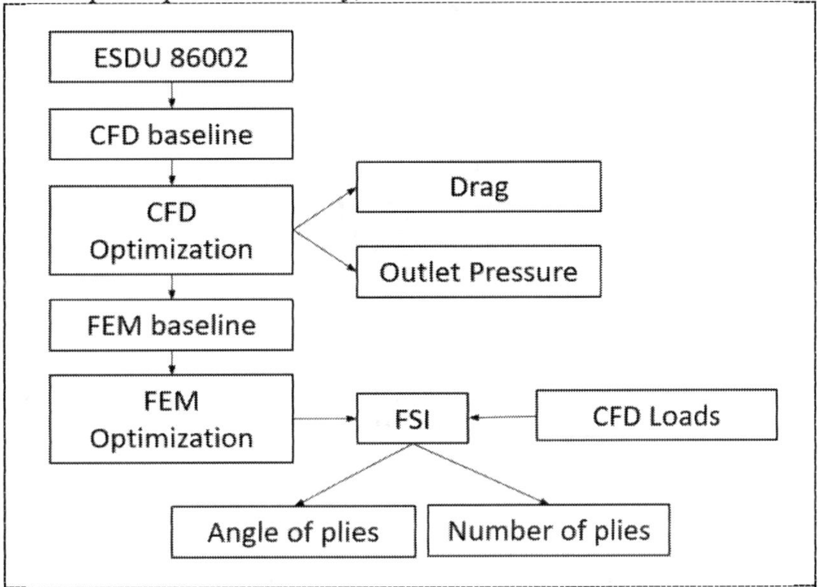

Figure 5. Design workflow

3.1. ESDU 86002

The ESDU86002 [17] standard was used for the preliminary sizing of a Scoop air intake according to defined operative working conditions. Such a geometry was used as baseline on which to generate the computational domain and to implement the shape parameterization. The intake was dimensioned to operate with a mass-flow rate of 0.86 kg/s and cruise conditions (angle of attack of 0 degrees, Mach number of 0.52).

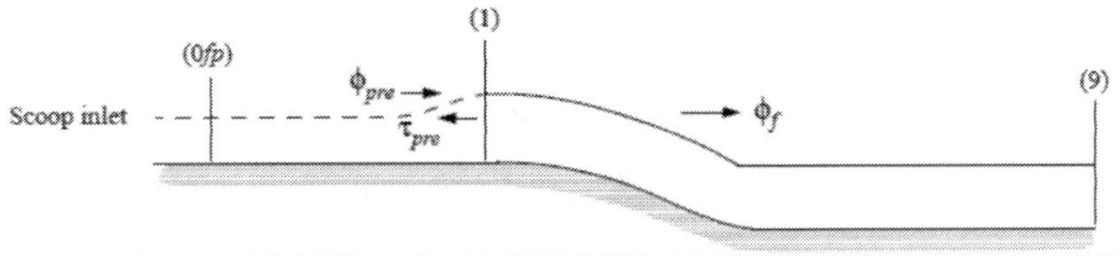

Figure 6. Figure extracted from the standard

IOP Conf. Series: Materials Science and Engineering 1306 (2024) 012025 doi:10.1088/1757-899X/1306/1/012025

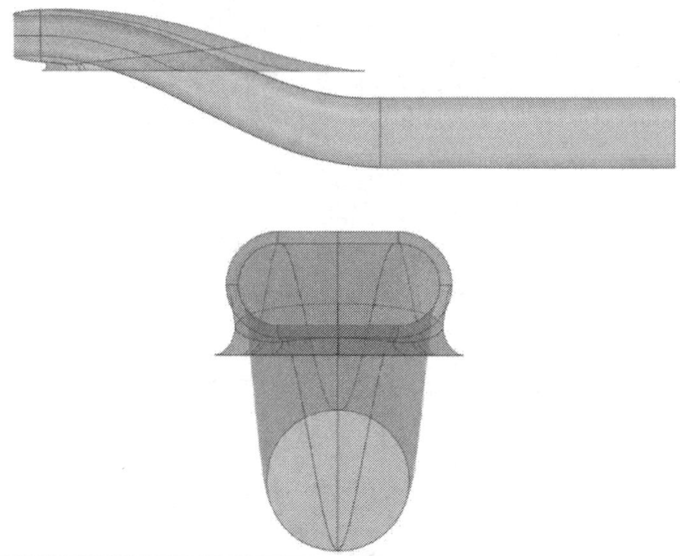

Figure 7. Geometry baseline

3.2. CFD Optimization

For shape optimisation, six shape parameters were defined acting in the most influential zones using a mesh morphing technique based on RBFs [18] [19]. Drag and output pressure were evaluated as output parameters. In fact, minimising drag leads to a reduction in consumption, while a higher outlet pressure leads to a lower pressure drop and less work for the compressor connected downstream of the air intake. To evaluate drag, all walls of the air intake (both external and internal) were considered. CFD optimisation is not the focus of this article, but for the sake of completeness, a comparison of initial and optimised geometry, and a comparison of pressure ranges has been reported. It can be seen that the separation zone on the lip of the air intake is less developed (drag reduction) and the flow within the duct is better guided. The optimization allows to obtain a reduction of drag about 32% and an increasing of outlet pressure about 86% .

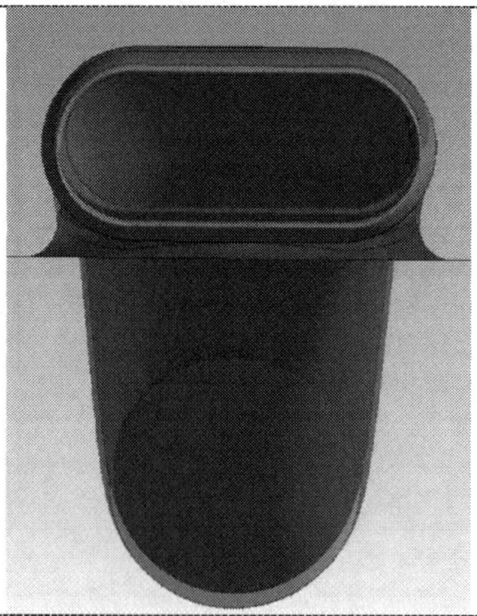

Figure 8. Comparison of baseline (red) and optimized shape (blue)

IOP Conf. Series: Materials Science and Engineering 1306 (2024) 012025 doi:10.1088/1757-899X/1306/1/012025

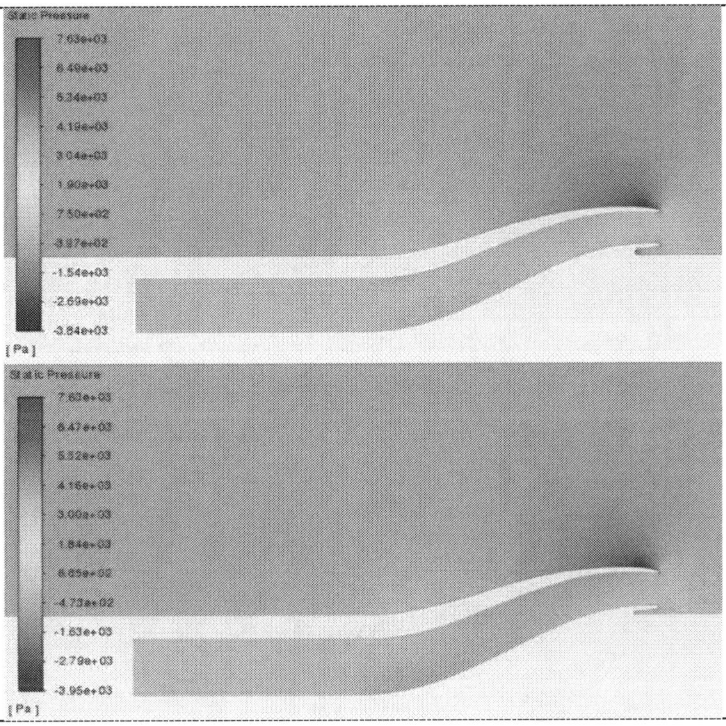

Figure 9. Pressure field of baseline (up) and optimized shape (below)

3.3. FEM Baseline

Aerodynamically optimised geometry was used as the FEM baseline. A carbon-fibre-reinforced composite material for aeronautical applications [20] was used, the main characteristics of which are shown. The material and stacking sequence used are often referred to wings, thus to much more deformable components than the one considered in this application. The Nastran solver was used for the FEM analysis. Table 2 summarises the main mesh statistics and boundary conditions used. The most severe condition was considered for the aerodynamic loads, i.e. Ma = 0.6 and angle of attack of 4 degrees. The orientation of the material follows the main flow direction.

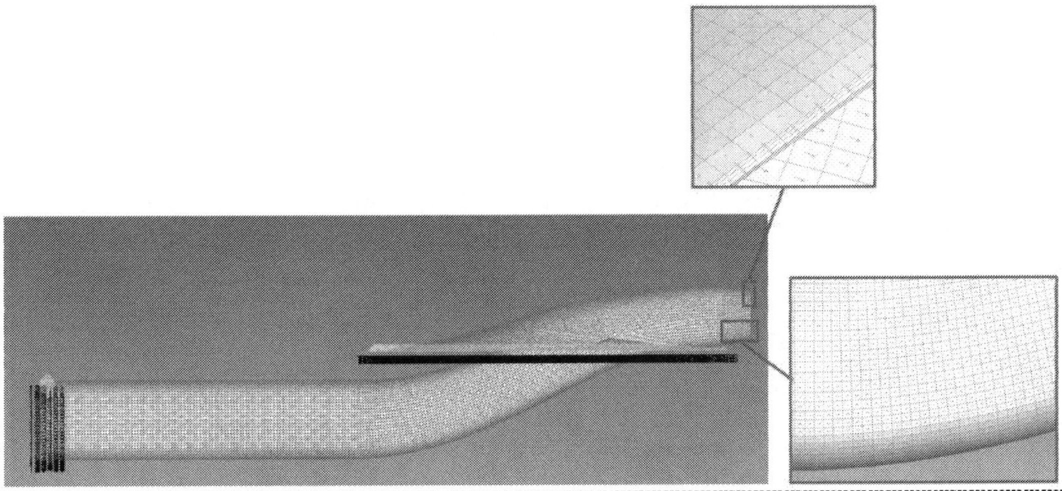

Figure 10. Mesh FEM with material orientation focus

IOP Conf. Series: Materials Science and Engineering 1306 (2024) 012025 doi:10.1088/1757-899X/1306/1/012025

Table 1. Material property

Density [kg/m^3]	*1530*
E11 [Pa]	*1.83 E+11*
E22 [Pa]	*9 E+9*
G12 [Pa]	*8 E+9*
G1z [Pa]	*8 E+9*
G2z [Pa]	*4 E+9*
X1t [Pa]	*1.5 E+9*
X2t [Pa]	*9.5 E+8*
X1c [Pa]	*7 E+7*
X2c [Pa]	*2 E+8*
S12 [Pa]	*8 E+7*
nu	*0.35*

Table 2. Main mesh statistics and Border Condition resume

Laminate Plate, Linear elements	48874
Failure Theory	Tsai-Wu
Constraints	Fixed curves
Loads	Pressure field of the most critical point
Parameters:	Number and angle of plies

Finally, an iterative process was used to fix the number of plies. The starting point was the literature configuration: 24 symmetric plies with stacking sequence $[45/-45/02/90/02/45/-45/02]_s$.
The maximum deformation was identified as the output parameter. In fact, the stresses for all the configurations studied are below the limit value and the Tsai-Wu coefficient is always less than 0.5. Conversely, with too high deformations there is a change in the fluid-dynamic behaviour. Therefore, the number of plies below which the structural deformations have a significant influence was identified. In this way, the initial configuration with 24 symmetrical plies with a thickness of 0.25 mm was changed to the final configuration with 4 symmetrical plies with a thickness of 0.125 mm. Reducing the number of plies leads to a reduction in mass of around 92%. As mentioned earlier, it should be noted that the initial configuration is generally used for other more deformable components, which explains the large gain
In the graphs shown, it can be seen that the final configuration provides good stiffness. In fact, the resulting deformations are not such as to induce significant changes in the fluid-dynamic behaviour and the FSI workflow converges to the first step. Therefore, a simple mapping of the aerodynamic loads (one way method) is sufficient to characterise the structural behaviour.
The last step considers the optimisation of the stacking sequence to minimise the maximum deformation.

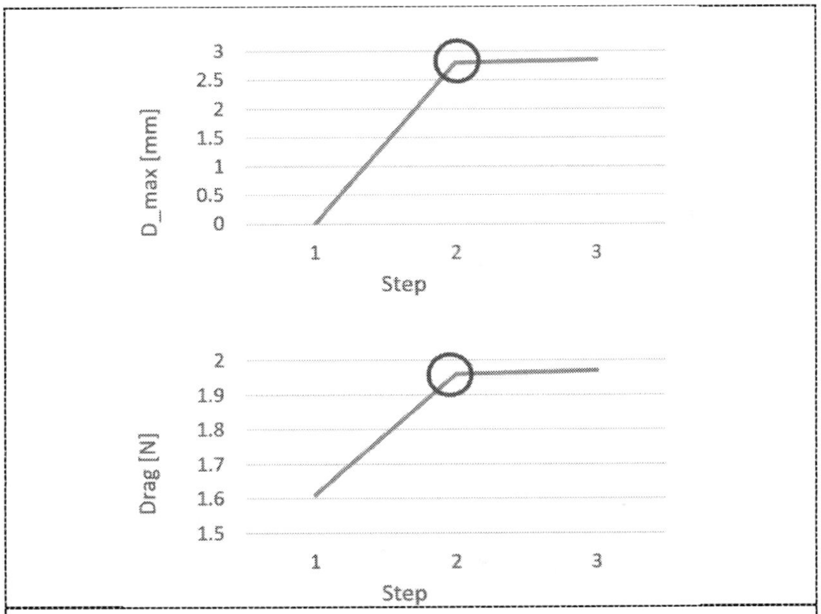

Figure 11. Convergence of FSI workflow for configuration with 4 x 0.125 mm plies [0/90]s

Table 3. Variation of CFD performances for configuration with 4 x 0.125 mm plies [0/90]s

Drag	+0.3 N
Outlet Pressure	+0.3%

4. Results

The performed analyses were used to create a ROM based DT. Ansys Twin Builder was utilized as the commercial code, and the obtained results were compared with a MATLAB code employing a neural network whose weights were optimized through LM backpropagation.

Eighty DPs were employed, with the angle of the 2 plies ranging from -90 to 90. Forty DPs were used for training, while the remaining forty were used as a test set. A comparison of the results reveals that the deformation field predicted by the ROM obtained through Ansys Twin Builder is highly reliable, with a relative error of 6%.

The ROM can be exported as a .fmu (functional mock-up unity), and it can be imported into Simulink as a black-box that takes input values and provides the desired field quantity as output. This enables the creation of a dashboard in the Matlab-Simulink environment, where all the involved physics can be visualized in a unified setting, or different components can be connected to build a more complex model.

Using the same procedure described, the CFD pressure ROM was also built. Both geometric parameters (the 6 shape parameters used for optimization) and physical parameters (cruise velocity and mass flow outlet) were considered as input parameters. In this case as well, the results are highly accurate, with a maximum relative error of 7%.

Consequently, an optimization platform was created within the Matlab-Simulink environment, which utilizes response surfaces to identify the optimum solution and employs ROMs to visualize real-time quantities of interest, both from CFD and FEM analyses, all integrated within the same interface. This allows for a better understanding of the involved physics, real-time exploration of different design points, and identification of the most interesting configurations.

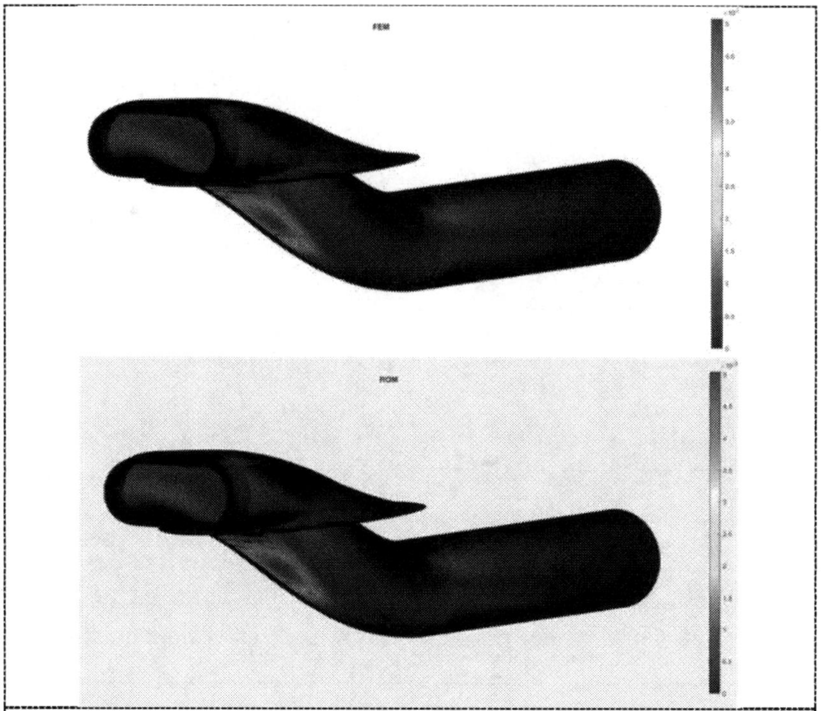

Figure 12. Comparison of FEM (up) and ROM (below) displacements for a random point of the test set

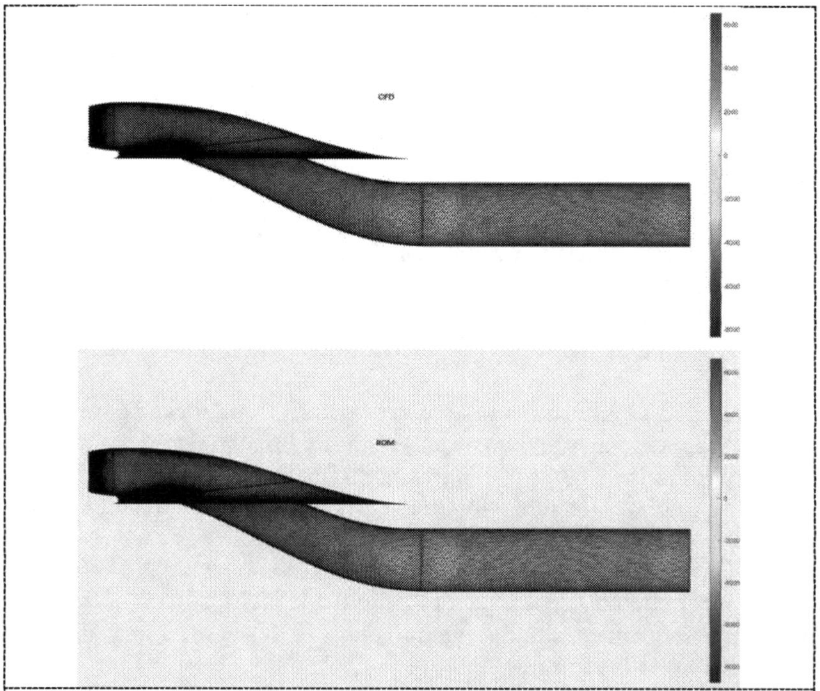

Figure 13. Comparison of CFD (up) and ROM (below) pressure field for a random point of the test set

Finally, the previously described Matlab code was used. The results are consistent with those obtained using commercial software. In particular, the following two graphs depict the displacement values for each node of a random design point from the test set and the ROM weights for the same point, respectively. It can be observed that the predicted and expected values overlap almost entirely. Furthermore, a comparison of the displacement distribution also shows a good correspondence.

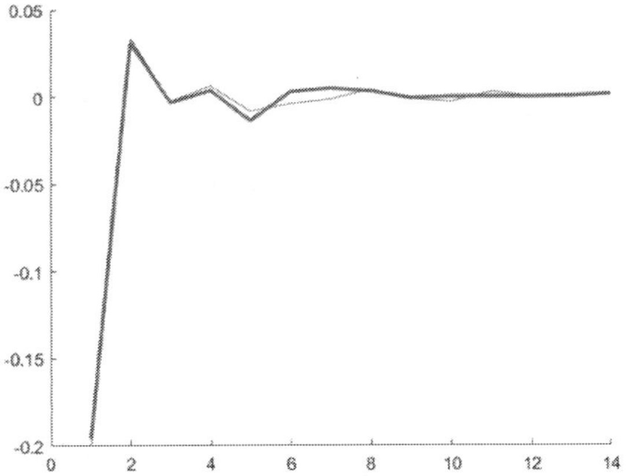

Figure 14. Weight of modes real (blue) and calculated by the neural network (orange)

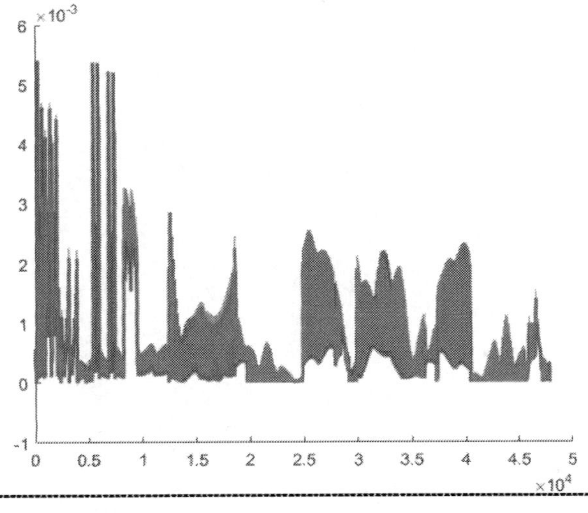

Figure 15. Displacement value for each node real (blue) and calculated by the neural network (orange)

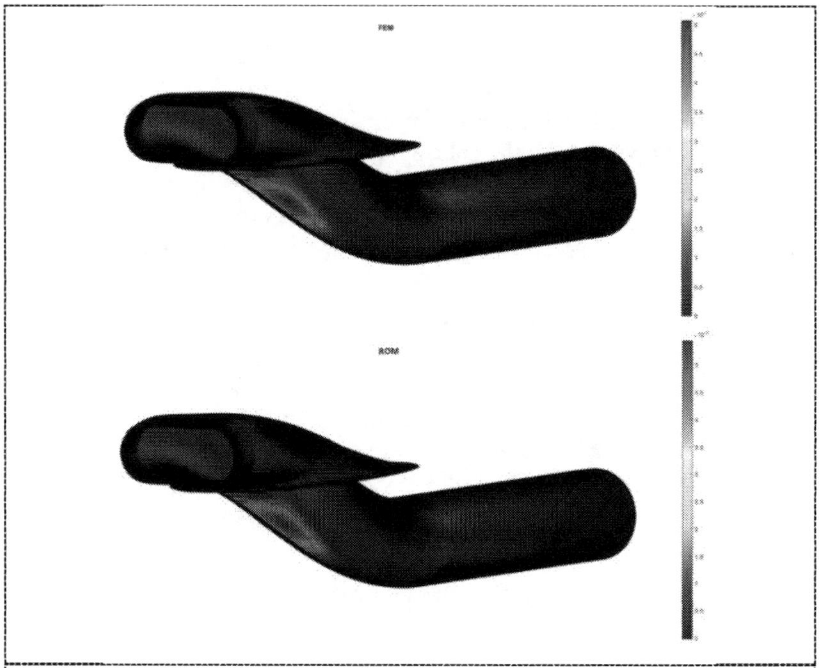

Figure 16. Comparison of FEM (up) and ROM (below) displacements for a random point of the test set

The methods described were used to identify the best configuration of ply angles. The optimal sequence is [-90/0]s and results in a 36% reduction in maximum deformation.

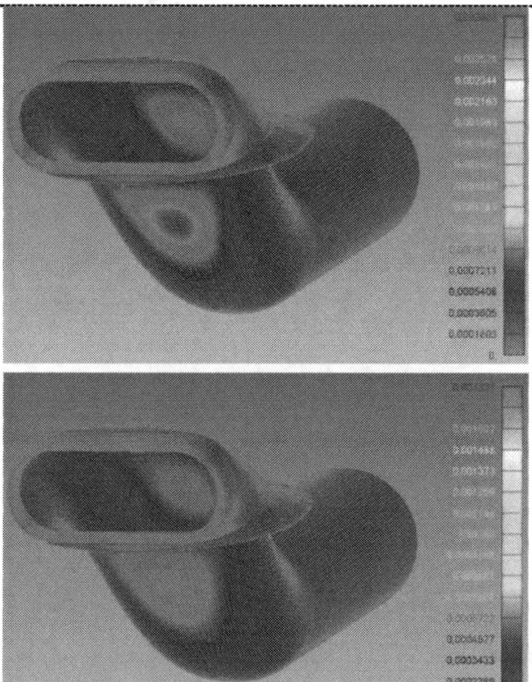

Figure 17. Displacements baseline (above) and optimized (below)

Table 4. Variation of max displacements

Initial	2.88 mm
Optimized	1.83 mm

5. Conclusions

This work has two main objectives. The first one is to provide a robust and flexible optimization method for aeronautical components. Specifically, a scoop air intake was considered, but the proposed design and optimization method are general and applicable to many other components. Compared to the geometry obtained by following reference standards, the proposed design and optimization method result in a significant improvement in fluid dynamic and structural behavior.

The second objective is to create a smart optimization dashboard that not only identifies the optimal point but also provides real-time visualization of results, exploration of parameter space while visualizing field quantities of interest to understand the physics of the problem, and the integration of all relevant physics into a single platform, all this in real time. For this purpose, DT ROM-based tools have proven to be a highly flexible and sufficiently accurate instrument.

References

[1] M. E. Biancolini, U. Cella, C. Groth, S. Porziani, Tool and methodologies for generating digital twins in medical research, 2020, Engisoft Newsletter

[2] L. Geronzi, E. Gasparotti, K. Capellini, U Cella, C Groth, S. Porziani, A. Chiappa, S. Celi, M. E. Biancolini , High fidelity fluid-structure interaction by radial basis functions mesh adaption of moving walls: A workflow applied to an aortic valve, 2021, Journal of computational science

[3] M. E. Biancolini, K. Capellini, E. Costa, C. Groth, S. Celi, Fast interactive CFD evaluation of hemodynamics assisted by RBF mesh morphing and reduced order models: the case of aTAA modelling, 2020, International Journal on Interactive Design and Manufacturing

[4] E. Kardampiki, E. Vignali, D. Haxhiademi, D. Federici, E. Ferrante, S. Porziani, A. Chiappa, C. Groth, M. Cioffi, M. E. Biancolini, E. Costa, S. Celi, The Hemodynamic Effect of Modified Blalock–Taussig Shunt Morphologies: A Computational Analysis Based on Reduced Order Modeling, 2022, Electronics

[5] M. Kasztelnik, P. Nowakowski, J. Meizner, M. Malawski, A. Nowak, K. Gadek, K. Zajac, A. Amedeo La Mattina, M, Bubak, Digital Twin Simulation Development and Execution on HPC Infrastructures, 2023, Computational Science – ICCS 2023

[6] M. E. Biancolini, Fast Radial Basis Functions for Engineering Applications, 2017, Springer

[7] C. Groth, A. Chiappa, S. Porziani a, M. E. Biancolini, E. Jacoboni, E. Serioli, F. Mastroddi, Multiphysics numerical investigation on the aeroelastic stability of a Le Mans Prototype car, 2019, Procedia Structural Integrity

[8] M. E. Biancolini, U. Cella, Flexible Engineering Toward Green Aircraft CAE Tools for Sustainable Mobility: CAE Tools for Sustainable Mobility, 2020, Springler

[9] RBF Morph User Guide

[10] D. H. Kapsoulis, V. G. Asouti1, E. M. Papoutsis-Kiachagias, K. C. Giannakoglou1, S. Porziani, E. Costa, C Groth, M. E. Biancolini, M. Andrejašič, D. Eržen, M. Bernaschi, A. Sabellico, G. Urso, G, AIRCRAFT & CAR SHAPE OPTIMIZATION ON THE RBF4AERO PLATFORM, 2016, HSTAM International Congress on Mechanics

[11] C. Groth, Adjoint-based shape optimization workflows using RBF, 2017, PhD Thesis

[12] A. S. Rowayda, SVD Based Image Processing Applications: State of The Art, Contributions and Research Challenges, 2012, International Journal of Advanced Computer Science and Applications

[13] S. L. Brunton, B. R. Noack, P. Koumoutsakos, Machine learning for fluid mechanics, 2020

Annual Review of Fluid Mechanics
[14] J. N. Kutz, Data-driven modeling & scientific computation: methods for complex systems & big data, 2013, Oxford University Press
[15] D. Chen, X. Gao, C. Xu, FlowGAN: A Conditional Generative Adversarial Network for Flow Prediction in Various Conditions, 2020, 32nd International Conference on Tools with Artificial Intelligence (ICTAI 2020)
[16] R. E. Chaal, M. Othman, A comparative study of back-propagation algorithms: Levenberg-Marquart and BFGS for the formation of multilayer neural networks for estimation of fluoride, 2022, Communications in Mathematical Biology and Neuroscience
[17] ESDU 86002. Drag and pressure recovery characteristics of auxiliary air inlets at subsonic speeds, 2004
[18] M. E. Biancolini, E. Costa, U. Cella, C. Groth, G. Veble, M. Andrejašič, Glider fuselage-wing junction optimization using CFD and RBF mesh morphing, 2016, Aircraft Engineering and Aerospace Technology
[19] U. Cella, C. Groth, M. E. Biancolini, Geometric Parameterization Strategies for shape Optimization Using RBF Mesh Morphing, 2016, Advances on Mechanics, Design Engineering and Manufacturing (Springer)
[20] C. Santiuste, X. Soldani, M. H. Miguélez, Machining FEM model of long fiber composites for aeronautical components, 2010, Composite Structures

IOP Conf. Series: Materials Science and Engineering 1306 (2024) 012026 doi:10.1088/1757-899X/1306/1/012026

Smart design of a high-performance seat frame for a luxury car

D Milone[1], L Dusini[2], A Nicoletta[2], D Santonocito[1]* and G Risitano[1]

[1]University of Messina, Department of Engineering, Contrada di Dio, 98166 Messina Italy
[2]Ferrari S.p.a., via Abetone, Maranello, 41053 Modena Italy

*email: dsantonocito@unime.it

Abstract. The design of mechanical components in recent years has been experiencing a series of innovative processes which, in addition to the classic methodologies, also support topological and structural optimization considering the product life cycle (LCA). These innovative aspects make it possible to reduce the weight of the components and the impact on the environment, both during production and during operation. The purpose of this work is to create a tool for the "intelligent design" of automotive components through optimization algorithms called "Smart Design". The final intent is to create a multidisciplinary methodology for mechanical design that combines topological optimization techniques and product life cycle analysis. The single studies, from the creation of the CAD and the finite element analysis, up to the topological optimization and the LCA analysis, are managed by a single code generated in the Matlab® environment. The work is carried out in collaboration with Ferrari S.p.a.

1. Introduction

The design of mechanical components in recent years has been experiencing a series of innovative processes which, in addition to the classic methodologies, also support topological and structural optimization considering the product life cycle (LCA) [1] These innovative aspects make it possible to reduce the weight of the components and the impact on the environment, both during production and during operation [2]

The purpose of this work is to create a tool for the "intelligent design" of automotive components through optimization automatic method called "Smart Design". Smart Design was presented for the first time in [3] by Previti et al. This new way of design, verifying and re-designing mechanical components has been called "Smart Design". Indeed, in the modern era, it is impossible to think that structural design does not make use of virtual, digital, rapid and secure prototyping methods. In fact, the final intent is to create a multidisciplinary methodology for mechanical design that combines topological optimization techniques and product life cycle analysis.

The single studies, from the creation of the CAD and the finite element analysis, up to the topological optimization and the LCA analysis, are managed by a single original optimized algorithm generated in the Matlab® environment by authors on a rear seat frame for a luxury car. The work is carried out in collaboration with Ferrari S.p.a.

2. Case study of a rear seat of a luxury car

The Smart Design algorithm was applied to the design of a rear seat of a luxury car. This kind of mechanical component must respect several prescriptions regarding mechanical performances and safety conditions. At the same time, it must be the lighter as possible, reducing its mass. This can be accomplished introducing in the design process also the analysis of the life cycle of the component, leading to a reduction of the environmental impact of all the production cycle of the component, starting from the extraction of raw materials. The goal of the Smart Design algorithm is to minimize an objective function which considers the aspects of mass reduction and equivalent greenhouse gases reduction. In the following sections the Smart Design algorithm will be analysed in all its aspects.

2.1. Design space preparation

To reduce the mass of the component, the Smart Design algorithm must work on a design space which represent the maximum and minimum boundary for topology optimization. The design space may also contain other objects which are related to the functionality of the component and cannot be of interest in the design space but allow to bound the component with the surrounding environment or allow to load it.

In this case study, we consider a rear seat of a luxury car, represented in Figure 1 as a series of box which represent the design region for the Smart Design algorithm. This design space was provided by Ferrari together with the coordinates of the body attachment place. The requirement is that the link between the rear seat and the metallic bracket must be ensure.

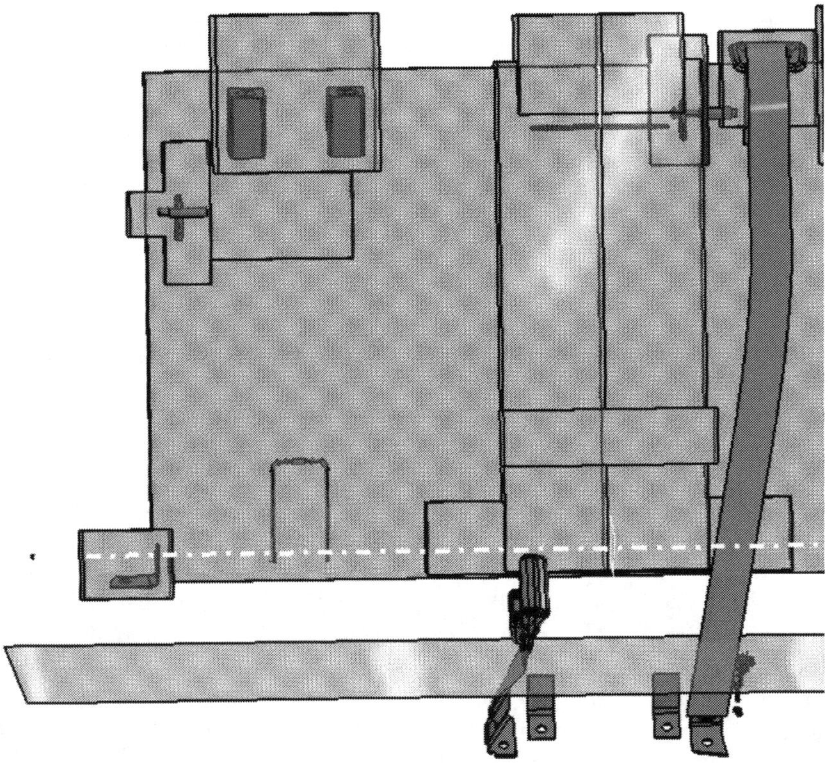

Figure 1. Mechanical component under study: rear seat of a luxury car.

All the mechanical connection within the body and the seat are not of interest for the algorithm and must be removed by boolean subtraction operations to provide a clear geometry with simplified constraints (Figure 2).

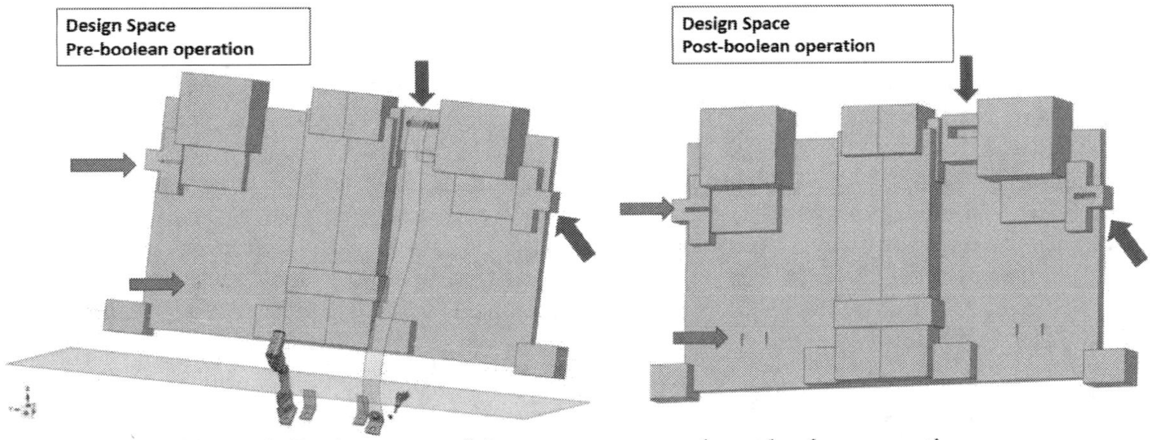

Figure 2. Design space of the rear seat, pre and post boolean operation.

2.2. Loadcases

The component must withstand to mechanical loads typical for its functions; indeed, it must assure the safety of the passengers. To achieve this, several loading conditions as prescribed by automotive standards were considered. The Federal Motor Vehicle Safety and Security (FMVSS) standard specifies the requirements relating to seats, equipment, and assembly to minimize the possibility of failure due to forces acting on the seat in the event of a vehicle impact. A load is applied to the centre of mass of the seat equal to 20 times its weight, in the horizontal direction (FMSVSS 207, Figure 3 case 1). The standard also provides to consider a safety margin, i.e. due care, which consists in applying a force equal to 24 times the weight of the seat, always in a horizontal direction. According to FMVSS301, to simulate rear impact, the rear of the vehicle must be subjected to 80 km/h ± 1 km/h with a movable deformable barrier of 1,386 kg. It will result in a static load of 10kN on the apex of the belt attachment, in the horizontal direction (Figure 3 case 2) [4]. the FMVSS 207/210 requires to perform a static anchorage test, by applying a load equal to 24 times (20 times + 20% of due care) the weight of the seat in its centre of gravity, and to apply a seat belt load of 13.35 kN plus a 20% of due care to shoulder and abdominal region with a 10° angle respect the horizontal plane (Figure 3, case 3) [5]. The ADR34/03 standard considers the presence of seats used for sitting infants and the presence of brackets called Top Tether (Figure 3, case 4), while the ECER-17 prescribes that the seat frame and head restraint system shall not exhibit any structural failure when subjected to a load of 1068 N (ECE + due care, Figure 3 case 5). It is designed to reduce damage to people due to the impact of luggage in the bonnet with the head of the seat.

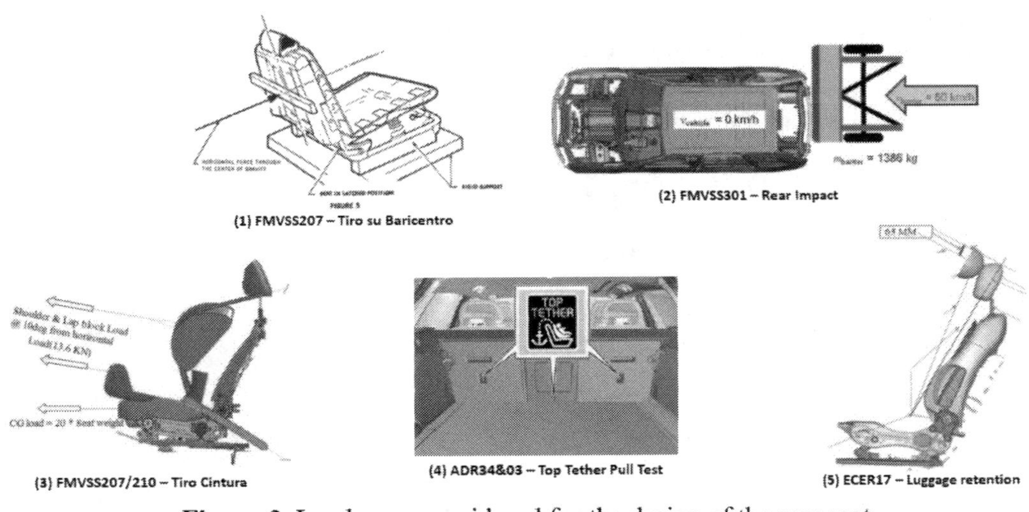

Figure 3. Loadcases considered for the design of the rear seat.

IOP Conf. Series: Materials Science and Engineering 1306 (2024) 012026 doi:10.1088/1757-899X/1306/1/012026

2.3. Finite Elements analysis and Topology optimization

All the previous load cases were replied with static structural finite elements analysis in Ansys Workbench. The design space was meshed with a SOLID186 hexa mesh, considering a minimum edge size of 5 mm and activating the option "mesh defeaturing" and "capture curvature" (Figure 4). After a convergence analysis, the number of elements was 137816. The seat was bonded considering the loading conditions as prescribed by the standards. A total number of 5 linear elastic static finite element analysis were created. To properly design the seat and to consider the combined effects of the 5 different load conditions, the superposition principle has been exploited. The results coming from the different load cases are the input for the topology optimization algorithm. In the FE simulation process several parameters were considered that allow to variate the mechanical properties (Young's Modulus, Poisson's ratio and density) of the material under study.

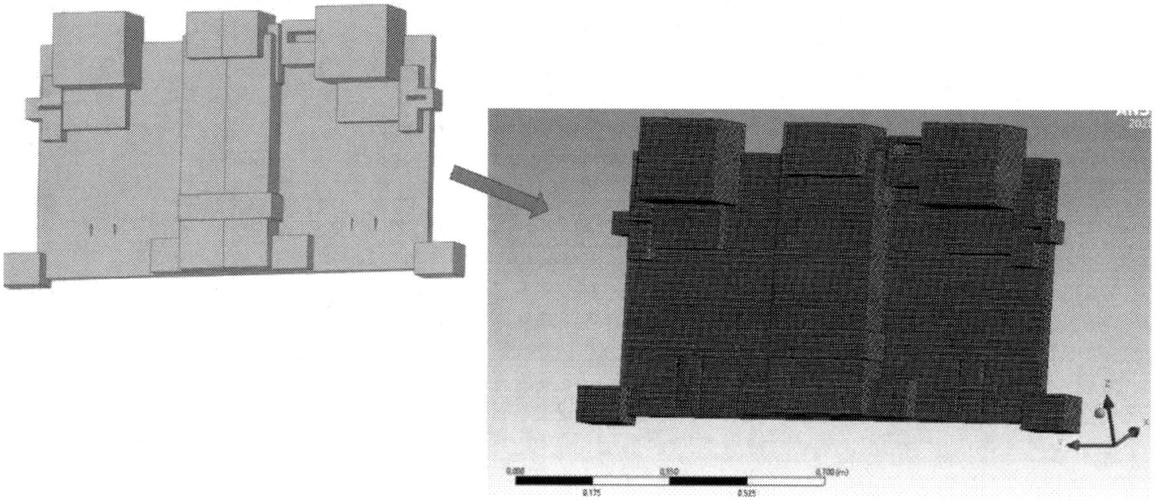

Figure 4. Mesh of the rear seat.

The topology optimization process consists in minimizing the mass of the components maintaining the same compliance, i.e. the stiffness of the component. All the five inputs load cases have a weight of 1 into the optimization process.

2.4. Life Cycle Assessment

The novelty of the Smart Design algorithm is to consider, in the design process, also the life cycle of the component starting from the raw material up to the introduction to the market of the component. This aspect was modelled setting all the parameters for a cradle-to-gate life cycle assessment (LCA) in the OpenLCA software. The scoping of the LCA was to analyse the seat production, considering a cradle-to-gate boundary condition. The impact category was selected as the mass of carbon dioxide emission ($kgCO_2$) in air or equivalent mass of carbon dioxide emission to consider the impact of several greenhouse gases on the environment ($kgCO_{2\ eq}$). The inventory considers the input and output in terms of total $kgCO_2$ equivalents relating to each process: extraction of materials, production of the component and related technological process, transport to the site. The transportation impact was considered with a raw material producer 100 km far away from the Ferrari production site. The evaluation of the environmental impact was assessed through the Climate Change Intergovernal Panel on Climate Change – Global Warning Potential on 20 years (IPCC 2013 GWP 20y). The results were retrieved to produce a single rear seat.

2.5. Control process and objective function

All the aspects of the Smart Design algorithm are controlled by a Matlab® script that guide the flux of the process by writing and reading the input and output parameters. The script takes as input the material properties (Young's Modulus E, Poisson' ratio υ, and density ρ) and the percentage of mass that we want to retain. Starting from an Ansys journal file, the script set the input variables and then run the FE simulation in batch mode. After that, if the FE simulations are verified (material failure stress are not reached), the script move to the LCA of the component and then to its topology optimization process. The output variables are the maximum, minimum and mean stress for each loadcases, the deformation on X axis, the initial and final mass (kg), the initial and final volume (m^3) and the initial and final carbon dioxide emissions ($kgCO_{2\ eq}$).

The goal of the algorithm is to minimize an objective function with requires as input the mass and the emission of carbon dioxide. The analytical formulation of the objective function is:

$$Obj\left(m,Kg_{CO2eq}\right)=a\frac{m_f}{m_i}+b\frac{Kg_{CO2eq\,f}}{Kg_{CO2eq\,i}}+c\frac{m_f}{m_{max}} \quad (1)$$

With a,b,c arbitrary weights where a+b+c= 1. The terms m_i and m_f are the initial and final mass; the $kgCO_{2\ eq\ i}$ and $kgCO_{2\ eq\ f}$ are initial and final results of the LCA analysis, while m_{max} is the maximum mass obtained in all the iterations.

The first ratio considers the reduction of mass, while the second one considers the reduction of emissions in the production cycle. The last ratio considers the reduction of mass in absolute terms considering all the tested materials.

Regarding the arbitrary weights, they allow to change the importance of each of the three contributions to the objective function. In our case study, we consider three different scenarios: Ferrari statement (a= 0.3; b= 0.3; c= 0.4); LCA dominant (a= 0.1; b= 0.8; c= 0.1); Mass dominant (a= 0.1; b= 0.1; c= 0.8).

3. Results

The Smart Design algorithm allows to analyse the effect of several materials in the production process of a mechanical component in terms of mass reduction, strength and environmental impact. After that the material has been chosen, the algorithm starts a simulation process where, at the end, the initial mass and initial greenhouse gas emissions have been compared with their final values, after that the optimized geometry has been reached.

The materials involved in this analysis are metals, plastics, carbon polymers and aluminium alloys. The input parameters considered are the mechanical properties of the material (E, υ, ρ) and the percentage of retained mass. In Table 1 are reported the results of the simulations performed with carbon polymers, while in Table 2 the result of the aluminium alloys. It is evident the reduction of the mass of the rear seat which also reflect a reduction of the emission of greenhouse gases.

Table 1. Results of the Smart Design algorithm for carbon balsa.

	Initial mass (kg)	Final mass (kg)	Initial LCA (kgCO2eq)	Final LCA (kgCO2eq)
Carbon balsa	66.76	39.4	120	89.4
Carbon PP	66.5	39.34	121	88.17

Table 2. Results of the Smart Design algorithm for aluminium alloys.

	Initial mass (kg)	Final mass (kg)	Initial LCA (kgCO2eq)	Final LCA (kgCO2eq)
Al 5005	128.8	76.05	208.3	132
Al 5754	125.8	76.05	208.27	131.97
Al 6061	125.8	76.05	208.28	131.8
Al Alloys Series 6000	127.80-139.25	75.49-82.25	209-210	130-132

The adoption of different materials lead to different final configuration of the optimized component. In Figure 5, it is possible to observe the difference between the topology optimization performed with AISI 316L and carbon balsa. There are small differences between the two final geometries, particularly near some critical points represented by the fixing site.

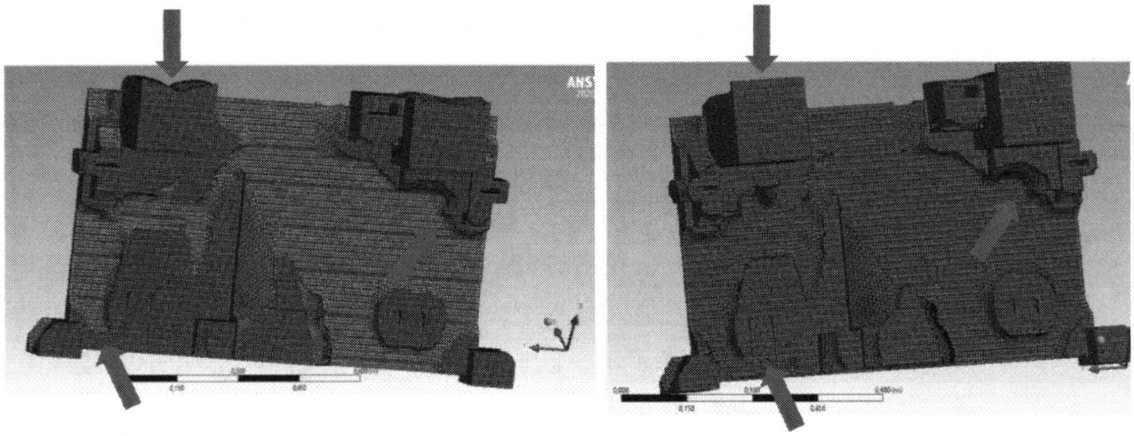

(a) (b)

Figure 5. Topology optimization of the rear seat: a) AISI 316L; b) Carbon Balsa.

The following graphs report the value of the objective function respect the adopted material for the three scenarios. For the Ferrari statement (Figure 6a), the materials which deeply reduce the weight and the environmental impact are the carbona balsa and the aluminium alloys. For this scenario, the mass reduction and the greenhouse gas emission aspects are considered with the same importance. For the LCA dominant scenario (Figure 6b), more attention has been paid in minimizing the contribution of the production cycle on the final component. In this case, carbon balsa has the best position within the material, followed by plastics. In the Mass dominant scenario (Figure 6c), it is important to drastically reduce the mass of the final component compared to all the other materials. Plastic materials allow to do that, obtaining very lightweight component.

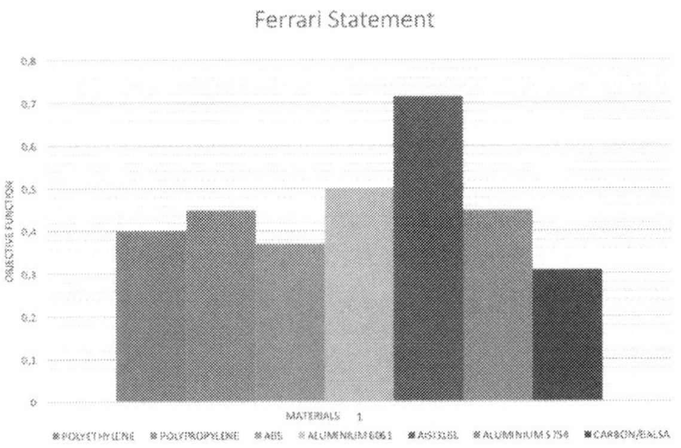

(a)

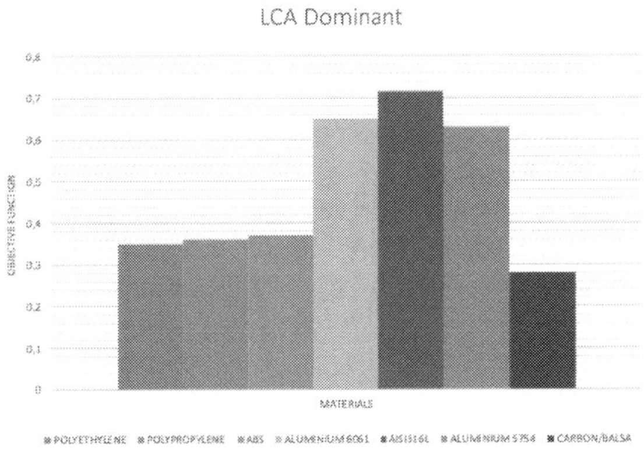

(b)

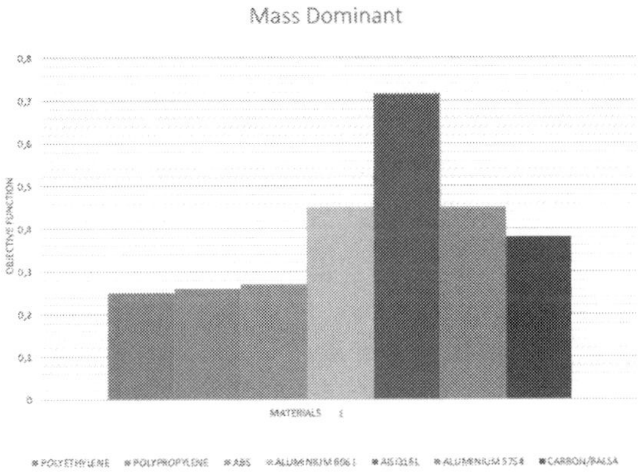

(c)

Figure 6. Results of the Smart Design algorithm in terms of Objective function for the three scenarios.

The component, which is obtained from the topology optimization process, must be redesigned to retrieve the final geometry for the production process. Raw geometries were redesigned considering the median plane by interpolating the points linearly and with quadratic splines (minimum guaranteed thickness 2 mm); in this way, clear shape is obtained. It is important to observe how the final component is not symmetrical due to the non-symmetry of the load cases (Figure 7).

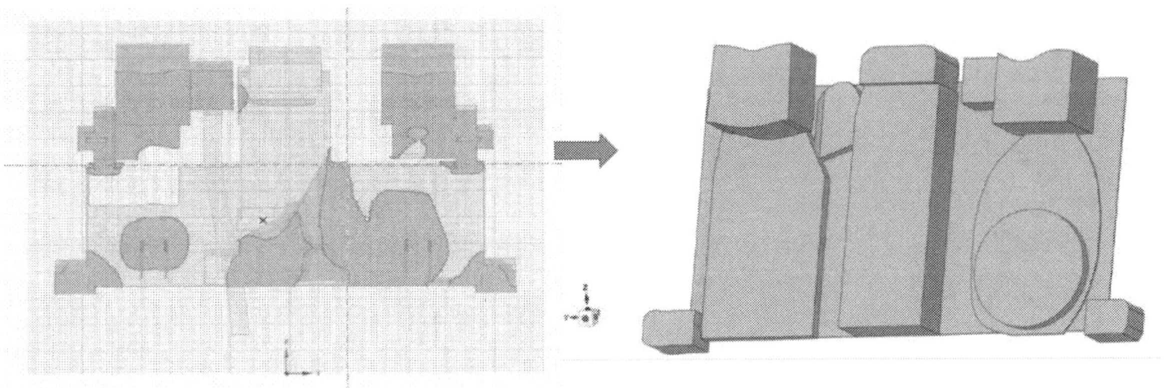

Figure 7. Design obtained from the topology optimization process and its cleaned version.

IOP Conf. Series: Materials Science and Engineering 1306 (2024) 012026 doi:10.1088/1757-899X/1306/1/012026

4. Conclusion

The aim of this work is to create a new tool for the "intelligent design" of seat frame for a luxury car through optimization algorithms. This new way of design, verifying and re-designing mechanical components has been called "Smart Design". The single studies, from the creation of the CAD and the finite element analysis, up to the topological optimization and the LCA analysis, are managed by a single original optimized algorithm generated in the Matlab® environment.
The design considered the aspects of mechanical performance, safety and environmental impact. This new approach allows you to carry out a custom-made design, leaving the user to choose the weights of the individual aspects.

The effects of mass reduction and CO2 production were analysed for different types of material, thus allowing an appropriate choice of the same.

Smart Design thus provided an optimized geometry which was the starting point for the final design of the component.

The individual phases (CAD, FEM and LCA) were managed by a single code, created by the authors. The most satisfactory results were obtained from aluminium alloys and carbon composite materials with balsa cores.

References

[1] AD La Rosa, G Cozzo, A Latteri, G Mancini, A Recca and G Cicala 2013 *A comparative life cycle assessment of a composite component for automotive* (AIDIC)
[2] Gonçalves M, Monteiro H and Iten M 2022 Life Cycle Assessment studies on lightweight materials for automotive applications - An overview *Energy Reports* **8** 338–45
[3] Previti U, Galvagno A, Risitano G and Alberti F 2022 Smart Design: Application of an Automatic New Methodology for the Energy Assessment and Redesign of Hybrid Electric Vehicle Mechanical Components
[4] Żuchowski A 2018 Analysis of the influence of the car impact velocity on the loads of the dummies in the front and rear seats *The Archives of Automotive Engineering* **81**
[5] Mulla Salim H, Yadv Sanjay D, Shinde D and Deshpande G 2013 Importance of Federal Motor Vehicle Safety Standards 207/210 in Occupant Safety - A Case Study *Procedia Eng* **64** 1099–108

IOP Conf. Series: Materials Science and Engineering 1306 (2024) 012027 doi:10.1088/1757-899X/1306/1/012027

An injury risk-based comprehensive framework for testing and assessing ADAS functions in critical road scenarios

Michelangelo-Santo Gulino[1], Giulio Vichi[1], Luigi Di Lillo[2], Alessandro Gianfelici[2], Dario Vangi[1]

[1]Department of Industrial Engineering, Università degli Studi di Firenze
[2]Property & Casualty Solutions, Swiss Reinsurance Company Ltd.

E-mail: michelangelosanto.gulino@unifi.it

Abstract. Assisted driving is currently considered a key aspect for improving road safety, and automakers and OEMs are working to achieve higher levels of vehicle automation by introducing new technologies and Advanced Driver Assistance Systems (ADAS) in the circulating fleet. This trend requires test protocols for vehicle safety assessment to be frequently reviewed and updated, considering the latest advances in the state of the art regarding ADAS functions and systems. As of today, performance assessment programs (such as NCAP) mainly evaluate how an ADAS behaves in terms of crash avoidance in specific critical scenarios, which represent the most frequent crash constellations among real-world impacts. However, enhanced safety can be also obtained in case the impact is not avoided if a decrease in Injury Risk (IR) for the involved road users is achieved by ADAS intervention, compared to the case of no intervention.

The purpose of this work is to propose an overall framework to draft or update test protocols for ADAS performance assessment based on real car-to-car impact observations, representative of impact scenarios in terms of both occurrence frequency and IR. First, the in-depth accident database IGLAD is analyzed to identify the most relevant car-to-car accident scenarios based on a relevance indicator, i.e., the risk level being the multiplication of the occurrence frequency and IR for a specific scenario. For each relevant scenario, a risk level-based strategy to identify one significant closing speed between vehicles for the tests is defined; the test collision speed for the two vehicles is determined analogously, and the risk level for each combination of speeds in a scenario represents the maximum achievable score by the ADAS if the collision is averted. Considering the well-established Euro NCAP framework as a relevant starting point for the definition of test protocols, two examples are highlighted regarding the proposal of a new test protocol and an update of an already existing one. Finally, a method is proposed for ADAS performance assessment if the impact is not avoided, scaling the maximum achievable score based on the IR reduction consequent to the ADAS intervention.

1. Introduction

In the rapidly evolving landscape of automotive technology, Advanced Driver Assistance Systems (ADAS) have emerged as a pivotal force in enhancing road safety, driving convenience, and the prospect of autonomous mobility. Continuing technological advances are driven by ADAS performance evaluation methodologies capable of highlighting points of possible weakness at the functional or logical level, to increase their flexibility towards critical situations and result in a higher user acceptability. ADAS performance evaluation methods are divided into predictive

IOP Conf. Series: Materials Science and Engineering 1306 (2024) 012027 doi:10.1088/1757-899X/1306/1/012027

(*a priori*) and retrospective (*a posteriori*) methods. If the ADAS system – or function – to be tested is implemented on a limited portion of the circulating fleet or is in the prototype phase, the method is called predictive and aims primarily at identifying conditions of highest efficiency [1, 2]. If, on the other hand, the implementation involves a large number of vehicles (as in the case of an established function such as Autonomous Emergency Braking, AEB), the method is referred to as retrospective: "what if" analyses are in this case typically performed, regarding how the outcome of a road impact would change in case of the function implementation onboard one or multiple vehicles [3].

Among the retrospective methods, protocols developed in the context of frameworks like the New Car Assessment Programme (NCAP) are an important reference for technicians and consumers. However, assessment methods are heterogeneous across continents. In the United States, NCAP is limited to assessing if a specific system is available in the vehicle and whether it performs according to a set of minimum requirements in specific situations; however, the ADAS performance scoring method is currently under definition[1]. Its European counterpart, Euro NCAP conversely introduced in 2018 a series of tests for various ADAS functions in car-to-vulnerable road users (car-to-pedestrian, car-to-bike, etc.) and car-to-car imminent collision scenarios. The latter[2] are aimed at representing the most frequent accident scenarios and those in which already established systems like AEB are able to operate at maximum efficiency in terms of crash avoidance. Nevertheless, increasing road safety does not only depend on crash avoidance but also on limiting the Injury Risk (IR) for the involved parties when an impact cannot be averted [4, 5].

An analysis of the Euro NCAP scoring criteria for an AEB system highlights that the fundamental performance evaluation parameter lies in the closing speed at collision V_r between the ego vehicle (or Vehicle Under Test, VUT) and the opponent vehicle (or Global Vehicle Target, GVT). Although V_r is an important factor influencing IR in collisions, its relationship with IR is a complex function that is often expressed by means of an S-shaped curve, derived by logistic regression on accident data [6]. As an example, a model of IR for frontal, centred impacts is reported in Figure 1, which is based on the injury metric Maximum Abbreviated Injury Scale higher than 3 (MAIS 3+) [1]. On the right in Figure 1 is reported the EuroNCAP scoring for a CCRs scenario of AEB testing with the VUT moving at 80 km/h and stationary opponent; starting from the IR value in the case of no intervention by the AEB (V_r=80 km/h, IR=53%) also known as Reference Scenario RS, the absolute reduction in IR is also reported on the right in Figure 1. This latter curve is directly derived from the IR model in the left of Figure 1. It can be seen that in the zone at $75<V_r\leq80$ a score of 0% is applied to the ADAS because the impact occurs at almost the same speed as the RS. In the $30<V_r\leq75$ zone, where the system has the possibility of significantly reducing IR with reduced changes to V_r, Euro NCAP provides a score of 0.25 or 0.50. Similarly, the score varies significantly between $0<V_r\leq30$, although IR in this range is almost constant and close to 0% – IR(V_r=30 km/h)=3%. Such a gap between the IR reduction trend and Euro NCAP scores emphasises that the representativeness of real accidents in terms of IR is not currently prioritized in the definition of ADAS performance assessment protocols. It should also be considered how, in more complex impact scenarios such as intersection crashes, different activations may result in changing the impact type (e.g., from front-to-front to front-to-side) with substantial changes to IR for the occupants of the involved vehicles [7]. For these reasons, it is essential to propose a set of tools for the definition of test protocols and scenarios that can effectively represent IR in real road impacts.

This work aims to propose a framework for the definition of car-to-car testing protocols and scenarios for ADAS performance assessment that are representative of real-world impacts

[1] Docket No. NHTSA-2021-0002

[2] European New Car Assessment Programme (Euro NCAP) – Assessment Protocol – Safety Assist Collision Avoidance, Implementation 2023, v10.2, November 2022

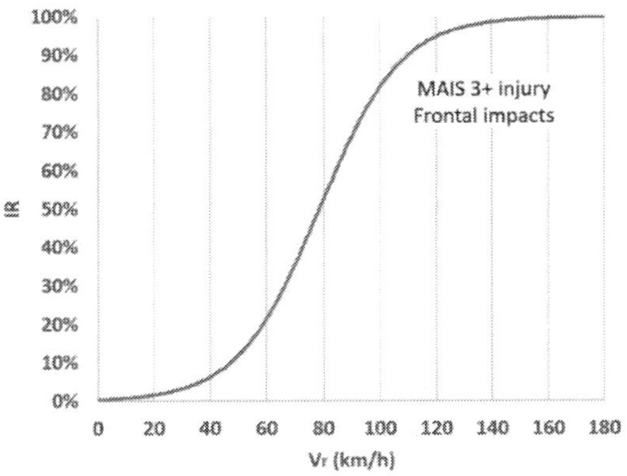

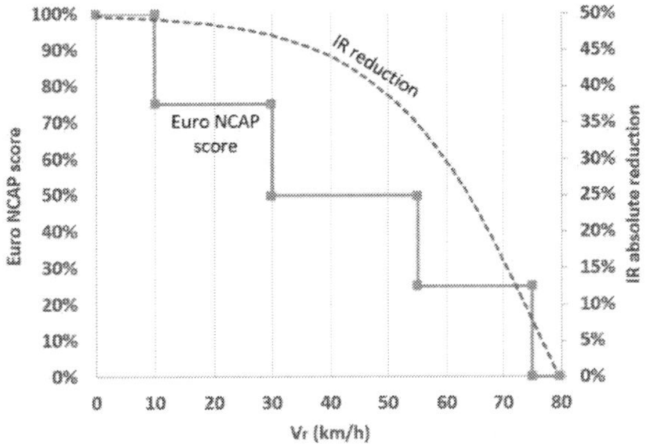

Figure 1. IR model for MAIS 3+ injuries in frontal, centred impacts (left [1]), comparison between Euro NCAP scoring method for a CCRs test at 80 km/h and achievable absolute reduction in IR as a function of V_r (right).

in terms of both frequency and IR for the vehicle occupants; this involves the provision of a methodology for evaluating performances in both cases where the onboard system is able to avoid the impact or not. The framework is applicable to both the development of new protocols and the updating of any existing protocols. From the latter perspective, the EuroNCAP protocols available to date are taken as a reference to illustrate the use of the proposed method.

The activities performed can be sequentially listed as follows:

(i) An in-depth accident database is employed to extract data on real car-to-car impact scenarios in terms of frequency and IR for the involved occupants so that the most relevant scenarios for ADAS testing can be identified;
(ii) For the selected scenarios, the closing speed and collision speed to be applied to the VUT and the GVT have been chosen based on frequency and IR-related criteria;
(iii) The test protocols and the maximum score achievable by an ADAS are proposed, for test procedures still to be developed or already available inside the Euro NCAP program;
(iv) An ADAS scoring methodology is introduced to provide an assessment of ADAS performance in both cases of avoided and unavoided impacts.

2. Data extraction from an in-depth accident database and selection of relevant scenarios

2.1. IGLAD database

The IGLAD database (Initiative for the Global Harmonization of Accident Data) includes a classification of relevant scenarios for ADAS assessment, initially proposed by a team of experts from Continental AG in 2019 [8]. All IGLAD data coded after 2019 are hence classified in terms of scenario ("Scenario Type" variable). Each participant in an accident is characterized by a label among 42 possible labels. The 42 possible scenarios are divided into 6 classes: "Driving", "Longitudinal", "Oncoming", "Turning", "Crossing", and "Other". Some of these scenarios are already considered inside the Euro NCAP program, as synthetically reported in Table 1, evidencing all available test scenarios for AEB, car-to-car EuroNCAP protocols and the related equivalent from the IGLAD classification. Based on these highlights, the Euro NCAP tests feasible for an update with the proposed framework are represented by seven protocols.

IOP Conf. Series: Materials Science and Engineering 1306 (2024) 012027 doi:10.1088/1757-899X/1306/1/012027

Table 1. AEB test protocols already implemented in the EuroNCAP framework (left column) with the corresponding scenario of the IGLAD classification (right column).

	EURONCAP	**IGLAD "Scenario Type"**
1	CAR-TO-CAR REAR STATIONARY (CCRs)	Running up
	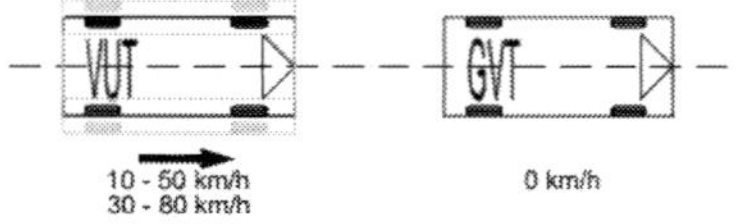	
2	CAR-TO-CAR REAR MOVING (CCRm)	Running up
	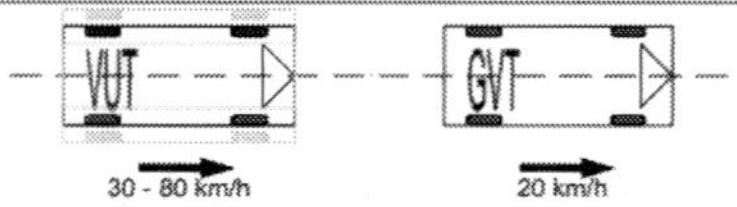	
3	CAR-TO-CAR REAR BRAKING (CCRb)	Running up
	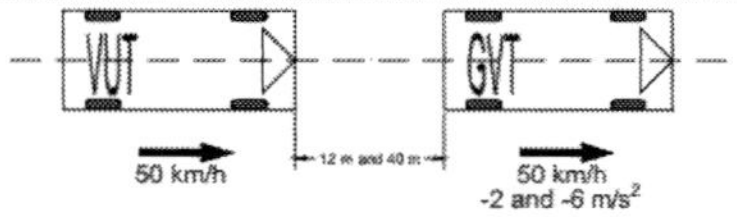	
4	CAR-TO-CAR FRONT TURN ACROSS PATH (CCFtap)	Turning farside (object oncoming)
	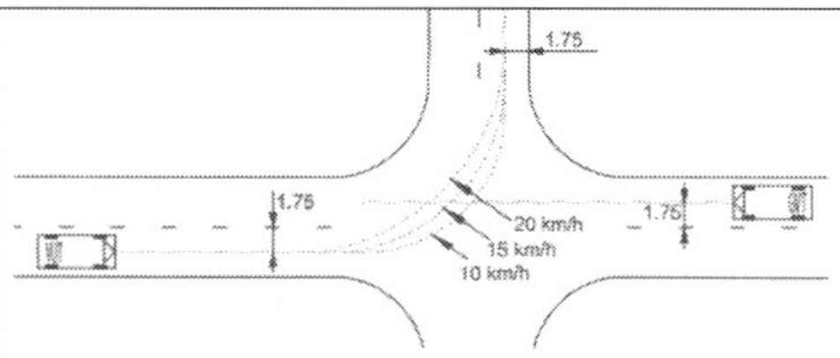	
5	CAR-TO-CAR CROSSING STRAIGHT CROSSING PATH (CCCscp)	Crossing from farside
	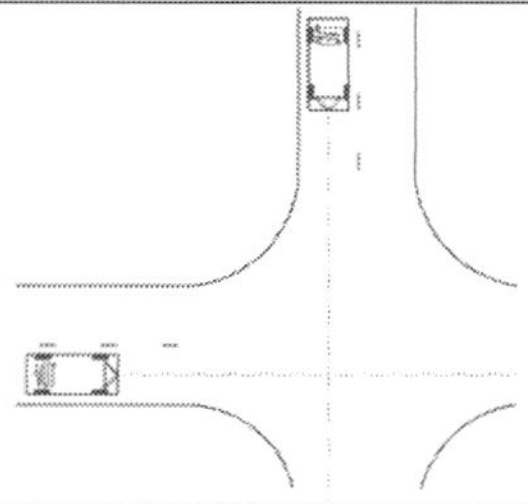	
6	CAR-TO-CAR FRONT HEAD-ON STRAIGHT (CCFhos)	Oncoming on the same lane
	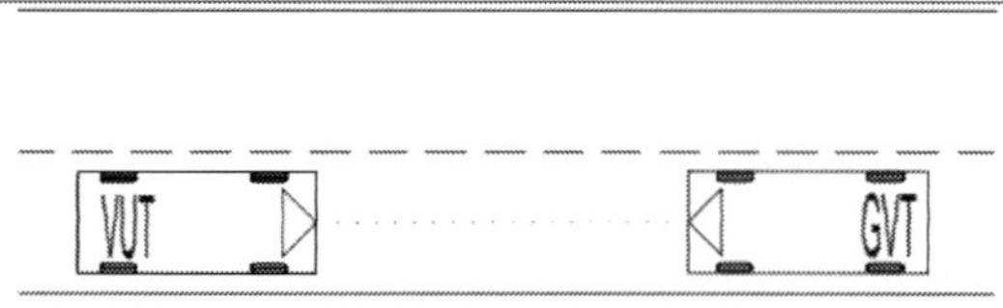	
7	CAR-TO-CAR FRONT HEAD-ON LANE CHANGE (CCFhol)	Oncoming on the same lane
	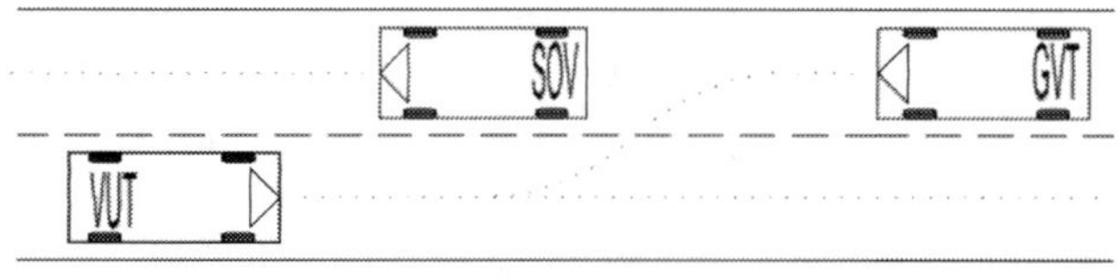	

2.2. Criterion for scenario selection
Limiting the interest to car-to-car impacts with accident dates starting from 2019, the IGLAD cases available for subsequent processing are 572. Data for these cases are extracted both in terms of scenario ("Scenario Type" variable) and impact-related features, i.e., velocity change of the single vehicle during the impact (ΔV [9]) and "Impact Type" (front, back, side, near-side). In particular, ΔV is an indicator of the impact forces (based on momentum conservation) to which the vehicle and the occupants are subjected. Accident type is conversely derived based on the position of the main intrusion on the vehicle (coded as "Collision Deformation Classification", CDC); if a lateral impact involves the compartment on the same side of an occupant's seating position, the impact is classified as a "near side". Using an established IR model [7], IR can be directly derived for the occupants of a specific vehicle based on ΔV and "Impact Type". At this point, to properly select scenarios that are representative from both standpoints of frequency and IR, an overall risk level has been applied to a specific scenario by multiplying its frequency (number of occurrences) with the average IR for the ego's occupants among all impacts included in such a "Scenario type". As an example, the highest risk level is equal to 104 and is associated with the scenario "Crossing from nearside" of the IGLAD classification. The scenarios were chosen considering only those with a risk level at least equal to one-third of the maximum. Table 2 summarizes the selected IGLAD scenarios, matched with the respective EuroNCAP test (if applicable). The scenario "Other Oncoming Accident" is highlighted in this step as a relevant scenario for the tests in terms of risk level; nevertheless, this category is not treated in the remainder of the paper, being comprehensive of a broad range of critical scenarios that cannot be classified otherwise based on the IGLAD coding scheme.

3. Test speed selection
Speed selection is a fundamental element for the proposal of a new testing protocol. To this end, a method is proposed to select one significant closing speed for the tests to be performed. One closing speed is chosen here for the sake of simplicity. To obtain the sought representative closing speed, a value of V_r is associated with a value of IR for each impact [6]. Considering a specific V_r value, the frequency of such a V_r value in the population has been multiplied by the IR for each impact, resulting in a risk level according to what prescribed in Section 2.2. Subsequently, a risk curve for each scenario is built derived from such a risk level by a second-order, least-square method polynomial fit, as shown in Figure 2 for the IGLAD "Crossing from near side" scenario. The speed value that has the highest risk level is considered in the proposal of new tests. This risk level at the specific V_r value is considered as the maximum score obtainable by the VUT in a test, in case it is able to avoid the impact. The results for identification of relevant V_r in all scenarios employing such a procedure are reported in Tables 3, dividing the tests between scenarios not covered and covered by EuroNCAP protocols.

IOP Conf. Series: Materials Science and Engineering 1306 (2024) 012027 doi:10.1088/1757-899X/1306/1/012027

Table 2. Relevant IGLAD scenarios matched with the respective Euro NCAP test (if applicable) and related risk level.

SCENARIO	IGLAD "Scenario Type"	**RISK LEVEL**	EURONCAP TEST
Crossing from nearside	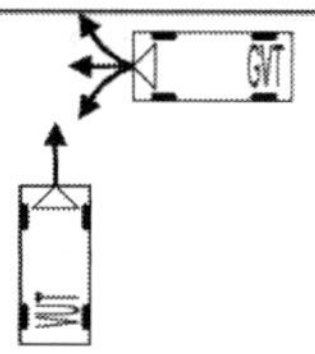	104	*Not available*
Turning farside and object oncoming	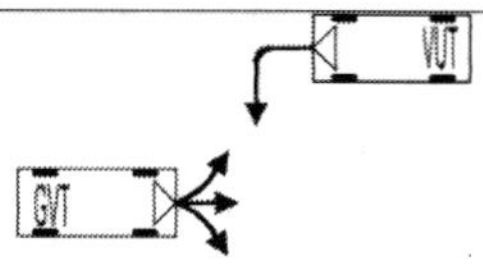	98	CCF_{tap}
Oncoming on the same lane	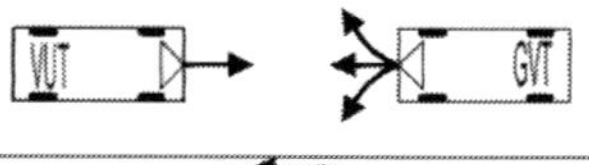	88	$CCFhos/CCFhol$
Turning farside and object from farside	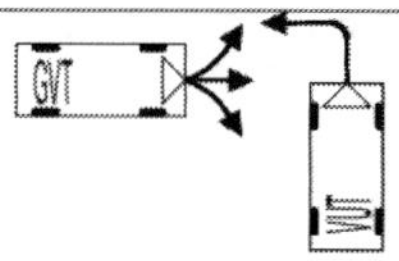	83	*Not available*
Crossing from farside	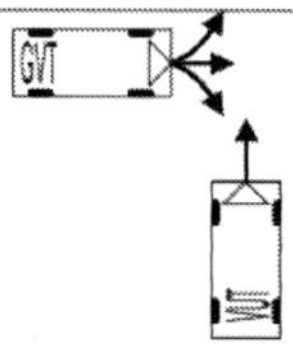	75	$CCCscp$
Other Oncoming accident	/	70	*Not available*
Running up	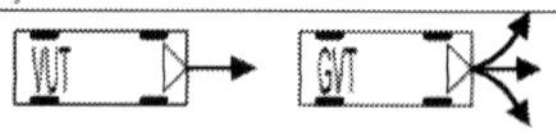	48	CCR_s/CCR_m
Lane changing to offside and oncoming	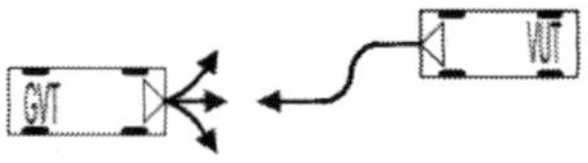	42	*Not available*

In addition, the identified representative V_r values must be distributed between both vehicles participating to the specific scenario; to this end, the most frequent collision speed of the ego in each considered scenario has been identified starting from the IGLAD database information. The histogram of collision speeds related to the scenario "Crossing from nearside" is reported in Figure 3 as an example. The most frequent collision speed values of the VUT are 35 km/h and 45 km/h. Then the VUT collision speed values for the tests related to scenario "Crossing from nearside" will be 35 km/h and 45 km/h. The results for the other scenarios are summarized in Table 4. The opponent collision speed is easily computable knowing the ego collision speed, the closing speed, and the impact configuration. Based on this, it is possible to propose new test procedures to evaluate ADAS functions, scenario by scenario.

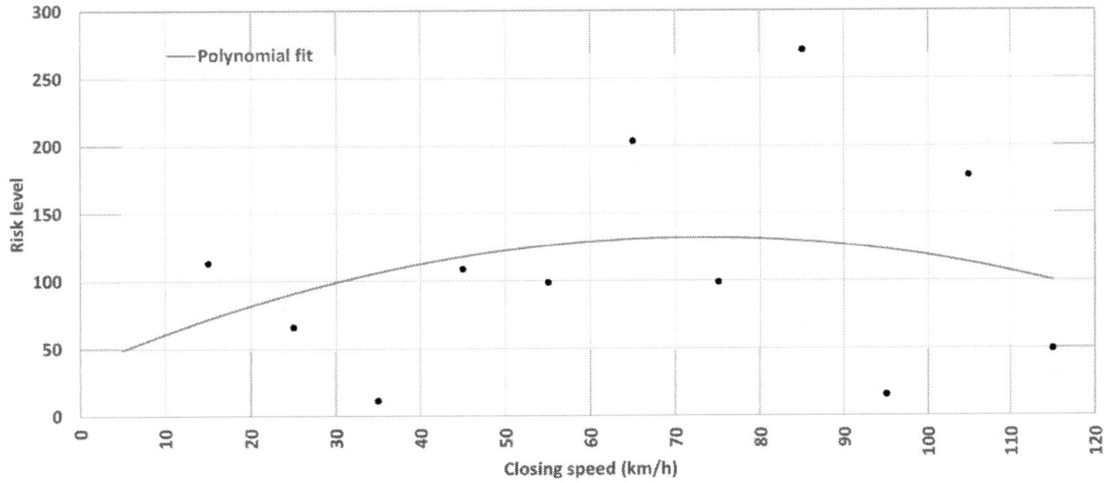

Figure 2. Risk level distribution for each V_r value in the IGLAD "Crossing from nearside" scenario.

Table 3. Closing speed and score (both for cases covered by Euro NCAP and those not).

SCENARIO	CLOSING SPEED (km/h)	RISK LEVEL
UNCOVERED CASES		
Crossing from nearside	75	140
Turning farside and object from farside	85	153
Lane changing to offside and oncoming	150	148
COVERED CASES		
Turning farside and object oncoming	90	312
Oncoming on the same lane	140	192
Crossing from farside	75	130
Running up	90	122

Table 4. Table of ego collision speed in the Euro NCAP covered and uncovered scenarios.

UNCOVERED CASES	Ego collision speed (km/h)
Crossing from nearside	35; 45
Turning farside and object from farside	15
Lane changing to offside and oncoming	0; 35; 65; 70
COVERED CASES	**Ego collision speed (km/h)**
Oncoming on the same lane	35; 40
Crossing from farside	40; 45
Turning farside and object oncoming	20
Running up	25; 35

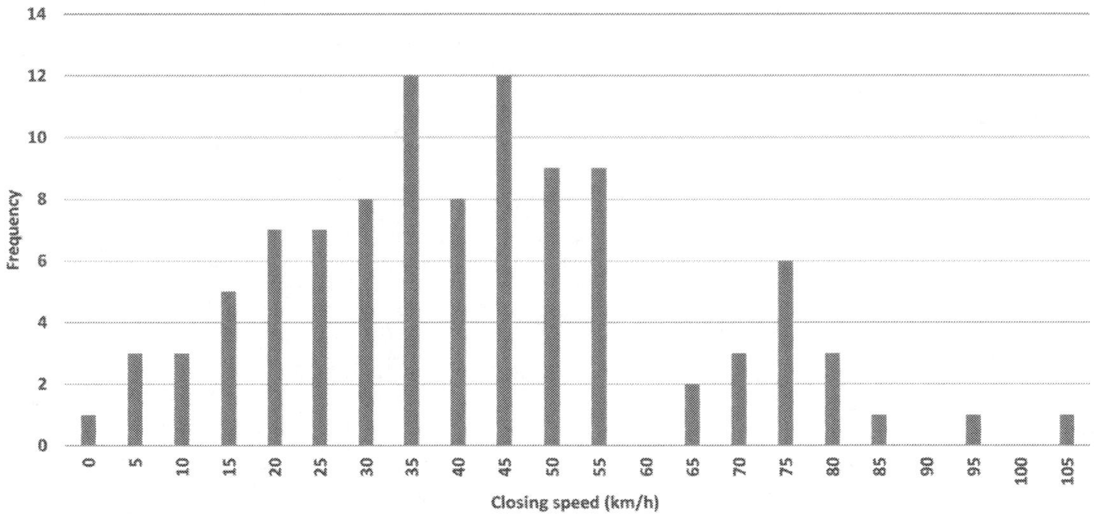

Figure 3. Histogram of collision speeds (Scenario "Crossing from nearside").

4. New proposal or update of test protocols

Once the scenarios for the tests, the VUT collision speed, and V_r have been selected, a test protocol must be proposed. As an established consumer program, Euro NCAP framework can be considered a solid reference for the proposal of new protocols or updates of already available testing procedures. For this reason, the present Section highlights two examples, one a new protocol and one regarding the update of an already available EuroNCAP protocol. In both instances and in the case of impact avoidance, the ADAS can obtain a maximum score equal to the risk level highlighted in Section 3 for the specific scenario and V_r under investigation.

4.1. New proposal: "Crossing From Nearside"

Figure 4 reports a possible configuration of the "Crossing from Nearside", identified as the most relevant scenario in terms of risk level (Table 2). Compared to the "CCCscp" test by Euro NCAP, rather then the front, the GVT impacts the side of the VUT (opposite side with respect to the driver's position). This difference in impact type enables diverse possibilities in terms of IR to be explored compared to the "CCCscp" alternative (IR in a side impact is higher than in the case of a front impact, ΔV being the same). The test must be performed two times at two different GVT and VUT collision speed combinations, in accordance with Table 5. Based on the V_r values in Table 3 and collsion speed values in Table 4, the GVT collision speed is computed as:

$$V_{GVT} = \sqrt{V_r^2 - V_{VUT}^2} \tag{1}$$

where V_{VUT} and V_{GVT} are respectively the speed of the VUT and the speed of the GVT.

Table 5. Test parameters for a newly proposed "Crossing from nearside" scenario.

TEST	Closing speed (km/h)	Ego collision speed	Opponent collision speed	MAX SCORE
1	75	35	65	140
2	75	45	60	140

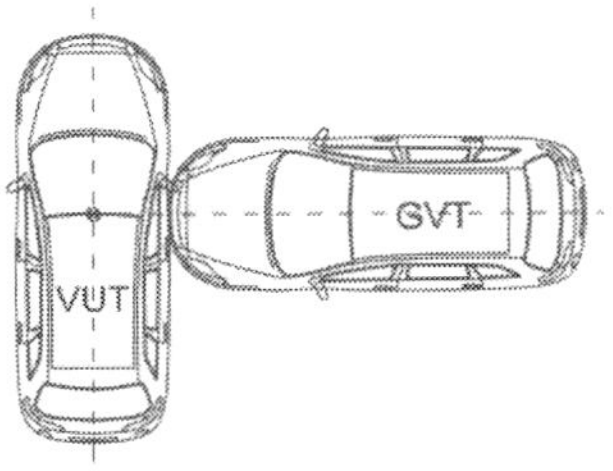

Figure 4. Impact configuration for a newly proposed "Crossing from nearside" scenario.

4.2. Update: "Turning Farside and Object Oncoming" (CCFtap)

Based on Table 1, the second most relevant scenarios in terms of risk level is the "Turning Farside and Object Oncoming" corresponding to the "CCFtap" scenario by Euro NCAP. Figure 5 reports the test path already implemented inside the Euro NCAP framework. From the data analysis in Table 3, the closing speed of interest is in this case equal to 90 km/h and the before proposed approach would suggest assigning a VUT collision speed of 20 km/h and performing only one test, with maximum score equal to 312 in case of avoided impact. Nevertheless, a solution accounting also for the already available methodology can be considered: the speed values prescribed by Euro NCAP can be employed, applying a maximum score based on the risk level corresponding to the test closing speed. For this scenario, the polynomial fit curve obtained as described in Figure 2 for the "Crossing from Nearside" scenario is used, reported in Figure 5. Referring to the EuroNCAP "CCFtap", the tests should hence be represented by 9 combinations of VUT speed values of 10, 15 and 20 km/h combined with GVT (opponent) speed values of 30, 45 and 55 km/h, as reported in Table 7.

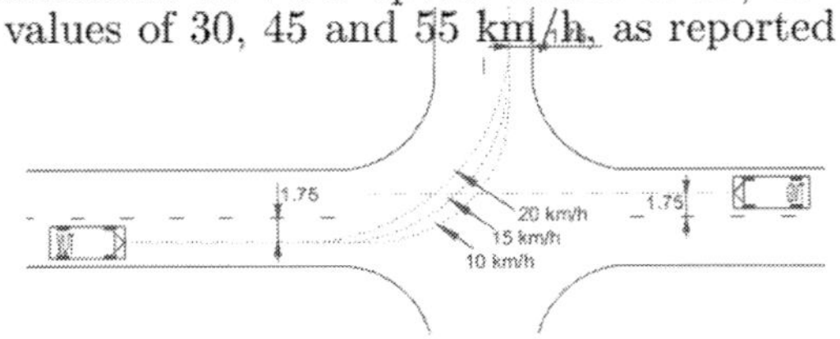

Figure 5. Test path for the already available "Turning Farside and Object Oncoming" scenario (Euro NCAP "CCFtap").

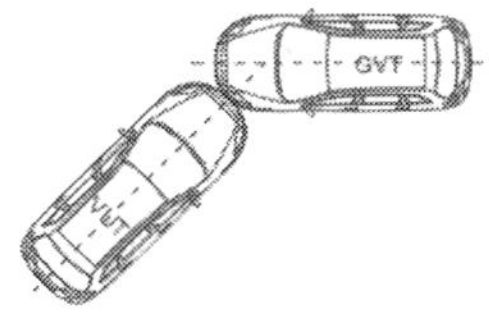

Figure 6. Impact configuration for the already available "Turning Farside and Object Oncoming" scenario (Euro NCAP "CCFtap").

The paths adopted ensure an impact configuration like the one shown in Figure 6. Based on the V_r values in Table 3 and collision speed values in Table 4, the GVT collision speed is computed according to Carnot's theorem as:

$$V_{GVT} = V_{VUT} \cos\alpha + \sqrt{V_{VUT}^2 \cos^2\alpha + V_r^2 - V_{VUT}^2} = V_{VUT} \cos\alpha + \sqrt{V_r^2 - V_{VUT}^2 \sin^2\alpha} \quad (2)$$

with α the angle between V_{VUT} and V_{GVT} collision speed of the closing speed vector in the velocity triangle. The value of α can be determined from the test paths shown in Figure 5. The total score for the scenario is the mean of all the scores achieved in each test.

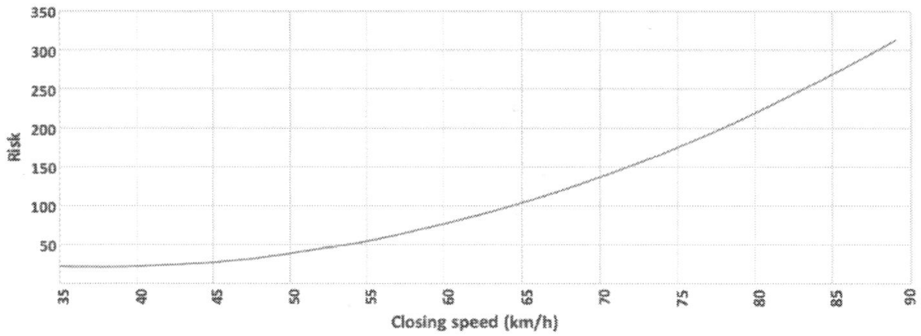

Figure 7. Risk curve for the definition of the maximum score at different V_r values in the already available "Turning Farside and Object Oncoming" scenario (Euro NCAP "CCFtap").

Table 6. Test parameters table ("Turning farside and object oncoming").

Ego speed [km/h]	Opponent speed [km/h]	Corresponding closing speed	MAX SCORE
10	30	35	22
10	45	50	38
10	60	65	100
15	30	40	22
15	45	55	54
15	60	70	130
20	30	45	27
20	45	60	76
20	60	75	170

5. SCORE SCALING CRITERIA

The previously introduced maximum score applies if the impact is avoided by the ADAS. If the ADAS intervention does not avert the impact in the test, the score is assigned by scaling the maximum score available with the IR value for the VUT as follows:

$$Score = MaxScore \cdot (IR_{reference} - IR_{modified})/IR_{reference} \tag{3}$$

where $IR_{reference}$ and $IR_{modified}$ are the IR values in the RS and in the scenario modified by the ADAS intervention, respectively. IR can be calculated, starting from the impact configuration and collision speed of the vehicles, by running a simulation with dedicated software (e.g., like that described in [10]) or by employing the so-called CMI-V_r approach (described by the authors in previous work [1, 9]): the vehicles do not actually sustain an impact because the GVT does not have inertial properties; therefore, the value ΔV for the identification of the IR cannot be directly measured. The previously reported equation accounts for the possibility that the ADAS intervention leads to an increase in IR because of a change in the impact configuration (like moving from a frontal and eccentric crash to a lateral and centred crash): in this case, the score for the ADAS in the analysed scenario becomes negative. The overall score for the ADAS is the sum of all scores achieved in each test scenario.

6. Conclusions

A framework has been highlighted for the proposal of testing protocols for ADAS performance assessment in car-to-car imminent impact scenarios. Compared to the previous literature, the methodological approach is based on occurrence probability (frequency) and injury consequences for the occupants (Injury Risk, IR) associated with real-world impact scenarios. These two factors are combined into a comprehensive risk level, representing an indicator of the specific scenario's relevance among all possible impact scenarios. The risk level allows also for the selection of proper closing speed at collision to be considered in the test development, as well as the definition of a maximum score obtainable by the ADAS if the impact is avoided in such relevant scenarios. A scaling criterion for such maximum score has been also highlighted in case of unavoided collision, based on IR values that can be assessed starting from the impact configuration and closing speed at collision for the vehicles.

The framework represents a valuable tool for both the proposal of new test protocols and the update of already available protocols. It also enables synthesizing the most relevant situations where ADAS function and devices can be tested, so that testing time (and cost) can be diminished. The described methodology can be further integrated by considering IR also for the occupants of the opponent vehicle, or by considering IR values associated with different injury metrics (i.e., other than MAIS 3+). In the end, a procedure able to evidence the strengths and weaknesses of active safety technology has been developed, which can efficiently contribute to the decrease in the time to market of new generation ADASs.

References

[1] Michelangelo S Gulino, Anita Fiorentino, and Dario Vangi. Prospective and retrospective performance assessment of advanced driver assistance systems in imminent collision scenarios: The cmi-vr approach. *European transport research review*, 14(1):1–14, 2022.

[2] Stefan Smit, Ernst Tomasch, Harald Kolk, Michael Alois Plank, Jürgen Gugler, and Hannes Glaser. Evaluation of a momentum based impact model in frontal car collisions for the prospective assessment of adas. *European transport research review*, 11(1):1–9, 2019.

[3] Ulrich Sander and Nils Lubbe. Market penetration of intersection aeb: Characterizing avoided and residual straight crossing path accidents. *Accident Analysis & Prevention*, 115:178–188, 2018.

[4] Anders Kullgren. Dose-response models and edr data for assessment of injury risk and effectiveness of safety systems. In *Proc of Int. IRCOBI Conf., Bern, Switzerland*, pages 3–14, 2008.

[5] Nico Kaempchen, Bruno Schiele, and Klaus Dietmayer. Situation assessment of an autonomous emergency brake for arbitrary vehicle-to-vehicle collision scenarios. *IEEE Transactions on Intelligent Transportation Systems*, 10(4):678–687, 2009.

[6] Michelangelo-Santo Gulino, Leonardo Di Gangi, Alessio Sortino, and Dario Vangi. Injury risk assessment based on pre-crash variables: The role of closing velocity and impact eccentricity. *Accident Analysis & Prevention*, 150:105864, 2021.

[7] Chris Jurewicz, Amir Sobhani, Jeremy Woolley, Jeff Dutschke, and Bruce Corben. Exploration of vehicle impact speed–injury severity relationships for application in safer road design. *Transportation research procedia*, 14:4247–4256, 2016.

[8] Antonio Lara, Jeffrey Skvarce, Harald Feifel, Michael Wagner, and Toshihisa Tengeiji. Harmonized pre-crash scenarios for reaching global vision zero. In *Proc. 26th Int. Tech. Conf. Enhanced Saf. Veh.*, pages 1–19, 2019.

[9] Dario Vangi, Michelangelo-Santo Gulino, Anita Fiorentino, and Antonio Virga. Crash momentum index and closing velocity as crash severity index. *Proceedings of the Institution of Mechanical Engineers, Part D: Journal of automobile engineering*, 233(13):3318–3326, 2019.

[10] Dario Vangi, Filippo Begani, Florian Spitzhüttl, and Michelangelo-Santo Gulino. Vehicle accident reconstruction by a reduced order impact model. *Forensic science international*, 298:426–e1, 2019.

IOP Conf. Series: Materials Science and Engineering 1306 (2024) 012028 doi:10.1088/1757-899X/1306/1/012028

Open ECU EPS System, preliminary design using a Hardware-in-the-loop approach

L Veneroso[a], A Anticaglia[a, b], L Lelli[a], A Mosci[a], C Annicchiarico[b] and R Capitani[a]

[a] DIEF, Università degli Studi di Firenze, Via Santa Marta 3, 50139 Firenze (FI), Italy
[b] Meccanica 42 srl, Via Ezio Tarantelli 15, 50019 Sesto Fiorentino (FI), Italy

E-mail: luca.veneroso@unifi.it

Abstract. The introduction of mechatronic systems in vehicles opens the possibility to a wide range of vehicle functionality. These systems allow active control of vehicle enabling a new level of driving comfort and safety. The development of these complex technological systems, however, requires a long time to integrate mechanical, electrical and software components. Innovative development strategies and devices are hence crucial to accelerate the integration and deployment of these mechatronic systems.

In order to fulfil these needs, this paper presents the process to size a development electronic power steering which aims to support vehicle industry during testing and validation of innovative steering-based control strategies, including autonomous driving, steer-by-wire and all-wheel steering applications. Using an hardware-in-the-loop test bench for a wide range of vehicle maneuvers a stock power steering unit has been characterized taking into account both driver and assistance motor torque contributions.

1. Introduction

In recent years, technological evolution in the automotive industry has led to a continuous development of the most advanced vehicle features, with particular attention to safety and dynamic behaviour.

One of the key areas where a central role has been observed is the active intervention on the steering system thanks to the introduction of electric power steering (EPS). Compared with hydraulic steering systems, the use of EPS systems has brought significant advantages, including greater energy efficiency (lower fuel consumption and reduced CO_2 emissions) and improved steering response and precision. More over they offer increased flexibility regarding system control, allowing for optimized steering response in different driving conditions [1]. Apart from providing the necessary assist torque and the aforementioned benefits, electro-actuated steering systems, enable the accurate execution of Advanced Driver Assistance Systems (ADAS) functionalities. These advanced driving assistance technologies leverage the electronic control capability of the steering system to offer features such as Lane-Keeping Assistance (LKA), active blind-spot monitoring, and automatic parking, contributing to driving safety and efficiency. In addition, EPS steering systems are more easily integrable with future autonomous driving systems [2]. Despite the growing interest in these topics, the development of customized control strategies and the implementation on existing experimental vehicles have proven to be challenging, as the programming of control units is often restricted to the specific system manufacturer. From these considerations arises the need to equip prototypal steering systems with a fully customizable open ECU

control logic, whose operation is controllable, to support research and development activities in the automotive field.
This article examines the design process of an Electric Power Steering (EPS) system with an open ECU, starting from the characterization of an existing EPS system installed on a commercial vehicle and its control logic. The examined EPS system is a parallel-axis electric motor system, which is commonly used in modern vehicles. The electric motor is controlled by an ECU that analyses data from sensors detecting the steering wheel angle, steering torque and vehicle speed. Based on this information, it provides the appropriate amount of assistance to facilitate steering manoeuvres. This analysis aims to identify the operating points of the electric servomotor, enabling the sizing and selection of a Brushless DC (BLDC) motor and a suitable drive, along with other electronic components necessary for system implementation.
The introduction of an open ECU EPS will expedite the development and calibration stages of advanced control logics based on steering management, both for ADAS systems and autonomous vehicles. To characterize the control logic, it was necessary to identify the transmission ratios and its geometric properties.
An experimental test campaign was conducted, integrating the steering system in the Hardware-in-the-loop test bench, provided by Meccanica 42 S.r.l. [3], [4], [5]. The tests were designed to investigate the EPS response to various driving conditions, including Open-loop and Closed-loop steering maneuvers.
The implementation of an open ECU steering system represents a significant step in automotive research, allowing greater flexibility in customizing driving functionalities and opening new opportunities for the development of advanced technologies for road safety and autonomous driving.

2. Methodology

For the first part of the activity, a complete disassembly of the steering rack was performed. Mechanical measurements were collected using a caliber and a coordinate-measuring machine (CMM). These dimensions were used for the evaluation of basic properties of the steering rack, such as the rack-pinion gear ratio.
After this phase, we were able to reproduce the geometry of every needed component using typical solid modelling software; this phase is required for detailed analysis (kinematic or stress analyses). In future, the whole mechatronic part of the rack (electric motor, belt, electronic board) will be redesigned and integrated with the existing components such as the case, the rack and the gear. The knowledge of these measurements was hence mandatory for the aim of this investigation, future activities included.
Similarly, this information is necessary to implement a simple, yet effective, numerical model to calculate the forces acting on the system.
To validate this model and to evaluate the EPS assist curves of this steering system, we decided to select a specific set of key manoeuvres and adopt a semi-experimental Hardware-in-the-Loop method. This test campaign was performed using an advanced Hardware-in-the-Loop simulator that integrates the entire steering system of a real vehicle in a static driving simulator. The layout of this HiL simulator is briefly explained in the scheme in Figure 1. In this simulator, two electric motors are used to reproduce the forces acting on the tie rods in every step of the real-time simulation; for each motor, one rocker is used to replicate the kinematic movement of the tie-rod as in the real vehicle. The hardware installed includes the entire steering system from the steering wheel up to the tie rod. The vehicle is modelled using the VI-CarRealTime model. All the signals needed for the functionality of the Electric Power Steering (EPS) are provided using the Controller Area Network (CAN) protocol, just like in the real vehicle.
Data used for the steering performance characterization are extracted directly from the CAN network of the steering unit. The most relevant signals are the steering column torque and steering wheel angle.
The steering column torque signal is sensed at the EPS torque sensor; the signal has a quantization of $8mNm$ and its value is sampled every $10ms$.
The steering wheel angle signal is measured at the EPS sensor; the signal has a resolution of $0.1deg$ and its value is sampled every $10ms$.

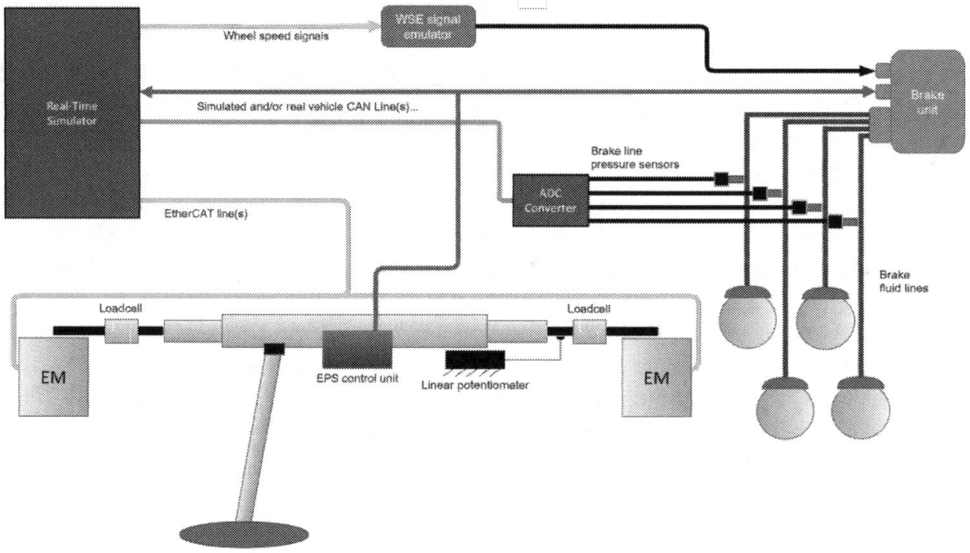

Figure 1. HiL Static simulator scheme

In this test bench, the stock tie rods have been modified to integrate two load cells. The transducer is certified for 0.2% Full Scale (FS) accuracy. These measurements are used for the test bench control logic and these data were used for the evaluation of this study. In addition, the steering rack position is quantified using a linear potentiometer. Figure 2 shows the simulator described and the more relevant components for the aim of this activity.

3. Maneuvers and test procedure

In literature the assistance torque is defined as the sum of the contribution of the essential functions[1], [6].

- **Power-assistance:** It aims to support the driver reducing the steering effort, especially during parking and low speed manoeuvre.
- **Friction compensation:** Reduces the effect of friction in the steering system, to provide a more responsive behaviour.
- **Inertia compensation:** Reduces the effect of EPS motor inertia to improve the dynamic performance of the steering.
- **Active damping:** Regulates system damping to improve stability and steering feel [7]. This function prevents steering wheel overshoot and unwanted oscillation around the center position during release.
- **Active return:** Generate an adequate and satisfactory runback response, particularly at low vehicle speed.

Therefore specifics, open and closed-loop manoeuvres were chosen to highlight all the aforementioned functions effects.
These manoeuvres were executed in a wide vehicle speed range, from 30 km/h up to 180 km/h. In Open-Loop maneuvers the driver provides vehicle inputs independently from vehicle response. On the contrary during Closed Loop maneuvers, the driver continuously adjusts inputs based on the feedback received from the vehicle. Each manoeuvre performed is described as follow:

IOP Conf. Series: Materials Science and Engineering 1306 (2024) 012028 doi:10.1088/1757-899X/1306/1/012028

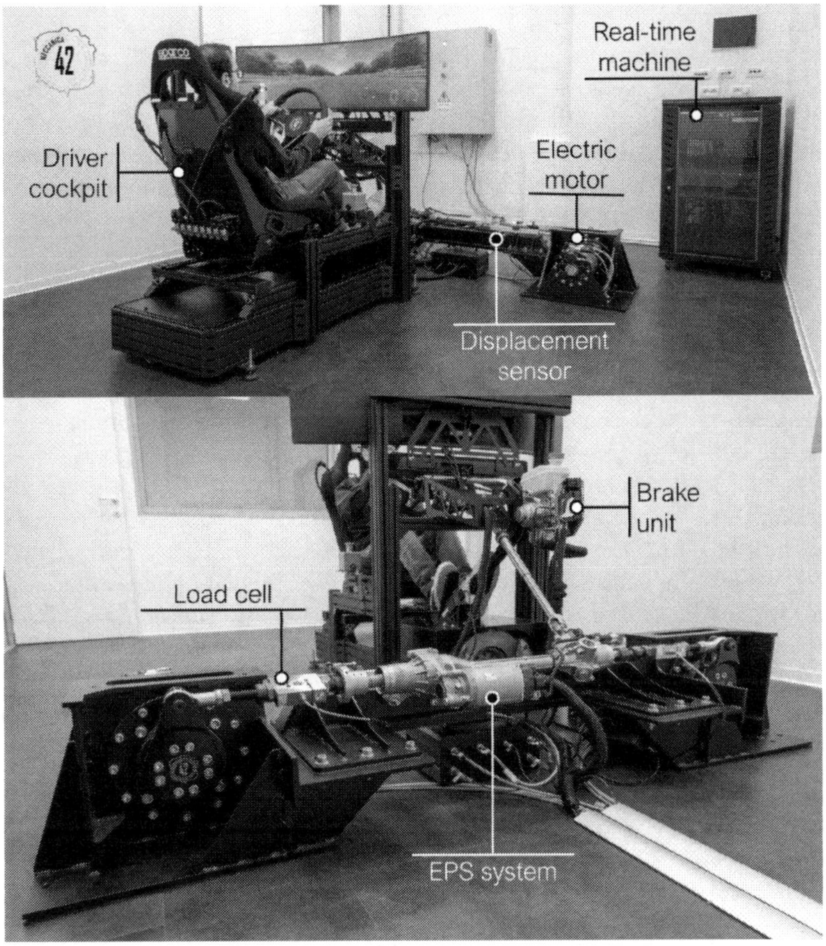

Figure 2. HiL static simulator: layout and main components

- **Sine steer:** A constant frequency and amplitude sine wave steering input is applied. The test set is composed by an high frequency (2 Hz) and low frequency ($0.5Hz$) sine both with $45deg$ steering angle amplitude.
- **Ramp input:** The ramp input test is performed increasing the steering angle with a constant steering velocity until the target angle is reached. The selected target angle was $\pm 90deg$ with a steering velocity of $50deg/s$.
- **Step steer:** In this test the driver is supposed to turn the steering wheel as quickly as possible to the chosen angular position, which must be held for at least $3s$ so that the measured variables reach a steady state. Every step manoeuvre was performed both clockwise and anticlockwise. The initial speed of the vehicle did not deviate from the nominal speed by more than $2km/h$. A $90deg$ steer step amplitude was chosen.
- **Release:** During steady state cornering the driver is supposed to suddenly release the steering wheel while maintaining a constant throttle pedal position. The initial steering wheel angle has been previously determined so that the initial lateral acceleration value is $2m/s^2$.
- **Track test:** To reproduce real driving steering input several lap were taken around Calabogie Motorsport Park track alternating low and high speed driving cycles in dry condition.

All the aforementioned manoeuvres were performed with an initial vehicle speed of $30km/h$ and then repeated again by increasing the initial speed by $30km/h$ until the limit value of $180km/h$ was reached. Lower speed values were not investigated because they are irrelevant for the automobile handling evaluation. Higher speeds, on the flip side, are difficult to perform with this vehicle due to its limited power output. Results evaluation and significance will hence not be trustworthy. During the execution of the tests throttle position was adjusted to maintain the vehicle speed with a tolerance of 5%. Prior to the execution of each test, a straight section was driven while monitoring the yaw rate of the vehicle so that this was at a value of 0 $\pm 0.5rad/s$.

4. Discussion

The analysis of the data collected during the test manoeuvres described before allows to extrapolate the assistance map of the EPS and individuate the motor operative map. The steering system has been modelled with the help of the one degree of freedom model as the one presented by Thomas Weiskircher et Al [8],[9]. The schematic representation of the single DOF model is shown in figure 3.
A more complex five degrees of freedom steering model, such as the one proposed by Harror et Al, [1] needs a detailed characterization of the steering system and all its components properties such as stiffness and damping coefficients. The single DOF model only requires the knowledge of the steering system kinematic properties, which have previously been identified, and its equivalent inertia. This dynamic model is depicted by equations 1, while a brief description of each equation parameter is provided in table 1. The non-linear friction behaviour is represented using the LuGre Friction model [10]. This model is one of the most popular approaches to treat friction in modelling and compensation algorithms thanks to its capacity to describe both the Stribeck effect and the stick-slip motion. The identification of the four static and two dynamic LuGre friction parameters has been carried out using dedicated manoeuvres as the one suggested by C. Canudas De Wit and P. Lischinsky [11]. LuGre model properly predicts friction phenomena in steering systems as it can accurately describe its dependence not only from steer velocity, but also from the applied torque [12].

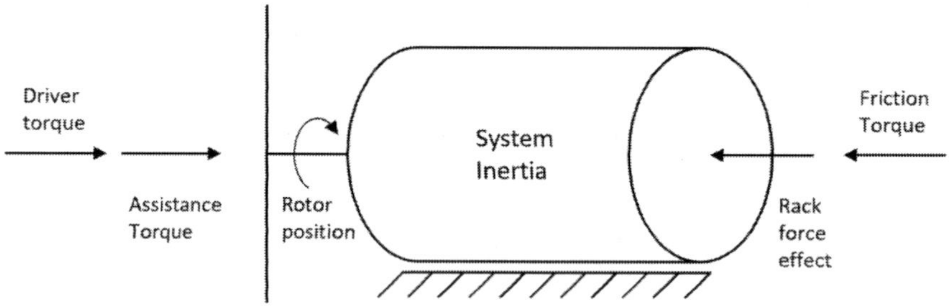

Figure 3. Single DOF system scheme, as described in [8]

$$\begin{cases} T = T_{em} + \dfrac{i_{bs}}{\tau_{pr}\tau_p} T_d & (1) \\ J\dot{\omega}_{rot} = T + \dfrac{F_r}{i_{bs}\tau_p} - T_f & (2) \end{cases}$$

.

The assistance map extrapolation has been carried out using the approach suggested by Pfeffer et Al.[13], which enable to define the steering assistance torque with two parameters only per vehicle velocity. This analytical approach allows to efficiently determine a harmonious increase of the assistance torque in relation to the steering wheel torque, avoiding the obsolete and time-consuming point-by-point definition of the assistance map. This method is based on the definition of the assistance torque ratio,

equation 3, which defines the support provided by the EPS motor to the driver to balance the steering torque generated by the tire action in steady state corners.
To enable proper steering feel and smooth boost curves, the assistance ratio should linearly increase with lateral acceleration as shown in equation 4. Given the kinematic properties of the steering system and analysing the moments acting on the steering axis, it is possible to express the required assistance torque as function of the applied steering torque as in equation 5. Considering this set of equation 3, 4, 5 to characterize the assistance boost curve it is sufficient to define D_a and K_a. All equation parameter are synthetically described in table 1. .The extrapolated assistance map is shown in figure 4a, where the steering torque, vehicle velocity and EPS assistance torque have been normalized with their respective maximum values. The assistance torque is a function of driver torque and vehicle velocity. In particular, at low vehicle speed, the EPS system supports the driver with high assistance torque while it decreases at high velocity to transmit a more authentic feeling to the driver who can perceive the vehicle state.
The total rack force is the result of the contribution of the driver action and the assistance motor torque multiplied by their respective transmission ratio. Figure 4b, shows how in recent vehicles the generation of the rack forces is mostly demanded by the assistance actuator.

$$\begin{cases} A_s = \dfrac{T_s}{i_s T_d} & (3) \\ A_s = \dfrac{m_f r}{i_s}(D_a + K_a a_y) & (4) \\ T_{em} = \dfrac{T_d(m_f r D_a + i_s T_d K_a - i_s)}{1 - k_a T_d} \dfrac{i_{bs}}{\tau_{pr} \tau_p i_s} & (5) \end{cases}$$

Table 1. Equation parameter

Symbol	Description
T	Total input torque
T_{em}	Electric motor torque
T_d	Steering wheel torque
τ_{pr}	Pinion to rack ratio
τ_p	Pulley transmission ratio
i_{bs}	Ball screw pitch
J	System total inertia
ω_{rot}	Electric motor velocity
F_r	Rack force
T_f	Friction torque
A_s	Assistance ratio
T_s	Steering axis torque
i_s	Steering ratio
m_f	Vehicle front mass
r	Sum of mechanical and pneumatic trail
D_a	Gradient factor
K_a	Degressivity factor
a_y	Lateral acceleration

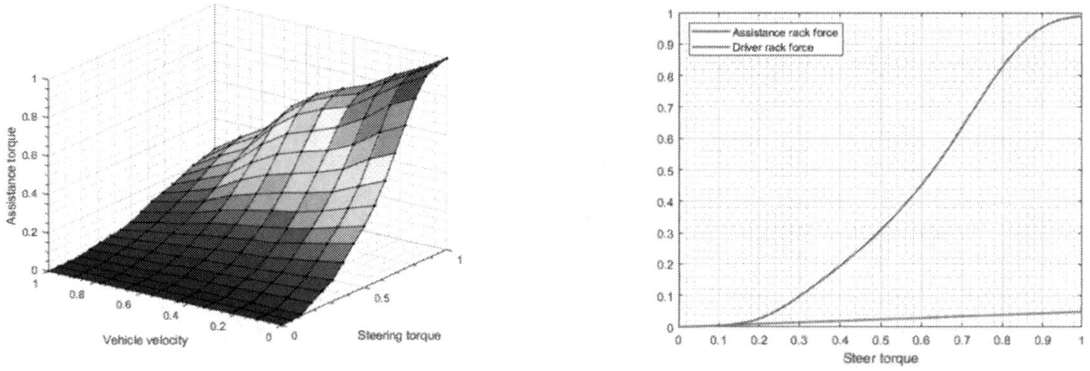

(a) EPS assistance torque map **(b)** Rack force contribution from EPS and Driver torque

Figure 4. Assistance and driver effort

Figure 5a, illustrates a comparison during a ramp steer manoeuvre, between the actual assistance torque signal recorded by the EPS control unit and the assistance torque generated using the extrapolated assistance map. The signal generated by the assistance map closely reproduces the original signal, proving the fidelity of the numerical model. In figure 5b is shown the error distribution.

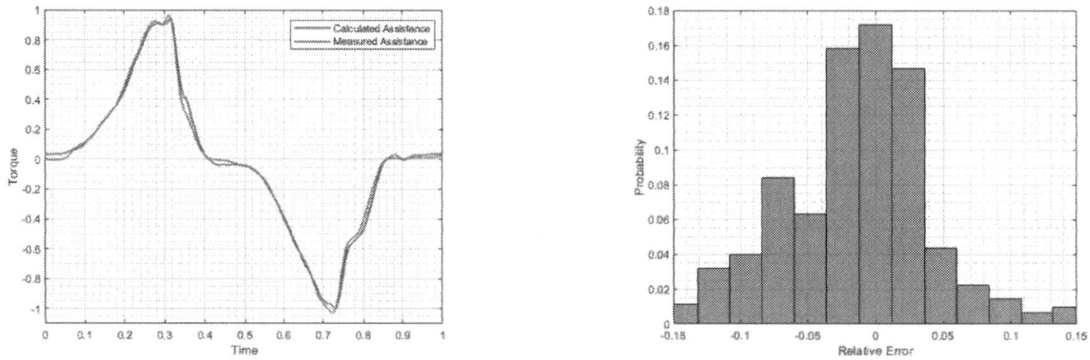

(a) Model and Measured assistance torque **(b)** Error distribution

Figure 5. Ramp steer test results

Figure 6, shows the operative zones of the BLDC motor, where torque and Velocity are normalized against their respective maximum values. For the EPS motor, the high-frequency sine manoeuvre proves to be the most demanding in terms of speed and power. Being a closed-loop test, the track drive is the most representative of real driving conditions, as shown in the figure motor velocity is significantly lower if compared to the one reached during the high-frequency sine manoeuvre. However, high torque levels are observed during these driving conditions, particularly during high-speed corners. The working points of the low-frequency sine and ramp/step manoeuvre are substantially included in those of track drive and high-frequency sine. The above considerations permit the selection of the appropriate commercially available BLDC motor capable of operating in the highlighted area.

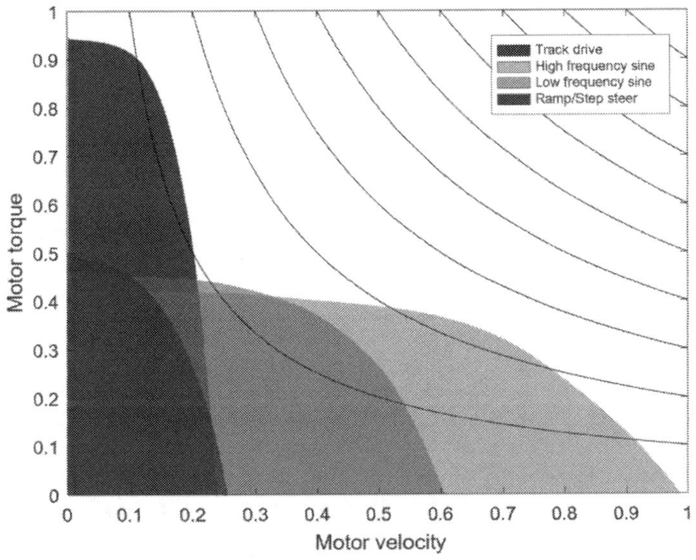

Figure 6. Electric motor operative zones

5. Conclusion

The possibility to actively control the steering system has played a central role both in the development of steering-based ADAS systems and in the control of autonomous driving vehicles. The introduction of these complex technologies has led to changes in the development process of vehicle functions, making fundamental the use of Hardware-in-the-loop simulators combined with development devices.
In this paper is proposed a novel testing method to determine the realistic working conditions of an electric power assist motor, in terms of maximum torque and velocity.
Through the execution of specific manoeuvres the validation of a simple and effective 1 DOF steering model was carried out along with the development of a basic steering assistance system control plant. With the aim to accelerate testing procedures, a steering actuator with customizable control logic will be developed starting from these findings.
To support vehicle industry in all the steering research fields, this development device could be coupled not only with classical assistive EPS systems but also with innovative actuation steering systems such as Steer-by-wire, all-wheel steering and autonomous driving.

Nomenclature

$BLDC$	Brushless Direct Current
CAN	Controlled Area Network
CMM	Coordinate-Measurement Machine
DOF	Degrees of freedom
ECU	Electronic Control Unit
EPS	Electric Power Steering
FS	Full Scale
HiL	Hardware-in-the-Loop
LKA	Lane-Keeping assistance

IOP Conf. Series: Materials Science and Engineering 1306 (2024) 012028 doi:10.1088/1757-899X/1306/1/012028

6. References

[1] Harrer M and Pfeffer P 2016 *Steering handbook* ISBN 9783319054490

[2] Winner H, Hakuli S, Lotz F and Singer C 2016 *Handbook of Driver Assistance Systems* (Springer International Publishing) ISBN 978-3-319-12351-6 URL http://link.springer.com/10.1007/978-3-319-12352-3

[3] Veneroso L, Alfatti F, Annicchiarico C and Capitani R 2023 *IOP Conference Series: Materials Science and Engineering* **1275**(1) 012041 ISSN 1757-8981

[4] Vinattieri F, Wright T, Capitani R, Annicchiarico C and Danisi G 2016 *SAE Technical Papers* ISSN 01487191

[5] Alfatti F, Annicchiarico C and Capitani R 2023 *IOP Conference Series: Materials Science and Engineering* **1275**(1) 012040 ISSN 1757-8981

[6] Brocker M 2006 *Vehicle System Dynamics* **44**(SUPPL. 1) 759–769 ISSN 00423114

[7] Chugh T, Chen W and Klomp M 2018 Design and control of model based steering feel reference in an electric power assisted steering system pp 43–49

[8] Weiskircher T, Fankem S and Ayalew B 2015 *Proceedings of the ASME Design Engineering Technical Conference* **3**

[9] Fankem S, Weiskircher T and Müller S 2014 *IFAC Proceedings Volumes* **47**(3) 8469–8474 ISSN 14746670 URL https://linkinghub.elsevier.com/retrieve/pii/S1474667016429502

[10] de Wit C C, Lischinsky P, Åström K J and Olsson H 1995 *IEEE Transactions on Automatic Control* **40**(3) 419–425 ISSN 15582523

[11] de Wit C C and Lischinsky P 1996 *IFAC Proceedings Volumes* **29**(1) 2078–2083 ISSN 1474-6670

[12] Li Y, Shim T, Wang D and Offerle T 2019 *Journal of Dynamic Systems, Measurement, and Control* **141**(11) ISSN 0022-0434

[13] Pfeffer P E and Harrer M 2008 *SAE Technical Papers* ISSN 0148-7191 URL https://www.sae.org/publications/technical-papers/content/2008-01-0502/

[14] Isermann R 2022 URL https://link.springer.com/10.1007/978-3-642-39440-9

[15] Peng Q, Liu Y, Gu P and Fan Z 2013 *Development of an Open-Architecture Electric Vehicle Using Adaptable Design* pp 79–90

[16] Isah A, Mohammed A and Hamza A 2019 *2019 2nd International Conference of the IEEE Nigeria Computer Chapter, NigeriaComputConf 2019*

[17] Talarico E M, Raimondi G, Alfatti F, Vitaliti D, Annicchiarico C, Talarico E M, Alfatti F, Vitaliti D and Annicchiarico C 2022 454–478 URL https://link.springer.com/chapter/10.1007/978-3-662-64550-5_27

[18] Certosini C, Vinattieri F, Capitani R and Annicchiarico C 2019 Development of a real-time steering system model for driving simulators vol 233 pp 2701–2713 ISSN 0954-4070 URL http://journals.sagepub.com/doi/10.1177/0954407018804389

[19] Galko C, Rossi R and Savatier X 2014 Vehicle-hardware-in-the-loop system for adas prototyping and validation pp 329–334

[20] Vinattieri F, Morrocchi L, Capitani R and Annicchiarico C 2017 Sviluppo di un' apparecchiatura sperimentale per l'integrazione del sistema di sterzo in un simulatore di guida statico (Associazione Italiana per l'Analisi delle Sollecitazioni) pp 6–9

[21] Gonter M and Seiffert U W 2013 *Integrated Automotive Safety Handbook* (SAE International) ISBN 978-0-7680-8029-2 URL https://saemobilus.sae.org/content/R-407/

[22] Winter J C F D, Leeuwen P M V and Happee R 2012 Advantages and disadvantages of driving simulators: A discussion URL http://ec.europa.eu/transport/roadsafety_library/publications/trainer_deliverable_2_1.pdf

[23] van Zanten A T, Erhardt R and Pfaff G 1995 Vdc, the vehicle dynamics control system of bosch URL https://www.sae.org/content/950759/

IOP Conf. Series: Materials Science and Engineering 1306 (2024) 012029 doi:10.1088/1757-899X/1306/1/012029

Experimental investigation on the potential of passive prechambers for use in 2S engines

Edoardo Risaliti[1,3], Giovanni Ferrara[2], Luca Romani[2], Francesco Del Pero[1] and Gabriele Arcidiacono[1]

1Guglielmo Marconi University, via Plinio 44, Rome 00193, Italy

2 Department of Industrial Engineering, University of Florence, Via di Santa Marta 3, 50139, Florence, Italy

3 Corresponding Author

Contacts: e.risaliti@unimarconi.it; giovanni.ferrara@unifi.it; luca.romani@unifi.it; f.delpero@unimarconi.it; g.arcidiacono@unimarconi.it

Abstract. Design improvement of Internal Combustion Engines (ICEs) is required by the continuous update of European exhaust emission standards and vehicle registration protocols. This has pushed research and industry efforts towards the development of Low Temperature Combustion (LTC) systems. Among various LTC technologies, the so called "jet ignition" is claimed to be well-suited for light engines, since it provides a more uniform and rapid combustion while ensuring the mixture ignition process. In this context, the adoption of prechambers was found well-suited for two-stroke engines, for which low-pressure direct injection technologies have been developed in the last decade to reduce the fuel short circuit phenomenon. The present paper experimentally investigates the use of different passive combustion prechambers in a Low-Pressure Direct Injection (LPDI) 2-stroke engine, with the main purpose to understand the relationship between prechamber geometrical parameters and engine performance. A 50-cc single cylinder LPDI motorcycle engine is chosen as the test case and it is re-arranged to run in jet ignition mode. The experimental analysis focuses on the evaluation of benefits provided by jet ignition combustion compared to the baseline LPDI propulsion unit at different engine operating points. Design criteria for prechambers development and the tuning process of engine combustion parameters (such as start of injection, ignition time and throttle valve opening) are described in detail. The results of experimental activity are finally shown and critically discussed, highlighting advantages and disadvantages of this application in terms of engine performance, efficiency and cycle-to-cycle variation.

1. Introduction

The evolving legislation on polluting emissions and vehicle type-approval procedures has posed challenges for car manufacturers in recent years. With the advent of electric and hybrid vehicles, the automotive market aims at reducing emissions on the tank-to-wheel side. Hybrid cars still incorporate Internal Combustion Engines (ICEs), both to power the vehicle and to assist the electric motor. REEV (Range Extender Electric Vehicle) technology represent an innovative solution to address the electrification challenge in transportation. These vehicles enable an extension of the electric-only range, reducing dependence on fossil fuels and helping mitigate direct CO2 emissions. However, they are not exempt from emission-related challenges, as they often feature ICEs that serve as generators to recharge the batteries. These ICEs can still produce pollutant emissions, even though at lower levels compared to traditional vehicles. Therefore, while REEVs represent a step in the right direction towards more sustainable mobility, it is essential to develop strategies for minimizing emissions from ICEs in such vehicles to maximize the environmental benefits of hybridization. To address these challenges, the sole adoption of advanced exhaust treatment systems is not sufficient, necessitating the introduction of efficient and clean combustion technologies. Low Temperature Combustion (LTC) holds significant potential in resolving issues related to ICEs. LTC involves the use of lean and homogeneous mixtures to reduce fuel consumption and emissions of nitric oxides (NOx) and particulate matter (PM) [1]. That said, LTC faces challenges such as ignition instability, cycle-to-cycle variation, and occasionally increased carbon monoxide (CO) and unburned hydrocarbon (UHC) emissions. Various LTC technologies [2] including Homogenous Charge Compression Ignition (HCCI), are targeted to overcome these issues. Among them, Jet Ignition (JI) stands out for its potential to stabilize the ignition process without significantly increasing system complexity. Simultaneously, two-stroke (2S) engines have experienced a decline in the motorcycle industry due to high fuel consumption and emissions, especially UHC emissions. These issues are primarily attributed to the inefficiencies of the scavenging process of the cylinder, leading to fuel and lube oil losses and increased cycle-to-cycle variation at low loads. Nevertheless, the 2S engines offer compelling advantages, such as high-power density, low weight, minimal maintenance requirements, and high reliability, making it suitable for motorsport and other applications. Such positive characteristics are also described by Mattarelli et al. [3], for both range extender cars and 2S engine cars. Since reducing vehicle mass can lead to lower fuel consumption [4], the use of 2S engines in replacement of 4S engines could be a viable option from an energy and environmental point of view. Similarly, Mattarelli et al. [5] compares the design of a 4-stroke and a 2S, for application in a range extender vehicle, stressing that the 2S engine allows a weight saving of 15 kilograms and a 6% improvement in efficiency. Serrano et al. [6] studies the application of a 2S engines with opposed pistons for range extender vehicles, achieving a 75% reduction in CO2 emissions compared to the non-hybrid counterpart. To further enhance the reduction of emissions, Direct injection technology, in particular Low-Pressure Direct Injection (LPDI), has the potential to mitigate fuel short-circuiting issues in 2S engines. LPDI is found effective in reducing UHC emissions in small 2S engines, as the injection timing is optimized to avoid fuel short-circuiting, while ensuring at the same time an adequate mixture homogenization. The combination of LPDI and LTC combustion shows great potential to improve the efficiency and to reduce pollutant emissions in 2S engines.

In this context, passive prechambers, which do not require significant engine modifications, have emerged as a feasible solution for small motorcycle engines, arousing interest among automotive manufacturers [7] [8] [9].

The design process of prechambers is primarily guided by considerations of the flow field within the two-stroke engine, as it plays a crucial role in the filling process of passive prechambers, as confirmed by Ciampolini at al. [10]. Furthermore, practical aspects are carefully considered to minimize any necessary modification to the existing engine configuration. As regards criteria for Applying Prechambers to 2S Engines, the paper refers to the findings of Ryu and Asanuma [8], that suggests placing the prechamber close to the center of the cylinder head as much as possible. For simplicity, the existing filleted spark plug hole is used as the prechamber housing. It should be noted that this hole is slightly offset towards the transfer ports, due to the asymmetrical shape of the cylinder head. Additionally, it has to be taken into account that this choice necessitates a small prechamber volume, as the original hole is designed to accommodate a M14 spark plug.

Compact prechambers require smaller volumes compared to elongated ones, as largely provided by literature [11] [12] [13]. Notably, Chiodi et al. [14] achieves promising results with a compact prechamber volume of 0.3 cc, corresponding to a prechamber-to-cylinder volume ratio (Vp/Vc) of 0.075%, in a four-stroke direct injection engine. The adoption of a small volume is justified by the fact that the fuel quantity within the prechamber does not directly contribute to produce useful power. However, Yamaguchi et al. [15] and Roethlisberger et al. [16] reveal that a larger prechamber volume may improve the combustion process within the main chamber. Therefore, a volume is chosen that would allow for the utilization of the existing cylinder top hole without significantly increasing the height of the prechamber, thus maintaining a reasonable prechamber-to-cylinder volume ratio. The volume of 0.13 cc is deemed a suitable baseline parameter, and an investigation is also conducted with a volume of approximately 0.16 cc, to assess the impact of prechamber volume on the combustion process of the specific engine under investigation.

As regards orifice inclination, Roethlisberger and Favrat [16] [17] recommended tilting the orifices as much as possible with respect to the horizontal plane. This consideration appears to be even more applicable in 2S engines, as the airflow from the transfer channels tends to assume an almost horizontal direction on due to the vertical inclination of the channels. However, in the case of the Piaggio "bell-like" cylinder head shape, direct impingement of hot jets onto the walls would reduce the adiabatic efficiency. To prevent such impingement, an inclination angle of 38° is chosen for the orifices, directed towards the transfer ducts, while an angle of 45° is selected for those aimed towards the exhaust port. A swirl angle is not adopted because the swirl motion inside the prechamber may lead to the stagnation of exhaust gases, as shown by Roethlisberger and Favrat [16].

When considering prechambers employed in Port Fuel Injection (PFI) engines, the existing literature presents a wide range of values for both the orifice area (At) and the At to prechamber volume ratio (Vp). This variability makes it challenging to establish a definitive baseline value for these parameters. Conversely, the literature offers limited data regarding Direct Injection (DI) engines. For instance, Fu et al. [18] uses a three-hole prechamber with a 12mm diameter, resulting in an At of 37.7 mm^2, for a four-stroke engine application. Bosi et al. [19] experiments a six-hole prechamber featuring a 1mm diameter, initially set at 4.7 mm^2 and later increased to 10.6 mm^2, in the context of a small two-stroke engine. The approach of Fu et al. [18] is characterized by a relatively large At, notably enhanced scavenging and facilitated efficient fuel filling. However, this strategy may exhibit an excessively high At for optimal combustion processes within the primary combustion chamber. Existing literature [15] suggests that smaller overall section areas tend to enhance combustion within the main chamber. Hence, values used by Bosi et al. [19] are adopted as baseline. Given that the prechamber volume is smaller than the one investigated by Bosi et al. [19] a reduced flow area is selected. Specifically, the baseline configuration is characterized by an overall flow passage area of 3.14 mm^2. Additionally, an investigation is carried out using a value of 7.1 mm^2 to assess the impact of flow passage area on engine performance. These values correspond to At/Vp ratios of 0.018 mm^{-1} and 0.040 mm^{-1}, equating to diameters of 1.2 mm and 1.5 mm for four-orifices prechambers, and 1.2 mm for six-orifices prechambers. It is worth noting that these surface areas may appear insufficient if the intention is to directly inject fuel into the prechamber, as exemplified by Fu[18]. Conversely, the objective is to swiftly vaporize the fuel within the cylinder, aligning with the principles of High-Pressure Direct Injection (HPDI). Subsequently, the prechamber is filled by capitalizing on the tumble vortex and compression motion within the engine. However, it is crucial to acknowledge that while the tumble vortex aids in scavenging the prechamber from exhaust gases and guiding fuel into it, it may also expel a portion of fuel that has entered before ignition. Consequently, the precise configuration of the injection window assumes paramount importance to mitigate the risk of misfiring.

This article aims at assessing the technical feasibility of adopting a passive prechamber in a small 50cc LPDI Piaggio engine, with the perspective of potentially applying this technique in REEV. Several passive prechamber designs are tested on the engine at laboratory scale. The study focuses on establishing the relationship between prechamber geometrical characteristics and the combustion process, to identify the optimal prechamber configuration for the selected engine. Combustion parameters are finely tuned, to achieve optimal engine performance and efficiency across a range of operating points which is typical for motorcycle operation.

The structure of the paper begins with the design criteria for passive prechambers based on a review of existing literature. The role of direct injection and LTC in improving performance follows. The paper then presents the prechambers tested in the project, along with the configuration of the 2S engine used for experimentation. Results from tests conducted at various engine speeds, exhaust CO percentages, and prechamber configurations are discussed, focusing on indicated efficiency and IMEP values for comparison between JI and the Standard engine (STD). Finally, the paper reports key findings of the research and outlines potential areas for future investigation.

2. Materials and method

2.1. Test Case

The experimental study took place at the LINEA Laboratory's engine testing facility, situated within the Department of Industrial Engineering at the University of Florence. The primary objective was to assess the practicality of implementing passive prechambers on small 2S engines. A prototype LPDI engine (**Figure** 1), derived from the Piaggio 2S engine – a single-cylinder, air-cooled motor with a displacement of 49.2 cm^3, was used for the tests.

Figure 1 *Test-bench with Piaggio 2S engine installed. The arrow indicates the location of the engine's spark plug, thus indicating the placement of the prechamber.*

The baseline LPDI engine features a bore-stroke of 40x39.3mm and a compression ratio of 10.5:1. Similar to previous investigations conducted by Bosi et al. [20] [19] an IC02NIj Continental injector was installed on the cylinder liner. This injector operated at an injection pressure of 5 bar and it was positioned a few millimeters above the central transfer duct. Gasoline was supplied to the injector via an external electric pump, while a Dell'Orto reciprocating pump, driven by electricity, introduced separately lubricating oil into the intake manifold. The oil pump's operation was managed by the Engine Control Unit (ECU), following a specific map that adjusted the oil quantity based on engine speed and load conditions.

To optimized engine performance, the Spark Timing (IGA) and Start of Injection (SOI) for the baseline LPDI engine underwent meticulous calibration. The aim was to achieve peak Indicated Mean Effective Pressure (IMEP) levels under full-load conditions. Conversely, during the calibration phase at lower loads, the focus shifted towards maximizing engine efficiency. In the case of the prechamber configuration, the baseline engine head accommodated a BR8ES M14 spark plug, while the prechamber itself was threaded into the existing aperture (as depicted in the figure). The prechamber was designed to host an ER9EH-6N M10 spark plug.

The engine was securely mounted on a 110-kW eddy current test bench, capable of reaching speeds of up to 13,000 rpm. Torque and power measurements were carried out at the gearbox output, connected to the wheel axle, simplifying the measurement setup. To measure the average fuel mass flow rate, the AVL PLUTRON CLASSIC system was employed. Additionally, CO and O2 concentrations were analyzed using the ASSEMBLAD INFRAGAS 309 gas analyzer (prior to the silencer). These measurements facilitated an indirect evaluation of the average air-fuel ratio within the combustion chamber and exhaust duct, respectively. Temperature readings were collected using K-type and T-type thermocouples (**Figure** 2)., monitoring the average temperature of the crankcase and exhaust manifold. Furthermore, the test cell's average temperature and pressure were continually monitored to standardize the output data. Data acquisition during the experiments is carried out using the CoBa software,), using a NI Compact RIO 9024 device [21]. The dynamic in-cylinder pressure is measured using an AVL GR12D piezoelectric sensor, and the instantaneous angular position of the crankshaft is recorded with an AVL 365X optical encoder. Measured data are acquired by means of the AVL IndiMicro measurement system and processed using the AVL Indicom software [22].

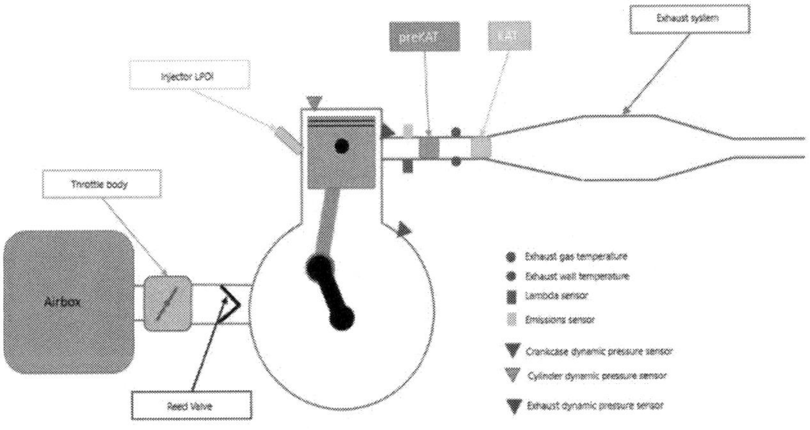

***Figure** 2 Scheme of sensor system*

3. Prechambers used in experiments and engine configuration

The experimental investigation encompassed the meticulous examination of twelve distinct prechamber configurations to discern their notable impact on engine performance. These configurations introduced variations in orifice diameter and count, prechamber volume, and the extent of penetration into the cylinder head (Table 1).

Table 1. Details of the several prechamber configurations tested during the experimental campaign.

Configuration	N° Orifices	Diameter Orifices (mm)	Prechamber Volume (mm^3)	Total Area per Volume Ratio (mm^{-1})	Vp/Vc (%)
C-1.1	4	1	128.43	0.035	2.44
C-1.2	4	1	128.43	0.035	2.39
C-1.3	4	1	161.61	0.028	3.07
C-1.4	4	1	161.61	0.028	3.01
C-2.1	4	1.5	132.62	0.053	2.52
C-2.2	4	1.5	132.62	0.053	2.47
C-2.3	4	1.5	165.8	0.043	3.15
C-2.4	4	1.5	165.8	0.043	3.09
C-3.1	6	1.2	133.46	0.051	2.54
C-3.2	6	1.2	133.46	0.051	2.49
C-3.3	6	1.2	166.65	0.041	3.17
C-3.4	6	1.2	166.65	0.041	3.1

Notably, the latter characteristic had an impact on the volumetric compression ratio of the engine. Among the configurations, C-#.2 uniquely managed to preserve an unaltered volumetric compression ratio, aligning with the STD. This was achieved by reducing the volume of the main combustion chamber to compensate for the additional prechamber volume. In contrast, configurations C-#.1 and C-#.2 exhibited the most compact prechamber volumes since the spark plug has been threaded until it reached the end of the internal prechamber thread. Configurations C-#.3 and C-#.4 featured a tactically inserted dedicated washer beneath the mechanical stop of the spark plug to amplify the prechamber's volume Visual representations delineating these diverse prechamber assemblies are thoughtfully presented in **Figure 3**. Noteworthy, the prechambers' construction material was AMPCOLOY 972 Cu-Cr-Zr alloy, meticulously machined via sophisticated five-axis milling techniques (**Figure 4**).

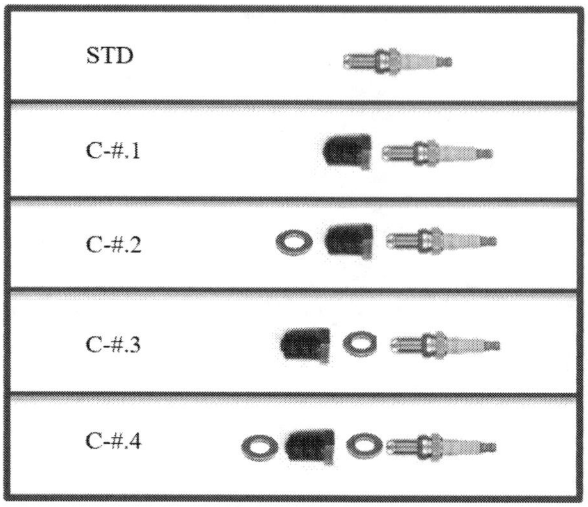

Figure 3 *Diagram of prechamber configurations. Thanks to the installation of the washer, it is possible to increase the internal volume of the prechamber and protrude the prechamber less into the cylinder.*

Figure 4 *On left, prechamber with 4 holes, 1mm diameter. On right, prechamber with 6 holes, 1.2 mm diameter*

This choice of material was guided by the requirements of high thermal conductivity and resistance to high-temperature oxidation, two pivotal attributes indispensable for prechambers. Previous analyses [19] demonstrated the tendency of aluminum to undergo thermal failure when utilizing JI combustion without a dedicated cooling system.

The experimental campaign evaluated engine performance at three specific operating points, simulating real-world driving conditions. These points included engine speeds of 5000 rpm and 2 bar BMEP (OP1), 6500 rpm at full charge (OP2) where maximum torque is achieved, and 7750 rpm at full charge (OP3) where maximum vehicle speed is attained. Before conducting the actual experiments, a series of preliminary tests were conducted to discern the optimal points characterized by maximum efficiency and IMEP. The tests were based on a specific degree of the start of injection (SOI) and spark timing

(IGA) for the C-1.2 prechamber configuration. For all operating points and prechamber configurations, the initial focus was on ascertaining the IGA that yielded the best efficiency for partial loads and the highest IMEP for full loads. Once the optimal IGA values were determined, the search for the best SOI values commenced. Subsequently, various tests were performed using the baseline prechamber configuration C-1.2, with IGA variations, to achieve specific CO percentages in the exhaust gas. The reported results highlight the achieved CO percentages of 2% relative to the exhaust gas mass, providing valuable insights into the equivalent air-to-fuel ratio.
However, it's noteworthy that no analysis was conducted on lower CO percentage values. For both the prechambers and the standard engine, no discernible enhancement in performance materialized in alignment with the predetermined performance criteria.

4. Results and discussion

This paragraph reports results of the experimental activity on the 49.2 cc Piaggio LPDI engine in the JI configuration. The various trends of the engine's significant output parameters (IMEP, indicated efficiency, cycle-to-cycle variation, and so on) are presented depending on the operating point and the pre-chamber configuration. It is worth to note that values coming from the analysis (dynamic pressure sensor inside the cylinder head) are collected using the AVL Indicom software by averaging 100 engine cycles. As already mentioned, the first part of the experimental campaign concerns the study of the best SOI and IGA values for the JI engine. For operating points below 5000 RPM, it is observed that the prechamber requires a larger throttle opening compared to the standard engine. This increase is aimed at enhancing volumetric efficiency while maintaining a 2 % CO level, consequently leading to a reduction in indicated efficiency and an increment in fuel consumption. As explained in [20] to achieve this enhanced efficiency, a delay in the ignition advance (IGA) is required. Such a delay, in turn, shifts the point of peak pressure away from the ideal timing, thus resulting in a reduction of the useful work provided by the piston and an increase in the temperature of exhaust gases. That said, it is important to note that these effects should not be interpreted as discrediting the use of passive prechambers. From the perspective of implementing this technology in REEV, it is crucial to consider operating conditions where partial load is not a concern, typically with a constant engine speed. Therefore, when assessing the tested engine, the data obtained from operating points above 5000 RPM become more pertinent and valuable.
Below are the selected operating points for the tests:
1 - At 5000 rpm and 2% of CO, the C-1.2 achieves the highest efficiency at SOI 220 IGA 30, while the STD engine at IGA 22. So, the operating points at IGA 22 and IGA 30 will be analyzed.
2 - From 6500 rpm the IMEP is selected as benchmark parameter: the C-1.2 leads to the maximum IMEP SOI 240 IGA 18 and the STD engine at IGA 22. The chosen operating points are IGA 18 and IGA 22
3 - At 7750 and CO 2%, both prechambers and STD engines have maximum IMEP at SOI 240 IGA 34. In this case, only the operating point at IGA 34 will be analyzed.
Values related to the STD engine are represented by black bars, the engine equipped with C-1.# prechambers by red bars, C-2.# prechambers by blue bars, and C-3.# prechambers by yellow bars in the upcoming graphs.

4.1. OP1:5000 rpm, 2 bar BMEP and CO 2% set.

This section reports the results of the OP1 configuration, which provides that the engine is set to run at 5000 rpm, with a BMEP set at 2 Bar and an exhaust CO of 2% compared to the mass measured at the exhaust. Such a case is borderline: the efficiency results to be not sensitive to the change of IGA. Anticipating IGA involves that the prechamber work at high peak pressures but also at low pressures during the expansion phase while postponing IGA does not allow to exploit the optimum of the peak point position and there is an efficiency loss due to the high exhaust temperatures.

Results show that the JI engine requires in general a lower throttle opening than the STD engine to achieve the same BMEP. The STD engine needs an opening of 38.5 % while the pre-chamber with the best efficiency, C-1.3 IGA 30, needs an opening of 34 % and the one with the worst efficiency (C-2.3 IGA 22) needs an opening of 36.5 %. Compared to partial loads, the higher BMEP setpoint leads to an

increase in the IMEP (as shown in **Figure** 5), and thus to an increase in the fuel required. This feature corresponds to an increase in the opening of the throttle valve. In particular, the C-1.3 prechamber, which has an indicated efficiency of 31.52% (**Figure** 6) has the lowest throttle opening and fuel consumption, measured at 0.524 kg/hr. The prechamber with the worst efficiency, C-2.3, consumes 0.572 kg/hr and shows an efficiency of 28.32%. The STD engine, with an indicated efficiency of 30.87%, consumes 0.576 kg/hr of fuel. The C-2.# and PC-3.# prechamber series show the lowest efficiencies, which are around 29%, but have a fuel consumption of about 0.550 kg/hr.
The graph in the Figure 7 reports specific CO_2 emissions per hour in relation to the effective power output by the engine. Consequently, we will have CO_2 emissions per unit of emitted energy. In the graph it can be observed that the prechamber with the highest indicated efficiency also has the lowest specific CO_2 emissions, with 2.051 kg/(kWh) against 2.211 kg/(kWh) of STD .

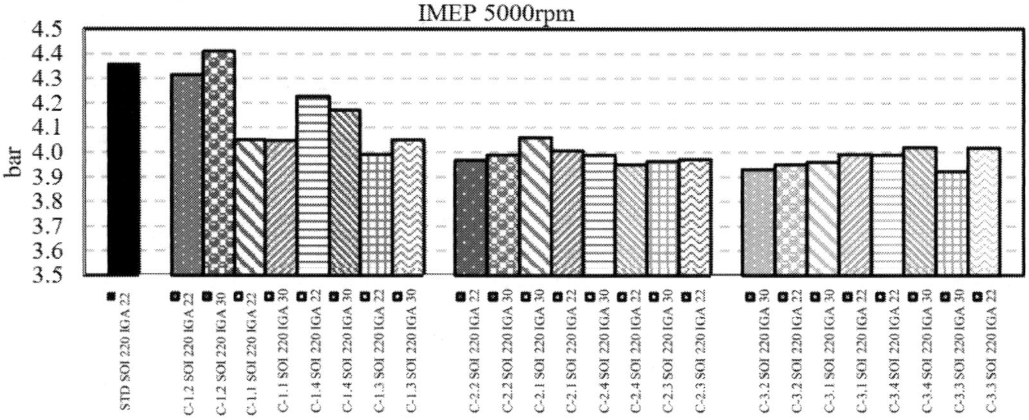

***Figure** 5 IMEP at 5000 rpm, BMEP set 2 bar, CO target 2%*

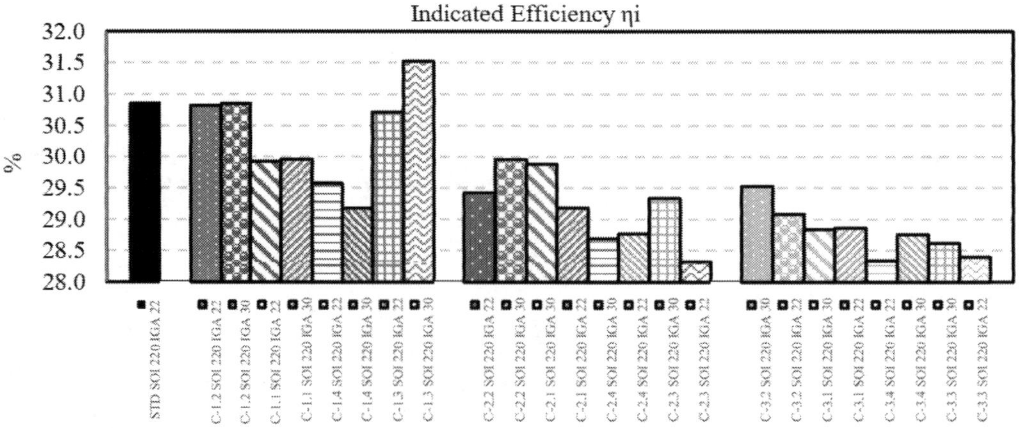

***Figure** 6 Indicated efficiency at 5000 rpm, BMEP set 2 bar, CO target 2%*

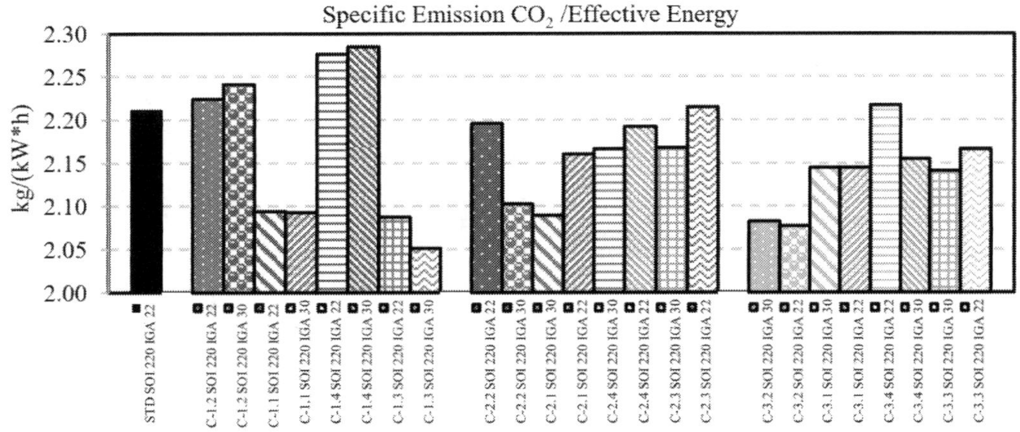

***Figure** 7 Specific CO2 emissions relative to the effective energy output of the Piaggio engine, at 5000 rpm, BMEP set 2 bar, CO target 2%,.*

Despite the low consumption of the prechambers, the low efficiency is due to the not optimal position of the pressure peak: in **Figure A** 1 and **Figure A** 2 (show in Appendix A) reveal that C-2.# and C-3.# series present pressure peak values around 21 bar, which occur on average at 16°-17° Crank Angle (CA) .The STD engine develops a higher pressure peak (24.32 bar), with a Standard Deviation (SD) of 3.02 bar, positioned at 15.5° CA, these latter with a SD of 1.63°. The C-2.3 pre-chamber has a peak pressure of 19.92 Bar, with a SD of 2.72 bar, located at 18.5° CA (SD of 3.57°). Finally, the C-1.3 prechamber shows a peak pressure of 22.22 bar, with 3.25 bar SD, positioned at 16 °CA (SD of 1.60°).
Consequently, the prechamber with the highest efficiency is the one that can ensure better combustion stability, reducing the SD of the peak pressure position and therefore decreasing cycle-to-cycle variability. In this case, the C-1.3 prechamber exhibits an SD comparable to that of the STD engine As a confirmation, measurements do not reveal any misfire in case of the C-1.3; that said, in the lower left of **Figure** 8 some points of the prechamber C-2.3 are found, thus introducing a standard deviation on the position then wider and therefore a greater cycle-to-cycle variability

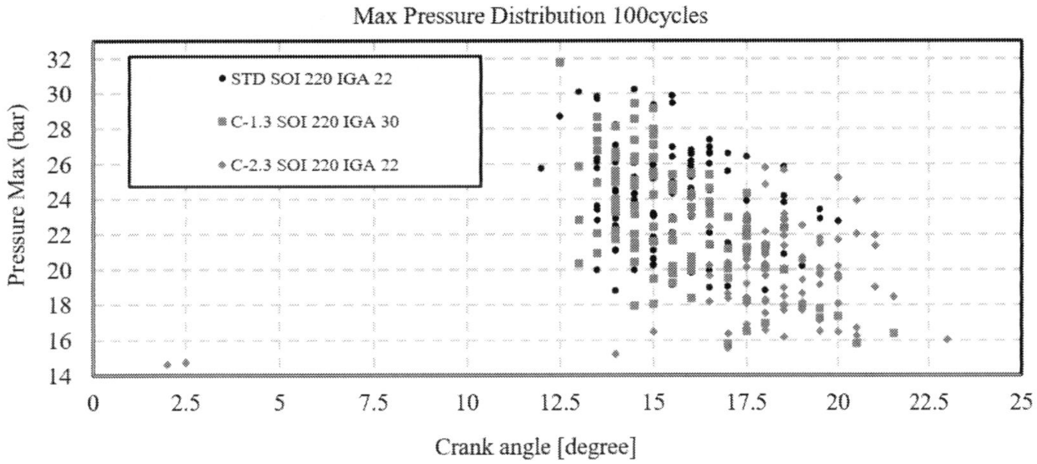

***Figure 8** Maximum pressure distribution over 100 cycles for standard engine (black), best efficiency prechamber (red) and worst efficiency prechamber (blue) 5000 rpm,, BMEP set 2 bar, CO target 2% , O° = top dead centre (TDC)*

Figure A 3 shows the heat release of the STD engine, C-1.3 and C-2.3. The opener with the highest efficiency is the one characterized by a similar heat release as the STD engine, despite the STD engine has a higher IGA value. Such an outcome substantially confirms that the prechambers can provide a faster burning process. The lowest efficiency prechamber has a more delayed start of combustion, causing lower and more delayed peak pressures. In **Figure A** 4 provides the P-V diagram, which shows how the C-1.3 expansion stage has a lower pressure, even than the C-2.3 prechamber, thus involving on one hand a decrease in the useful work of the piston and on the other hand a lower IMEP.
As regards the C-2.3, during the expansion phase it almost overlaps with the STD engine, but due to a lower peak pressure and a bigger delay it does not provide optimal work, as well as it causes higher exhaust temperatures (the STD engine with C-2.3 has a temperature of 618 C° while the STD engine with C-1.3 has an exhaust temperature of 578 C°).

4.2. OP2:6500 rpm, WOT and CO 2% set.
This section reports the results of the OP2 configuration, which provides that the engine is set to run at 6500 rpm, with a WOT set and an exhaust CO of 2% compared to the mass measured at the exhaust. In such a configuration the prechambers with the highest IMEP are considered, instead of those with the highest efficiency since the analysis is targeted at investigating motor points at full load. Looking at **Figure** 9 shows that all the prechambers have approximately the same IMEP as the STD engine. The STD engine shows an IMEP of 7.30 bar, the prechamber with the best IMEP is again the C-1.2 IGA 18 °CA with 7.53 bar, while the worst configuration is the C-1.4 IGA 22 °CA with 6.89 bar. **Figure** 10 shows that the efficiency trend is almost the same as in the IMEP graph: the STD engine has an efficiency of 25.55 % with the C-1.2 (26.96 %), while the C-1.4 is characterized by the lowest value (24.62 %). On the other hand, the prechamber that turns out to be the most efficient in terms of indicated efficiency is the C-3.2 IGA 18°, with an efficiency of 27 %.

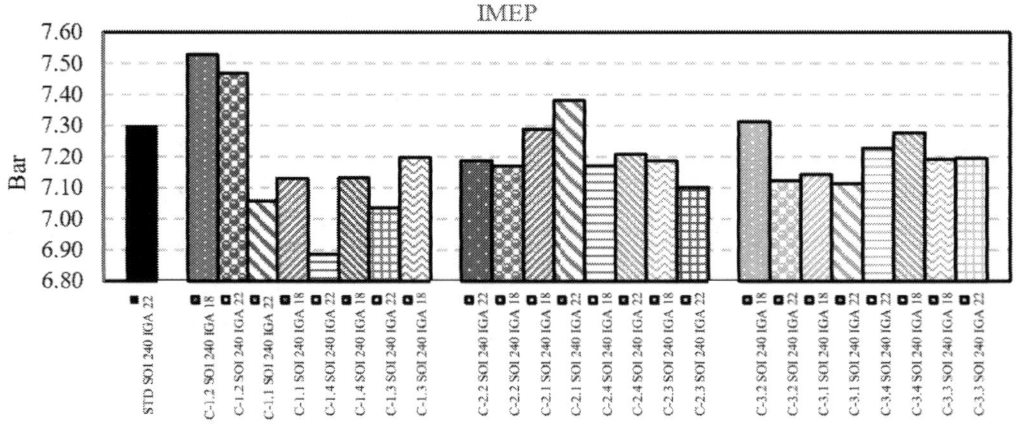

***Figure 9** IMEP at 6500rpm, WOT set, CO target 2%*

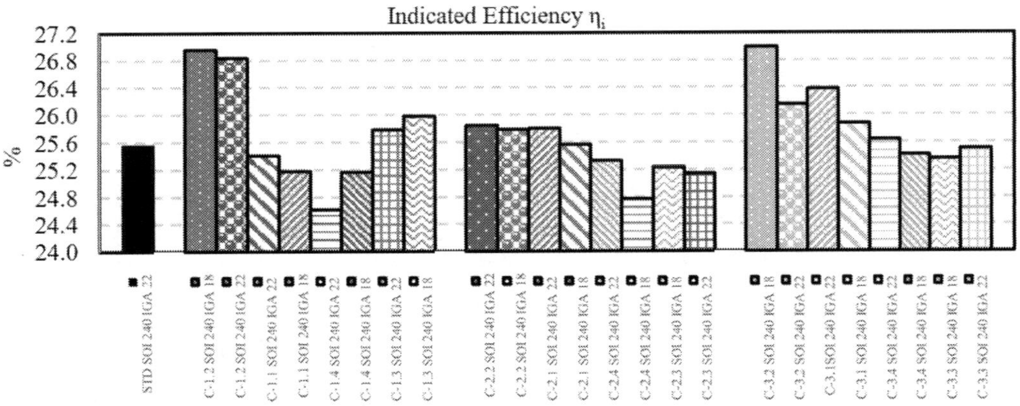

Figure 10 *Indicated efficiency at 6500rpm, WOT set, CO target 2%*

Such a configuration slightly improves the IMEP if compared to the standard engine, enhancing it by 0.18 %. That said, **Figure** 11, shows an improvement in emissions with reduced fuel consumption. As a confirmation, the consumption in the C-1.2 configuration is 1.14 kg/hr, and 1.105 kg/hr in the C-3.2 configuration, while the STD engine consumes 1.16 kg/hr. Tests show that the C-1.4 pre-chamber consumes 1.14 kg/hr despite its lower efficiency, since it operates with less fuel demand and higher combustion speed, but it fails to exploit combustion at the optimal point. The diagrams in **Figure** 12 and **Figure A** 5 reveal that the C-1.4 configuration has an extremely faster heat release than the STD engine, the C-1.2 and the C-3.2; consequently, the C-1.4 prechamber has an earlier pressure peak. More specifically, the C-1.4 prechamber shows the pressure peak at 12 °CA, with a SD of 1.63°, while the STD engine reaches the pressure peak at 15.5 ° CA with a SD of 1.7° (the C-1.2 at 15.5 °CA with a SD of 1.43°, the C-3.2 at 16.5°CA with a SD of 1.30°).

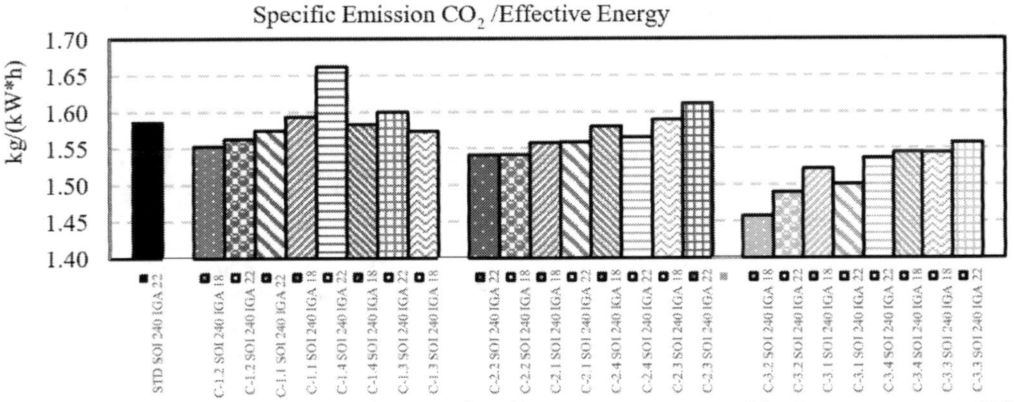

Figure 11 *Specific CO2 emissions relative to the effective energy output of the Piaggio engine at 6500 rpm, WOT set, CO target 2%*

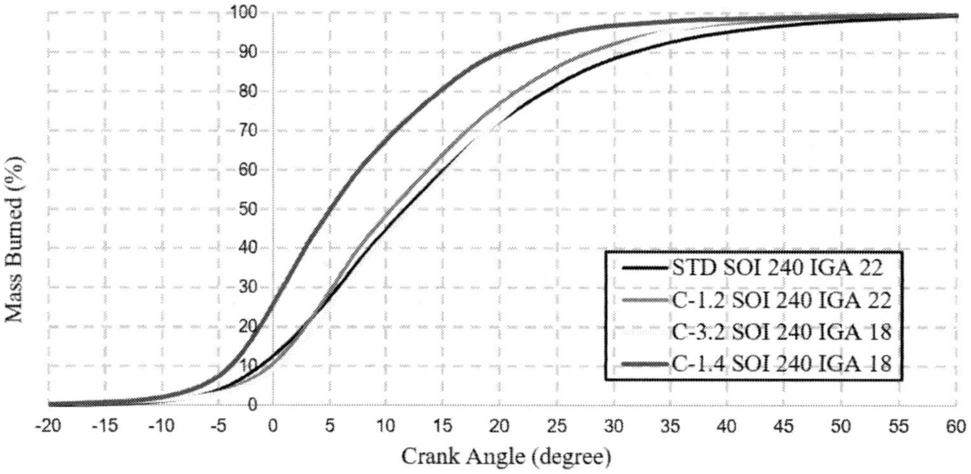

Figure 12 *Heat release for standard engine (black), best IMEP prechamber (red), worst IMEP prechamber (purple) and best efficiency prechamber (yellow) at 6500rpm, WOT set, CO target 2%, 0°=TDC*

The lower value of SD of the JI engine is indicative of a decrease in the cycle-to-cycle variation. From the above, it can be said that prechamber performs better at high engine loads: this is due to the improved flushing of internal volume, which can be associated with the stronger flow motion and the wider compression variation, which also increases the combustion stability.

Despite the higher peak pressure, the C-1.4 prechamber presents a pressure line in the expansion phase (see **Figure A** 5), which is below the STD engine , C-1.2 and C-3.2, thus enabling to obtain less useful work than the others, and therefore reducing the overall efficiency. This result is in line with **(19)**, which states that the use of a prechamber provides a lower cycle-to-cycle variation and a greater combustion stability. At the same time, as the combustion speed is higher, the JI engine needs a delay in the ignition: indeed, in case of anticipated ignition, the pressure peak occurs too early, resulting in sub-expansion effects or blow-by effects.

Figure 12 shows results related to the C-3.2 prechamber, with a heat release similar as the STD engine: the combustion begins later, but it is so fast that there is a faster heat release 5 ° of crank angle after the TDC. The C-3.2 has a mean peak pressure of 37.8 bar, with a 3.31 bar SD, lower than the C-1.4 (43.20 bar and 3.96 bar SD) and C-1.2 (40.1 bar with a 4.48 bar SD), but higher than the STD engine (37.55 bar and 5.2 bar SD). A sufficient pressure peak centered at an optimal point also provides more useful work during the expansion phase: in **Figure A** 5 shows that the C-1.2 and C-3.2 curves have the highest pressures during the expansion phase. This is also confirmed by the fact that the C-1# and C-3# ensure a lower exhaust temperature than the STD and therefore a better exploitation of the fuel potential.

4.3. OP3:7750 rpm, WOT and CO 2% set.

In this subchapter the results of the OP3 configuration are presented: the engine is set to run at 7750 rpm, with a WOT set and an exhaust CO of 2% compared to the mass measured at the exhaust. During preliminary experiments, it was noted that both the STD and the JI engine had the highest IMEP performance with an IGA of 34 °CA. Therefore, the results reported below are all at the same IGA. **Figure** 13 and **Figure** 14 report the trend of IMEP and indicated efficiency. As expected, by increasing the engine rpm and moving away from the area where the engine delivered maximum torque, the IMEP decreases in all configurations. The STD engine goes from an IMEP of 7.30 bar at 6500 rpm and 2% CO to an IMEP of 5.61 bar. Similarly, the PC-1B pre-chamber, which is also the best at 6500 rpm, passes from an IMEP of 7.53 bar to an IMEP of 5.64 bar. However, the different prechamber configurations analysed present an IMEP value which is comprised within a relatively small range (5.15-5.40 bar).

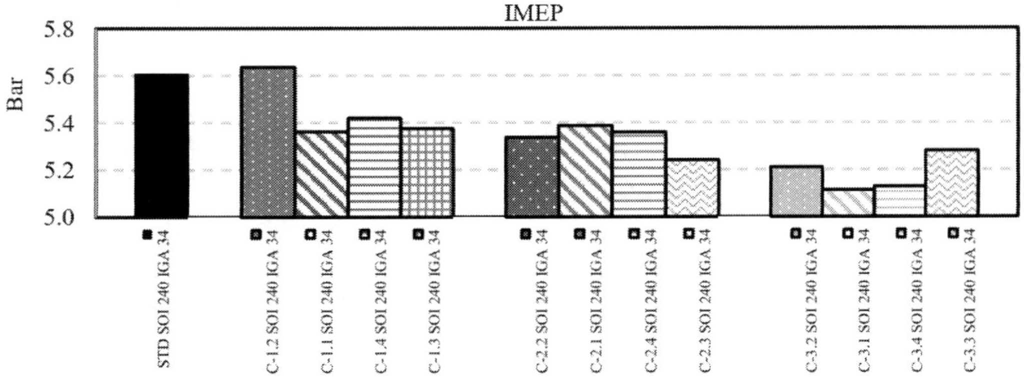

***Figure 13** IMEP at 7750rpm, WOT set, CO target 2%*

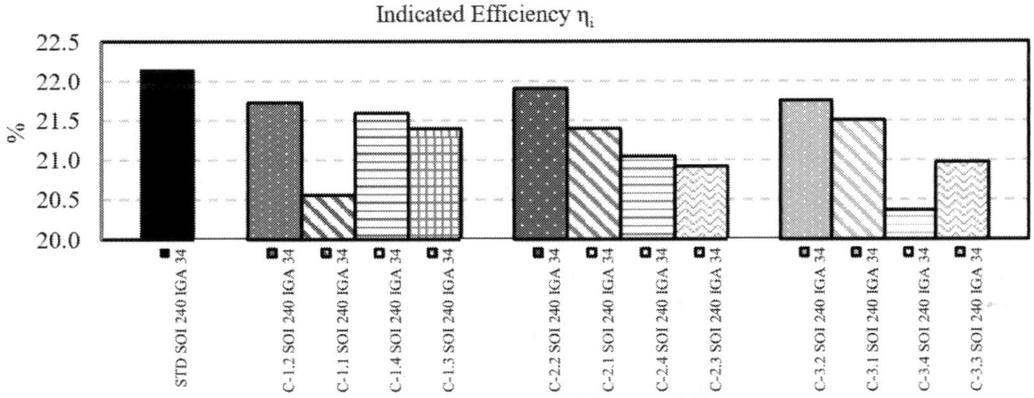

***Figure 14** Indicated efficiency ηi at 7750rpm, WOT set, CO target 2%*

The worst case is represented by the C-3.1, with an IMEP value of 5.11 bar. With regards to the indicated efficiency, the STD engine is found to achieve the best results, (22.14 %), consuming 1.03 kg/h of fuel, while the C-1.2 shows an efficiency of 21.73 %, with a consumption of 1.05 kg/h. Finally, the worst configuration is found to be the C-3.1, achieving a 21.50 % efficiency, with a 0.97 kg/h fuel consumption. The configuration with the highest efficiency is the C-2.2 (21.9 % efficiency), with a consumption of 0.99 kg/h. This is confirmed by CO2 emissions data reported in **Figure** 15: C-2.2 emits 1.93 kg/(kW*h), C-1.2 emits 2.04 kg/(kW*h), C-3.1 emits 2.00 kg/(kW*h) and STD emits 2.01 kg/(kW*h). The higher amount of fuel consumption of the C-1.2 leads on the one hand to a higher IMEP, and on the other hand, to a lower indicated efficiency. On the contrary, the C-3.1 shows a lower IMEP, but also a lower efficiency, despite the lower fuel consumption.

The prechambers are found to have a lower IMEP since the peak pressure is not at the optimum point. The JI engine at 7750 rpm shows more or less the same issues as in the lower operating point (engine speeds below 5000 rpm): prechambers, other than C-1.2, show a pressure peak position that is delayed than the STD engine, as it can be seen in **Figure A** 6 , **Figure A** 7 and **Figure A** 8

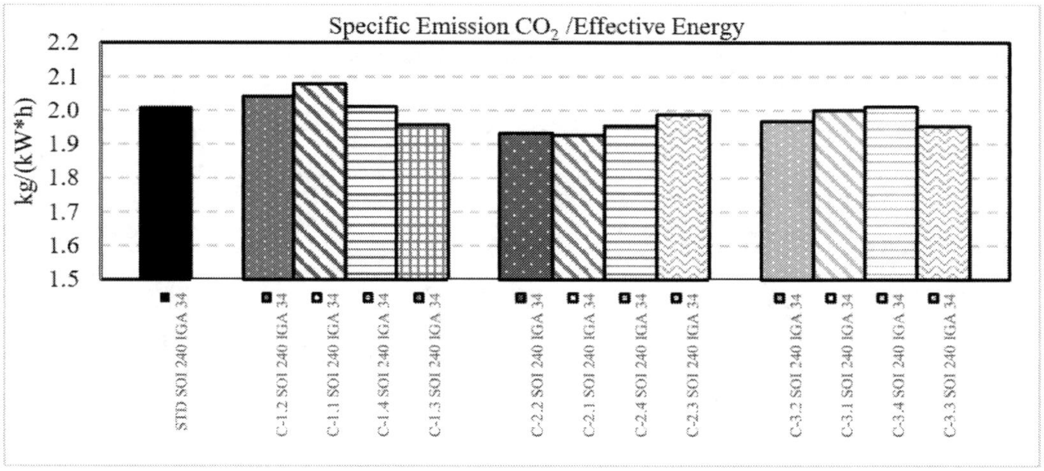

Figure 15 *Specific CO2 emissions relative to the effective energy output of the Piaggio engine at 7750 rpm, WOT set, CO target 2%*

In addition, all the JI configurations show a lower pressure peak than the STD engine. The STD engine has a peak pressure of 33.47 bar, with a 4.22 bar standard deviation, positioned at 11.5° crank angle, with a 1.64 °CA standard deviation. The C-1.2 has a peak pressure of 31.20 bar, with 3.96 bar SD, positioned at 13° crank angle, with a 1.57 °CA standard deviation, and the C-3.1 27.03 bar, with 3.55 bar, positioned at 14.5°, with a 1.50° SD. Despite the lower efficiencies, the prechambers under consideration have a higher combustion stability than the STD engine, as they present lower SD. The prechamber with the highest efficiency (C-2.2) has a peak pressure of 27.89 bar with 3.45 bar SD positioned at 14.5°, with 1.45° SD. Due to the delayed peak position than the STD, the JI configurations are found to have a delayed start of combustion. **Figure** 16 shows that the combustion of the STD engine starts earlier, and JI configurations have almost the same combustion rate of the STD, the only exception being the C-2.2, which instead is characterized by a higher rate. In the curve of the C-2.2, which substantially has the same shape of the STD engine, except for the lower pressure peak, is 2.27 bar lower.

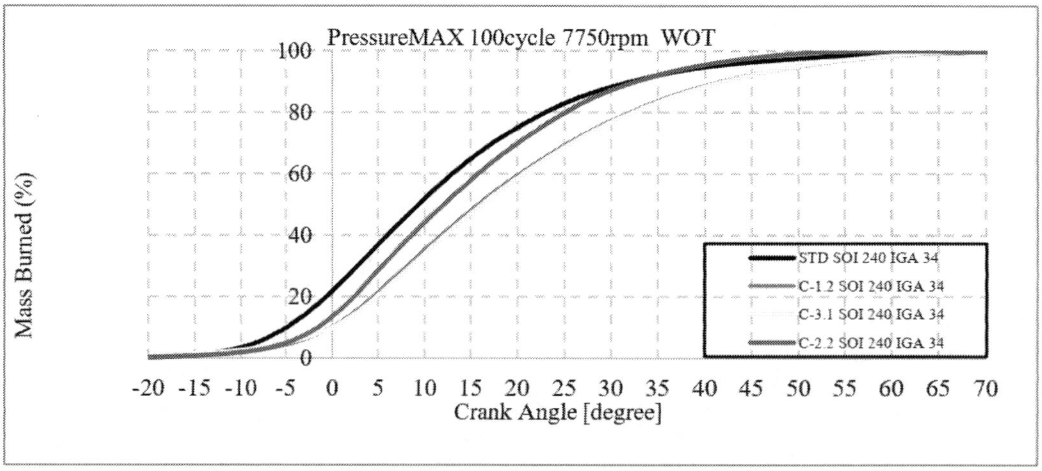

Figure 16 *Heat Release for STD (black), best IMEP prechamber (red), worst IMEP prechamber (yellow) and best efficiency prechamber (blue) at 7750rpm, WOT set, CO target 2%, 0°=TDC*

It can also be noted that during the expansion phase the pressure line of the C-1.2 remains above the curve of the STD, thus providing a higher IMEP. It can be also noted that all the prechambers have a higher exhaust temperature than the STD engine, due to the delayed pressure peak. From the above considerations, it can be stated that running the prechambers at higher engine speeds than the maximum torque point (6500 rpm) involves high total area-to-volume ratio and big orifice diameters. However, results highlight that this affects negatively prechamber efficiency at lower speeds, by postponing the peak pressure point.

5. Conclusion

The paper presents a preliminary study on the application of different passive prechambers to the LPDI Piaggio 2S 49.2 cm3 engine, as literature review highlights the suitability of "stumpy" multi-orifice prechambers for use in 2S engines, in particular GDI ones. Three different prechamber geometries are tested: C-1# features four holes with a diameter of 1.2 mm, C-2# features four holes with a diameter of 1.5 mm, C-3 #features six holes with a diameter of 1.2 mm. The volume and sinkage within the main combustion chamber are changed during tests, to assess the variation in terms of indicated efficiency and IMEP.

The analysis stresses that the JI combustion allows increasing efficiency and IMEP when specific pressure conditions occur within the main combustion chamber. For engine speed lower than 5000 rpm, the JI engine cannot reach the same performance as the STD engine; this is due to the fact that the engine fails to fire properly, and it is necessary to increase the throttle valve opening to maximize the amount of fuel injected. The C-1# prechamber proves to be the best JI setup for the vast majority of tests carried out, exceeding the performance of the STD engine, especially in terms of IMEP, but still failing to reach the efficiencies of the STD motor at low loads. To achieve higher efficiencies, a greater number of orifices should be used for 6500 rpm, to achieve better combustion by reducing the cycle-to-cycle variation. Meanwhile, for 7750 rpm, a larger orifice diameter is needed to counteract the ignition delay inside the combustion chamber. The other prechambers tested can reach no more than the performance of the STD engine.

Results above show that a prechamber with a volume as small as possible is preferable, as big volumes generate more energy jets, thus involving problems of heat exchange and consequent efficiency loss. On the other hand, the deepness of the prechamber inside the main combustion chamber has no consequences on efficiency and IMEP, since the scavenging occurs due to the compressions (not the charge motions), and therefore there are no benefits by sinking more the prechamber.

In the light of previous considerations, it can be concluded that prechambers tested do not show a significant increase in performance and efficiency, since they provide better performances only under specific conditions that make favorable the scavenging of the prechamber volume; in all the other cases, the prechambers worsen the performance of the engine, unless a dedicated design is carried out to make that the prechamber operate at speeds other than those of the engine maximum efficiency. Nevertheless, the adoption of prechambers for 2S ICEs in application to REEV can be deemed a path worth exploring: since ICEs are used under constant operating conditions (solely tasked with recharging batteries), the adoption of a prechambers to enhance combustion efficiency and performance is a valid option.

Future research can be focused on finding the optimal configuration of SOI and IGA for each prechamber, to achieve the best performance also at partial loads. Another interesting pint is exploring the possibility of using active/passive scavenging (e.g., designing a pigtail connected to the exhaust to promote prechamber scavenging), as well as applying other types of materials with low thermal conductivity (such as ceramic materials to achieve better results in terms of heat exchange).

6. Appenix A

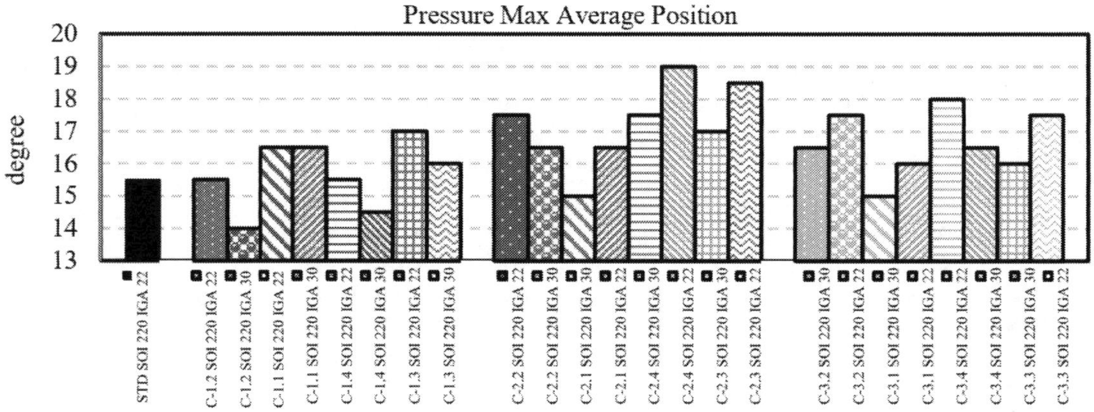

Figure A 1 *Pressure Max Average Position in cylinder at 5000 rpm, BMEP set 2 bar, CO target 2%*

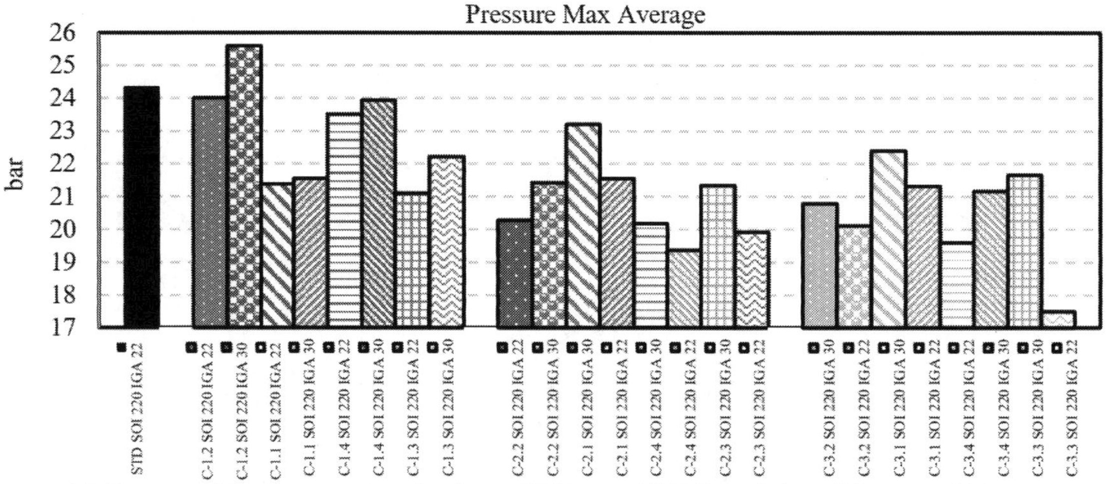

Figure A 2 *Pressure Max Average in cylinder at 5000 rpm, BMEP set 2 bar, CO target 2%*

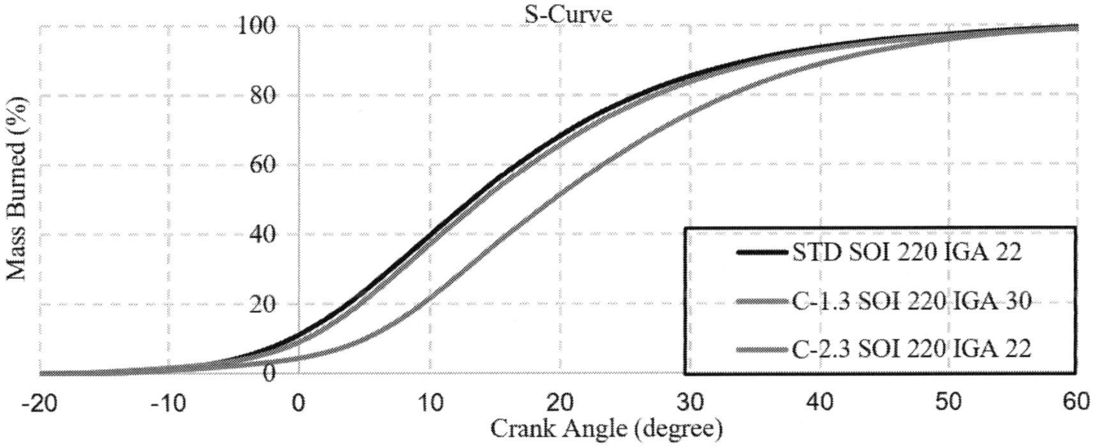

Figure A 3 *Heat release for standard engine (black), best efficiency prechamber (red) and worst efficiency prechamber (blue) 5000 rpm, BMEP set 2 bar, CO target 2%, 0°= TDC*

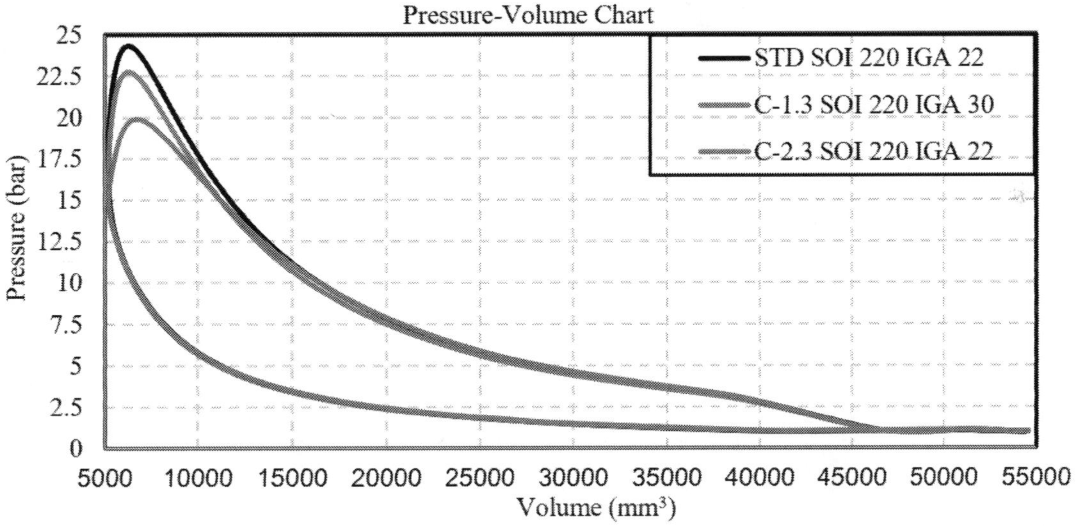

Figure A 4 *Indicated cycle for standard engine (black), best efficiency prechamber (red) and worst efficiency prechamber (blue) 5000 rpm, BMEP set 2 bar, CO target 2%*

IOP Conf. Series: Materials Science and Engineering 1306 (2024) 012029 doi:10.1088/1757-899X/1306/1/012029

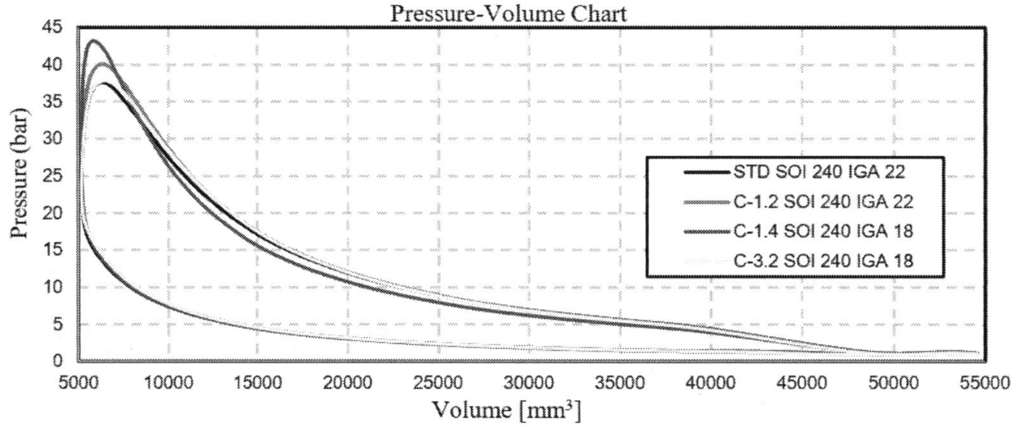

Figure A 5 *Indicated cycle for STD engine (black), best IMEP prechamber (red), worst IMEP prechamber (purple) and best indicated efficiency prechamber (yellow) at 6500rpm, WOT set, CO target 2%*

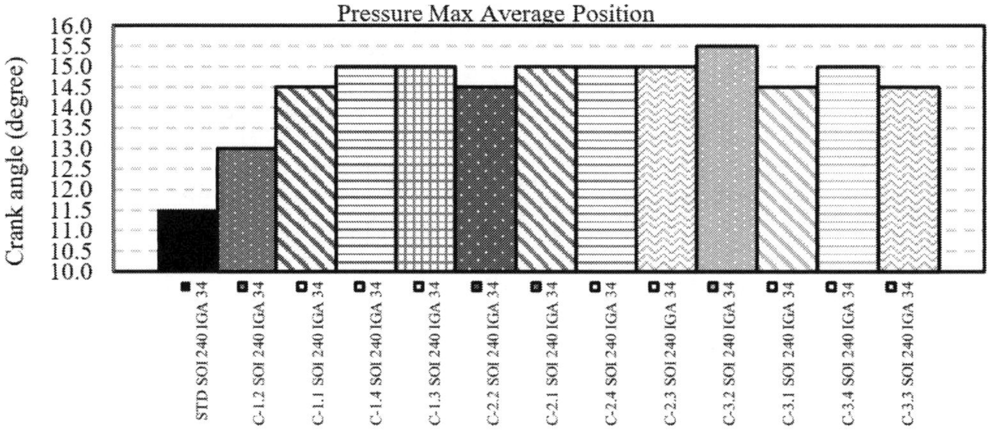

Figure A 6 *Average Position Pressure Max in cylinder at 7750rpm, WOT set, CO target 2%*

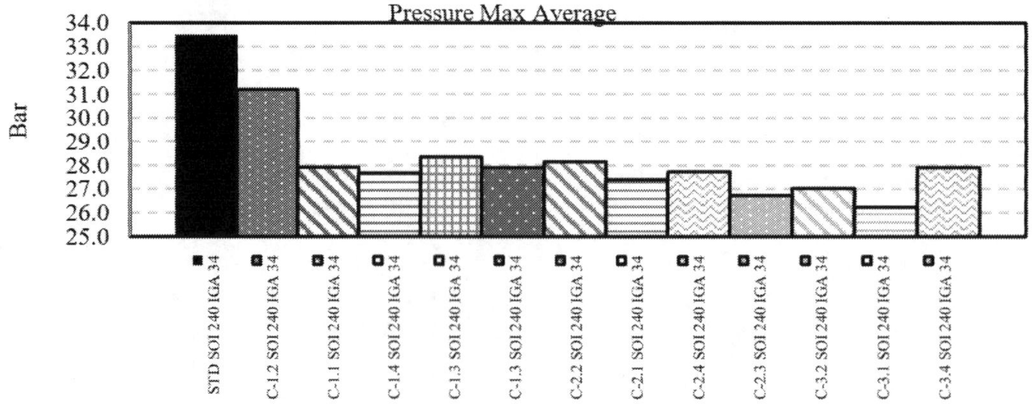

Figure A 7 *Pressure Max Average in cylinder at 7750rpm, WOT set, CO target 2%*

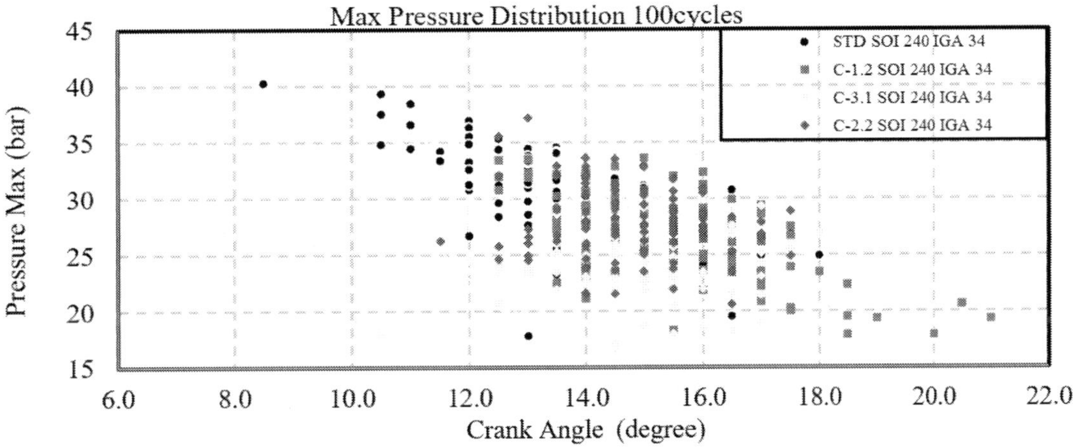

Figure A 8 *Maximum pressure distribution over 100 cycles for STD (black), best IMEP prechamber (red), worst IMEP prechamber (yellow) and best efficiency prechamber (blue) at 7750rpm, WOT set, CO target 2%, 0°=TDC*

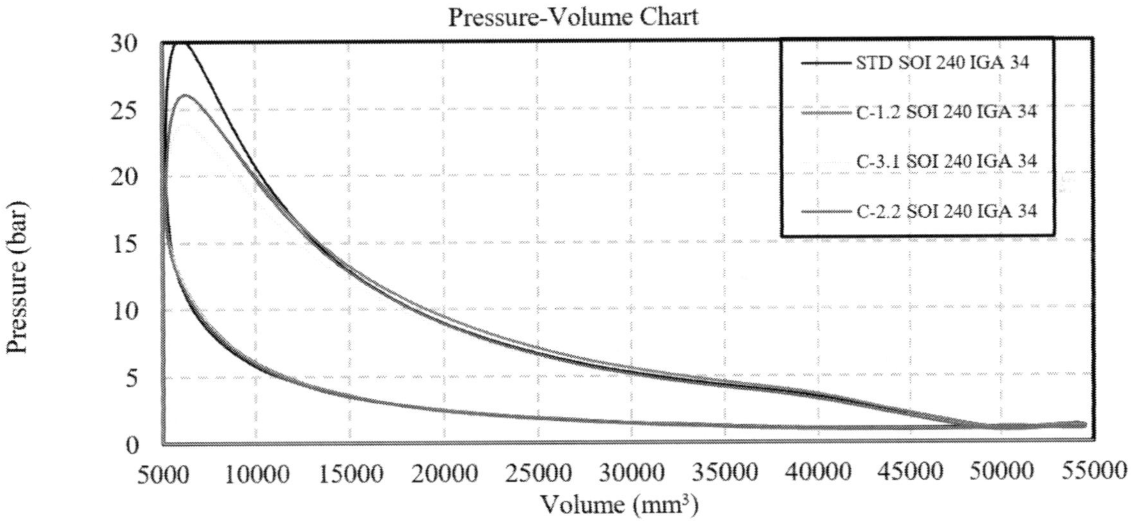

Figure A 9 *Indicated cycle for for STD (black), best IMEP prechamber (red), worst IMEP prechamber (yellow) and best efficiency prechamber (blue) at 7750rpm, WOT set, CO target 2%, 0°=TDC*

7. References

[1] Olmeda P, García A, Monsalve-Serrano J, Lago Sari R 2018 Experimental investigation on qRCCI heat transfer in a light-duty diesel engine with different fuels: Comparison versus conventional diesel combustion., *Appl. Therm. Eng.* **144** 424-436.

[2] Pachiannan T, Zhong W, Rajkumar S, He Z, Leng X, Wang Q 2019 A literature review of fuel effects on performance and emission characteristics of low-temperature combustion strategies. *Appl. Energy* **251** 11380

[3] Mattarelli E, Rinaldini C A, Cantore G, Baldini P, 2012 2-stroke externally scavenged engines for range extender applications. *SAE Technical Paper* 2012-01-1022,

[4] Czerwinski F 2021 Current trends in automotive lightweighting strategies and materials *Materials* **14.21** 6631

[5] Mattarelli E, Rinaldini C A, Cantore G, Agostinelli E 2015 Comparison between 2 and 4-stroke engines for a 30 kW range extender. *SAE Int. J. of Alternative Powertrains*, **4(1)** 67-87

[6] Serrano J R, García A, Monsalve-Serrano J, Martínez-Boggio S 2021 High efficiency two stroke opposed piston engine for plug-in hybrid electric vehicle applications: Evaluation under homologation and real driving conditions. *Applied Energy* **282** 116078

[7] Benajes J, Novella R, Gomez-Soriano J, Martinez-Hernandiz P J, Libert C, Dabiri M 2019 Evaluation of the passive pre-chamber ignition concept for future high compression ratio turbocharged spark-ignition engines *Applied Energy* **248** 576-588

[8] Frasci E, Novella R, Moreno B P,Arsie I, Jannelli E. 2023, Impact of prechamber design and air–fuel ratio on combustion and fuel consumption in a SI engine equipped with a passive TJI. *Fuel* **345** 128265

[9] Posch S, Winter H, Zelenka J,Pirker G, Wimmer A 2021 Development of a tool for the preliminary design of large engine prechambers using machine learning approaches *Applied Thermal Engineering* **191** 116774

[10] Ciampolini M, Bigalli S, Balduzzi F, Bianchini A, Romani L,Ferrara G 2020 CFD Analysis of the Fuel–Air Mixture Formation Process in Passive Prechambers for Use in a High-Pressure Direct Injection (HPDI) Two-Stroke Engine *Energies* **13** 11: 2846,

[11] Sandoval M H B , Alvarez C E C , Roso V R et al. 2020, The influence of volume variation in a homogeneous prechamber ignition system in combustion characteristics and exhaust emissions *J Braz. Soc. Mech. Sci. Eng.* **42** 72

[12] Moreira, T A A 2014 "Análise e caracterização de um sistema de ignição por lança chamas operando com carga homogênea"

[13] Gombosuren N, Ogami Y, and Asada H 2020 A charge possibility of an unfueled prechamber and its fluctuating phenomenon for the spark ignited engine *Energies*, **13.2** 303

[14] Chiodi M, Kächele A, Bargende M, Koch D, Wachtmeister G, Wichelhaus D 2018 Development of an innovative combustion process: Spark-assisted compression ignition *Int. Stuttgarter Symposium* **18**

[15] Yamaguchi S, Ohiwa N, Hasegawa T 1985 Ignition and burning process in a divided chamber bomb *Combustion and flame* **59** 177-187

[16] Roethlisberger R P, Favrat D 2003 Investigation of the prechamber geometrical configuration of a natural gas spark ignition engine for cogeneration: part II. Experimentation *Int. J. Thermal Science* **42** 239-253.

[17] Roethlisberger R P, Favrat D 2003 Investigation of the prechamber geometrical configuration of a natural gas spark ignition engine for cogeneration: part I. Numerical Simulation *Int. J. Thermal Science* **42** 223-237.

[18] Fu L, Ishima T, Long W Q, Tian J P 2009 Research on the ignition-chamber GDI engine combustion system *J. of Thermal Sci. and Tech.* **4(1)** 53-62.

[19] Bosi L, Ciampolini M, Romani L, Balduzzi F, Ferrara G 2020 Experimental Analysis on the Effects of Passive Prechambers on a Small 2-Stroke Low-Pressure Direct Injection (LPDI) Engine *JSAE 2020*

[20] Bosi L, Ciampolini M, Raspanti S, Romani L, Ferrara G 2021 Jet ignition in small two-stroke engines: an experimental survey on benefits and challenges *E3S Web of Conferences* **312** 07012

[21] https://www.ni.com/en/shop/compactrio.html.

[22] https://www.avl.com/en/testing-solutions/all-testing-products-and-software/advanced-measurement-technologies.

[23] Gentz G, Thelen B, Gholamisheeri M, Litke P, Brown A, Hoke J, Toulson E 2015 A study of the influence of orifice diameter on a turbulent jet ignition system through combustion visualization and performance characterization in a rapid compression machine *Applied Thermal Eng.* **81** 399-411

[24] Ryu H, Asanuma T 1985, Combustion analysis with gas temperature diagrams measured in a prechamber spark ignition engine *Twentieth Symposium (Int.) on Combustion* **20** 195-200

IOP Conf. Series: Materials Science and Engineering 1306 (2024) 012030 doi:10.1088/1757-899X/1306/1/012030

Retrofitting pedestrian and cyclist collision warning for improved safety of heavy trucks employed in urban areas: a case study in Florence and Pistoia, Italy

Cosimo Lucci[1], Marco Pierini[1], and Giovanni Savino[1,2]

[1] Dept. of Industrial Engineering, University of Florence, Via di Santa Marta 3, 50139 Florence, Italy

[2] Monash University Accident Research Centre, 21 Alliance Ln, Clayton VIC 3800, Australia

email: cosimo.lucci@.unifi.it

Abstract. The increasing use of heavy vehicles for services such as public transport, waste collection and logistics in densely populated urban areas causes numerous collisions with serious consequences every year, particularly for pedestrians, cyclists and motorcyclists. Advanced Driver Assistance Systems (ADAS) have proven to be a solution because they can significantly reduce the occurrence and consequences of road crashes. Although many ADAS technologies will be mandatory on new trucks in 2024, it will take more than a decade before the benefits are realised due to the average fleet renewal time. To accelerate this process, the most promising solution is therefore the introduction of aftermarket ADAS (also referred to as 'retrofit' ADAS) on the circulating fleet. This study aims to identify a tailored solution for retrofitting ADAS capable of providing pedestrian collision warnings on heavy vehicle fleets currently circulating in urban areas. The study involves a sample of vehicles and drivers operating in the metropolitan areas of Florence and Pistoia for waste collection services. The company owner of the vehicles integrated a commercial camera-based ADAS system with three additional cameras installed on the frontal bumper and edges of the vehicle, linked with three screens positioned in the vehicle cabin. In this study, a field test campaign was performed integrating trials in a proving ground closed to traffic with testing and monitoring in urban areas. This allowed to define a set of recommended parameters depending on vehicle speed that represent a good balance between alert frequency, minimizing false positives and ensuring that the alerts indicate real potential hazards to the driver. The recommendations provided in this study can also be extended to other vehicle fleets so that rapid and substantial improvements in the safety of heavy vehicles operating in cities can be achieved.

1. Introduction

Over the past decade, there has been a significant increase in the number of trucks and heavy vehicles required for essential services such as public transportation, deliveries, logistics, and urban services [1]. The presence of these vehicles in urban areas, which are densely populated by vulnerable road users like pedestrians, cyclists, and motorcyclists, was indicated as one of the primary causes of severe crashes resulting in life-changing injuries and fatalities in urban areas, as documented in [2-3].

This issue has specific relevance in historical city centres of major European cities, where heavy trucks drive on narrow roads densely populated by pedestrians and cyclists, creating a complex interaction between vulnerable road users and large vehicles. The confined spaces and limited visibility make it difficult for truck drivers to manoeuvre safely, increasing the risk of crashes, especially when turning corners or negotiating tight streets. In such settings, the potential for serious injuries and fatalities is worsened due to the close proximity of pedestrians imposed by narrow pathways, and limited visibility of trucks provided with high cabins. Different approaches can be employed to reduce the risk of these crashes, including route and timing planning, drivers' training, pedestrian awareness, and the implementation of Advanced Driver Assistance Systems (ADAS) like blind-spot detection and collision avoidance systems.

ADASs have been proven to significantly decrease the likelihood and severity of crashes and provide technological support for an efficient and safe transportation system in many traffic scenarios [4-5]. In Europe, most ADAS technologies are set to become mandatory for new trucks starting in 2024 [6].

However, the expected benefits may not become evident until the next decade, due to the typical fleet renewal time ranging from 10 to 15 years in European countries [7].

Prior research using Italy as a case study, has highlighted that retrofitting currently operational trucks [8] with Advanced Driver Assistance Systems (ADAS) has the potential to save a huge number of lives (up to 250 every year in Italy) and prevent crashes causing severe injuries [9]. Also, the introduction of ADAS on all circulating trucks would be economically beneficial, as demonstrated by a cost-benefit analysis of the Italian truck fleet, which has indicated a saving of societal costs by more than 350 million euros [9]. The expense of their installation would be outweighed by the crash expenses they would prevent. Furthermore, the long-term benefits of retrofitting ADASs could be even more substantial, thanks to the enhancement of users' driving behaviour facilitated by ADAS support and their remarkable potential for improving vehicle efficiency, thereby reducing environmental impact [9].

However, the complexity of the traffic scenarios typical of historical city centres (see for example Figure 1, the setting of this study), mixing high density of vulnerable road users and narrow roads, force heavy vehicles to driving conditions for which common retrofit ADAS like pedestrian collision warning and blind spot detection systems are not designed for, fostering issues related to missing detection of pedestrian and cyclists in dangerous conditions and annoyance of drivers due to increased frequency of the alerts. Despite the increasing attention to this specific type of crash and the chances to retrofit ADASs on trucks at reasonable costs offered by different commercial solutions, these issues jeopardize the successful implementation of this technology and frustrate their potential benefits for crash prevention and vulnerable road users' protection.

The objective of this study was testing an innovative approach for retrofitting pedestrian and cyclist collision warning systems for trucks and heavy vehicles and identify recommendations to improve its efficacy in historical city centres characterized by a high density of vulnerable road users and limited visibility.

2. Methods

2.1. ADAS device

A standard retrofit Advanced Driver Assistance System (ADAS) technology (Mobileye shield [11]) was selected by the company owner of the vehicles providing an appropriate range of ADAS functionalities. This device is based on a unit which includes a frontal master camera installed on the windshield (see Figure 2), a GNSS receiver and internal sensors such as accelerometers and gyrometers. Mixing the data collected by cameras and sensors, it offers a range of ADAS functionalities including Forward Collision Warning (FCW), Urban Forward Collision Warning (UFCW), Lane Departure

Warning (LDW), Headway Monitoring and Warning (HMW), Speed Limit Indicator (SLI), and Pedestrian and Cyclist Collision Warning (PCW). The FCW function relies on Time-to-Collision (TTC) calculations, triggering an alert when the TTC falls below 2.4 s, 2.7 s, or 3 s, depending on the configuration. It operates within a speed range of 30-180 km/h. UFCW functions similarly to FCW but operates at lower speeds within the 0-30 km/h range. LDW becomes active above 65 km/h and utilizes a front camera to detect lane boundaries, issuing an alert when the vehicle leaves its lane without using turn signals. HMW continuously monitors the time-based distance from the vehicle in front and generates a warning if this distance drops below 0.7 seconds. SLI uses the front camera to track road speed limit signs and alerts the driver if the vehicle exceeds the posted speed limit. PCW comes into play when the vehicle is travelling below 50 km/h and provides a warning if the TTC with a pedestrian or cyclist reaches 2.5 s.

Figure 1. Vehicles and urban environment object of the study

Starting from the standard configuration, three more cameras were installed by the owner of the vehicles in collaboration with a commercial partner specialized in providing ADAS: one on the frontal side of the vehicle (just above the frontal bumper), and two on the frontal edges (A-pillar) of the vehicle - see Figure 2. These cameras have a wide field of view and a possible range for detecting obstacles, pedestrians and cyclists up to 30 m. For each camera an internal screen was installed inside the vehicle cabin (see Figure 3), providing two levels of warning: i) a yellow alert with or without sound based on the settings, when a pedestrian or cyclist is detected within the field of view of the ADAS system; ii) a red alert with repeating sound when a pedestrian or cyclist is estimated as in collision with the vehicle. Each screen reports the warnings coming from only one of the additional cameras.

Figure 2. Position of the four cameras installed on the trucks (Yellow: Master camera, Light green: Frontal camera, Dark green: Lateral cameras)

Figure 3. Internal view of the truck and position of the three screens for Pedestrian Danger Zone and Warning

2.2. *Testing procedure*

The test procedure, which was developed based on previous work [9], consisted of two phases: first, a test session was designed and performed in a proving ground, to assess the capability of ADAS to prevent crashes in realistic crash scenarios, using a moving mannequin. After that, test sessions were performed in urban areas during the normal use of the vehicles in collecting waste around the city centre of Florence, Italy, to assess the frequency of the alerts and their relevance to hazard situations.

Proving ground

The testing procedure was designed to evaluate the effectiveness of the ADAS functionalities, specifically pedestrian collision warning and blind spot detection, in a controlled environment reproducing relevant traffic conditions for the system. This assessment aimed to simulate real-world scenarios and assess the ADAS system's capabilities accurately.

A dedicated testing space suitable for reproducing actual traffic conditions was selected among those owned by the utility company (ALIA spa) owner of the vehicles and partner of the project. Within this controlled area, a variety of pedestrian collision scenarios were designed and replicated using simplified mannequins to mimic typical pedestrian movements and behaviours.

Two models of the truck employed for waste collection retrofitted with the ADAS system were subjected to these simulated scenarios. This included different driving conditions like straight lines and curves, speeds, and pedestrian behaviours (crossings with different angles and walking speeds). The ADAS system's response was closely monitored and evaluated in each situation, focusing on its ability to detect pedestrians and issue timely alerts to the driver. There was no opportunity to test the system under different environmental conditions, such as varying lighting, weather, and road surface conditions, to ensure its reliability across all real-world scenarios. The first, qualitative, assessment of the performance of the system was based on the capability of the driver to avoid the crash against the mannequin, braking just in response to the alert.

The data collected during these tests allowed the team of engineers and the provider of the ADAS system to fine-tune the settings. By replicating typical crash scenarios involving pedestrians, this testing procedure aimed to assess the capability of the system to provide alerts timely enough to the driver to brake, and to cover all the relevant blind spots and crash situations.

Urban area

To validate the effectiveness of pedestrian collision warnings, a second test phase was performed in densely populated urban areas of Florence city centre. This procedure involved conducting field tests within the operational environment of these vehicles. Trucks equipped with ADAS systems were driven by operators through the area during their normal use for waste collection, and data (videos and vehicle data) were recorded. A researcher participated in the data collection as a passenger of the vehicle, taking notes about hazardous situations. During short breaks between each test session, qualitative analyses of the warning systems were performed to assess their responsiveness and relevance to potential crash conditions by a team including an engineering expert in camera parameters and a road safety engineer. Further analysis was then performed analysing recorded data to assess the system's ability to detect and respond to potential collisions accurately. A further validation phase consisted of using the vehicle for their usual trip around the city at different times of the day and days of the week, collecting data only about the warning (position and type) in the different trips. The validation procedure focused on achieving a balance between warning frequency and minimizing false positives, ensuring that the warnings genuinely indicate potential hazards. Iterative testing and refinement of ADAS and camera parameters were necessary to achieve optimal results. This testing process ensured that ADAS and collision warning systems for trucks were optimized in these specific urban environments.

2.3. Recording instrumentation

During the tests, a datalogger (Race Technology DL1club) recorded data from internal sensors and GNSS position, providing vehicle speed, acceleration and position. Such data was recorded and synchronized with a video from a camera installed inside the cabin of the vehicle, recording driver actions and the view from the windscreen of the area in front of the vehicle, including collision warning alerts produced by the ADAS on the three screens. At the end of each test session, subjective data from an interview were collected from drivers. For tests performed in proving ground closed to traffic, a high-resolution stationary camera, a laser scanner and a drone provided with a camera and positioned above the collision area were employed to record the relative position between the vehicle and the mannequin.

3. Field Tests and Results

To achieve a clear comprehension of truck driving patterns within the urban area of major interest for pedestrian and cyclist collision warning systems, we undertook a data analysis of their driving speeds during waste collection trips. The distribution of the velocities divided for the two areas of the city of Florence (inside and outside UNESCO) is displayed in Figure 4. The results indicate that about 30% of the time is spent stationary, while inside the UNESCO area, most of the driving time is spent in the range of 5-30 km/h, while outside the UNESCO area, about 10% of the time is spent also in the range 30-40 km/h. Based on this data and the analysis of previous studies available in the literature regarding reaction time [12-13], the working parameters of the ADAS were set for the specific urban scenario.

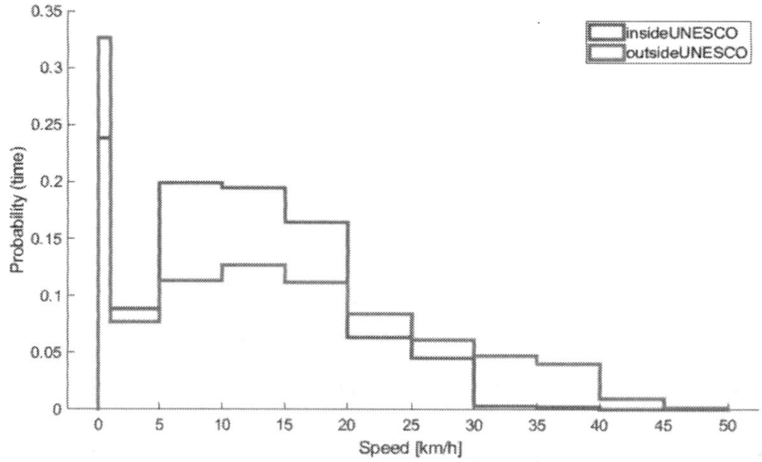

Figure 4. Speed distribution of vehicles operating in urban areas, inside and outside UNESCO area

3.1. *Field test results*

The primary objective of the field test campaign was to evaluate the effectiveness of the retrofitted Advanced Driver Assistance System (ADAS) functionalities, with a specific focus on pedestrian collision warning provided by the additional cameras presented in previous paragraphs and related settings. The initial phase of the assessment was based on a qualitative evaluation of the ADAS system's performance. This evaluation was made on the vehicle's ability to avoid a collision with the simulated pedestrian represented by a mannequin, braking just in response to the alert. The measure of the distance between the vehicle and the mannequin at the different speeds and the timing at which the alert is issued were employed as parameters to fine-tune the warning system.

The analysis of the data collected during these tests provided relevant information on the precision and responsiveness of the ADAS system, allowing the team of engineers and the ADAS system provider to fine-tune and optimize the system's settings.

A further analysis was performed to confirm whether the ADAS system including the additional cameras effectively covered all relevant blind spots of the vehicle and potential crash situations, ensuring comprehensive protection in typical scenarios.

The observation of ADAS system behaviour performed during the normal usage of the vehicle in the city centre aimed to validate the effectiveness of pedestrian collision warnings within the densely populated urban areas like Florence city centre, in conditions which are the usual driving scenarios. The qualitative analysis performed during breaks between test sessions focused on assessing the responsiveness and relevance of the warning systems in potential crash conditions that occurred during the session. This allowed to identify driving conditions in which the system reported alerts which were not required, like people walking on narrow footpaths without crossing their trajectory with the vehicle, in order to adjust the triggering criteria of the alerts and limit false positives. At the same time, the ADAS system's ability to accurately detect potential collisions was monitored and discussed with the driver, who indicated in the data events of potential risk for which he/she expected an alert. As part of the validation procedure, the trucks were also employed for their usual trips around the city at different times of the day and days of the week, in order to evaluate the frequency of the alert and the influence of the changes of parameters.

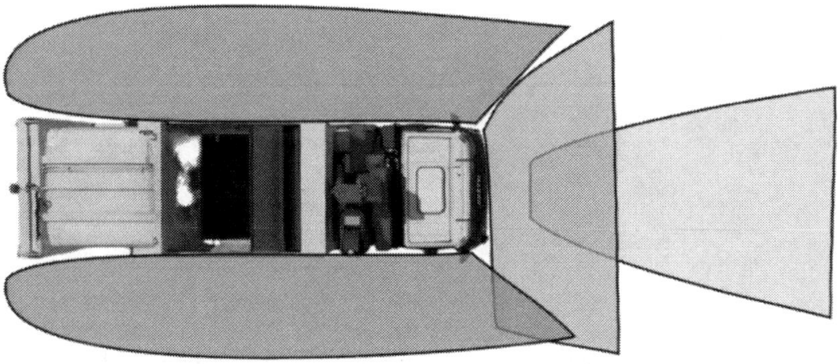

Figure 5. Qualitative representation of the field of view of cameras installed in the vehicles

3.2. *Parameters design and selection*

The working parameters were set based on vehicle speed, and were divided into three different conditions for the three cameras (see Figure 5): Stationary vehicle, Low speed (< 10 km/h), and High speed (> 10 km/h).

Stationary vehicle: in order to inform the driver of the presence of pedestrians and cyclists in close proximity of the vehicle, an alert consisting of a low-volume constant sound and a yellow light on the central screen was implemented. The alert is issued when a pedestrian is at a distance lower than 1-2 m from the frontal bumper, according to the frontal blind spot of the vehicle. This alert is designed to inform the driver of the presence of pedestrians in the blind spots of the vehicle, which are located in

the frontal and lateral part due to the height of the vehicle cabin and the presence of A-pillars (see Figure 6), in order to prevent crashes occurring when the vehicle takes off again after stops.
Vehicle at low speed (< 10 km/h): this range of speed is representative of driving conditions in which the vehicles move slowly in pedestrian zones or extremely busy areas. In these conditions warning must be issued to inform the driver of the presence of pedestrians and cyclists in the blind spots of the vehicle, or in a lateral position when the driver moves towards a pedestrian while turning. Similarly to stationary vehicle conditions, a frontal warning is issued when a pedestrian is at a distance lower than 2-5 m in front of the vehicle. The triggering distance is proportional to the vehicle speed and possible braking deceleration, thus allowing the driver to stop the vehicle before the crash considering a reaction time for braking of 1 s. An alert consisting of a high-volume blinking sound and a red blinking light on one of the two lateral screens is issued when the Time to Collision (TTC) between the vehicle and the pedestrian or cyclist is estimated below 1.5-2.5 s, proportional to driving speed.
Vehicle at high speed (> 10 km/h): for driving speed above 10 km/h (but lower than 30 km/h, for which the standard pedestrian collision provided by the system is fully effective), the alert is issued based on the Time to Collision for both frontal and lateral cameras. The TTC to issue the alert from the frontal camera is in the range of 2-2.5 s and for the lateral camera in the range 2.5-3 s. The exact value is selected in the range proportional to driving speed and considering the lower visibility and complexity for the driver to recognise hazards coming from a lateral direction. All the values of TTC in the speed range 10-30 km/h are defined considering an average reaction time of the driver of 0.8-1 s as indicated in literature for attentive drivers [12-13], thus allowing the driver to brake up to a full stop before the crash.

Figure 6. Blind spots of the vehicle due to cabin height (orange) and A-pillars (red)

4. Discussion
In this paper, an innovative approach for retrofitting pedestrian and cyclist collision warnings for trucks operating in narrow spaces of historical city centres is presented, in order to provide a tailored solution and recommendations to address crashes involving heavy trucks used for urban services like waste collection and vulnerable road users.

First, a field test procedure was designed and used in a proving ground closed to traffic to validate the effectiveness of pedestrian collision warnings in preventing crashes by providing alerts that can allow the driver to react and brake before the crash, confirming that timing and reaction times indicated in literature are suitable also for this application [12-13]. Then, a testing process was performed in real-world driving scenarios. This, combining qualitative assessments with data analysis, contributed to the optimization of the parameters and aimed at providing a system effective in warning the driver while minimizing false warnings in these specific urban settings.

The approach presented allowed to validate via iterative experimental testing a new hardware setup specifically designed for heavy trucks operating in urban areas densely populated by vulnerable road users. The setup consists of one frontal master camera with a narrow field of view for the detection of high-distance obstacles at high velocities and three low-range cameras with a wide field of view for the detection of vulnerable road users at low speeds. The hardware setting allows monitoring the surroundings of the vehicle with good precision and informs the driver of the presence of pedestrians and cyclists, with a specific focus on the blind spots produced by the geometry of the vehicle (see Figure 6). The hardware setting was combined with software settings and alert triggering parameters tailored for the specific condition in which operate the vehicles, characterized by narrow roads and footpaths, with pedestrians and cyclists moving in close proximity to the vehicle. The final setup was tested in the operating conditions and worst-case scenarios, obtaining recommendation to be used in similar settings. The frequency of the warning was considered easily tolerable for drivers, resulting in not annoying and with limited false positives, while the alerts were issued in response to circumstances of realistic potential risk.

Despite the encouraging results, this study did not consider a systematic analysis and definition of the triggering parameters, thus not allowing to reach an optimized setting and triggering strategy, which will require further analytical studies. Also, the data acquired is limited to repeated sessions of field testing using two prototype vehicles, which cannot be considered as extensive testing of the proposed setup. However, within a forthcoming research project, the current setup will be implemented on a wider number of vehicles and monitored for longer timeframes, providing a wide dataset to develop a more detailed and robust setting.

The testing approach and the recommendations achieved in this study can also be adapted to various types of vehicle fleets, thereby enabling rapid and significant enhancements in the safety of large vehicles operating within urban areas.

5. Conclusions

This paper presented an integrated methodology to tailor a retrofitted ADAS system for heavy trucks employed for waste collection to be effective in preventing crashes involving pedestrians and cyclists in busy urban areas and historical city centres common to many European countries.

Throughout a testing and validation process, iterative testing and refinement of the ADAS system and camera parameters were conducted. The validation process was focused on achieving a balance between alert frequency and minimizing false positives, in order to ensure that the alerts provided truly indicated potential hazards to the driver.

The combination of hardware performances employed in this study and software settings is already capable of providing useful support for drivers to prevent crashes involving pedestrians and cyclists, and will serve as a foundation for further refinement and optimization of the ADAS system for the specific traffic scenario of historical city centres.

Acknowledgement

The authors would like to acknowledge the department "Direzione Progettazione e Trasformazione Servizi" of Alia Servizi Ambientali spa for their support and collaboration in this project.

Funding

This paper is part of a project that has received funding from the Fondazione Cassa di Risparmio di Pistoia e Pescia under the founding scheme "Giovani@Ricerca Scientifica", project ARTeMiS (Advanced driver assistance systems: Retrofitting Trucks making Mobility Safe and sustainable in Pistoia).

Disclosure statement: All authors have read and agreed to the submitted version of the manuscript. The authors declare no conflict of interest.

References

[1] Eurostat, "Road freight transport by vehicle characteristics", 2021.
[2] European Commission, "Statistics on road fatalities", 2020.
[3] European Transport Safety Council, "How to improve the safety of goods vehicles in the EU?", PIN Flash Report 39, 2020.
[4] Li, X., Lin, K.Y., Meng, M., Li, X., Li, L., Hong, Y. and Chen, J., 2022. A Survey of ADAS Perceptions with Development in China. IEEE Transactions on Intelligent Transportation Systems.
[5] Kuehn, M., Hummel, T., & Bende, J., "Advanced driver assistance systems for trucks–benefit estimation from real-life accidents." In 22st International Technical Conference on the Enhanced Safety of Vehicles, 2011.
[6] Regulation (EU) 2019/2144 on type-approval requirements to ensure the general safety of vehicles and the protection of vulnerable road users. 2019.
[7] Automobile Club d'Italia, "Autoritratto 2019", Roma, 2020.
[8] J. Scholliers, M. Tarkiainen, A. Silla, M. Modijefsky, R. Janse, G. van den Born, "Study on the feasibility, costs and benefits of retrofitting advanced driver assistance to improve road safety", Final Report, European Commission, 2020.
[9] M. Pizzicori, S. Piantini, C. Lucci, M. Pierini, and G. Savino, "Studio sull'accettabilità di sistemi avanzati di assistenza alla guida (ADAS) retrofit per veicoli pesanti tramite prove di guida in condizioni di traffico reale", AIAS conference, Padova, 2022.
[10] Wu, C., Cao, J., Du, Y., "Impacts of advanced driver assistance systems on commercial truck driver behaviour performance using naturalistic data." IET Intell. Transp. Syst. 1–10 2022.
[11] Mobileye, Mobileye Shield+ Connect User Manual, 2023.
[12] Johansson, G., & Rumar, K. (1971). Drivers' Brake Reaction Times. Human Factors, 13(1), 23–27. https://doi.org/10.1177/001872087101300104
[13] Abe, G., Richardson, J., The influence of alarm timing on braking response and driver trust in low speed driving, Safety Science, Volume 43, Issue 9, 2005, https://doi.org/10.1016/j.ssci.2005.04.006.

IOP Conf. Series: Materials Science and Engineering 1306 (2024) 012031 doi:10.1088/1757-899X/1306/1/012031

Suitable solutions and EU regulatory framework of electric light mobility vehicles for last-mile delivery: an overview

A Raimondi[1], G Savino[1], J Lagrimino[2], G Biagioni[3] and N Baldanzini[1]

[1] Department of Industrial Engineering (DIEF), University of Florence, Via di Santa Marta 3, 50139, Florence, Italy.
[2] Department of Architecture (DIDA), University of Florence, Via della Mattonaia 8, 50121, Florence, Italy.
[3] Department of Legal Sciences (DSG), University of Florence, Via delle Pandette 32, 50127, Florence, Italy.

E-mail: andrea.raimondi@unifi.it

Abstract. Due to current environmental European regulations in terms of pollutant emissions, electric cargo cycles are receiving growing interest compared to heavy-duty vehicles as a cheaper and healthier alternative for last-mile delivery of goods in urban and suburban areas. Within the PNRR Next Generation EU program, the Sustainable Mobility Center aims to design innovative e-cargo bikes considering technical features along with rider behaviour and infrastructure characteristics. The market offers plenty of solutions, which may differ in technical and design features. Nowadays a standardized terminology, widely shared among researchers for those light mobility vehicles does not exist. Several authors employ different terms or introduce classifications based upon the vehicle type of use or even on the vehicle layout (e.g., payload position or number of wheels). Concurrently, EU regulatory framework is undergoing rapid changes and appears unclear. Certain categories of light mobility vehicles might be ruled out from the type-approval system established by the regulation EU 168/2013. For those vehicles, harmonised standards are under drafting or approval. This situation requires a considerable effort among manufacturers in the identification of the proper product directives or regulations and the associated mandatory requirements and testing procedures for the introduction of a new vehicle model in the market. Within this context, an overview of the EU regulatory framework for these light mobility vehicles is presented. A literature evaluation with a structured and coded methodology of scientific databases, company websites and commercial reports is performed, evaluating commercial and prototype solutions of e-cargo bikes from a technical and legislative perspective. Furthermore, a case study is addressed, deepening the national legislation which rules the use of e-cargo bikes in Italy. The information provided in this work may be useful among manufacturers as a starting point for the design process of these types of vehicles. Moreover, this paper lays the foundations for a widely recognized taxonomy of e-cargo bikes based upon distinctive features.

IOP Conf. Series: Materials Science and Engineering 1306 (2024) 012031 doi:10.1088/1757-899X/1306/1/012031

1. Introduction

Transport accounts for nearly a quarter of Greenhouse Gas (GHG) emissions in the EU (European Union) and stands as the primary contributor to urban air pollution [1]. Within this sector, road transport is the largest emitter, responsible for over 70% of EU GHG emissions from transport in 2019 [2] and entails negative impacts as accidents, traffic congestion, and noise pollution [3]. Furthermore, road transport contributes significantly to the production of pollutant emissions, as sulphur dioxide (SO2), nitrogen oxides (NOx), carbon monoxide (CO), unburned hydrocarbons (HC) and particulate matters (PM) [3]. Passenger cars and vans, including light commercial vehicles, are responsible for approximately 12% and 2.5% of the total EU carbon dioxide (CO2) emissions, which is the primary GHG [2]. The European Commission proposes ambitious targets for new vehicles, requiring a minimum 55% reduction in net GHG emissions for cars and vans by 2030 [4], as part of its efforts to attain climate neutrality by 2050, in alignment with the European Climate Law [5].

Urban mobility of people and goods has a substantial impact on air quality of urban and suburban areas with high population density and, together with the problems related to traffic congestion, parking issues as parking space and parking time, air pollution and noise pollution, has a relevant effect on the quality of life [6]. Moreover, the accessibility to city centres for conventional vehicles is restricted or prohibited (e.g., low emission areas, pedestrian areas, or time windows). In this sub mentioned framework, electric light mobility vehicles appear as a viable solution to mitigate or partially solve the negative effects of the transportation systems [7]. Among these vehicles, the most popular are e-bikes, also known as pedelecs (pedal electric cycles) or EPACs (Electric Pedal Assisted Cycles). However, a sustainable solution that is increasingly appearing in the urban environment, particularly in European countries, is the so-called e-cargo bike [8]. This light mobility vehicle is environmentally friendly and has the potential to replace private cars or commercial vehicles, further than other cycle-like vehicles [9]. Indeed, it does not result in any local emissions and has a limited urban surface usage. Moreover, it reduces costs compared to using a van [10], [11], [12] and in several situations is competitive to a conventional vehicle in terms of travel kilometres or time [13]. Within this context, e-cargo bikes are receiving growing interest for last-mile delivery of goods in urban and suburban areas [12], [14]. The expression "last-mile delivery" refers to the final delivery service route that covers a range of a few kilometres in a city environment, after large volumes of items are unloaded from trucks at strategical logistic points (e.g., distribution centres or logistic hubs) near or just outside cities. To date e-cargo bikes are particularly suitable for densely populated urban and suburban areas when logistic hubs are close to the city center [15].

Several models are available in the market or under development, mainly classified in macro-categories (two-, three- and four-wheeled). Technical details for each existing or feasible layout are not fully exposed and a standardised taxonomy does not exist. Moreover, relevant features from a regulatory perspective or related to the usage of these light mobility vehicles within the EU member states are not highlighted. European standards are yet to be established [16], [17]; the regulatory framework is not properly defined, and the presence of regulatory gaps is observed. In addition, current literature does not explore country-specific restrictions on the use of e-cargo bikes within the EU member countries [8].

This work focuses on a specific e-cargo bike category, similar to an EPAC, with the following characteristics:

- maximum continuous rated power is less than or equal to 250 W;
- electric assistance switches off when the cyclist stops pedalling;
- electric assistance is progressively reduced and finally switched off before the vehicle reaches the velocity of 25 km/h.

Furthermore, the presence of a trailer as and additional equipment to increase the transported payload is not considered in this paper.

As explained in detail in Section 3, e-cargo bikes that meet these conditions can be placed on the market without European type-approval. However, the CE marking process is mandatory and the manufacturer is required to follow a conformity assessment procedure.

Within the PNRR Next Generation EU program, the Sustainable Mobility Center [18], Light and Active Mobility Spoke aims to design innovative e-cargo bikes, considering technical features along with rider behaviour and infrastructure characteristics; a target is the introduction of a small fleet of e-cargo bikes for last-mile delivery of mails and packages in Italy. This work is functional for the project design, and aims to:

- identify commercial or prototype e-cargo bikes layouts, to determine the most relevant features from a regulatory perspective or related to the vehicles usage and to introduce a classification based upon distinctive characteristics;
- investigate the EU regulatory framework to identify European directives, regulations, and standards, along with the homologation workflow, and to present the existing issues;
- determine and explore the national legislation in Italy, to ascertain whether and how usage rules of e-cargo bikes are applied within the national territory.

The information provided in this work, beside constituting a solid foundation for the reference project, may be useful among manufacturers as a starting point for the design process of these light mobility vehicles.

2. Materials and methods

A literature evaluation with a structured and coded methodology of scientific databases, company websites and commercial reports is performed, reviewing commercial and prototype solutions of e-cargo bikes in the EU-27 (i.e., the twenty-seven member states of the European Union) from a technical and legislative perspective. Several keywords Three scientific databases were consulted: Scopus, ScienceDirect and Google Scholar. The desired data were mainly obtained by visiting the websites of e-cargo bikes manufacturers and component manufacturers, as powertrain components. The consultation of the user manual, often available in the "support" or "download" section of the manufacturer website, was frequently required. For each identified layout, the following data were collected and further analysed, as described in section 3:

- name of the e-cargo bikes manufacturer;
- name of the EU member country;
- name of the e-cargo bike model;
- vehicle layout;
- type of use;
- vehicle price;
- vehicle dimensions;
- unloaded vehicle weight;
- admissible payload;
- cargo box dimensions;
- cargo box capacity (i.e., weight and volume);
- powertrain features;
- vehicle range;
- tyres size;
- European product directives and regulations identified by the manufacturers;
- standards applied to guarantee the product conformity with the EU directives and regulations.

EU directives, regulations and standards are listed in the declaration of conformity, which is usually available in the “support” or “download” section of the manufacturer website, as a part of the user manual or as a single document.

Concurrently, the EU regulatory framework is explored by employing similar keywords in two primary sources: the official website of EU law and other public documents of the EU (EUR-Lex) [19] and the official website of the European Commission (EC) [2]. To find out relevant European directives, regulations, and standards for the product “e-cargo bike” it was necessary to identify the "entity" of the product, which included the intended use and the list of components along with their functionalities.

Each relevant directive, regulation and standard is identified by reviewing the title and the "scope" section.

Finally, the Italian legislative acts for e-cargo bikes usage within the national territory are identified and a review is performed, focusing on several topics:

- vehicle classification;
- vehicle requirements;
- infrastructure characteristics;
- infrastructure compatibility and usage rules.

3. Results

E-cargo bikes may lead researchers and manufacturers to ambiguity under several topics. The market offers numerous solutions, which differ from each other in a wide range of technical features; only overall data are available, and a standardised classification for these light mobility vehicles is absent.

Concurrently, EU regulatory framework is not properly defined and is undergoing rapid changes; European standards are still under drafting or approval. Moreover, no standard exists for the definition of an e-cargo bike within EU member states. Therefore, it is up to the individual countries to establish a proper definition. Within this context, Italian legislation provides sufficiently concise requirements and usage rules for e-cargo bikes within the national territory.

3.1. Layout identification of e-cargo bikes

Based on the several studies analysed, the keyword "e-cargo bike" is the main term for denoting this category of vehicles, although it is frequently used in a general manner rather than being specific to a particular layout. The existing or under development solutions can be broadly divided in three macro categories, based on the number of wheels: bicycles or bikes (i.e., two-wheeled/r layouts), tricycles or trike (i.e., three-wheeled/r layouts) and quadricycles or quads (i.e., four-wheeled/r layouts) [8].

Approximately seventy companies were identified and more than a hundred of e-cargo bike models, mainly two- wheeled and three-wheeled layouts, while only a few numbers of four-wheeled layouts are available in the market. Figure 1 represents a map chart with a sequential two colours gradient, providing information on EU manufacturers of e-cargo bikes; among the twenty-seven member countries the manufacturing is active across eleven, mainly in central and northern Europe. Germany is the country with the highest number of manufacturers, followed by the Netherlands and Denmark. Figure 2 shows the number of the models identified for each macro category.

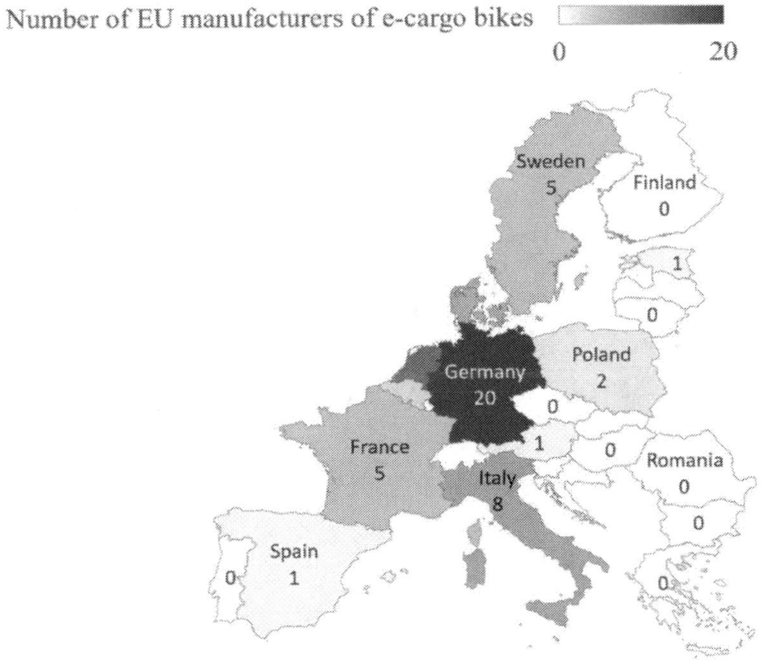

Figure 1. Map chart of EU e-cargo bike manufacturers. Adapted using graphical data powered by GeoNames, Microsoft, OpenStreetMap and TomTom.

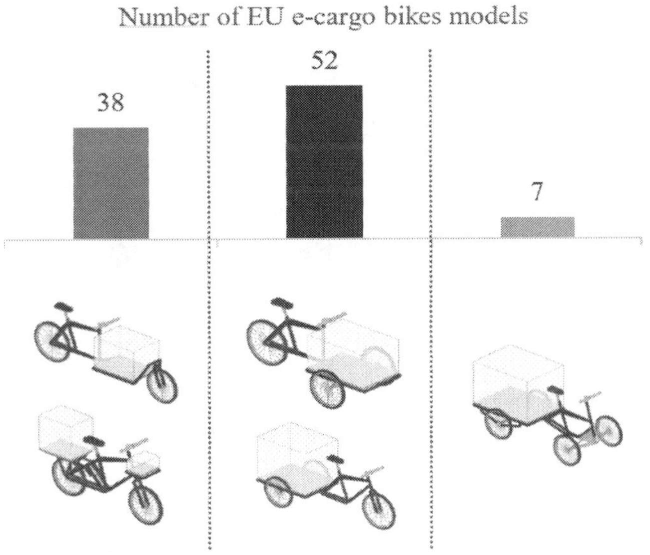

Figure 2. Number of EU e-cargo bike models and design representation of the macro categories.

Among the three macro categories, e-cargo bikes differ from each other by several overall features [9], [13], [20], [21], as shown in Figure 3:

- curb weight [kg]: this parameter represents the weight of the assembled vehicle without considering the rider, the cargo box, and the payload;

IOP Conf. Series: Materials Science and Engineering 1306 (2024) 012031 doi:10.1088/1757-899X/1306/1/012031

- admissible payload [kg];
- cargo box capacity [l];
- range [km]: this parameter is declared by the manufacturers and represents the average km travelled with a single battery. If a dual battery is implemented, it theoretically doubles;
- price [€].

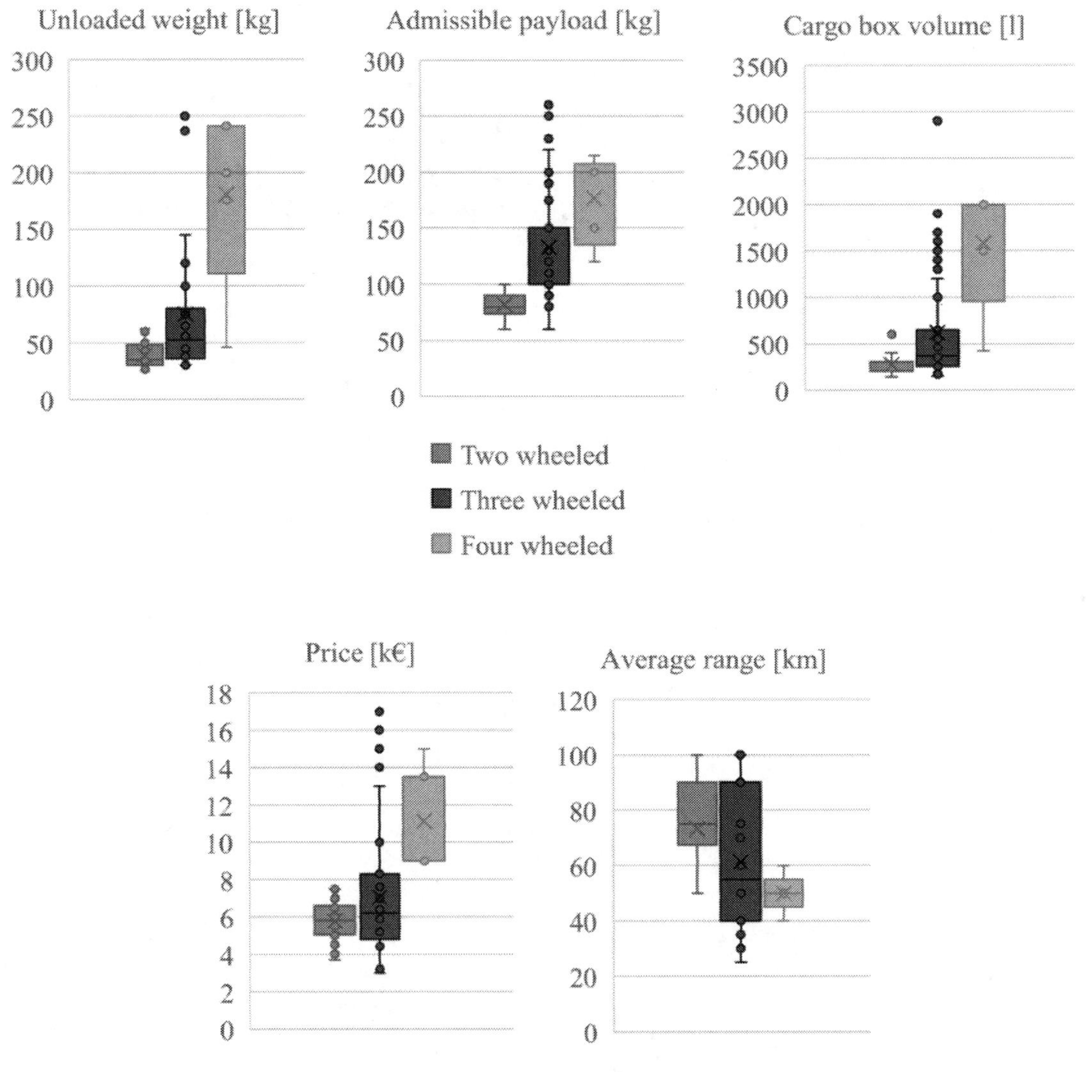

Figure 3. Overall features of e-cargo bikes.

Each macro category comprises several layouts and a different terminology is used among researchers to identify each layout. These terms are summarized in Table 1.

Table 1. E-cargo bikes layouts: terminology

Category	Load position	Layout	Terminology	Reference
Two wheeled	Front		Long Jhon bike Front-load(ing) cargo bike	[13] [12], [21]
	Rear	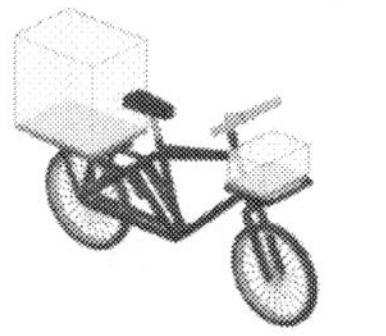	Delivery bike Messenger bike Conventional bike	[13] [12] [22]
	Rear		Long Tail bike Rear-load(ing) cargo bike	[13] [12], [21]
Three wheeled	Front		Front load tricycle Front-load(ing) cargo trike Tadpole tricycle	[13] [12], [21] [16]
	Rear	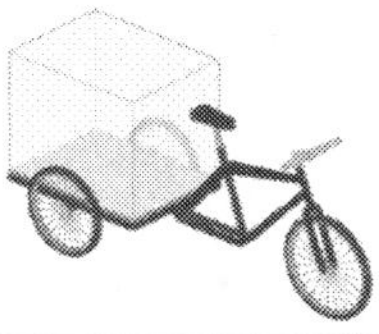	Heavy-load tricycle Heavy-duty cargo bike Freight tricycle Rear-load(ing) cargo trike Delta tricycle	[13] [23] [24] [12], [21] [20]
Four wheeled	Rear	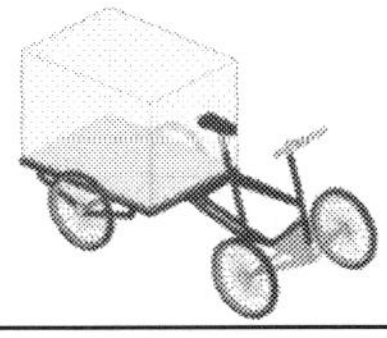	Four-wheel cargo cycle Quad or quadricycle Cargo Quadracycle	[25] [22], [26] [21]

The existing layouts differ by several features, as listed below:

- number and position of the wheels: in the case of a two-wheeled and a four-wheeled layout, these are similar, respectively, to a bicycle and a quadricycle. Otherwise, a three-wheeled layout may have two different configurations, commonly known as tadpole and delta layouts [16], [20]; these are distinguished by the position of the two-wheeled axle, which is frontal for the tadpole layout and rearward for the delta layout;
- position of the cargo box: respect to the rider, the position of the payload can be forward (front load) and backward (rear load);
- dynamics: four different configurations were identified: Tilting (T), Tilting All Frame (TAF), Tilting Front Frame (TFF) and Not Tilting (NT). An e-cargo bike with a tilting all frame is designed with a tilting mechanism that allows the tilting of the entire frame, while

IOP Conf. Series: Materials Science and Engineering 1306 (2024) 012031 doi:10.1088/1757-899X/1306/1/012031

a tilting front frame implies the tilting of the front frame respect to the rear side of the vehicle, mainly consisting of the rear frame connected to the rear wheels. Instead, a not tilting layout has a rigid frame that prevent the rolling motion;

- driving wheel/s: three different options were identified: Front-Wheel Drive (FWD), Rear-Wheel Drive (RWD) and All-Wheel Drive (AWD);
- steering wheel/s: two different options were identified: Front-Wheel Steering (FWS) and Rear-Wheel Steering (RWS).

Despite each of these features may contribute to the classification of individual layouts, it was observed that the main distinctive and useful features for defining the taxonomy of e-cargo bikes are those related to the axles of the vehicle. The proposed classification considers the number of wheels of each axle and the dynamics of the axles. Nevertheless, the identification of the axle that is most heavily loaded due to the payload (i.e., the load position), adds details. In the case of a two-wheeled layout, there are only two possible configurations (i.e., front load and rear load), while a four-wheeled layout has the cargo box always positioned above the rear axle. In the case of a three-wheeled layout, the payload is often positioned above the two-wheeled axle, and only in special cases it can be placed on the one-wheeled axle. The nomenclature used for the identification of each layout consists of two pairs of two characters, separated by a horizontal dash. The first two characters provide information about the front axle, while the second two characters refer to the rear axle. For each pair of characters, the first indicate the number of wheels on the axle, while the second is related to dynamics of the axle; T and R stand, respectively, for Tilting and Rigid. As further integration, the information related to traction and steering are included in the nomenclature. The proposed classification is presented in Table 2.

Table 2. E-cargo bikes layouts: classification and taxonomy

Loaded axle	Dynamics	Layout	Nomenclature
Two wheeled layout – E-cargo bike or bicycle			
Front	Tilting		1T – 1T (RWD – FWS)
Rear	Tilting		1T – 1T (RWD – FWS)
Rear	Tilting		1T – 1T (RWD – FWS)
Three wheeled layout – E-cargo trike or tricycle			
Front	Not tilting		2R – 1R (FWD – RWS)
Front	Not tilting		2R – 1R (RWD – FWS)

Front	Tilting All Frame		2T – 1T (RWD – FWS)
Rear	Not tilting		1R – 2R (RWD – FWS)
Rear	Tilting front frame		1T – 2R (RWD – FWS)
Rear	Tilting all frame		1T – 2T (RWD – FWS)
Four wheeled layout – E-cargo quad or quadricycle			
Rear	Not tilting	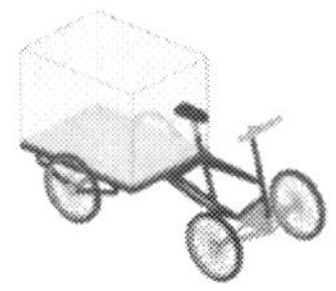	2R – 2R (F/R/AWD – FWS)

3.2. *EU regulatory framework of e-cargo bikes*

3.2.1. *Regulation EU 168/2013*

E-cargo bikes do not have their own definition within EU regulatory framework, although they can be considered part of the broader category of “cycles”, primarily covered by regulation EU 168/2013 on the approval and market surveillance of two- or three-wheel vehicles and quadricycles [27]. This regulation replaces directive CE 2002/24 [28], that in article 1, letter h, explicitly excluded "cycles with pedal assistance which are equipped with an auxiliary electric motor having a maximum continuous rated power of 0.25 kW, of which the output is progressively reduced and finally cut off as the vehicle reaches a speed of 25 km/h, or sooner, if the cyclist stops pedalling", and their components or technical units unless they were intended to be fitted to vehicles covered by the directive. Regulation EU 168/2013 broadens the scope of exclusion, given that it does not apply to "pedal cycles with pedal assistance which are equipped with an auxiliary electric motor having a maximum continuous rated power of less than or equal to 250 W, where the output of the motor is cut off when the cyclist stops pedalling and is otherwise progressively reduced and finally cut off before the vehicle speed reaches 25 km/h," as specified in article 2, paragraph 2, letter h. Exemption from the application of the regulation generally means that vehicles can circulate without being subject to type-approval, thus without requiring a helmet, a license plate, a driver license, and a mandatory insurance for damages resulting from the operation of the vehicles. However, individual EU member states may still introduce further limitations, especially those aimed at ensuring road safety, both in terms of additional product requirements or controls and vehicle circulation rules, which should be considered in the design phase of each individual vehicle. As a result, the EPAC is exempt from the approval procedure, provided that it complies with the requirements imposed by regulation EU 168/2013. This is further supported by harmonised standard EN 15194: 2017, that defines an EPAC as a "cycle, equipped with pedals and an auxiliary electric motor, which cannot be propelled exclusively by means of this auxiliary electric motor, except in the start-up assistance

IOP Conf. Series: Materials Science and Engineering 1306 (2024) 012031 doi:10.1088/1757-899X/1306/1/012031

mode." Within this context, e-cargo bikes, which fall within the category of "pedal cycles" as described by regulation EU 168/2013, are exempt from type-approval.

3.2.2. Safety obligations and conformity assessment procedure

E-cargo bikes not covered by regulation EU 168/2013 are nevertheless encompassed within the framework of a well-defined and comprehensive set of European directives and regulations aimed at ensuring the safety of products entering the European single market. Within this context, the New Legislative Framework (NLF) is in force since 2008 [29]. It comprises a series of measures designed to enhance market surveillance and improve the quality of the conformity assessment procedure, that is the process undertaken by a manufacturer to demonstrate whether specific requirements related to a product are met before it can be sold. This process may involve a conformity assessment body, as required by the applicable legislations, known as notified body, responsible for conducting tasks related to the conformity assessment procedure. EU product legislation outlines the conformity assessment procedure for each product and manufacturers are legally obligated to comply with EU directives and regulations.

However, in some cases they are allowed to self-certify the product compliance. To simplify this process, the manufacturers may use harmonised standards, which are drawn up specifically as a useful tool for the manufacturer to demonstrate product compliance with EU legislation. This means that if the product complies with harmonised standards, the presumption of conformity with the reference EU directives and regulations is ensured. On the other hand, as these are voluntary standards, the manufacturer has the option to select an alternative technical approach to establish compliance with the obligatory legal standards, even if it might result in higher expenses to ensure the presumption of conformity. Harmonised standards are developed by the European Committee for Standardisation (CEN), the European Committee for Electrotechnical Standardization (CENELEC), and the European Telecommunications Standards Institute (ETSI), officially designated by the European Union and the European Free Trade Association as responsible entities [30]. The references to harmonised standards must be published in the Official Journal of the European Union (OJEU) [31]. Manufacturers are responsible to carry out the conformity assessment procedure, set up the technical file, issue the Declaration of Conformity and affix the CE marking to a product. Within this context, the binding EU directives and regulations for e-cargo bikes are listed below; the main harmonised standards that may be applicable are also reported. As already mentioned, the list cannot be neither complete nor exhaustive.

1) Machinery directive (2006/42/EC) [32], as the e-cargo bike falls under the scope of the directive as a "system powered by a force other than direct human or animal force, composed of multiple parts, one of which is mobile", pursuant to article 2. This directive includes a comprehensive list of essential health and safety requirements concerning the design and construction of machinery. E-cargo bikes can only be placed in the market if compliant with these requirements. This directive has already been replaced by a new community regulation [33], [34] that, however, will not be effective until forty-two months after its publication (i.e., 2027), although some rules are already effective.

   The main harmonised standard is EN 15194:2017, specifically designed for EPACs. Compliance with this standard does not allow the product to be compliant with the machinery directive regarding the essential safety requirements relating to hand-arm and whole-body vibrations transmitted by the vehicle to the user. The proposed solutions to overcome this issue are listed below:

   - identify harmonised standards under the machinery directive that cover safety requirements on user vibrations, as EN 1032:2003+A1:2008;
   - identify not harmonised standards, with the awareness that compliance with those documents does not give presumption of conformity to the machinery directive (e.g., EN ISO 5349-1 and EN ISO 5349-2).

IOP Conf. Series: Materials Science and Engineering 1306 (2024) 012031 doi:10.1088/1757-899X/1306/1/012031

2) Electromagnetic compatibility directive (2014/30/EU) [35], as there are electric devices that may influence each other when interconnected or close to each other. The purpose is to keep all those side effects under reasonable control; most of these requirements are covered by EN 15194: 2017.
   This directive has recently been modified, although only regarding the electromagnetic compatibility of control circuit devices, actuating elements, and multimedia equipment, which must currently comply with implementing decision 2021/455 [36].
   An example of a harmonised standard is EN IEC 60947-5-2020.
3) Low voltage directive (2014/35/EU) [37], as there is electrical material (i.e., the battery charger) intended to be used within the limits indicated by the directive, specifically with a nominal voltage in the range 50 - 1000 V in AC and between 75 - 1500 V in DC.
   An example of a harmonised standard is EN 60335-2-29: 2021.
4) Battery directive (2006/66/CE) [38], that rules the collection, recycling, treatment and disposal of batteries and accumulators. Since batteries may contain metals such as zinc, copper, manganese, lithium, and nickel, which may be dangerous for the environment and human health, this directive prohibits the placing on the market of most batteries and accumulators with a certain mercury or cadmium content. It has been already subject to revision through the introduction of a new regulation, which will take effect, for most of its content, next year [39].
5) WEEE Directive (2012/19/EU) [40] on WEEE (Waste Electrical and Electronic Equipment), that is aimed at improving the collection and recycling of the said waste.
6) REACH Regulation (EC 1907/2006) [41], relative to the REACH (Registration, Evaluation, Authorisation and Restriction of Chemicals); through the consultation of the available declarations of conformity of EU e-cargo bikes manufacturers it emerges that only in a few cases it has been deemed appropriate to comply with this regulation, without normally assuming any relevance.
7) Display Regulation (EU 2019/2021) [42], as there may be a display intended to show driving data to the user; however, it only applies when the display area is greater than 100 square cm.
8) RoHS Directive (2011/65/EU) [43], that lays down rules on the RoHS (Restriction of the use of Hazardous Substances in electrical and electronic equipment (EEE) for the protection of human health and the environment, including the environmentally sound recovery and disposal of EEE waste. E-cargo bikes may not contain any lead, mercury, cadmium, hexavalent chromium, polybrominated biphenyls (PBB) or polybrominated diphenyl ethers (PBDE), that can be found in several components (e.g., battery or frame).
   An example of a harmonised standard is EN IEC 63000: 2018.
9) Radio Equipment Directive (2014/53/EU) [44], as there may be radio equipment (e.g., Bluetooth or Wi-Fi devices).

The identified EU directives and regulations do not express obligation for the manufacturers to rely on notified bodies, hence they can carry out individual tests and checks independently. Nevertheless, many manufacturers turn to professional testing organizations to test e-cargo bikes. As already mentioned, it should be emphasized that the listed standards do not adapt completely to e-cargo bikes, as to date a standard definition for e-cargo bikes do not exist. Within this context, the European Committee for Standardisation has established the working group "carrier cycles" to develop European standards for e-cargo bikes, which are still under drafting or approval. The status of the European standardization process is documented in the working programme on the CEN webpage, in the section "Technical Work" [45]. At national level, two member states have issued ad-hoc standards for safe design and construction of e-cargo bikes. France released NF R30-050 standard in 2018 and Germany released DIN 79010 standard in 2020, both aimed at providing rules and tests for e-cargo bikes with defined maximum transportable weight, length, electric motor power (i.e., 250 W) and speed (i.e., 25 km/h). Within this context, these two national standards lay the foundation for the European standardisation process under development by CEN.

3.3. *E-cargo bikes in Italy: a case study*

E-cargo bikes, as extensively explained, belong to the broader category of “cycles”. Following the Italian legislation, the reference standard is the legislative decree 285/1992 [46] and subsequent amendments. Article 50, paragraph 1, first part of the legislative decree defines the "cycles" as vehicles with two or more wheels, operated exclusively by human power and by means of pedals. Thanks to a recent amendment, article 50 also includes in this category electrically pedal assisted cycles, equipped with an auxiliary electric motor with a maximum continuous rated power of 0.25 kW, whose power supply is progressively reduced and eventually interrupted when the vehicle reaches 25 km/h, or earlier if the cyclist stops pedalling. Therefore, the national legislation is in line with the regulation EU 168/2013. Furthermore, the law 178/2020 [47] allows to equip these vehicles with a button for activating the motor while the pedals are stationary, although in this mode the vehicle must not exceed a velocity of 6 km/h. Moreover, national legislation has recently introduced provisions for electrically pedal assisted cargo cycles, as specified in article 50, paragraph 1, second part of the legislative decree 285/1192, according to which cargo cycles can be equipped with an auxiliary electric motor with a maximum continuous rated power of 0.5 kW. However, the provision regarding the increased motor power prescribed for the cargo version of the cycle appears to be in contrast with EU regulation 168/2013, which limits the maximum continuous rated power at 0.25 kW. Therefore, only the cargo cycle with a 0.25 kW power should be considered exempt from EU type- approval. Paragraph 2 specifies the maximum cycle dimensions, prescribing that it cannot exceed 1.3 m in width, 3.5 m in length, and 2.2 m in height. Moreover, it establishes that cycles must have an approximately flat and horizontal loading platform, open or closed, whose loading surface complies with the following criteria:

$$Surface_{loading\,platform} \leq 0{,}3 \cdot Length_{cycle} \cdot Width_{cycle} \quad (1)$$

For the design of e-cargo bikes or any other cycle-like vehicles, it is also necessary to assess the regulations of public roads. To date EU regulations are absent, except for directive EU 2019/1936 [48] on road infrastructure safety, which applies to the trans-European transport network (TEN-T).

In Italy, traffic regulations for these vehicles and the provisions related to the different parts of the infrastructure are of particular importance, as listed below:

- cycle lanes: currently there are two types, as defined by article 4 of ministerial decree 557/1999 [49]: cycle lane “on site” and “on reserved lane”. The former is physically separated from the road or the sidewalk by physical barriers and can be one-way or two-way, while the latter is almost one-way and can be obtained as a part of the road or the sidewalk, separated from vehicular or pedestrian traffic by means of a longitudinal delineation stripe. Article 7 specifies the lanes dimensions, which must be at least 1.5 m wide (including markings) for a one-way cycle lane and 1.25 m wide per lane in the case of a two-way cycle lane, resulting in a total width of 2.5 m. Although these values are the standard dimensions for cycle lanes, the width can be reduced to 1 meter per lane;
- cycling and pedestrian area: this is primarily dedicated to pedestrian traffic. According to article 189, cyclists are not obliged to use this shared path;
- pedestrian area: this is defined as the “area prohibited to the circulation of vehicles, except for those in emergency service, bicycles, and vehicles serving individuals with limited or impaired motor abilities, as well as possible exceptions for zero-emission vehicles with dimensions and speeds comparable to bicycles”. Unless otherwise specified by the local authority through the placement of road signs, cycles are allowed to travel within pedestrian areas. However, article 182 recommends pushing the cycle by hand whenever, due to traffic conditions, it may obstruct or represents a danger to pedestrians;
- pedestrian pathway: this is exclusively for pedestrians. Therefore, there is no allowance for cycles circulation, except when they are being pushed by hand;

- cycling and vehicular pathway: this represent the most dangerous scenario for cyclists and can only be implemented when there is insufficient space to create a dedicated cycle path. However, the legislation purpose is to promote shared spaces, as confirmed by the legislative decree 67/2020 [50].

Considering the results of the market analysis described in Section 3 in terms of cycle size and loading platform area dimensions, an assessment with the requirements imposed by Italian legislation is shown in Figure 4 and Figure 5. As a result, only one model exceeds the admissible length, while for the loading platform area of several models does not respect the condition imposed by Equation 1.

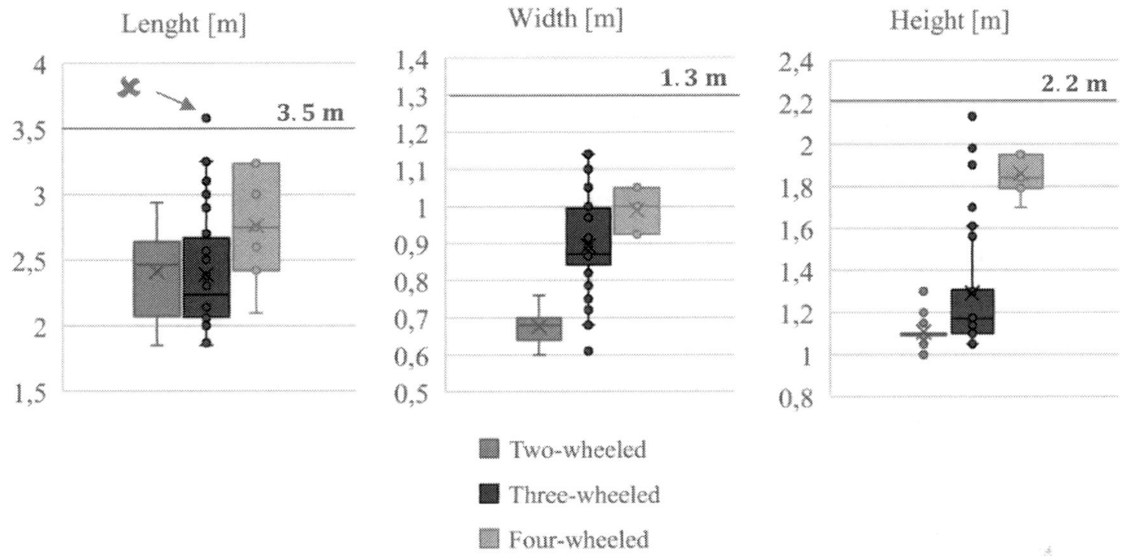

Figure 4. Comparison of e-cargo cycles models dimensions from commercial analysis with the size limitations imposed by Italian legislation for cycles dimensions.

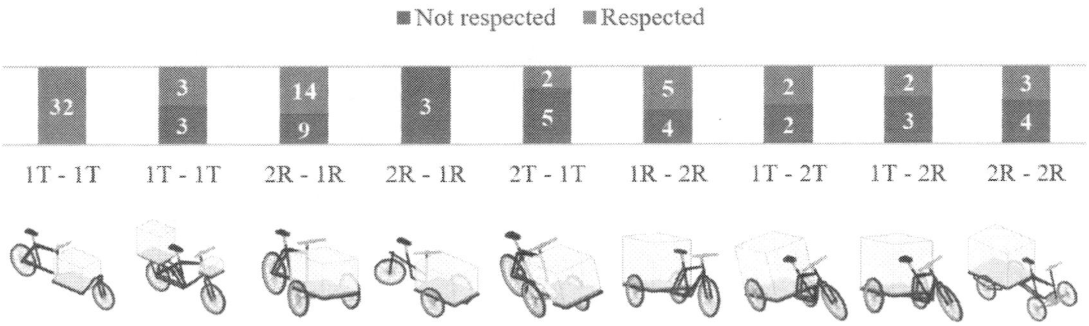

Figure 5. Graphical representation of e-cargo cycles models from commercial analysis which respect or not the condition imposed by Italian legislation regarding the loading platform surface of cargo cycles.

4. Conclusions

The work presented in this paper focused on the framework of e-cargo bikes considering existing or under development layouts from a technical and legislative perspective. In addition, EU regulatory framework is outlined. Furthermore, a case study is addressed, deepening the Italian legislation on the use of these light mobility vehicles within the national territory.

A literature review with a structured and coded methodology aimed to identify the main existing issues related to this topic is performed. The results show that, to date, a standardised terminology for e-cargo bikes is absent. The taxonomy of e-cargo bikes is explored, and a classification based upon distinctive features is presented. The European regulatory framework was also analysed. The work highlighted a lack of an e-cargo bike definition and, consequently, of specific standards, although there is ongoing work in the standardization committees. The current EU regulatory framework is explained and the feasible European product directives and regulations, along with harmonised or not harmonised standards were identified. As, for the design phase, the assessment of national legislation on the use of these vehicles has to be considered, the Italian relevant laws were also identified and presented.

The study performed in this paper represents a baseline for to project performed within the Sustainable Mobility Center, Light and Active Mobility Spoke, and may be useful among researchers and manufacturers as a preliminary step for the design phase of these light mobility vehicles.

Acknowledgements

This work is funded by the European Union - Next Generation EU. "Sustainable Mobility Center (Centro Nazionale per la Mobilità Sostenibile - CNMS)", M4C2 - Investment 1.4, Project Code CN_00000023, CUP B13C22001000001.

References

[1] European Environmental Agency *Greenhouse gas emissions from transport in Europe* Accessed 20 September 2023 https://www.eea.europa.eu/ims/greenhouse-gas-emissions-from-transport#:~:text=Preliminary%20estimates%20of%20emissions%20in,their%201990%20level%20in%202029

[2] European Commission *Road transport: Reducing CO_2 emissions from heavy-duty vehicles* Accessed 20 September 2023 https://climate.ec.europa.eu/eu-action/transport/road-transport-reducing-co2-emissions-vehicles/reducing-co2-emissions-heavy-duty-vehicles_en

[3] European Environment Agency *Transport and environment report 2022: digitalisation in the mobility system: challenges and opportunities* 2022 Luxembourg Publications Office of the European Union pp 25-35

[4] Eur-Lex *REGULATION (EU) 2019/631 OF THE EUROPEAN PARLIAMENT AND OF THE COUNCIL of 17 April 2019 setting CO2 emission performance standards for new passenger cars and for new light commercial vehicles, and repealing Regulations (EC) No 443/2009 and (EU) No 510/2011* Accessed 20 September https://eur-lex.europa.eu/legal-content/EN/TXT/?uri=CELEX%3A02019R0631-20230905

[5] Eur-Lex *Regulation (EU) 2021/1119 of the European Parliament and of the Council of 30 June 2021 establishing the framework for achieving climate neutrality and amending Regulations (EC) No 401/2009 and (EU) 2018/1999 ('European Climate Law')* Accessed 20 September 2023 https://eur-lex.europa.eu/eli/reg/2021/1119/oj

[6] Sperling D 2018 *Three Revolutions* Washington, D.C. Island Press pp 55-76

[7] Shaheen S, Cohen A, Chan N and Bansal A 2020 *Sharing strategies: carsharing, shared micromobility (bikesharing and scooter sharing), transportation network companies, microtransit, and other innovative mobility modes* Transportation, Land Use, and Environmental Planning pp 237-62

[8] Narayanan S and Antoniou C 2022 Electric cargo cycles – A comprehensive review *Transport Policy* vol **116** pp 278-303

[9] Carracedo D and Mostofi H 2022 Electric cargo bikes in urban areas: A new mobility option for private transportation *Transportation Research Interdisciplinary Perspectives* vol **16** 100705
[10] Giglio C, Musmanno R and Palmieri R 2021 Cycle Logistics Projects in Europe: Intertwining Bike-Related Success Factors and Region-Specific Public Policies with Economic Results *Applied Sciences* vol **11** 1578
[11] Arnold F, Cardenas I, Sörensen K and Dewulf W 2018 Simulation of B2C e-commerce distribution in Antwerp using cargo bikes and delivery points *Eur. Transp. Res. Rev.* vol **10** pp 1-13
[12] Verlinghieri E, Itova I, Collignon N and Aldred R 2021 *The Promise of Low-Carbon Freight* Possible.Inspiring climate action
[13] Gruber J and Narayanan S 2019 Travel Time Differences between Cargo Cycles and Cars in Commercial Transport Operations *Transportation Research Record* vol **2673** pp 623–37
[14] Caggiani L, Colovic A, Prencipe L P and Ottomannelli M 2021 A green logistics solution for last-mile deliveries considering e-vans and e-cargo bikes *Transportation Research Procedia* **52** 75-82
[15] Naumov V 2021 Substantiation of Loading Hub Location for Electric Cargo Bikes Servicing City Areas with Restricted Traffic *Energies* vol **14** 839
[16] D'hondt J, Slaets P, Demeester E and Juwet M 2022 Effects of a Torsion Spring Used in a Flexible Delta Tricycle *Applied Mechanics* vol **3** pp 1040–51
[17] Farman G 2020 *CARGO BIKE ANALYSIS AND FIELD TESTING* Topsector Logistiek
[18] MOST Centro Nazionale per la Mobilità Sostenibile *Home* 2023 Accessed 20 September 2023 https://www.centronazionalemost.it/
[19] EUR-Lex *EU Law* Accessed 20 September 2023 https://eur-lex.europa.eu/homepage.html
[20] D'Hondt J, Juwet M, Demeester E and Slaets P 2022 DEVELOPMENT OF AN ELECTRIC TRICYCLE FOR SERVICE COMPANIES AND LAST-MILE PARCEL DELIVERY *TRANSPORT PROBLEMS* **17** 175–87
[21] Court B and Lane T *on behalf of the Central London Sub- Regional Transport Partnership* Element Energy Limited
[22] Cairns S and Sloman L 2019 Potential for e-cargo bikes to reduce congestion and pollution from vans in cities Transport for Quality of Life Ltd
[23] Bogdanski R, Cailliau C, Seidenkranz M, Bayer M and Reed M 2021 Development of a General Specification Sheet for Heavy-Duty Cargo Bikes *2021 Sixteenth International Conference on Ecological Vehicles and Renewable Energies (EVER)* pp 1–11
[24] Tipagornwong C and Figliozzi M 2014 Analysis of Competitiveness of Freight Tricycle Delivery Services in Urban Areas *Transportation Research Record* vol **2410** pp 76–84
[25] Last Mile Logistics Life Cycle Assessment: A Comparative Analysis from Diesel Van to E-Cargo Bike *Energies* vol **15** 7817
[26] Schier M, Offermann B, Weigl J D, Maag T, Mayer B, Rudolph C and Gruber J 2016 Innovative two-wheeler technologies for future mobility concepts *2016 Eleventh International Conference on Ecological Vehicles and Renewable Energies (EVER)* pp 1–7
[27] Eur-Lex *Regulation (EU) No 168/2013 of the European Parliament and of the Council of 15 January 2013 on the approval and market surveillance of two- or three-wheel vehicles and quadricycles* Accessed 20 September 2023 https://eur-lex.europa.eu/eli/reg/2013/168/2020-11-14
[28] Eur-Lex *Directive 2002/24/EC of the European Parliament and of the Council of 18 March 2002 relating to the type-approval of two or three-wheel motor vehicles and repealing Council Directive 92/61/EEC)* Accessed 20 September 2023 https://eur-lex.europa.eu/eli/dir/2002/24/2013-12-11
[29] European Commission *New legislative framework* Accessed 20 September 2023 https://single-market-economy.ec.europa.eu/single-market/goods/new-legislative-framework_en

[30] European Commission 2023 *Harmonised Standards* Accessed 20 September 2023 https://single-market-economy.ec.europa.eu/single-market/european-standards/harmonised-standards_en
[31] Eur-Lex *Access to Official Journal* Accessed 20 September 2023 https://eur-lex.europa.eu/oj/direct-access.html
[32] Eur-Lex *Directive 2006/42/EC of the European Parliament and of the Council of 17 May 2006 on machinery, and amending Directive 95/16/EC* Accessed 20 September 2023 https://eur-lex.europa.eu/eli/dir/2006/42/2019-07-26
[33] Eur-Lex *Regulation (EU) 2023/1230 of the European Parliament and of the Council of 14 June 2023 on machinery and repealing Directive 2006/42/EC of the European Parliament and of the Council and Council Directive 73/361/EEC* Accessed 20 September 2023 https://eur-lex.europa.eu/eli/reg/2023/1230/2023-06-29
[34] Eur-Lex *Corrigendum to Regulation (EU) 2023/1230 of the European Parliament and of the Council of 14 June 2023 on machinery and repealing Directive 2006/42/EC of the European Parliament and of the Council and Council Directive 73/361/EEC (Official Journal of the European Union L 165 of 29 June 2023)* Accessed 20 September 2023 https://eur-lex.europa.eu/eli/reg/2023/1230/corrigendum/2023-07-04/oj
[35] Eur-Lex *Directive 2014/30/EU of the European Parliament and of the Council of 26 February 2014 on the harmonisation of the laws of the Member States relating to electromagnetic compatibility* Accessed 20 September 2023 https://eur-lex.europa.eu/eli/dir/2014/30/2018-09-11
[36] Eur-Lex *Commission Implementing Decision (EU) 2021/455 of 15 March 2021 amending Implementing Decision (EU) 2019/1326 as regards harmonised standards for electromagnetic compatibility of control circuit devices and switching elements, and multimedia equipment* Accessed 20 September 2023 https://eur-lex.europa.eu/eli/dec_impl/2021/455/oj
[37] Eur-Lex *Directive 2014/35/EU of the European Parliament and of the Council of 26 February 2014 on the harmonisation of the laws of the Member States relating to the making available on the market of electrical equipment designed for use within certain voltage limits* Accessed 20 September 2023 https://eur-lex.europa.eu/eli/dec_impl/2021/455/oj
[38] Eur-Lex *Directive 2006/66/EC of the European Parliament and of the Council of 6 September 2006 on batteries and accumulators and waste batteries and accumulators and repealing Directive 91/157/EEC* Accessed 20 September 2023 https://eur-lex.europa.eu/eli/dir/2006/66/oj
[39] Eur-Lex *Regulation (EU) 2023/1542 of the European Parliament and of the Council of 12 July 2023 concerning batteries and waste batteries, amending Directive 2008/98/EC and Regulation (EU) 2019/1020 and repealing Directive 2006/66/EC* https://eur-lex.europa.eu/eli/reg/2023/1542/oj
[40] Eur-Lex *Directive 2012/19/EU of the European Parliament and of the Council of 4 July 2012 on waste electrical and electronic equipment (WEEE)* Accessed 20 September 2023 https://eur-lex.europa.eu/eli/dir/2012/19/2018-07-04
[41] Eur-Lex *Regulation (EC) No 1907/2006 of the European Parliament and of the Council of 18 December 2006 concerning the Registration, Evaluation, Authorisation and Restriction of Chemicals (REACH), establishing a European Chemicals Agency, amending Directive 1999/45/EC and repealing Council Regulation (EEC) No 793/93 and Commission Regulation (EC) No 1488/94 as well as Council Directive 76/769/EEC and Commission Directives 91/155/EEC, 93/67/EEC, 93/105/EC and 2000/21/EC* Accessed 20 September 2023 https://eur-lex.europa.eu/legal-content/EN/TXT/?uri=CELEX%3A02006R1907-20230806
[42] Eur-Lex *Commission Regulation (EU) 2019/2021 of 1 October 2019 laying down ecodesign requirements for electronic displays pursuant to Directive 2009/125/EC of the European Parliament and of the Council, amending Commission Regulation (EC) No 1275/2008 and*

IOP Conf. Series: Materials Science and Engineering 1306 (2024) 012031 doi:10.1088/1757-899X/1306/1/012031

repealing Commission Regulation (EC) No 642/2009 Accessed 20 September 2023 https://eur-lex.europa.eu/eli/reg/2019/2021/2021-05-01

[43] Eur-Lex *Directive 2011/65/EU of the European Parliament and of the Council of 8 June 2011 on the restriction of the use of certain hazardous substances in electrical and electronic equipment* Accessed 20 September 2023 https://eur-lex.europa.eu/legal-content/EN/TXT/?uri=CELEX%3A02011L0065-20230901

[44] Eur-Lex *Directive 2014/53/EU of the European Parliament and of the Council of 16 April 2014 on the harmonisation of the laws of the Member States relating to the making available on the market of radio equipment and repealing Directive 1999/5/EC* Accessed 20 September 2023 https://eur-lex.europa.eu/eli/dir/2014/53/2022-12-27

[45] CEN 2023 *CEN/TC 333/WG 9 – Carrier cycles* Accessed 22 September 2023 https://standards.cencenelec.eu/dyn/www/f?p=205:7:0::::FSP_ORG_ID:2370473&cs=192170B6219BE3069ED0717D8C7DD0E01

[46] GAZZETTA UFFICIALE *TESTO AGGIORNATO DEL DECRETO LEGISLATIVO 30 aprile 1992, n. 285* 1992 Accessed 20 September 2023 https://www.gazzettaufficiale.it/eli/id/1994/03/22/093A6158/sg

[47] GAZZETTA UFFICIALE *LEGGE 30 dicembre 2020, n. 178* 2020 Accessed 20 September 2023 https://www.gazzettaufficiale.it/eli/id/2021/01/18/21A00174/sg

[48] Directive (EU) 2019/1936 of the European Parliament and of the Council of 23 October 2019 amending Directive 2008/96/EC on road infrastructure safety management Accessed 20 September 2023 https://eur-lex.europa.eu/eli/dir/2019/1936/oj

[49] GAZZETTA UFFICIALE *DECRETO 30 novembre 1999, n. 557* 1999 Accessed 20 September 2023 https://www.gazzettaufficiale.it/eli/id/2000/09/26/000G0315/sg

[50] GAZZETTA UFFICIALE *LEGGE 5 giugno 2020, n.67* 2020 Accessed 20 September 2023 https://www.gazzettaufficiale.it/eli/id/2020/06/26/20G00083/sg

IOP Conf. Series: Materials Science and Engineering 1306 (2024) 012032 doi:10.1088/1757-899X/1306/1/012032

Analysis of Energy Consumption for electric motorcycles depending on vehicle configuration and driving cycles

A Niccolai*, **L Berzi, F Barone, N Baldanzini**

Department of Industrial Engineering (DIEF), University of Florence, Via di S. Marta, 3, 50139, Florence, Italy.

* e-mail: adelmo.niccolai@unifi.it

Abstract. This article presents an analysis of energy consumption for battery electric motorcycles (BEM) based on vehicle configuration and driving cycles. The study investigates the impact of two architectures: rear-wheel regenerative (RWR), and two-wheel regenerative (2WR) on consumption. For the analysis, the study considered the effect of various parameters on the longitudinal vehicle dynamics such as speed profiles (WMTC, real-world profile), longitudinal road slope (from -5° to 5° with 0.5° step increment), and 5 different brake distributions. The research employs simulation techniques using the Matlab/Simulink environment to develop a simplified longitudinal vehicle dynamics model (1 DOF) and for the implementation of a serial brake-blending control strategy. The results of the analysis provide energy potential regeneration and insight into the sensitivity of energy consumption to road slope variations, powertrain working points, and potential energy harvesting. The findings contribute to a valuation of the factors that influence energy consumption in electric motorcycles and have implications for the development of vehicle architectures through accurate range assessment on real-world riding conditions and provide valuable information for powertrain components right-sizing.

1. Introduction

The electrification of road transport is one of the strategies adopted to reduce or mitigate mainly the CO_2 emission related to road vehicles [1]. In the last years, also due to the nudging by the policy maker, an increment of electric vehicles in terms of units produced and sold has occurred [2].Furthermore, analysis and study predict the growth of these types of vehicles in the next year [2]. The characteristics of the electric powertrain such as the high torque offered at zero speed of the electric machine, and the lower energy density of the battery than the fossil fuels influence the vehicle design. This latter type of vehicle needs/allows different architectures than the conventional vehicles. Thus, a rethinking of the vehicle design is necessary, especially for motorcycles due to the smaller volume or size than the car. The battery capacity and vehicle mass are closely related, and an increment of the vehicle mass means also an increment on consumption [3]. A reduction of the battery capacity without changes the drive range, for example via regenerative braking, would be ideal. The aim of this study is to investigate *unusual* configurations that could play a pioneering role in further developments. A big difference between conventional vehicles and electric vehicles is the possibility of recovering energy by operating the electric machine as a generator, although the presence and use of a mechanical braking system is still needed for technical and safety reasons [4]. The two architectures investigated by the authors were the rear-wheel regenerative configuration (RWR) and the two-wheel regenerative configuration (2WR). The braking regeneration was achieved by a single electric machine for the RWR and by two, one for each

IOP Conf. Series: Materials Science and Engineering 1306 (2024) 012032 doi:10.1088/1757-899X/1306/1/012032

axle, on the 2WR to allow better braking behaviour [5]. The traction was always provided by the rear motor. In this work, many aspects of the 2WR configuration were not investigated such as the increment of the unsprung masses, wheel inertia, and steering inertia. The focus was to determine the influences on consumption which is obviously a single aspect of the entire powertrain puzzle, aimed to help in the choice of the vehicle architecture.

2. Materials and Method

This section describes the materials and methods used in the work. In the first subsection, the vehicle models and speed profiles were introduced, and then the braking distribution and braking control strategy were explained in the second subsection. In the last subsection, the methodology and simulation outputs were defined.

2.1 Vehicle Model and Speed Profiles

For the analysis of the vehicle architectures and an estimation of the consumption, a simplified motorcycle model was developed. The simulation environment Matlab/Simulink was utilized for this scope. The vehicle model, based on in-plane vehicle dynamics (1 DOF), was comprehensive of the mechanical and electrical system (powertrain), the rider (speed profiles controller), the environment (road grades), and the braking strategy. The two configurations (RWR, 2WR) shared the same component for the rear wheel powertrain which consists of a permanent magnet synchronous machine, transmission (single gear ratio), final chain transmission, and battery. For the 2WR architecture, an additional wheel hub motor was implemented in the front wheel, increasing also the vehicle mass of 15 kg in this configuration. The total mass, including the rider mass, was 370 kg for the RWR architecture. A schematic representation of the 2WR architecture is shown in Figure 1.

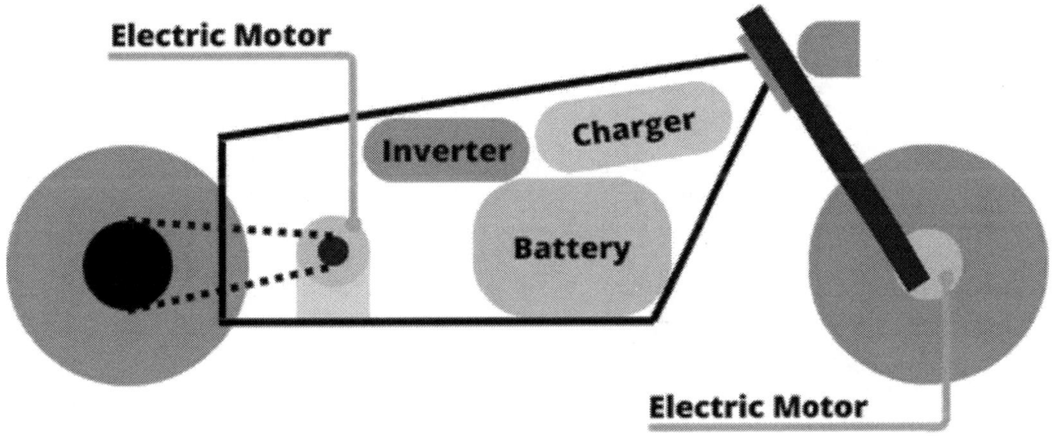

Figure 1: A simple scheme of two-wheel driven electric motorcycles. The rear motor torque is supplied to the rear wheel via a chain transmission, while the front motor is implemented directly in the front wheel hub.

All the vehicle parameters for the calculation of driving resistance were kept constant for the whole simulation and based on literature values. The product of front cross section area and drag coefficient is 0.35, and rolling coefficient is 0.013. The vehicle model neglects many aspects like the tire slip, the influence of suspension, the lateral dynamics, and other aspects that might influence the consumption; the efficiency of the powertrain group is based on a typical PMSM machine, a technology typically suitable for motorcycle application. Despite these limitations, a control on adhesion longitudinal coefficient and wheelie was implemented to avoid possible unphysical phenomena. The two speed profiles are shown in Figure 2. The first one is the WMTC speed profiles while the second one was obtained by processing of naturalistic riding in a mixed urban and rural trip (Firenze-Bologna on open

IOP Conf. Series: Materials Science and Engineering 1306 (2024) 012032 doi:10.1088/1757-899X/1306/1/012032

roads) using an electric motorcycle whose size is comparable to the simulated one; therefore, such real-world data are considered a typical case study for the vehicle class under analysis.

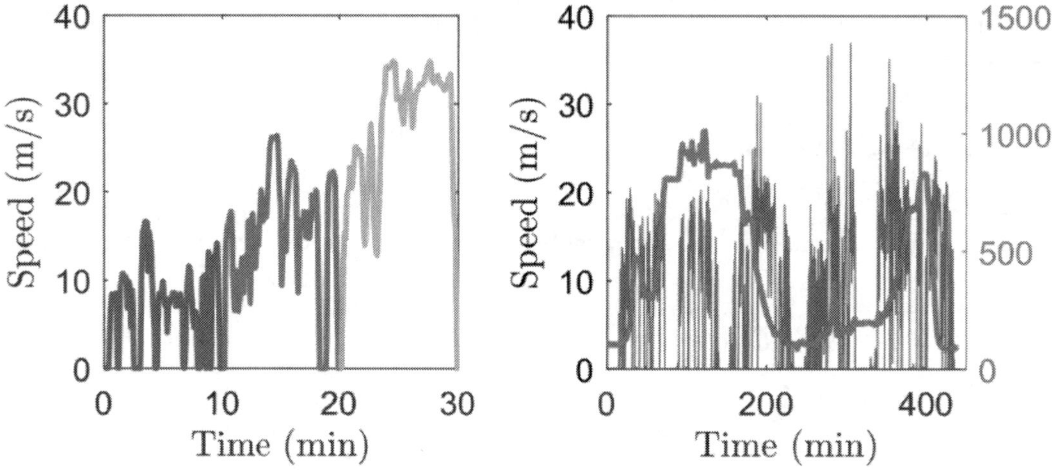

Figure 2: shows the vehicle class 3 WMTC speed profile on the left subplot. Only part 2 was used for the analysis (colored in orange). On the right subplot is displayed the speed and altitude signals recorder during a motorcycle trip and used for the creation of the experimental driving cycle.

The WMTC part 2, class 3, subclass 2 was chosen because is the standard for test homologation of motorcycles with a maximum vehicle speed bigger or equal to 140 km/h [6]. The average driving speed is 58.8 km/h (WMTC) and 48.6 km/h (experimental), and the maximum speed is 94.9 km/h (WMTC) and 133 km/h for the other speed profile. The two cycles have a standard deviation of acceleration value of 0.68 m/s^2 (WMTC) and 0.76 m/s^2 (experimental). Finally, the WMTC has a difference in altitude of zero meters while the other speed profile has a maximum altitude variation of 926 meters.

2.2 Brake distribution and Brake-blending

Electric motors can also work as a generator to recover the vehicle's kinetic energy during the braking manoeuvres. The two main aspects that influence the braking performance in the simplified model were the braking distribution (ρ), and the brake-blending strategy. These latter variables describe the braking force axle repartition, and the brake axle force splitting between the friction brakes and the regenerative brakes, respectively. The braking distribution (ρ) was defined in equation (1), where X_R is the rear axle brake force, and X_F is the front axle brake force.

$$\rho = \frac{X_R}{X_R + X_F} \#(1)$$

The variable (ρ) ranges from 0 (no rear brake) to 1 (only rear brake). The braking distribution was kept constant from the beginning to the end of the simulations, but it was changed between different runs, and it belongs to the following set: $\boldsymbol{\rho} = \{0, 0.25, 0.5, 0.75, 1\}$. In order to increase the energy regeneration, a serial control strategy was implemented for each axle [7], albeit it means a worse and less safety adherence coefficient allocation with potential issues at high deceleration [4]. This latter strategy first splits the braking torque request between the two axles, then aims to demand the maximum regenerative torque allowed by the motor, and then the remaining braking torque request is added by the frictional/dissipative brakes as done in previous work [8]. The feasibility of the braking system or the possibility of implementing wasn't considered in this work, but the focus was the outcomes in terms of energy and regenerative braking. The brake-blending strategy, for the RWR architecture, was defined by equation (2), while the graphic representation of the logic for the 2WR architecture is shown in Figure 3.

$$F_{brake} = \{F_{reg}, \quad F_{brake\_request} \leq F_{reg\_max} \; F_{reg\_max} + F_{friction}, \quad F_{brake\_request} > F_{reg\_max} \; \#(2)$$

Where F_{brake} is the braking force required by the rider, F_{reg} is the regenerative force applied by the motor, F_{reg_max} is the maximum force value allowed by the motor due to mechanical and operational limits, and $F_{friction}$ is the force supplied by the frictional brakes.

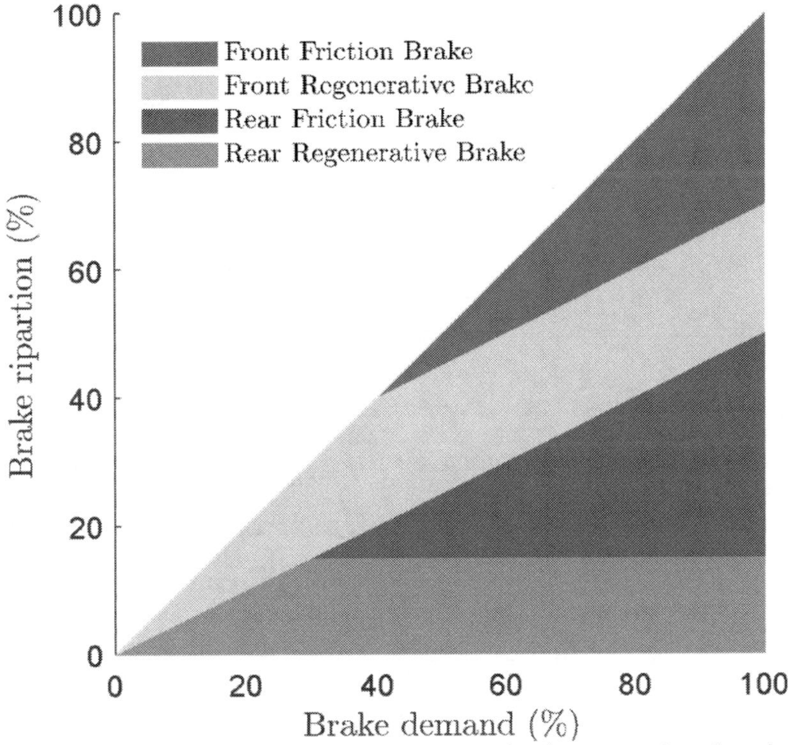

Figure 3: shows the brake serial control strategy for a two-wheel regenerative electric motorcycle (ρ = 0.5). Highlighted areas represent the cumulative braking force for each type of brake device.

The action of braking was implemented via blending the two types of braking (regenerative, and dissipative), independently of the rider's actions. The brake lever feedback of a hypothetical rider will not be affected during the braking manoeuvres because the rider brake demand was guaranteed with different combinations of the brakes without changing the output (total brake force).
Finally, for both architectures, the torque needed for the traction was supplied only by the rear motor for simplicity given the aim of studying the braking energy and regeneration.

2.3 Method
Multiple vehicle simulations were performed to evaluate how the different powertrains, driving cycles, and braking parameters have an influence on consumption. Investigation of the powertrain components was conducted by switching from the two architectures (RWR, 2WR), changing the braking distribution value (for a total of five values), and changing the drive profile (WMTC, experimental). In addition, an analysis of the road slope influence on operation point and energy flow was conducted, adding altitude profiles at the WMTC speed profile to evaluate the influence of the road slope on consumption [9]. The altitude profiles were obtained considering a constant road slope angle (α) for the whole distance travelled by the vehicle. This latter variable ranges from minus five to five degrees with incremental steps of a half degree. The modified speed profiles are shown in Figure *4*.

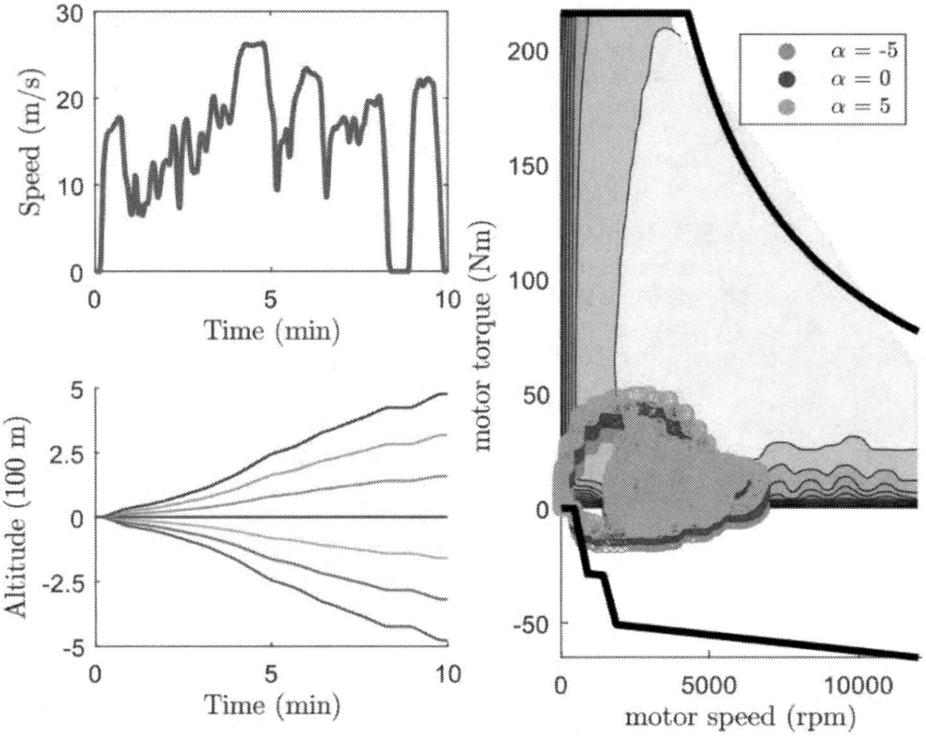

Figure 4: In the top right subplot is plotted the WMTC speed vs. time signal. In the bottom left subplot, different altitude profiles are shown for different constant road slope values. On the right subplot the operation points for three different road slope angles.

The net energy, average motor efficiency, and equivalent energy were used for each simulation to compare the different vehicle configurations and speed profile results. Net Energy was defined as the difference between supplied battery energy and the regeneration energy, at the battery, normalized by the distance travelled by the vehicle as described in equation (3).

$$\underline{E} = \frac{E_{battery_{out}} - E_{battery_{in}}}{distance} \#(3)$$

Average motor efficiency was defined as the mean efficiency of the electric motors (only rear, to a fair comparison between architectures) achieved during all the operations (traction and braking), using a look-up table depending on RPM and torque, equivalent to a typical PMSM efficiency, as said. *Equivalent energy* is the net energy minus the normalized potential energy by the distance, and it is the energy spent by the vehicle considering a zero net altitude but with a different efficiency of the powertrain. With these parameters were possible to compare results obtained by different boundary and initial condition, input data, and vehicle configuration.

3. Results

In the first subsection, how the braking distribution and braking-blending control strategy influence consumption is shown. The influence of road slope on the motor operative points and consumptions is provided in the second subsection.

3.1 Regenerative braking for vehicle configurations

The results obtained for the two driving cycles (WMTC, experimental) and the architectures (RWR, 2WR) are shown in Figure 5.

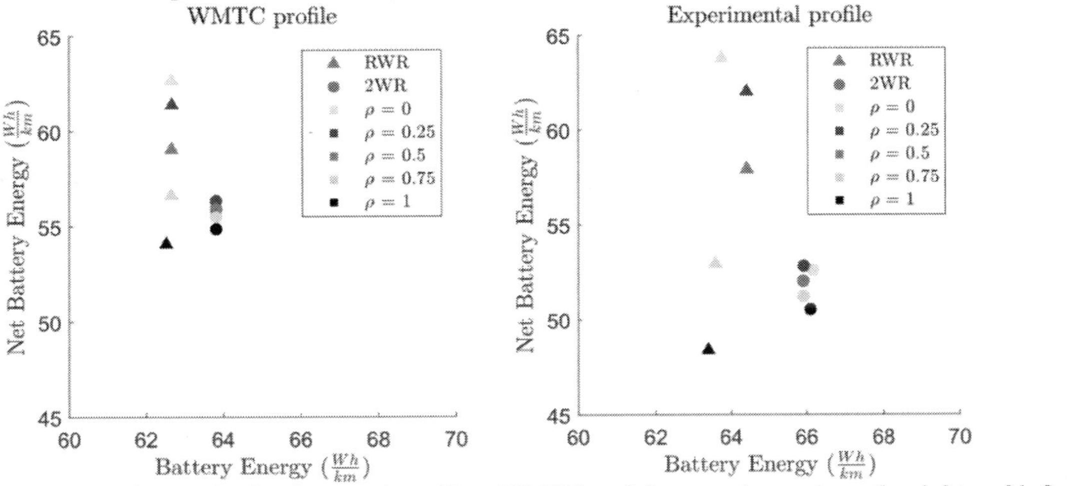

Figure 5: shows the results for the speed profiles, WMTC and the experimental, on the right and left subplots respectively. The geometric dot shape (triangle, circle) means the vehicle configuration (RWR, 2WR). Different colors mean different braking distribution values (see the legend for the details).

The RWR configuration had a normalized energy request of 62.6 Wh/km and 63.9 Wh/km for the WMTC and experimental profiles respectively. The 2WR energy demand was 63.8 Wh/km and 66 Wh/km for the same cycle list. When the braking distribution was equal to one the lowest consumptions were obtained for all the vehicle architecture (RWR, 2WR) and speed profiles (WMTC, experimental). The minimum value of the 2WR was always higher than the minimum value of the RWR configuration. The maximum value for the RWR configuration was obtained always for ρ equal to zero (no regeneration at all). The 2WR maximum was obtained for ρ equal to 0.25 in both cycles ($\rho = 0$ means only front regeneration). The RWR consumption ranges from 54.1 to 62.7 Wh/km and from 48.4 to 63.8 Wh/km for the WMTC and experimental profiles respectively. While the 2WR configuration ranges from 54.9 to 56.3 Wh/km (WMTC profile) and 50.5 to 52.8 Wh/km (experimental profile).

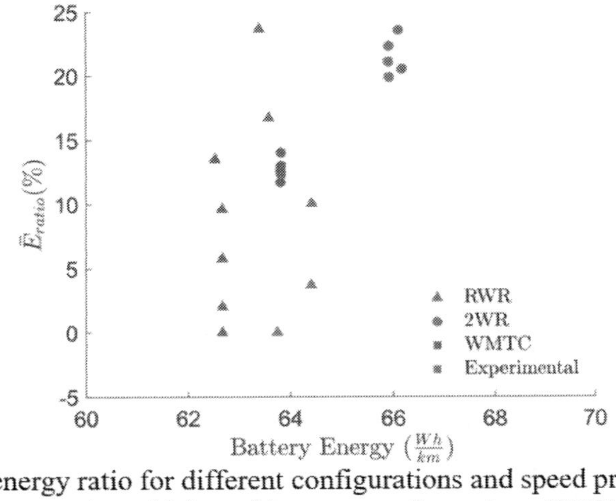

Figure 6: shows the energy ratio for different configurations and speed profiles. The geometric dot shape (triangle, circle) means the vehicle architecture configurations (RWR, 2WR) while the dot color (blue, orange) means the speed profile (WMTC, experimental).

The energy ratio, described as the ratio between the regenerated energy and the delivered energy, is shown in Figure 6. The minimum ratio value (0%) was obtained for the RWR in the case of no regeneration ($\rho = 0$) independently of the driving cycle. The 2WR configuration has a minimum of 11.7 % (WMTC profile) and a maximum of 23.6 % (experimental profile), while the maximum ratio was 23.7 % for the RWR-WMTC combination.

3.2 Road slope influence

As described in subsection 2.3, an analysis of the road slope influence on consumption was performed by adding altitude profiles at the WMTC speed profiles (constant road slopes, see Figure 4 for some examples). The net energy results for the two configurations are shown in Figure 7, RWR on the left subplot and 2WR on the right subplot.

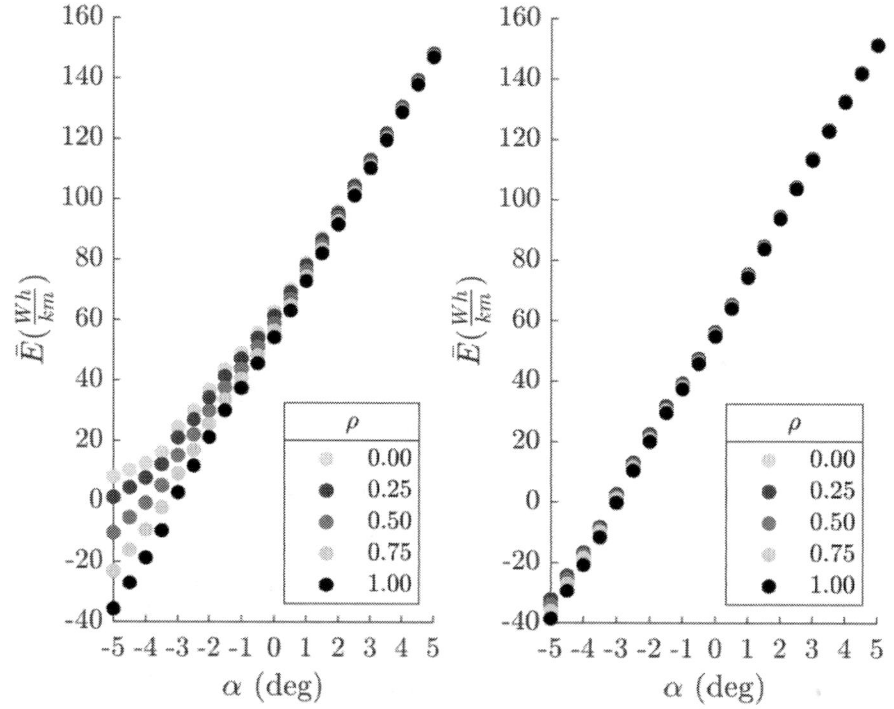

Figure 7: shows the net energy for the RWR on the left subplot and the 2WR on the right subplot for five values of the braking distribution (ρ).

For positive road slope values (uphill direction) the net energy was less influenced by the braking distribution value. In the downhill direction (negative alpha values), variation of ρ affects more the net energy, and in the case of 2WR, it was always negative for all the ρ values. At minus five degrees, for the WMTC profile, the net energy range was 43.7 Wh/km (RWR) and 5.1 Wh/km (2WR). The maximum consumption was 148.3 (RWR) and 151.6 Wh/km (2WR) at alpha equal to five degrees and changing the braking distribution, reduce consumption of 1.3 and 0.3 Wh/km (RWR, 2WR) at maximum. With the addition of the road slope, there is also an additional demand for energy (potential energy) and so on consumption. Thus, to evaluate the architecture performance at the net energy, the normalized potential energy was subtracted from it (called as equivalent energy in subsection 2.3). This latter *new energy* and the average rear motor efficiency are displayed in Figure 8 for the modified WMTC profile. Higher was the road angle and higher was the average rear motor efficiency for all the configurations and braking distribution values. Furthermore, looking at the RWR results, the difference between the net energy and the normalized potential energy decreased for ρ equal to 0 and 0.25 and increases for the other values. While for the 2WR with positive road grades, the increment was always bigger than with zero road angle. The efficiency of the rear motor increased with positive angles in both vehicle configurations, while it had a non-monotonic trend for negative alpha values (increasing and decreasing).

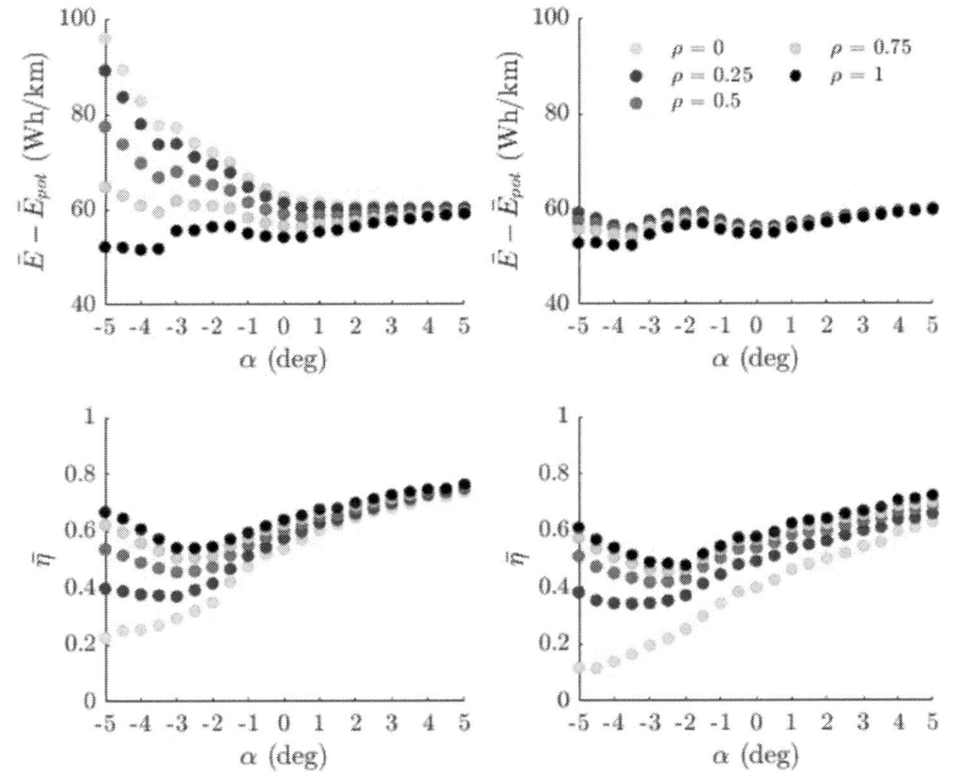

Figure 8: the two top subplots show the net energy minus the potential energy normalized by the distance for the two vehicle configurations (RWR on the left and 2WR on the right). In the two bottom subplots, the average rear motor efficiency is displayed.

4. Discussion

In this section, a discussion of the results is provided. In the first subsection, the diversity of consumption for the two architectures is threatened. The second subsection focuses on the road slope's influence on consumption.

4.1 Architectures influence

The 2WR had always the highest energy demand (63.8 and 66 Wh/km) in both cycles (WMTC, experimental) than the RWR, due to the mass increment given by the front motor. The RWR had the lowest consumptions (54.1, 48.4 Wh/km) in each driving cycle when ρ was equal to one (only rear braking force). Setting $\rho = 0$ could also be seen, energetically speaking, as a strategy where the regenerative rear brake is saturated, and then the other braking power is supplied by frictional brakes (front or rear). The worst consumption results (56.3, 52.8 Wh/km) for the 2WR were obtained for rho equal to 0.25 but they were lower than the second better result of the RWR (at $\rho = 0.75$). The 2WR with $\rho = 0$ (only front braking regeneration) performed a little better than 0.25 braking distribution. With ρ equal to one the 2WR vehicle had higher consumption than the RWR with the same braking distribution value, due to the energy losses inducted by the mass increment such as the tire rolling resistant moment and the not unitary powertrain efficiency (impossibility of recovering all the energy). This result was still the better result of the 2WR even if the front motor does not recover energy. The RWR consumption

was reduced, compared to the non-regenerative case (RWR, $\rho = 0$) of 13.6% (WMTC, $\rho = 1$) and 24.1% (experimental, $\rho = 1$). The 2WR reductions were of the 12.4% (WMTC) and 21% (experimental). While the worst results were 2% (WMTC, $\rho = 0.25$) and 2.7% (experimental, $\rho = 0.25$) for the RWR. The 2WR architecture obtained 10.1% (WMTC, $\rho = 0.25$) and 17.14% (experimental, $\rho = 0.25$) as worst consumption reductions, which were more than three times bigger than the RWR worst results. The lower energy recovery results on WMTC than the experimental profile was related to the lower energy dissipated via braking, rather than air drag and rolling resistance, so the braking regeneration influenced less the consumption due to the speed profile nature. Looking at the energy ratio, the conclusions are the same as made for the net energy, in fact, the energy ratio is equal to one minus the ratio between the net energy and the supplied energy.
The final energy consumption of the 2WR was less sensitive to the braking distribution in comparison with the RWR; due to the increased mass however, the absolute value was comparable or higher. Considering braking distribution with the prevalence on the front axle the potential energy consumption reduction due to regeneration was estimated to be remarkable, from 2.7% (experimental profile, RWR, rho = 0.25) to 17.1% (experimental, 2WR, rho = 0.25). Therefore, for these speed profiles (moderate deceleration, coherently with naturalistic riding on open roads and with WMTC values) a solution with only a rear braking regeneration is assessed to be the optimal solution for energy regeneration and consumption but it might be not for a safety allocation of the longitudinal adherence coefficient (road slippery condition or influence the ABS activation).

4.2 Road slope influence
An increment of the net energy by adding a positive road slope was trivial because, for the same speed profile, it was added a request of more energy (potential energy). As well as a decrement with negative α values. Interesting was the relative increment/decrement of net energy (+136 %/-87.3%, $\alpha = 5/-5$) which was more than the potential energy itself because of the non-unitary efficiency of the powertrain. With positive slope angles, the influence of the braking distribution on consumption was neglectable for both architectures because of the same deceleration (same speed profile) the braking force was mainly supplied by the force of gravity component parallel to the road (decrement of the braking force). While a lower longitudinal road angle meant a higher difference in net energy for different ρ, especially for the RWR configuration. The average rear motor efficiency (Figure 8) increased with the increment of positive α values, thus the motor worked at an overall higher average efficiency (higher in traction and lower in braking) than the zero-altitude profile. Instead with a negative road slope, the efficiency decreases until a local minimum between -2 and -3 degrees and then increased again due to the limit of braking regeneration (motor cut-off). Positive α values reduced the ρ independency on the equivalent energy because the tire forces were mainly positive (traction). For a certain value of α or bigger there are no braking tire forces at all, thus the equivalent energy convergence to a unique value, independently by the ρ variable. Vice versa for negative road slope values the braking tire forces increase and ρ influences strongly the results, especially for the RWR configuration. In fact, even if the average rear motor efficiency had similar behaviour in both vehicle configurations (Figure 8, two bottom subplots), the equivalent energy (Figure 8, two top subplots) and net energy (Figure 7) were less affected for 2WR vehicle by the choice of ρ because of the front motor efficiency increment.

5. **Conclusion**
In this work, the influence on the consumption of two fully electric motorcycle architectures was investigated via two different speed profiles, five braking distribution values, and a serial control braking strategy. In addition, the influence of the road slope was analysed by adding altitude profiles at the WMTC speed profile. Results suggested that the rear wheel regenerative (RWR) architecture is preferable (less consumption, much simpler, and cheaper) if it is possible to apply a logic where the rear regenerative torque is prioritized ($\rho = 1$). The 2WR benefit of an almost braking distribution independence on energy reduction. This latter configuration could be better in situations where high decelerations and high adhesion coefficient are required, thus the investigation via harsh driving cycle

could be worthy. The energy dissipated by the brakes (regenerative and frictional) was the same for the same cycle and vehicle configuration. Thus, the only rear wheel regeneration must be preferred with low deceleration and not harsh application such as moped, due to the less vehicle mass, complexity, and cost. The 2WR, despite the complexity increment, could be a solution for "sport" driving cycles because the ρ independence allows it to regenerate with the front or rear wheel proportional to the braking request. Furthermore, two motors could allow for the optimization of the traction phases that were not studied in this work. On the other hand, adding mass and moment of inertia to the front assembly influences the vehicle handling so a different setup (caster angle, normal trail, etc.) could be necessary for this type of vehicle. An analysis of the road slope was conducted, even with all the limitations of adding a posteriori altitude profile on a speed profile. The results showed that consumption obviously increased (uphill direction) and decreased (downhill direction) but road slope has also an influence on the operative points. The efficiency (not unitary) of the powertrain even for an altitude profile with zero potential energy implicates an increment of consumption and neglected the road slope at all is a simplification, especially for the RWR configuration. The road slope acts by transferring the weight from one axle to the other (uphill or downhill direction) shifting the braking force consequentially, so a 2WR configuration could be preferable to avoid wheel locking. More detailed studies for harsh cases and more architectures, and considering the legal constraint, should be made to clarify if the increment of mass and complexity fulfilment the energy reduction otherwise is better to invest in vehicle development of other solutions such as mass reduction, tire improvement, and powertrain efficiency optimization.

References

[1] European Commission, 2017 European Roadmap Electrification of Road Transport Status: final for publication
[2] - International Energy Agency I 2023 Global EV Outlook 2023: Catching up with climate ambitions
[3] Weiss M, Cloos K C and Helmers E Energy efficiency trade-offs in small to large electric vehicles
[4] Oleksowicz S, Ruta M, Burnham K, Curry E and Garces H 2013 Legal, safety and practical regenerative braking control challenges *Measurement and Control (United Kingdom)* **46** 283–8
[5] Spichartz P, Bokker T and Sourkounis C 2017 Comparison of electric vehicles with single drive and four wheel drive system concerning regenerative braking *2017 12th International Conference on Ecological Vehicles and Renewable Energies, EVER 2017*
[6] European Commission, 2019, Regulation No 134 of the Economic Commission for Europe of the United Nations (UN/ ECE) — Uniform provisions concerning the approval of motor vehicles and their components with regard to the safety-related performance of hydrogen-fuelled vehicles (HFCV) [2019/ 795]
[7] Qiu C, Wang G, Meng M and Shen Y 2018 A novel control strategy of regenerative braking system for electric vehicles under safety critical driving situations
[8] Yuan Y, Zhang J, Li Y and Li C 2018 A Novel Regenerative Electrohydraulic Brake System: Development and Hardware-in-Loop Tests *IEEE Trans Veh Technol* **67**
[9] Björnsson L-H and Karlsson S 2016 The potential for brake energy regeneration under Swedish conditions

IOP Conf. Series: Materials Science and Engineering 1306 (2024) 012033 doi:10.1088/1757-899X/1306/1/012033

Advancing Structural Optimization of an Electric Motor Rotor through Mesh Morphing Techniques

S Porziani[1], C Groth[2], H Zhang[3] and M E Biancolini2

[1]RBF Morph srl, Via Rosmini 4, 00077 Monte Compatri, Italy
[2]Department of Enterprise Engineering, University of Rome Tor Vergata, 00133 Rome, Italy
[3]ANSYS 2600 Ansys Dr, Canonsburg, PA 15317

E-mail: `stefano.porziani@rbf-morph.com`

Abstract. The electric motor has been one of the most important inventions of the last two centuries and has become increasingly important in the last 10 years thanks to its efficiency and its ability to reduce pollution. Given this increasing importance of electrical motors, an interest in design and optimization approaches for motor components is arising. Given the different physics involved in the transformation from electrical to mechanical energy, a valuable and reliable approach has to be applied in the optimization process. In past years, Mesh Morphing based on Radial Basis Functions (RBF) has largely proved its validity in generating shape modification for Finite Element Method (FEM) models. In this work, authors will demonstrate the advantages in applying two opposite approaches for shape optimization using RBF Mesh Morphing applied to electrical motors rotors. These components need to be accurately designed in order to guarantee an adequate life duration by reducing stresses in the rotating components. The two approaches that will be described are the parameter based one, in which a set of design parameters will be varied to generate shape modification used to feed a meta-model that will be used to identify an optimal configuration, and a parameter-less approach, in which the shape modification will be driven by FEM analysis results, with the aim to reduce stress hot-spots.

1. Introduction

The Electric Motor (EM) has been one of the most important inventions of the last two centuries. It was invented almost two hundred years ago and since then it has undergone many evolution. In the last years, the electric motor has become increasingly important in motor vehicles such as electric cars thanks to its efficiency and its ability to reduce pollution. This is not only due to its ability to operate without producing harmful emissions for the environment, but also to the fact that the electric motor is much more efficient than internal combustion engines because it converts electrical energy into mechanical energy with very high efficiency.

In the last decade, the automotive industry put a lot of effort in transitioning from internal combustion engine vehicle to electrical motor vehicles. The increased interest is obviously pushed by increasing environmental concerns, but the adoption of this new engine for everyday mobility will introduce also other benefits. EMs can produce higher torque and power with respect to a thermal engine of equal size or weight, emit lower noises, have lower maintenance and operating costs.

There are several typology of EM, but recently Interior Permanent Magnet (IPM) motors have become a predominant choice for driving modern electric vehicles (EVs). Capitalizing on their high efficiency and torque capabilities, IPMs serve as an ideal response to the automotive sector's transition from internal combustion engines to electric drivetrains. Contemporary IPM rotor designs are the result of extensive research spanning several decades, striking a careful equilibrium between performance, efficiency, affordability, and reliability. At high speeds up to 20000 RPM, rotors experience considerable centrifugal forces. Thus, it's crucial for the rotor's laminates to firmly anchor the magnets, ensuring they remain intact and undamaged. Proper rotor engineering is key to mitigating such potential challenges.

Numerical tools developed to support and push further design of mechanical components can be employed in the EM rotors design too. In particular Finite Element Method (FEM) can provide a reliable way to investigate both thermal and structural behaviour of such components. As a matter of facts, each FEM model can represent the behaviour of a single rotor configuration, and in order to identify the optimal configuration, a set of models have to be prepared to be analysed. Each analysis can be very time consuming, and, to obtain results faster, each configuration can be generated using Radial Basis Functions (RBF) mesh morphing [1, 2]. This mesh-less technique allows to generate modified configuration using shape modifications without the need to generate the underlying modified geometry. RBF mesh morphing has been proved to be reliable and time-saving in several advanced workflows, such as Fluid-Structure Interaction (FSI) [3, 4, 5, 6, 7, 8], or crack front evolution [9, 10], but also in stress recovery in composite plates [11, 12] in fracture mechanics applications [13], FEM results upscaling [14] and shape optimization [15].

The shape modification imposed by RBF mesh morphing to the numerical model can be defined according to user input or can be driven by numerical model itself, exploiting, for example evaluated surface stress. This approach, which can be defined parameter-less, is at the basis of the Biological Growth Method (BGM) [16], which mimics the behaviour of natural tissues when stress and forces are applied on surfaces. The application of this parameter-less approach has already proven its effectiveness even compared with parameter based optimizations methods and other parameter-less ones ([17] and [18]).

In the present work, both the parameter based and the parameter-less procedures are used to optimize stresses on a EM rotor. The numerical tools used are the ones included in the ANSYS Workbench Finite Element Analysis (FEA) framework [19], including RBF Morph Structures [20], that has been used to apply mesh morphing in FEM models.

It is worth to remark that, given the complexity of an electric motor design, the numerical simulations should include all physics involved. In this work, the attention has been focused on the optimization of the structural model, in order to give a prompt and clear view of the benefits achievable with the proposed approaches. Due to the meshless nature of the RBF based mesh morphing, the introduction of multi-physics analysis in the optimization workflow will not invalidate it, since the same shape modification can be applied to each different numerical grid.

In the next sections a brief introduction on RBF based mesh morphing and BGM will be given (section 2), then the numerical model will be described (section 3), finally the optimizations results and the conclusion will be reported (sections 4 and 5).

2. Mesh Morphing and Biological Growth Method

In this section an overview of the RBF mesh morphing and of the BGM will be given, together with an insight of how these two techniques can be used and combined to perform both Parameter-less and Parameter Based optimizations.

2.1. Background on RBF Mesh Morphing

RBF were used as an interpolation tool to fit data with no prescription about the number of dimensions on which this data is defined [21]: they are a set of scalar functions capable to interpolate data defined in the so called source points, in which known values are defined. Generally speaking, the interpolation function can be written as:

$$s(\boldsymbol{x}) = \sum_{i=1}^{N} \gamma_i \varphi\left(\|\boldsymbol{x} - \boldsymbol{x}_{k_i}\|\right) + h(\boldsymbol{x}) \tag{1}$$

In equation (1) $\boldsymbol{x}_{k_i}$ are the source points defined in the space $\mathbb{R}^n$ and $\boldsymbol{x}$ are the points at which the function is evaluated, called also target points. φ is the radial function. A detailed description of the RBF theory and its application was given in [16].

2.2. Background BGM

BGM was inspired by how biological tissues growth is influenced by external actions and forces: layers are added when an activation stress is reached. The studies from which this method has been developed are those by Heywood [22], Mattheck and Burkhardt [23] and Waldman and Heller [24].

In the present work another formulation for BGM is used, which has been presented in [16] and in [25]. The displacement applied to each target node (S_{node}) is aligned with the surface normal direction and can be both inward and outward. To evaluate it, equation (2) is used, in which σ_{node} is the stress value for each node, σ_{th} is a threshold value for stress chosen by user, σ_{max} and σ_{min} are the maximum and minimum value for stress evaluated in the current set of source nodes. d is the maximum offset between the nodes on which the maximum and the minimum stress are evaluated. This parameter is defined by the user to control the nodes displacement whilst limiting the possible distortion of the mesh:

$$S_{node} = \frac{\sigma_{node} - \sigma_{th}}{\sigma_{max} - \sigma_{min}} \cdot d \tag{2}$$

Equation (2) allows to impose a displacement for nodes on the surface to be optimized that can be either inward, in case the stress for the current target node is lower than the stress chosen by user as threshold, or outward, in case stress on target node is higher than threshold one.

2.3. Parameter-less Based Optimisation

Combining RBF based mesh morphing (section 2.1) and BGM (section 2.2) can result in an automatic surface sculpting procedure driven by FEM analysis results and needing no shape modification parameters. The numerical procedure is described in [16] and summarized in the following steps:

(i) stress from baseline configuration are evaluated by means of FEM simulation;
(ii) BGM routines retrieve surface stresses and evaluate each target node displacement using equation (2);
(iii) RBF problem is set up using BGM evaluated displacement and, if needed, additional RBF source points displacements defined by user;
(iv) baseline model is then morphed and FEM solution evaluated again;
(v) updated stresses are analyzed to decide if further optimization is needed; if so, the workflow can be iterated form point (ii), otherwise, the optimized configuration is reached.

In the described automatic surface sculpting procedure, the analyst has only to define two parameters: σ_{th} and d of equation (2).

IOP Conf. Series: Materials Science and Engineering 1306 (2024) 012033 doi:10.1088/1757-899X/1306/1/012033

2.4. Parameter based Shape Optimisation
The parameter based optimization can be performed with RBF mesh morphing by prescribing actions to specific groups of nodes (scaling, translation, surface offset, ...) so that the shape of surfaces and of the volume mesh is updated accordingly. The entity of such actions (scaling factors, component of translation, amount of offset ...) are then combined and controlled so that a certain number of new shapes is generated by Design of Experiment (DoE) and optimal performances are then computed on the response surface [17].

3. FEM Model
In this section a description of the numerical model used to perform optimization is given. Numerical analyses were performed in the framework of Ansys Workbench, since it has all the numerical tools needed for these tasks: a FEM modeler and solver (Ansys Mechanical) an RBF based mesh morphing tool (RBF Morph Structural) which also has implemented the BGM routines, and a surface response based optimization tool (otpiSLang).

3.1. Numerical Model Description
Due to the peculiar geometry of the electric motor rotor, the modelling has taken advance of two simplifications based on the symmetry conditions applicable to the model. Only a sector of the rotor has been modelled, corresponding to a 45° span of the rotor (see Figure 1), rotor external radius is $70mm$ whilst internal radius is $23.5mm$. The other symmetry condition applied is the planar strain of the analysed component, which is acceptable considering the electric motor rotor configuration and the boundary conditions applied. Due to this, the thickness of the model is $5 \cdot 10^{-4}m$, which is required to generate 1 element through the thickness. In particular all FEM analyses were performed considering a constant rotational velocity equal to $1885 rad/s$ and all symmetry faces are constrained using frictionless support condition. Contact between rotor and the four magnets has been modeled using bonded contact condition.

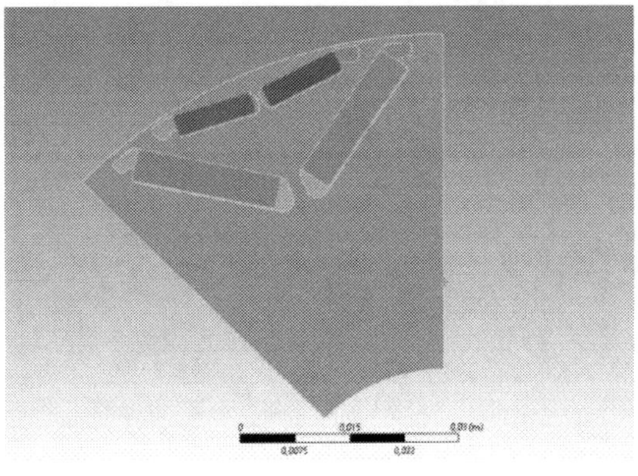

Figure 1. Rotor sector geometrical model

After the meshing process (i.e. the discretization of the solids) the FEM model was composed by 342875 nodes and 47756 elements. Mesh refinements rules were applied in order to have more resolution in the areas in which high surface stress levels are located (hot-spots), as can be seen in Figure 2.

The identified hot-spots were three, located in the fillet area of the three magnet housing, as depicted in Figures 3, 4 and 5, respectively. Maximum equivalent stress values are reported in Table 1.

IOP Conf. Series: Materials Science and Engineering 1306 (2024) 012033 doi:10.1088/1757-899X/1306/1/012033

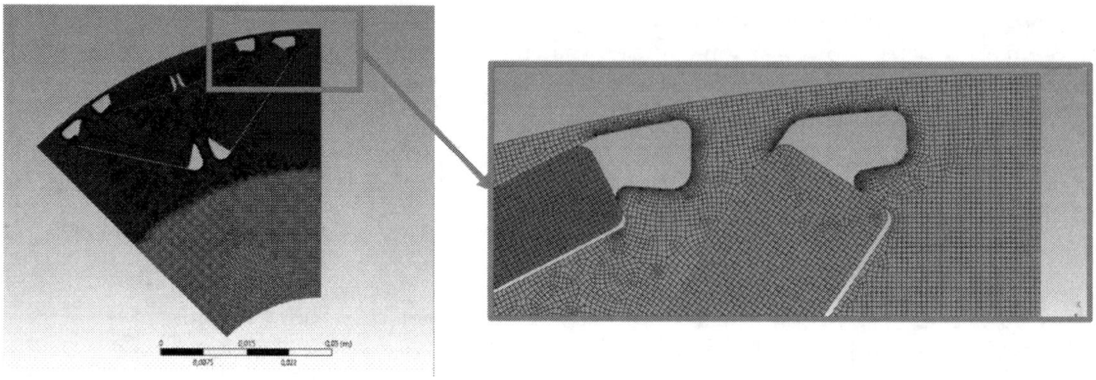

Figure 2. Model discretisation levels

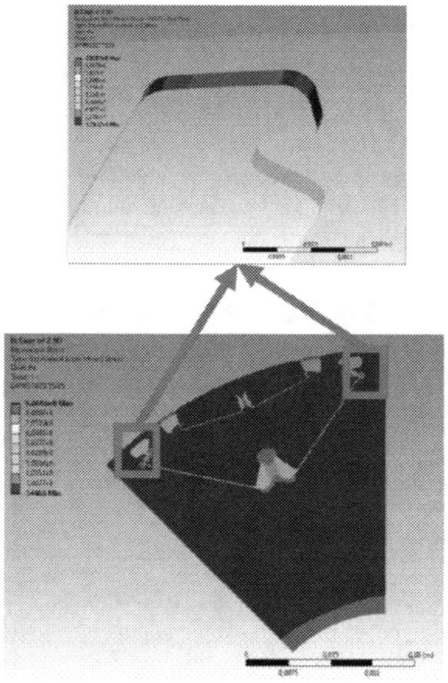

Figure 3. Surface stress at hot-spot 1

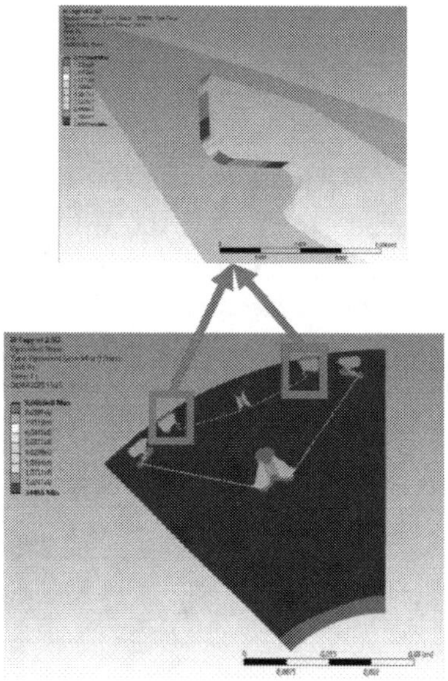

Figure 4. Surface stress at hot-spot 2

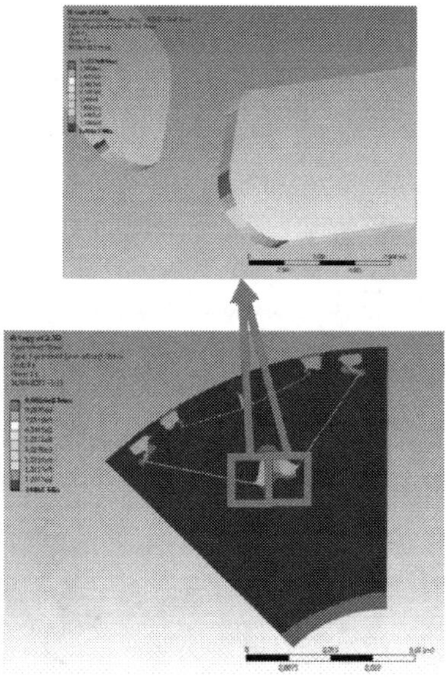

Figure 5. Surface stress at hot-spot 3

Table 1. Maximum equivalent stress in identified hot-spots

Location	Maximum Equivalent Stress
Hot-spot 1	208 MPa
Hot-spot 2	215 MPa
Hot-spot 3	319 MPa

3.2. Parameter Based Setup

In order to perform the parameter based optimization a particular RBF set-up has been defined. The RBF set-up is referred to the source points definition, i.e. the points on which displacement (null or not) is imposed and on which the displacement field is evaluated.

It has been decided to define the shape of the magnet housing fillet using three parameters. On the circular arc portion of the fillet, three points (at start, middle and end of the arc) have been selected and moved radially using a cylindrical local reference system (Figure 6). Exploiting the Hierarchical morph features of RBF Morph Structures, the RBF field obtained with these three point displacement is then used to drive the shape modification of the entire circular arc fillet portion, as depicted in Figure 7. Then, maintaining fixed the points of the fillet near the magnet contact area, the magnet housing fillet is morphed accordingly, as previewed in Figure 8.

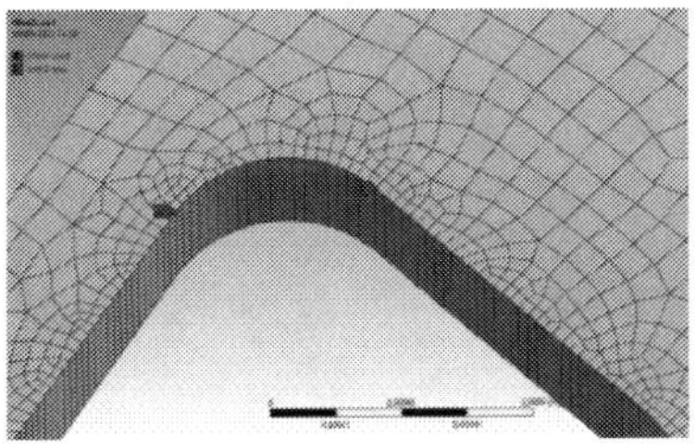

Figure 6. Circle arc points displacement

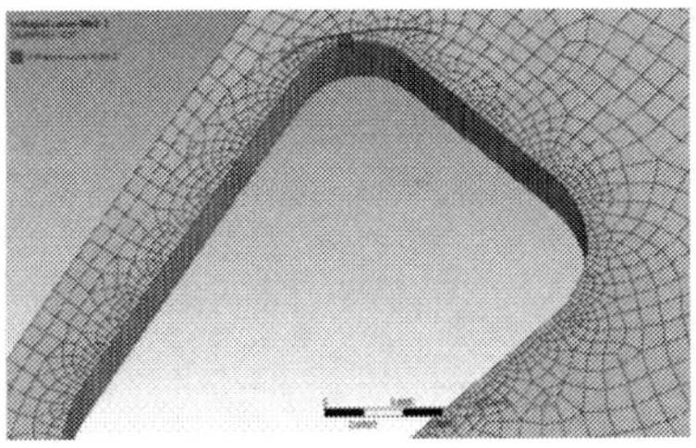

Figure 7. Circle arc shape modification

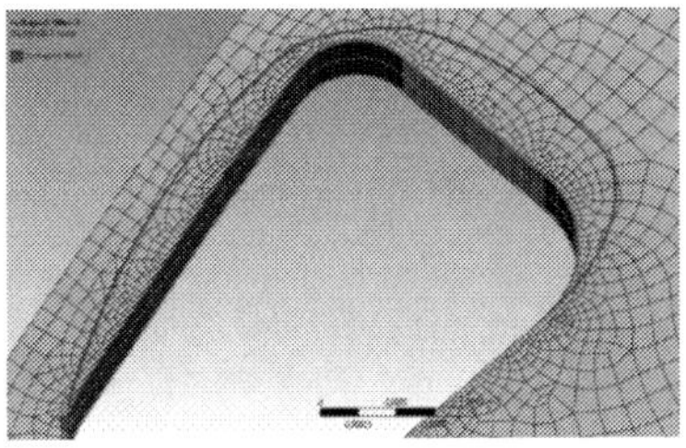

Figure 8. Magnet housing fillet shape modification

In Figure 9 the mesh in the baseline configuration is reported and compared with the same mesh portion in the morphed configuration (Figure 10) after the modification of the three shape parameters (i.e. points displacement depicted in Figure 6).

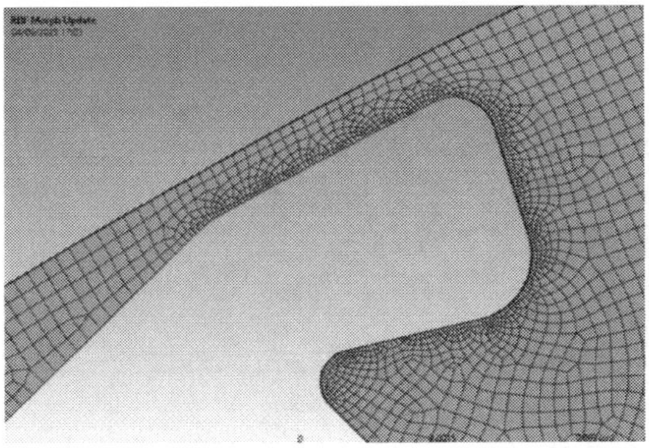

Figure 9. Mesh before morphing

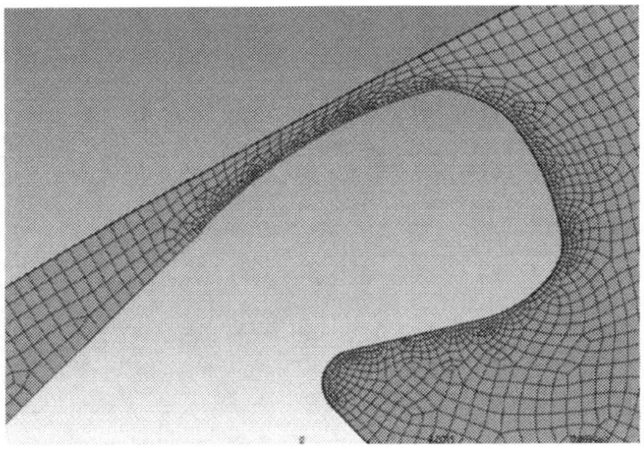

Figure 10. Mesh after morphing

To finalize the parameter based optimization setup, three Response Surface Optimization tasks were executed action on a single hot-spot separately (Figure 11). The Design of Experiment (DoE) tables were set up using the Latin Hypercube Sampling with CCD Samples algorithm, which generated 15 Design Points (DP) for each hot-spot. Each DP is generated by prescribing a different combination of the radial displacement of the three control points defined at each fillet. The Response Surface built on the outcome of the DoE was the Genetic Aggregation one and the optimization method used was the Multi-Objective Genetic Algorithm (MOGA). Each candidate point identified was finally re-evaluated using FEM, obtaining the final optimized shape for the three hot-spots after 48 FEM simulations.

IOP Conf. Series: Materials Science and Engineering 1306 (2024) 012033 doi:10.1088/1757-899X/1306/1/012033

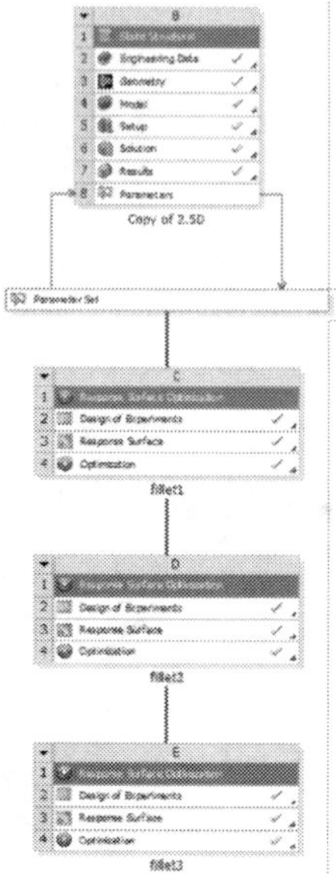

Figure 11. Response Surface Optimization Tasks

3.3. Parameter-Less Setup

The RBF Set-up used for the parameter-less optimization involved only the selection of the hot-spot surfaces (Figure 12), using the surface offset driven by surface equivalent stress levels (i.e. BGM). The parameters (see Section 2.2) used were $1 \cdot 10^{-5}m$ for the maximum growth per iteration d and $200MPa$ for the threshold stress σ_{th}.

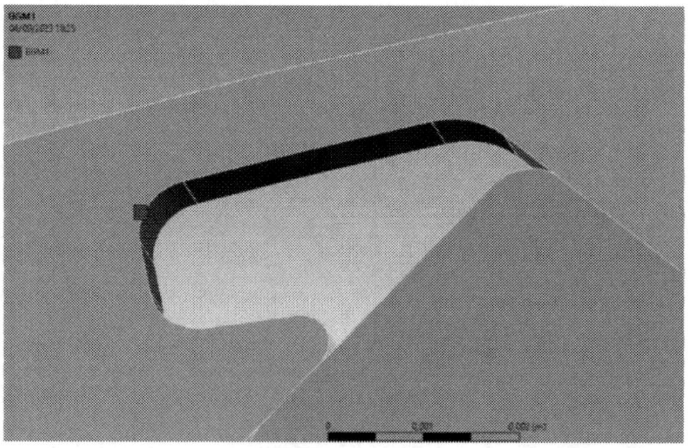

Figure 12. Hot-spot surfaces selected for BGM sculpting

To complete the parameter-less set-up the sequential morphing procedure was prepared, using Ansys Workbench parameters to instruct RBF Morph Structure to use previous FEM results to morph the current model. Letting the code to sculpt the hot-spot surfaces using surface equivalent stress levels, the procedure required 17 steps to reach the minimum value for hot-spot 1 and 2 and 7 steps to reach the minimum value for hot-spot 3 (Figure 13).

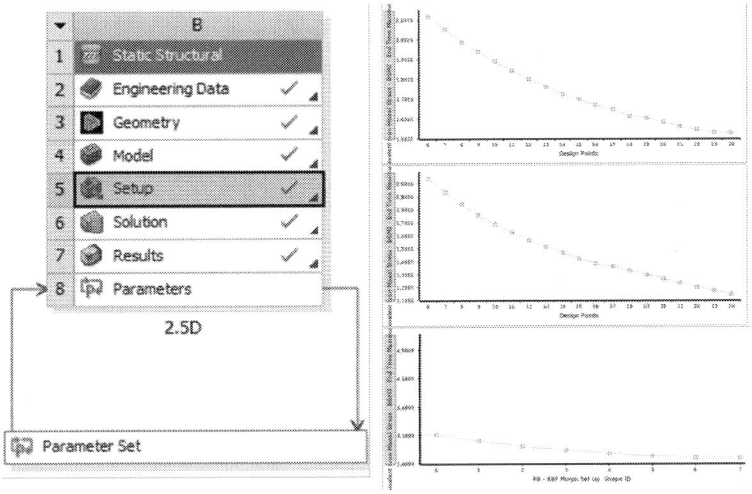

Figure 13. Sequential morph set-up and maximum equivalent stress evolution for the three hot-spots

4. Results

4.1. Parameter based optimization

Results of parameter based optimization are depicted in Figures 14 to 19. In particular, maximum equivalent stress in hot-spot 1 is $143MPa$ (Figure 14), in hot-spot 2 is $181MPa$ (Figure 16) and in hot-spot 3 is $294MPa$ (Figure 18), obtaining maximum equivalent stress reduction of 31.25%, 18.78% and 7.84% respectively.

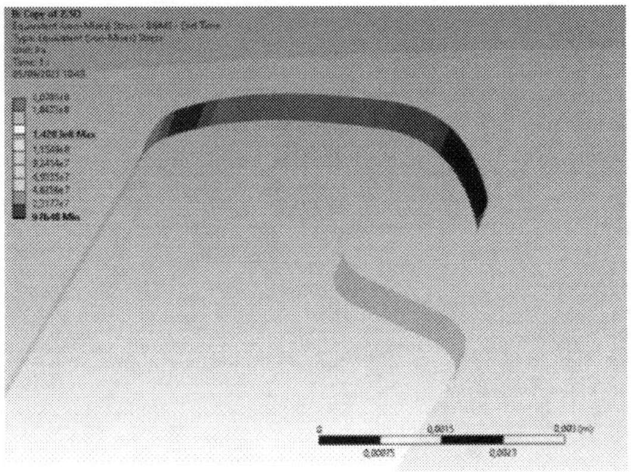

Figure 14. Hot-spot 1 equivalent stress

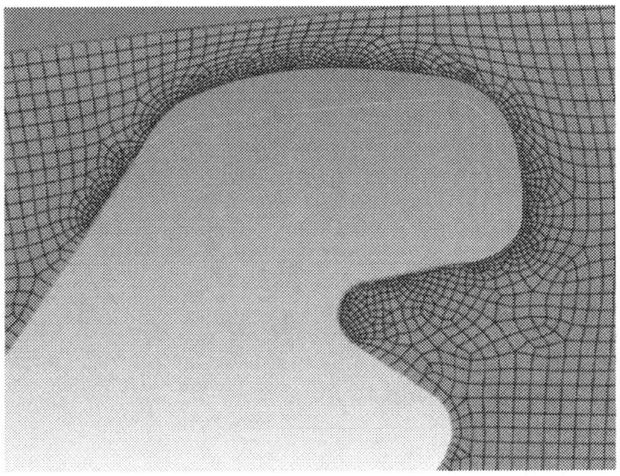

Figure 15. Hot-spot 1 optimized shape

4.2. Parameter-less optimization

Results of parameter based optimization are depicted in Figures 20 to 25. In particular, maximum equivalent stress in hot-spot 1 is $121MPa$ (Figure 20), in hot-spot 2 is $156MPa$ (Figure 22) and in hot-spot 3 is $278MPa$ (Figure 24), obtaining maximum equivalent stress reduction of 41.82%, 27.44% and 12.85% respectively.

In Table 2 all optimizations results are presented and compared.

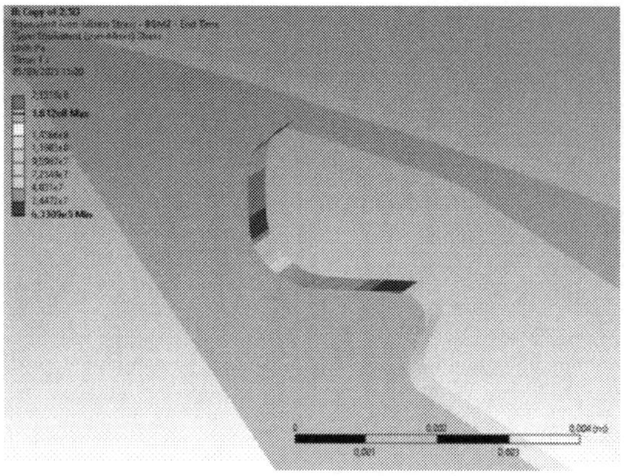

Figure 16. Hot-spot 2 equivalent stress

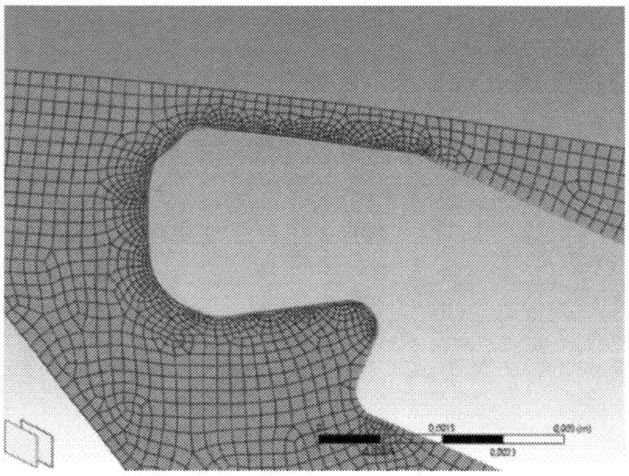

Figure 17. Hot-spot 2 optimized shape

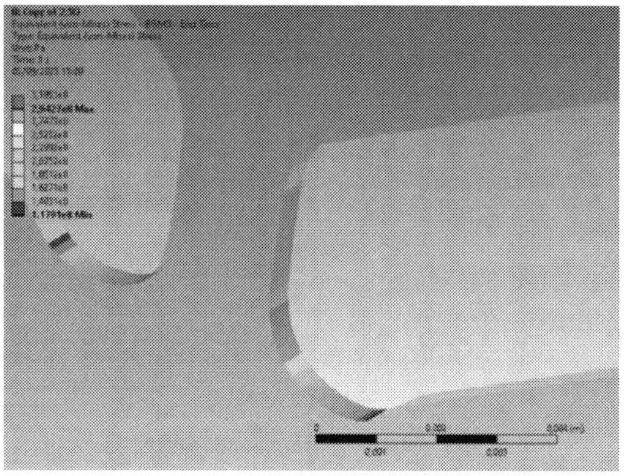

Figure 18. Hot-spot 3 equivalent stress

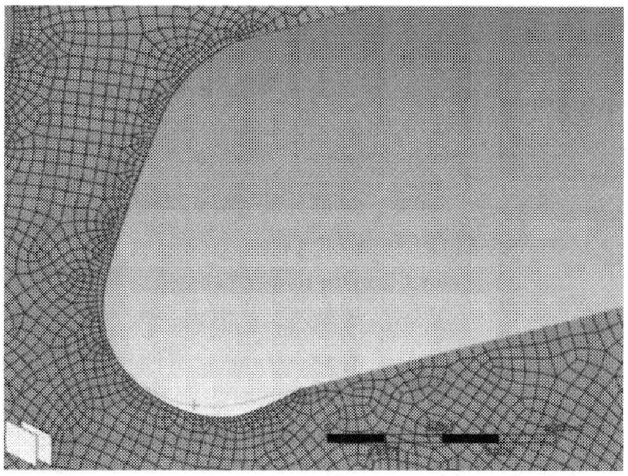

Figure 19. Hot-spot 3 optimized shape

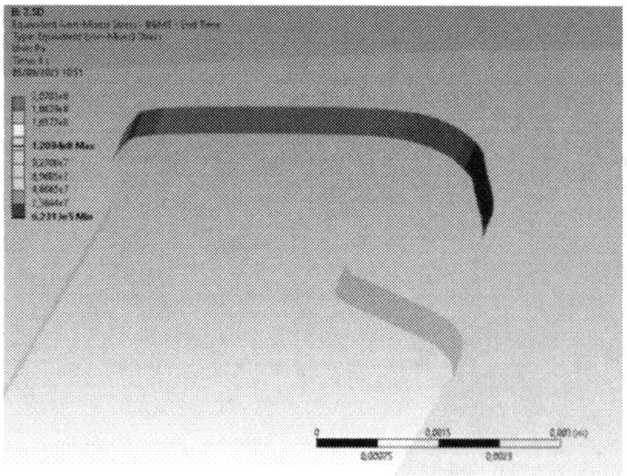

Figure 20. Hot-spot 1 equivalent stress

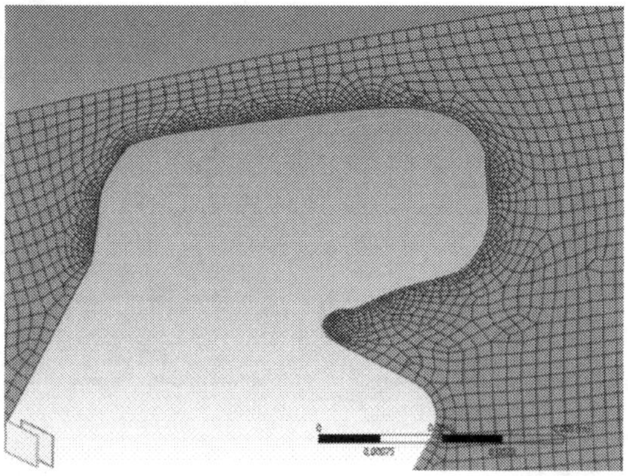

Figure 21. Hot-spot 1 optimized shape

IOP Conf. Series: Materials Science and Engineering 1306 (2024) 012033 doi:10.1088/1757-899X/1306/1/012033

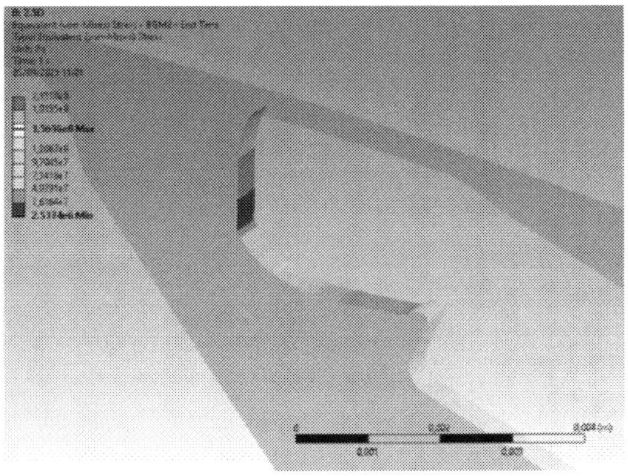

Figure 22. Hot-spot 2 equivalent stress

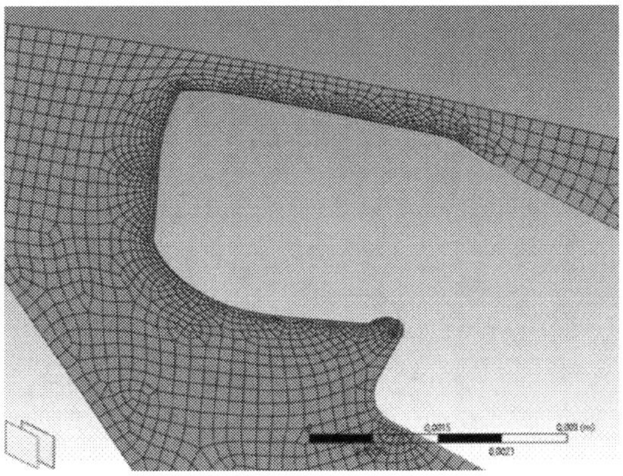

Figure 23. Hot-spot 2 optimized shape

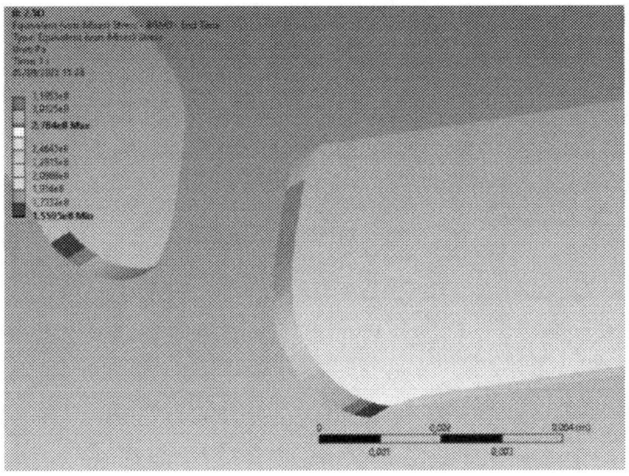

Figure 24. Hot-spot 3 equivalent stress

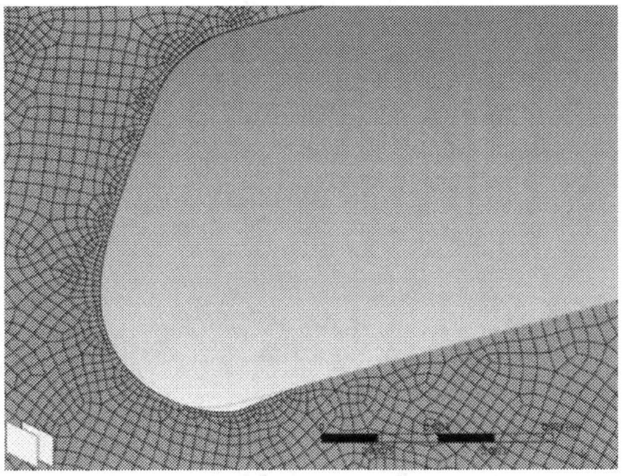

Figure 25. Hot-spot 3 optimized shape

Table 2. Optimizations results comparison

Location	Baseline	Parameter Based Max Eq. Stress	Reduction	Parameter-less Max Eq. Stress	Reduction
Hot-spot 1	208 MPa	143 MPa	31.25 %	121 MPa	41.82 %
Hot-spot 2	215 MPa	181 MPa	18.78 %	156 MPa	27.44 %
Hot-spot 3	319 MPa	294 MPa	7.84 %	278 MPa	12.85 %

5. Conclusions

In the present work two optimization approaches for EM rotors are presented, both based on RBF mesh morphing. The two optimization methods were tested on a generic rotor, which presented three regions in which high equivalent stress levels were identified. These regions (hot-spots) were located in the fillets of magnet housing.

The first approach presented is the Parameter based one: the numerical model mesh in the hot-spots areas are made parametric by means of RBF mesh morphing by properly chose the fixed and moving source points of the RBF problem. The parametric model is then used generate all the FEM models in morphed configurations corresponding to the DP of a DoE on which the Response Surface Optimization was executed.

The second approach presented, the Parameter-Less one, combined the RBF based mesh morphing with the BGM, creating an automatic surface sculpting tool driven by surface stress levels

Both approaches proven to be effective in reducing stresses in the identified hot-spots. Stress reduction obtained with the Parameter Based approach on the three hot spots were respectively $31,25\%$, $18,78\%$ and $7,84\%$. Stress reduction obtained with the Parameter-Less approach on the three hot spots were respectively $41,82\%$, $27,44\%$ and $12,85\%$.

The Parameter Based approach optimization reached lower stress reduction rate at a higher computational cost: 48 FEM simulation were run, besides the response surface calculation cost. However optimized shapes results smoother if compared with ones obtained with the other approach.

The Parameter-Less approach required a lower computational cost (17 FEM simulation were run) to obtain optimized configuration.

The present work focused attention on the structural analysis only, but it is worth to remark that both approaches can be used in a complex multi-physics workflow EM design, since mesh morphing field evaluated to obtain the optimized shape can be used to morph other physic numerical models, so that all numerical model shape can be synchronized and, eventually, forces and interaction can be updated and optimization workflow iterated to increase accuracy of results.

References

[1] De Boer A, Schoot M, and Bijl H. Mesh deformation based on radial basis function interpolation *Computers And Structures* **85**, 784-795

[2] Biancolini M 2011 Mesh morphing and smoothing by means of radial basis functions (RBF): a practical example using Fluent and RBF Morph *Handbook Of Research On Computational Science And Engineering: Theory And Practice* pp. 347-380

[3] Lombardi M, Parolini N and Quarteroni A 2013 Radial basis functions for inter-grid interpolation and mesh motion in FSI problems *Computer Methods In Applied Mechanics And Engineering*

[4] Groth C, Cella U, Costa E and Biancolini M 2019 Fast high fidelity CFD/CSM fluid structure interaction using RBF mesh morphing and modal superposition method *Aircraft Engineering And Aerospace Technology*

[5] Biancolini M, Cella U 2010 An advanced RBF morph application: coupled CFD CSM aeroelastic analysis of a full aircraft model and comparison to experimental data *MIRA International Vehicle Aerodynamics Conference. Grove, United Kingdom.* 243-258

[6] Biancolini M E, Cella U, Groth C and Genta M 2016 Static aeroelastic analysis of an aircraft wind-tunnel model by means of modal RBF mesh updating *Journal of Aerospace Engineering* **29**(6), 04016061

[7] Groth C, Cella U, Costa E and Biancolini M E 2019 Fast high fidelity CFD/CSM fluid structure interaction using RBF mesh morphing and modal superposition method *Aircraft Engineering and Aerospace Technology* **91**(6), 893-904

[8] Groth C, Porziani S, Biancolini M E 2021 Radial basis functions vector fields interpolation for complex fluid structure interaction problems *Fluids* **6**(9) 314

[9] Giorgetti F, Cenni R, Chiappa A, Cova M, Groth C, Pompa E, Porziani S and Biancolini M 2018 Crack Propagation Analysis of Near-Surface Defects with Radial Basis Functions Mesh Morphing *Procedia Structural Integrity* **12** 471-478

[10] Biancolini M, Chiappa A, Giorgetti F, Porziani S and Rochette M 2018 Radial basis functions mesh morphing for the analysis of cracks propagation. *Procedia Structural Integrity* **8** 433-443

[11] Chiappa A, Groth C, Reali A and Biancolini M E 2020 A stress recovery procedure for laminated composite plates based on strong-form equilibrium enforced via the RBF Kansa method *Composite Structures* **244** 112292

[12] Chiappa A, Groth C and Biancolini M E 2023 A two-scale RBF meshless method for the interface stress retrieval in simply bended and torqued long-fibres laminates *Composite Structures* **306** 116600

[13] Chiappa A, Groth C, Brutti C, Salvini P and Biancolini M E 2019 Post-processing of 2D FEM Q1 models for fracture mechanics by radial basis functions and balance equations *Int. J. Mech* **13** 104-113

[14] Chiappa A, Groth C and Biancolini M E 2019 Improvement of 2D finite element analysis stress results by radial basis functions and balance equations *Int. J. Mech* **13** 90–99
[15] Groth C, Chiappa A and Biancolini M E 2018 Shape optimization using structural adjoint and RBF mesh morphing *Procedia Structural Integrity* **8** 379-389
[16] Porziani S, Groth C, Waldman W and Biancolini M E 2021 Automatic shape optimisation of structural parts driven by BGM and RBF mesh morphing *International Journal of Mechanical Sciences* **189**, 105976
[17] Porziani S, Groth C and Biancolini M E 2018 Automatic shape optimization of structural components with manufacturing constraints *Procedia Structural Integrity* **12** 416–428
[18] Porziani S, Groth C, Mancini L, Cenni R, Cova M and Biancolini, M. 2019 Optimisation of industrial parts by mesh morphing enabled automatic shape sculpting *Procedia Structural Integrity* **24** 724-737
[19] ANSYS, Inc., http://www.ansys.com/products/structures. *Accessed on 2023-09-20*
[20] RBF Morph srl, http://www.rbf-morph.com/act-module/. *Accessed on 2023-09-20*
[21] Davis P 1963 Interpolation and Approximation (Blaisdell Publishing Company, New York)
[22] Heywood R 1969 Photoelasticity for designers (Pergamon Press,1969)
[23] Mattheck C and Burkhardt S 1990 A new method of structural shape optimization based on biological growth *International Journal Of Fatigue* **12**, 185-190
[24] Waldman W and Heller M 2015 Shape optimization of holes in loaded plates by minimization of multiple stress peaks. Technical Report, *Defence Science and Technology Organisation* Fisherman Bend, Australia, Aerospace Div
[25] Biancolini M E 2018 Fast Radial Basis Functions for Engineering Applications, Springer, Berlin.

IOP Conf. Series: Materials Science and Engineering 1306 (2024) 012034 doi:10.1088/1757-899X/1306/1/012034

Influence of car front-end designs on motorcyclists' trajectory in head-on and side-on-head crashes

A Perticone[1], V Nardomarino[1], N Baldanzini[1]
[1] Department of Industrial Engineering, University of Florence, Via di Santa Marta 3, 50139 Firenze, Italy

E-mail: alberto.perticone@unifi.it

Abstract. Motorcyclists are highly vulnerable road users, and cars are one of their primary crash opponents. This study investigates the influence of car front-end designs on motorcyclist trajectory in head-on and side-on-head crashes. The analysis is based on a dataset of 120 multi-body crash simulations conducted using MADYMO and post-processed with MATLAB. An analysis of 1412 real-world Powered Two-Wheeler (PTW) to car accidents was conducted to determine the most common crash configurations and the associated ranges of the variables, such as vehicle speeds and contact points. Three PTW styles (sport-touring, scooter, and sport) and four car front-end designs (Sport utility vehicle (SUV), Family Car/Sedan (FCR), Roadster (RDS), and Multi-purpose vehicle (MPV)) were considered.
The study examined the riders' thrown distance in both collision types. It was observed that, regardless of the collision type, the head was identified overall as the primary body region coming into contact with the opposing vehicle, followed by the chest and neck. In frontal collisions, an augmented bonnet height corresponded to an increased incidence of head contact, whereas a lower bonnet height resulted in a higher frequency of chest contact. Moreover, the thrown distance depended also on PTW speed, particularly for sport and sport-touring motorcycles. Notably, contact with the car windscreen was only observed at velocities exceeding 60 km/h, whereas impact with the bonnet leading edge occurred exclusively below this threshold. Due to the shielding effect of their PTW's fairing, scooter riders predominantly experienced no contact with the opposing vehicle. Sport-touring motorcycles exhibited a more vertical trajectory upon ejection, leading to a greater likelihood of overturning and subsequent rearward head impact with the vehicle. In contrast, sport motorcycles tended to forward projections with a high likelihood of chest contact. In the case of lateral impacts, it was observed that vehicles with a more prominent profile, such as SUVs and MPVs, equipped with protruding bumpers, effectively restrained riders. In this case, vehicle speed did not exert a significant influence on the thrown distance. Additionally, the presence of a conspicuous fuel tank and the initial posture of the rider on the PTW played a crucial role in determining the final thrown distance. Due to their upright postures and the absence of a pronounced fuel tank, scooter dummies were thrown further than others, thus causing head contact with the windscreen.

IOP Conf. Series: Materials Science and Engineering 1306 (2024) 012034 doi:10.1088/1757-899X/1306/1/012034

These findings highlight the importance of car front-end design and PTW fairings in mitigating riders' injuries and provide valuable insights to vehicle manufacturers for developing tailored safety measures for riders.

1 Introduction

Motorcyclists represent a significant group of road users whose safety is of paramount concern due to their inherent vulnerability on the road. The exposure to potential risks and the limited protection available on their Powered-Two-Wheelers (PTWs) make riders particularly vulnerable in accidents. Statistics concerning riders' fatalities and injuries in Europe underscore the urgent need for adequate safety measures and motorcycle consumer rating programs [1].
Although road deaths in 2021 fell by 10% (2,000 fatalities) compared with pre-pandemic 2019, in 2022, fatalities rose by 3% to the previous year [2]. Furthermore, the share differs greatly among road users, with vulnerable road users (pedestrians, cyclists, and PTW riders) representing almost 70% of total fatalities [2]. The road safety context is also different across Europe, and it is recognized the significance of tailored approaches to address the vulnerabilities faced by motorcyclists in each country. In this regard, Italy and France still had the highest number of rider fatalities in 2020, even though this number has decreased more than the EU average over the past decade [3]. In addition to fatalities, non-fatal injuries among motorcyclists also present a concerning picture. In 2018, the EU set a 50% reduction target by 2030 for road deaths and, for the first time, also for serious injuries. This objective was defined in the Commission's Strategic Action Plan on Road Safety and EU road safety policy framework 2021-2030 [4], which also lays out road safety plans aiming to reach zero road deaths by 2050 within the "Vision Zero" program. Serious injuries can lead to long-term disabilities, medical costs, and a diminished quality of life for affected individuals. As evidenced by the statistical data, motorcyclists are indeed vulnerable road users (VRUs) in Europe, facing a disproportionately high risk of fatalities and injuries compared to that faced by other road users, such as car drivers.
Addressing this issue necessitates understanding accident events in detail, as well as the injuries and their causes, to develop countermeasures for the protection of PTW riders. In this regard, it is already known that impact speed [5–7], impact angle [8,9], impact location [10,11] and PTW style [1,12–15] have great influences on the riders' risk of injuries. The present study wants to answer the research question of the car front-end design influence on the motorcyclist's trajectory during the two most representative types of crashes: head-on and side-on-head. Similar studies on the influence of vehicle shapes and stiffnesses have been common in recent years for pedestrian and cyclist safety, but the research efforts on PTW users were scarce. It was demonstrated that bicyclists are exposed to lower impact speed and are less frequently and severely injured than pedestrians, with longer head contact time and shorter longitudinal throws [16]. Furthermore, it was shown that the geometry of the car largely determines the contact points of the different body regions with the bonnet [17] and that vehicles with a high bonnet leading edge are likely to cause higher head ground contact speeds at 30 km/h [18], and consequently, low-fronted cars provide better containment of the pedestrian on the bonnet. SUVs and vans appear to be more aggressive for head–ground contacts, while at 40 km/h, a low front might not give any benefit in reducing the severity of head–ground impacts. Later, relationships between passenger car front shape and pedestrian injury risk derived from accident data were quantified, showing that the bumper geometry is statistically significant for AIS2+ leg injury risk and femur/pelvis injury occurrences [19]. On

the other hand, the authors concluded that the passenger car front shape did not show statistical significance for AIS3+ thorax and head injuries and, in general, the passenger car front shape parameters were less influential than impact speed and pedestrian age for pedestrian injury risk. Recently, a couple of papers have been published about the influence of vehicle front-end design on the head injury risk of motorcyclists (Xiao et al., 2022, 2020), demonstrating that the bonnet length has high influences on motorcyclist head injuries and that one-box vehicles can be more likely to cause head injuries than the other four vehicle types: i.e., Sedan, MPV, SUV, and EV (Electric Vehicle). The authors also noted that the effects of vehicle shape parameters on motorcyclist and pedestrian head injuries are quite different, and the trend of decreasing the bonnet angle and increasing the bonnet length can potentially reduce the head injury risk of motorcyclists.

To answer the research question, the present study investigates the effect of four car front-end designs on the riders' thrown distance in two collision types, i.e., head-on and side-on-head, in conjunction with three different PTW styles: sport-touring, scooter, and sport.

2 Method

2.1 Data collection

Configurations and speed ranges were defined based on a dataset of 1412 real PTW-to-car accidents collected from ISO 13232, MAIDS, and GIDAS databases. The dataset included only accidents involving L3 vehicles, i.e., PTWs with a maximum design speed exceeding 45 km/h and an engine capacity exceeding 50 cm^3. The dataset exclusively consisted of accidents with a two-wheeled motor vehicle (MC) without a pillion rider and rider seated on an upright motorcycle at the time of impact. The dataset was sorted to identify the most representative crash configurations coded with the three-digit system provided by ISO 13232. The code identifies the contact points (CPs) and relative heading angle (RHA) between MC and opposite vehicle (OV), based on the following coding scheme: the first two digits of the code indicate the CPs of the OV and MC, respectively, while the third digit corresponds to the RHA between the two vehicles; seven contact points are identified for the OV in a clockwise direction, starting from the center line of the frontal bumper (CP=1) and ending at the center line of the rear bumper (CP=7); for the MC, the first contact point (CP=1) represents the front wheel, the third contact point (CP=3) represents the rear wheel, and the second and fourth contact points (CP=2,4) indicate the left and right sides of the MC, respectively; the third digit in the code is the multiple 1 to 8 of a 45-degree range that returns the RHA measured clockwise from the front.

The 1412 collected accidents were sorted according to occurrence frequency (Figure 2.1). Since this study was devoted to evaluating the influence of car front-end designs, head-on-side (42%), head-on-rear (9%), and rear-end (4%) accidents were not considered. Configurations 114, 115, and 143 were investigated, representing head-on (21%) and side-on-head accidents (10%), respectively. The data was used to define the conditions for the simulations.

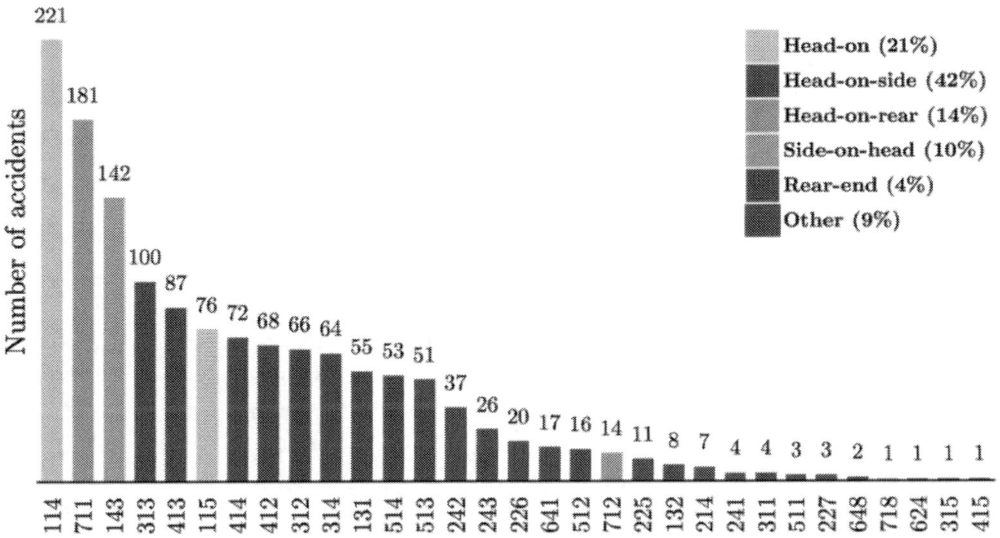

Figure 2.1 – Crash configuration distributions across data collected from ISO 13232, MAIDS, and GIDAS.

2.2 *Simulation set-up*

For the head-on crashes, the OV was chosen to be stationary since it is crucial to emphasize that this study focuses exclusively on evaluating trajectories rather than injuries. Specifically, the assessment considers trajectories within a reference system integral to one of the two vehicles, namely the OV, in this study. Consequently, the cinematic characteristics of the system, unaffected by the forces generating it, remain consistent for a given relative speed, regardless of which vehicle possesses a higher speed or greater mass. Thus, under fixed relative speed conditions, the accelerations measured by the dummy or the OV are identical, irrespective of whether the speed is ascribed to the heavier or lighter vehicle. However, it is essential to note that when considering an external reference system and accounting for injury assessment, discrepancies in speed attribution yield divergent outcomes. The speed of the MC was defined between 30 and 90 km/h with five steps of 15 km/h each. Even though the original dataset reported the range for the MC down to 0 km/h in "114" and "115" accidents, 30 km/h was considered the minimum significant speed of the MC to appreciate a thrown trajectory since no ejections were noticed by simulating a few sample configurations below this threshold with the OV stationary. Conversely, the MC was kept stationary for the side-on-head crashes to maximize the chance that the PTW rider could hit the OV, and the OV speed range was defined between 30 km/h and 90 km/h, with increments of 15 km/h, for uniformity with the other crash type, although its maximum speed across the databases was 65 km/h. The simulations were set up in MADYMO and performed with version 2020.2. A total of 120 simulations were run, considering the five speeds, three PTWs, four OV profiles, and two collision types. The RHA between the two vehicles was maintained at its nominal value, i.e., 180 degrees in the first collision type and 90° in the second one. This study did not consider angled crashes, and their effect on dummy trajectories was not evaluated. Every parameter (speed, angle, OV type, MC type) was managed by HyperStudy to automatically generate the simulation files with a run-time of 500 ms. As this study only evaluated primary impacts, a 500 ms window was considered adequate to capture these events in agreement with other studies [23–29]. The coefficient of friction between the dummy and the car was demonstrated to have a remarkable effect on trajectories and head impact time and was therefore set to 0.3, which is in accordance with

IOP Conf. Series: Materials Science and Engineering 1306 (2024) 012034 doi:10.1088/1757-899X/1306/1/012034

several studies [6,18,20,30,31]. A friction coefficient of 0.8 was assumed for the tire-ground contact to represent a dry asphalt road [32–34].
The thrown trajectories were recorded until the first interaction between the rider's head and the OV. The visual representation of the trajectories is provided in Appendix A (Figure A.2, Figure A.3, Figure A.4, Figure A.5). Shaded stickmen represented the riders in their initial positions on the PTWs, excluded from the visual representation; the solid colors indicated their final position when the acquisition stopped.

2.3 Vehicle models

Three vehicles, representative of three PTW styles, and four front-end designs of car models were used in this study. The three PTWs were developed during previous research at the University of Florence and were already used in multi-body crash simulations [35,36]. The computer-model fleet comprehended a sport-touring, a sport, and a scooter style PTWs, Figure 2.2, and four Generic Vehicles (GVs) for cars, Figure 2.3 (left). GV models are generic replications of current car fronts. The car fronts were developed within the TB024 EuroNCAP Protocol [37] for kinematic comparisons only and should not be used for evaluations of injury metrics as they do not meet the UN-R127 requirements. The GVs represented Sport Utility Vehicle (SUV), Family Car/Sedan (FCR), Roadster (RDS), and Multi-Purpose Vehicles and Superminis (MPV). In a previous study [20], three-dimensional median geometries were used to obtain representative vehicle shapes. The four FE models retrieved from that work, Figure 2.3 (left), were converted in MADYMO as facet surfaces and contact characteristics, Figure 2.4, were attributed to the five main front-end sub-groups: roof/windscreen, bonnet, bonnet leading edge (BLE), bumper, and spoiler, Figure 2.3 (right). These contact characteristics were obtained numerically by propelling a rigid cylindrical impactor against their FE counterparts in RADIOSS. The procedure was repeated at eight locations on the spoiler, bumper, bonnet lead, and bonnet at the vehicle centerline. The output curves (Figure 2.4) complied with the corridors supplied in the previous studies [20] to generate reasonable contact forces in the pedestrian-vehicle interaction study. However, they did not account for the stiffness out of the centerline of each car model since, in this study, all the simulated crashes were centered and ignored potential regions of localized stiffness along the centerline.
The vehicle curb weights were increased by 150 kg for the driver and front passenger as specified in the Euro NCAP pedestrian testing protocol for a total amount of: 1775 kg for the SUV, 1690 kg for FCR, 1590.5 kg for MPV, and 1462.5 kg for RDS. The vehicles have only one degree of freedom: the X-direction in the vehicle coordinate system. The global coordinate system was defined as shown in Figure 2.3: X-direction was the driving direction of the OV (longitudinal axis), and x=0 at the foremost point of the vehicle at the beginning of the acquisition (t=0). Negative values referred to the acquisition points in front of the OV when the riders were still on their PTW. Y-direction was the lateral vehicle axis with y=0 at the vehicle centerline, whereas Z-direction was vertical and upwards with z=0 at the ground level.

Figure 2.2 - PTW fleet, helmeted rider dummy, and BSJ.

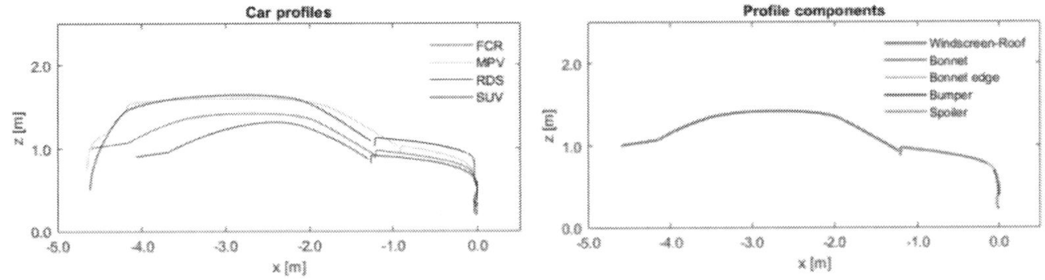

Figure 2.3 - The four OV profiles (left) and the five parts of the front-end design (right).

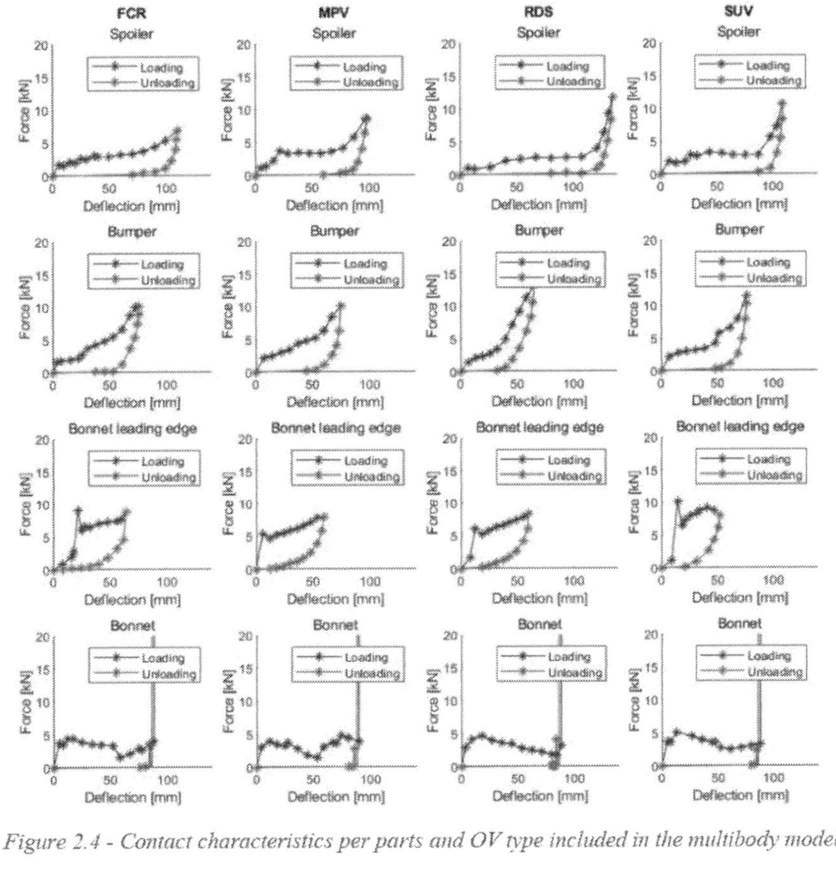

Figure 2.4 - Contact characteristics per parts and OV type included in the multibody models.

IOP Conf. Series: Materials Science and Engineering 1306 (2024) 012034 doi:10.1088/1757-899X/1306/1/012034

The Motorcycle Anthropometric Test Dummy (MATD) used in this study is based on the Hybrid III Anthropometric Test Dummy design, and it was specifically developed to model riders. The MATD was fitted with a helmet, and a state of equilibrium was achieved for the entire model by applying a gravitational load to the dummy and the PTW (e.g., pre-compressed suspensions). It is important to note that a rider's sitting position can vary significantly, and it is influenced by numerous factors, including human factors (such as physical characteristics, riding experience, duration of riding, cognitive load), motorcycle design characteristics, and environmental variables (such as road conditions, weather, and traffic conditions) [38]. When comparing the actual comfortable position of a real rider with the anthropometric measurements of the MATD, several disparities were observed [39]. Hence, the positioning of the dummy was carried out following the recommended procedures outlined by ISO [40], and the joint angles were extrapolated from relevant literature to achieve a natural ergonomic posture that corresponds to the optimal riding position [41].

3 Results

The results are categorized into frontal and lateral crashes, Figure A.1. Each collision type was subjected to a comprehensive analysis of two influential parameters: the style of the PTW and the type of front-end design. The output under control was the trajectory of the head, henceforth denoted in figures as "Distance," which represented the thrown distance along the X-axis measured at the head's center of gravity in the reference system described in section 2.3.

3.1 Frontal Crash

In frontal crashes, the thrown distance of the dummy increased with a magnitude dependent on speed for both the PTW styles and the OV front-end designs, Figure 3.1. Every collision causing head contact was plotted in the chart, identifying the car part involved. A trendline minimizing the sum of squares of a power function was drawn for each PTW, Figure 3.1 (left), and car profile, Figure 3.1 (right). Every line showed a clear dependence on the closing speed, primarily upon different PTW styles: the sport PTW had the highest slope and vertical shift, while the scooter reported the lowest values, with a weak dependence on speed. The sport-touring showed intermediate values of slope and vertical shift. With reference to car profiles, the dependence on speed was more accentuated for RDS and less for FCR. Similar behaviors were noticed for SUV and MPV. It was worth noting that contacts on the windscreen happened only at and above 60 km/h, whereas those on the bonnet edge were only below 60 km/h. Contacts on the bonnet were largely spread across the whole speed span.

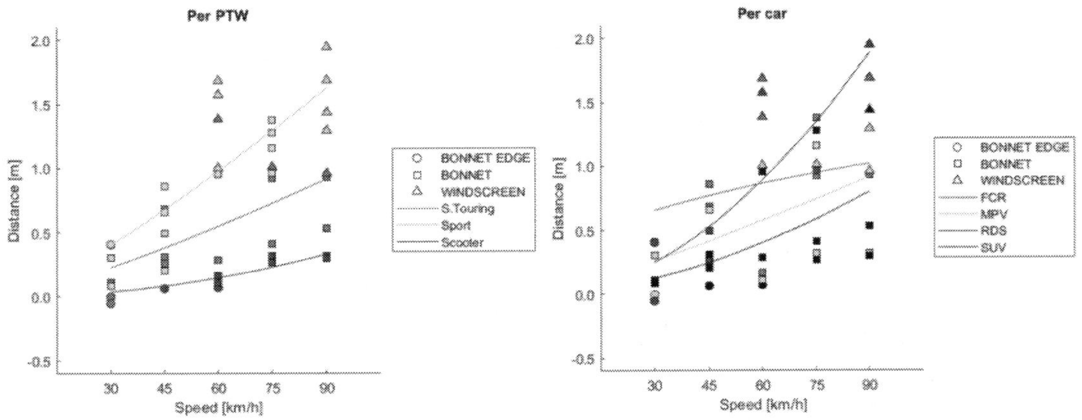

Figure 3.1 - Influence of speed in frontal crash per PTW (left) and per car (right). Every collision causing head contact was plotted in the scatter with a geometric symbol and a trendline was drawn over this data.

3.1.1 Influence of PTW

The boxplot analysis, Figure 3.2, and ANOVA tests confirmed that the scooter (M=0.27 m, SD=0.106 m) had the shortest thrown distance among the PTWs, significantly different (p-value < 0.01) from that of the other styles. Also, the sport (M=0.98 m, SD=0.550 m) and the sport-touring (M=0.41 m, SD=0.425 m) differed significantly from one another (p-value < 0.01). The total number of all the speed pairs per PTW was 20, but each boxplot did not count for the same number of simulations due to the presence of some no-contact instances ($N_{S.Touring}=15$; $N_{Sport}=20$; $N_{Scooter}=10$).

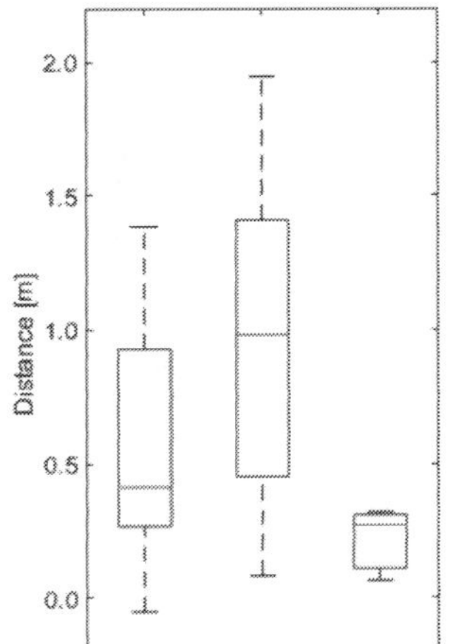

Figure 3.2 - Collision distance per PTW in frontal crashes.

A further investigation interested the contact points during the crash. The distribution per car parts, Figure 3.3 (left), showed that between 50% and 75% of the collisions involved the three PTWs with direct contact between the dummy and the car bonnet or bonnet leading edge. Other contacts involved the windscreen or even no contact at all, in the case of scooter and sport-touring. However, a Chi-squared test with the null hypothesis that one car part is independent of an assigned PTW style could not be rejected on a 5% significance level for any PTW style,

Figure A.6 (a). Based on the distribution per dummy's body regions, Figure 3.3 (right), the sport PTW dummy consistently came into contact with the OVs, with a notable occurrence of chest contacts (40%), which was deemed significantly related to the type of PTW by the Chi-squared test, Figure A.6 (b). On the other hand, sport-touring motorcycles and scooters exhibited instances where no contact with the OV was observed in 25% and 50% of cases, respectively, and the sport-touring motorcycle dummy experienced the highest head contact percentage (75%). Notably, the no-contact instances, despite not being below the 5% significance level, Figure A.6 (b), showed a marked variability in the PTW style.

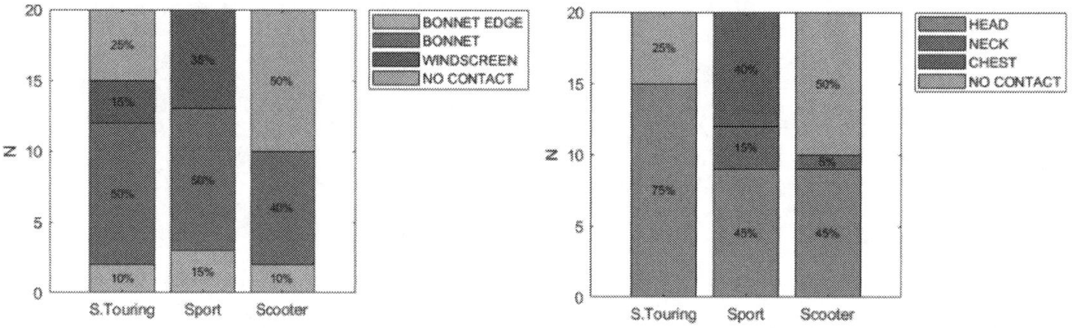

Figure 3.3 - Contact points per car parts (left) and dummy's body regions (right) across PTWs in frontal crashes.

3.1.2 *Influence of car profile*

The boxplot analysis and ANOVA tests confirmed that the SUV (M=0.29 m, SD=0.450 m) had the lowest thrown distance among the OVs, significantly different (p-value < 0.01) from that of the other car profiles, Figure 3.4. FCR scored higher values (M=0.89 m, SD=0.546 m), while MPV (M=0.49 m, SD=0.446 m) and RDS (M=0.49 m, SD=0.730 m) reported similar median values, with wider variance for the latter. Each boxplot did not count for the same number of simulations (*N=15*) due to the presence of some no-contact instances ($N_{FCR}=12$; $N_{RDS}=7$; $N_{MPV}=12$; $N_{SUV}=14$).

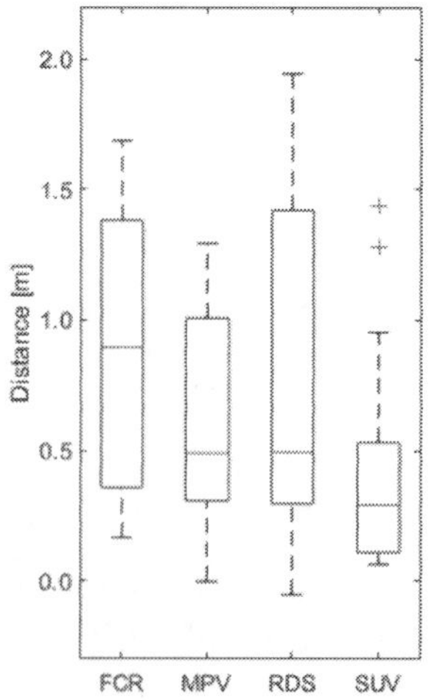

Figure 3.4 - Collision distance per car profile in frontal crashes.

The FCR and RDS exhibited an equal percentage of chest contacts (20%), Figure 3.5 (right). Notably, the SUV demonstrated the highest frequency of contact with the dummy, with the lowest occurrence of no-contact instances and contacts on the windscreen, both at 7%, Figure 3.5 (left). Most of the contacts with the SUV involved the head on the bonnet. In contrast, approximately half of the collisions involving the RDS did not directly result in contact (53%), while a moderate percentage (20%) of contacts occurred on the chest (a value shared with the FCR). Contacts on the head were the lowest reported (27%), as illustrated in Figure 3.5 (right). Eventually, the MPV exhibited the highest number of contacts on the windscreen (27%). From the Chi-squared tests, neither body regions nor car parts could reject the null hypothesis of independence from the front-end designs with a 5% significance level, Figure A.6 (a)-(b).

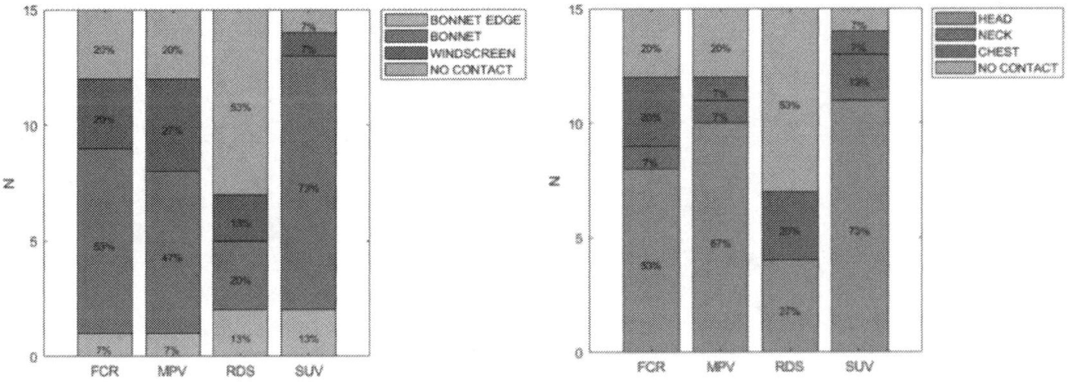

Figure 3.5 - Contact points per car parts(left) and dummy's body regions (right) across OVs in frontal crashes.

3.2 *Lateral Crash*

In this case, the scooter reported the longest distance, the sport style the shortest, and the sport-touring maintained its central position between the other two, Figure 3.6 (left). Only the scooter showed a moderate positive derivative between 30 and 90 km/h. Grouping the results per car profile, FCR and RDS performed similarly, with almost the same distances across the five-speed steps, while MPV and SUV reported slightly lower values, up to 0.6 m less than FCR and RDS, Figure 3.6 (right). No predominant match with specific car parts was highlighted in any speed range, and no substantial evidence was found between the impact speed at which the OV hit the MC and the thrown distance of the dummy in lateral crashes.

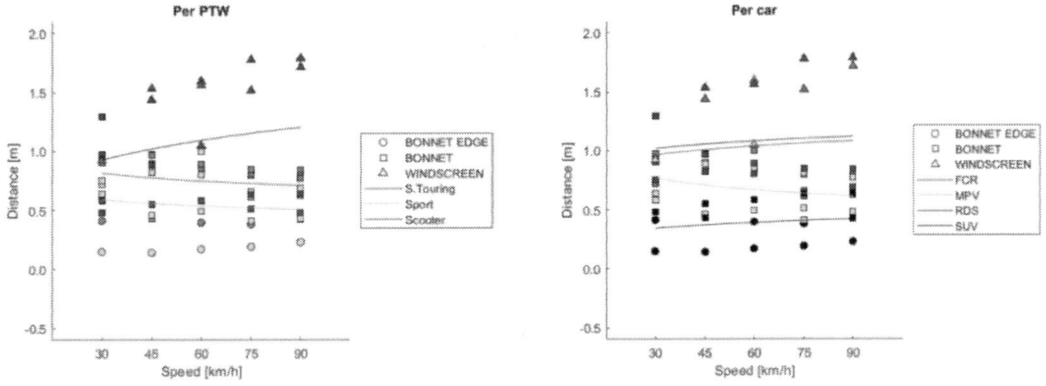

Figure 3.6 - Influence of speed in lateral crash per PTW (left) and per car (right). Every collision causing head contact was plotted in the scatter with a geometric symbol and a trendline was drawn over this data.

3.2.1 *Influence of PTW*

In this collision type, dummies ended their trajectories always with head contact on the OVs. The scooter maintained a significant (p-value < 0.01) higher thrown distance (M=1.01 m, SD=0.498m), Figure 3.7, hence, a higher percentage of contacts on the windscreen (45%), Figure 3.8, which the Chi-squared test attributed to the PTW style, Figure A.6 (c). Conversely, the sport-touring (M=0.84 m, SD=0.213 m) and the sport (M=0.62 m, SD=0.262 m) reported the shortest distances, not significantly different, turning into a comparable percentage of contacts on bonnet and bonnet edge. Each bar and boxplot counted for the same number of simulations (*N=20*).

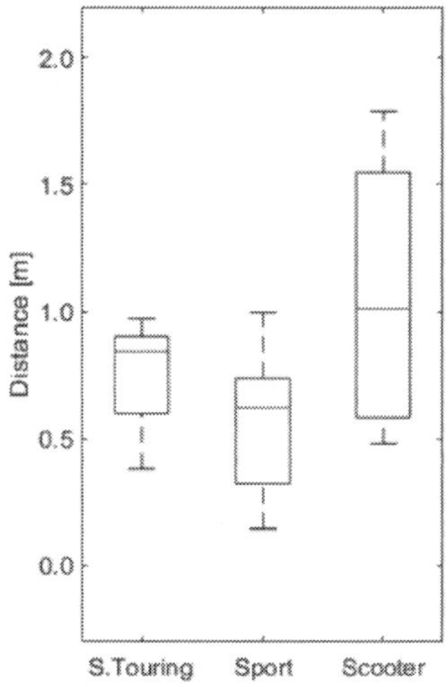

Figure 3.7 - Collision distance per PTW in lateral crashes.

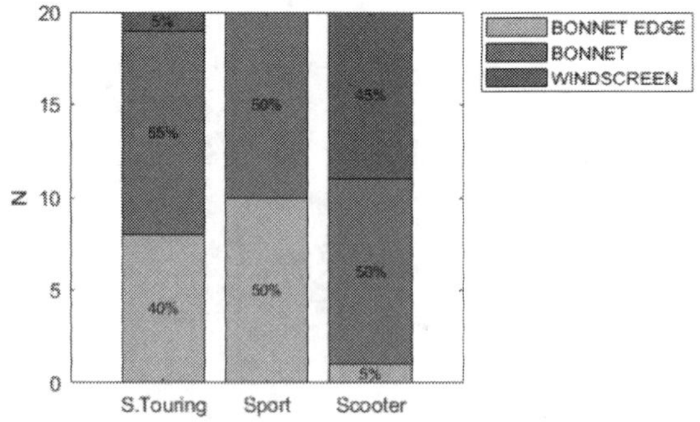

Figure 3.8 - Contact points per car parts across PTWs in lateral crashes.

3.2.2 *Influence of car profile*

A remarkable difference was noticed between car profiles: the SUV reported the shortest thrown distances among the four profiles (M=0.41 m, SD=0.174 m), significantly different from the others (p-value < 0.01), Figure 3.9. Examining the contact points on the car profiles, it became evident that the SUV was the only profile with a substantial percentage of contacts (53%) occurring on the bonnet edge, which depended on the front-end design from the Chi-squared test, Figure A.6 (c); while no contacts were observed on the windscreen, Figure 3.10. In contrast, the remaining three profiles exhibited a higher percentage of contacts on the bonnet and a smaller portion, approximately one-quarter (lower for the MPV at 13%), on the windscreen. None of these three profiles showed an interaction with the bonnet edge in any simulation. Each bar and boxplot counted for the same number of simulations (*N=15*).

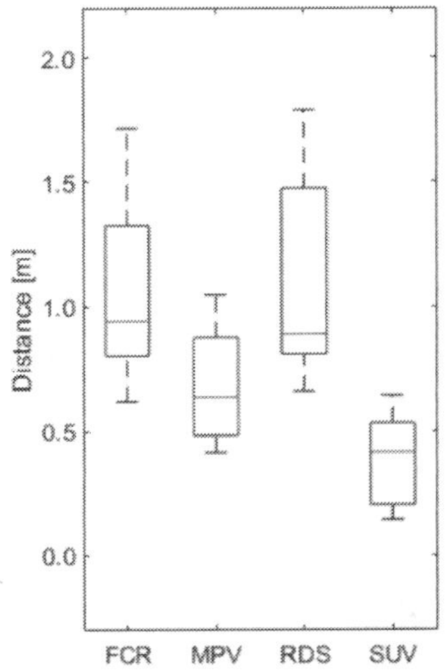

Figure 3.9 - Collision distance per car profile in lateral crashes.

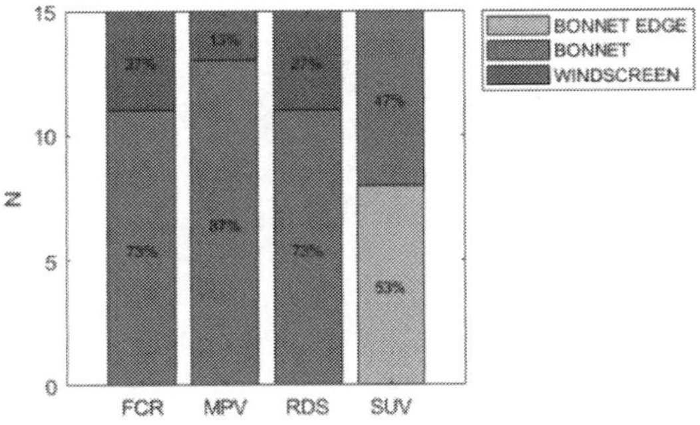

Figure 3.10 - Contact points per car parts across car profiles in lateral crashes.

4 Discussion

The frontal crash results showed that the rider's thrown distance depended on the MC speed, especially for sport and sport-touring PTWs. The contact on the windscreen happened only at speeds over 60 km/h, whereas the bonnet edge was prevalently hit at 30 and 45 km/h. As expected in frontal collisions, the bumper never hit the dummy, which collided solely with the PTW front wheel. Contact points on dummies and cars, indeed, depended on the trajectories developed during the collision, which in turn were determined by the PTW and car geometries. The data concerning rider-car contact points find evidence on a wide range of distinct trajectories, a phenomenon influenced by the inherent characteristics of PTWs, which highlights the unique dynamics at play during such collisions. Understanding these diverse trajectories is crucial for enhancing safety measures and accident prevention strategies in the context of PTW and car interactions. The rider on the scooter was prevalently restrained by the shielding of its vehicle, in agreement with other studies [8], with many no-contact instances against the OV and slight contacts on the bonnet after it turned around the PTW frontal headlamp. Only at the highest speeds was the dummy fully projected off the PTW, hitting the windscreen of the OV. Riders on the sport and sport-touring had more upward trajectories, maybe attributable to their different postures. The dummy on the sport-touring first hit the pelvis or lower limbs on the handlebars so that its flight was more vertical than forward, and it tended to overturn, with the head impacting on the OV in a rearward position, without any direct contact on the neck, as seen for cyclists in other studies, [9]. The dummy turned around the front of the MC, ending with head contact on the OV bonnet or even flying over the roof after a complete flip without any contact, in agreement with other studies [42]. Conversely, the sport PTW had a lower handlebar, promoting a free release of the dummy from the saddle with a high variance of end contacts and new body regions involved, such as the chest, in addition to the head and neck. From the point of view of cars, generally, as the height of the bonnet increased, there were more contacts on the head; conversely, with a lower bonnet height, there were more contacts on the chest, confirming the trend observed in pedestrians [43]. In the crashes against the OV with the highest bonnet, SUV, and MPV, the PTW pitch was lower, confirming that frontal structures may aim to control the energy transfer and prevent a rollover in frontals [44]. However, the dummy's motion was more upward than forward, which favored its rotation with direct head contact on the proximal part of the bonnet (short thrown distance). Conversely, FCR and RDS allowed extended thrown distances, and the riders hit the OV far away on the bonnet or directly on the windscreen. The RDS reported shorter distances and many no-contact instances, as the OV was more likely to move backward after the collision due to its lighter mass; hence, using a reference system integral to the car, its distance median appeared lower, and its dependence on MC speed was higher. The no-contact instances were dominated primarily by two mechanisms: at low speed, dummies on sport and sport-touring were thrown over the PTW, falling to the ground without touching the OV after a slight rebound of both the vehicles; at high speed, riders could easily pass over the OVs (RDS and FCR) without touching it.

Different from frontal crashes, in laterals, the thrown distance did not show dependence on the OV impact speed, neither when it was analyzed per style of PTW nor type of OV. Conversely, the distance was greatly influenced by the PTW style since, after the first contact, the dummy lays down on the bonnet and develops its trajectory only if the lower limb is not restrained between the two vehicles. The fuel tank assumes a critical role in determining two potential outcomes: whether the dummy is trapped and experiences head contact with the bonnet or whether it remains free and subsequently strikes the windscreen with its head. Besides, the initial posture the dummy took on the PTWs determines its final distance: the more upright the dummy is, the further it is thrown after the crash. Thus, these combined elements correlate well with the finding that scooters are associated with the longest thrown distance in side impact scenarios. In addition, it is evident that OVs characterized by high profiles, such as MPVs and SUVs, played a significant role in confining the riders and preventing head contact with the windscreen. From this perspective, this study corroborates the findings that decreasing bumper height and bonnet length can increase head injury risk for riders [21,22] and pedestrian protection [45].

Although the OV profiles presented in this study are intended for trajectory assessment only, a certain level of riskiness can be inferred, considering that the car parts hit by riders have different stiffnesses and that the kinematics of the human body during the primary impact influence the severity of the injuries, at least for pedestrians, [46]. According to a study involving 153 PTW riders [42], the windscreen region (made by glass, wiper, and frame) resulted in the most dangerous, with an average AIS reported of 2.2, followed by the rear half of the bonnet (2.0), the bonnet leading edge (1.9), the bumper (1.7) and, eventually, the front half of the bonnet (1.2). Moreover, despite its low frequency (5%), cervical spine injuries were found to be extremely important, with an average AIS of 3.0, followed by injuries on the head (2.4, 60%), chest (2.2, 30%), and legs (1.9, 79%). The windscreen region was found likely to result in severe injuries not only for PTW riders [47] but also for head contact of pedestrians [9,48] and cyclists [9,16]. Based on the previous findings, RDS should represent a considerable threat for riders, given its high percentage of head contacts in lateral crashes and chest involvement in frontals, prevalently on the windscreen and rear bonnet. Conversely, the SUV may be more tolerant of riders, considering its contacts on the first half of the bonnet and their absence on the windscreen. Nevertheless, while certain vehicle variables, such as the inclination angle of the windscreen, appear to have no direct correlation with injuries [46], other studies propose that certain vehicles, such as SUVs, might possess bonnets that are even stiffer than windscreens found in compact cars [49]. However, since Otte's findings [42], many improvements interested the bonnet region, such as the pop-up bonnet ("Active Hood Lifting System") [50] and the pedestrian airbag, found effective also for cyclists [51]. Hence, assessing the influence of specific car components on the risk of injuries is a complex task [52]. It was emphasized that estimates of countermeasure efficacy cannot be based on the assumption that all injuries are solely caused by vehicle-related factors, but monitoring post-vehicle impact kinematics is crucial to address injuries during crashes completely [53].

This study showed that a higher bonnet height in frontal collisions was associated with an increased likelihood of head contact, while a lower height led to a higher chance of chest contact. In lateral impacts, the SUV effectively restrained riders with its bumper, avoiding contact with the windscreen. Statistical tests, such as the ANOVA and the Chi-squared tests, confirmed a significant correlation between the SUV front-end design and the thrown distances (or contact points). However, these tests also suggested that FCR, RDS, and MPV front-ends were not significantly different from a trajectory perspective, and, thus, they may be interchangeable for studies on car-to-PTW crashes. Differently, the PTW style significantly affected the outcome of frontal collisions: scooter riders rarely came into contact with the opposing vehicle; sport-touring motorcycles tended to have a higher likelihood of overturning with rearward head impact on the car; sport PTWs typically projected forward, increasing the likelihood of chest contacts. In lateral collisions, the presence of a conspicuous fuel tank on the PTW and the initial posture of the rider also played a significant role in determining the distance the rider was thrown: scooter riders were thrown further, often leading to head contact with the windscreen. These findings can serve as a valuable resource for future research aiming to design car profiles more forgiving towards PTW riders, ultimately enhancing their safety during crashes.

IOP Conf. Series: Materials Science and Engineering 1306 (2024) 012034 doi:10.1088/1757-899X/1306/1/012034

A. Appendix

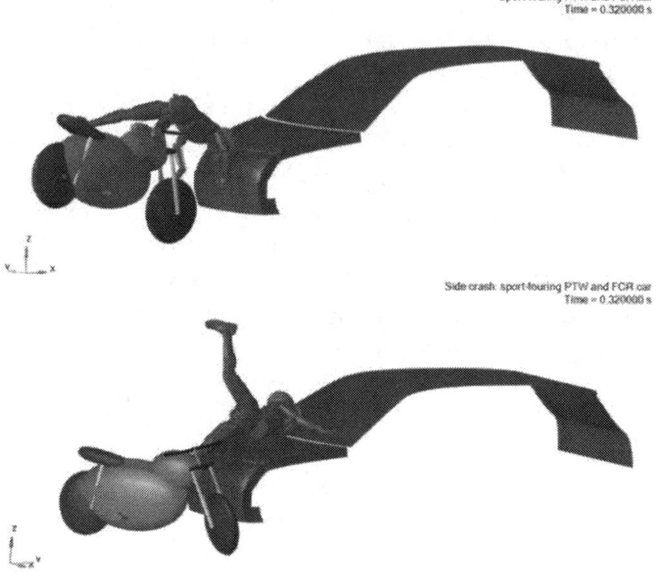

Figure A.1 - Frontal (upper) and side (lower) crashes reproduced in MADYMO.

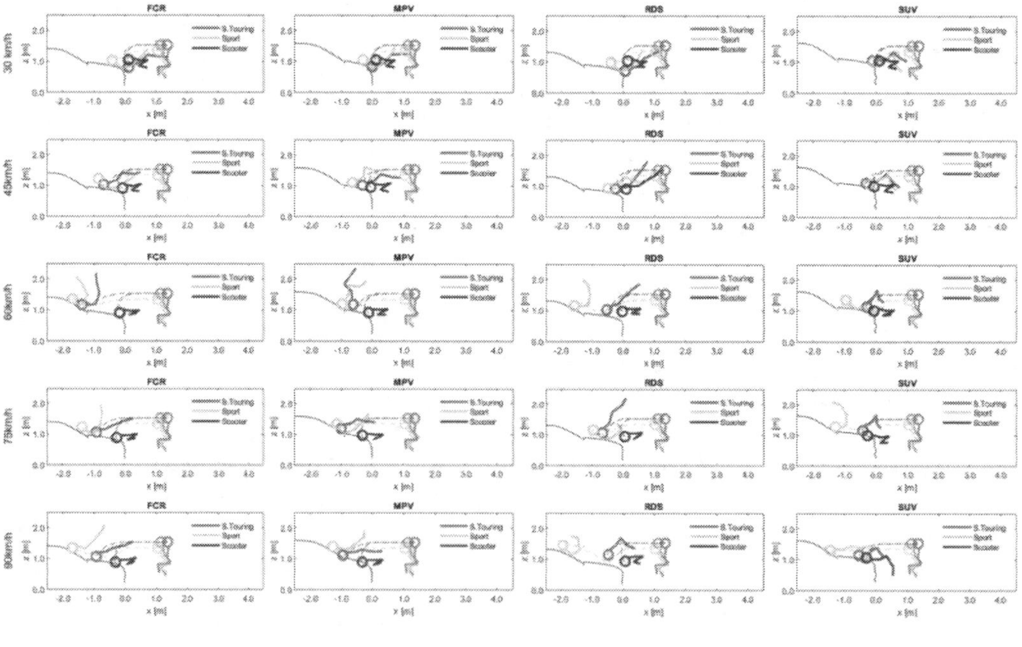

Figure A.2 - Overview of frontal collisions divided per OV across five-speed steps and grouped per PTW style.

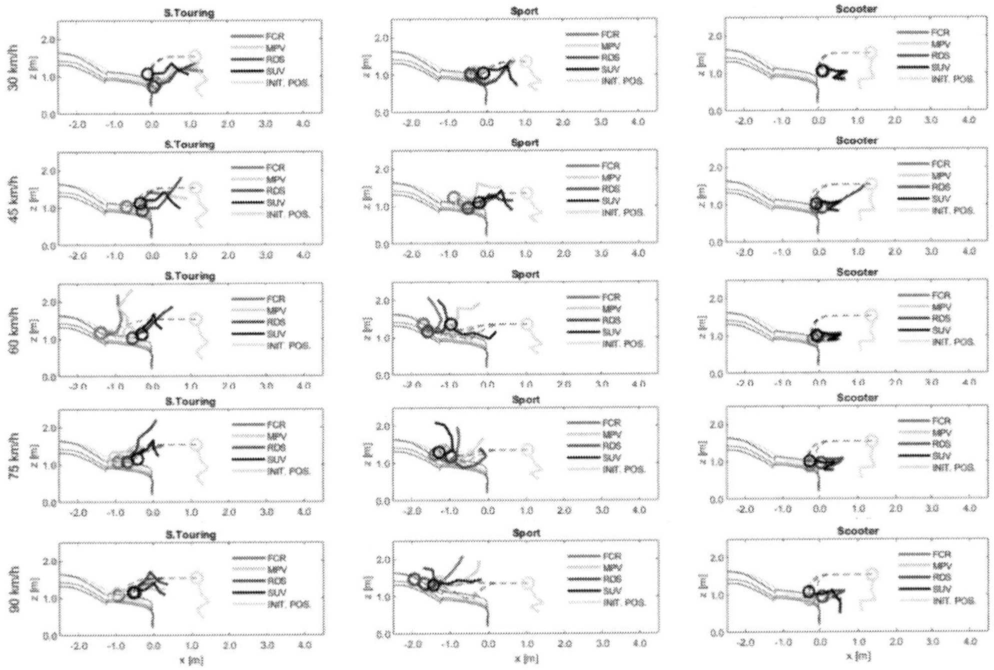

Figure A.3 - Overview of frontal collisions divided per PTW across five-speed steps and grouped per OV style.

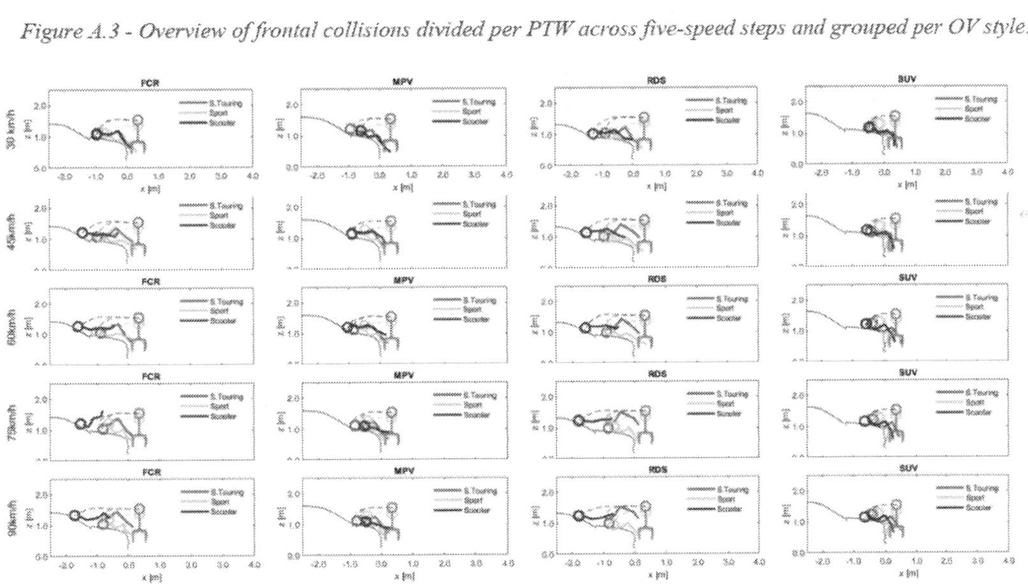

Figure A.4 - Overview of lateral collisions divided per OV across five-speed steps and grouped per PTW style.

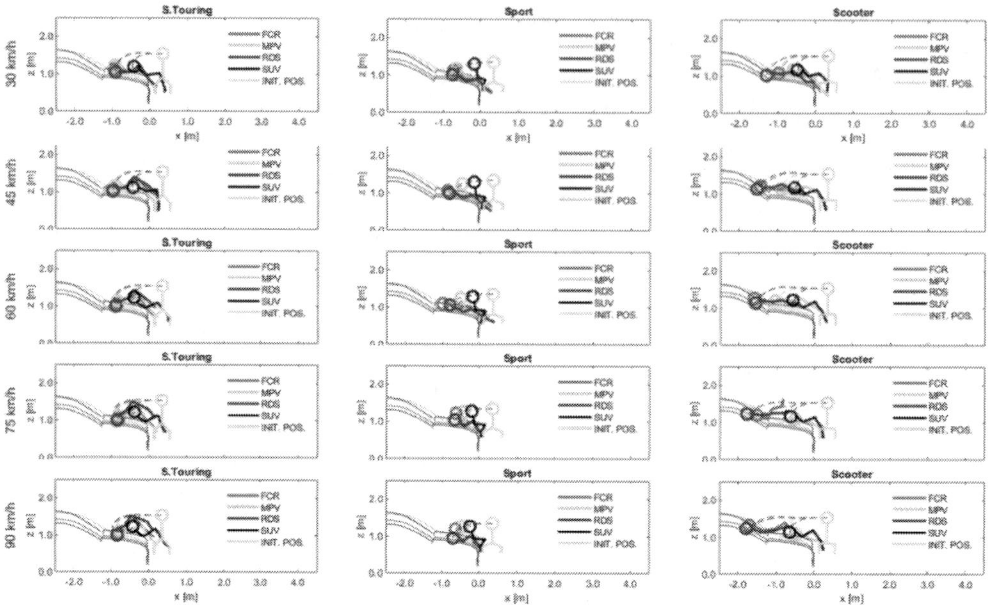

Figure A.5 - Overview of lateral collisions divided per PTW across five-speed steps and grouped per OV style.

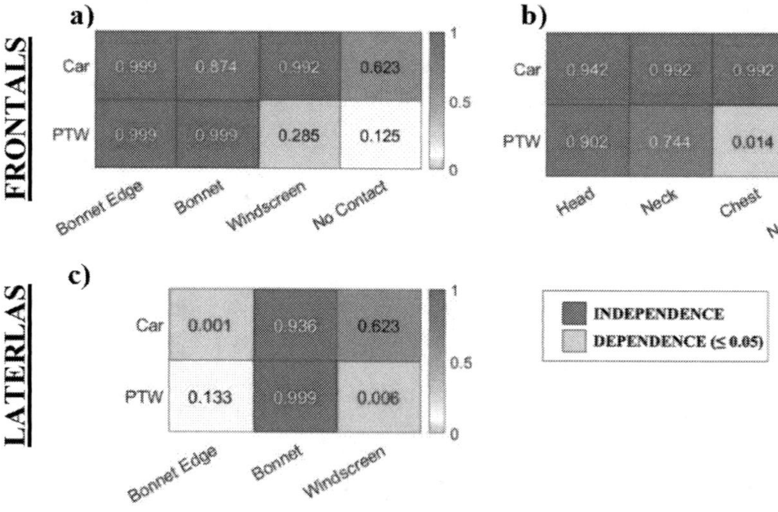

Figure A.6 - Chi-squared test at the 0.05 significance level for the independence of the variables (car parts and body regions) from the categorical groups (types of front-end and PTW) in head-on (first row) and side-on-head (second row) crashes.

REFERENCES

[1] Rizzi M 2016 *Towards a safe system approach to prevent health loss among motorcyclists* (Göteborg: Chalmers University of Technology)

[2] Directorate General for Mobility and Transport 2023 *Road safety in the EU: fatalities below pre-pandemic levels but progress remains too slow*

[3] Slootmans F 2023 *Facts and figures motorcyclists and moped riders*

[4] European Commission 2018 *Communication from the commission to the European parliament, the council, the European economic and social committee and the committee of the regions* (Brussels)

[5] Lubbe N, Wu Y and Jeppsson H 2022 Safe speeds: fatality and injury risks of pedestrians, cyclists, motorcyclists, and car drivers impacting the front of another passenger car as a function of closing speed and age *Traffic Safety Research* **2** 1–26

[6] Simms C K and Wood D P 2006 Effects of pre-impact pedestrian position and motion on kinematics and injuries from vehicle and ground contact *International Journal of Crashworthiness* **11** 345–55

[7] Li J, Fang S, Guo J, Fu T and Qiu M 2021 A Motorcyclist-Injury Severity Analysis: A Comparison of Single-, Two-, and Multi-Vehicle Crashes Using Latent Class Ordered Probit Model *Accid Anal Prev* **151**

[8] Serre T and Llari M 2010 Numerical Analysis of the Impact between a PTW Rider and a Car in Different Accident Configuration *IFMBE Proceedings* vol 31 pp 521–4

[9] Nie J, Li G and Yang J 2015 A Study of Fatality Risk and Head Dynamic Response of Cyclist and Pedestrian Based on Passenger Car Accident Data Analysis and Simulations *Traffic Inj Prev* **16** 76–83

[10] Mukherjee S, Chawla A, Mohan D, Singh M, Sakurai M and Tamura Y 2001 Motorcycle-car side impact simulation *Conference proceedings International Research Council on the Biomechanics of Injury, IRCOBI* (Isle of Man (UK): IRCOBI)

[11] Serre T, Masson C, Perrin C, Martin J L, Moskal A and Llari M 2012 The motorcyclist impact against a light vehicle: Epidemiological, accidentological and biomechanic analysis *Accid Anal Prev* **49** 223–8

[12] Chawla H, Karaca I and Savolainen P T 2019 Contrasting Crash- and Non-Crash-Involved Riders: Analysis of Data from the Motorcycle Crash Causation Study *Transp Res Rec* **2673** 122–31

[13] Carroll J, Gidion F, Rizzi M and Lubbe N 2022 Do motorcyclist injuries depend on motorcycle and crash types? An analysis based on the German In-Depth Accident Study *International Motorcycle Conference (IFZ)*

[14] Watson P M 1979 Features of the Experimental Safety Motorcycle - ESM1 *International Technical Conference on Experimental Safety Vehicles (ESV)* (Paris (France): NHTSA)

[15] Hurt H H Jr, Ouellet J V. and Thom D R 1981 *Motorcycle accident cause factors and identification of countermeasures* vol 1 (Los Angeles, California)

[16] Peng Y, Chen Y, Yang J, Otte D and Willinger R 2012 A study of pedestrian and bicyclist exposure to head injury in passenger car collisions based on accident data and simulations *Saf Sci* **50** 1749–59

[17] Van Rooij L, Bhalla K, Meissner M, Ivarsson J, Crandall J, Longhitano D, Takahashi Y, Dokko Y and Kikuchi Y 2003 Pedestrian crash reconstruction using multi-body modeling with geometrically detailed, validated vehicle models and advanced pedestrian

injury criteria *International Technical Conference on the Enhanced Safety of Vehicles (ESV)*

[18] Crocetta G, Piantini S, Pierini M and Simms C 2015 The influence of vehicle front-end design on pedestrian ground impact *Accid Anal Prev* **79** 56–69

[19] Li G, Lyons M, Wang B, Yang J, Otte D and Simms C 2017 The influence of passenger car front shape on pedestrian injury risk observed from German in-depth accident data *Accid Anal Prev* **101** 11–21

[20] Klug C, Feist F, Raffler M, Sinz W, Petit P, Ellway J and Van Ratingen M 2017 Development of a procedure to compare kinematics of human body models for pedestrian simulations *Conference proceedings International Research Council on the Biomechanics of Injury, IRCOBI* vol 2017-Septe (Antwerp (Belgium): IRCOBI) pp 508–30

[21] Xiao Z, Wang L, Mo F, Lv X and Yang C 2020 Influences of impact scenarios and vehicle front-end design on head injury risk of motorcyclist *Accid Anal Prev* **145** 1–11

[22] Xiao Z, Wang L, Zhang Y and Yang C 2022 A study on motorcyclist head reponses during impact against front end of vehicle *International Journal of Crashworthiness* **27** 147–59

[23] Namiki H, Nakamura T and Iijima S 2005 A computer simulation for motorcycle rider injury evaluation in collision *International Technical Conference on the Enhanced Safety of Vehicles (ESV)* (Washington DC (USA): NHTSA)

[24] Barbani D, Baldanzini N and Pierini M 2014 Development and validation of an FE model for motorcycle-car crash test simulations *International Journal of Crashworthiness* **19** 244–63

[25] Yettram A L, Happian-Smith J, Mo L S M, Macaulay M A and Chinn B P 1994 Computer simulation of motorcycle crash tests *International Technical Conference on the Enhanced Safety of Vehicles (ESV)* (Munich (Germany): NHTSA) pp 1227–40

[26] Grassi A, Barbani D, Baldanzini N, Barbieri R and Pierini M 2018 Belted Safety Jacket: a new concept in Powered Two-Wheeler passive safety *Procedia Structural Integrity* **8** 573–93

[27] Xu S, Jin X, Luo C and Fu S 2021 Modeling method of rider-two-wheeler-road coupled system in two-wheeler traffic accident reconstrucitons *International Journal of Crashworthiness* 1–9

[28] Cherta B O, Llari M, Afquir S, Martin J L, Bourdet N, Honoré V, Masson C and Arnoux P J 2019 Analysis of trunk impact conditions in motorcycle road accidents based on epidemiological, accidentological data and multi-body simulations *Accid Anal Prev* **127** 223–30

[29] Deguchi M 2005 Simulation of motorcycle-car collision *International Technical Conference on the Enhanced Safety of Vehicles (ESV)* (Washington (USA): NHTSA)

[30] Wang Q, Lou Y, Jin X, Kong L, Qin C and Hou X 2022 Reverse reconstruction of two-wheeled vehicle accident based on Facet vehicle model and hybrid human model *International Journal of Crashworthiness* **27** 661–76

[31] Himmetoglu S, Acar M, Bouazza-Marouf K and Taylor A J 2010 Whiplash protection by energy absorbing car-seat concepts *Biennial Conference on Engineering Systems Design and Analysis, ESDA2010* vol 3 (Istanbul, Turkey: ASME) pp 311–20

[32] Shokouhfar S 2017 *A Virtual Test Platform for Analyses of Rolling Tyres on Rigid and Deformable Terrains*

[33] Shokouhfar S, Rakheja S and El-Gindy M 2015 Verification of the Part-Composite Approach for Modeling the Multi-Layered Structure of a Rolling Truck Tire *LS-DYNA Users Conference* (Würzburg (Germany): DYNAmore GmbH)
[34] Wong J Y 1978 *Theory of Ground Vehicles* (John Wiley & Sons Inc.)
[35] Perticone A, Barbani D and Baldanzini N 2023 Enhancing rider safety through Personal Protective Equipment: investigating the influence of Powered Two-Wheeler styles on the performance of the Belted Safety Jacket (Reviewed: waiting for approval) *Traffic Inj Prev*
[36] Perticone A, Barbani D and Baldanzini N 2023 Preliminary evaluation of a passive safety device concept for the protection of riders *International Journal of Crashworthiness*
[37] Klug C and Ellway J 2021 *TB 024 - Pedestrian Human Model Certification*
[38] Arunachalam M, Mondal C, Singh G and Karmakar S 2019 Motorcycle riding posture: A review *Measurement (Lond)* **134** 390–9
[39] Smith T, Zellner J and Rogers M N 2006 A three dimensional analysis of riding posture on three different styles of motorcycle *International Motorcycle Safety Conference* (Long Beach (USA))
[40] ISO 13232-6 2005 Motorcycles — Test and analysis procedures for research evaluation of rider crash protective devices fitted to motorcycles — Part 6: Full-scale impact-test procedures
[41] Arunachalam M, Singh A K and Karmakar S 2021 Perceived comfortable posture and optimum riding position of Indian male motorcyclists for short-duration riding of standard motorcycles *Int J Ind Ergon* **83** 1–17
[42] Otte D 1980 A Review of Different Kinematic Forms in Two-Wheel- Accidents-Their Influence on Effectiveness of Protective Measures *Stapp Car Crash Conference* (Troy (USA): SAE International) pp 561–605
[43] Longhitano D, Henary B, Bhalla K, Ivarsson J and Crandall J 2005 Influence of Vehicle Body Type on Pedestrian Injury Distribution *SAE World Congress & Exhibition* (Detroit (USA): SAE International)
[44] Maier S and Fehr J 2023 Accident simulations of a novel restraint safety concept for motorcyclists *International Technical Conference on Experimental Safety Vehicles (ESV)* (Yokohoma (Japan): NHTSA)
[45] Sankarasubramanian H, Göhlich D, Mukherjee S and Chawla A 2012 A bio-mechanics based methodology to optimize vehicle front profile for pedestrian safety *Proceedings of the 10th International Symposium on Biomechanics and Biomedical Engineering* pp 705–10
[46] Otte D 2004 Use of Throw Distances of Pedestrians and Bicyclists as Part of a Scientific Accident Reconstruction Method *SAE World Congress & Exhibition* (Detroit (USA): SAE International) p 246
[47] Piantini S, Pierini M, Delogu M, Baldanzini N, Franci A, Mangini M and Peris A 2016 Injury analysis of powered two-wheeler versus other-vehicle urban accidents *Conference proceedings International Research Council on the Biomechanics of Injury, IRCOBI* (Malaga (Spain): IRCOBI) pp 840–53
[48] Otte D 1999 Severity and mechanism of head impacts in car to pedestrian accidents *Conference proceedings International Research Council on the Biomechanics of Injury, IRCOBI* (Sitges (Spain): IRCOBI)

IOP Conf. Series: Materials Science and Engineering 1306 (2024) 012034 doi:10.1088/1757-899X/1306/1/012034

[49] Ballesteros M F, Dischinger P C and Langenberg P 2004 Pedestrian injuries and vehicle type in Maryland, 1995–1999 *Accid Anal Prev* **36** 73–81

[50] Oh C, Kang Y and Kim W 2008 Assessing the safety benefits of an advanced vehicular technology for protecting pedestrians *Accid Anal Prev* **40** 935–42

[51] Rodarius C, Mordaka J and Versmissen T 2008 *Bicycle safety in bicycle to car accidents*

[52] Li G, Yang J and Simms C 2017 Safer passenger car front shapes for pedestrians: A computational approach to reduce overall pedestrian injury risk in realistic impact scenarios *Accid Anal Prev* **100** 97–110

[53] Kendall R, Meissner M and Crandall J 2006 The Causes of Head Injury in Vehicle-Pedestrian Impacts: Comparing the Relative Danger of Vehicle and Road Surface *SAE World Congress & Exhibition* (Detroit (USA): SAE International)

Robotization of the IR-Thermographic technique – impact on the visualisation quality and considerations on the data workflow

N Montinaro [1], C Mineo [2], A Pantano [1] and D Cerniglia [1]

[1] Department of Engineering, University of Palermo, 90128 Palermo, Italy;

[2] Institute for High-Performance Computing and Networking, National Research Council of Italy, 90146 Palermo, Italy

Corresponding author email: nicola.montinaro@unipa.it

Abstract. Quality control automation is becoming increasingly popular in industrial production lines. Active thermography techniques are well-regarded for their adaptability, providing fast, non-contact, and full-field non-destructive evaluation. Automating thermographic evaluation can increase assessment speed and repeatability without sacrificing inspection accuracy. By using a robot arm to manipulate the thermographic setup, it becomes possible to inspect large components and refine scans on suspicious zones, even in parts with complex geometries. In this study, the performance of a new thermographic inspection platform is compared with a conventional setup to showcase the potential improvements. A plastic curved-shaped sample with artificial flat bottom hole defects is used as a benchmark for the comparison. The advantages and disadvantages of robotizing the infrared non-destructive setup are analyzed, and the impact of the data workflow and future research activities are also discussed.

1. Introduction

Non-destructive tests (NDTs) are the sum of examinations, tests, and surveys accomplished through methods that do not alter the material and do not require the destruction or removal of samples from the structure, aimed at detecting and characterizing flaws. This type of test makes it possible to assess the damage in the examined object without compromising its continued usability [1].

Conventional manual NDT techniques are insufficient in some scenarios since they could cause a manufacturing process bottleneck, and call for skilled experts [2]. Consequently, there exist underlying incentives for augmenting automation in nondestructive testing. These days, it is possible to quickly produce large and complicated geometries by combining cutting-edge additive manufacturing techniques with conventional subtractive methods thanks to computer-aided design (CAD) and computer-aided manufacturing (CAM) [3]. As a result, NDT automation will need to handle the examination of these complex structures. Although manual scanning remains suitable for specific areas of a structure, the industry emphasizes creating reliable automated solutions to cut down on inspection times and costs. An industrial robot is a versatile manipulator that can be autonomously operated and programmed in three or more axes [4]. Some modern robot solutions fall within the range of robot manipulators that are appropriate for creating automated NDT systems. They offer precision mechanics, the ability to master every joint precisely, and the capacity to output positional data at rapid update rates. The following are the primary challenges in implementing a robotic NDT system:

- coupling the robotic manipulator with the NDT instrumentation;
- developing an NDT software for data collection and visualization;
- developing an appropriate robot inspection route for the component under inspection.

To get over these challenges, some six-axis robotic arm applications for automating ultrasonic nondestructive testing (NDT) inspection have been developed [5–11]. Other than these uses, automation has not been attained in other inspection kinds; thermographic infrared (IR) testing is one example of this. Infrared thermography is a non-destructive method used to obtain and visualize thermal images. In active infrared thermography, an IR camera is used to track the surface temperature's temporal evolution while a heat flow is applied to the component being tested. The idea behind defect identification is that various materials possess distinct thermo-physical characteristics, which result in unique thermal signatures that an infrared sensor, like an IR camera, can detect [12–14]. Even though a thermographic setup in reflection mode, with a heat excitation source and an IR camera on the same side of the part under inspection, is not ideal for detecting deep defects, it offers some advantages over ultrasonic-based inspections. Infrared thermography (IR-NDT) is a contactless and full-field technique. It allows for the remote inspection of the entire area of a component within the field of view of an IR camera at once. Over the last few decades, IR-NDT techniques have become widely popular due to the availability of accurate and affordable devices in the market. The primary benefit of robotizing IR thermography is the ability to inspect large, intricately shaped components by capturing multiple thermographic images from specific positions. Additionally, critical parameters for defect detection, including spatial resolution, energetic resolution, thermal energy input, and sampling rate, can be selected without any trade-offs and with greater flexibility.

Initial research on active thermography examination employing industrial robots as manipulators can be found in [15–16]. Detecting and characterizing flaws requires accurate and consistent pixel blending, which makes thermographic picture alignment difficult for NDT examination. Due to incorrect alignment and blending, the composite thermography image may contain erroneous artifacts, which could result in a false-positive defect diagnosis. The authors of [17] tackled the problems of thermographic picture fusion by creating a process for automatically transforming, aligning, and blending thermographic acquisition data. This work demonstrates the improvements that can be achieved with the new robotized IR-thermographic inspection platform. Investigating artificial flat bottom hole flaws on a curved plastic sample allows for a comparison. The benefits and drawbacks of automating the IR-NDT setup are examined, followed by a discussion of the data workflow's impact and next research projects.

2. IR-thermography

Infrared thermography is a non-contact optical imaging technique used to detect infrared radiation. In the field of non-destructive testing (NDT), it is used to assess the structure, monitor the distribution of infrared radiation, and convert the measurements into a temperature scale. The two primary categories of the IR-NDT are often classified as passive and active. In passive ones, anomalies in the temperature distribution point to areas of concern. This method has important applications in several domains, including manufacturing processes, predictive maintenance, medicine, and building thermal efficiency assessment programs. It is mostly utilized for qualitative inspection to identify anomalies.

Active thermography was developed to lessen the susceptibility to environmental factors and produce more accurate findings. In active thermography, a thermal contrast is created in the thermographic image of the part's surface by an external heat source. Since an object with internal anomalies (such as voids, delamination, or inclusions) will need to be excited into a thermal disequilibrium to produce a characteristic thermal signature that can be detected with an IR camera, the latter method is especially well-suited to nondestructive testing (NDT). Various methods have been created based on the kind of external stimuli that are being used. Pulsed (PT), step heating (SHT), and lock-in (LIT) are the most often used thermal stimulation techniques. In Figure 1 a schematic representation of the PT and SHT is shown. In [18] a detailed comparison of the LIT and PT approach are compared to quantitatively characterize artificial flaws in steel and carbon fiber-reinforced plastic. With the same aim to quantitatively characterize flaws in composite material, the quadrupole method is used in [19] in comparison to the PT and thermal signal reconstruction method (TSR).

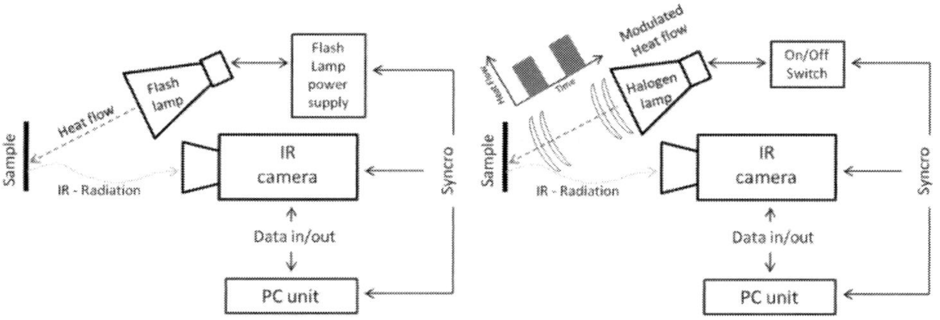

Figure 1. Schematic representation of the traditional active IR thermographic setup for pulsed thermography (on the left) and step heating (on the right).

The geometry, material, and size of the flaw of the being inspected part usually determine the best approach and configuration to be used. Research has indicated that not all defects can be easily identified using the same method; for instance, not all materials react noticeably to the same heating source [20]. PT is the most promising method for robotization because of its simplicity and speed. An alternative IR-thermographic system that holds potential for robotization utilizes a small laser for heat injection instead of a bulkier flash lamp. The use of laser excitations to detect vertical cracks in metal parts and delamination/debonding in composite materials has been experimentally applied in [21–23]. The active IR techniques are also profitably used to inspect the typical flaws in metal additively manufactured parts [24-25].

To minimize environmental effects and enhance the thermal contrast of the defect, PT data is often post-processed. In pulsed phase thermography (PPT), the phase and frequency of the PT acquisitions are extracted through post-processing employing the same principles used in lock-in thermography [14, 26–28]. In particular, flashlights emit quick and high-intensity bursts of heat while the following temperature decline is recorded by the IR camera over a truncation window. Once the raw data has been collected, the Discrete Fourier Transform (DFT) is calculated to find the frequency content of the temperature response. Lastly, the phase of a certain harmonic content can be extracted and shown as a phasegram. The value assigned to each pixel in the image represents the phase of the object. Any sudden change in phase indicates a potential issue or is a result of the object's shape. In the PPT method, deeper anomalies are more clearly visible in the low-frequency phasegrams, while the high-frequency phasegram is better suited for sub-cortical flaws. It is also expected that normalizing the signal when assessing the phase will reduce the negative effects of uneven heat distribution and ambient reflections.

3. Methods

3.1. Data workflow

The robotization of the IR-NDT technique enables precise adjustment of critical inspection parameters (e.g., integration time, sample rate, focal distance) without the typical trade-offs of the conventional technique. This enhanced flexibility leads to improved defect detection and characterization performance on samples of various shapes and materials.

The key parameters to be configured in a PT inspection setup are summarized in Table 1. Figure 2 illustrates the data workflow of the proposed robotized inspection method. The data workflow of the robotized inspection platform begins with defining the input parameters. These parameters are selected based on the anticipated defect type, its shape, and position. Typically, an expert adjusts the IR camera acquisition parameters and the experimental setup geometry to define these critical settings.

Table 1. Main parameters to be set in the IR-thermography robotized setup.

Feature	Physical phenomena involved	Parameter to set
Part material	Thermal wave velocity	Length of acquisition, sample rate (frame per second)
Position of the defect concerning the scanned surface	Thermal wave travel distance	Length of acquisition, integration time
Planar dimension of the defect	Field of view	Spatial resolution, focal distance, lens
Type of defect/inclusion	Wave scattering	Integration time, sample rate (frame per second)
Size of the part	Field of view	Number of poses, scanning path

After defining the input parameters, they are inserted into the offline programming user interface (2) created in MATLAB (refer to [17] for more information). The graphical user interface enables the creation of a raster inspection tool path for the sample, based on the user-specified maximum spacing between consecutive image acquisition poses and offset from the sample edges.

After obtaining the IR videos (one for each pose), the data needs to undergo post-processing for accurate visualization and interpretation. After completing step 3 in Figure 2, the data can follow either the 4A or the 4B workflow direction. The acquired data, which consists of thermographic temperature map evolutions, is processed individually using a PPT algorithm or any alternative method. Subsequently, the data is aligned and corrected in step 5A using the Fine Pixel-based Alignment Method (FiPAM), the details of which can be found in [17]. Alternatively, the acquired temperature map evolutions for each pose can be merged with the FiPAM and then post-processed with a thermographic algorithm such as the PPT. In both cases, the final outcome is a map of the scanned sample surface in step 6, where the defect signature can be easily detected and eventually geometrically characterized.

3.2 Thermographic images fusion

In order to collect data optimally during automated inspections, significant effort is made to ensure that the robot's tool path is accurately aligned with the sample reference frame [29]. Despite these efforts, there will always be a difference between the actual tool path and the ideal one due to several factors:

- Physical tolerances in the robot joints
- Geometric deviations in the mounting support of the sensing instrumentation
- Residual inaccuracies in the calibration of the part's position
- Discrepancies between the actual sample geometry and the digital model of the part.

The resulting data often shows imperfect alignment when encoded through robot positional feedback and plotted on a single map. This issue leads to a clear misalignment of thermographic images in robotic thermographic inspections. In a practical solution presented in [17], each picture's plotting position and potential aberrations are corrected to provide a misalignment-free full-field view of the sample being examined. This method is suitable when the rough position of the camera (the shooting pose of each image) is known. The direct method finds the best alignment between a pair of images through an iterative search, where one image is transformed in relation to the other using one of the five planar transformations (see Figure 3).

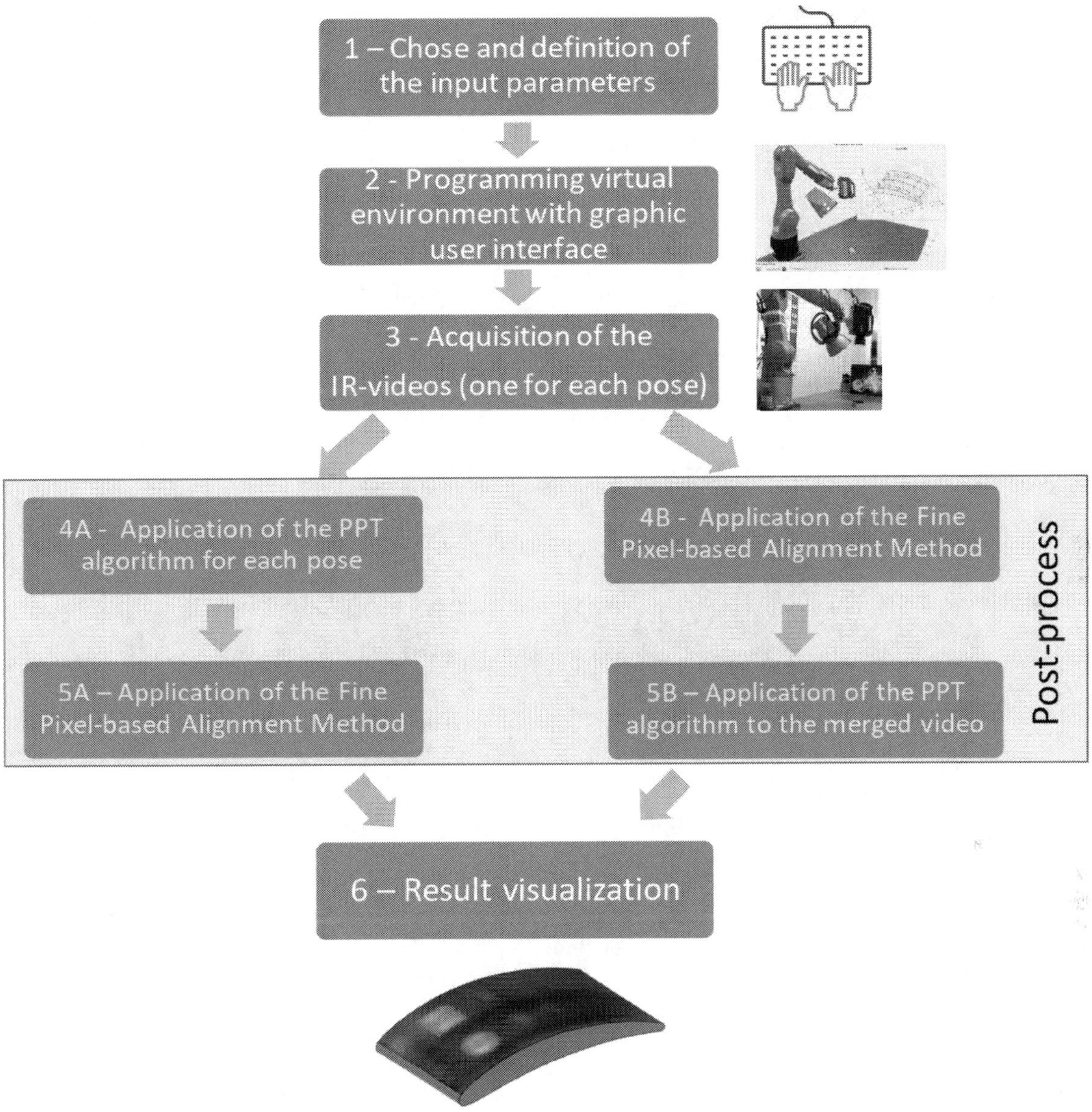

Figure 2. Robotised IR-thermography data workflow.

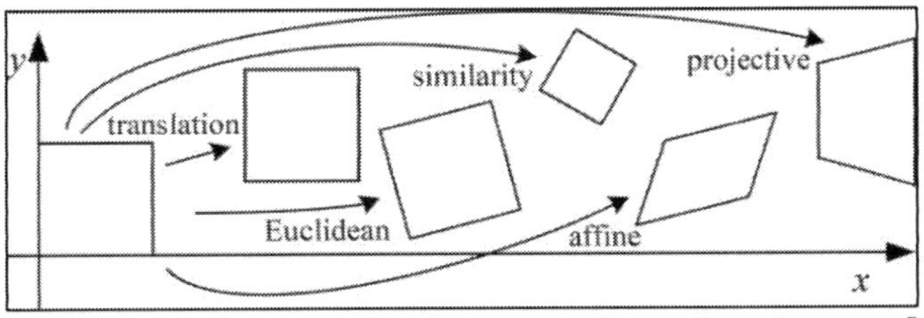

Figure 3. The basic set of 2D planar transformations (source: [30]).

Simply aligning all images in a dataset is not enough to merge them into a single composite image. There may be noticeable variations in pixel brightness across numerous aligned photos where they overlap. To address this, a technique has been developed to preserve the NDT data. The final composite image is created using Laplacian pyramid blending, which allows for a smooth transition between pictures without causing any visual artifacts.

4. Experimental activity

4.1. Sample

The benchmark sample is a piece of bulk epoxy that replicates the curved geometry of a compressor blade with variable thickness. It has one convex side and one concave side, both with a single direction of curvature. On the concave side of the sample, six flat bottom holes (FBHs) were machined as artificial flaws - three with square portions and three with circular sections. As a result, the FBHs are not visible from the sample's convex side. The sample's shape, primary dimensions, and the location and size of the FBHs are shown in Figures 4a and 4b. Additionally, the sample is painted with an acrylic-based, high-temperature resistant matte black paint to enhance the surface emissivity uniformly.

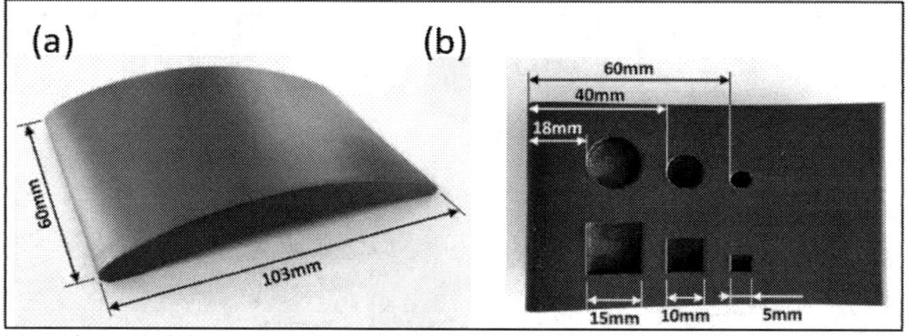

Figure 4. (a) a picture of the curved sample with the indication of its dimensions; (b) a picture of the back side of the sample showing the artificial FBHs location and size;

4.2. Setups

The thermography platform used in this study is illustrated in Figure 5a. The robotic manipulator could carry up to 11.1 kg and had a maximum reach of 110 cm (KUKA KR10 R1100-2 arm [31]). Both the flashlight and the IR camera were positioned on the same side of the component being inspected, as this setup was designed for PT examination in reflection mode. The IR camera and flash lamp were attached to the robot using a custom-made mounting bracket, maintaining a fixed relative position throughout the examination. The support allows for adjusting the orientation of the flash lamp, which enables modification of the angle between the camera axis and the flash lamp's illumination axis. An Elinchrom Twin X4 Lamphead emitting a pulse with a power output of 4800 W/s and a duration of 1/180 s served as a heat source [32]. The adopted IR camera is a cooled FLIR X6540sc equipped with a 640 x 512 (horizontal x vertical) pixel detector [33] and a 50 mm F/2.0 lens. The camera is connected to the PC with a Gigabit Ethernet link for full-bandwidth data acquisition. The MATLAB graphic user interface was used to create a tool path for inspecting a sample. For this particular application a spacing of 25 mm between consecutive poses and a 10 mm offset from the sample edges was set. The inspection path consisted of 15 poses distributed across three passes with five poses each. The Interfacing Toolbox for Robotic Arms (ITRA) [34] was used to control the connection between the computer and the robot.

To evaluate the results obtained by the robotized setup, the same sample has been inspected with a conventional PT experimental setup shown in Figure 4b. The traditional PT setup used the same IR camera and flash lamp as the robotized setup. However, in the robotized setup, the acquisition is sub-windowed to 10% of the full frame while maintaining equal data flow. This allows for a potential increase in sample rate of up to 5.5 times. It's important to note that the focal distance and lens are consistent across both setups. The experimental parameters for both setups are summarized in Table 2.

IOP Conf. Series: Materials Science and Engineering 1306 (2024) 012035 doi:10.1088/1757-899X/1306/1/012035

Table 2. Experimental setup parameters.

	Robotised	Conventional
Int. time	3192 μs	3192 μs
Frame per second	100 Hz	18 Hz
Length acquisition	10 s	10 s
Field of view H x V	224 x 192 pixel	640 x 512 pixel
Focal distance	55 cm	55 cm
Type of post-process	PPT	PPT
Resolution	150 μm/pixel.	150 μm/pixel.

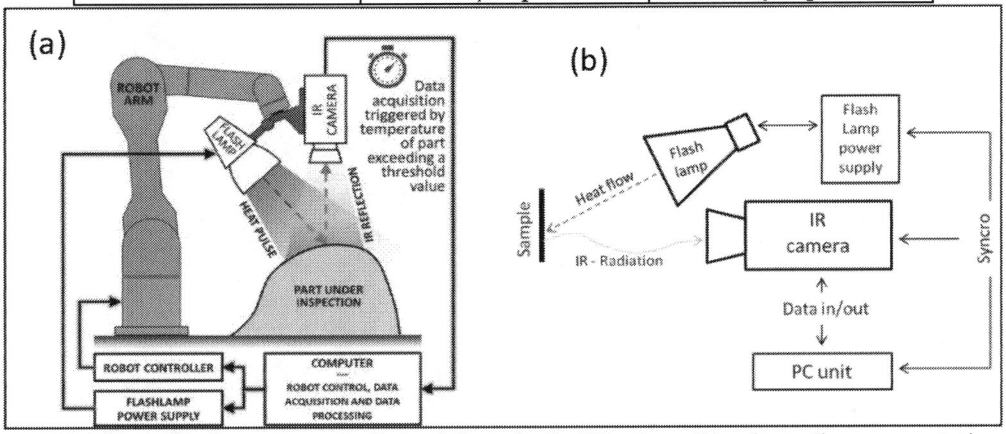

Figure 5. (a) schematic reppresentation of the robotic PT setup; (b) schematic representation of the conventional PT setup used as a comparison.

5. Results and discussion

To assess the platform's performance, we examined the benchmark sample using the same instruments as the traditional PT setup (see Figure 5b) and the robotized setup (see Figure 5a). In both cases, we analyzed the surface temperature's temporal evolution with the PPT algorithm in mind. The data workflow path through 4A and 5A (see Figure 2) is used for this comparison. Figure 6 displays the phasegrams at 0.05 Hz obtained by the PPT algorithm after step 4A of the workflow in Figure 2, but before the FiPAM application 5A. As shown in Figure 6, the original 15 images encoded through the robot poses need alignment and contrast equalization.

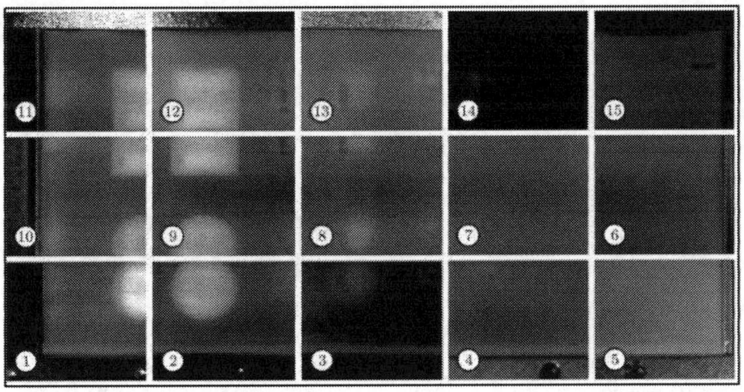

Figure 6. Set of 15 phasegrams at 0.05 Hz before the application of the FiPAM algorithm.

IOP Conf. Series: Materials Science and Engineering 1306 (2024) 012035 doi:10.1088/1757-899X/1306/1/012035

In Figure 7a, the visualization of the composite phasegram obtained after the application of the FiPAM is shown. The similarity transformation, which allows four degrees of freedom (horizontal translation, vertical translation, rotation, and scaling), is proved to be sufficient to permit a fine alignment of the 15 images of the set. In Figure 7b, the post-processed phasegram extracted at 0.25 Hz with the PPT algorithm resulting from the conventional setup is compared.

In Figure 8, the same phasegram of Figure 7a is wrapped onto the CAD model sample surface to increase the clarity of the visualization to a potential NDT operator.

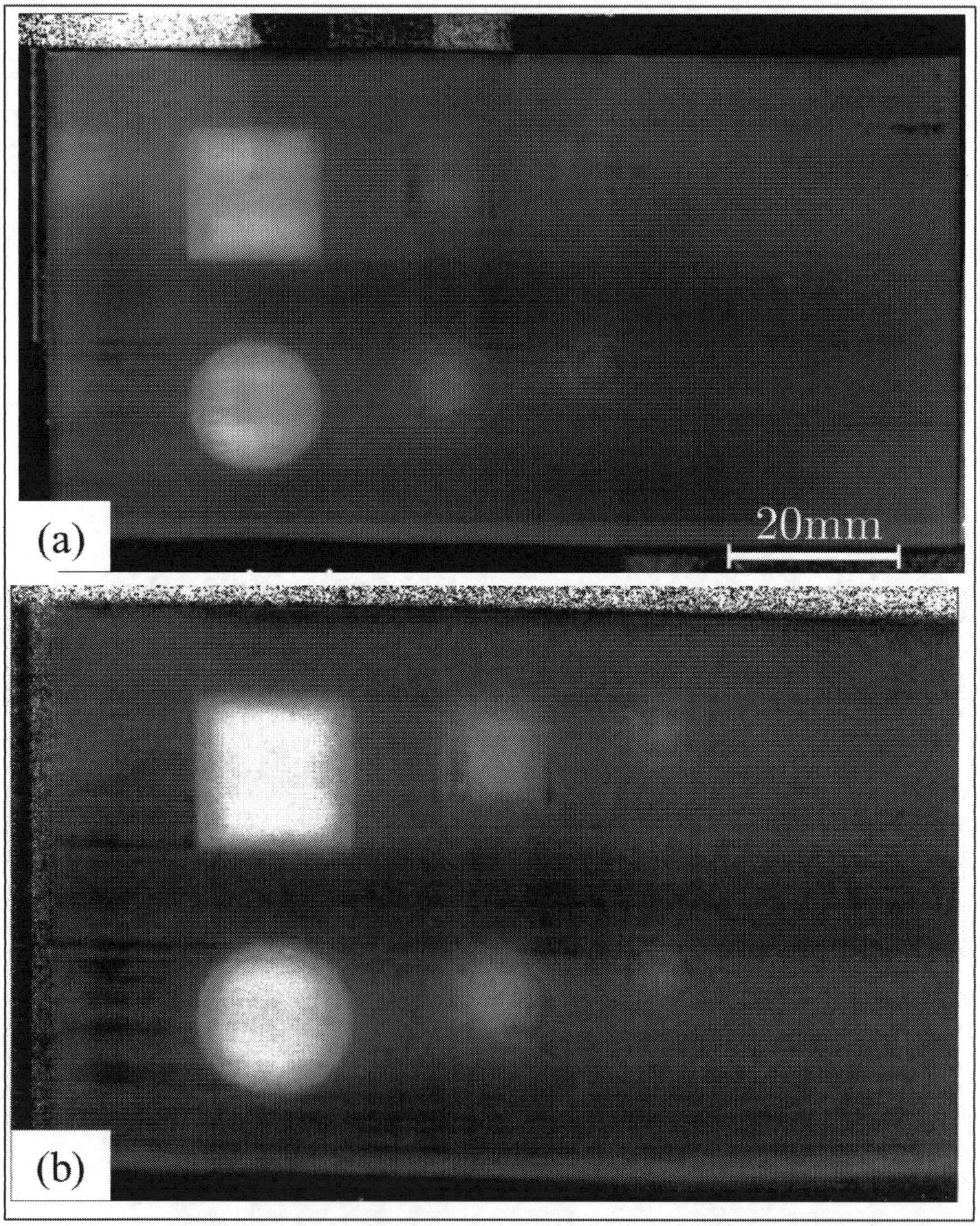

Figure 7. (a) composite phasegram obtained with the robotized setup; (b) phasegram obtained with the conventional setup and same instrumentation.

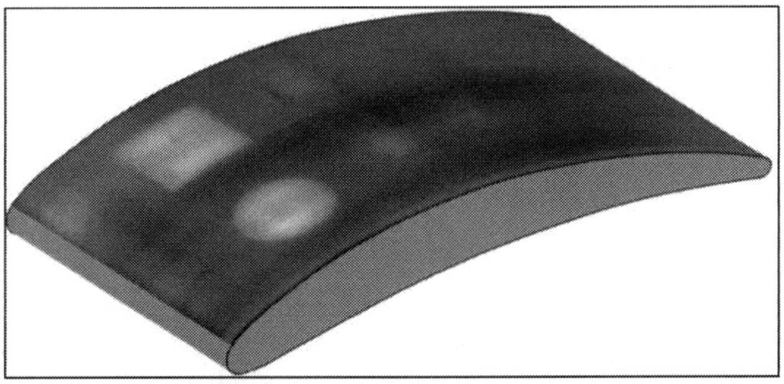

Figure 8. Phasegram of Figure 7a wrapped onto the CAD model of the sample.

The comparison shown on Figure 7 between the two experimental setups shows a clear improvement in the fidelity of reproduction of the position and geometry of the defect. The increased fidelity becomes clearer in the case of smaller defects (both circular or squared). The phasegram coming from the conventional setup with the same PPT algorithm post-process does not allow any geometry characterization but only detection of the defect. Contrariwise, all defects' sizes and positions are faithfully reproduced using the dataset acquired through the robotic platform.

Conventional setup image is then suffering from blurring. This issue is probably due to the fact that only a focal plane can be clearly in focus on the inspection of curved surfaces, thus resulting in a trade-off set. Instead, the robotized inspection allows for reducing the focalization problem, partitioning the whole image, and maintaining the focus in a specific tolerance range. The focalization error decreases with the number of partitions tending to infinity.

The application of the robotic setup does not bring only advantages; in fact, it must be considered that the scanning time for the given area increases linearly with the number of poses into which it is partitioned. Furthermore, the image processing time due to the FiPAM algorithm must be considered. In [17], the computational time for applying the FiPAM was studied.

It must be noted that the presented comparison is performed following the data workflow 4A → 5A (see Figure 2) but different scenarios could be expected in the case of 4B → 5B workflow. In this latter workflow, the temporal evolution of the temperature acquired in the *n* poses will be aligned and merged in a single file and subsequently processed using a PPT algorithm (or another one). This alternative approach would reduce the number of pixels to be processed with the PPT algorithm (which is the most computationally expensive) since the overlapping portions of images are processed only once. The analysis in terms of computational cost and quality of this approach's results will be studied soon. Moreover, future work will investigate on the optimization of the processing algorithms allowing to control the inspection in real time. A data-driven control of the platform would increase the capability of the system to adapt to the part geometry and material acquiring high-quality thermographic data.

References

[1] Hull, JB, Vernon J 2012 Non-Destructive Testing *Springer Nature*

[2] Sattar, TP 2010 Robotic Non-Destructive Testing *Industrial Robot: An International Journal* **37**(5)

[3] Gibson I, Rosen DW, Stucker B, Khorasani M 2021 Additive Manufacturing Technologies *Springer*

[4] Appleton E, Williams DJ 2012 Industrial Robot Applications *Springer Science & Business Media*

[5] Mineo C, Pierce SG, Wright B, Cooper I, Nicholson PI 2015 Paut Inspection of Complex-Shaped Composite Materials through Six Dofs Robotic Manipulators *Insight-Non-Destructive Testing and Condition Monitoring* **57** 3 161-66

[6] Mineo C, Pierce SG, Nicholson PI, Cooper I 2016 Robotic Path Planning for Non-Destructive

Testing–a Custom Matlab Toolbox Approach *Robotics and Computer-Integrated Manufacturing* **37** 1-12

[7] Mineo C, Cerniglia D, Ricotta V, Reitinger B 2021 Autonomous 3D geometry reconstruction through robot-manipulated optical sensors *International Journal of Advanced Manufacturing Technology* **116** (5-6) 1895-1911

[8] Mineo C, Cerniglia D 2023 Adding Autonomy to Robotic Enabled Sensing *Lecture Notes in Civil Engineering* **270** LNCE 338 - 347

[9] Bosse J, Thaler B, Ilse D, Bühling L Automated Air-Coupled Ultrasonic Technique for the Inspection of the Ec145 Tail Boom *NDT in Aerospace 2012* - Tu.2.B.3

[10] Cooper I, Nicholson I, Yan D, Wright B, Liaptsis D, Mineo C 2013 Development of a Fast Inspection System for Aerospace Composite Materials-the Intacom Project *5th International Symposium on NDT in Aerospace 2013-11-13 - 2013-11-15*

[11] Mineo C, Cerniglia D, Poole A 2022 Autonomous Robotic Sensing for Simultaneous Geometric and Volumetric Inspection of Free-Form Parts *Journal of Intelligent and Robotic Systems: Theory and Applications* **105** (3) 54

[12] Meola C 2012 Infrared Thermography Recent Advances and Future Trends Bentham Science Publishers

[13] Holst GC 2000 Common Sense Approach to Thermal Imaging Vol. 1 SPIE Optical Engineering Press Washington

[14] Maldague X 2001 Theory and Practice of Infrared Technology for Nondestructive Testing John Wiley & Sons Inc

[15] Schmidt T, Dutta S 2012 Automation in Production Integrated Ndt Using Thermography *NDT in Aerospace, Augsburg* **8**

[16] Massaro A, Galiano A 2020 Infrared Thermography for Intelligent Robotic Systems in Research Industry Inspections: Thermography in Industry Processes *Handbook of Research on Advanced Mechatronic Systems and Intelligent Robotics*, 98-125 IGI Global

[17] Mineo C, Montinaro N, Fustaino M, Pantano A, Cerniglia D 2022 Fine Alignment of Thermographic Images for Robotic Inspection of Parts with Complex Geometries *Sensors* **22** (16) 6267

[18] Maierhofer C, Röllig M, Krankenhagen R, & Myrach P 2016 Comparison of quantitative defect characterization using pulse-phase and lock-in thermography *Applied optics* **55**(34).

[19] Müller J P, Dell'Avvocato G, & Krankenhagen R 2020 Assessing overload-induced delaminations in glass fiber reinforced polymers by its geometry and thermal resistance. *NDT & E International* **116** 102309

[20] Shull PJ 2002 Nondestructive Evaluation: theory, techniques, and applications New York: Marcel Dekker Ed

[21] Montinaro N, Cerniglia D, Pitarresi G 2017 Flying Laser Spot Thermography technique for the NDE of Fibre Metal Laminates disbonds *Composite Structures* **171** 63–76

[22] Montinaro N, Cerniglia D, Pitarresi G 2018 Evaluation of interlaminar delaminations in titanium-graphite fibre metal laminates by infrared NDT techniques *NDT and E International*, **98** 134–146

[23] Montinaro N, Cerniglia D, Pitarresi G 2019 Evaluation of Vertical Fatigue Cracks by Means of Flying Laser Thermography *Journal of Nondestructive Evaluation*, **38**(2)

[24] D'Accardi E, Krankenhagen R, Ulbricht A, Pelkner M, Pohl R, Palumbo D, & Galietti U 2022 Capability to detect and localize typical defects of laser powder bed fusion (L-PBF) process: An experimental investigation with different non-destructive techniques. *Progress in Additive Manufacturing* **7**(6), 1239-1256

[25] Montinaro N, Cerniglia D, Pitarresi G 2018 A numerical and experimental study through laser thermography for defect detection on metal additive manufactured parts *Frattura ed Integrità Strutturale* **12** (43) 231-240

[26] Maldague X, Marinetti S. 1996 Pulse Phase Infrared Thermography *Journal of applied physics*

79(5) 2694-98

[27] Maldague X, Galmiche F, Ziadi A 2002 Advances in Pulsed Phase Thermography *Infrared physics & technology* **43**(3-5) 175-81

[28] Ibarra-Castanedo C, Maldague X 2004 Pulsed Phase Thermography Reviewed *Quantitative Infrared Thermography Journal* **1**(1) 47-70

[29] Mineo C, Cerniglia D, Poole A 2022 Autonomous Robotic Sensing for Simultaneous Geometric and Volumetric Inspection of Free-Form Parts *Journal of Intelligent & Robotic Systems* **105**(3) 54

[30] Paragios N, Chen Y, Faugeras OD 2006 *Handbook of Mathematical Models in Computer Vision: Springer Science & Business Media*

[31] KUKA Kr 10 R1100-2. https://www.kuka.com/-/media/kuka-downloads/imported/6b77eecacfe542d3b736af377562ecaa/0000290003_en.pdf?rev=3e82b095d46c4a86b7e195cabdb980cb&hash=A4F0F7D35D9485B612F8BC455B5805E4 (accessed 02/04/2024)

[32] Elinchrom Digital Rx. https://www.elinchrom.com/wp-content/uploads/download-center/73256_digital_rx_manuel--en-de-fr.pdf (accessed 02/04/2024)

[33] FLIR X6540sc http://www.flirmedia.com/MMC/THG/Brochures/RND_088/RND_088_US.pdf (accessed 02/04/2024)

[34] Mineo C et al.. 2020 Enabling Robotic Adaptive Behaviour Capabilities for New Industry 4.0 Automated Quality Inspection Paradigms *Insight-Non-Destructive Testing and Condition Monitoring* **62**(6) 338-44

IOP Conf. Series: Materials Science and Engineering 1306 (2024) 012036 doi:10.1088/1757-899X/1306/1/012036

Impact of sand feed rate on the damage of railway wheel steels

N Zani[1,*], K Shu[2], L Ghidini[1], C Petrogalli[1] and A Mazzù[1]

[1] Department of Mechanical and Industrial Engineering, University of Brescia, Brescia, 25123, Italy
[2] College of Mechanical Engineering, Hunan Institute of Science and Technology, Yueyang, 414006, China

* E-mail: nicola.zani@unibs.it

Abstract. The global railway industry plays a pivotal role in economic development, offering efficient transportation solutions. However, railways operating in desert environments face unique challenges due to windblown sand. This study investigates the influence of sand feed rates on wheel-rail wear in desert conditions. Experimental tests were conducted using a bidisc apparatus to simulate sand feed rates. Results indicate that low feed rates lead to spalling and pitting, while high rates increase abrasive and fatigue wear. A critical transition point at 0.4 g/min suggests sand-induced abrasion of wheel surfaces. Moreover, the research highlights the crucial role of sand feed rates not only in wear but also in surface roughness, further emphasizing the complex interplay between sand transport rates, adhesion, and wear mechanisms. These insights provide valuable guidance for mitigating wear-related challenges in desert railway operations and optimizing maintenance strategies.

1. Introduction

The global railway industry has played a pivotal role in driving industrialisation and economic development for well over a century. Railway networks have provided unmatched advantages, such as substantial transportation capacity, cost-efficiency, energy conservation, environmental sustainability, and safety. This enduring legacy has led to continual expansion, resulting in the establishment of extensive railway networks. Among the challenging environmental conditions and extreme landscapes, deserts stand out as the most significant technical challenge that new railways must address.

A significant concern in these regions is the pronounced damage and wear experienced by railway materials, surpassing what is observed in dry environments [1–4]. The gravity of this issue often becomes evident only after extended periods of service. This complexity arises from the dynamic and multifaceted nature of windblown sand environments, subject to constant changes driven by factors such as landform features and weather fluctuations. Furthermore, various factors, including the accumulation of sand on rail surfaces following sandstorms, issues affecting rail ballast due to dynamic windblown sand phenomena and sand infiltration can influence the wear and tear on wheels and rails.

A particular area of research has been dedicated to studying the influence of sand feed rate on wear, damage, and adhesion at the interface between the railway wheel and rail. Several papers showed that, when there is no liquid lubricant at the contact interface, the increase of sand feed rate has been associated with decreasing adhesion, primarily due to solid lubrication effects [2,4–10]. This phenomenon is caused by sand, which notably reduces the overall friction coefficient by reducing tangential forces. This reduction results from both the separation of wheel and rail surfaces and the sand grains' rolling motion.

Experiments conducted using small-scale bidisc apparatus have elucidated that the sand feed rate notably affects damage patterns and wear regimes. Under conditions of low sand feed rates (up to approximately 2.5 g/min) [6,9], predominant surface damage phenomena manifest as spalling and pitting, with the former exhibiting decreased prominence as transport rate conditions escalate. It is crucial to emphasise that slow feeding rates are directly linked to adhesive and oxidative wear. The acceleration of black oxide formation is attributed to an augmented quantity of pits induced by sand particles infiltrating the wheel-rail interface. Noteworthy findings include cross-sectional analyses revealing a thin oxide layer hosting entrapped sand particles within.

Conversely, higher sand feed rates are unequivocally correlated with escalated wear rates for both rail and wheel materials, regardless of adhesion conditions [2]. The wear mechanisms observed encompass abrasive and fatigue phenomena [1.11], resulting in a surface layer characterised by pronounced ratcheting due to the presence of sand fragments and wear debris, as well as a subsurface layer exhibiting less severe plastic deformation [2,12–13]. It is essential to highlight that these damage features become significantly more pronounced as sand feeding rates increase, emphasising the complex interplay between sand, adhesion, and wear mechanisms [14]. Moreover, the influence of sand extends beyond the realms of adhesion and wear, introducing erosive wear. This adds an additional layer of complexity to the dynamic interplay. As such, it is imperative to comprehensively examine how sand transport rates exert influence over wheel and rail wear and damage in windblown sand environments.

This work embarks on an exhaustive exploration of these challenges within the context of desert windblown sand environments. This study seeks to closely examine how varying rates of sand feed rate impact the damage characteristics and wear of the materials. We will achieve this objective by employing a two-disc testing machine. Our aim is to clarify the points where the shift occurs between oxidative and abrasive wear patterns.

2. Experimental setup

The experimental tests described here were carried out using an advanced bi-disc apparatus with high performance capabilities. Figure 1 provides an illustration of the setup used in these experiments. The apparatus was equipped with two mandrels, one of which could move along linear slides. These mandrels were powered by two separate alternating current motors (33 kW), capable of achieving speeds of up to 1200 rpm. A servo-hydraulic actuator with a capacity of 75 kN was employed to apply the contact load. Information regarding the normal load and torque was gathered at a frequency of 5 Hz using a data acquisition system from National Instruments.

The cylindrical test specimens were manufactured from the railhead and wheel-tread surfaces, with a thickness of 15 mm and a diameter of 60 mm. Given their cylindrical shape, the contact track width matched the specimen thickness. In this work, the wheel and rail steels chosen were UNI EN ER8 and 900A, respectively. Their mechanical properties and chemical compositions are listed in Table 1. The tests were carried out under the theoretical Hertzian pressure (1100 MPa), creepage (0.0314 m/s) and average rolling speed (500 r.p.m.). All the tests were performed for 40×10^4 cycles. The experiments were conducted under both clean and solid contaminated conditions. Silica sand was used in these tests; the primary component of the sand is SiO_2. The maximum granulometry of the particles was 0.500 mm. The tested sand feed rate ere 0 (clean condition), 0.1, 0.2, 0.4, 1 and 3 g/min.

During the tests, the friction coefficient of every test was determined from torque measurement. After the experiments, the samples were ultrasonically cleaned and weighed by a precision balance with a resolution of 0.001 g and the mass loss was calculated. The surface analysis was used by a laser scanning confocal microscope (LSCM, OLS5000-SAF, Japan) and a scanning electron microscope (SEM, TESCAN MIRA LMS, Czech Republic) equipped with an energy disperse spectroscopy (EDS). The wheel samples were also sectioned along the mid plane orthogonally to their axis; the section was ground, polished and etched with 2% Nital and examined with a LEICA DM6000M light optical microscope.

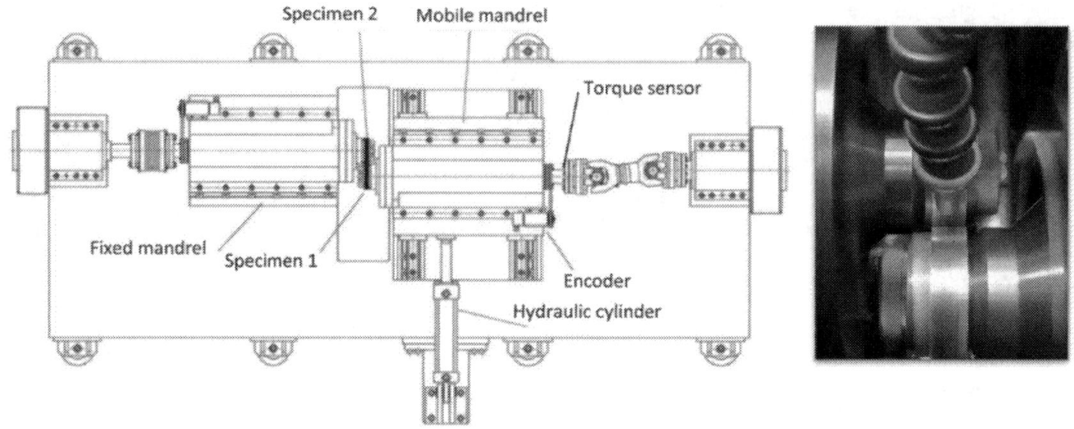

Figure 1 – Bi-disc testing machine.

Table 1 – Mechanical properties and chemical composition of the samples' steels.

		ER8	900 A
Chemical composition [wt%]	**C**	0.52	0.71
	Mn	0.74	1.07
	Cr	0.22	0.03
	Ni	0.16	0.02
	Mo	0.031	0.01
	Si	0.37	0.27
Ultimate tensile strength [MPa]		940	930
Yield strength [MPa]		590	470
Elongation [%]		17	14
Vickers hardness		298	263

3. Results

3.1. Friction coefficient and wear rate

The experimental friction coefficient was calculated by dividing the measured torque by the normal load and the sample's radius. Figure 2 shows the stabilised friction coefficients according to the sand feed rates. Compared to the value around 0.55 under clean conditions, the friction coefficient shows a descending trend to about 0.21 under the highest tested feed rate (3 g/min).

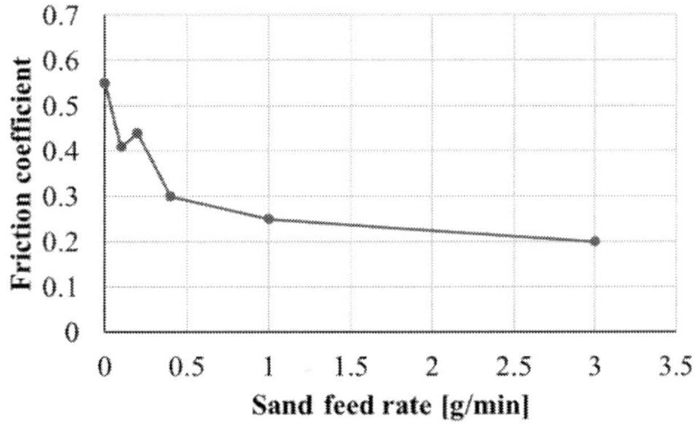

Figure 2 – Friction coefficients with respect to sand feed rate.

Figure 3 depicts the change in wear rate, measured as the amount of mass loss per unit distance, concerning the sand feed rate. There are two distinct phases observed. When the sand feed rate is below 0.4 g/min, there is a transient pattern characterised by an increase in the wear rate of the wheel and a decrease in the wear rate of the rail. On the other hand, when the sand feed rate equals or exceeds 0.4 g/min, the wheel wear rate begins to decline, while the rail wear rate starts to rise, eventually reaching a stabilised level. This change in behaviour at higher feed rates can be attributed to sand particles embedded in the wheel sample and to abrasion on the surface of the rail.

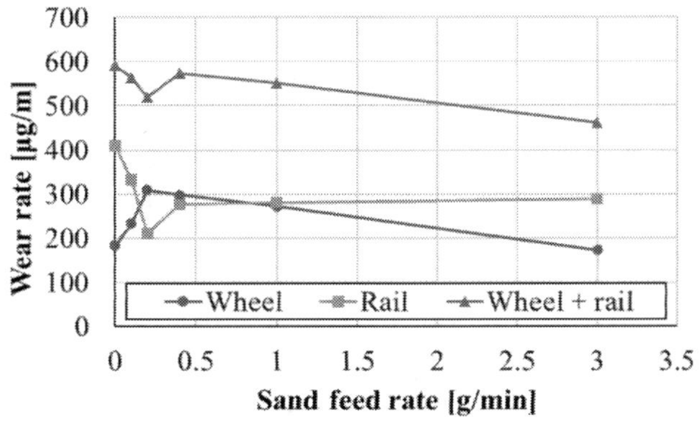

Figure 3 – Wear rate as a function of sand feed rate.

3.2. Surface damage and roughness

Figure 4 shows the surface roughness of worn wheel and rail specimens respect to sand feed rate. Both the surface roughness of rail and wheel kept relatively stable between 0 and 0.4 g/min, and subsequently presented a sharply increasing trend. It also indicated that the sand feed rate of 0.4 g/min was a critical point when the sand started to affect the surface roughness of wheel and rail evidently. Furthermore, the surface roughness of wheel was also very close to that of rail.

IOP Conf. Series: Materials Science and Engineering 1306 (2024) 012036 doi:10.1088/1757-899X/1306/1/012036

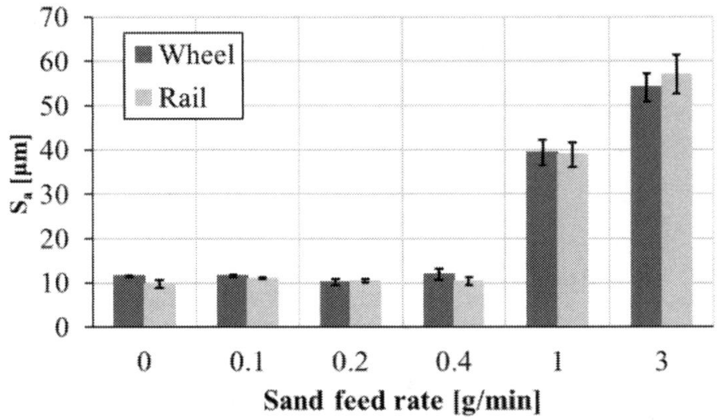

Figure 4 – Arithmetical mean height S_a as a function sand feed rate for wheel and rail samples.

Figure 5 illustrates the macroscopic damage patterns on the wheel surface under varying sand feed rates. Specifically, in Figure 4a, it's noticeable that when the sand feed rate was below 0.1 g/min, the predominant surface damage feature on the wheel was clearly fatigue cracks. However, as the sand feed rate increased, the prominence of fatigue cracks diminished, giving way to the formation of grooves. To gain insight into the dimensions of these macroscopic damage patterns, a three-dimensional profile of the surface morphology was measured and is presented in Figure 6. The data reveal a clear trend: as the sand feed rate rose, both the depth and width of the grooves increased. Moreover, it became increasingly evident that the overall surface condition deteriorated with higher sand feed rates.

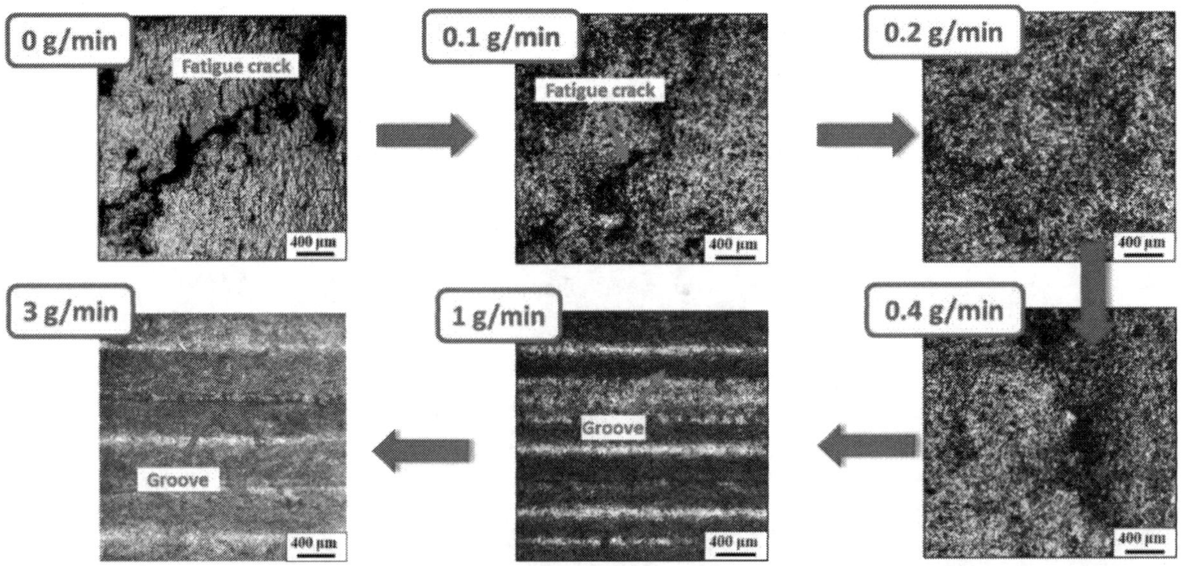

Figure 5 – Optical micrographs of the worn wheel surfaces.

Figure 7 displays the SEM images of the wheel surfaces after wear tests with three different sand feed rates (0, 0.2, 3 g/min). In Figure 6a, when no sand was introduced, aside from noticeable fatigue cracks, the wheel surface damage also featured clear furrows caused by abrasive wear from debris. Figures 7b and c illustrate that as the sand feed rate increased, the surface damage included not only fatigue cracks and furrows but also unevenly distributed oxides. Interestingly, the level of oxidation at the sand feed rate of 0.2 g/min was more severe than that at the sand feed rate of 3 g/min. This suggests

IOP Conf. Series: Materials Science and Engineering 1306 (2024) 012036 doi:10.1088/1757-899X/1306/1/012036

that the oxidative wear of the wheel initially intensified and then diminished as the sand feed rate increased. Concurrently, fatigue wear decreased as the sand feed rate rose. Notably, abrasive wear was consistently present under all conditions.

Figure 8 shows the EDS analysis of worn wheel surfaces under three grades of sand feed rate (0, 0.2, 3 g/min). The elements Fe, O, and Si were detected on the worn wheel surfaces under all conditions. However, the proportion of these three elements varied significantly with different sand feed rates. The Si content on the worn wheel surfaces exhibited an upward trend as the sand feed rate increased. Additionally, it indicated that the quantity of sand remaining on the worn wheel surface increased with the rise in sand feed rate, which could potentially impact the measurement of the actual mass loss of the metal material.

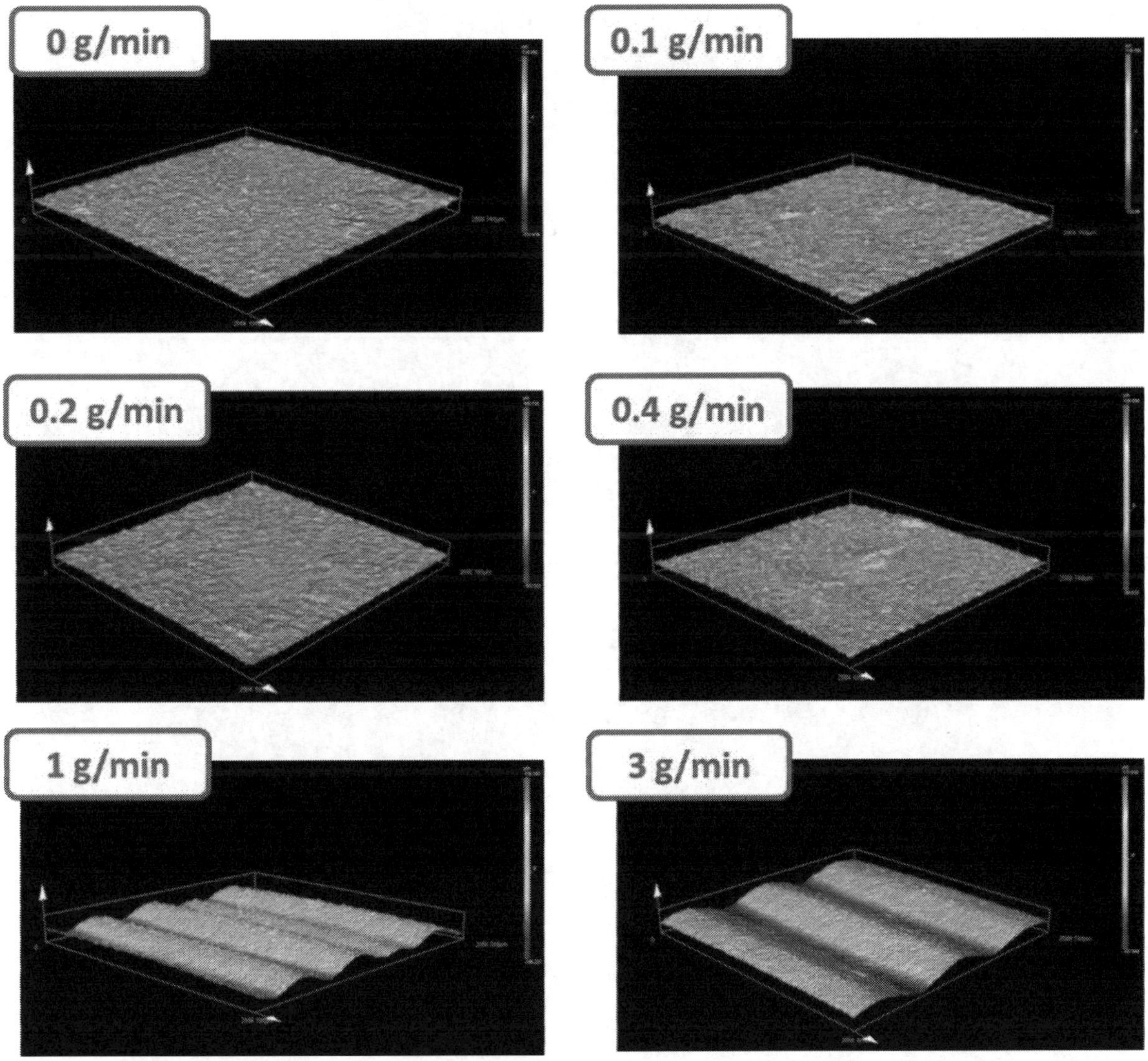

Figure 6 – Macroscopic damage morphology and three-dimensional profile of wheel surfaces with respect to sand feed rate.

IOP Conf. Series: Materials Science and Engineering 1306 (2024) 012036 doi:10.1088/1757-899X/1306/1/012036

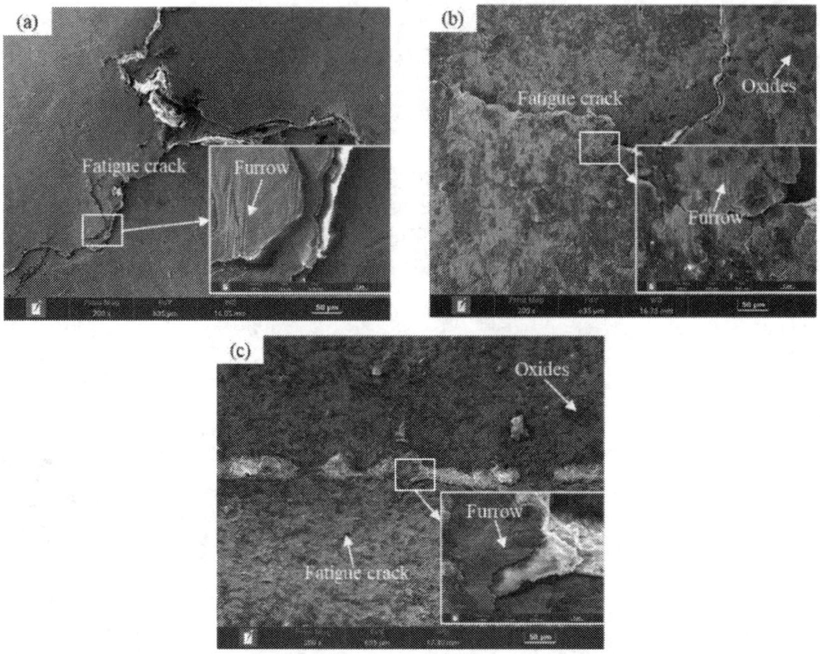

Figure 7 – SEM morphology of the wheel surfaces under different sand feed rates: (a) 0 g/min; (b) 0.2 g/min; (c) 3 g/min.

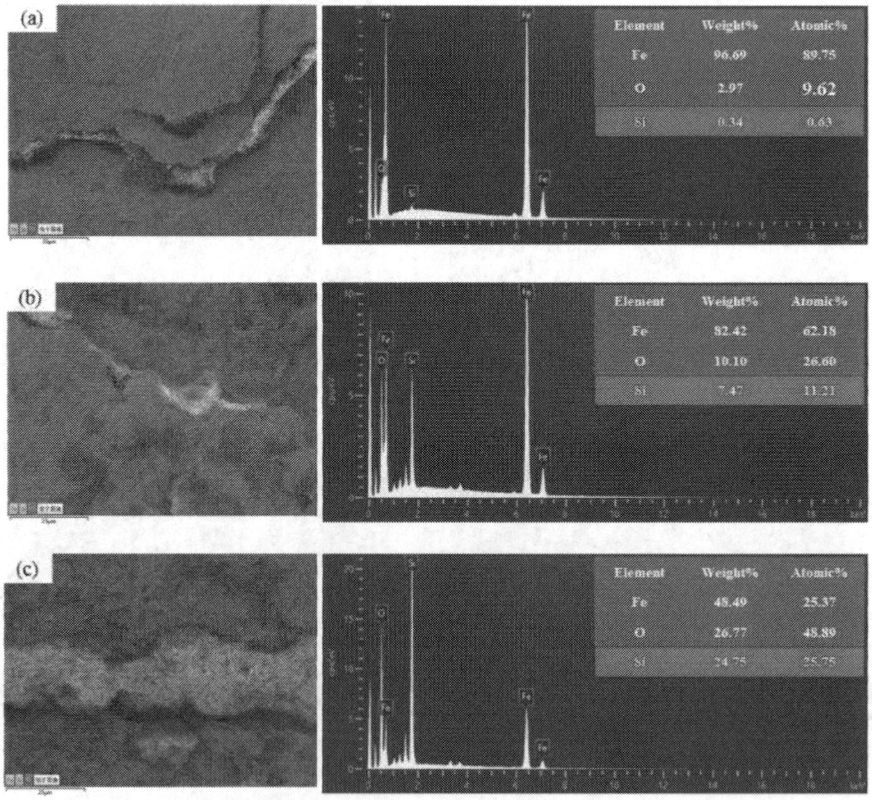

Figure 8 – EDS analysis of wheel surfaces under different sand feed rates: (a) 0 g/min; (b) 0.2 g/min; (c) 3 g/min.

3.3. Subsurface damage

When contact friction is present, the steel microstructure exhibits elongation along shear bands aligned with the friction direction. The plasticized depth is commonly defined as the distance from the surface to the point where the grains in the subsurface have not experienced deformation (see Figure 9). Figure 10 shows the depth of plasticisation in the wheel samples as it changes with varying sand feed rates. Up to 0.2 g/min, this parameter experiences a reduction. This finding could be somewhat deceptive if examined in isolation, as it suggests increased resistance to ratcheting when small amounts of sand are introduced into the contact area. However, it is worth noting that the wear rate increases under these conditions, leading to a reduction in deformed layer depth. When the feed rate exceeds 0.4 g/min, the depth of the plasticised layer increases, even though the wear rate decreases. This could be related to different wear damage phenomena when higher feed rates are applied: two distinct plastically deformed layers can be observed. The first is a surface layer exhibiting intense plastic flow and ratcheting caused by grooves resulting from wear debris and sand trapped inside the contact patch. The second is a deeper surface layer with less intense ratcheting. Figure 11 illustrates a cross-section of the sample tested at 3 g/min, with SEM-EDS analysis results showing crushed particles embedded in the surface disc.

Figure 9 – Example of optical micrograph and detail of the plasticised layer (wheel sample, 0.2 g/min).

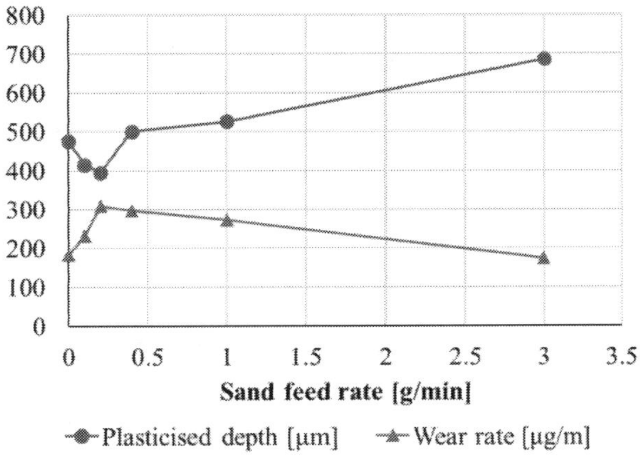

Figure 10 – Plasticised depth correlated to wear rate according to sand feed rate.

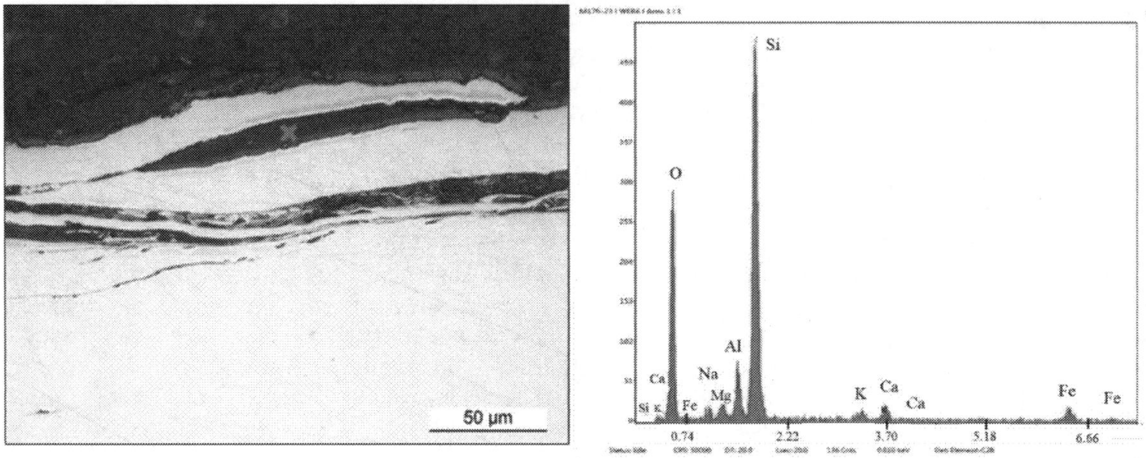

Figure 11 – On the left, optical micrograph of cross section wheel sample; on the right, EDS analysis of the material entrapped in the surface layer (ER8, 3 g/min).

4. Conclusions

The study examined the impact of sand feed rate on the damage and wear characteristics of railway wheel steel using bi-disc tests. Comparative tests were also conducted without sand. The study's findings lead to the following conclusions:

1. The friction coefficient displayed a consistent decreasing trend and reached an asymptotic value when the sand feed rate exceeded 1 g/min.
2. Fatigue wear was the main damage phenomenon in clean condition. Fatigue wear and oxidative wear were the primary wear mechanisms observed when the sand feed rate was less than 0.4 g/min. However, above this threshold, abrasive wear became the predominant mode of wear.
3. The oxide layers and the surface cracks contained randomly distributed sand particles.
4. Sand feed rates exceeding 0.4 g/min worsened surface damage on the samples, leading to the formation of grooves with increasing depth and width as the feed rate increased.
5. The plasticized depth exhibited a significant increase when the feed rate exceeded 0.4 g/min.

More work will be required to investigate whether these outcomes might be influenced by the sand particles granulometry and the geographical provenance of the sand.

References

[1] Arias-Cuevas O, Li Z and Lewis R 2011 A laboratory investigation on the influence of the particle size and slip during sanding on the adhesion and wear in the wheel-rail contact *Wear* **271** 14–24.

[2] Faccoli M, Petrogalli C, Lancini M, Ghidini A and Mazzù A 2018 Effect of desert sand on wear and rolling contact fatigue behaviour of various railway wheel steels *Wear* **396–397** 146–161.

[3] Wang W J, Lewis R, Yang B, Guo L C, Liu Q Y and Zhu M H 2016 Wear and damage transitions of wheel and rail materials under various contact conditions *Wear* **362–363** 146–152.

[4] Zani N and Petrogalli C 2022 Predictive maps for the rolling contact fatigue and wear interaction in railway wheel steels *Wear* **510–511** 204513.

[5] Gutsulyak D, Stanlake L J E and Qi H 2021 Twin Disc Evaluation of Third Body Materials in the Wheel / Rail Interface *Tribol. - Mater. Surfaces Interfaces* **15:2** 339–349.

[6] Shu K, Wang W, Ding H, Lin Q, Meli E, Guo J, Mazzù A and Liu Q 2022 Influence of sand transport rate on rolling wear and damage behaviors of wheel/rail in Gobi and desert windblown sand environments *Tribol. Int* **172** 107584.

[7] Gallardo-Hernandez E A and Lewis R 2008 Twin disc assessment of wheel/rail adhesion *Wear* **265:9–10** 1309–1316.

[8] Omasta M, Machatka M, Smejkal D, Hartl M and Křupka I 2015 Influence of sanding parameters on adhesion recovery in contaminated wheel–rail contact *Wear* **322–323** 218–225.

[9] Wang W J, Zhang H F , Wang H Y, Liu Q Y, and Zhu M H 2011 Study on the adhesion behavior of wheel/rail under oil, water and sanding conditions *Wear* **271:9–10** 2693–2698.

[10] Zani N, Chaise T, Ghidini A, Faccoli M and Mazzù A 2021 Numerical study about the effect of bainitic traces on plasticity in ferritic-pearlitic railway wheels *Proc. Institutions Mech. Eng. Part F J. Rail Rapid Transit* **235(6)** 726–740

[11] Grieve D G, Dwyer-Joyce R S, and Beynon J H 2001 Abrasive wear of railway track by solid contaminants *Proc. Inst. Mech. Eng. Part F J. Rail Rapid Transit* **215:3** 193–205.

[12] Mazzù A, Ghidini A, Zani N and Faccoli M 2021 A simplified numerical study of wheel/rail material coupling in presence of solid contaminants *Tribol. - Mater. Surfaces Interfaces* **15:2** 102–114.

[13] Mazzù A, Ghidini A, Zani N and Faccoli M 2018 Study of wheel/rail material coupling in presence of solid contaminants *Proc. 11th Int. Conf. Contact Mech. Wear Rail/wheel Syst (CM2018)* (Delft: the Netherlands) pp 701–710.

[14] Huang W, Cao X, Wen Z, Wang W, Liu Q, Zhu M and Jin X 2017 A Subscale Experimental Investigation on the Influence of Sanding on Adhesion and Rolling Contact Fatigue of Wheel/Rail under Water Condition *ASME J. Tribol.* **139:1** 011401.

IOP Conf. Series: Materials Science and Engineering 1306 (2024) 012037 doi:10.1088/1757-899X/1306/1/012037

Parametric Analysis on the Static and Modal Response of Folded Metamaterials

P.M. Santucci and A. Baldi

University of Cagliari, Department of Mechanical, Chemical and Materials Engineering, Via Marengo 2, 09123 Cagliari, Italy

E-mail: pietrom.santucci@unica.it

Abstract. Metamaterials have been studied and analyzed in the past three decades because of their outstanding properties. Generally speaking, a metamaterial is a material that exhibits a mechanical behavior that does not depend only on the bulk material but also on the geometrical configuration in which it lies. This aspect leads to the possibility of tuning and engineering the structural response. One of the most interesting properties is the auxetic behavior of metamaterial. An auxetic material shows a global negative Poisson's ratio. Shock absorption, acoustic dissipation, and shape morphing are some of the most popular employment for auxetic materials. In this article, we focus on the response of folded material under static and dynamic load conditions. Folded materials consist of folding a sheet under specific geometrical constraints. One of the most famous is the Miura-ori pattern, which comes from the origami-folding technique. The geometrical parameters, such as folding angles and edge lengths, play a fundamental role in achieving the desired auxetic behavior. These geometrical parameters define a unit cell that can be stacked into a periodic structure. This article proposes an experimental parametric study of the thickness impact on the auxetic behavior while edge dimensions and folding angles are fixed. The geometrical complexity of the pattern forced us to use additive manufacturing for the specimen fabrication. In particular, we choose Fused Filament Fabrication (FFF) using polymers like ABS and PLA. Digital Image Correlation (DIC) is used for monitoring the displacement and strain fields onto the Miura-ori surface under tensile load. Finally, Time Averaged Speckle Interferometry is employed for evaluating the modal response by using a quasi-full out-of-plane sensitivity setup.

1. Introduction

Metamaterials have been the main focus of a solid mechanics branch for several decades. Their name suggests the potential for engineering application: the challenge is the mechanical response control by working on the geometrical configuration rather than the bulk material. In this landscape, the auxetic behavior is still one of the most searched and studied [1–4]. Examples of "naturally" auxetic materials are biological tissues [5–7], cubic metals [8] and several crystals [9, 10]. The auxetic response can be achieved using several strategies, such as two or three-dimensional lattices [11, 12], reentrant structures [13–16], structures with cut slots [17–20], or custom-shape inclusions [21, 22], instability [23–25] and many others [26–28].

Within the family of auxetic materials, the folded ones have been regarded as groundbreaking since their possible applications in several engineering fields [29]. These materials come along with the Japanese origami folding paper technique. The first reference of such a structure is due to Koryo Miura[30], who developed the Miura-ori folding (Figure 1a): it is obtained by surface tessellation using parallelograms to form the faces. The light weight of these structures and the capacity to deploy made them quite interesting in the aerospace[31] and automotive [32] fields. The energy absorption properties [33, 34] and the possibility of stacking Miura-ori layers in sandwich core composites [35, 36] are other remarkable applications for this geometry.

Thus, the Miura-ori structural response, focusing on the auxetic behavior, has been studied from a kinematic perspective [37, 38]. The kinematical model considers the face skeleton as a rigid articulate parallelogram, concentrating the elastic behavior on the creases [39]. Therefore, the faces rotate relatively around creases. This article proposes an experimental evaluation of thickness influence regarding the auxetic response of a Miura-ori sample under static tensile load. In addition, the modal response of a thick Miura-ori specimen is evaluated by employing interferometric experimental analysis. The article is organized as follows. Section 2 introduces the Miura-ori unit cell, with its geometrical parameters and manufacturing techniques. Section 3 presents the experimental techniques used in the investigation process. Section 3.1 reports more details about the Digital Image Correlation, while section 3.2 focuses on the Time-Averaged Speckle Interferometry. Finally, section 4 presents the experimental results of the parametric analysis on thickness impact and the modal response.

2. Model and Methods

The Miura-ori is a folded metamaterial that exhibits an auxetic behavior under specific geometrical prescriptions.

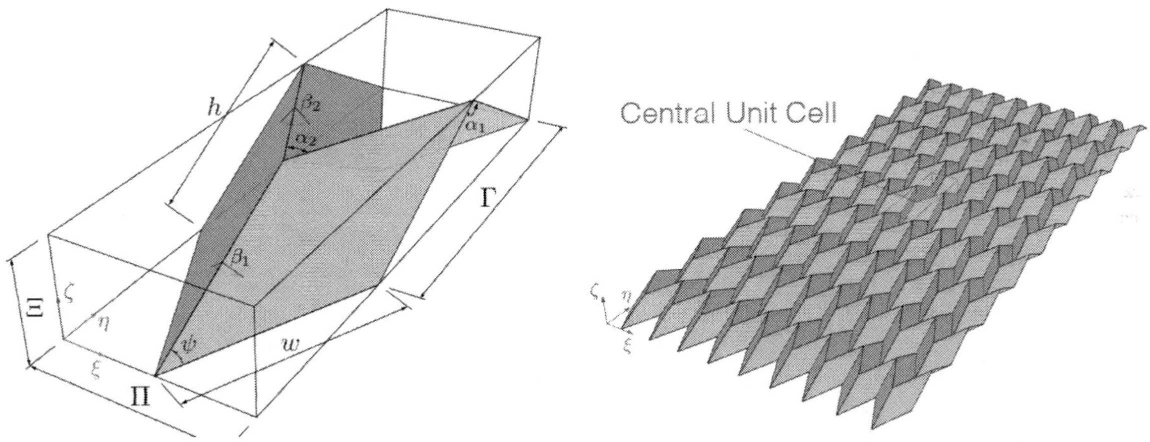

(a) *Miura-ori unit cell.* **(b)** *Miura-ori periodic pattern.*

Figure 1: *Miura-ori geometry. a) Fundamental parameters of a Miura-ori unit cell. b) Periodic disposition of* 9×9 *Miura-ori unit cell.*

Figure 1a shows the fundamental geometrical parameters that describe the Miura-ori unit cell: it can be defined using the length of the edges w and h, and the folding angles α_1 and α_2. However, angles β_1, β_2, and ψ represent another parameter set, expressed by the following relationships [39]:

$$\cos(\psi) = \sin\left(\frac{\alpha_1}{2}\right)\cos\left(\frac{\alpha_2}{2}\right) \tag{1}$$

$$\cos(\beta_1) = \frac{\cos(\alpha_1) + \cos(\alpha_2)\left[\cos(\alpha_1) + 3\right] - 1}{\cos(\alpha_1) + \cos(\alpha_2)\left[\cos(\alpha_1) - 1\right] + 3} \tag{2}$$

$$\cos(\beta_2) = \frac{\cos(\alpha_2) - \cos(\alpha_1)\left[\cos(\alpha_2) - 3\right] + 1}{\cos(\alpha_1) + \cos(\alpha_2)\left[\cos(\alpha_1) - 1\right] + 3} \tag{3}$$

The unit cell outer dimensions Γ, Π and Ξ are defined as:

$$\Gamma = (2w)\sin\left(\frac{\alpha_1}{2}\right) \tag{4}$$

$$\Pi = (2h)\sin\left(\frac{\alpha_2}{2}\right) \tag{5}$$

$$\Xi = (w)\cos\left(\frac{\alpha_1}{2}\right) \tag{6}$$

The replication of the unit cell along the ξ and η axis leads to a periodic pattern of 9×9 cells, as reported in Figure 1b.

The article of Schenk et al. [37] shows a strong relation, limited to the in-plane analysis, between the Poisson's ratios $\nu_{\eta\xi}$ ($\nu_{\xi\eta}$) and the angle α_2, which holds for a kinematical analysis of the deformation mode (eqs. (7) and (8)).

$$\nu_{\eta\xi} = -\frac{1}{\tan^2\left(\frac{\alpha_2}{2}\right)} \tag{7}$$

$$\nu_{\xi\eta} = -\tan^2\left(\frac{\alpha_2}{2}\right) \tag{8}$$

The aforementioned kinematical approach is based on modeling the edges as discontinuity between adjacent faces (Figure 2a), which can relatively rotate around them [37, 39].

In this article, we investigate, from the experimental side, the mechanical response of a Miura-ori structure whose thickness is not negligible (Figure 2b).

A first experimental campaign is oriented to study the influence of thickness on the in-plane auxetic response of Miura-ori periodic structure.

Figure 3a shows a specimen designed for tensile test. The main periodic pattern is composed by 9×9 thick unit cell, with additional two rows at the top and bottom of the structure: those are used as a connection to the gripping ends. The manufacturing process of these patterns is remarkably challenging because of the sharp elements that discourage conventional manufacturing techniques. Additive manufacturing proved to be a good solution for Miura-ori manufacturing. In particular, Fused Filament Fabrication (FFF) has been employed: the specimens have been printed along ξ axis to avoid any material deposition at cell vertices (Figure 3b). The printer used for manufacturing the samples is a Prusa i3 MK2. The materials employed in the experimental campaign are Polylactic Acid (PLA) and Acrylonitrile Butadiene Styrene (ABS), commonly used polymers in 3D printing. The gripping ends must be stiffer than the rest of the pattern because of the polymeric material: this requirement avoids deformation localization on the ends. Thus, a prismatic slot with reference surfaces guarantees an easier alignment in the tensile test machine (Figure 3a and 5a). In addition, M6 nuts are inserted in the specimen heads,

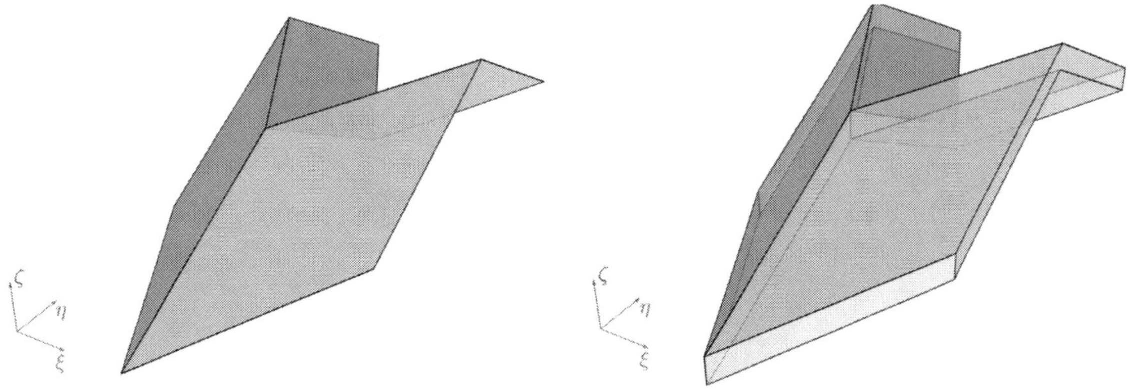

(a) *Kinematical model.* (b) *Continuum model.*

Figure 2: *Comparison between the two unit cell in terms of facets thickness. a) Miura-ori unit cell with negligible facets thickness: the in-plane response is only kinematics related. b) Miura-ori unit cell with a not negligible facets thickness: the in-plane mechanical response is modeled using the continuum mechanics theory.*

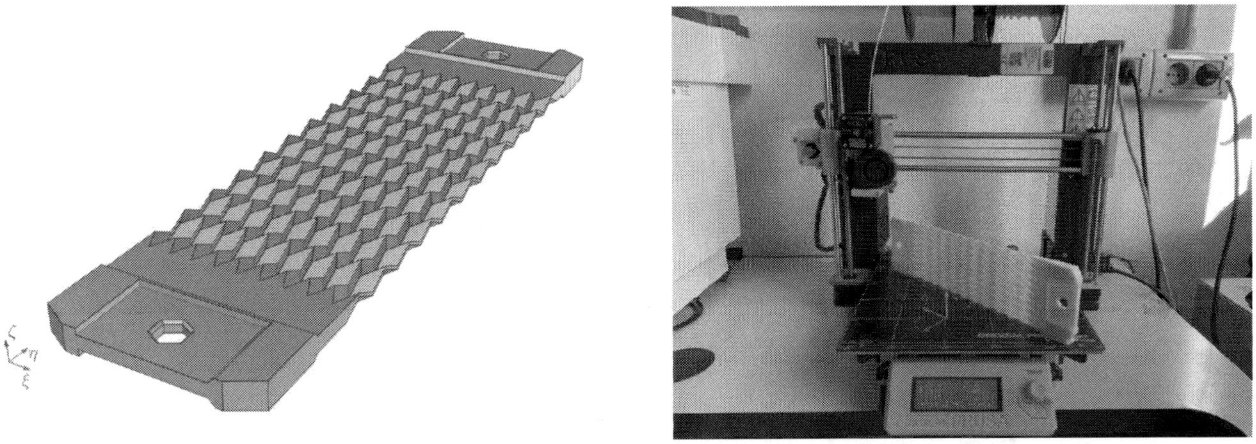

(a) *Specimen geometry.* (b) *3D printing process.*

Figure 3: *Specimen design and fabrication. a) Characteristics of a custom Miura-ori specimen made of* 9×9 *active cells. b) Manufacturing of a specimen using a Prusa i3 MK2 3D printer.*

where the grips exert their pressure: the nuts act as stroke-end for the grips, avoiding dangerous stress concentration, and preventing the loss of structural integrity due to material failure.

Then, a second campaign focuses on the modal response of such structures. The specimen has the same geometrical characteristics as the one used for the tensile test. It only differs from the previous by the number of cells, which is reduced, for room reasons, to 7 cells along η axis and 9 cells along ξ axis.

3. Experimental Techniques and Setup

Two different experimental techniques have been used for investigating the mechanical response under tensile tests and the modal response. Digital Image Correlation (DIC) has been employed

IOP Conf. Series: Materials Science and Engineering 1306 (2024) 012037 doi:10.1088/1757-899X/1306/1/012037

Figure 4: *Specimen geometry that has been used for testing the modal response.*

for the tensile tests: this technique gives remarkable advantages in displacement and strain field measurements. The modal response has been evaluated using Time Averaged Speckle Interferometry, a specifically developed interferometric approach for acquiring the vibration modes of objects.

3.1. Digital Image Correlation in Tensile Tests
Digital Image Correlation is an optical technique that estimates the displacement field over an object's surface by correlating two images acquired before and after the motion. In its standard formulation (local approach), the images are partitioned into small sub-regions, called subsets. Thus, DIC applies to bi-dimensional (DIC-2D) and three-dimensional (DIC-3D) displacement fields on surfaces, as well as volumetric fields [40]. The fundamental hypothesis that holds the DIC approach is the optical flow constancy [40], defined as:

$$I(x,y,t) = I(x+u, y+v, t+\Delta t) \tag{9}$$

where $I(x,y,t)$ is the intensity field at point, defined by coordinates (x,y), and at the time instant t. The terms u and v that are present in eq. (9) are, respectively, the x and y components of displacement, while Δt is a discrete-time increment.

$$\frac{\partial I}{\partial t} + \frac{\partial I}{\partial x}\underbrace{\frac{\partial x}{\partial t}}_{\dot{u}} + \frac{\partial I}{\partial y}\underbrace{\frac{\partial y}{\partial t}}_{\dot{v}} = 0 \tag{10}$$

Eq. (10) is the first-order expansion of the relationship (9) second member, under the hypothesis of small displacements. The eq. (10) is a single equation with two unknowns ($\dot{u}$ and $\dot{v}$), which leads to an ill-posed problem. Several solutions have been proposed [41, 42], but the most used is the so called *Lucas-Kanade* approach [43]. This approach minimizes an error function, in which the displacements are modeled as an affine transformation.

$$u = a_0 + x_s + a_1 x_s + a_2 y_s + \overbrace{a_3 x_s^2 + a_4 x_s y_s + a_5 y_s^2}^{\text{parabolic terms}} \tag{11}$$

$$v = b_0 + y_s + b_1 x_s + b_2 y_s + \underbrace{b_3 y_s^2 + b_4 x_s y_s + b_5 x_s^2}_{\text{parabolic terms}} \tag{12}$$

Eqs. (11) and (12) are an example of parabolic shape functions that model an affine transform [40]. The coordinates x_s and y_s belong to a reference system attached to the subset center, while the parameter sets $\boldsymbol{a} = \{a_0, a_1, \dots a_n\}$ and $\boldsymbol{b} = \{b_0, b_1, \dots b_n\}$ represent the minimization aim.

$$\chi^2 = \sum_i \sum_j [R(i,j) - T(i+u, j+v)]^2 \tag{13}$$

The eq. (13) is the cost function, in which R and T are the reference and the target image, while i and j are indexes that define the pixel coordinates inside a subset whose center is in (i_0, j_0). The cost function minimization requires posing to zero the derivatives, which are calculated with respect to the $\boldsymbol{a}$ and $\boldsymbol{b}$ parameter sets. This procedure leads to solve a system of the form $\boldsymbol{Ax} = \boldsymbol{c}$, in which $\boldsymbol{A} = \sum_k \boldsymbol{h}_k \boldsymbol{h}_k^T$ and $\boldsymbol{c} = \sum_k (R_k - T_k)\, \boldsymbol{h}_k$, where the vector $\boldsymbol{h}$ is defined as:

$$\boldsymbol{h} = \left\{ \frac{\partial T}{\partial x}, \frac{\partial T}{\partial y}, \frac{\partial T}{\partial x} x_s, \frac{\partial T}{\partial y} x_s, \dots \right\}$$

by assuming that parameters follows a sorting rule like $a_0, b_0, a_1, b_1, \dots, a_n, b_n$. The index k moves on the whole image points involved in the calculation.

Therefore, expanding in series the second member of eq.(9) is not the only possible choice. Expanding the reference image [44, 45] R gives remarkable advantages since it does not change during calculation history, as the intensity derivatives. In addition, this strategy is not sensitive to noise bias [46, 47]. An alternative solution is the expansion of both target and reference intensity fields at midpoint [48, 49].

The aforementioned local approach is the most common but is not the only one possible. The so-called global method uses the same course as the local one, but it differs by introducing a global function that describes the displacement field, instead of using the subset sampling [50, 51]. The global function is continuous, belonging to the C_0 class, although other choices have been proposed in literature [52, 53]. In this case, the region of interest (ROI) is discretized by a Finite Element-like mesh, where displacements depend on the nodal degrees of freedom, and are shared between neighboring elements [54]. Thus, the error function (13) can still be employed, but the parameter sets ($\boldsymbol{a}$ and $\boldsymbol{b}$) act now as nodal degrees of freedom [54].

Another remarkable advancement in the DIC world is the Integrated approach (iDIC) [55, 56]. The iDIC differs from the standard local method because of the shape functions: in this case, they are designed to match the problem-related displacement field, instead of using the general purpose ones. Problem-related shape functions make the domain partitioning useless: thus, the analysis procedure employs the whole pixels inside the ROI. Several examples of iDIC employment are residual stress [57] and fracture mechanics [58]: the outputs are problem-related, leading to great advantage because the data do not need additional post-processing.

In this campaign, we use 2D-DIC software with a local approach developed by one of the authors (A. Baldi). The specimens are tested under tensile load using an MTS Landmark 370 universal test machine (Figure 5b), with a load capacity of ±100 kN and a load cell with the same measurement range. The tests are conducted in displacement control. The images have been acquired using an Allied Vision Pike F421b (Figure 5c), with a resolution of 2048 × 2048 and a pixel size of 7.4 µm, and it is equipped with a Schneider macro lens. The synchronization

IOP Conf. Series: Materials Science and Engineering 1306 (2024) 012037 doi:10.1088/1757-899X/1306/1/012037

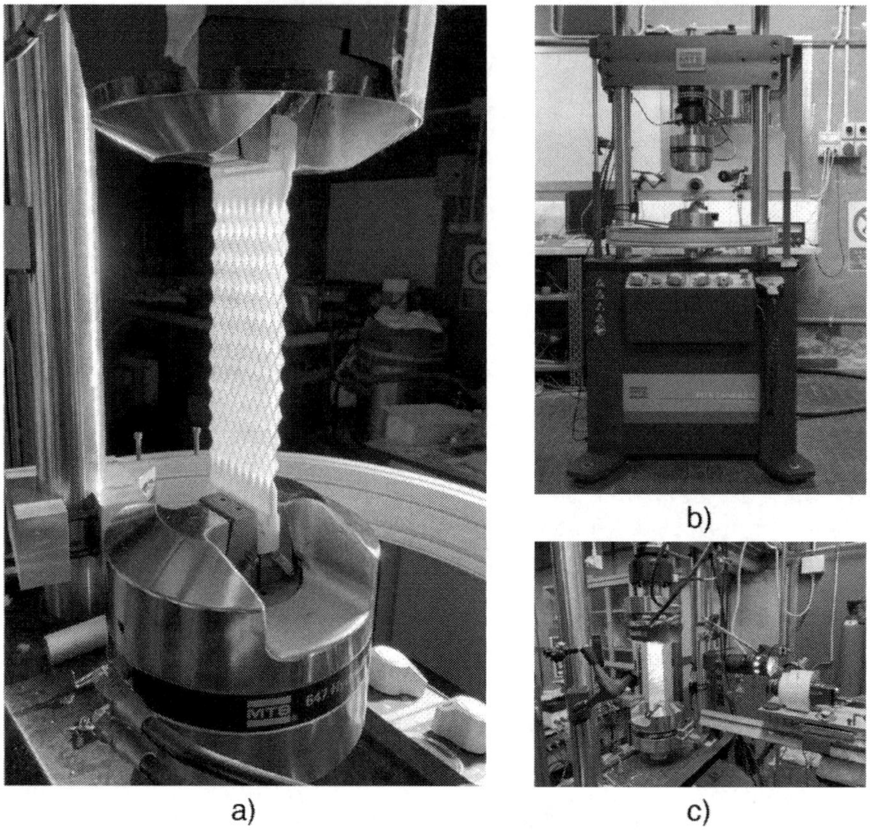

Figure 5: *Details of experimental setup. a) Sample of Miura-ori gripped in the test machine. b) MTS Landmark 370 universal testing machine. c) Overview of the camera setup with the lighting system.*

between the tensile test and the image acquisition is accomplished through a National Instrument board that sends a trigger signal to the camera at a fixed sampling rate.

3.2. Time-Averaged Speckle Interferometry for Modal Response

Time-Averaged Speckle Interferometry is an interferometric technique developed for investigating the modal response of a vibrating object [59–63].

Figure 6a shows the interferometric setup employed for the mode measurement. The small angle between the object beam I_O and the CCD axis ($\sim 10°$) leads to an almost out-of-plane displacement sensitivity. The laser source is a Helium-Neon tube with a wavelength of 632.7 nm and 50 mW power. The CCD is an Allied Vision Pike F421b, with the same characteristics reported in the DIC section, and it is equipped with a Schneider macro optics. The setup differs from conventional ones only by the reference branch: the reference beam is focused through a lens on the iris aperture plane of the CCD optics. Although this feature raises the alignment complexity (i.e., the CCD sensor must be fully enlightened), it has the remarkable advantages of using an high-quality lens [64] and simplifying the speckle field focusing. The phase-shifting is pursued using the PZT-actuated mirror on the reference branch. The specimen gripping system is a fixed-center vice (Figure 6b), with one harm modified for embedding the excitation system. The latter is composed by a solid-state elastic spring, actuated by PZT stack element.

The interference field is expressed by the following relationship:

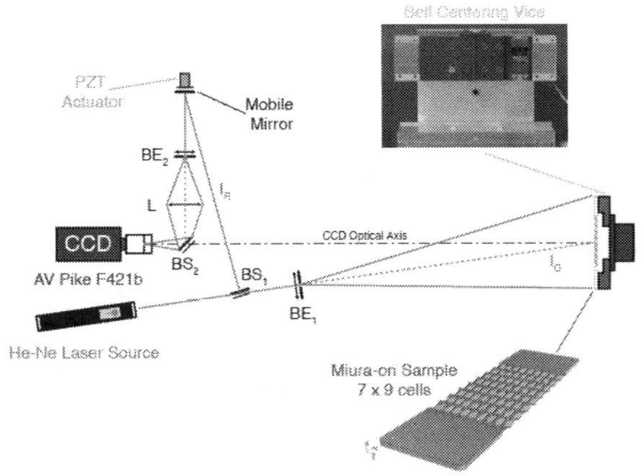

(a) *Interferometric setup.*

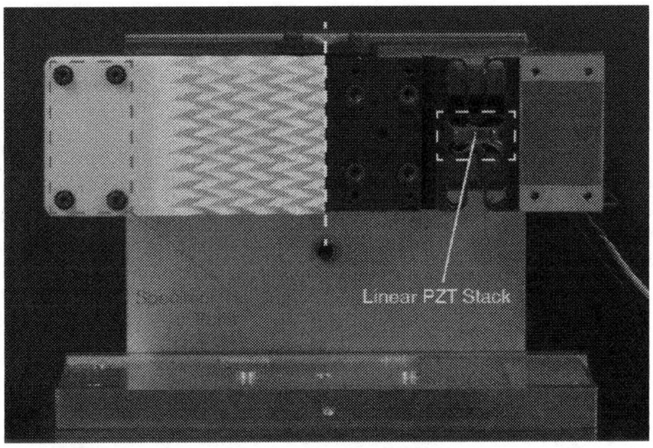

(b) *Gripping/excitation system.*

Figure 6: *Time-Averaged Speckle setup. a) Scheme of the interferometric setup (I_R: Reference beam. I_O: Object Beam. BS: Beam Splitter. BE: Beam Expander. L: Lens). b) Fixed center vice, modified to embed the excitation system.*

$$I(x,y,t) = I_O + I_R + 2\sqrt{I_O I_R}\cos(\phi + \Delta) \tag{14}$$

Where I_O and I_R are respectively the object and the reference beam, $\phi = \phi(x,y)$ is the phase field. The term $\Delta = \Delta(x,y,t)$ needs special attention: it depends on either vibration amplitude A and angular frequency ω, illumination angle θ, and the coherent source wavelength λ.

$$\Delta(x,y,t) = \frac{2\pi\,[1+\cos(\theta)]\,[A\cos(\omega t)]}{\lambda} \tag{15}$$

The intensity (14) is integrated by the CCD over the shutter aperture time τ, leading to an output voltage V_{CCD} (16), where the term Γ depends on the CCD gain and chromatic sensitivity [64].

$$V_{\mathrm{CCD}} = \Gamma \int_{t_0}^{t_0+\tau} I_O + I_R + 2\sqrt{I_O I_R}\cos(\phi + \Delta) \quad \text{where} \quad \tau = nT + \delta \tag{16}$$

In eq. (16), the term τ is defined as an integer number n of vibration period $T = \frac{2\pi}{\omega}$ and a remainder δ that arises from the division $\frac{\tau}{T}$. The CCD output voltage (16) can be expanded in a convenient series form [63].

$$\begin{aligned} V_{\mathrm{CCD}} \propto \frac{2n\pi}{\omega}\left[I_O + I_R + 2\sqrt{I_O I_R} J_0(k)\cos(\phi)\right] + \\ + \delta\left\{I_O + I_R + 2\sqrt{I_O I_R}\left[\cos(\phi) - \frac{k\sin(\delta\omega)}{\delta\omega}\sin(\phi) - \frac{k^2\sin(2\delta\omega)}{\delta\omega}\cos(\phi) + \dots\right]\right\} \\ \text{where} \quad k = \frac{2\pi A\,[1+\cos(\theta)]}{\lambda} \end{aligned} \tag{17}$$

The second member of eq. (17) is modulated by δ: as the vibration frequency increases, the term δ decays[1], leading to the loss of relevance of the eq.(17) second member [63].

[1] The term δ is, therefore, the time delay between the shutter opening time, and the temporal interval the system needs to accomplish an integer number of oscillations.

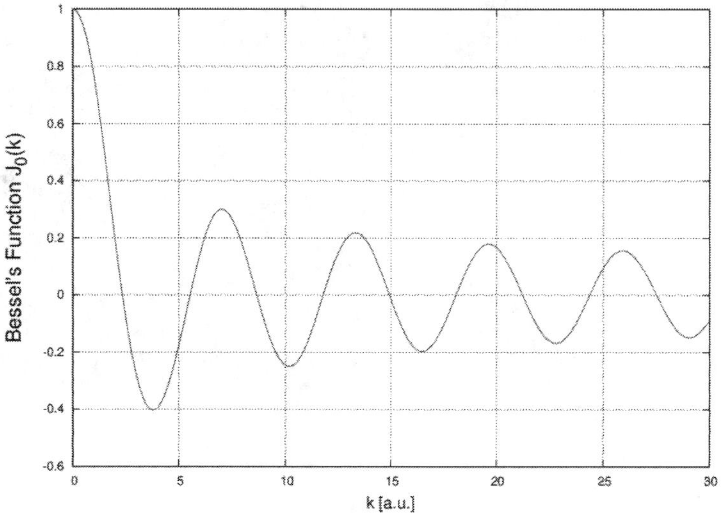

Figure 7: J_0 *Bessel's function.*

Thus, the first member of the (17) contains the 0th-order Bessel's function (Figure 7) and the term k, which essentially depends on the vibration amplitude A because the illumination angle and the wavelength are intrinsic parameters of the setup. The fringes intensity field is modulated by J_0 Bessel's function (eq.(17)): thus, the nodes, where the amplitude A is null, are characterized by the brightest fringe (if $A = 0$, then $J_0 = 1$). In addition, as the frequency ω increases, the amplitude A decreases, with the consequent loss of contrast between fringes.

In Time-Averaged Speckle Interferometry, the acquisition method can follow two different strategies. The first consists of applying a phase-shifting approach [65] when the sample under analysis is at rest and then under the desired frequency excitation. The interferogram is obtained by subtracting the modulo 2π phase fields, leading to an almost binary fringe pattern [66]. A second possibility is to discard the phase modulo 2π [64] of the phase-shifting post-processing and focus only on the modulation (fringe intensity field). If the illumination is constant, the points with the same vibration amplitude have the same modulation. In this perspective, for each fringe that comes from the modulo 2π approach, there will be two Bessel's fringe: this is a consequence of the J_0 modulation (Figure 7), where at each peak corresponds two nodes. The advantages of this second method are the absence of a reference field and the immediate visibility of the fringe pattern [64], which comes from the modulation field only.

We employ the second approach because of its convenience. A Miura-ori specimen, reshaped as reported in Figure 4 for fitting in the gripping system, is excited through a frequency sweep with a constant amplitude. The specimen response has been monitored via a PZT transducer, bonded on the specimen surface. The PZT signal is acquired for identifying the resonant conditions, which are the main focus of the investigation. The interference field is then acquired and processed through the 5-frame Hariharan approach [65], focusing on the fringe visibility to obtain Bessel's fringe pattern.

4. Results

In this section, the results from the experimental tests are reported for the parametric analysis under tensile load and the modal response.

4.1. Parametric Analysis on Thickness under Tensile Tests

The experimental campaign for assessing the effect of thickness on the auxetic behavior of the Miura-ori geometry has been conducted on two sets of samples: one made of PLA and the other of ABS. The specimen geometry follows the specification of section 2, and the thickness is defined as the distance between the top and bottom surfaces of the faces. The images for the DIC analysis have been acquired in the linear elastic regime only, and the Poisson's ratio $\nu_{\eta\xi}$ is evaluated by the ratio of the ε_ξ and ε_η, averaged on a mask of 7×7 cells, centered on the central one.

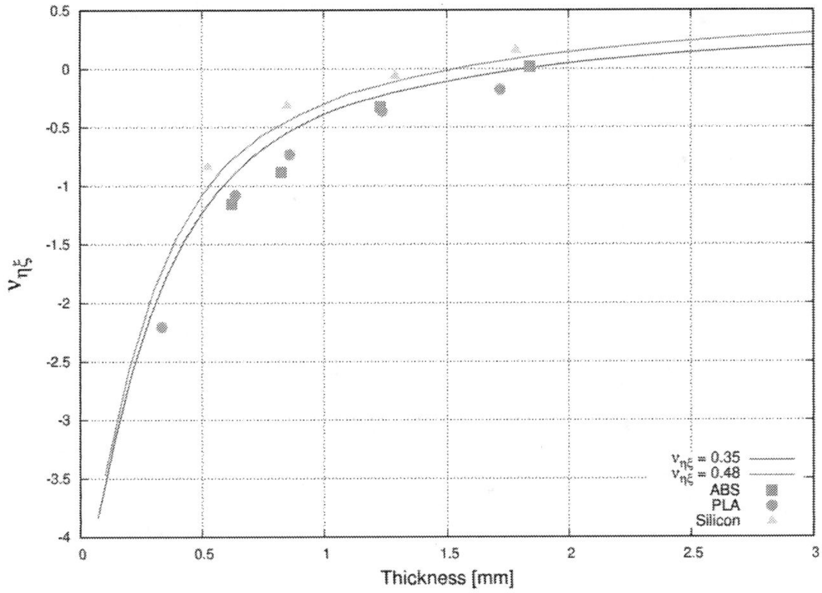

Figure 8: *Parametric analysis on thickness: comparison between experimental data an numerical simulation*

Figure 8 presents the data from the experimental observations, besides numerical simulation curves, used as a comparison. Both PLA and ABS have a Poisson's ratio of 0.35. Although the experimental data follow the numerical curve trend, a deviation occurs. During the fabrication, the specimens have been printed by filament deposition, which is oriented in the same direction for each layer. This consideration leads to the introduction of an anisotropic behavior of the samples. The validation has been pursued by developing a new Miura-ori specimen set made of silicon rubber: the latter has been poured into a mold to ensure an isotropic behavior of the samples. Therefore, the data relative to the silicon specimens, reported in Figure 8, are closer to the relative numerical curve, suggesting that the fabrication-induced anisotropy plays a not negligible role in mechanical response.

Figure 9 shows the displacement and strain fields of a PLA specimen with a thickness of 0.853 mm, that exhibits a Poisson's ratio of −0.73. The load has been applied in the negative direction of η axis. Therefore, Figure 9a reports the ξ component of displacement, which clearly shows the auxetic behavior. The data analysis and post-processing have been conducted using the software ABc developed by A. Baldi.

4.2. Modal Response

The modal response has been evaluated following the procedures explained in section 3.2. The shutter aperture time has been 50 ms. The specimen has been made of PLA and is 0.34 mm

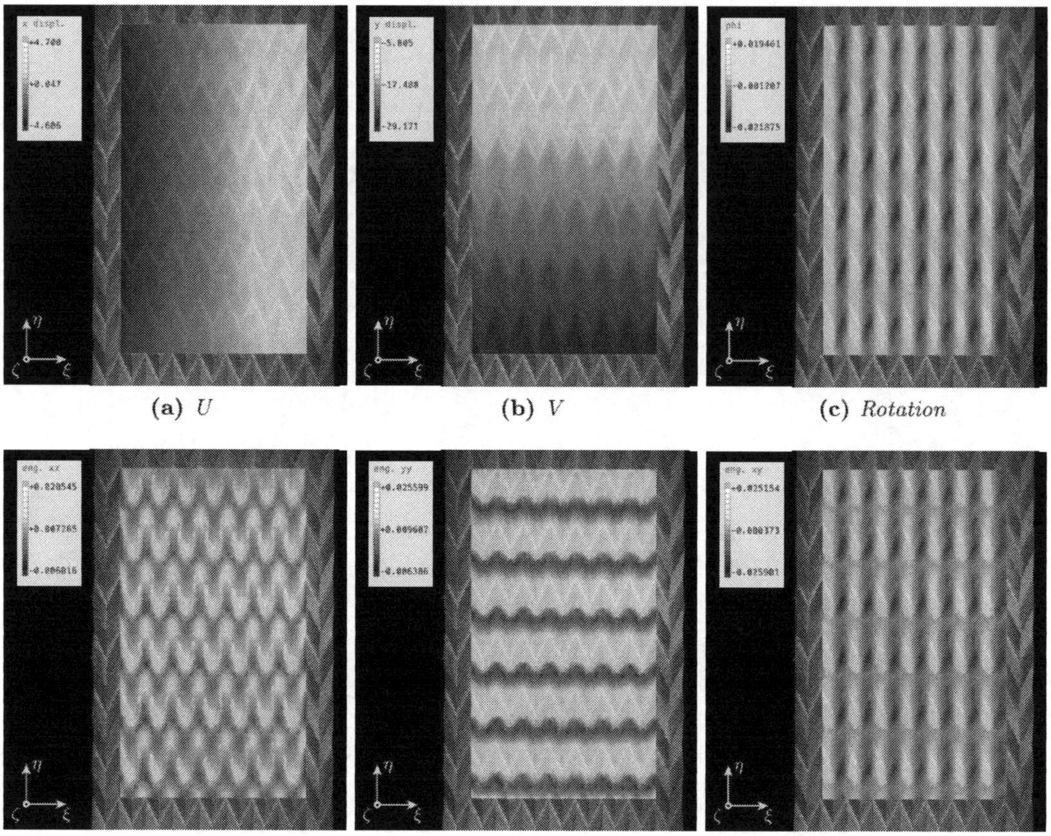

(a) U **(b)** V **(c)** *Rotation*

(d) $\varepsilon_{\xi\xi}$ **(e)** $\varepsilon_{\eta\eta}$ **(f)** $\varepsilon_{\xi\eta}$

Figure 9: *Results of DIC analysis on a PLA specimen 0.853 mm thick.*

thick. A thin layer of varnish has been applied on the sample surface to avoid any transparency effect.

Figure 10 reports the experimental results that come from the Time-Averaged Speckle Interferometry. The numerical resonant modes, calculated with the Finite Element, are used as a comparison. The interferograms give qualitative information about the shape of resonant modes, which agree with the numerical predictions. The small deviation observed between the experimental and numerical resonant frequencies can be addressed to the varnish layer or to the boundary conditions.

5. Discussion

The parametric analysis of the thickness influence on the auxetic response of Miura-ori geometry shows that the global Poisson's ratio $\nu_{\eta\xi}$ strongly depends on the thickness. The experimental data from DIC analysis fits the numerical simulations trend, even though the deviation is due to fabrication anisotropy. The silicon rubber specimen plays a fundamental role in identifying the source of the bias between numerical and experimental data. DIC proved to be a consolidated technique for measuring displacement and strain field, even for challenging geometry such as the Miura-ori one.

Time-averaged speckle interferometry is a remarkable technique for analyzing the modal

IOP Conf. Series: Materials Science and Engineering 1306 (2024) 012037 doi:10.1088/1757-899X/1306/1/012037

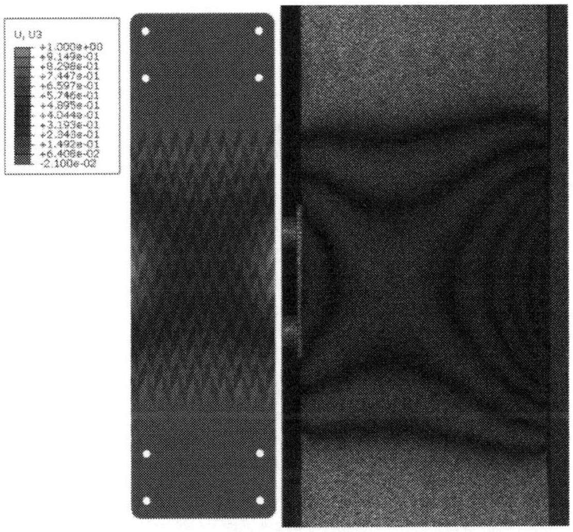

(a) 1st *mode. Numerical eigenfrequency: 130.6 Hz. Experimental eigenfrequency: 133.8 Hz*

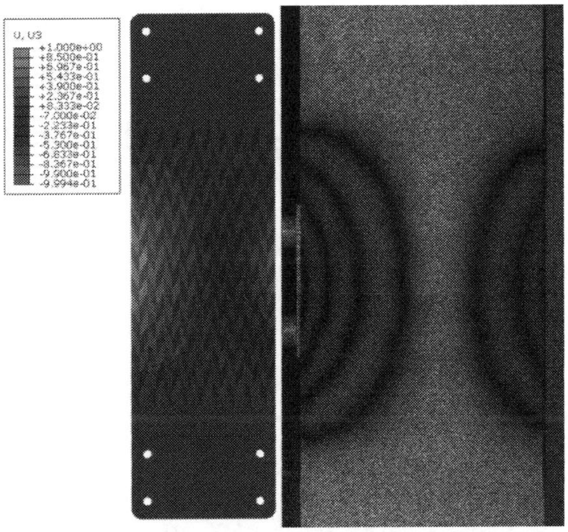

(b) 2nd *mode. Numerical eigenfrequency: 135.5 Hz. Experimental eigenfrequency: 145.5 Hz*

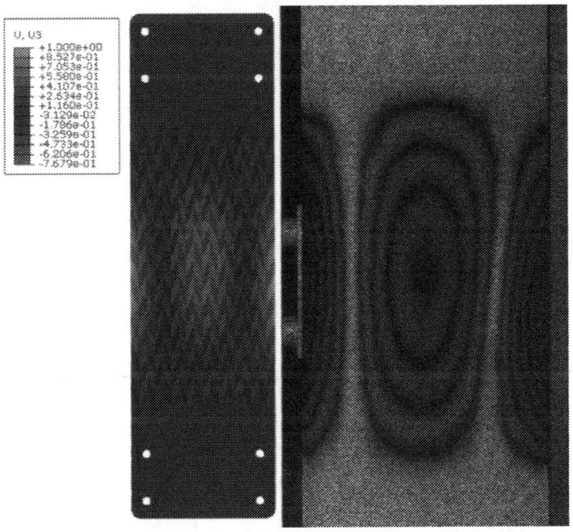

(c) 3rd *mode. Numerical eigenfrequency: 294.5 Hz. Experimental eigenfrequency: 321.0 Hz*

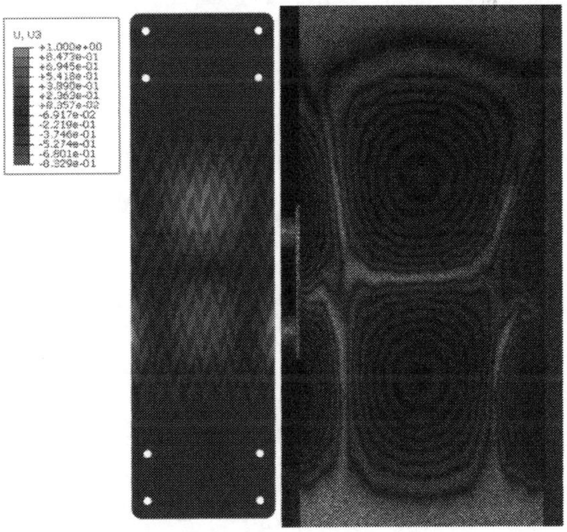

(d) 4th *mode. Numerical eigenfrequency: 558.7 Hz. Experimental eigenfrequency: 602.0 Hz*

Figure 10: *Comparison between the resonant modes of a Miura-ori structure acquired with the Time-Averaged Speckle Interferometry and numerical simulations.*

response of a vibrating object. Thus, it avoids the more expensive high-speed cameras or laser vibrometers. The experimental interferograms exhibit a closer shape with respect to the numerical predictions: this poses the basis for further investigation on the Miura-ori modal properties.

References

[1] Evans K E 1991 *Endeavour* **15** 170–174

[2] Lakes R S and Elms K 1993 *Journal of Composite Materials* **27** 1193–1202

[3] Evans K and Alderson K 2000 *Engineering Science & Education Journal* **9** 148–154

[4] Alderson A and Alderson K L 2007 *Proceedings of the Institution of Mechanical Engineers, Part G: Journal of Aerospace Engineering* **221** 565–575

[5] Veronda D R and Westmann R A 1970 *Journal of Biomechanics* **3** 111–124

[6] Lees C, Vincent J F V and Hillerton J E 1991 *Bio-Medical Materials and Engineering* **1** 19–23

[7] Timmins L H, Wu Q, Yeh A T, Moore J E and Greenwald S E 2010 *American Journal of Physiology-Heart and Circulatory Physiology* **298** H1537–H1545

[8] Baughman R H, Shacklette J M, Zakhidov A A and Stafström S 1998 *Nature* **392** 362–365

[9] Goldstein R V, Gorodtsov V A and Lisovenko D S 2010 *Mechanics of Solids* **45** 529–545

[10] Lubarda V A and Meyers M A 1999 *Scripta Materialia* **40** 975–977

[11] Cabras L and Brun M 2014 *Proceedings of the Royal Society A: Mathematical, Physical and Engineering Sciences* **470** 20140538

[12] Cabras L and Brun M 2016 *Journal of the Mechanics and Physics of Solids* **91** 56–72

[13] Lakes R 1987 *Science* **235** 1038–1040

[14] Choi J B and Lakes R S 1992 *Journal of Materials Science* **27** 4678–4684

[15] Peng X L, Soyarslan C and Bargmann S 2020 *Extreme Mechanics Letters* **35** 100641

[16] Zhang X Y, Ren X, Zhang Y and Xie Y M 2022 *Thin-Walled Structures* **174** 109162

[17] Grima J N and Gatt R 2010 *Advanced Engineering Materials* **12** 460–464

[18] Carta G, Brun M and Baldi A 2016 *Mechanics of Materials* **97** 67–75

[19] Carta G, Cabras L and Brun M 2016 *Journal of the European Ceramic Society* **36** 2183–2192

[20] Baldi A, Brun M and Carta G 2022 *Mechanics of Materials* **164** 104114

[21] Theocaris P S, Stavroulakis G E and Panagiotopoulos P D 1997 *Archive of Applied Mechanics* **67** 274–286

[22] Mizzi L, Mahdi E M, Titov K, Gatt R, Attard D, Evans K E, Grima J N and Tan J C 2018 *Materials & Design* **146** 28–37

[23] Bertoldi K, Reis P M, Willshaw S and Mullin T 2010 *Advanced Materials* **22** 361–366

[24] Shen J, Zhou S, Huang X and Xie Y M 2014 *physica status solidi (b)* **251** 1515–1522

[25] Khan K A, Alshaer M H and Khan M A 2021 *Advanced Engineering Materials* **23** 2001041

[26] Saxena K K, Das R and Calius E P 2016 *Advanced Engineering Materials* **18** 1847–1870

[27] Ren X, Das R, Tran P, Ngo T D and Xie Y M 2018 *Smart Materials and Structures* **27** 023001

[28] Prawoto Y 2012 *Computational Materials Science* **58** 140–153

[29] Miura K and Pellegrino S 2020 *Forms and Concepts for Lightweight Structures* (Cambridge University Press)

[30] Miura K 1985 *The Institute of Space and Astronautical Science report* **618** 1–9

[31] Zhang C, Yang Q and Tao R 2021 *Smart Materials and Structures* **30** 075004

[32] Lv C, Krishnaraju D, Konjevod G, Yu H and Jiang H 2014 *Scientific Reports* **4** 5979

[33] Tolman S S, Delimont I L, Howell L L and Fullwood D T 2014 *Smart Materials and Structures* **23** 094010

[34] Liu S, Lu G, Chen Y and Leong Y W 2015 *International Journal of Mechanical Sciences* **99** 130–142

[35] Pydah A and Batra R C 2017 *Thin-Walled Structures* **115** 311–322

[36] Xiang X M, You Z and Lu G 2018 *Composite Structures* **195** 359–374

[37] Schenk M and Guest S D 2013 *Proceedings of the National Academy of Sciences* **110** 3276–3281

[38] Wei Z Y, Guo Z V, Dudte L, Liang H Y and Mahadevan L 2013 *Physical Review Letters* **110** 215501–

[39] Moradweysi P, Santucci P M, Carta G, Goudarzi T, Aghdam M M, Baldi A and Brun M 2022 *Mechanics of Advanced Materials and Structures* 1–19

[40] Schreier H, Orteu J J, Sutton M A *et al.* 2009 *Image Correlation for Shape, Motion and Deformation Measurements* vol 1 (Springer New York, NY)

[41] Horn B K P and Schunck B G 1981 *Artificial Intelligence* **17** 185–203

[42] Bruhn A, Weickert J and Schnörr C 2005 *International Journal of Computer Vision* **61** 211–231

[43] Lucas B D and Kanade T 1981 An iterative image registration technique with an application to stereo vision *IJCAI'81: 7th international joint conference on Artificial intelligence* vol 2 (Vancouver, Canada) pp 674–679

[44] Shum H Y and Szeliski R 1998 Construction and refinement of panoramic mosaics with global and local alignment *Proceedings of the IEEE International Conference on Computer Vision* pp 953–958

[45] Baker S and Matthews I 2004 *International Journal of Computer Vision* **56** 221–255

[46] Shao X, Dai X and He X 2015 *Optics and Lasers in Engineering* **71** 9–19

[47] Baldi A, Santucci P M and Bertolino F 2022 *Optics and Lasers in Engineering* **154** 107012

[48] Davis C Q and Freeman D M 1998 *Optical Engineering* **37** 1290–1298

[49] Tong W 2013 *Strain* **49** 313–334

[50] Sun Y, Pang J H L, Wong C K and Su F 2005 *Applied Optics* **44** 7357–7363

[51] Hild F and Roux S 2006 *Strain* **42** 69–80

[52] Cheng P, Sutton M A, Schreier H W and McNeill S R 2002 *Experimental Mechanics* **42** 344–352

[53] Roux S, Hild F and Berthaud Y 2002 *Applied Optics* **41** 108–115

[54] Baldi A and Bertolino F 2016 *Meccanica* **51** 979–991

[55] Roux S and Hild F 2006 *International Journal of Fracture* **140** 141–157

[56] Gao J and Shang H 2009 *Applied Optics* **48** 1371–1381

[57] Baldi A 2014 *Experimental Mechanics* **54** 379–391

[58] Baldi A and Santucci P M 2022 *Engineering Fracture Mechanics* **259** 108082

[59] Tiziani H J and Klenk J 1981 *Applied Optics* **20** 1467–1470

[60] Joenathan C 1991 *Applied Optics* **30** 4658–4665

[61] Li X and Tao G 2002 *Optics & Laser Technology* **34** 259–264

[62] Romero G, Alvarez L, Alanís E, Nallim L and Grossi R 2003 *Optics and Lasers in Engineering* **40** 81–90

[63] Wang W C, Hwang C H and Lin S Y 1996 *Applied Optics* **35** 4502–4509

[64] Baldi A, Jacquot P and Ginesu F 2004 Una applicazione della tecnica speckle all'analisi di uno statore per piezo-motors *Atti del XXXIII Convegno AIAS*

[65] Creath K, Crennel K M, Gåsvik K J, Halliwell N A, Pickering C J, Hariharan P, Kujawinska M and Yatagai T 1993 *Interferogram analysis: digital fringe pattern measurement technique* (IOP Publishing)

[66] Borza D N 2004 *Optics and Lasers in Engineering* **41** 515–527

On the Compositional Step of the IC-GN Digital Image Correlation Algorithm

Antonio Baldi[1] **and Pietro Maria Santucci**[1]

[1]Dipartimento di Ingegneria Meccanica, Chimica e Nucleare, Universitá degli Studi di Cagliari, Via Marengo 2, 09123 Cagliari, IT

E-mail: `antonio.baldi@unica.it`

Abstract. Digital Image Correlation (DIC) is a well-known experimental optical technique for measuring the displacement field based on the assumption that pixel intensity does not change with motion.

DIC can be implemented using various alternative approaches. The most used in the Experimental Mechanics field are the Forward-Additive Gauss-Newton (FA-GN) and the Inverse-Compositional Gauss-Newton (IC-GN) formulations. The former corresponds to the original formulation proposed by Lucas and Kanade, while the latter was proposed by Baker and Matthews twenty years later. Although both formulations give the same results at the first order, their speed, convergency characteristic and noise robustness differ considerably. Nowadays, the IC-GN method is usually preferred because of its lower computational load and the significantly better noise sensitivity. However, the Inverse Compositional approach, as its name states, requires the inversion and composition of the displacement field, thus enforcing the use of invertible displacement fields. This can be a significant limitation because it may introduce an under-matching error in the solution.

This work shows that a simple modification of the viewpoint makes the compositional step simpler, thus giving a DIC formulation as fast as that of the IC-GN, with the same noise-bias characteristic and without requiring an invertible displacement field.

1. Introduction

Digital Image Correlation is a well-known optical, non-interferometric, experimental technique [1]. It works on the assumption that the intensity of a single pixel remains constant during motion. Since the detected intensity includes both the diffuse and reflected components, this apparently simple assumption implies two constraints (uniform illumination and Lambertian surface) [2], which are usually not fully satisfied; however, "standard" experimental setups are usually close enough to the theoretical requirements to make the algorithm work.

The equation resulting from the constant intensity hypothesis, known as the *optical flow equation*,

$$\frac{\partial I}{\partial x}\frac{\partial u}{\partial t} + \frac{\partial I}{\partial y}\frac{\partial v}{\partial t} + \frac{\partial I}{\partial t} = 0 \tag{1}$$

where I is the image intensity and u and v are the displacement components, clearly shows that the displacement identification problem is ill-posed because it involves two unknowns (u and v); without further assumptions, we can only estimate the motion in the direction of the image intensity gradient [3, 4]. Thus, to find a solution, one has either to add a stabilization term [3]

IOP Conf. Series: Materials Science and Engineering 1306 (2024) 012038 doi:10.1088/1757-899X/1306/1/012038

to the differential equation (1) or search for the best *local* solution in the sense of least-squares. The latter is the approach most commonly used in the field of Experimental Mechanics which will be discussed below.

Following the article by Lucas and Kanade [5], several modifications have been proposed [6] to improve solution reliability and computational efficiency. However, they can all be described within the framework of the original algorithm: the *local* mapping between the reference image $f(\mathbf{x})$ and the current image $g(\mathbf{x})$ is assumed to be a simple function of a limited number of parameters, and the solution is found in the sense of least squares. Thus, by considering a sufficiently large number of pixels around the point of interest, it is possible to find the best fit with respect to a given error function. In the following, the Parametric Sum of Squared Differences (PSSD)

$$\chi^2_{PSSD} = \sum_{k=1}^{N} \left[a + bf(x_k, y_k) - g(x_k + u_k, y_k + v_k)\right]^2 \tag{2}$$

is assumed [7], although we will sometimes refer to the Sum of Squared Differences (SSD) criterion

$$\chi^2_{SSD} = \sum_{k=1}^{N} \left[f(x_k, y_k) - g(x_k + u_k, y_k + v_k)\right]^2 \tag{3}$$

to simplify the discussion if the modifications can be easily inferred.

Whatever the criterion (either (2) or (3)), the summation is over all the N pixels of the active area (usually a small subset of the image around the point of interest), while the a and b parameters in (2) take into account the avarage intensity and contrast correction between $f(\mathbf{x})$ and $g(\mathbf{x})$.

A simple polynomial mapping is usually accepted for the u and v displacement components:

$$\begin{aligned} u &= s_0 + s_1\xi + s_2\eta + [s_3\xi^2 + s_4\xi\eta + \ldots] \\ v &= t_0 + t_1\xi + t_2\eta + [t_3\xi^2 + t_4\xi\eta + \ldots] \end{aligned} \tag{4}$$

where ξ and η are the *local* coordinates of the subset ($\xi = x - x_0$, $\eta = y - y_0$, where x_0, y_0 are the *global* coordinates of the point of interest), s_i and t_i are the set of parameters controlling the mapping (the unknowns of the problem), and the terms in brackets appear only when parabolic shape functions are used.

The solution is identified by a Gauss-Newton approach: the function $g(x + u, y + v)$, which describes the intensity of the current image at point $(x + u, y + v)$, is expanded in Taylor's series truncated to the first order, allowing the calculation of a linear approximation of the local behavior. The χ^2 function (3) thus becomes

$$\chi^2_{SSD} = \sum_{k=1}^{N} \left\{ f_k - \left[g_k + \sum_{i=1}^{m} \left(\frac{\partial g_k}{\partial x}\frac{\partial u}{\partial s_i} ds_i + \frac{\partial g_k}{\partial y}\frac{\partial v}{\partial t_i} dt_i \right) \right] \right\}^2 \tag{5}$$

where we have noted that the displacement component u does not depend on the t_i (v does not depend on s_i) and $du \equiv d\xi \equiv dx$ ($dv \equiv d\eta \equiv dy$).

Computing and setting to zero the derivatives of (5) with respect to ds_i and dt_i (plus a and b in the case of the PSSD criterion) gives a system of $2m$ linear equations ($2(m + 1)$ for PSSD) that can be easily solved using the Cholesky decomposition [8, 9] (the matrix is symmetric and positive definite), thus allowing the updating of the current set of parameters.

Given the new set of parameters, the entire process can be iterated to further refine the solution, until convergence is reached.

IOP Conf. Series: Materials Science and Engineering 1306 (2024) 012038 doi:10.1088/1757-899X/1306/1/012038

The algorithm outlined is known as the Forward-Additive Gauss-Newton (FA-GN) approach, because the displacement components map $f(\mathbf{x})$ to $g(\mathbf{x})$ and the solution is obtained by summing a series of progressively smaller corrections, each obtained by performing one step of the iterative Gauss-Newton algorithm.

Some points are to be noted:

- the least squares fitting approach requires a sufficiently large number of independent data points; so the subset must be large enough (a practical minimum size is around 9×9 pixels) and the surface must have a texture (either natural or artificial);
- we have neglected all second order terms in the Hessian matrix. Although their inclusion may improve convergence near the minimum [10], it does not affect the final result because the second-order derivatives are identically zero when convergence is achieved. In addition, the inclusion of second-order terms makes the algorithm slower and more complex to implement; finally, it can instabilize the algorithm far from the minimum if the model fits the real data poorly or experimental data are contaminated by outlier [9];
- the formulation described assumes a forward mapping and a Taylor series expansion for the $g(\mathbf{x})$ field. In fact, this is not the only possible choice, and all four possible combinations have been studied [6]. Furthermore, *consistent* [11, 12] and *symmetric* [13, 14] formulations have also been proposed; of all these, the Inverse-Compositional Gauss-Newton approach described in the following section seems to be the most widely used because of its noise robustness and numerical efficiency;
- the displacement functions u and v map a point in one of the images to the other one. Thus, the proposed formulation is sometimes described in the literature in terms of *warping functions* (hereafter referred to as $W(\mathbf{x}, \mathbf{p})$ or $\zeta(\mathbf{x}, \mathbf{p})$, where $\mathbf{p}$ is the full set of parameters controlling the function, i.e. $\mathbf{p} = \mathbf{s} \cup \mathbf{t}$).

Regardless of the formulation, DIC requires the comparison of two intensity fields; since the displacements are not restricted to integer values, the intensity of at least one of the images must be interpolated at non-integer locations. In addition, the algorithm requires the calculation of intensity derivatives in both the x and y directions at the same locations. Thus, intensity interpolation and derivative estimation are by far the most computationally intensive steps of DIC. The aim of this work is to identify an efficient and reliable formulation that minimizes these requirements.

2. The Inverse Compositional Approach

The most computationally intensive step of a DIC algorithm is the assembly of the Hessian matrix. Indeed, this step requires the evaluation of the intensity errors and derivatives at each point and for each iteration. Since $u \wedge v \in \mathbb{R}$, at least one of the intensity evaluations of each pair of points requires an interpolation (of either the field $f(\mathbf{x})$ or $g(\mathbf{x})$); however, if the intensity derivatives are computed at fixed locations, it is possible to factor-out this step by pre-evaluating the derivatives before entering the iterative loop: considering that the computation of the derivatives requires approximately the same number of operations of the interpolation, this results in a speed-up by a factor of two.

This objective is achieved by the Inverse Compositional approach [15, 16]. In this algorithm, an incremental parameter update is computed as

$$\Delta\mathbf{u}^{(n)} = \operatorname{argmin} \sum_k \left\{ g\left[W\left(\mathbf{x}_k, \mathbf{p}^{(n)}\right)\right] - f\left[\zeta\left(\mathbf{x}_k, \mathbf{u}_k^{(n)}\right)\right]\right\}^2 \tag{6}$$

where the warping function ζ, related to the reference image, is always an identity, i.e. the shape of the reference subset is reset at the beginning of each iteration. The use of this assumption in combination with the exchange of roles between the reference and the current image leads to a significant computational improvement. Since the shape functions of the reference image never change, the sampled points are always the same; moreover, due to the exchange of the roles, the computation of intensity derivatives is performed on $f(\mathbf{x})$ (at locations that never change). Thus, the Hessian matrix can be pre-computed before entering the iterative loop[1].

The solution process is as follows:

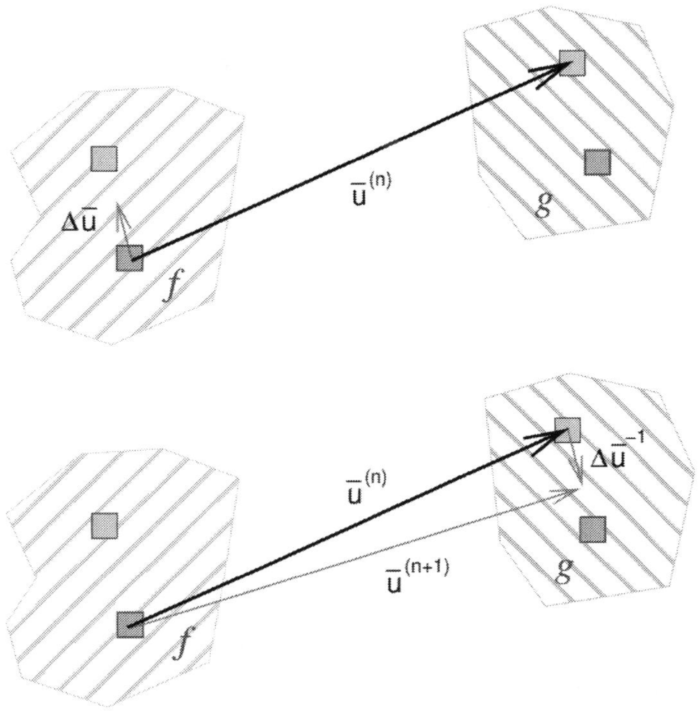

Figure 1: Inverse Compositional mapping. Top: Starting from the reference image f, the green point is coupled with the pink one in the test image g using a forward mapping (should be green). The result of the Gauss-Newton refinement step is the *backward* correction vector $\mathbf{\Delta u}$. Bottom: The backward correction is inverted and composed with the previous vector to produce an improved *forward* estimate (the brown vector).

(i) compute the x and y derivatives of the reference image at the center of the pixels of the subset (i.e. at integer locations);

(ii) compute an initial estimate of the corresponding points in the current image g; thus, the set of parameters $\mathbf{p}$ is related to a *forward* mapping (from f to g);

(iii) using the current forward displacement field, estimate intensity of the pixels of the subset in the current image g (this step requires polynomial interpolation);

(iv) assemble the solution matrix using the pre-computed intensity and derivative of the reference image in combination with the new values of intensities of corresponding points in the current image;

[1] A further significant advantage is that it is possible to use an ad-hoc differential operator working on integer locations.

IOP Conf. Series: Materials Science and Engineering 1306 (2024) 012038 doi:10.1088/1757-899X/1306/1/012038

(v) solve the system. Sincethe roles of f and g are swapped, the solution vector is a *backward* displacement-correction set of parameters;

(vi) since the current set of parameters is related to a forward mapping while the correction set is related to a backward mapping and we are warping both the current and the reference images, it is impossible to use the standard approach and add the correction to $\mathbf{p}$. Instead, the correction must be inverted and composed with the current one, i.e.

$$W\left(\mathbf{x},\mathbf{p}^{(n+1)}\right) \leftarrow W\left(\mathbf{x},\mathbf{p}^{(n)}\right) \circ \zeta^{-1}\left(\mathbf{x},\Delta\mathbf{u}\right)$$

(vii) check for convergence; on failure, go back to step 3, otherwise exit the loop.

Note that $\zeta\left(\mathbf{x},\Delta\mathbf{u}\right)$ is never updated, so the intensities $f(\mathbf{x}_k)$ and their x and y derivatives $f_x(\mathbf{x}_k)$ and $f_y(\mathbf{x}_k)$, once computed do not need to be updated. The whole process is illustrated in figure 1. In principle, we should move the point in f (figure 1-top); instead, we move the starting point of the "virtual" backward displacement (figure 1-bottom) in such a way that the mapped point of f is always the same. This change of viewpoint takes place in step 6, which is therefore the key point of the algorithm. In fact, the inversion of the incremental displacement and its composition with the current one is a relatively simple operation if a linear mapping is used, because in this case $W(\mathbf{x},\mathbf{p})$ can be expressed in matrix form as

$$\mathbf{u} = \begin{bmatrix} 1+s_1 & s_2 & s_0 \\ t_1 & 1+t_2 & t_0 \\ 0 & 0 & 1 \end{bmatrix} \cdot \begin{pmatrix} \xi \\ \eta \\ 1 \end{pmatrix} \text{ i.e. } \begin{bmatrix} 1+u_x & u_y & u_0 \\ v_x & 1+v_y & v_0 \\ 0 & 0 & 1 \end{bmatrix} \cdot \begin{pmatrix} \xi \\ \eta \\ 1 \end{pmatrix} \tag{7}$$

Thus, the inversion of the incremental displacement corresponds to a matrix inversion and the composition to a matrix multiplication. However, the former operation constitutes a significant limitation, not so much because of the computational load[2] but because the assumed displacement functions must be invertible. This is a relatively straightforward requirement in the linear case, but only recently has an invertible parabolic function been suggested [18]; to the author's knowledge, there is no formulation for cubic shape functions.

3. From Inverse Compositional to Backward Subtractive

The key point to an efficient implementation of DIC is the pre-computation of the Hessian matrix to avoid evaluating the x and y derivatives of the intensities at all points of the subset for each iteration. As shown in the previous section, this can be achieved by using a backward formulation in combination with the inversion of the incremental displacement[3]. However, this implies the inversion of the incremental displacement matrix and its composition with the current one. The objective of this section is to remove this limitation.

The first step towards this goal is to observe that the displacement function is only used at the beginning of each iteration to identify coupled points in the test and reference images. Thus, we can formally state that the displacement vector starts from g and arrives at f, even though we are actually storing the parameters of the vector from f to g.

The second observation is the fact that the polynomial functions used in (4) form the basis of an m-dimensional real space—$m = 3$ for linear shape functions, 6 for parabolic, 10 for cubic, and so on. Thus, a set of parameters (either s or t) constitutes the coordinate of a point in this space, so that the inverted direction vector can easily be obtained by reversing the sign of all components.

[2] Both matrix multiplication and Cholesky decomposition require $\mathcal{O}(n^3)$ multiplications and a similar number of sums [17], but $n = 3$ in the linear case,

[3] Note that the simple use of a backward formulation is not sufficient, as with this formulation the intensity-evaluation points in f would move, thus requiring a re-computation of $f(\mathbf{x}_k)$ and its derivatives.

Thus, assuming that the roles of reference and current images are swapped (i.e. to adopt a backward formulation), the SSD χ^2 error function becomes

$$\chi^2_{SSD} = \sum_{k=1}^{N} [g(x_k, y_k) - f(x_k + u_k, y_k + v_k)]^2 \tag{8}$$

where u_k and v_k are the *backward* displacement components ($-u_k$ and $-v_k$ are the forward displacement components). Note that

- unlike the Inverse Compositional approach, we apply the warp function only to the reference image;
- the radiosity compensation has to be applied to g;
- we are working in a purely additive framework.

Expanding $f(x_k + u_k, y_k + v_k)$ in Taylor series, equation (8) becomes

$$\chi^2_{SSD} = \sum_{k=1}^{N} \left\{ g(\mathbf{x}_k) - \left[f(\mathbf{x}_k) + \sum_{i=1}^{m} \left(\frac{\partial f(\mathbf{x}_k)}{\partial x} \frac{\partial u}{\partial s_i} ds_i + \frac{\partial f(\mathbf{x}_k)}{\partial y} \frac{\partial v}{\partial t_i} dt_i \right) \right] \right\}^2$$

Thus, the normal-equation solution matrix $\mathbf{A}$ and the known term vector $\mathbf{b}$ become $\mathbf{A} = \sum_k \mathbf{h}_k \mathbf{h}_k^{\mathrm{T}}$ and $\mathbf{b} = \sum_k (f_k - g_k)\mathbf{h}_k$, where

$$\mathbf{h}_k^{\mathrm{T}} = \{-f_x, -f_x\xi, -f_x\eta, \ldots, -f_y, -f_y\xi, -f_y\eta, \ldots\}$$

and the unknown vector is

$$\mathbf{x}^{\mathrm{T}} = \{ds_0, ds_1, \ldots, ds_n, dt_0, dt_1, \ldots, dt_n\}$$

Note that (8) assumes a backward formulation, i.e. u and v are backward displacements and the incremental parameters[4] resulting from solving the system $\mathbf{A\,x} = \mathbf{b}$ are backward values that should be added to the current backward displacement components to make the point in the reference image f equal to the start value in the current image g. But this would require an update of the target point in the reference frame (i.e. a new calculation of all the f_k of the subset and their derivatives f_x and f_y at non-integer locations). Instead, following the IC-GN approach, we move the starting point in the current image g; but because we are using an additive formulation, and in force of our observation on the vector space, we simply add the sign-reversed incremental vector (figure 2, bottom) instead of computing and composing the inverse.

The resulting algorithm is identical to the Inverse Compositional algorithm, except that step 6 now reads

v ...

vi add $-\Delta\mathbf{p}$ to the current set of parameters, i.e. *subtract* the incremental vector $\Delta\mathbf{p}$ from p (hence the name of the algorithm).

vii ...

Thus, although the algorithm is formulated in terms of backward displacement, we start with (and refine at each step) a set of parameters that relate to forward displacement. The Hessian matrix is computed only at the beginning of the algorithm (before entering the iteration), whereas to perform a new iteration step we only need to interpolate intensity values in g (the current image).

[4] Note that when radiosity compensation is used, the first and second entries in the $\mathbf{x}$ vector are actually total values; indeed, $\mathbf{x}^{\mathrm{T}} = \{a, b, ds_0, ds_1, \ldots, ds_n, dt_0, \ldots, dt_n\}$.

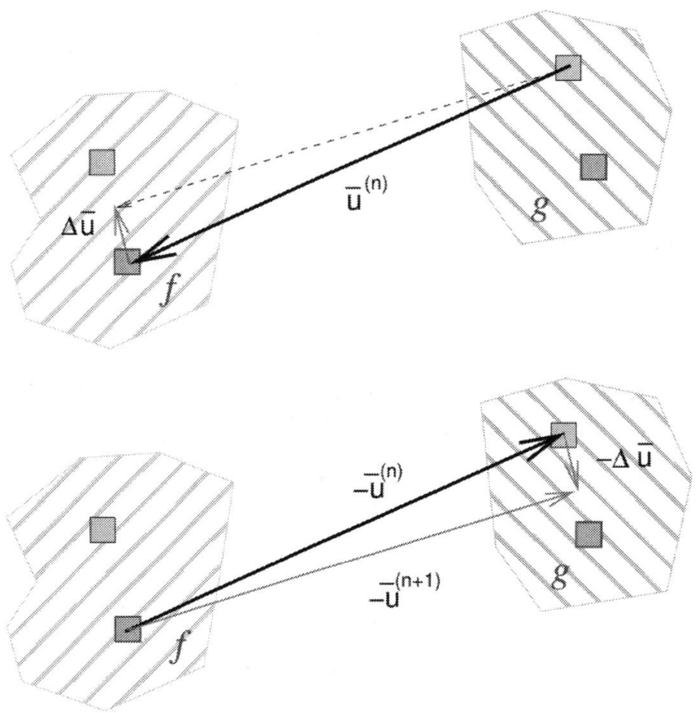

Figure 2: Backward Subtractive algorithm. Above: the algorithm is formulated as a "pure" backward additive algorithm (thus, the subset on image f should move), but (below) we actually store the forward displacement ($-\mathbf{u}$) and add $-\Delta\mathbf{u}$.

4. Equivalence of BS-GN and IC-GN

In [1] the authors show that $\Delta\mathbf{u} = -\Delta\mathbf{p}$ for rigid body motion. In this section we want to show that this is true up to the first order in general and not only for rigid body motion. It is worth noting that near the minimum the second derivatives go to zero, so this condition is sufficient to prove the equivalence.

Assuming that Δu_0, Δu_x, ..., Δv_y (Δs_0, Δs_1, ..., Δt_2) are the incremental values computed using the backward formulation, the (linear) shape functions associated with a generic step of the Backward Subtractive methods are

$$\mathbf{u} = \begin{bmatrix} 1+u_x-\Delta u_x & u_y-\Delta u_y & u_0-\Delta u_0 \\ v_x-\Delta v_x & 1+v_y-\Delta v_y & v_0-\Delta v_0 \\ 0 & 0 & 1 \end{bmatrix} \cdot \begin{pmatrix} \xi \\ \eta \\ 1 \end{pmatrix} \tag{9}$$

where u_0, u_x, ..., v_y are the total *forward* values of the displacement components at the start of the step and ξ and η are local coordinates with respect to the center of the subset.

Computing the corresponding matrix for the Inverse Compositional algorithm requires the inversion of the incremental matrix and composition of the result with the current displacement function:

$$\mathbf{u} = \begin{bmatrix} 1+u_x & u_y & u_0 \\ v_x & 1+v_y & v_0 \\ 0 & 0 & 1 \end{bmatrix} \cdot \begin{bmatrix} 1+\Delta u_x & \Delta u_y & \Delta u_0 \\ \Delta v_x & 1+\Delta v_y & \Delta v_0 \\ 0 & 0 & 1 \end{bmatrix}^{-1} \cdot \begin{pmatrix} \xi \\ \eta \\ 1 \end{pmatrix} \tag{10}$$

The inverse of the incremental matrix can be computed analytically and then multiplied by the current matrix. The first and second rows of the composed matrix $\mathbf{C} = \mathbf{P}\,\Delta\mathbf{u}^{-1}$, thus result:

$$C(1,1) = \frac{(1+u_x)(1+\Delta v_y)}{d} - \frac{u_y \Delta v_x}{d}$$
$$C(1,2) = \frac{u_y(1+\Delta u_x)}{d} - \frac{(1+u_x)\Delta u_y}{d}$$
$$C(1,3) = u_0 + \frac{u_y(\Delta u_0 \Delta v_x - \Delta u_x \Delta v_0 - \Delta v_0)}{d} + \frac{(1+u_x)(\Delta u_y \Delta v_0 - \Delta u_0 \Delta v_y - \Delta u_0)}{d}$$
$$C(2,1) = \frac{v_x(1+\Delta v_y)}{d} - \frac{(1+v_y)\Delta v_x}{d}$$
$$C(2,2) = \frac{(1+v_y)(1+\Delta u_x)}{d} - \frac{v_x \Delta u_y}{d}$$
$$C(2,3) = v_0 + \frac{(1+v_y)(\Delta u_0 \Delta v_x - \Delta u_x \Delta v_0 - \Delta v_0)}{d} + \frac{v_x(\Delta u_y \Delta v_0 - \Delta u_0 \Delta v_y - \Delta u_0)}{d}$$

where $d = 1 + \Delta u_x - \Delta u_y \Delta v_x + \Delta v_y + \Delta u_x \Delta v_y$.

Matrix $\mathbf{C}$ can be simplified by neglecting all products of incremental factors. Furthermore, by noting that u_x, u_y, v_x and v_y are more than one order of magnitude smaller than displacements, all mixed products (e.g. $u_y \Delta u_x$, $u_x \Delta u_y$, ...) can also be neglected. The first order approximation of $\mathbf{C}$ thus is

$$\hat{C}(1,1) = \frac{1+u_x+\Delta v_y}{1+\Delta u_x+\Delta v_y} \qquad \hat{C}(1,2) = \frac{u_y - \Delta u_y}{1+\Delta u_x+\Delta v_y} \qquad \hat{C}(1,3) = u_0 - \frac{\Delta u_0}{1+\Delta u_x+\Delta v_y}$$
$$\hat{C}(2,1) = \frac{v_x - \Delta v_x}{1+\Delta u_x+\Delta v_y} \qquad \hat{C}(2,2) = \frac{1+v_y+\Delta u_x}{1+\Delta u_x+\Delta v_y} \qquad \hat{C}(2,3) = v_0 + \frac{\Delta v_0}{1+\Delta u_x+\Delta v_y}$$

where $1 + \Delta u_x + \Delta v_y$ is the first order approximation of d.

$\hat{C}(1,1)$ and $\hat{C}(2,2)$ can be further simplified by summing and subtracting to the numerator Δu_x and Δv_y respectively, thus obtaining

$$\hat{C}(1,1) = \frac{1+u_x+\Delta v_y+\Delta u_x-\Delta u_x}{1+\Delta u_x+\Delta v_y} = 1 + \frac{u_x - \Delta u_x}{1+\Delta u_x+\Delta v_y}$$

and

$$\hat{C}(2,2) = \frac{1+v_y+\Delta u_x+\Delta v_y-\Delta v_y}{1+\Delta u_x+\Delta v_y} = 1 + \frac{v_y - \Delta v_y}{1+\Delta u_x+\Delta v_y}$$

Finally, we can observe that the factors Δu_x and Δv_y in the denominator of all the matrix entries are very small and go to zero with iterations, so the denominator is almost unitary, i.e. the (IC-GN) matrix $\mathbf{C}$ is identical to the BS-GN matrix (9) to the first order.

5. Experimental Validation

To validate the new formulation, we processed one of the experimental datasets related to the polynomial bias analysis described in [19] with the new algorithm. As the experimental apparatus and its calibration are fully detailed in the aforementioned paper, we will only summarize the relevant data in this article: each data set consists of 43 images acquired during a rigid body translation test. The experiment simulates real-use conditions, so the images were acquired using a telecentric lens (Edmund Optics 0.9× Large-Format Telecentric Lens). Consequently, the displacement corresponding to one pixel of our camera (AVT Pike F-421

IOP Conf. Series: Materials Science and Engineering 1306 (2024) 012038 doi:10.1088/1757-899X/1306/1/012038

Figure 3: Experimental setup. Left: during calibration; note the two Twyman–Green Interferometers used to accurately measure displacements. Right: experimental configuration during DIC tests. One of the interferometers has been removed to allow the installation of a 45-degree mirror in front of the telecentric lens. The second interferometer is still installed, but is not used during the acquisition: displacements are recorded using the strain gauge bridges installed on piezo actuators.

FireWire-B 1.2 CCD camera; 2048 × 2048 pixels; pixel size 7.4 μm) is less than 10 μm. To accurately control displacements, a monolithic translation stage actuated by two low-voltage piezo crystals (ThorLabs PZS001, nominal displacement 11.6 $\pm$ 2 μm at 100 V) was built. The PZTs are instrumented using strain gauges (full-bridge configuration) to allow closed-loop control of the system.

Figure 3 shows the experimental setup during calibration (left) and during actual test (right). To correlate the strain gauge bridge signal with the actual displacements, two Twynam-Green interferometers were built taking care to install one of the mirrors on the translation stage. Thus, an accurate calibration was performed by monitoring the phase signal[5]. Figure 3-right shows the experimental setup during analysis. To reduce vibrations, the camera was installed horizontally on the optical bench: the specimen is thus imaged by a 45 degree tilted mirror. For installation, one of the two interferometers (no longer required) was removed.

Figure 4 shows the results of the analysis. Each curve plots the difference between the mean displacement value and the strain-gauge measured translation. Specifically, we sampled the image surface over a regular grid, taking care to obtain independent results (i.e. the sampling step is larger than a subset, so that each subset does not share pixels with its neighbors). For each image, we then computed the mean and standard deviation of the displacements. Finally, we plotted the difference between the normalized displacement (as acquired by strain-gauge signals) and the mean of the DIC-estimated displacements. In order to perform the normalization, we had to identify pixel size. This should be 8.22 μm (the nominal pixel size is 7.4 μm and the (nominal) primary magnification of the telecentric lens is 0.9×), but the actual size depends on the alignment of the lens and the actual primary magnification. Instead of carrying out a difficult experimental analysis, we simply examined the error curves associated with different interpolation polynomials. As none of them shows bias for displacement equals to $n/2$ pixel, it suffices to look for their intersection points: the second corresponds to a shift of one-pixel.

Analysis of fig. 4 shows that the BS-GN method behaves in the same way as the IC-GN. It is well known that the IC-GN algorithm is less affected by polynomial bias than FA-GN because

[5] For measurement, strain gauges are much more convenient than interferometers because they do not require any post-processing (apart from the scale factor) and provide absolute data.

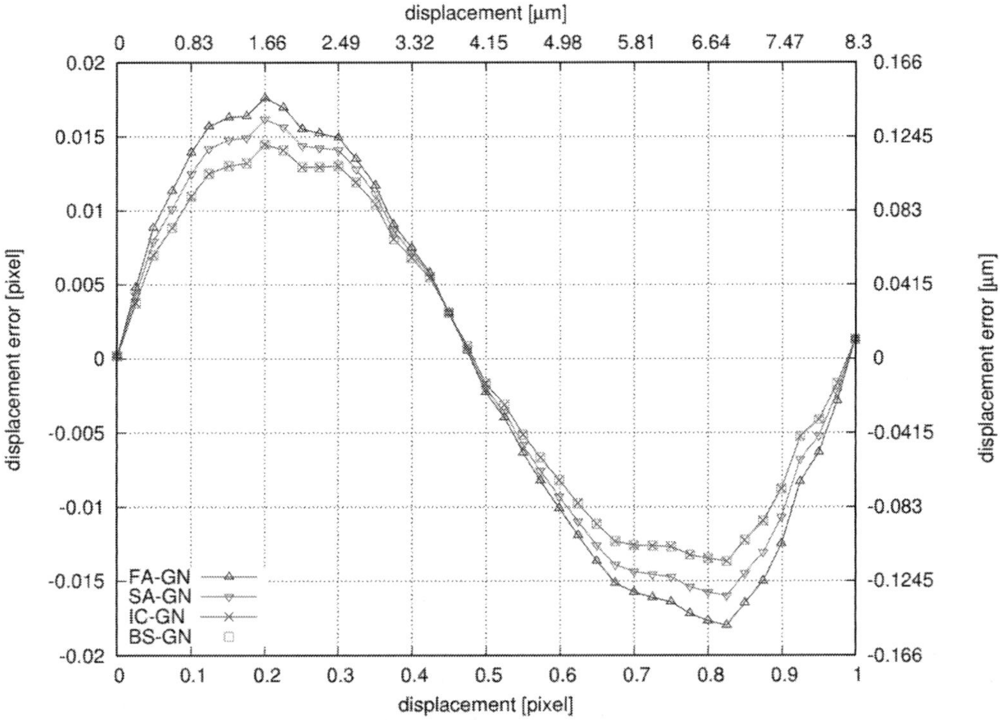

Figure 4: Bias of the c_4 polynomial. The Forward-Additive Gauss-Newton method (FA-GN) shows the largest bias whereas the Inverse-Compositional (IC-GN) and Backward-Subtractive (BS-GN) methods the smallest. The Symmetric-Additive approach stays in the middle.

the intensity derivatives are calculated at integer locations. But the Backward-Subtractive algorithm also uses pre-computed image intensity derivatives estimated at integer locations, so, we expect a similar performance. Indeed, the results of the two approaches are so similar that they appear as a single curve in the figure. The Forward additive shows a somewhat larger bias and the symmetric approach falls in the middle.

Figure 5 shows the results of the analysis of sample 14 of the DIC Challenge dataset. The 2D–DIC Challenge initiative [20] supplies a set of sample images and a set of analysis guidelines for standardized reporting of 2D–DIC results. Of all the datasets, Sample 14 is becoming the de facto standard for testing DIC algorithms and for this reason we show here the results of the BS-GN algorithm when dealing with this data set.

We analyzed the L_1, L_3 and L_5 load cases using a subset size of 21×21 pixels, linear shape functions and the BS-GN formulation. The image surface was sampled every two pixels and radiosity compensation was enabled to mimic real-use conditions (actually, radiosity compensation is not required as the images are numerically generated, so there is no illumination variation). Figure 5-top shows the mean displacement versus the x-location of the three load cases. As the displacements are only along the x axis, it is possible to statistically analyze each column of the image to extract the mean and standard deviation of the estimated displacements. The translucent areas around the solid lines correspond to ± 1 standard deviation error. We have also reported the expected displacements: apart from a few local errors, a fairly good agreement with the expected results was observed. Figure 5-bottom shows the density plot and contour lines associated withthe L_5 load case to visually support the analysis: the blue regions correspond to maximum negative displacements while the yellow areas correspond to maximum positive values

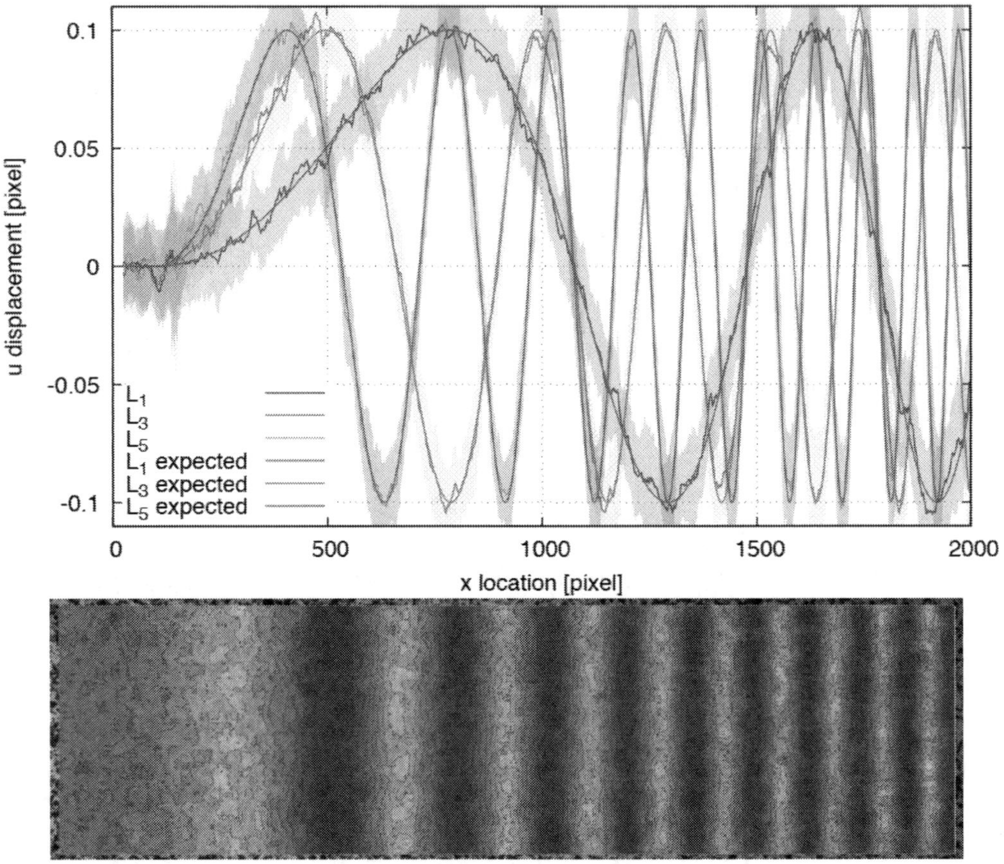

Figure 5: DIC Challenge Sample 14. Top: mean displacements related to load levels L_1, L_3 and L_5. The translucent regions correspond to ± 1 standard deviation. Bottom: density plot of u displacements related to step L_5.

(respectively -0.1 and $+0.1$ pixel).

6. Discussion and Conclusions

This paper discusses a modification of the Inverse Compositional DIC formulation. In particular, the inversion and composition steps are replaced by a simple subtraction. Since the Hessian matrix is evaluated in the template image at integer locations, it can be pre-evaluated at the beginning of the computation exactly as in the IC-GN method. For the same reason, the proposed approach is not affected by noise-bias when appropriate differential operators are used [21]. Furthermore, the inversion and matrix multiplications are replaced by a simple subtraction, so the resulting algorithm is theoretically somewhat faster[6].

A significant advantage of the proposed formulation is the fact that it removes the constraint of using invertible shape functions. Although a parabolic formulation has recently been proposed [18], there is no solution for higher order shape functions, making it impossible to correctly describe complex strain gradient regions. Finally, although the displacement field should always be invertible when affine shape functions are used [6], the matrix may be ill-conditioned and

[6] The speed gain is almost unnoticeable if linear shape functions are used, but may become significant if higher order shape functions are used.

introduce significant numerical errors. This problem obviously does not exist in the proposed formulation.

References

[1] Michael A. Sutton, Jean-José Orteu, and Hubert Schreier. *Image Correlation for Shape, Motion and Deformation Measurements: Basic Concepts,Theory and Applications.* Springer, New York, USA, 2009. ISBN 978-0-387-78746-6. doi: 10.1007/978-0-387-78747-3.

[2] A. Del Bimbo, P. Nesi, and J.L.C. Sanz. Analysis of optical flow constraints. *IEEE Trans. Image Process.*, 4(4):460–469, Apr 1995. ISSN 1057-7149. doi: 10.1109/83.370674.

[3] Berthold Horn, Berthold Klaus, and Paul Horn. *Robot vision.* MIT press, 1986.

[4] RB Lehoucq, PL Reu, and DZ Turner. The effect of the ill-posed problem on quantitative error assessment in digital image correlation. *Experimental Mechanics*, pages 1–13, 2017.

[5] Bruce. D Lucas and Takeo Kanade. An iterative image registration technique with an application to stereo vision. In *Proceedings of Imaging Understanding Workshop*, volume 130, pages 121–130, 1981.

[6] W Tong. Formulation of lucas–kanade digital image correlation algorithms for non-contact deformation measurements: a review. *Strain*, 49(4):313–334, 2013.

[7] B. Pan, H. Xie, and Z. Wang. Equivalence of digital image correlation criteria for pattern matching. *Applied Optics*, 49(28):5501–5509, 2010.

[8] André-Louis Cholesky. Sur la résolution numérique des systèmes d'équations linéaires. *Bulletin de la Sabix. Société des amis de la Bibliothèque et de l'Histoire de l'École polytechnique*, (39):81–95, 2005.

[9] W. H. Press, S. A. Teukolsky, W. T. Vetterling, and B. P. Flannery. *Numerical Recipes. The Art of Scientific Computing.* Cambridge University Press, 3rd edition, 2007. ISBN 0-521-88068-8.

[10] J Neggers, B Blaysat, JPM Hoefnagels, and MGD Geers. On image gradients in digital image correlation. *International Journal for Numerical Methods in Engineering*, 105(4): 243–260, 2016.

[11] Zhengyou Zhang. Motion and structure from two perspective views: from essential parameters to euclidean motion through the fundamental matrix. *JOSA A*, 14(11):2938–2950, 1997.

[12] Gary E Christensen and Hans J Johnson. Consistent image registration. *IEEE transactions on medical imaging*, 20(7):568–582, 2001.

[13] C. Quentin Davis and Dennis M. Freeman. Statistics of subpixel registration algorithms based on spatiotemporal gradients or block matching. *Optical Engineering*, 37(4):1290–1298, 1998.

[14] Brian B Avants, Charles L Epstein, Murray Grossman, and James C Gee. Symmetric diffeomorphic image registration with cross-correlation: evaluating automated labeling of elderly and neurodegenerative brain. *Medical image analysis*, 12(1):26–41, 2008.

[15] Simon Baker and Iain Matthews. Lucas-kanade 20 years on: A unifying framework. *International Journal of Computer Vision*, 56(3):221–255, 2004.

[16] Heung-Yeung Shum and Richard Szeliski. Construction of panoramic image mosaics with global and local alignment. *International Journal of Computer Vision*, 48(2):151–152, 2002.

[17] Lloyd N Trefethen and David Bau III. *Numerical linear algebra*, volume 50. Siam, 1997.

[18] Ruixiang Bai, Hao Jiang, Zhenkun Lei, and Weikang Li. A novel 2nd-order shape function based digital image correlation method for large deformation measurements. *Optics and Lasers in Engineering*, 90:48–58, 2017.

IOP Conf. Series: Materials Science and Engineering 1306 (2024) 012038 doi:10.1088/1757-899X/1306/1/012038

[19] Antonio Baldi and Filippo Bertolino. Experimental analysis of the errors due to polynomial interpolation in digital image correlation. *Strain*, 51(3):248–263, 2015. doi: 10.1111/str.12137.

[20] Phillip L Reu, Eveline Toussaint, Elizabeth Jones, Hugh A Bruck, Mark Iadicola, Ruben Balcaen, Daniel Z Turner, Thorsten Siebert, Pascal Lava, and M Simonsen. Dic challenge: developing images and guidelines for evaluating accuracy and resolution of 2d analyses. *Experimental Mechanics*, pages 1–33, 2017. URL `http://sem.org/dicchallenge/`.

[21] Xinxing Shao, Xiangjun Dai, and Xiaoyuan He. Noise robustness and parallel computation of the inverse compositional gauss–newton algorithm in digital image correlation. *Optics and Lasers in Engineering*, 71:9–19, 2015.

IOP Conf. Series: Materials Science and Engineering 1306 (2024) 012039 doi:10.1088/1757-899X/1306/1/012039

Homogenized Stiffness Matrix of Two-Node Elements through Experimental Flexibility tests

C Iandiorio[1], F Cafolla[1], E Marotta[1] and P Salvini[1]

[1] University of Rome "Tor Vergata", Department of Enterprise Engineering, Via del Politecnico 1 – 00133 Rome, Italy

Corresponding author: christian.iandiorio@uniroma2.it

Abstract. Homogenization is a numerical technique to obtain the equivalent mediated response of a complex material. However, there are some scenarios in which the complexity of the physical phenomena present in the real structure makes it difficult to generate a faithful Representative Volume Element (RVE). In these cases, it may turn easier and more reliable to perform a homogenization directly by experimental test results. For this reason, we present a procedure to build-up the whole stiffness matrix starting from experiments. This approach is here discussed to get a homogenised two-node beam element. The procedure requires some flexibility measurements, thus allowing fewer measurements if compared to stiffness-based approach. The method is verified with some experiments carried on a frame structure. The comparisons with a Finite Element model build up with stiffness matrix assembly demonstrates the validity and robustness of the proposed procedure.

1. Introduction

Homogenization is a method used in the field of structural analysis which allows the determination of the effective (mediated) mechanical properties of complex material made up of different phases (constituents), combining the properties of the individual constituents and their geometry [1,2]. Some typical applications of this method regard composite [1], cellular, granular or porous materials [3]; the mechanical properties of them could significantly vary at a microscopic level, but homogenization provides the macroscopic behaviour for structural design.

There are two main approaches for homogenization. The first one uses variational approaches based on asymptotic developments [4,5] and it is useful to obtain an analytical understanding of upper/lower homogenization bounds. Unfortunately, this approach is limited to a few elementary geometries and phase distributions, as the case of ellipsoidal inclusions in a structural bulk. The second approach is numerical homogenization [6,7]. This last is the most used for its helpfulness when addressing engineering problems. The technique requires the generation of a fully detailed CAD model, called Representative Volume Element (RVE) to extract the equivalent elastic directional response and material-stiffness coefficients. Some parameters could significantly affect the final results using this numerical process, i.e. the reliability of the substitution: the RVE dimension, the orientation and geometry of the constituent phases [8] and the level of discretization.

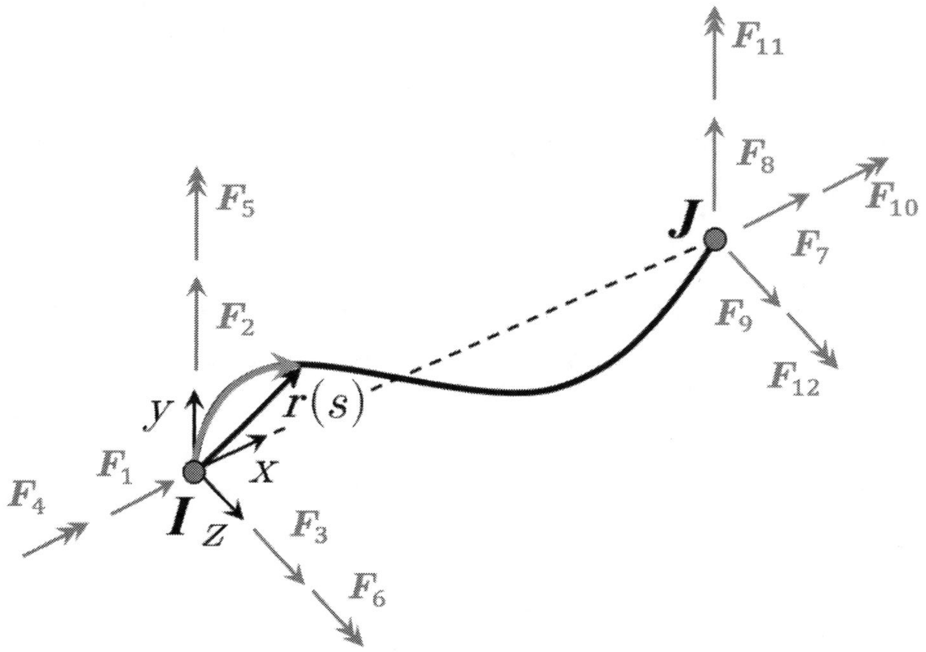

Figure 1. Forces and geometry notation of the two-node reference element

Even if homogenization regards elastic material response these properties can vary with temperature, pressure, humidity and other environmental factorsa so that aprecise and careful modelling is necessary. In some particular cases numerical homogenization is difficult to apply. Examples are: biological materials which may present a pseudo-random phases distribution (e.g. a bones) [9,10]; damaged elements affected by cyclic loading or ageing (e.g. reinforced beam/column concrete) [11,12]; elements in which the stiffness evaluation requires a multiphysics approach (e.g. a hollow pipe with closed caps with internal gas, subjected to bending). In these cases, it may be better to perform the homogenization directly through experimental results.
This paper proposes an experimental procedure applied to two-nodes elements (i.e., straight or curved beam element).
The presented procedure uses a flexibility approach to build-up the stiffness matrix [13-15] since flexibility tests are much easier to carry on than stiffness measurements.

2. Stiffness matrix from the minimum number of Flexibility measurements

The stiffness matrix of a two-node element with six degrees of freedom (d.o.f.'s) per node (figure 1) is the following:

$$[\boldsymbol{K}]_{12\times 12} = \begin{bmatrix} k_{0101} & \cdots & k_{0112} \\ \vdots & \ddots & \vdots \\ k_{1201} & \cdots & k_{1212} \end{bmatrix} = \begin{bmatrix} [\boldsymbol{K}_{II}] & [\boldsymbol{K}_{IJ}] \\ [\boldsymbol{K}_{JI}] & [\boldsymbol{K}_{JJ}] \end{bmatrix} \tag{1}$$

where in the right side of the equation 1 it is represented the partitioned form according to the end nodes $\boldsymbol{I}$ and $\boldsymbol{J}$.
It is well-known that a structural Finite Element model works with stiffness matrices instead of flexibility ones [16], but for the following developments it is useful to define the symmetric element flexibility matrix as follow:

$$[\boldsymbol{C}]_{12\times12} = \begin{bmatrix} c_{0101} & \cdots & c_{0112} \\ \vdots & \ddots & \vdots \\ c_{1201} & \cdots & c_{1212} \end{bmatrix} = \begin{bmatrix} [\boldsymbol{C}_{II}] & [\boldsymbol{C}_{IJ}] \\ [\boldsymbol{C}_{JI}] & [\boldsymbol{C}_{JJ}] \end{bmatrix} \quad (2)$$

Note that it is not possible to compute the whole stiffness matrix by simple inversion of the flexibility one (and vice-versa) inasmuch, for a free-free element, they are singular matrices.
The aim is to build up the homogenized stiffness matrix (1) by experimental measurements. Fulfill this mission with a directl approach, i.e. by performing all the 78 measures (not 144 if symmetry is used) is too onerous, as well as gradually less reliable due to experimental uncertainties involved in every measurement. Therefore, the best approach is to explore if exist a procedure that guarantees the minimum number of measurements.
Furthermore, perform stiffness measurements could be very difficult; the definition of the generic stiffness component $k_{ij} = F_i/u_j$ imply to impose to measure the force F_i explicated along the d.o.f. "i" when a displacement u_j is imposed along the "j" d.o.f. (a 2D example in figure 2) while all other displacement being null through the action of constraints. Therefore, a direct stiffness measure requires to face a statically undetermined structure, which is very difficult to conduct experimentally.

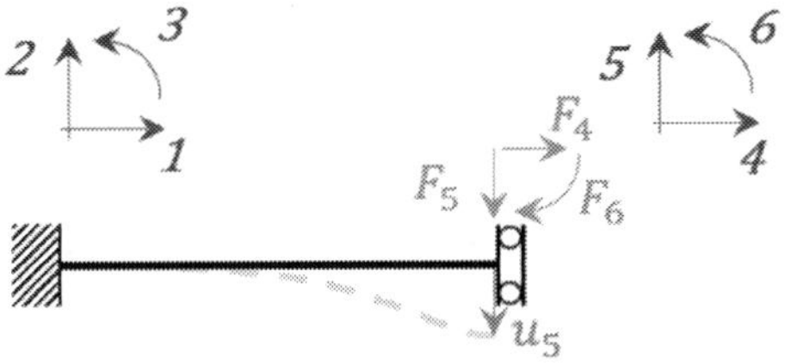

$$k_{45} = \frac{F_4}{u_5} \qquad k_{55} = \frac{F_5}{u_5} \qquad k_{65} = \frac{F_6}{u_5}$$

Figure 2. Stiffness coefficients obtained from a statically undetermined structure

Differently, the definition of the generic flexibility $c_{ij} = u_i/F_j$ imply to measure the displacement u_i along the d.o.f. "i" when a force F_j is applied along the "j" d.o.f., while all the other forces are null, except the reactive forces that make statically determined the structure (a 2D example in figure 3). Therefore, measure a flexibility (i.e. work with a statically determined structure) is experimentally much easier that works with stiffness measurements.

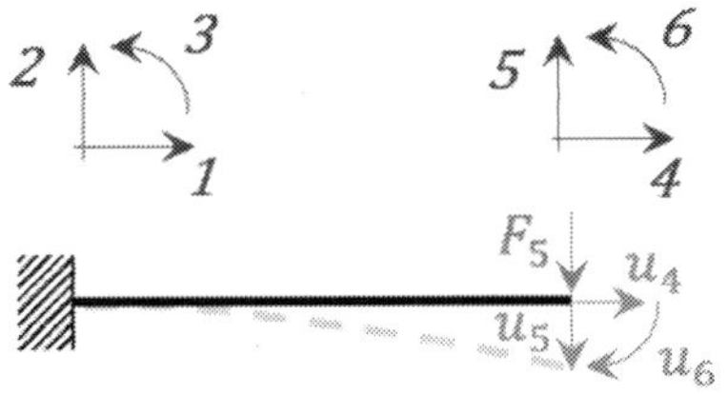

$$c_{45} = \frac{u_4}{F_5} \qquad c_{55} = \frac{u_5}{F_5} \qquad c_{65} = \frac{u_6}{F_5}$$

Figure 3. Flexibility coefficients obtained from a statically determined structure

Since one does not know a-priori either the material behaviour or the element geometry, the specimen results as a sort of black box. Despite this, the mechanical equilibrium must hold. The concentrated forces and moment are applied only at the nodes; therefore, we can impose the fundamental equations of statics:

$$\sum_i \boldsymbol{F}_i = \boldsymbol{0} \tag{3}$$

$$\sum_i \boldsymbol{M}_i + \boldsymbol{r}_i \times \boldsymbol{F}_i = \boldsymbol{0} \tag{4}$$

Where $\boldsymbol{F}_i$ are the *i-th* force, $\boldsymbol{M}_i$ are the *i-th* concentrated moment and $\boldsymbol{r}_i$ the radius vector *i-th* force, starting from the local element reference system with origin on the node $\boldsymbol{I}$.
It is convenient, for the sake of compactness, to define the following transformation matrices:

$$[\boldsymbol{T}_I(s)] = \begin{bmatrix} [\boldsymbol{I}]_{3\times3} & [\boldsymbol{0}]_{3\times3} \\ -[\boldsymbol{r}(s)]_\times & [\boldsymbol{I}]_{3\times3} \end{bmatrix} \tag{5}$$

$$[\boldsymbol{T}_J(s)] = \begin{bmatrix} [\boldsymbol{I}]_{3\times3} & [\boldsymbol{0}]_{3\times3} \\ [\boldsymbol{r}_{IJ}]_\times - [\boldsymbol{r}(s)]_\times & [\boldsymbol{I}]_{3\times3} \end{bmatrix} \tag{6}$$

in which $[\boldsymbol{I}]_{3\times3}$ is the 3×3 unitary matrix, $[\boldsymbol{0}]_{3\times3}$ the 3×3 zeros matrix and

$$[\boldsymbol{r}(s)]_\times = \begin{bmatrix} 0 & -z(s) & y(s) \\ z(s) & 0 & -x(s) \\ -y(s) & x(s) & 0 \end{bmatrix} \tag{7}$$

$$[\boldsymbol{r}_{IJ}]_\times = \begin{bmatrix} 0 & -z_J & y_J \\ z_J & 0 & -x_J \\ -y_J & x_J & 0 \end{bmatrix} \tag{8}$$

are the skew-symmetric matrices generated by the axial vector $\{\boldsymbol{r}(s)\} = \{x(s)\ y(s)\ z(s)\}^T$ that is the radius vector (function of the curvilinear abscissae) of a generic point along the element axis, while $\{\boldsymbol{r}(L)\} = \{\boldsymbol{r}_{IJ}\} = \{x_J\ y_J\ z_J\}^T$ is the position vector of the $\boldsymbol{J}$ node.
Using definitions (5-8), the equilibrium equations (3-4) can be written in the follow compact notation:

$$\{\boldsymbol{F}_J\} = -[\boldsymbol{T}_I(L)]\{\boldsymbol{F}_I\} \tag{9}$$

$$\{\boldsymbol{F}_I\} = -[\boldsymbol{T}_J(0)]\{\boldsymbol{F}_J\} \tag{10}$$

where $\{\boldsymbol{F}_I\} = \{F_1\ F_2\ F_3\ F_4\ F_5\ F_6\}^T$ and $\{\boldsymbol{F}_J\} = \{F_7\ F_8\ F_9\ F_{10}\ F_{11}\ F_{12}\}^T$.
Furthermore, other that the equilibrium relations between the nodal forces, it is possible to perform the equilibrium at a generic point along the curvilinear abscissae "*s*", to obtain the expression of the internal forces as a function of forces $\{\boldsymbol{F}_I\}$ or $\{\boldsymbol{F}_J\}$:

$$\{\boldsymbol{F}(s)\} = -[\boldsymbol{T}_I(s)]\{\boldsymbol{F}_I\} \tag{11}$$

$$\{\boldsymbol{F}(s)\} = -[\boldsymbol{T}_J(s)]\{\boldsymbol{F}_J\} \tag{12}$$

These latter relations will be useful in the section §4 to obtain the element length scaling.

The partitioned form of the fundamental Finite Element relation of the element in figure 1 appear as:

$$\begin{Bmatrix} \{\boldsymbol{F}_I\} \\ \{\boldsymbol{F}_J\} \end{Bmatrix} = \begin{bmatrix} [\boldsymbol{K}_{II}] & [\boldsymbol{K}_{IJ}] \\ [\boldsymbol{K}_{JI}] & [\boldsymbol{K}_{JJ}] \end{bmatrix} \begin{Bmatrix} \{\boldsymbol{U}_I\} \\ \{\boldsymbol{U}_J\} \end{Bmatrix} \tag{13}$$

where $\{\boldsymbol{U}_I\}, \{\boldsymbol{U}_J\}$ are the nodal displacements, and due to the stiffness matrix symmetry, the following relations hold:

$$[\boldsymbol{K}_{II}] = [\boldsymbol{K}_{II}]^T \tag{14}$$

$$[\boldsymbol{K}_{IJ}] = [\boldsymbol{K}_{JI}]^T \tag{15}$$

$$[\boldsymbol{K}_{JJ}] = [\boldsymbol{K}_{JJ}]^T \tag{16}$$

If equation (9,14-16) are applied in (13), the stiffness matrix can be written as follow:

$$[\boldsymbol{K}] = \begin{bmatrix} [\boldsymbol{K}_{II}] & -[\boldsymbol{K}_{II}][\boldsymbol{T}_I(L)]^T \\ -[\boldsymbol{T}_I(L)][\boldsymbol{K}_{II}] & [\boldsymbol{T}_I(L)][\boldsymbol{K}_{II}][\boldsymbol{T}_I(L)]^T \end{bmatrix} \tag{17}$$

Therefore, equation (17) shows that the complete stiffness matrix can be obtained only by the knowledge of the submatrix $[\boldsymbol{K}_{II}]$, that is a 6×6 matrix for 3D analysis or 3×3 for 2D ones. Furthermore, it is worth to pointing out that the matrix $[\boldsymbol{T}_I(L)]$ depends only on the position of the node $\boldsymbol{J}$, that is a trivial data.

At this point, it is useful to recover the flexibility convenience about experiments, hence it is possible to compute the stiffness submatrix associated to the node $\boldsymbol{I}$ with the inversion of the flexibility matrix associated at the same node, i.e. $[\boldsymbol{K}_{II}] = [\boldsymbol{C}_{II}]^{-1}$.

In other words, we can build up the entire stiffness matrix of a two-node element performing only more convenience flexibility measurements. Furthermore, due to the symmetry of the flexibility sub-matrix $[\boldsymbol{C}_{II}]$, are necessary 21 measurements to build up the entire 144×144 3D stiffness matrix, or only 6 measurements for the 36×36 2D stiffness matrix.

It is possible to obtain a similar result of (17) that instead use the stiffness and flexibility matrices associated to the node $\boldsymbol{J}$ if equation (10,14-16) are applied in (13):

$$[\boldsymbol{K}] = \begin{bmatrix} [\boldsymbol{T}_J(0)][\boldsymbol{K}_{JJ}][\boldsymbol{T}_J(0)]^T & -[\boldsymbol{T}_J(0)][\boldsymbol{K}_{JJ}] \\ -[\boldsymbol{K}_{JJ}][\boldsymbol{T}_J(0)]^T & [\boldsymbol{K}_{JJ}] \end{bmatrix} \tag{18}$$

where $[\boldsymbol{K}_{JJ}] = [\boldsymbol{C}_{JJ}]^{-1}$.

3. Experimental tests to build-up the Stiffness matrix

In this section we apply the procedure previously exposed to build-up the stiffness matrix of a 2D beam element. The specimen taking into account is a three-layered hollow and straight pipe (figure 4); the three layers are made by different type of polypropylene. The section is not perfectly circular but presents a non-regular ovalization along the axis.

Six components of the flexibility matrix must be computed; the second and third rows of $[\boldsymbol{C}_{JJ}]$ could be deduced experimentally by the schemes shown in figure 5, therefore with two different tests that involves a simple cantilever beam. However, to avoid the uncertainty about the realization of a perfect clamping, it is better to proceed with a three-point bending test (figure 6) for the two components on the second row, and a four-point bending test (figure 7, constant bending moment) for the components in

IOP Conf. Series: Materials Science and Engineering 1306 (2024) 012039 doi:10.1088/1757-899X/1306/1/012039

the third row. The measured quantities referring therefore to half specimen, inasmuch the configurations in figure 6 and 7 can be viewed as two facing cantilever beams.
There are no difficulties to execute these simple tests, that for the test in figure 6 requires only to measure the middle displacements and the ends rotation, this latter measured by laser pointer. The four-point bending instead, requires only the measure of the angle variation.
Three tests have been repeated three times for each loading scheme, for three different loads, to obtain a sort of secant matrix.

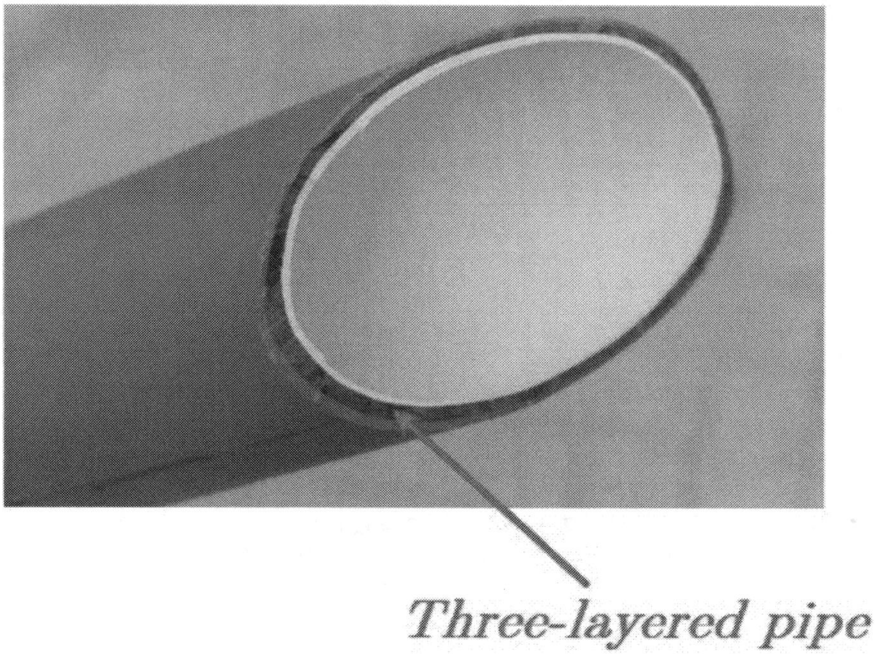

Figure 4. Three-layer polypropylene pipe specimen

The mean measured flexibility are: $\overline{c_{55}} = 2.194 \times 10^{-3}\, m/N$, $\overline{c_{56}} = 4.083 \times 10^{-3}\, N^{-1}$, $\overline{c_{66}} = 7.036 \times 10^{-3}\, N^{-1}m^{-1}$. Due to the fact that the interesting behaviour for this hollow pipe is the bending one, we did not perform any tests about the first row (axial behaviour) of the flexibility matrix $[\boldsymbol{C_{JJ}}]$. Being the pipe straight the components c_{45} and c_{46} can be assumed null, while the first component can be estimated as $\overline{c_{44}} \cong \left(\frac{I}{A}\right) c_{66} = 1.762 \cdot 10^{-3}\, c_{66}$.
Therefore, the evaluated flexibility submatrix is:

$$[C_{JJ}] = 10^{-3} \times \begin{bmatrix} 1.239 \times 10^{-5} & 0 & 0 \\ 0 & 2.194 & 4.083 \\ 0 & 4.083 & 7.057 \end{bmatrix} \tag{19}$$

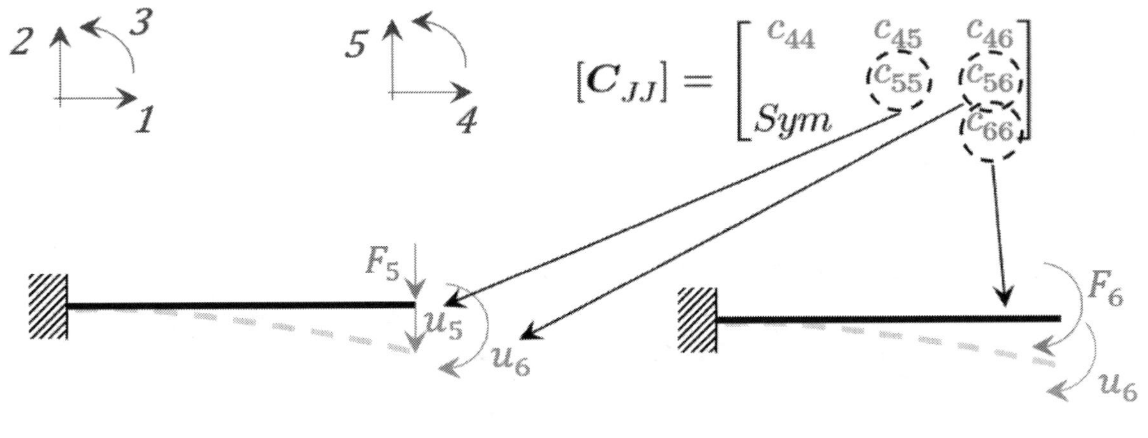

Figure 5. Ideal scheme to obtain the bending components of the partitioned flexibility matrix

From which, using the equation (18) and the length $L = 0.652\ m$ it is immediately compute the entire 2D element stiffness matrix:

$$[\boldsymbol{K}] = \begin{bmatrix} 5.2 \cdot 10^{-8} & 0 & 0 & -5.2 \cdot 10^{-8} & 539.77 & 1057.3 \\ & 1182.4 & 55.613 & -564.88 & -13.19 & -585.36 \\ & & 2819.6 & -1239.2 & 644.36 & 24.627 \\ & & & 1271.2 & 0.67342 & -2.9158 \\ & Sym & & & 1252.5 & -2.3348 \\ & & & & & 1100 \end{bmatrix} \tag{20}$$

4. Homogenized Section Flexibility and Element length scaling

Previous section described the experimental step to obtain the stiffness matrix from flexibility measurements, but a problem still arises. The stiffness matrix in equation (20) is valid only for an element length equal to $L = 0.652\ m$, and this reduce its usefulness. It is essential to find a method to compute the homogenized stiffness matrix of an element made by the same material and section, but with a different length. To achieve this result, the strain energy of the generic curvilinear one-dimensional element is considered:

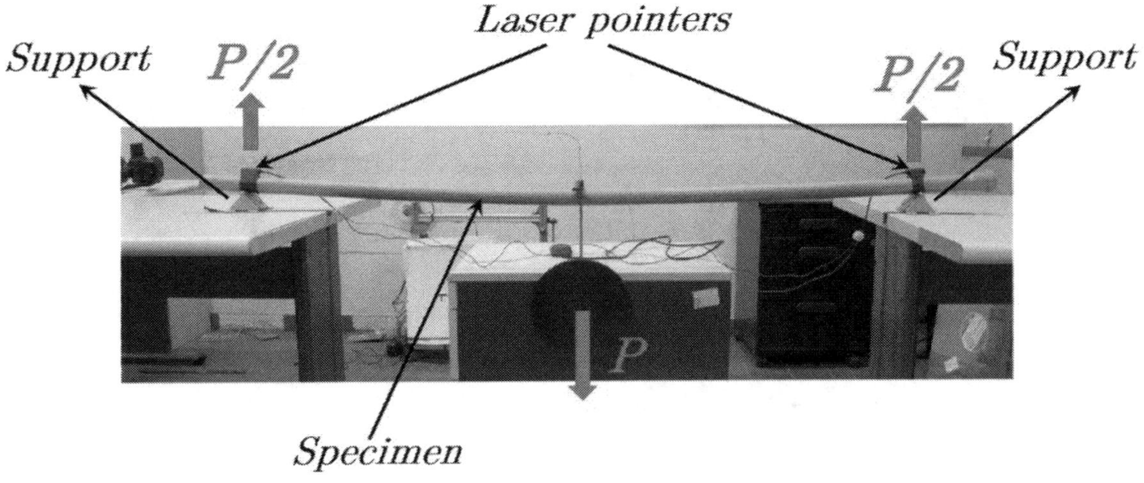

Figure 6. Set-up of the three-point bending test

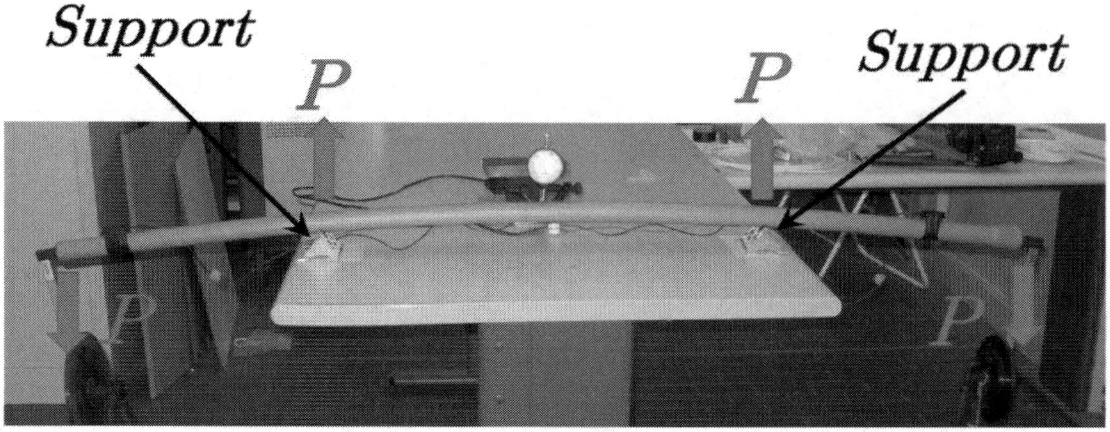

Figure 7. Set-up of the four-point bending test

$$U = \int_V \int_0^{\varepsilon} \boldsymbol{\sigma} : d\boldsymbol{\varepsilon} \; dV = \frac{1}{2} \int_0^L \{\boldsymbol{\varepsilon}(s)\}^T \{\boldsymbol{F}(s)\} \, ds \tag{21}$$

Follow the well-known beam element, under the hypothesis of linearity, it is possible to assume a linear relationship between the strain vector $\{\boldsymbol{\varepsilon}(s)\}$ and the internal force vector:

$$\{\boldsymbol{\varepsilon}(s)\} = [\boldsymbol{C}^{sec}]\{\boldsymbol{F}(s)\} \tag{22}$$

in which:

$$[\boldsymbol{C}^{sec}] = \begin{bmatrix} c_{11}^{sec} & c_{12}^{sec} & c_{13}^{sec} & c_{14}^{sec} & c_{15}^{sec} & c_{16}^{sec} \\ & c_{22}^{sec} & c_{23}^{sec} & c_{24}^{sec} & c_{25}^{sec} & c_{26}^{sec} \\ & & c_{33}^{sec} & c_{34}^{sec} & c_{35}^{sec} & c_{36}^{sec} \\ & & & c_{44}^{sec} & c_{45}^{sec} & c_{46}^{sec} \\ & & & & c_{55}^{sec} & c_{56}^{sec} \\ Sym & & & & & c_{66}^{sec} \end{bmatrix} \tag{23}$$

for a two-node element with out of plane deformations (12 × 12 stiffness matrix), and:

$$[\boldsymbol{C}^{sec}] = \begin{bmatrix} c_{11}^{sec} & c_{12}^{sec} & c_{13}^{sec} \\ & c_{22}^{sec} & c_{23}^{sec} \\ Sym & & c_{33}^{sec} \end{bmatrix} \tag{24}$$

for an element with a planar deformation (6 × 6 stiffness matrix).

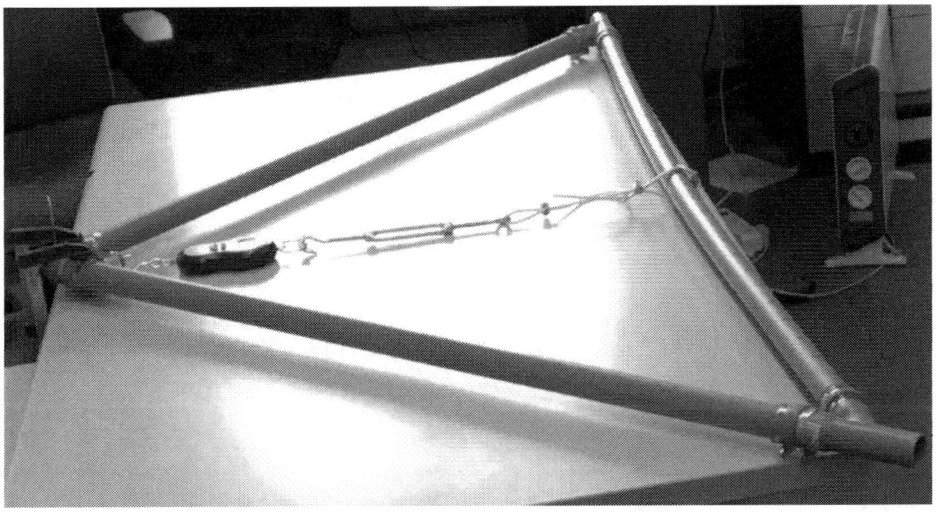

Figure 8. Unloaded triangular frame

We call $[\boldsymbol{C}^{sec}]$ "*sectional flexibility matrix*", inasmuch it contains information only about material and section's moments of inertia (all order, static moment included).
Applying equation (22) and (12) into (21) turn out:

$$U = \frac{1}{2}\int_0^L \{\boldsymbol{F}_J\}^T [\boldsymbol{T}_J(s)]^T [\boldsymbol{C}^{sec}][\boldsymbol{T}_J(s)]\{\boldsymbol{F}_J\}\, ds \tag{25}$$

The generic flexibility components c_{ij} can be analytical obtained from the energy expression using the Theorem of Complementary Energy (second Castigliano's Theorem) as follow:

$$c_{ij} = \frac{\partial^2 U}{\partial F_i \partial F_j} = \frac{1}{2}\frac{\partial^2}{\partial F_i \partial F_j}\int_0^L \{\boldsymbol{F}_J\}^T [\boldsymbol{T}_J(s)]^T [\boldsymbol{C}^{sec}][\boldsymbol{T}_J(s)]\{\boldsymbol{F}_J\}\, ds = \frac{\alpha_{ij}}{2}\int_0^L [\boldsymbol{T}_J(s)]^T [\boldsymbol{C}^{sec}][\boldsymbol{T}_J(s)]\, ds \tag{26}$$

in which the term $\alpha_{ij} = 2$ if $i = j$, and $\alpha_{ij} = 1$ if $i \neq j$.
For a straight element with a planar deformation (6×6 stiffness matrix), the analytical integration of equation (25) gives:

$$\begin{Bmatrix} c_{44} \\ c_{45} \\ c_{46} \\ c_{55} \\ c_{56} \\ c_{66} \end{Bmatrix} = \begin{bmatrix} L & 0 & 0 & 0 & 0 & 0 \\ 0 & L & \frac{1}{2}L^2 & 0 & 0 & 0 \\ 0 & 0 & L & 0 & 0 & 0 \\ 0 & 0 & 0 & L & L^2 & \frac{1}{3}L^3 \\ 0 & 0 & 0 & 0 & L & \frac{1}{2}L^2 \\ 0 & 0 & 0 & 0 & 0 & L \end{bmatrix} \begin{Bmatrix} c_{11}^{sec} \\ c_{12}^{sec} \\ c_{13}^{sec} \\ c_{22}^{sec} \\ c_{23}^{sec} \\ c_{33}^{sec} \end{Bmatrix} \tag{27}$$

The left side of equation (26) contains the experimental measured flexibility coefficients, on the right there is a matrix that contains only information about the geometry, and the vector of unknown coefficients associated to the sectional flexibility. Therefore, find the vector of coefficients c_{ij}^{sec} is immediately by computing the inversion of the matrix in equation (26).

If the coefficients c_{ij}^{sec} are known, it is possible to apply the equation (26) to compute the flexibility matrix $[C_{JJ}]$ of an element with a whatever length; from the already computed $[C_{JJ}]$ it is possible to build-up the new stiffness matrix using again the equation (18). This new stiffness matrix is computed on a different length, but it refers to an element with homogenized section properties (material and moments of inertia) equal to the tested specimen. In other words, these coefficients are extensive quantities that allow the length scaling of the element with respect to the stiffness matrix. It is worth pointing out that this result can be applied also to curvilinear elements, performing a numerical integration.

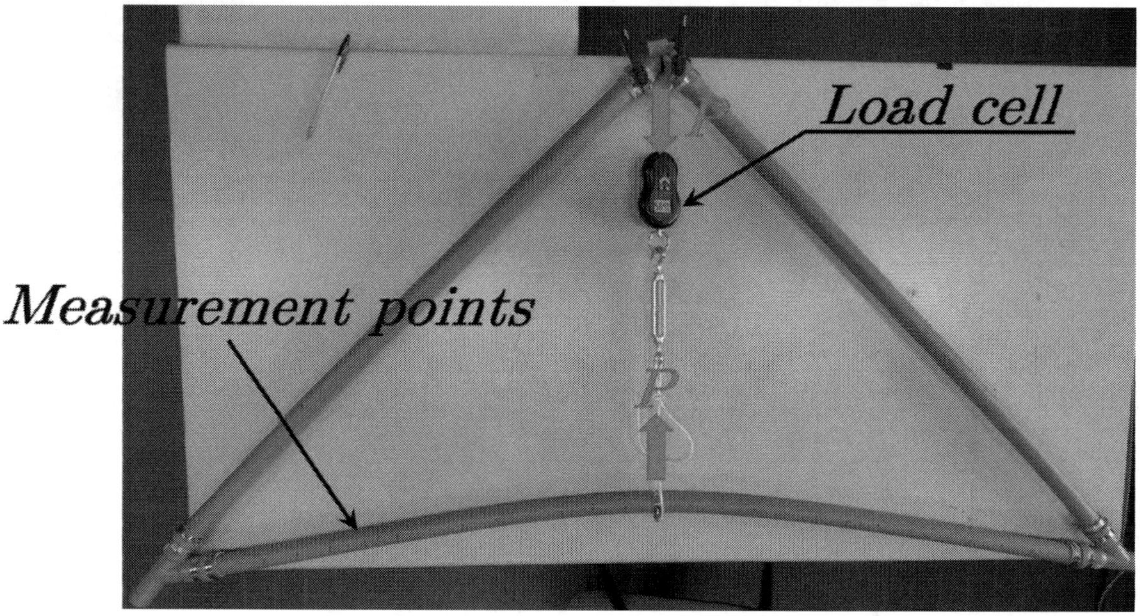

Figure 9. Deformed configuration of the triangular frame with an imposed displacement of 64 mm

5. Tests on a Frame

To validate the previous described length-scaling procedure other experimental tests have been conducted. We assembled a triangular frame formed by three pipes made with the same material of the already tested in §3, but different lengths. The base of the triangle took a curved shape (figure 8) due to the tightening operation. In view of the viscoelastic effect of the polypropylene, we perform the loading applying three imposed displacements (figure 9). The base of the triangle contains 27 measurement points, and the deformed shape was deduced using a laser distance meter (figure 10).

To compare these experimental results and to validate the effectiveness of the length-scaling procedure, we have implemented the identified element in the software Ansys® APDL, using the user defined element MATRIX27 (figure 11) and performing a linear analysis. The experimental and Finite Element results are compared in figures 12, 13 and 14. As expected, the error between them increase if the load increases, but it keeps sufficiently small (above all in figure 12) to validate the proposed approach. Probably, a high improvement can be obtained if the non-linear geometric effect is considered, i.e. implementing an incremental analysis by updating each element stiffness matrix for each iteration according to its current configuration.

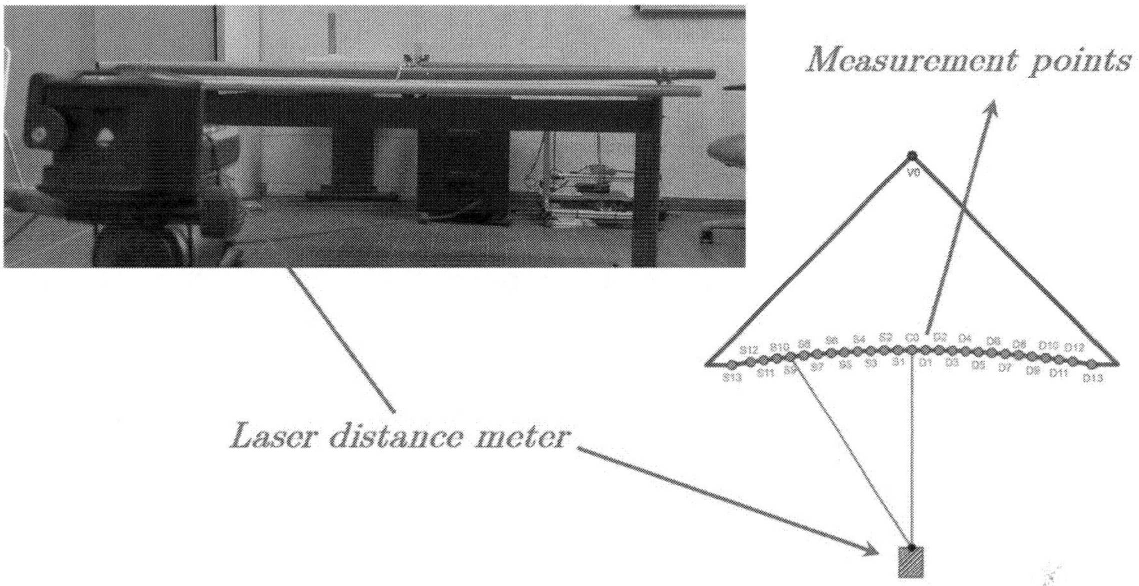

Figure 10. Measurement set-up of the deformed configuration of the triangle's base

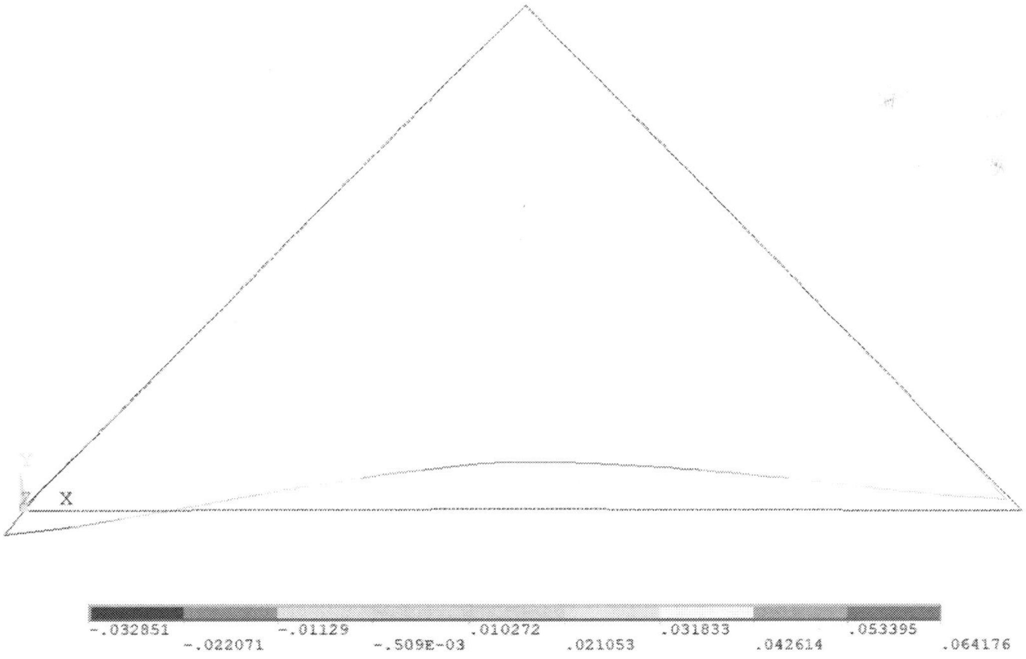

Figure 11. Finite Element computation of the deformed shape triangular frame with an imposed displacement of 64 mm

IOP Conf. Series: Materials Science and Engineering 1306 (2024) 012039 doi:10.1088/1757-899X/1306/1/012039

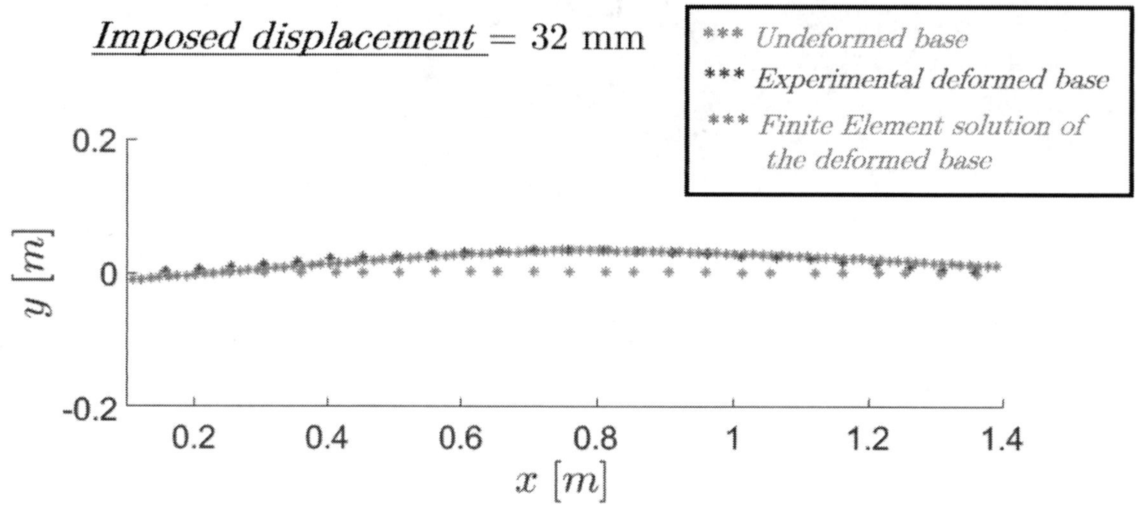

Figure 12. Finite Element computation of the deformed shape triangular frame with an imposed displacement of 32 mm

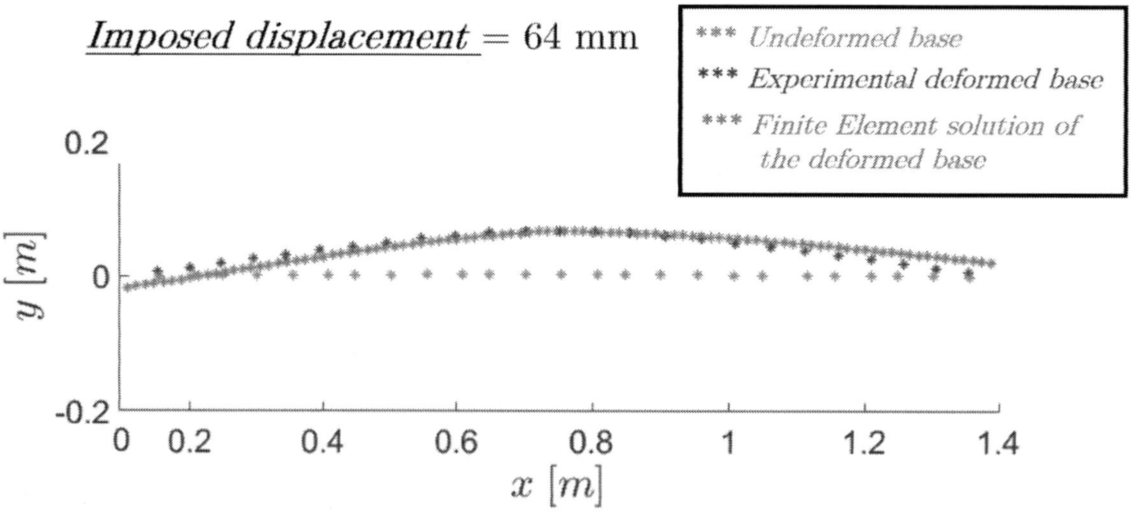

Figure 13. Finite Element computation of the deformed shape triangular frame with an imposed displacement of 64 mm

IOP Conf. Series: Materials Science and Engineering 1306 (2024) 012039 doi:10.1088/1757-899X/1306/1/012039

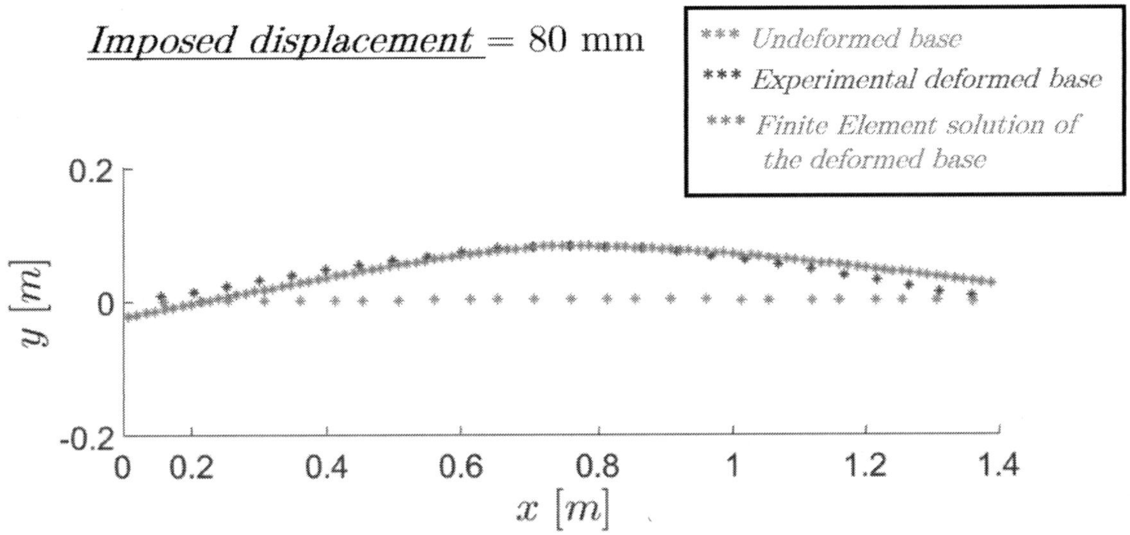

Figure 14. Finite Element computation of the deformed shape triangular frame with an imposed displacement of 80 mm

6. Conclusions

In this paper we present a method that allows to perform a structural homogenization of two-node elements through experimental test. The result of this experimental homogenization directly provides the entire stiffness matrix of the element, therefore differently from the classical homogenization the results do not contain only materials information but also on geometry and inertia properties.
The paper shows the advantage to execute flexibility tests instead then stiffness ones, since flexibility works with statically determined structure while stiffness requires the solving of statically undetermined structures. The use of flexibility allows the minimization of the necessary number of tests. A mathematical procedure to obtain the full stiffness matrix from the flexibility matrix associated only to one node is analytically deduced, setting the equilibrium condition.
The experiments are performed on a straight pipe, made by three different layers of polypropylene. Only two three-point bending tests and a four-point test are necessary to fully characterize the bending behaviour of a stiffness matrix in a 2D analysis.
Another mathematical procedure to perform the length-scaling of the element is presented. This last is based on the strain energy of the element and allows to obtain the "sectional" flexibility matrix, that is a matrix of extensive quantities that depends only on section information (material and inertia). This procedure allows to build-up the stiffness matrix whatever is the element length.
The length-scaling procedure was tested on a triangular frame structure made by three beams that differ from the length of the reference specimen. The results of experiments are compared with a Finite Element Analysis performed using the constructed stiffness matrix. Finite Element results and experimental tests show a similar trend; obviously, the error increase if the imposed displacements increase, inasmuch we perform a linear analysis. To obtain better results one should take into account the geometric non-linearity, addressing the problem with an incremental analysis and updating the global stiffness matrix of each element according to its current attitude. The presented experiments are conducted on a straight structure, but the entire presented procedure works also with a generic curvilinear element (also non-planar), whatever is its shape.

References

[1] Andreassen E, Andreasen CS 2014 How to determine composite material properties using numerical homogenization *Computational Materials Science* **83** pp 488–495

[2] Boso DP Lefik M Schrefler BA 2009 Recent developments in numerical homogenization. *Computer Assisted Mechanics and Engineering Sciences* **16** 161–183

[3] Hollister SJ Kikuchi N 1992 A Comparison of Homogenization and Standard Mechanics Analyses for Periodic Porous Composites *Computational Mechanics* **10** 73–95

[4] Fang Z Sun W Tzeng JT 2004 Asymptotic Homogenization and Numerical implementation to Predict the Effective Mechanical Properties for Electromagnetic Composite Conductor *Journal of Composite Materials* **38**

[5] Hashin Z, Shtrikman S 1963 A variational approach to the theory of composite elastic materials. *Journal of the mechanics and physics of solids* **11**(2) pp 127-140

[6] Hassani B, Hinton E 1998 A Review of Homogenization and Topology Optimization I – Homogenization *Theory for Media with Periodic Structure. Computer & Structures* **69** pp 707–717

[7] Hassani B, Hinton E 1998 A Review of Homogenization and Topology Optimization II – Analytical and Numerical Solution of Homogenization Equations *Computer & Structures* **69** pp 719–738

[8] Berryman JG, Milton GW 1988 Microgeometry of random composites and porous-media *Journal of Physics D-Applied Physics* **21** pp. 87-97

[9] Chen H, Kubo KY 2014 Bone three-dimensional microstructural features of the common osteoporotic fracture sites *World J Orthop* **5** pp. 486-495

[10] Lu Y 2021 Computational modelling of bone microstructure *Computational Modelling of Biomechanics and Biotribology in the Musculoskeletal System* pp. 251-256

[11] Almansour H, Mohammed A, Lounis Z 2018 Modeling the resilience of aging concrete bridge columns subjected to corrosion and extreme climate events *Life Cycle Analysis and Assessment in Civil Engineering: Towards an Integrated Vision* pp. 8

[12] Coronelli D, Gambarova P 2004. Structural assess-ment of corroded reinforced concrete beams: modelingguidelines *Journal of Structural Engineering* **130** pp.1214–1224

[13] Iandiorio C, Salvini P 2023 Elastic-plastic analysis with pre-integrated beam finite element based on state diagrams: Elastic-perfectly plastic flow *European Journal of Mechanics - A/Solids* **97** pp.1214–1224

[14] Iandiorio C, Salvini P 2023 Updated Lagrangian Curvilinear Beam Element for 2D Large Displacement Analysis *Lecture Notes in Mechanical Engineering* Springer

[15] Marotta E, Massimi L, Salvini P 2020 Modelling of structures made of filiform beams: Development of a curved finite element for wires *Finite Elements in Analysis and Design* **170**

[16] Zienkiewicz OC, Taylor R L 2000 The Finite Element Method, Solid Mechanics (Volume 2) *Wiley*

IOP Conf. Series: Materials Science and Engineering 1306 (2024) 012040 doi:10.1088/1757-899X/1306/1/012040

Material Characterization Augmented with Artificial Intelligence.

Matteo Vettori[1] , Adriano Marchi[1,4], Enrico Bellocchio[2], Alessandro Devo[2], Davide Belfiori[2,5], Francesco Soncini[3], Francesco Musiari[4], Fabrizio Moroni[4], Alessandro Pirondi[4]

[1] MaCh3D Srl, Parma, ITA
[2] Red Lynx Robotics Srl, Perugia, ITA
[3] 3DPR Srl, Langhirano, PR, ITA
[4] Dept. of Engineering and Architecture, University of Parma, Parma, ITA
[5] Dept. of Mathematics and Computer Science, University of Perugia, Perugia, ITA

m.vettori@mach3d.it

Abstract. The present paper investigates the application of artificial intelligence to improve the results from simple, non-instrumented, tensile tests, performed with a desktop-size MaCh3D smart universal testing machine. Non-instrumented tensile tests, performed on any testing machine, are affected by both deterministic and random factors that introduce errors in the test results. Specific features of the MaChh3D tester minimize random factors effects on test results while introducing a larger effect of deterministic factors. Artificial intelligence is identified as a novel approach to correct errors in non-instrumented tensile test, capable of simulating a direct strain measure onto the test, replacing traditional contact or non-contact instrumentations (like strain-gages, extensometer and optical measures) that introduce complexity into test procedure and require time for setup. The resulting AI model implementation is described and its performance evaluated in comparison with instrumented tests, also comparing different training strategies. The developed *AI-extensometer* (artificial intelligence virtual extensometer), is capable of a precise mechanical properties evaluation, with errors from 0 to 10% depending on the specific parameter.

1. Introduction

The traditional supply-chain in manufacturing is undergoing a deep transformation pushed by scaling technologies and socio-political and economic trends. The consolidation and consequent growing adoption of new manufacturing technologies, like the full set of additive manufacturing technologies, together with the emerging market trends related to circularity (recycling and re-use of materials and resources), sustainability (development and application of natural or bio-friendly materials) and reduced emissions (metal replacement strategies) are pushing a growing demand of material testing at different levels: for new materials development, for qualification and control of manufacturing processes and to push circularity approaches at higher levels. Thus, material testing activities need to move closer and become more accessible directly into industrial productions [1], by simplifying procedures and downscaling machines.

IOP Conf. Series: Materials Science and Engineering 1306 (2024) 012040 doi:10.1088/1757-899X/1306/1/012040

The MaCh3D smart universal testing machine [2, 3] addresses these needs with a compact sized hardware and drop-shaped fixture that allows easier, safer and faster testing procedure and avoids operator errors.

Like in traditional universal testing machines, MaCh3D is also subjected to test errors that affect test results when performing a simple non-instrumented test. Machine stiffness and slips occurring across the kinematic chain from the specimen to the actuator, affect the displacement measure that results in a non-precise strain measure. The specific MaCh3D fixture (Figure 1) transmits load to the specimen "by shape" and not "by force" like in traditional clamps. Due to this, slip occurs between the specimen and the fixture, depending on the applied load and the friction coefficient between specimen material and fixtures materials. Although being larger than in mechanical or hydraulic clamps, the slip occurring in the MaCh3D clamps is repeatable between tests, between test sessions, between test devices and even between different operators. The same cannot be said about both mechanical and hydraulic clamps because of the tightening force or pressure, affecting the specimen slips in fixtures, that may vary from machines, sessions and operators.

In order to avoid errors, a common solution is to perform instrumented tests, applying a contact instrument like a strain-gage or an extensometer, or adopting non-contact techniques like optical strain gauges. All of these options are introducing complexity to the test procedure, requiring a specific expertise from the operator and time-consuming operations for setup. More simplified strategies need to be developed to maintain ease of test and allow at-line material testing in industrial productions.

The work described follows the goal to improve the precision of a non-instrumented test combining the simplicity and compact size of the MaCh3D hardware.

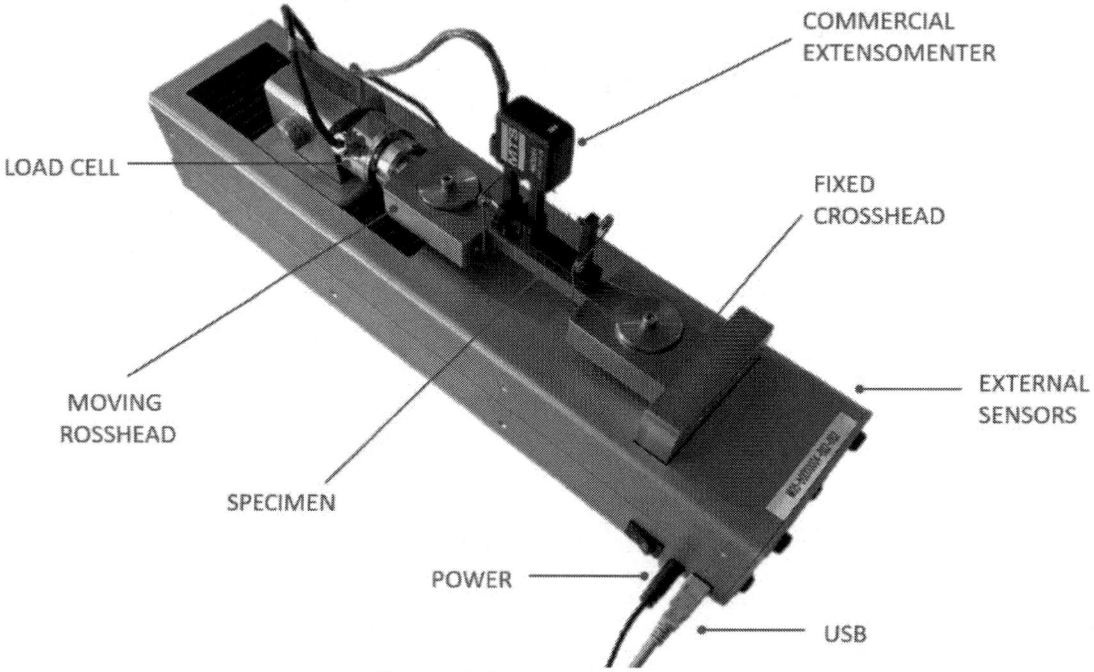

Figure 1. Experimental setup

2. Non-physical correction of non-instrumented tests

Two different approaches were presented in literature for the correction of test errors occurring when mechanical stress-strain behavior is calculated from load-displacement data from a non-instrumented test: an (i) analytical approach [3] and a (ii) numerical approach [4,5].

In the analytical approach the system is modeled by a combination of the single relevant elements of the system, each one described by a mathematical model. Properly combining together the single lumped-elements the overall system is described. Precise characterization of each element (calibration) is necessary for an accurate model prediction. The numerical approach consists in the modeling of the system through an accurate finite element model describing the system itself and the specimen. Precise modeling and calibration of FE parameters is necessary too.

Both the models from the two different techniques, starting from the load-displacement curve measured from a non-instrumented test, can estimate the load-displacement curve of the specimen gauge length (uniform section length), or directly the mechanical properties of the material under test. A certain error is intrinsic in both the approaches, higher fidelity can be achieved the more the model is tuned for a specific application (i.e. a specific material under test).

Artificial intelligence is experiencing growing applications in engineering and even in the fields of experimental mechanics and materials modelling [6,7]. In order to reduce the error from non-instrumented tests, artificial intelligence can be used to build a virtual strain measurement instrument (*AI-virtual extensometer*), returning a stress-strain curve starting from the simple load-displacement curve of the non-instrumented test. A properly architected and trained AI model can be accurate according to the type of errors introduced in the test, deterministic errors (structural compliance, friction, …) can usually be well identified and corrected, while random errors (misalignment, operator accuracy, …) generate a layer of uncertainty in the model that affect the prediction error.

The present paper covers the development of an AI-virtual extensometer to be used together with the MaCh3D device model MaCh5s. Execution and results of an industrial pilot experiment is also described.

3. Workflow

As previously introduced, the goal of the research is the development of an AI-virtual extensometer that receiving as input the load-displacement curve of a tensile test, together with other relevant parameters, is capable to return a stress-strain curve of the same tests, excluding all the errors deriving from all the factors that occurs outside the specimen.

Activity has been subdivided in 3 macro-activity regarding: (i) the development of the artificial intelligence model, (ii) the building of a dataset of tests for training the model and, for validation purposes, (iii) the execution of an industrial pilot experiment.

4. Artificial intelligence model

The AI-virtual extensometer tool can be actually described as a transfer function that from the non-instrumented load-displacement curve returns a virtually measured stress-strain curve. The complexity of the transfer function is due to the number of phenomena that the transfer function has to describe, their non-linearity and the non-linearity of the two input and output curves.

The final version of this AI consists in a deep learning model made of both vanilla neural networks and the more advanced recurrent neural networks. Model architecture is depicted in Figure 2. Model is fed with two inputs (highlighted in green):

- I_1 , a matrix with the dimension $[m_n, 2]$, made of the "*load*" and "*displacement*" time series, where m_n represents the length of the two time series for the nth test sample
- I_2 , a vector in the form [*W1, W2, W3, T1, T2, T3, LS, LU, BP, M0, M1, M2, M3*] containing relevant specimens dimensions and the *Mx* parameters consisting in an I/O input string (so called one-hot vector) currently based on four digits, which combination refers to the specimen's material family.

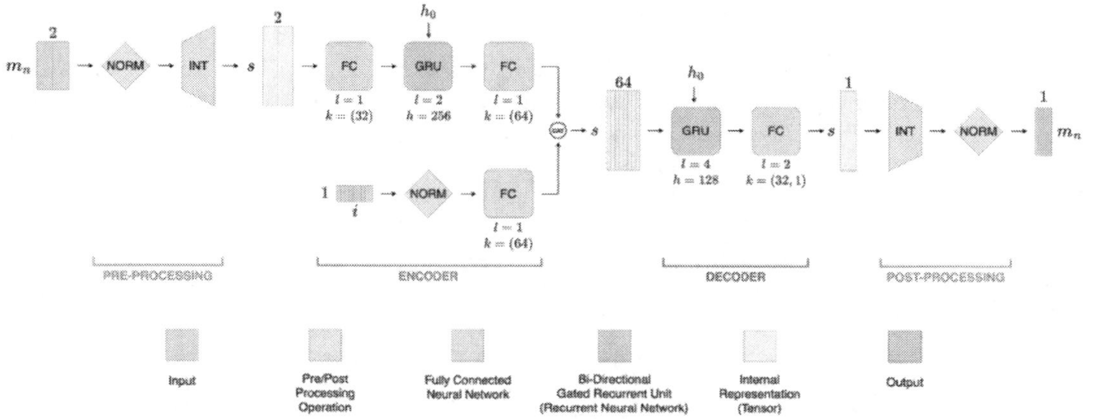

Figure 2. Schematic of the AI model architecture.

The values in the $I1$ matrix are normalized as follows:

$$LOAD_{norm} = \frac{LOAD}{LOAD_{max} - LOAD_{min}} \quad \text{and} \quad ENC1_{norm} = \frac{ENC1}{ENC1_{max} - ENC1_{min}}$$

then the length of each test sample is made identical by interpolation at a fixed number of points. This is a numerical strategy to allow faster model training by combining the single matrices into a 3D matrix. The length s=2000 was chosen arbitrarily, larger than the maximum length of the input load-displacement curves.

The conditioned data are then passed to a fully connected layer with 32 neurons, a Gated Recurrent Unit (GRU) with 2 layers and 256 neurons (hidden size h = 256) and 10% dropout, and a second fully connected layer with 64 neurons.

The output produced is then concatenated (CAT) with the encoding of I_2 , preliminary normalized. Encoding is made by a fully connected (FC) layer with 64 neurons. Resulting tensor, with dimension $[s, 64]$ is the input the input of a second GRU with 4 layers and hidden size $h = 128$, and 10% dropout. In both the GRUs, hidden state h is initialized with a random h_0 tensor.

The last fully connected network, made of two layers with 32 and 1 neurons each, ios needed to produce the final output. A *softplus* activation is defined as follows:

$$softplus(x) = log_{10}(1 + e^{\beta x + \alpha}), \text{ where } \alpha = \beta = 3$$

The final output is brought back to the original length m_n and restored to the non-normalized state to return the output series O.

Model training was performed with data batches of size b=32, and with epochs c=1000. Training is performed comparing the load-displacement series with the load-strain series with the goal to minimize the *Mean Absolute Percentage Error* (*MAPE*) of the predicted strain versus the measured strain. *MAPE* is expressed by the following loss function:

$$MAPE = \frac{1}{b}\sum_{n=1}^{b} \frac{|O_n - \frac{EXTS_n}{LG_n}|}{max(\epsilon, |\frac{EXTS_n}{LG_n}|)} .$$

It should be noticed that in order to be able to compute this loss function batch-per-batch both the O and the *strain* time series need to be of the same size for all samples, thus the comparison is performed onto the non-post-processed output O and the interpolated *strain* series

Although the interpolation introduces a small error in the overall process, it is negligible if compared against the model error. Therefore, it is an extremely good trade-off to use the interpolation when considering the tremendous increase in learning speed (higher than 10x).

IOP Conf. Series: Materials Science and Engineering 1306 (2024) 012040 doi:10.1088/1757-899X/1306/1/012040

5. Training dataset composition

The final dataset for model training is made of more than 450 tensile tests. Table 1 summarizes the dataset compositions, stored in three batches according to the periods of productions. Because the industrial pilot is performed into an additive manufacturing service, and because MaCh3D have wider application in this market, additive manufactured materials are represented only in the database.

Table 1. Training dataset composition.

Dataset	*Technology*	*Material*	*Printer*		*# samples*
D01	FFF	PLA	Various printers		66
D02	FFF	PLA, PETG, ABS	Creality Ender-3		47
	FFF	PETG	Prusa MK-I Witbox-I		90
	FFF	PLA	Creality CR10 Anycubic Vyper		56
D03	SLS	PA12	EOS model P396	J-1	130
				J-2	78
				Total	**467**

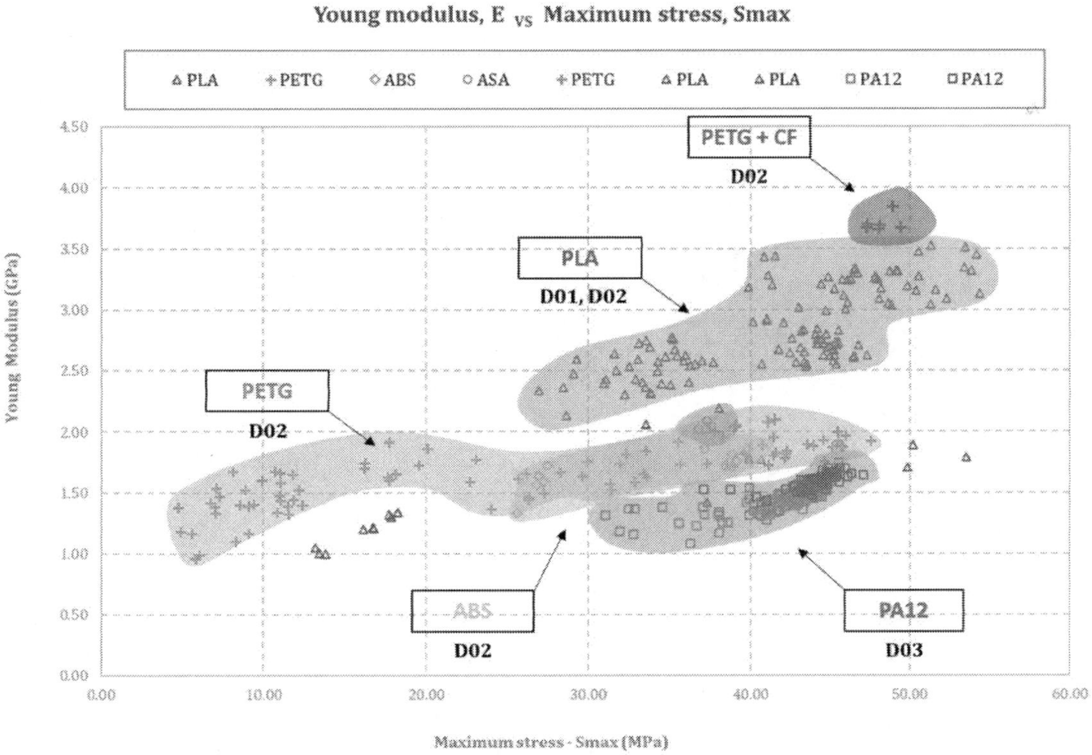

Figure 3. Ashby map of the mechanical properties from the dataset.
Areas are grouped according to material type.

6. Industrial pilot experiment

6.1. Experiment setup

The developed AI-virtual extensometer is tested into a relevant industrial workspace, a small additive manufacturing service. The experiment consists in the integration of the MaCh3D material testing solution in the AM service routinary workflow to evaluate the benefits from shopfloor material testing activity. The AI-virtual extensometer is used for each test. Over the pilot duration (5 weeks) three sessions of instrumented tests are scheduled to verify and validate the performance of the AI-virtual extensometer.

Specimens are produced with SLS technology using EOS model P396 systems. Two materials have been used: PA12 (commercial name "EOS PA2200") and glass filled PA12 (commercial name "EOS PA3200GF"). Specimens are randomly placed in each production batch together with other components, leading to different job density from different batches. Tests were usually performed within one day after job de-powdering and finishing. Because of orthotropy several orientations have been considered for specimen's production. Build volume temperature and exposure parameters were kept constant throughout the experiment period. Once produced, the specimens were sandblasted with glass beads as part of the standard finishing procedure, no other types of post-processing were applied.

A total of 12 production batches were investigated and 150 specimens were tested.

Regarding the AI-virtual extensometer, the model used in the experiment is performed with training based on the full dataset containing all the different types of materials. A second training based only upon the specimens manufactured from the same site (D03) is tested, in order to evaluate the impact of a site specific training versus a generalized training.

(a) (b) (c)

Figure 4. Industrial pilot project. (a) the 3DPR AM service production floor with the two EOS P396 systems used for the experiment; (b) a sample job nesting, with the highlighted specimens for testing; (c) the MaCh3D universal testing machine during an instrumented test.

6.2. Experiment results

Results of the AI-virtual extensometer compared to results from instrumented tests and those of non-instrumented tests are listed in the following table 2.

Tables list the MAPE value of each single mechanical property comparing the non-instrumented and the virtual instrumented tests. MAPE is calculated onto the full set of tests. The three sub-tables correspond to three evaluations performed on the full verification dataset, the PA12 tests only and the

IOP Conf. Series: Materials Science and Engineering 1306 (2024) 012040 doi:10.1088/1757-899X/1306/1/012040

PA12GF tests only, this last corresponding to a "blind" condition for the model, because PA12GF material is not covered by the training dataset.

The MAPE values for the stress related parameters are null because of direct load measure by the load cell. MAPE values for the strain related parameters are globally within the 10-11% of maximum error. The best results are coming from the model trained with the full dataset (including all materials and technology covered) that lead to the conclusion that a wider number of tests even if from different materials and technologies can help the AI to better perform in the prediction of the stress-strain curve.

Table 2. AI model validation results.

Full Experiment (Test Dataset)

MAPE is calculated on the full test dataset

	E	σ_y	ε_y	σ_{max}	ε_{max}	σ_u	ε_u	ε_r
Naked Test	68.41%	**4.94%**	188%	0%	124%	0%	118%	57.4%
Tr_D3	23.71%	8.6%	36.21%	0%	27.96%	0%	35.27%	62.96%
Tr_ALL	**11.15%**	7.58%	**15.63%**	0%	**4.74%**	0%	**5.64%**	**15.24%**

PA12

MAPE is calculated onto PA12 tests only

	E	σ_y	ε_y	σ_{max}	ε_{max}	σ_u	ε_u	ε_r
Naked Test	66.7%	**2.97%**	182.8%	0%	93.2%	0%	90.5%	54.2%
Tr_D3	12.85%	11.29%	19.11%	0%	8.58%	0%	9.38%	11.56%
Tr_ALL	**6.24%**	6.75%	**11.01%**	0%	**4.71%**	0%	**5.3%**	**7.9%**

PA12GF tests

MAPE is calculated onto PA12GF tests only

	E	σ_y	ε_y	σ_{max}	ε_{max}	σ_u	ε_u	ε_r
Naked Test	67.9%	**8.4%**	164.2%	0%	101%	0%	96.7%	56.6%
Tr_D3	21.35%	6.38%	29.81%	0%	7.22%	0%	11.28%	13.05%
Tr_ALL	**19.8%**	9.94%	**23.65%**	0%	**4.65%**	0%	**4.87%**	**8.43%**

Opportunity of performing periodic testing activity at the service bureau allowed the impact and benefits of such activity integrated in the shop everyday routine. Impact of performing tensile tests at production floor, onto specimens printed in batch with standard production is reported negligible in terms of the impact in everyday productivity. Simplicity of testing deriving from the MaCh3D fixtures

IOP Conf. Series: Materials Science and Engineering 1306 (2024) 012040 doi:10.1088/1757-899X/1306/1/012040

and performing non-instrumented tests allows tests to be performed quickly by the production department with no specific training.

Measurements performed by the service bureau during the experiment lead to the following outcomes.

Material properties declared by the material supplier/manufacturer are not consistent with those measured in the "real-world" production, resulting in about 10% lower than those declared.

Comparing test results over time (Figure 5) an oscillation of the mechanical properties is revealed, normally within the 5% of the parameter but showing peaks even higher than 10-15%. For sure a more accurate campaign should be performed to properly characterize the mechanical properties oscillation over time but results outlined the fact that monitoring mechanical performances of materials and components is an important activity in additive manufacturing production, especially when related to structural components.

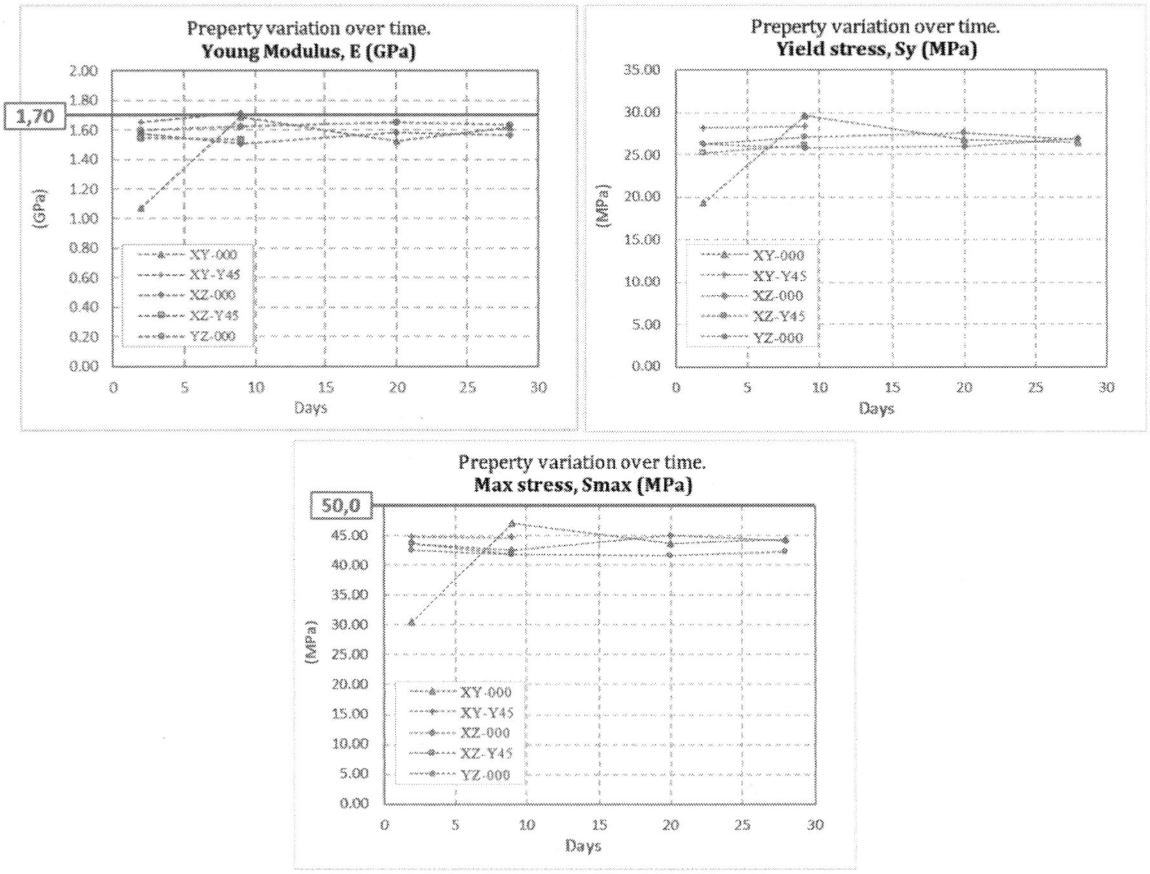

Figure 5. Results from the industrial pilot experiment. The three graphs represent Young modulus, Yield stress and Max stress mean values variations over the time, from different production batches, for several specimen orientations. The red horizontal lines in the graphs correspond to the material datasheet's declared values (Yield stress is not declared).

7. Conclusions

An AI-virtual extensometer is developed to reduce the error emerging from non-instrumented tests performed onto the MaCh3D compact universal testing machine. The tool is predicting the stress-strain behavior of the tested specimen starting from the load-displacement curve with an error limited within the 10% with respect to instrumented tests.

The AI-virtual extensometer and the MaCh3D testing device is adopted into a small additive manufacturing service bureau in their everyday workflow. Impact of MaCh3D tests onto daily activities/productivity is negligible while allowed firstly to identify deviation the declared material properties compared to the measured one, and subsequently to characterize properties variation according to specimen orientation and to detect and measure the oscillation in the same properties across different production batches. Results proved that datasheets are not always representative of real world mechanical properties confirming the need of material testing in additive manufacturing production for quality control and monitoring the mechanical properties of printed materials and parts.

Acknowledgments
Part of this work is co-financed by the European Union's Horizon 2020 action program for research and innovation, with a cascade funding mechanism promoted in the context of the KYKLOS4.0 European project (Grant Agreement n 872570).

References

[1] AAVV 2016 *Latest Trends in the Material Testing Market* (Frost & Sullivan) Frostperspective
[2] Bergonzi L, Vettori M, Pirondi A, Moroni F and Musiari F 2018 Numerical and experimental validation of a non-standard specimen for uniaxial tensile test *Procedia Structural Integrity* **12** pp 392–403
[3] Bergonzi L, Vettori M and Pirondi A 2019 Development of a miniaturized specimen to perform uniaxial tensile tests on high performance materials *Procedia Structural Integrity* **24** pp 213–24
[4] Hajy Akbary F, Santofimia M J and Sietsma J 2014 Elastic Strain Measurement of Miniature Tensile Specimens. *Exp Mech* **54** pp 165–173
[5] Liu H, Shen Y, Yang S, Zheng P, Zhang L 2017 A comprehensive solution to miniaturized tensile testing: Specimen geometry optimization and extraction of constitutive behaviors using inverse FEM procedure *Fusion Engineering and Design* **121** pp 188-197
[6] Kibrete F, Trzepieciński T, Gebremedhen HS and Woldemichael DE 2023 Artificial Intelligence in Predicting Mechanical Properties of Composite Materials *Journal of Composites Science*; **7**(9) 364
[7] Mianroodi J R, Siboni N. and Raabe D 2021 Teaching solid mechanics to artificial intelligence - a fast solver for heterogeneous materials *npj Comput Mater* **7** 99.

IOP Conf. Series: Materials Science and Engineering 1306 (2024) 012041 doi:10.1088/1757-899X/1306/1/012041

Predictive sustainability analysis applied to an automotive design case study.

A. Antonacci[1], F. Del Pero[2], E. Innocenti[1], A. Giraldi[1], M. Pierini[1], M. Delogu[1]

[1] Department of Industrial Engineering, University of Florence, Italy
[2] Department of Engineering Science, Guglielmo Marconi University, Italy

E-mail: andrea.antonacci@unifi.it

Abstract. The paper deals with a predictive sustainability analysis applied to the design of automotive components in lightweight perspective. The analysis is conducted through the integration of the traditional Life Cycle Assessment (LCA) methodology with tailored forecasting algorithms able to provide a predictive evaluation of Climate Change (CC) by elaboration data contained in commercial environmental datasets. The comparison is referred to the entire Life Cycle (LC) of the system (including production, use and End-of-Life) according to a "from cradle to grave approach".
A medium-class car rear crash management system is used as case study, assessing the potential benefits related to the substitution of conventional steel with 6000/7000 series aluminium alloys, along with other minor design changes. Particularly, the study compares the environmental profile of the two solutions based on the CC impact category in application to both an Internal Combustion Engine Vehicle (ICEV) and a Battery Electric Vehicle (BEV).
The results show the potentiality of the proposed methodology, highlighting possible improvements/worsenings: new materials and manufacturing technologies adopted in the lightweight rear crash management system entail contrasting environmental effects depending on LC phases, that is, increased CO2eq in production (around 125% - mainly due to the strong energy intensity of aluminium supply chain) and reduced burdens in use and EoL (primarily provided by component mass reduction). That said, the influence of different boundary conditions on the predictive models is significant only for the electricity produced to energize the BEV configuration, leading to an overall variability of comparison results ranging within 3-30% for the BEV case study.

1. Introduction

The current EU regulations are driving towards the reduction of Greenhouse Gas (GHG) emissions, setting a global reduction target of 60% by 2050 [1]. This issue is of great interest in the automotive field, where over 20% of GHG impacts in the European Union are attributed [2], with 50% of them originating from passenger vehicles [3]. Urgent measures are therefore needed to reduce the environmental burdens of vehicles. One of the most commonly used methodologies in scientific research to assess such impacts is the Life Cycle Assessment (LCA) methodology [4], which allows for quantitative sustainability evaluation of various subsystems. To date, the adoption of LCA is well-established in the automotive sector. Eco-design principles promote the use of recycled materials and the introduction of innovative design solutions to reduce consumption [5]. In support of these principles, customized tools can be found in literature that assess the sustainability of solutions selected by the designer [6]. More specifically, the attention to green solutions is heavily focused on vehicle electrification [7]. Numerous articles can be found that highlight the sustainability of electric vehicles compared to internal combustion cars. Helmers et al. [8] presents results for 18 impact categories obtained from the comparison of Battery Electric Vehicles (BEVs) and Internal Combustion Engine Vehicles (ICEVs) using the Recipe method. The results are reported in terms of break-even point and demonstrate a strong dependence on battery production scenarios. Del Pero et al. [9] shows that BEVs result in zero emissions during thee usage phase, but the production of the electric drivetrain poses a barrier to achieving maturity for this technology. In line with this assessment, Anderson et al. [10] demonstrates the potential of using a plug-in hybrid engine powered by renewable fuel to reduce the size of batteries, which are typically large for full electric vehicles. Batteries are also a critical element in the environmental assessment of End-of-Life (EoL) vehicle scenarios, which greatly impact the overall environmental effect [11]. The complete transition to electric mobility has also been discussed at the COP 26

[12], but numerous automakers have opposed the complete exclusion of biofuels. However, there is still a need for further increase in green energy sources for electricity generation and strengthening of the infrastructure system [13]. Several studies study the environmental impact of using alternative materials (such as polymeric, composites, or aluminium), with a view to reducing component weight or assessing the specific impact of the material itself [14-15]. Akhshik [16] et al. investigates the use of bio-based composites for the production of four components (battery tray, engine beauty shield, cam cover, and oil pan). The results demonstrate a reduction of over 2 times in GHG emissions. The use of such materials brings substantial benefits for car usage exceeding 200,000 km [17], a value that is expected to grow due to the increasing efficiency of propulsion systems over the years. Moreover, the classic approach of LCA does not take into account the influence of time, and the values of environmental impacts during the use and EoL stages are discounted when the analysis is performed. To account for this effect, it is possible to carry out forecasting analyses. Some examples of such applications can be found in literature. Memary et al. [18] discusses the temporal effect (time-trend) on the environmental impact associated with the extraction of metallic materials (copper). The results show that the effect in terms of SO2 and CO2 is expected to increase by up to 300% within the considered time horizon due to the anticipated decline in metal concentration in raw minerals (ore grade). Zhu1 et al. [19] introduces the temporal effect regarding electricity production in the Chinese scenario. The study provides a detailed forecast of the increasing use of renewable sources in the national energy mix throughout the analyzed period. The results predict a reduction of over 50% in greenhouse gas emissions by 2035 compared to 2017. Stranddorf et al. [20] implements forecasts regarding the environmental effect of the European energy mix. The proposed green scenario, which involves a significant increase in renewable energy sources, results in a Climate Change savings of over 80% by 2050.

In this work, the temporal trends' effect on the sustainability of an automotive component is evaluated using insights extracted from the work of Sacchi et al. [21]. In particular, a forecast scenario is implemented following the principles dictated by the Shared Socioeconomic Pathways (SSP) [22], which quantify the impacts on Integrated Assessment Models (IAMs) of economic, technological, and urbanization evolution. A predictive LCA is developed to evaluate the time-based effect of an innovative lightweighting solution. The analysis follows a comparative approach between the traditionally used solution and an innovative solution that presents a weight reduction. First, the paper introduces a re-engineering activity conducted on a rear Crash Management System (CMS) module. The study begins with the reference steel-based CMS and develops a lightweight version based on ultra-high-strength aluminium extrusions. Such a lightweight variant substantially reduces mass, preserving performance, safety, and functionality standards. Both solutions undergo an evaluation and comparison, examining its lightweight potential, design enhancements, and sustainability improvements, in the predictive LCA context and different propulsion system (BEV or ICEV). Moreover, a detailed sensitivity analysis is presented, based on different powertrain, as well as different vehicle segments and average annual mileage. The results are presented in terms of CO_{2eq}, and they evaluate the dependence of the Break Even Point (BEP) on the aforementioned parameters. The results can be used to inform decision-making regarding the adoption of innovative lightweight solutions and to guide the design of future vehicles.

2. Materials and Methods

2.1. Module Design

The main components of both the reference (REF) and lightweight (LW) versions of the CMS module are shown in **Figure 1** with an exploded view in a CAD environment. The presented assemblies are divided into sub-assemblies, numbered as follows: 1. Backplate; 2. Crash Box; 3. Beam; 4. Towing System.

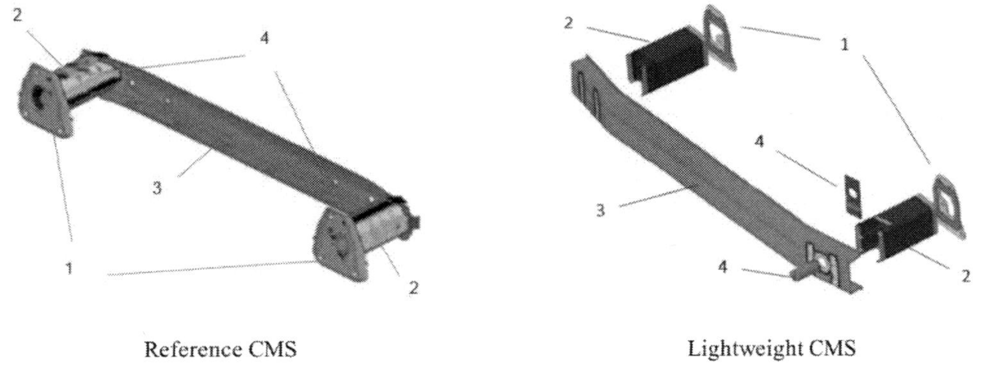

Figure 1 - CMS CAD - exploded view of reference (a) and lightweight (b) solutions.

Table 1 and **Table 2** present a comprehensive outline of the primary design characteristics of the REF and LW versions of CMS. Such a representation includes relevant information concerning each component, such as composition, thickness, and manufacturing process. The vehicle on which the CMS is installed is a C-class car, both in the ICEV and BEV configurations.

Table 1 - Overview of reference (REF) design solutions for the CMS assembly.

ID	Components	Material	Manufacturing Process	Mass [kg]
1	Backplate	HCT600X	Cold stamping	0.787
2	Crash Box	HCT600X	Cold stamping	1.370
3	Beam	HCT980X	Cold stamping	1.614
4	Towing System	HC420LA/HC260L	Cold stamping	0.218
			Total	3.989

Table 2 - Overview of lightweight (LW) design solutions for the CMS assembly.

ID	Components	Material	Manufacturing Process	Mass [kg]
1	Backplate	EN AW7003 LS	Extrusion, Punching	0.296
2	Crash Box	EN AW7003 LS	Extrusion, Milling	0.787
3	Beam	EN AW7003 HS	Extrusion, Forming, Punching	1.202
4	Towing System	EN AW6082	Extrusion, Milling, Punching	0.131
			Total	2.416

The REF design solution is made of cold-stamped steel, weighing approximately 4 kg. The selection of steel grades is guided by attaining optimal performance within the economically favorable range of cold stamping materials.

The LW version of CMS comprises the same components as the baseline version, it does not need for any modification to the surrounding parts, and it is engineered to fulfill the performance criteria of the baseline version concerning two specific factors:

1. maximum deformation and energy absorption in the RCAR low-speed structural crash test protocol for rear impact [23];
2. bending moment when subjecting the beam to deformation at its center.

An ultra-high-strength aluminum alloy is employed as the base material for the LW design, resulting in a total mass of approximately 2.4 kg. The backplate (1) and crash box (2) are extruded parts made of aluminum alloy 7003, with respective thicknesses of 4.0 mm and 2.1 mm. A highly ductile, high-strength version of the AW7003 in a slightly overaged temper (T7) is employed for the crash boxes, to meet the demanding ductility requirements of these components. Additionally, the chosen alloy provides enhanced performance in terms of crack resistance, deformability, and stability of mechanical properties in large-scale production, thereby combining high strength with good ductility and production efficiency. The novel developed beam (3) is formed using an open beam extruded profile made of the precipitation-hardenable high-strength aluminum alloy 7003, with a thickness ranging from 1.8 to 2.0 mm. This extruded profile design is coupled with a specialized forming concept, offering an optimal balance between mass reduction, cost-effectiveness, and crash resilience. The chosen alloy for the beam profile exhibits slightly higher strength compared to the other two components. The beam, crash box, and backplate components offer favorable characteristics in terms of mass reduction, strength, production efficiency, and design flexibility due to the utilization of a highly ductile aluminum extrusion alloy. Finally, the towing system (4) is crafted from 6082 aluminium.

Overall, the design choices implemented in the LW CMS version, such as material alteration, application of heat treatment, and use of tailored extruded blanks, result in a weight reduction of nearly 40 % compared to the REF CMS version, while simultaneously maintaining the performance, safety, and functionality requirements.
Additional requirements apply to both design alternatives, specifying that no breakage should occur at a 105 mm intrusion in the bending load case; furthermore, the CMS must meet the strength and durability standards set by OEM for the towing attachment.

2.2. *Predictive Sustainability Assessment*

This section aims at undertaking a comparative predictive analysis between the CMS reference and lightweight alternatives. To accomplish this, a baseline sustainability assessment is conducted using the LCA methodology [24-26]. Such an analysis is carried out by integrating the conventional LCA methodology with tailored forecasting algorithms, allowing a predictive assessment of environmental impacts through advanced data processing techniques.

The analysis comprehensively encompasses the entire LC of the CMS module, subdivided into production (i.e., raw material extraction and manufacturing process), use, and End-of-Life (EoL) stages. The environmental profile of both CMS solutions is assessed using the ReCiPe Midpoint (H) 1.13 impact assessment methodology [27] in terms of the Climate Change midpoint (CC) impact category. In this context, the Functional Unit (FU) is defined as the CMS module installed on the reference vehicle, with an average life cycle age of 12 years. [28]. It is assumed that both design options maintain unaltered both mechanical and functional performances. The assessment of Fuel Consumption (FC) and exhaust air emissions during operation is based on the Worldwide Harmonized Light-Duty Test Procedure (WLTP) [29]. Inventory data is collected for the above three main LC phases, encompassing materials/energy consumption, waste production, and environmental emissions. These factors are modeled using Life Cycle Inventory (LCI) processes and elementary flows sourced from the Ecoinvent 3.9 database.

Regarding production, the inventory is conducted separately for raw material acquisition and manufacturing sub-phases. Raw material acquisition accounts for the entire production chain, from primary material extraction to the production of semi-finished products. Manufacturing incorporates activities involved in converting semi-finished products into final module components. The inventory of both materials and manufacturing steps is conducted through a break-down approach [30], previously applied in ENLIGHT [31] and e-LCAr [32] projects. This approach assesses each mono-material component of the CMS system, with the overall impact of the production stage determined by aggregating contributions from the individual mono-material parts.

Data collection for raw material acquisition primarily consists of secondary data from scientific literature and the Ecoinvent database, based on composition and manufacturing processes of the CMS materials (see **Table 1** and **Table** *2*). The manufacturing considers energy consumption and material losses during production of finished parts. Processes such as joining, assembly, and transportation are excluded from the system boundaries, as a preliminary investigation indicates that their impact is negligible for the CC. Unlike material provision, data collection for the manufacturing phase relies on primary data obtained through direct measurements at supplier and OEM processes sites. Regarding the predictive analysis perspective, the production is modeled by considering the current year of the study (i.e., the year 2023), since the specific assessment is conducted in the present timeframe, thus excluding the use of forecast algorithms.

The use stage involves evaluating the impacts related to the operation of the CMS module, influenced by two factors: the energy/fuel transformation upstream to energy/fuel consumption itself (i.e., Well-to-Tank sub-phase, WTT) and the energy/fuel consumption for car driving, also considering the exhaust air emissions (i.e., Tank-to-Wheel, TTW), modeled through inventory framework reported in **Table 3** (see **Equations 1-9**).

Table 3. Use stage equations for predictive assessment.

Use Stage Equations			
Use		$CC^{Use}(y) = CC^{WTT}(y) + CC^{TTW}$	(1)
	ICEVs	$CC^{WTT}(y) = cc_{fuel}(y) \cdot EC_{comp}$	(2)
		$CC^{TTW} = (CO_{2veh} \cdot mil_{veh}(y) \cdot EC_{comp})/EC_{veh}$	(3)
		$EC_{comp} = (FRV \cdot \rho_{fuel} \cdot m \cdot mil_{veh}(y))/10000$	(4)
		$EC_{veh} = (EC_{100km} \cdot \rho_{fuel} \cdot mil_{veh}(y))/100$	(5)
		$EC_{100km} = (CO_{2veh} \cdot 100)/m_{CO_2}$	(6)
	BEVs	$CC^{WTT}(y) = cc_{mix}(y) \cdot EC_{comp}$	(7)
		$CC^{TTW} = 0$	(8)
		$EC_{comp_i} = (ERV \cdot m \cdot mil_{veh}(y))/10000$	(9)

$CC^{Use}(y)$ = climate change of the operational stage of specific year (y) [kg CO_{2eq}].
$CC^{WTT}(y)$ = climate change of WTT stage of specific year (y) [kg CO_{2eq}].
CC^{TTW} = climate change of TTW stage [kg CO_{2eq}].
m = mass of component – both REF and LW CMS module [kg].
$mil_{veh}(y)$ = vehicle mileage related to specific year (y) [km]
CO_{2veh} = vehicle CO_2 emissions [gCO_2/km]
ρ_{fuel} = density of fuel [kg/l]
m_{CO2} = CO_2 mass per liter of fuel [g/l]
EC_{veh} = vehicle energy consumption during use stage; [kg] for ICEVs, [kWh] for EVs.
EC_{comp} = component energy consumption during use stage; [kg] for ICEVs, [kWh] for EVs.
FRV = fuel reduction value [l/100km*100kg]
ERV = energy reduction value [kWh/100km*100kg]
$cc_{fuel}(y)$ = CC per mass unit due to fuel production of specific year (y) [kg CO_{2eq}/kg]
$cc_{mix}(y)$ = CC per anergy unit due to electrical mix grid production of specific year (y) [kg CO_{2eq}/kWh]

The assumptions of the predictive evaluation are described in the following bullet po ints, which refer to secondary data from the LCI Ecoinvent 3.9 database:

- the Well-To-Tank (WTT) phase is time-variant, as it depends on the environmental impact associated with the production of fuel/energy throughout the entire car operation, which varies over time (i.e., cc_{fuel} and cc_{mix}).
- the Tank-To-Wheel (TTW) phase does not depend on time, since vehicle emissions are not time-variant;
- according to the framework developed by Del Pero et al. [33-34], the mileage is based on the ACEA report [28], which provides an average lifespan of passenger vehicles of 12 years for the European context. Therefore, an annual mileage of 12000 km/year is assumed, resulting in a total mileage of 144000 km;
- the specific impacts of fuel and electricity production (cc_{fuel} and cc_{mix}) are calculated using predictive algorithms over 12 years within the European context (i.e., 2023-2035) [28], considering four predictive scenarios [21][22]:

  1. TRA scenario. It represents the traditional LCA scenario, where the use phase is calculated considering present-day impacts during the vehicle's operational years (TRADITIONAL SCENARIO);
  2. SSP2-BASE scenario. It assumes no climate policies, resulting in a temperature increase beyond 3.5°C compared to pre-industrial levels (WORST CASE SCANARIO);
  3. SSP2-RCP26 scenario. Aligned with the Paris Agreement, this predictive scenario anticipates a temperature increase between 1.8-2°C compared to pre-industrial levels (MIDDLE CASE SCENARIO);
  4. SSP2-RCP19 scenario. Also in line with the Paris Agreement, this predictive scenario projects a temperature increase of 1.5°C compared to pre-industrial levels (BEST CASE SCENARIO).

- the reduction values (i.e., FRV and ERV) are obtained on the basis of modeling framework developed by Del Pero et al. [33-34]

The inventory of EoL stage is modelled by according to the Directive 2000/53/EC [35]. The scenarios considered for steel and aluminium components are in line with ISO standard 22628:2002 [36], based on the prevailing technological capabilities in Europe. Given that the CMS module is not typically disassembled for reuse or recovery due to its low mass and unfavourable positioning within the vehicle, it is assumed that it is not disassembled at from the vehicle and it undergoes the shredding process. Subsequently, the scenario involves material separation and open-loop recycling. The assessment of recycling takes into account both the environmental impacts associated with material/energy consumption and the impacts caused by recycling activities. The inventory is modeled as the avoided production of primary material, based on the substitution factor approach [37]. Concerning the predictive perspective, the EoL phase is modeled by considering the last year of vehicle life-time (i.e., the year 2035), since the above processes occur in the future time-frame, and this is done by using forecast algorithms.

The LCI data for all LC stages are provided in **Table 4**.

Table 4. LCI (Life Cycle Inventory) data collection of all LC stages.

Inventory Data Collection				
		LCI Data		**Quality**
Cradle-to-Grave Approach	Production (2023)	CMS Material composition	See **Table 1** and **Table 2**	Primary
		Raw materials production	Ecoinvent 3.9 Database	Secondary
		CMS Manufacturing processes	See **Table 1** and **Table 2**	Primary
		Manufacturing processes	Ecoinvent 3.9 Database	Secondary
	Use (2023-2035)	Propulsion technology	ICEV Gasoline BEV	Primary
		Vehicle Class	C	Primary
		CO_{2veh} [gCO_2/km]	108*	Primary
		mil_{veh} [km]	144000**	Primary
		ρ_{fuel} [kg/l]	0.741	Primary
		m_{CO2} [g/l]	2370	Primary
		FRV [l/100km*100kg] ERV [kWh/100km*100kg]	Framework Modeling [36-37]	Secondary
		m [kg]	3.989 (REF CMS)	Primary

		2.416 (LW CMS)	
EoL (2035)	Electricity for shredding [kWh/kg]	0.05	Secondary
	Electricity for materials sorting [kWh/kg]	0.033	Secondary
	Sorting Efficiency [-]	98%	Secondary
	EoL processes	Ecoinvent 3.9 Database	Secondary
	Substitution Factor for Materials	0.25 - Steel 0.15 - Aluminium	Secondary

*Average value representative of generic C-class gasoline vehicle.
**Value calculated according to ACEA Report [28] and Modeling Framework

3. Results and Discussion

LCIA results for REF and LW design alternatives across each LC stage, considering both different propulsion technologies (ICEV, BEV) and predictive scenarios (TRA, SSP2-BASE, SSP2-RCP26, SSP2-RCP19), are reported in **Table 5**.

Subsequent paragraphs delve into the discussion regarding the impact variations resulting from the adoption of the forecasting scenario in comparison to the baseline approach, as well as the differences between the REF and LW design solutions. These discussions stress the implications that different methodological approaches and design choices have on the environmental performance of the CMS module.

Table 5. LCIA results for REF and LW design solutions, with different propulsion technologies during use stage (ICEV, BEV).

	CMS Solution	Propulsion Technology	Predictive Scenario	Production (2023)		Use (2023-2035)		EoL (2035)	LC (2035)
				Material	Manufacturing	WTT	TTW		
CC [$kgCO_{2eq}$]	REF	ICEV	TRA	10.30	0.096	4.422	23.143	-2.031	35.93
			SSP2-BASE	10.30	0.096	4.416	23.143	-1.952	36.003
			SSP2-RCP26	10.30	0.096	4.344	23.143	-1.596	36.288
			SSP2-RCP19	10.30	0.096	4.259	23.143	-1.411	36.388
		BEV	TRA	10.30	0.096	12.238	0.00	-2.031	20.604
			SSP2-BASE	10.30	0.096	11.120	0.00	-1.952	19.564
			SSP2-RCP26	10.30	0.096	9.399	0.00	-1.596	18.20
			SSP2-RCP19	10.30	0.096	5.637	0.00	-1.411	14.623
	LW	ICEV	TRA	22.20	1.18	2.678	14.017	-2.571	37.504
			SSP2-BASE	22.20	1.18	2.675	14.017	-2.586	37,486
			SSP2-RCP26	22.20	1.18	2,631	14.017	-2.53	37,498
			SSP2-RCP19	22.20	1.18	2.58	14.017	-2.483	37,494
		BEV	TRA	22.20	1.18	7.412	0.00	-2.571	28.221
			SSP2-BASE	22.20	1.18	6.735	0.00	-2.586	27.53
			SSP2-RCP26	22.20	1.18	5.693	0.00	-2.53	26.544
			SSP2-RCP19	22.20	1.18	3.414	0.00	-2.483	24.311

3.1. Contribution Analysis of Impacts: Influence of LC Stages on CMS module

Figure 2 presents the contribution analysis of environmental impact, categorized by LC stage, powertrain technology, and predictive scenarios. Regardless of the module design, the most impactful LC phases are production and use, representing approximately 90-95 % of total impact for both BEV and ICEV. Regarding production, it can be observed that CC shows a significant increase, reaching 125 % when passing from the REF to the LW design alternative, mainly due to the higher mass-specific impact associated with aluminium production compared to steel. The mass reduction achieved through aluminium (about 39 %) does not counterbalance the higher mass-specific CC.

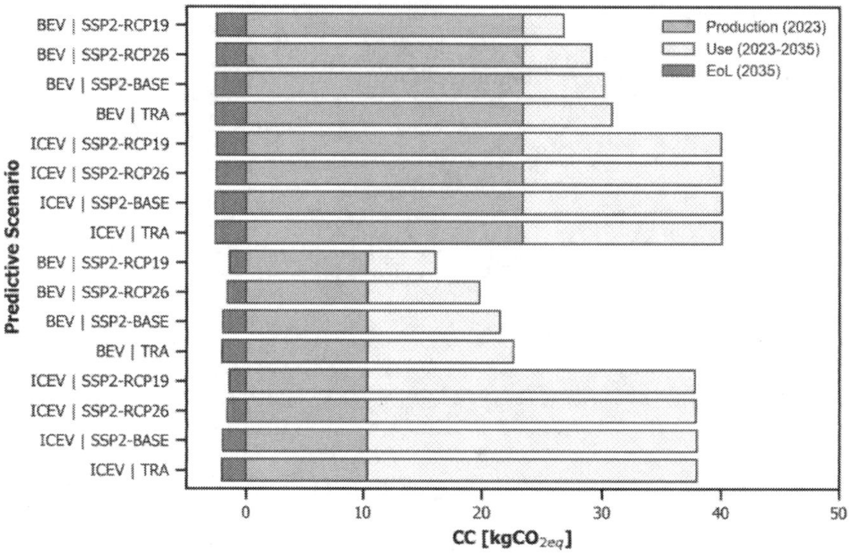

Figure 2. Contribution analysis of impact by LC stage, powertrain technology, and predictive scenarios for REF and LW CMS design solutions.

Concerning the REF CMS design, the use stage is the most significant for both powertrain types, except for the predictive scenarios SSP2-RCP26 and SSP2-RCP19 for BEV, for which the production becomes more impactful than the operational phase. This outcome can be attributed to production and procurement of electric energy consumed during operation, whose impact is expected to significantly decrease over time due to increasing use of renewables, as predicted by the employed forecasting algorithms.

Conversely, in the case of the LW CMS design, the operational stage remains the second most significant for all powertrain and predictive scenario configurations since the operational phase directly depends on module mass (see Eq. 1-9). Thus, mass reduction leads to an impact decrease of use stage impact. However, the production shows a higher CC due to the procurement of aluminium, which exhibits higher specific effects than steel.

Explicitly considering the use phase, it remains relatively constant across predictive scenarios for the ICEV configuration, primarily due to the fuel production required for the operational phase, where the impact remains relatively stable or experiences only a slight reduction over time (according to the prediction of the employed forecasting algorithms). On the other hand, for the BEV configuration, a significant reduction is observed, as the predictive scenarios change from the worst (i.e., traditional approach - TRA) to the best (i.e., SSP2-RCP19); such a reduction is mainly due to the decreasing environmental impact of electricity production, as previously specified.

The EoL phase represents the least significant LC step, accounting for approximately 5 % and10% of the total impact respectively for ICEV and BEV. That said, results show that EoL provides an environmental credit due to metallic composition of both design alternatives. However, the percentage shares associated with EoL are minimal: this is primarily due to the low substitution factor of steel and aluminum recycling, which derives from both the high energy demand of re-melting and the need for primary alloy elements to achieve the same quality of primary materials (See **Table 4**). Interestingly, the environmental credit associated with EoL of LW solution is higher in absolute terms than the one associated with the REF solution, as a result of combination of production and substitution factor. Furthermore, regardless of powertrain configuration and design solution, it is noteworthy to observe a reduction in EoL credit, as the predictive scenarios become more stringent because of stricter environmental policies (from the traditional approach - TRA - to the best approach - SSP2-RCP19). This reduction is not only due to the decreased environmental impact of the recycling process, but also to a reduction in the impact related to specific material procurement. Consequently, this leads to an overall reduction in the total credit.

In the light of previous points and results shown in **Table 5**, the solutions related to the ICEV configuration exhibit a slightly higher impact as the predictive scenarios become more stringent. These results are due to the reduction in the environmental credit associated with EoL, which is not compensated by use phase, despite this latter provides a decrease at forecast scenarios varying. Instead, the solutions associated with BEV configuration show a lowering impact with at predictive scenarios varying. In this case, the environmental credit associated with EoL is outweighed by a significant decrease in the use phase, since the decrease in electricity production impact is relevant.

Figure 3 illustrates the variation of CC impact for CMS module with respect to mileage over the vehicle life-span (i.e., 12 years with an annual mileage of 12000 km/year).

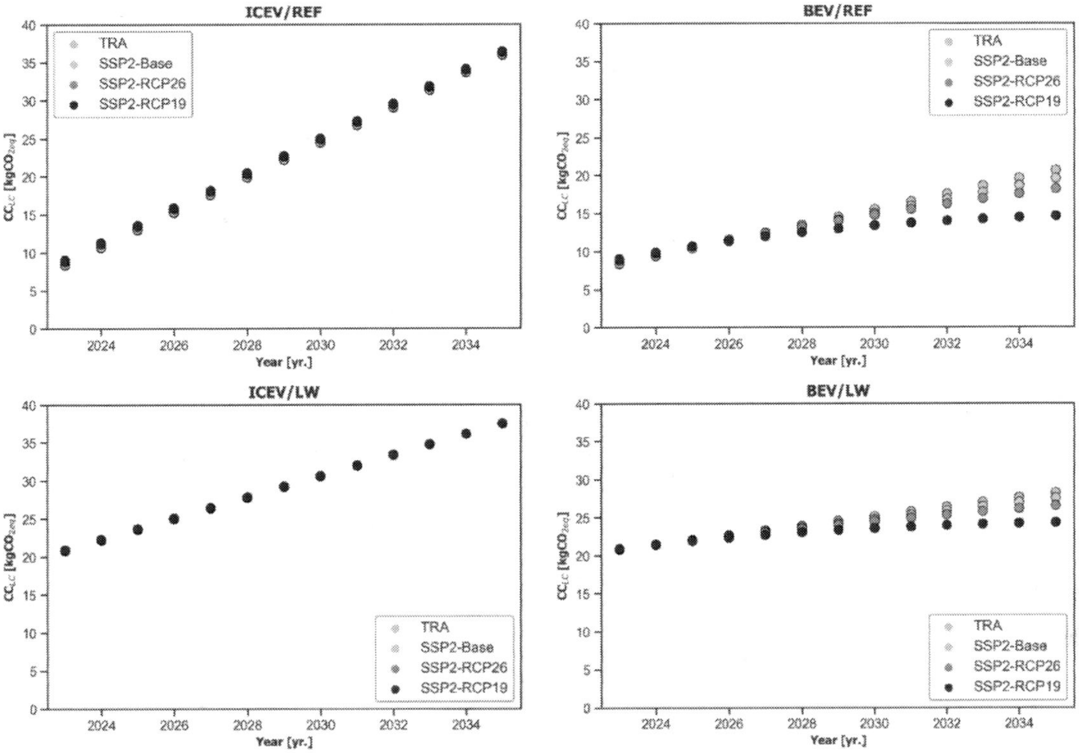

Figure 3. CC representation of CMS module respect to mileage for ICEV/BEV and REF/LW combinations.

All the previously defined predictive scenarios are presented in the analysis. **Table 6** and **Table 7** present the CC percentage variation between the traditional LCA scenario and predictive SSP2 scenarios for both ICEV/BEV and REF/LW combinations. It can be observed that:

1. in the case of ICEV/REF combination, the impacts of the predictive scenarios are slightly higher than the traditional approach, regardless of the year considered (ranging from 0.2% to 7.5%). This outcome can be explained through the reduction in EoL credit, which is not compensated by the impact decrease during use phase (insufficient to compensate for the credit reduction);
2. in the case of BEV/REF combination, the impacts of the predictive scenarios are slightly higher than the traditional approach only over the first 3/4 years of the operational phase, after which a significant reduction in environmental impact is observed (ranging from -0.6% to -29% for the SSP2-RCP19 scenario). These results are due to the reduction in the environmental credit associated with EoL, which is counterbalanced by a bigger impact reduction in use stage (concerning the decreasing CC from electricity production in the future years);
3. for ICEV/LW configuration, the percentage variations depend on scenario: the more stringent scenarios in line with the Paris Agreement (i.e., SSP2-RCP26 and SSP2-RCP19) exhibit slightly higher impacts than the traditional approach until the last operational year, where there is a reversal of this trend (ranging from -0.014% to -0.027%). On the other hand, the SSP2-BASE scenario shows lower percentage values regardless of the year (ranging from -0.047% to -0.068%). The reduction in EoL credit influences these variations for the first two scenarios. In contrast, the third scenario experiences an increase in credit due to the production and procurement of aluminum required for production and EoL. The impact of aluminum, according to the predictive algorithms, is expected to increase slightly over time for the SSP2-BASE scenario, thus increasing the credit;
4. in the case of BEV/LW configuration, the percentage variations also differ among scenarios: the SSP2-RCP26 and SSP2-RCP19 scenarios show slightly higher impacts than the traditional approach only during the first two years of the operational phase, after which there is a notable reduction in environmental impact (ranging from -0.6% to -13.9% for the SSP2-RCP19 scenario). On the other hand, the SSP2-BASE scenario exhibits lower percentage values regardless of the year (ranging from -0.068% to -2.45%).

The environmental credit associated with EoL is counterbalanced by a stronger reduction in use stage, mainly due to a greener electricity production in the future.

Table 6. CC percentage variation of CMS module between TRA scenario and predictive SSP2 ones, related to ICEV/BEV and REF combinations.

Year [yr.]	ICEV - ΔCC [%]			BEV - ΔCC [%]		
	SSP2-Base	SSP2-RCP26	SSP2-RCP19	SSP2-Base	SSP2-RCP26	SSP2-RCP19
2023	0.932%	5.197%	7.410%	0.932%	5.197%	7.410%
2024	0.731%	4.070%	5.786%	0.546%	3.912%	5.094%
2025	0.600%	3.335%	4.719%	0.011%	2.553%	2.346%
2026	0.508%	2.813%	3.959%	-0.603%	1.149%	-0.672%
2027	0.439%	2.422%	3.386%	-1.250%	-0.283%	-3.849%
2028	0.386%	2.116%	2.936%	-1.897%	-1.727%	-7.110%
2029	0.344%	1.869%	2.571%	-2.521%	-3.176%	-10.399%
2030	0.310%	1.664%	2.268%	-3.104%	-4.623%	-13.676%
2031	0.281%	1.491%	2.010%	-3.634%	-6.061%	-16.911%
2032	0.257%	1.342%	1.789%	-4.100%	-7.487%	-20.080%
2033	0.236%	1.211%	1.595%	-4.495%	-8.899%	-23.166%
2034	0.217%	1.096%	1.425%	-4.812%	-10.292%	-26.153%
2035	0.201%	0.994%	1.273%	-5.047%	-11.666%	-29.030%

Table 7. CC percentage variation of CMS module between TRA scenario and predictive SSP2 ones, related to ICEV/BEV and LW combinations.

Year [yr.]	ICEV - ΔCC [%]			BEV - ΔCC [%]		
	SSP2-Base	SSP2-RCP26	SSP2-RCP19	SSP2-Base	SSP2-RCP26	SSP2-RCP19
2023	-0.068%	0.200%	0.425%	-0.068%	0.200%	0.425%
2024	-0.064%	0.185%	0.390%	-0.142%	0.003%	0.012%
2025	-0.061%	0.170%	0.354%	-0.276%	-0.276%	-0.631%
2026	-0.058%	0.153%	0.316%	-0.455%	-0.628%	-1.472%
2027	-0.056%	0.136%	0.277%	-0.669%	-1.044%	-2.479%
2028	-0.054%	0.118%	0.238%	-0.905%	-1.517%	-3.627%
2029	-0.052%	0.099%	0.199%	-1.153%	-2.041%	-4.892%
2030	-0.051%	0.081%	0.160%	-1.404%	-2.609%	-6.252%
2031	-0.050%	0.062%	0.121%	-1.651%	-3.217%	-7.687%
2032	-0.049%	0.043%	0.083%	-1.885%	-3.858%	-9.181%
2033	-0.048%	0.024%	0.045%	-2.100%	-4.529%	-10.716%
2034	-0.048%	0.005%	0.009%	-2.290%	-5.226%	-12.278%
2035	-0.047%	-0.014%	-0.027%	-2.450%	-5.944%	-13.854%

3.2. Sensitivity and BEP Analysis

A critical aspect that must be considered is that all solutions related to LW CMS design are more impactful if compared to the REF version (see **Figure 2**). The rationale behind this outcome lies in selecting the average annual mileage (12000 km/year) used for calculating use stage, resulting in a total mileage at the end of the vehicle's operational life that does not reach the break-even point (BEP) between the two CMS design solutions.

Against this scenario, a sensitivity analysis is conducted, where the main parameters are powertrain (ICEV, BEV), vehicle segment (A/B, C, D/E), design solution (REF, LW), predictive scenario (TRA, SSP2), and annual mileage; the latter varies from 10000 km/year to 15000 km/year. **Figure 4** reports a Heatmap representation of the CC results (calculated in the year 2035), as the above parameters are varied.

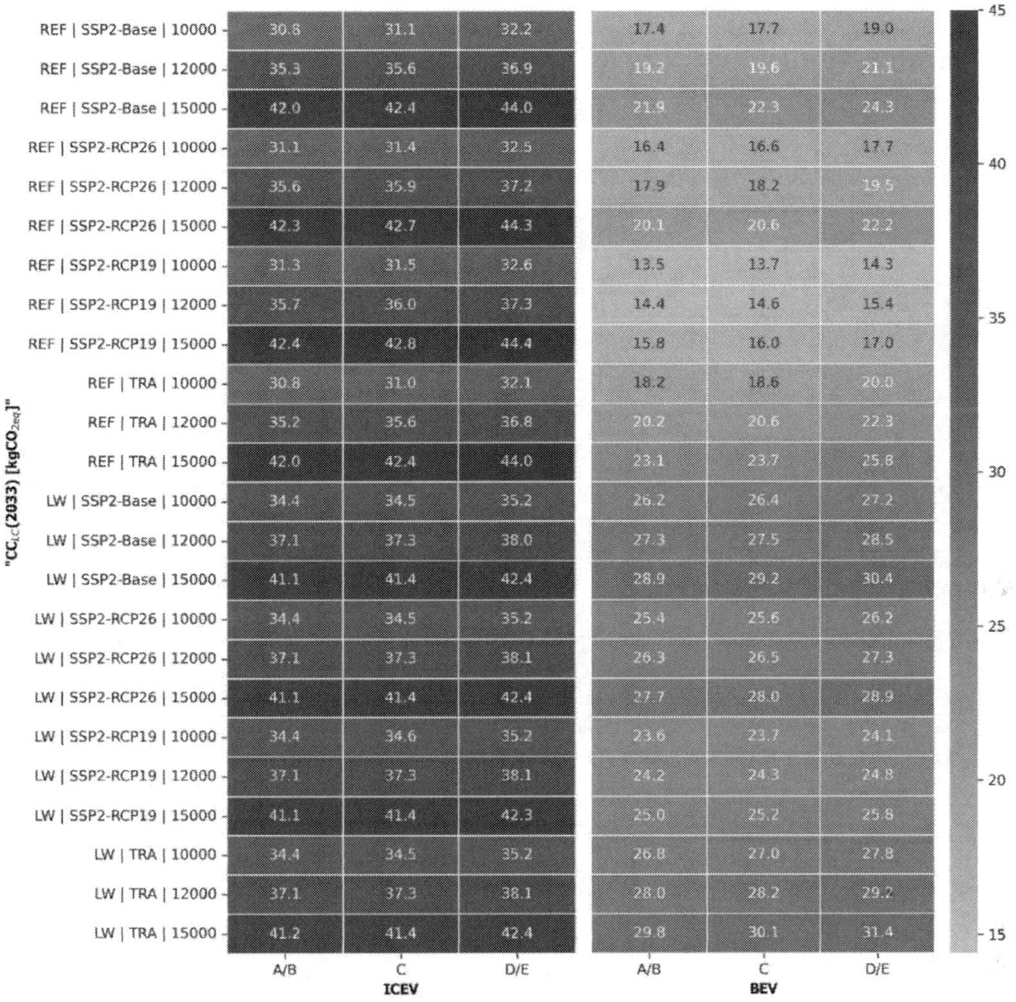

Figure 4. CC Heatmap representation of sensitivity analysis results.

Interestingly, regardless of the annual mileage, solution, and predictive scenario, the impacts associated with any BEV segment is lower than the ICE configuration (except for the last values associated with the TRA scenario for the LW solution with an avg=15000 km/year). Furthermore, for the BEV configuration, the total impact of the LW solutions fails to reach the CC equivalence with its ICEV counterparts. This is due to the reduction over time of the impact related to the production and sourcing of the electrical energy required for the operational phase, thus leading to a reduction in the overall impact for both REF and LW solutions. Consequently, reaching the BEP does not occur for mileage values which are incompatible with the operational life of passenger vehicles.

Concerning the ICE configuration, it is observed that the CC outcomes of LW alternatives exceed those of the REF solutions for all predictive scenarios (TRA, SSP2) and vehicle segments (A/B, C, D/E), considering the mileage of 15000 km/year. The mass reduction achieved in the LW solution and the slight impact reduction of fuel production lead to a significant CC decrease associated with use phase. As a result, the BEP is reached before that vehicle operational life is concluded. Errore. L'origine riferimento non è stata trovata. provides a comprehensive overview of BEP variation based on sensitivity analysis.

4. Conclusion

This paper focuses on a predictive LCA to assess the time-based effects on a generic automotive component, adopting a comparative approach between the traditionally used design solution and the innovative one, which involves a significant weight reduction.

Table 8. BEP Results obtained from sensitivity analysis.

BEP [km]			ICEV			BEV		
			A/B	C	D/E	A/B	C	D/E
REF - LW	SSP2-BASE	10000	NR	NR	NR	NR	NR	NR
		12000	NR	NR	NR	NR	NR	NR
		15000	168000	168000	159000	NR	NR	NR
	SSP2-RCP26	10000	NR	NR	NR	NR	NR	NR
		12000	NR	NR	NR	NR	NR	NR
		15000	165000	165000	156000	NR	NR	NR
	SSP2-RCP19	10000	NR	NR	NR	NR	NR	NR
		12000	NR	NR	NR	NR	NR	NR
		15000	165000	162000	156000	NR	NR	NR
	TRA	10000	NR	NR	NR	NR	NR	NR
		12000	NR	NR	NR	NR	NR	NR
		15000	171000	168000	162000	NR	NR	NR

NR = BEP Not Reached

The starting point is the re-engineering of the chosen case study (i.e., a rear CMS module), by replacing conventional steel with ultra-high-strength aluminum alloys, specifically using extruded AW 7003 for beam and crash box components. The innovative design results in a significant mass reduction of nearly 40 %. From a design perspective, adopting the 7000 series aluminum alloy strikes a good balance between strength, production efficiency, and design flexibility. Additionally, using an open beam profile with a customized extrusion design concept yields satisfactory outcomes for both lightweight potential and crashworthiness.

The comparative sustainability assessment, however, yields contrasting results across specific parameters, like powertrain, annual mileage, car segment, predictive scenarios, and design solutions. The environmental assessment outcomes can be summarized as follows:

- the contribution analysis of environmental impact, considering LC stages, powertrain technology, and predictive scenarios for the CMS design solutions, reveals that production and use stages are the most impactful, representing approximately 90 - 95% of total LC impact for both BEV and ICEV powertrains;
- the CC impact category shows a significant increase (125 %) when passing from REF to LW design alternative, mainly due to the higher mass-specific impact associated with aluminum production if compared to steel;
- for the REF CMS design, the use stage is the most significant for both powertrain types. However, for specific predictive scenarios of BEV, the production phase becomes more impactful due to the production and procurement of electric energy, which is expected to decrease over time due to the increasing use of renewables. On the other hand, in the case of the LW CMS design, the operational stage remains the second most significant stage for all powertrain and predictive scenario configurations, with weight reduction leading to a reduced impact of. However, the production phase shows a higher impact due to the procurement of aluminium;
- the use phase for ICEV remains relatively constant across predictive scenarios. At the same time, there is a significant reduction in BEV as predictive scenarios change from worst to best, mainly due to the decreasing environmental impact of electricity production;
- EoL represents the least significant stage, accounting for approximately 5% -and 10 % respectively for ICEV and BEV, resulting in an overall environmental credit;
- the percentage variation of CC with mileage demonstrates different behaviors passing from the traditional LCA scenario to the predictive SSP2 ones, considering ICEV/BEV and REF/LW combinations. The predictive results show that the ICEV solutions are almost aligned with the traditional approach, with a percentage reduction ranging from approximately 0 % to 1.2 %. Whereas for BEV solutions, significant impact reductions are obtained, with a percentage reduction ranging from 3 % to 30%;
- the LW design solutions are more impactful than the REF alternatives due to the chosen average annual mileage (12000 km/year), leading to a total mileage that does not provide a BEP between the two solutions;
- a sensitivity analysis is conducted considering several parameters, such as powertrain, vehicle segment, design solution, predictive scenario, and annual mileage. The analysis shows that the impacts associated with any BEV vehicle segment are consistently lower than their ICEV counterparts, except for one scenario. Furthermore, the LW solutions for BEV fail to reach CC equivalence with ICEV configuration, due to the reduced impact over time of electricity production. Thus, reaching the BEP is only feasible for mileage values which are not compatible

with passenger car life-time. For the ICEV configuration, the CC outcomes of REF alternatives become bigger than LW solutions for all scenarios and vehicle segments at a mileage value of 15000 km/year. The mass reduction achieved, coupled with a slight impact reduction in fuel production, leads to a significant CC decrease during use phase, enabling a BEP between the two alternatives.

Overall, the proposed methodology highlights the need for trade-offs between material and manufacturing choices, emphasizing the importance of a holistic approach to predictive sustainability assessment.

References

[1] European Commission. White Paper: Roadmap to a Single European Transport Area—Towards a Competitive and Resource Efficient Transport System; *European Commission: Brussels*, Belgium, 2011; /* COM/2011/0144 final */
[2] *European Environment Agency*. Greenhouse gas emissions by aggregated sector. 2019.
[3] *European Environment Agency*. Greenhouse gas emissions from transport in Europe. 2019.
[4] Michael Z. Hauschild, Ralph K. Rosenbaum, *Stig Irving Olsen* 2018 Life Cycle Assessment: Theory and Practice. DOI 10.1007/978-3-319-56475-3
[5] Staniszewska E, Klimecka-Tatar D, Obrecht M. Eco-design processes in the automotive industry. *Production Engineering Archives*. 2020 Dec 1;26:131–7.
[6] Schöggl J. P., Baumgartner R. J., Hofer, D. Improving sustainability performance in early phases of product design: A checklist for sustainable product development tested in the automotive industry. *Journal of Cleaner Production*. 2017,*140*, 1602-1617
[7] Neidhardt M, Mas-Peiro J, Schneck A, Pou JO, Gonzalez-Olmos R, Kwade A, et al. Automotive Electrification Challenges Shown by Real-World Driving Data and Lifecycle Assessment. *Sustainability*. 2022 Jan;14(23):15972.
[8] Helmers E, Dietz J, Weiss M. Sensitivity Analysis in the Life-Cycle Assessment of Electric vs. Combustion Engine Cars under Approximate Real-World Conditions. *Sustainability*. 2020; 12(3):1241
[9] Del Pero F, Delogu M, Pierini M. Life Cycle Assessment in the automotive sector: a comparative case study of Internal Combustion Engine (ICE) and electric car. *Procedia Structural Integrity*. 1 Jan. 2018;12:521–37.
[10] Andersson Ö, Börjesson P. The greenhouse gas emissions of an electrified vehicle combined with renewable fuels: Life cycle assessment and policy implications. *Applied Energy*. 1 May 2021;289:116621.
[11] Koroma MS, Costa D, Philippot M, Cardellini G, Hosen MS, Coosemans T, et al. Life cycle assessment of battery electric vehicles: Implications of future electricity mix and different battery end-of-life management. *Science of The Total Environment*. 20 Jul 2022;831:154859.
[12] Department for Transport. COP26 declaration on accelerating the transition to 100% zero emission cars and vans. *GovUk*; 2021. https://www.gov.uk/government/publications/cop26-declaration-zero-emission-cars-and-vans/cop26-declaration-on-accelerating-the-transition-to-100-zero-emission-cars-and-vans.
[13] Mamalis AG, Spentzas KN, Mamali AA. The impact of automotive industry and its supply chain to climate change: Somme techno-economic aspects. *Eur Transp Res Rev*. Mar 2013;5(1):1–10
[14] Mohammadi H, Ahmad Z, Mazlan SA, Faizal Johari MA, Siebert G, Petrů M, et al. Lightweight Glass Fiber-Reinforced Polymer Composite for Automotive Bumper Applications: A Review. *Polymers*. Jan 2023;15(1):193
[15] Gonçalves M, Monteiro H, Iten M. Life Cycle Assessment studies on lightweight materials for automotive applications - An overview. *Energy Reports*. 1 Jun 2022;8:338–45.
[16] Akhshik M, Panthapulakkal S, Tjong J, Sain M. The effect of lightweighting on greenhouse gas emissions and life cycle energy for automotive composite parts. *Clean Techn Environ Policy*. 1 Apr 2019;21(3):625–36
[17] Kawajiri K, Kobayashi M, Sakamoto K. Lightweight materials equal lightweight greenhouse gas emissions?: A historical analysis of greenhouse gases of vehicle material substitution. *Journal of Cleaner Production*. 20 Apr 2020;253:119805.
[18] Memary R, Giurco D, Mudd G, Mason L. Life cycle assessment: a time-series analysis of copper. *Journal of Cleaner Production*. 1 Sep 2012;33:97–108
[19] Zhu X, Wang S, Wang L. Life cycle analysis of greenhouse gas emissions of China's power generation on spatial and temporal scale. *Energy Science & Engineering*. 1 Mar 2022;10.
[20] Stranddorf LK, Clavreul J, Prieur-Vernat A, Ryberg MW. Evaluation of life cycle impacts of European electricity generation in relation to the Planetary Boundaries. *Sustainable Production and Consumption*. 2023 Jul 1;39:414–24.
[21] Sacchi R, Terlouw T, Siala K, Dirnaichner A, Bauer C, Cox B, et al. PRospective EnvironMental Impact asSEment (premise): A streamlined approach to producing databases for prospective life cycle assessment using integrated assessment models. *Renewable and Sustainable Energy Reviews*. 1 May 2022;160:112311.
[22] Riahi K, van Vuuren DP, Kriegler E, Edmonds J, O'Neill BC, Fujimori S, et al. The Shared Socioeconomic Pathways and their energy, land use, and greenhouse gas emissions implications: An overview. *Global Environmental Change*. 1 Jan 2017;42:153–68.

[23] *RCAR* Low-Speed Structural Crash Test Protocol. Available online: http://rcar.org/Papers/Procedures/RCAR%20Structure%20Test%20procedure%20Version%202_3.pdf#zoom=80%

[24] EC-JRC. International Reference Life Cycle Data System (ILCD) Handbook-Recommendations for Life Cycle Impact Assessment in the European Context; *Publications Office of the European Union*: Brussels, Belgium, 2011; ISBN 978-92-79-17451-3.

[25] ISO 14040:2006. Environmental management – Life cycle assessment – Principles and framework.

[26] ISO 14044:2006. Environmental management – Life cycle assessment – Requirements and guidelines.

[27] Huijbregts M, Steinmann Z, Elshout P, Stam G, Verones F, Vieira M, Hollander A, Zijp, M. ReCiPe 2016 v1. 1 - A Harmonized Life Cycle Impact Assessment Method at Midpoint and Endpoint Level: Report I. Charact. No RIVM Rep. 2016–0104a *Natl Inst Public Health Env*. Bilthoven Neth 2017.

[28] ACEA Report: VEHICLES IN USE EUROPE 2023 - JANUARY 2023

[29] Mock P, Kühlwein J, Tietge U, Franco V, Bandivadekar A, German J. TheWLTP: How a new test procedure for cars will affect fuel consumption values in the EU.Work. Pap.—*Int. Counc. Clean Transp*. 2014, 9, 35–47.

[30] Del Pero F, Delogu, M, Fernandez V, Ierides M, Seidel K, Thirunavukkarasu D. Lightweight Design Solutions in the Automotive Sector: Impact Analysis for a Door Structure. In Sustainable Design and Manufacturing; KES-SDM2019. *Smart Innovation. Systems and Technologies*; Springer: Singapore, 2019;

[31] Enhanced Lightweight Design, ENLIGHT Project. Available online: https://cordis.europa.eu/project/id/314567

[32] E-Mobility Life Cycle Assessment Recommendations, e-LCAr Project. Available online: http://www.elcarproject.eu/

[33] Del Pero F, Berzi L, Antonacci A, Delogu M. Automotive Lightweight Design: Simulation Modeling of Mass-Related Consumption for Electric Vehicles. *Machines*. 2020; 8(3):51.

[34] Del Pero F, Delogu M, Pierini M. The effect of lightweighting in automotive LCA perspective: Estimation of mass-induced fuel consumption reduction for gasoline turbocharged vehicles. *Journal of Cleaner Production*. 2017; *154*, 566-577

[35] Directive 2000/53/EC on End-of-Life Vehicles European Commission. Available online: https://eur-lex.europa.eu/legal-content/EN/ALL/?uri=CELEX%3A32000L005

[36] ISO 22628. Road Vehicles, Recyclability and Recoverability Calculation Method; ISO: Geneva, Switzerland. Available online: https://www.iso.org/standard/35061.htm

[37] *Sphera*, GaBi6 Software. 2020. Available online: http://www.gabi-software.com/italy/index/

IOP Conf. Series: Materials Science and Engineering 1306 (2024) 012042 doi:10.1088/1757-899X/1306/1/012042

Product Re-Design Towards Sustainability: a Case Study of a Mass-Production Device

M. Moreschini[1], A. Migliavacca[2], A. Mansutti[2], D. Ilare[1,2], M. Guagliano[1]

[1]Politecnico di Milano, Dip.to Meccanica, Via La Masa, 1 - 20156 Milano (Italy)
[2]Elettrotecnica Rold srl, Via della Merlata, 1 - 20024 Nerviano (MI), Italy

mario.guagliano@polimi.it

Abstract. Sustainability issues are nowadays embedded in society and organizations cannot avoid tackling them. Product design represents the phase in which 80% of the overall emissions are determined, thus the one on which changes should be realized. In some industrial sectors the environmental impact has never been accounted systematically during the conceptualization phase. The resulting products are consequently characterized by poor environmental performances, that have a deep negative effect if realized on a mass production scale. A sustainable re-design and process innovation approach is therefore required. This is the objective of this research, that uses *Rold* Washing Machine Doorlock as a case study example. After the realization of a cradle-to-gate Life Cycle Assessment (LCA) of the component under exam, design and process modifications are identified. The assessment is not limited to the proposal of environmentally improved versions of the product. A barrier towards sustainability adoption is represented by the common perception of its unaffordability. It is important for the identified solutions leading to improved ecological performances, but they should be sustainable for the company also in financial terms. At this purpose, for each proposed change, a technical assessment of the operational steps required is performed, together with an economic analysis. The resulting re-engineered product passes all the desired conditions, and the methodology applied can be extended to every other sector facing the same issue.

1. Introduction

Sustainability is nowadays a key concern for every business, for ethical, economic, and legislative reasons [1] [2]. In 2019, the European Commission (EC) published the European Green Deal (EGD), whose objective is providing an overall direction for the EU legislation and regional development strategies, where climate policy and emissions reduction are the top priority[3][4].The plan aims to make Europe climate neutral by 2050 and to become a global leader in tackling climate change. In the EGD, the industrial sector plays a key role, as it represents one of the most important industries in the continent and accounts for 20 percent of the EU's greenhouse gas emissions [5]. The EGD addresses product design as a fundamental practice towards more sustainable products, as it represents the phase in which 80% of the overall emissions are determined [6]. Industrial sectors in which sustainability concerns have never been integrated systematically in the design phase, need to incorporate this transition, pursuing sustainable design and process innovation. From this requirement derives the following research scope.

IOP Conf. Series: Materials Science and Engineering 1306 (2024) 012042 doi:10.1088/1757-899X/1306/1/012042

2. Research Objectives and Methodology

2.1. Research Gap and Objectives

Given the empirical approach of the project, the research gap is identified through an integration between the literature review, industry trends analysis and company's interviews. Products conceived when sustainability was not considered in design choices are likely to have high improvement margins in terms of environmental impact and they could lead to significant change to the project. Improvement actions are most of the time, focalised on the downstream phases of product lifecycle, as their consolidated design is given for granted. The present literature does not systematically tackle this criticality. What organizations need is a product re-engineering approach, aimed at integrating sustainability considerations in a structured way, thus achieving better results. This kind of approach could be useful for already existing devices, while for brand new projects the integration of sustainability issues should be included in the first stages of the design. Sustainability definition has a triple perspective: environmental, economic and social [7]. The objective of the study is performing an analysis of a consolidated industrial product and suggesting design and process re-engineering actions aimed at improving its sustainability performances. The most addressed tool for the evaluation of products' environmental impact is Life Cycle Assessment (LCA) which represents the starting point of the analysis. Finally, this study aims to evaluate whether the methodology applied could be generalized and useful for other applications or industrial sectors, facing the same problem. The previously described motivations are the foundation of the defined research questions:

R.Q.1: *Is it possible, starting from the LCA results of a consolidated product, re-engineering it (in terms of design, material, production process) to make it more environmentally sustainable?*
R.Q.2: Can the re-engineered product be sustainable also from an economic perspective?
R.Q.3: On the basis of the present case study, is it possible to draw some general guidelines that could be used also for other products and/or in other industrial sectors?

2.2. Research Methodology

For providing answers to the identified research questions, the work is structured in two principal phases:

- Methods and tools identification
- Case study analysis

Methods and tools identification

The first phase consists into acquiring knowledge of the theoretical concepts required to perform the work. Specifically, four topics are investigated:

- Sustainability and Sustainable Manufacturing;
- Design for X (DfX) Methodologies- characteristics and application rules
- LCA-steps and methodologies.

Scientific databases are consulted: Scopus, Emerald, Web of Science and Google Scholar. Articles are selected through: keywords insertion, filtering by abstract and then by content, snowballing. For the acquisition of DfX and LCA concepts, it is referred to professional handbooks, providing explanations about the steps required.

Case Study

The case study analysis phase is articulated as shown in the framework provided in **Figure 1**. In a preliminary step, knowledge regarding the company's reference sector and the related sustainability trends are acquired through papers, company's annual reports and business research reviews. During the solutions assessment, proposals not passing all the gates will not be completely discarded. Technology innovations or other contingent factors could offer the possibility to make them worthiness in the future, thus entering the cycle again.

3. Case Study Context

3.1. The Company

With the aim of understanding the current company's business model characteristics under the sustainability point of view, and to identify improvement directions, Rold's Sustainable Business Model Canvas has been realized.

Particularly, it is decided to use the Triple Layered Business Model Canvas (TLBMC) framework [8]. The final model obtained is reported in Appendix A. The TLBMC is a tool that allows to link business model innovation with sustainable business model development, by assisting organizations in overcoming barriers towards sustainability-oriented changes [8]. The tool analysis reveals that the company is committed in the sustainability transition and is aware of its need, but concrete actions have not been put in place yet.

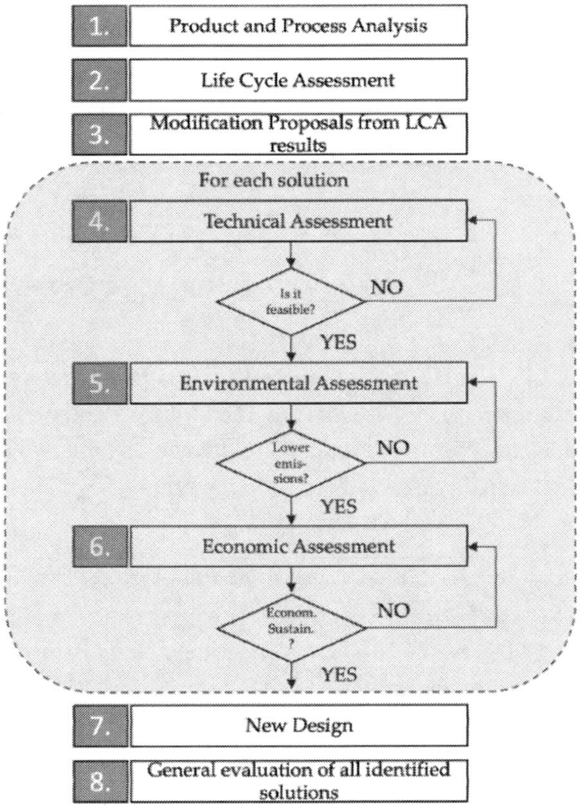

Figure 1. Case Study Analysis Methodology

This work provides a contribution in such direction, allowing to propose measures aimed at reducing the environmental costs of running the business and including sustainability concerns in the value proposition.

3.2. The Product and its Production Process

The product analysed for the study is a safety, magnetic doorlock for front loading washing machines, named *DKS08618* (**Figure 2**). It will be furtherly referred to it as *DK*. It is composed of two main parts: an internal one (that will be identified as *CORE*) and an external one, enabling the *CORE* connection with the washing machine.

Figure 2. 3D drawing of the DKS08618 magnetic doorlock (front view)

The *DK* realization is based on two main manufacturing processes, both performed internally by the company:

- Polymeric components are made through injection moulding technology;
- The *CORE* and the final product are assembled through a completely automatic assembly line.

The average annual demand is around 20 millions of pieces.

4. Life Cycle Assessment

The first phase of the work consists into realizing an LCA of the product under analysis. Life Cycle Assessment is a structured, comprehensive, and internationally standardised method. It quantifies all relevant emissions and resources consumed, and the related environmental health impacts and resource depletion issues that are associated with goods or services [9]
LCA method is defined by the following standards:

- EN ISO 14040:2006+A1:2020 - Environmental management — Life Cycle Assessment — Principles and framework
- EN ISO 14044:2006+A2:2020 - Environmental management — Life Cycle Assessment — Requirements and guidelines

The steps performed are presented accordingly with the LCA Framework [10] (**Figure 3**).

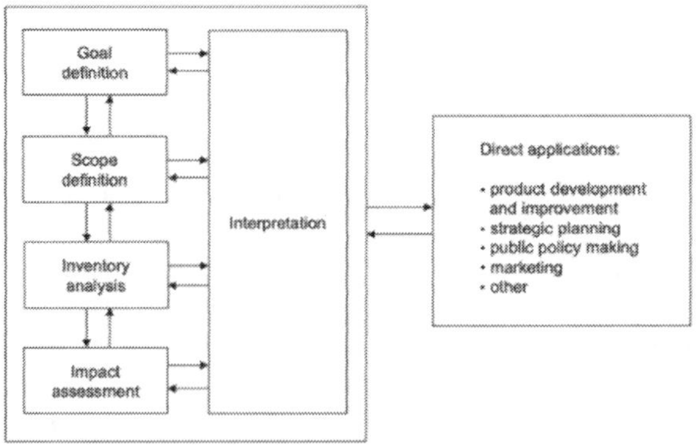

Figure 3. Life Cycle Assessment Framework [10]

4.1. *Goal and Scope Definition*

The goal of the study is the evaluation of the environmental impact from cradle till the company's gate, of the *DK*. It is not the aim of the analysis either to evaluate the product with respect to alternative items or to compare different End-of-Life scenarios. The obtained results are not intended for public declarations or audits.

4.2. *Inventory Analysis*

The data modelling phase is realized through the software *OpenLCA*. The used database is a free accessible one: *the elcd_greendelta_2122*. Through the product Bill of Materials and flow chart, all the inputs and outputs are identified, and the model is realized accordingly. LCI data of standard components and raw materials are not available in the database (because of the impossibility to access a professional one). Thus, the needed information is extracted from the software *Granta EduPack,* and each product system is realized from scratch. LCI data of transportation, heat and energy consumption are sourced from *OpenLCA* database.

4.3. *Life Cycle Impact Assessment*

Life Cycle Impact Assessment is applied at Midpoint level. Only mandatory features, according to ISO14044, are applied: Impact Category Selection, Classification, and Characterization. The adopted impact assessment method is IPCC GWP 100a. Therefore, the selected impact category is Global Warming Potential (GWP). General results obtained from the software are shown in **Figure 4**. One unit of *DK* is responsible for 0,39168 kg_{CO2eq} emissions.

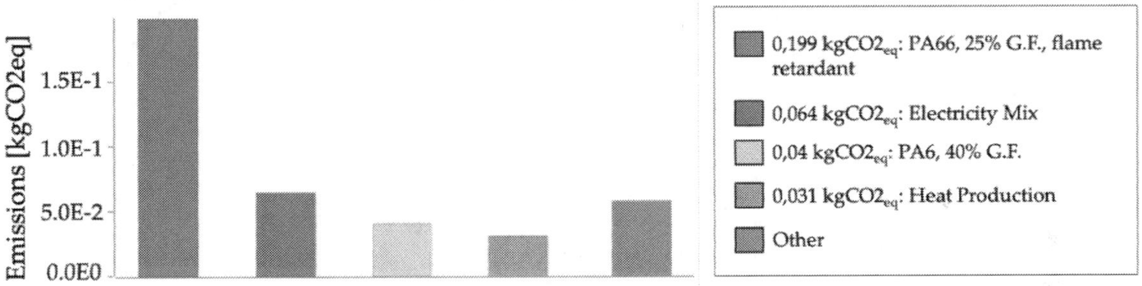

Figure 4. OpenLCA General Information Results

4.4. *Interpretation of the Results*

The results show that a relevant contribution to the product environmental impact is caused by the materials employed. Specifically, Polyamide 66 (PA66), 25% glass fibre (G.F.) with flame retardant (F.R.) accounts for more than a half of the total emissions. For the most impactful materials, the quantity needed for the product realization (expressed as a percentage of the total product weight) and the impact (in terms of percentage of emissions with respect to the total) are computed.
Results are provided in **Table 1**.

Table 1. Impact evaluation of main materials employed

	$Quantity = \frac{Q_{material}[kg]}{Q_{DK}[kg]}$	$Impact = \frac{I_{material}[kg_{CO2eq}]}{I_{DK}[kg_{CO2eq}]}$
PA66, 25%G.F., F.R.	54,35%	51%
PA6, 40%G.F.	11%	10%

IOP Conf. Series: Materials Science and Engineering 1306 (2024) 012042 doi:10.1088/1757-899X/1306/1/012042

5. Design Solutions

On the basis of the result interpretation, the work initially focuses on reducing the emissions caused by the usage of **PA66, 25% G.F. with flame retardant.**

Then the possibility to apply other modifications, to act on additional fronts, is evaluated. All the reasonings and the proposals taken into exam will be discussed in the following paragraphs. It is decided not to act on energy and heat consumption despite their relevance, because in order to measure evident improvements on such sides, acting on product design is not enough, and this would be out this research scope.

5.1. PA66 Intervention Solution

The solution to intervene on PA66, 25% G.F., F.R. negative environmental effects, consists in its replacement with a new material. At this purpose the software *Granta EduPack* is used. Through the software support, the following steps are performed:

- all the materials with a *Nearness Value* higher or equal 89% with respect to PA66, 25% G.F., flame retardant are isolated. *Granta EduPack* considers the Euclidean distance between the value of some parameters of the reference substance, and the assessed one. It is decided to set as reference parameters:
  a. *Material properties: Young 's Modulus, Yield strength, Tensile Strength, Melting Point, Thermal Conductivity*
  b. *Process and operation characteristics: Mould Temperature* and *Moulding Pressure Range, Maximum Service Level,;*
- the isolated materials are plotted on a graph having price on the X-axis and CO_2eq production emissions on the Y-axis;
- all the materials belonging to the quadrant where

$$X_{material} \leq X_{PA66,\ 25\%\ G.F.,F.R.} \text{ and } Y_{material} \leq Y_{PA66,\ 25\%\ G.F.,F.R.}$$

are considered, and the related properties are deeply compared against those of the reference one and with respect to the characteristics required by the component where it should be used **(Figure 5)**.

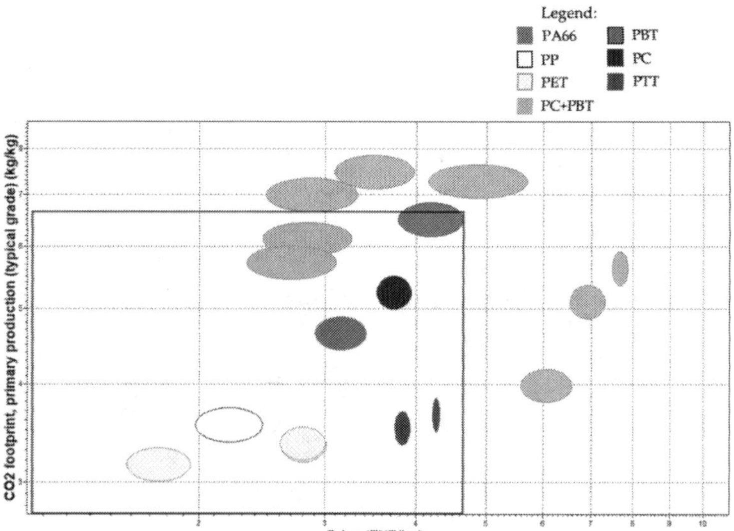

Figure 5. Granta EduPack materials comparison graph

The resulting substitutes are: *Polycarbonate (PC), 20-30% G.F., flame retardant; Polytrimethylene Terephthalate (PTT), 15% G.F.; PTT, 30% G.F.; Polyethylene Terephthalate (PET) 15% G.F., flame retardant; PET, 35% G.F., super toughened; Polypropylene (PP), 40% long G.F; Polybutylene terephthalate (PBT), 30% G.F., flame retardant; PC+PBT, 20% G.F. and PC+PBT, 30% G.F.*

After a deeper assessment (based on economic considerations and an analysis of the other physical, mechanical and electrical properties), the most suitable material is PP, 40% G.F. An advantage of choosing PP is that *Rold* employs it for other applications and has already a reference supplier. In order to make the product realized with the new material compliant with the legislation, a flame retardant addition would be necessary. This is not a criticality because available suppliers can provide the polymer with the required reinforcement. The higher environmental impact caused by F.R. is accounted in the environmental assessment.

A scenario analysis is run to assess the reduction in terms of CO_2eq emissions that material substitution would enable to reach. The project section of *OpenLCA* is used at this purpose. Firstly, the environmental footprint variation, associated to each modified component, is assessed. Then, the overall impact on the final product environmental performance is computed. The obtained results are shown quantitatively in **Table 2** and graphically in **Figure 6**. Only components that would be subject to the material substitution are reported in the analysis.

Table 2. Emissions variation achievable through material substitution

Component	Emissions caused with the usage of PA66, 25% G.F., F.R. [kg_{CO2eq}] *	Emissions caused with the usage of PP, 40% long G.F., F.R. [kg_{CO2eq}] *
DK Rast Body	0,038	0,028
E-Box Cover	0,059	0,042
Pull to Open Support Doorlock	0,105	0,072
DK	0,392	0,331

**Values rounded to the third decimal place*

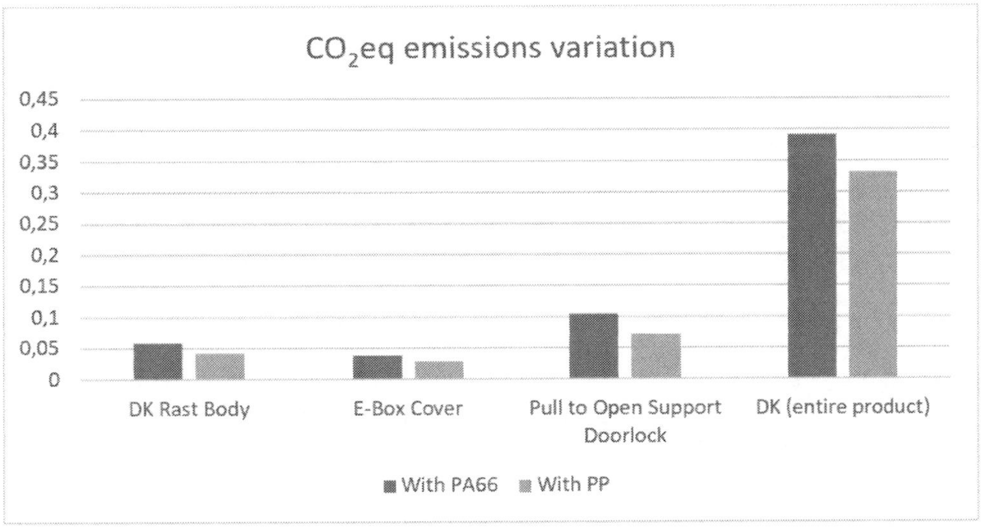

Figure 6. $CO2_{eq}$ emissions variation after material substitution

With this modification, the achievable reduction in terms of CO_2eq emissions is higher than 15% (considering the entire product). Specifically, 0,06 kg_{CO2eq}/*DK* can be saved. Taking into account that the *DK* annual production rate amounts to around 20 million pieces/year, adopting this solution could permit an emission reduction of more than 1140 $tons_{CO2eq}$/year.
The further economic assessment reveals that 950,000 €/year of savings can be generated. Indeed, the new material is less expensive and no cost of contracts with new suppliers nor training expenses would be required.

5.2. Alternative Design Solutions
The previously identified proposals are mainly based on the implementation of Design for Environment (DfE) Methodologies [13] An additional analysis is performed considering Design for Manufacturing (DfM) [13], Design for Assembly (DfA) [14] and Design for Disassembly (DfD) guidelines. Two further solutions emerge. They are deepened below.

5.2.1. Screw Removal
Both DfM and DfA pursue the reduction of the number of components (if not necessary) and the related costs. Elimination of not needed parts can also allow to decrease product's environmental impact. An analysis of the *DK* under this perspective allows to identify the presence of a screw (**Figure 7**), over the sensing layer, without a real stabilizing function.

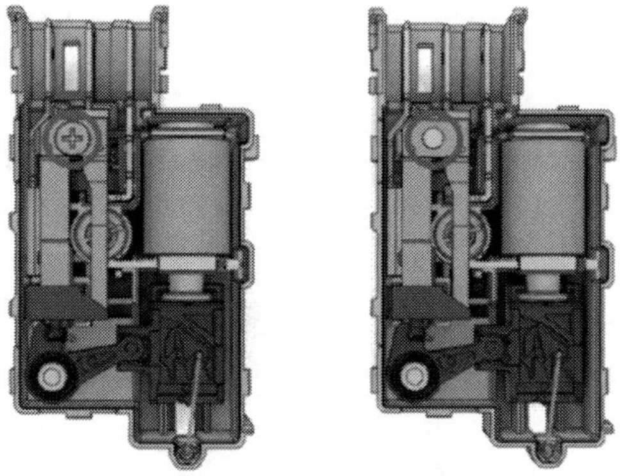

Figure 7. Internal part of DK with the screw (left image) and with the rivet (right image)

This role is largely assumed by an already present riveting system. The LCA results show that the reference screw has a not negligible impact on the entire product environmental performances (0,00042 kg_{CO2eq}/DK over a total of 0,39168 kg_{CO2eq}/DK).
The LCA results show that the reference screw has a not negligible impact on the entire product environmental performances (0,00042 kg_{CO2eq}/DK over a total of 0,39168 kg_{CO2eq}/DK).
The component presence is mainly due to the clients' past requests, who preferred to maintain it as additional warranty of the layer stability. The screw environmental footprint considers the effects of the constituent material and production process, together with the emissions caused by the transportation phase required for its supply. To reduce these contributions, the identified solution consists in its replacement with a riveting system. For the geometry of the new rivet, it is proposed to maintain the same dimensions as the hold located on the laminate supporting the screw. In this way no need to modify the quotes of the sensing layer (and consequently the related production process) could occur. **Figure 7** shows the CAD image of the product before and after the modification. The environmental impact of the new solution is computed and compared with the base case. The decrease in terms of CO_2eq emissions per *DK* is $3{,}968\ x\ 10^{-4} kg_{CO2eq}$ .The solution can reduce the product emissions by 0,1%. It is apparently a negligible improvement, but it corresponds to 7,5 $tons_{CO2eq}$ saved per year on the global production. To evaluate its relevance, it can be considered that it may allow to compensate the emissions caused by more than 2 roundtrip flights from Milan to Tokyo for a person. From the economic assessment evaluation, it results that no additional expenses would be required. Savings can be generated

by the missed need of purchasing the screws. Only efforts in terms of operational steps are necessary. Firstly, the substitution of the screw insertion phase in the *Assembly Line*, with the riveting. Secondly, the adaptation of the design of the body upon which inserting the laminate, to locate the rivet there. From the company feedback, both the changes result easily feasible.

5.2.2. Disassembly Improvement

The contribution tree results of the LCA show that the highest contributors to the overall environmental impact are represented by the E-BOX Cover and DK Body of the product's *CORE* (**Figure 8**).

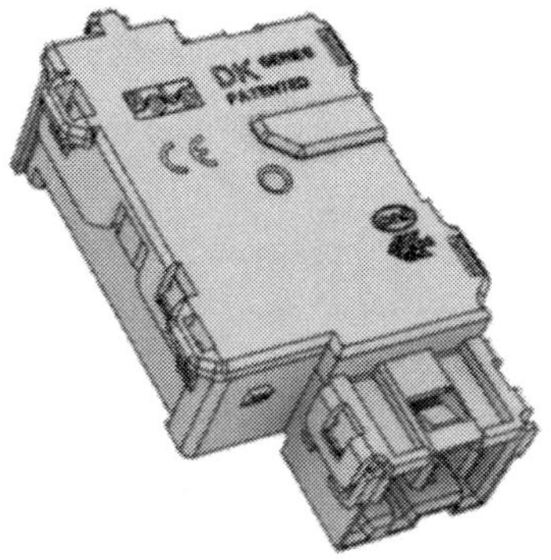

Figure 8. CAD image of the DK's CORE showing the E-BOX Cover and the DK Body

An analysis under the DfD point of view is therefore. It reveals that the *DK* is not designed for disassembly, as demonstrated by the presence of permanent snap-fits. Improving the disassembly performance of the components could enable their recovery at the product End-of-Life. The technical assessment of the solution, realized through recent literature review articles and company interviews, proves the impossibility of its realization. A proper reverse logistic network, able to collect home appliance components and supplying them back to the producing companies, has not found implementation yet, especially in Italy. When a development in such field will take place, the *DK* disassembly improvement will pass the technical assessment and it will be possible to realize the other needed considerations.

6. Critical discussion

The case study realized allows to answer the first and the second research questions.
R.Q.1: *Is it possible, starting from the LCA results of a consolidated product, re-engineering it (in terms of design, material, production process) to make it more environmentally sustainable?*

Three of the identified solutions permit to suggest new product and process design procedures, able to improve the product environmental impact. LCA proved essential in the analysis: firstly, to identify where acting on; secondly to quantitatively assess the performance enhancement. Thanks to the results provided, the company can increase its knowledge regarding the emissions released by its product and the main responsible factors, thus implementing focused countermeasures. The applied methodology can assist the organization into acquiring awareness of how introducing in a more systematic way the environmental aspect in the re-engineering phase of all its produced devices.
This additionally confirms the potential of LCA as a central tool to assess sustainability performances under the environmental aspect, and consequently drive product development and improvement. It revealed to be fundamental to evaluate quantitatively whether the proposals could be beneficial or not, without the risk of adopting burden shifting.

R.Q.2: Can the re-engineered product be sustainable also from an economic perspective?

The economic assessment proved essential to perform this step. what literature claims as one of the limitations of LCA is its pure environmental perspective, that disregards the economic and social analysis. The research method adopted permits to evaluate also the *"Profit"* side of the TBL, as a second step after the confirmation of the benefits provided by the proposals on the environmental aspect. Considering a project *"sustainable"* from the economic perspective does not always mean that no investment is required or that it allows to reduce costs. It depends on various factors, that are not univocal but vary from one company to another. Every organization may have a different perception of an *"economically sustainable "*change and it is up to them making such evaluation. For this reason, investment and efforts required from each suitable proposal have been computed and presented. Feedbacks have then been collected to establish the related feasibility. The result is that the *Material Substitution* and the *Screw Removal* can be positively evaluated under this perspective: their implementation on the *DK* will make the re-engineered product sustainable in both environmental and economic terms.

A critical analysis of the adopted methodology provides the answer to the third research question.

R.Q.3: On the basis of the present case study, is it possible to draw some general guidelines that could be used also for other products and/or in other industrial sectors?

The generality of the process developed make it suitable for all the products that, like *DK*, have been conceptualized without accounting systematically for the environmental aspect. Therefore, it can be used as a guideline in all the industrial sectors where the design approach has never included the sustainability performance as a main driver. The actual problems and developing legislations are pushing industries to revise the conceptualization approach of product development, to integrate environmental concerns since the design phase. Following a similar procedure, as the one adopted for the case study, will permit to identify viable re-engineering solutions. In performing the process, the LCA will play a fundamental role, firstly for the identification of the primary aspects on which acting on, then to verify if the proposals could improve the eco-friendliness of the output. Afterwards, the investment analysis will be useful to evaluate the sustainability of the solution in economic terms.

6.1. Study Limitations

The realized study has some limitations. The first one is related to how the LCA has been realized. As mentioned, it has not been possible to realize a complete assessment, because of the inaccessibility to professional libraries and updated data. For this reason, the model was built by using the available information, even obtained from different sources, and only Climate Change impact category could be measured. It is decided to tolerate these limitations for two reasons: the reached results can be anyway considered reliable, and the objective of the work is not limited to the LCA realization, but to the re-engineered solution proposals. Moreover, these limits are acceptable as it is not the scope of the environmental analysis the pursuit of public declarations, but it has been used only for internal research. If the company in the future will decide to perform a detailed LCA by asking a specialized organization for its realization, this limit will be overcome. In this way it will be possible to compare the results with those obtained in the study and implement further actions, if new hotspots would be revealed. The other limitation of the study consists into having considered only the environmental and economic dimension of sustainability, disregarding the social one. This decision has been adopted because of the difficulty in introducing social considerations at product development level, as demonstrated by the lack of standard indicators aimed at assessing SLCA. Anyway, the social layer of the TLBMC realized could represent a fundamental for the company to pursue improvements in such directions. Moreover, when more structured SLCA methodologies will be developed, it will be possible to integrate this assessment in the proposed procedure.

7. Conclusions

The actual environmental issues and the new governmental legislations are requiring the industrial sector to incorporate sustainability aspects in all the product lifecycle phases, especially during early stages.

IOP Conf. Series: Materials Science and Engineering 1306 (2024) 012042 doi:10.1088/1757-899X/1306/1/012042

Products designed when the environmental aspect was not considered so relevant, need a re-engineering approach, aimed at incorporating this dimension, without disregarding the others.
This paper represents a contribution in providing such a methodology, using a case study analysis to illustrate the steps required. Thanks to the empirical study, the company increases its knowledge regarding new design and process technologies. These have not been evaluated in the past because of the unawareness of the product environmental impact, and the attitude of maintaining a consolidated design. The work deepens traditional LCA studies based on providing environmentally improved solutions, without accounting systematically for their operational and economic feasibility. Its novelty is based on the introduction of two additional level of analysis: the technical and the economic assessment. A strength of the case study realized lies in the possibility to extend the performed steps to all the other industries facing the same problem as Rold and wishing to re-think their product conceptualization, with the objective of innovating sustainably.
At the end of the study, the company will result more aware about *DK* environmental impact and will be able to empirically act to improve the related sustainability. At the same time, industries facing the same issues, will have a reference regarding how approaching the re-engineering phase of their products. Possible future developments of the work have been distinguished between those suggested to the reference company and those regarding future integration studies.
Regarding suggested progresses proposed to Rold, in the future the company will be able to apply the same used methodology to other products belonging to its portfolio, in order to expand the sustainability improvement actions to different items. Moreover, concerning the proposed solutions, the suggestion is monitoring whether production and technology advancements will allow to modify the realized evaluations. This regards specifically the DfD product performance improvement: if in the future a supporting reverse-logistic network will be available, it will be possible to consider the option viable and testing the ecological and economic worthiness of its implementation.

8. References

[1] Lubin D A and Esty D C 2010 The sustainability imperative *Harvard Business Review Magazine*
[2] Deloitte & Touche (M.E.) 2010 Sustainability, a growing business imperative
[3] Eckert E and Kovalevska O 2021 Sustainability in the European Union: Analyzing the Discourse of the European Green Deal *Journal of Risk and Financial Management* **14** 80
[4] European Commission 2019 *Communication from the Commission: The European Green Deal* (Brussels)
[5] European Commission 2020 *Communication from the commission to the European Parliament, the European Council, the Council, the European Economic and Social Committee and the Committee of the Regions: A new Industrial Strategy for Europe* (Brussels)
[6] Bogue R 2014 Sustainable manufacturing: A critical discipline for the twenty-first century *Assembly Automation* **34** 117–22
[7] Elkington J 1994 Towards the Sustainable Corporation: Win-Win-Win Business Strategies for Sustainable Development *California Management Review 36,no. 2* 90–100
[8] Joyce A and Paquin R L 2016 The triple layered business model canvas: A tool to design more sustainable business models *J Clean Prod* **135** 1474–86
[9] European Commission. Joint Research Centre. Institute for Environment and Sustainability. 2010 *International Reference Life Cycle Data System (ILCD) Handbook general guide for life cycle assessment: detailed guidance.* (Publications Office of the European Union)
[10] Hauschild M Z, Rosenbaum R K and Olsen S I 2018 *Life Cycle Assessment*
[11] Aliyeva N, Sas H S and Saner Okan B 2021 Recent developments on the overmolding process for the fabrication of thermoset and thermoplastic composites by the integration of nano/micron-scale reinforcements *Compos Part A Appl Sci Manuf* **149**
[12] DuPont *Applying Teflon Coatings*
[13] Ulrich K T and Eppinger S D 2011 *Product design and development-Fifth Edition* (McGraw-Hill/Irwin)
[14] Boothroyd G (Geoffrey), Dewhurst Peter and Knight W A (Winston A 2002 *Product design for manufacture and assembly* (M. Dekker).

IOP Conf. Series: Materials Science and Engineering 1306 (2024) 012042 doi:10.1088/1757-899X/1306/1/012042

Appendix A - Rold Triple Layered Rold's Business Model Canvas

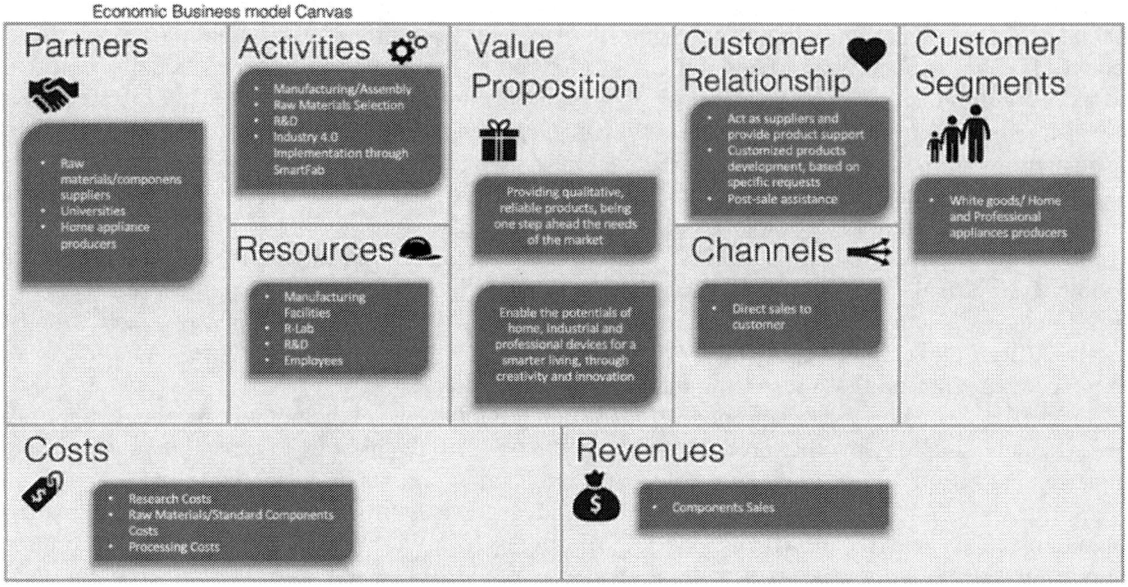

Figure A. 1: Rold Economic Business Model Canvas

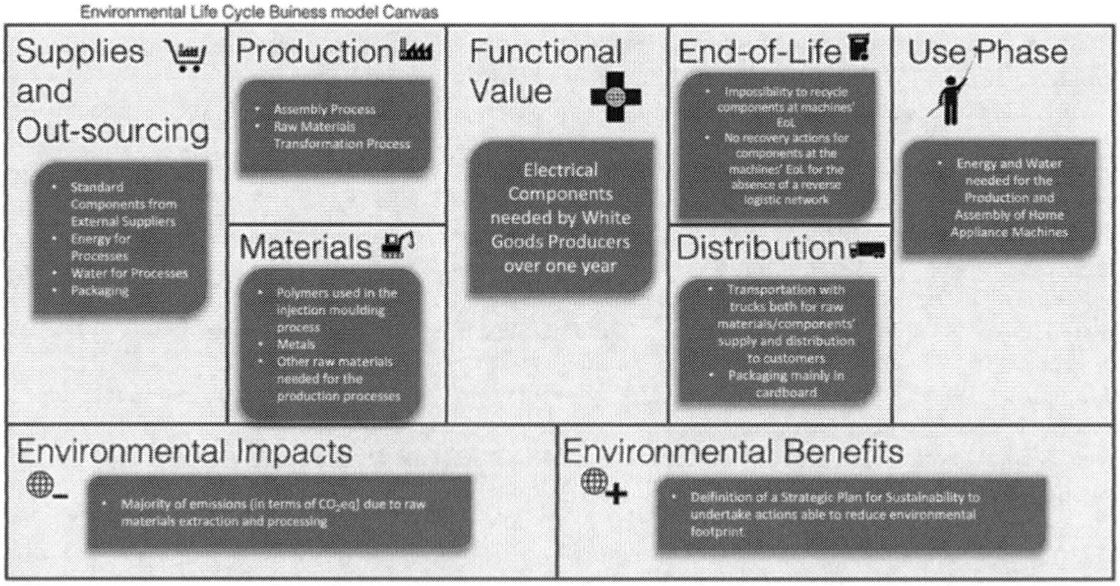

Figure A. 2: Rold's Environmental Business Model Canvas

IOP Conf. Series: Materials Science and Engineering 1306 (2024) 012042 doi:10.1088/1757-899X/1306/1/012042

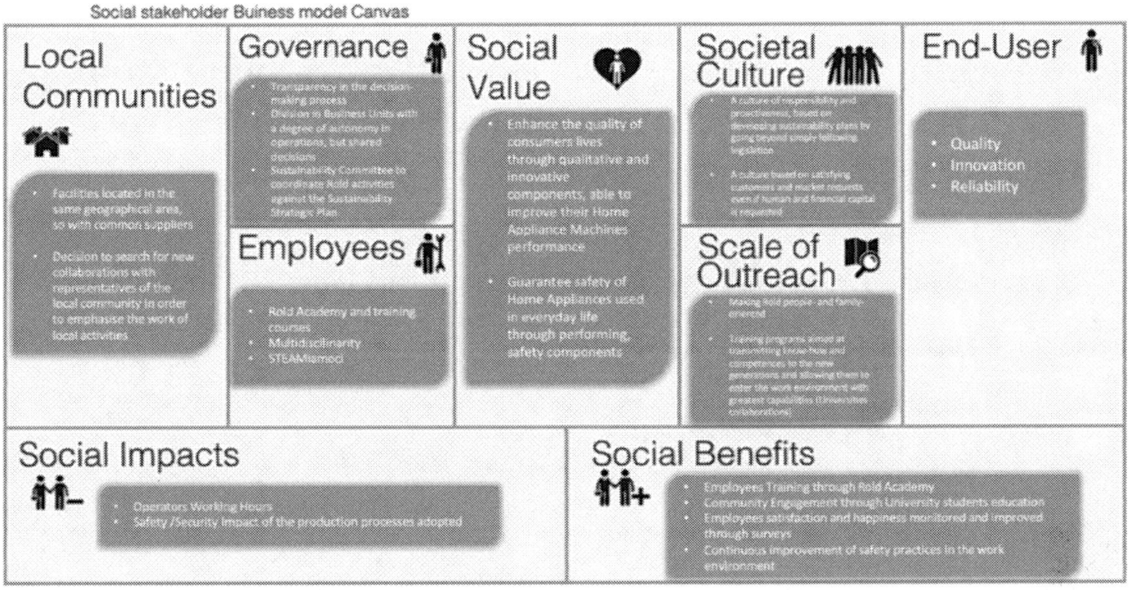

Figure A. 3: Rold's Social Business Model Canvas

IOP Conf. Series: Materials Science and Engineering 1306 (2024) 012043 doi:10.1088/1757-899X/1306/1/012043

Approaching the re-design of reusable packaging from an environmental perspective: a case study in the railway sector

Berzi L.*, Cocci M, Barbieri, R., Pierini, M., Delogu M.
Università Degli Studi di Firenze, Department of Industrial Engineering DIEF, 50139 Florence (FI), Italy

*email: lorenzo.berzi@unifi.it

Abstract.
The production of railway vehicles determines the handling of a large amount of materials which are transferred from suppliers to production site using industrial packaging. The activity here described is part of the MORIMB project, which dealt with the reduction of waste related to railway sector; wood, plastic, cardboard and metal fasteners are the typical materials which are adopted for such packaging.
The study is motivated from the evidence that, due to the peculiarities of the transported components, most of the packaging units were different from standardized formats therefore they were frequently used for a single transport and subsequently scrapped. An applicative methodology for the analysis, redesign, and evaluation of environmental and economic impacts was defined, with particular attention to packaging reusability. An adaptation of the DMAIC approach was used to analyze the various components and find solutions reusability and performance improvement, especially in terms of volume efficiency and component protection. Depending on the item to be transported, specific crates have been designed and structurally verified according to existing technical regulation, and their sustainability has been analyzed considering environmental and economic impact during production, transport and storage phase; appropriate scenarios regarding distance travelled and volume occupancy were defined. The case study presented in detail in this work concerns the concept of special crates for the transport of passenger seats – a key component for train production - for which an increase in packaging density and a reduction of transportation costs was achieved. Other case studies were also considered.

1. Introduction

The impact related to packaging is well known as a cause of relevant impact for consumer and industrial products. Due to the increasing sensitiveness to the environmental impact, since the last decade of the XXth century a tendency to the reduction and optimization of packaging waste is started. Various studies [1] highlight that the substitution of single-use packaging in favor of reusable packaging can support waste reduction, environmental indicators improvement and, under certain conditions, economic advantages; in other words, reusable packages are potentially more sustainable. In this context, our analysis is focused on the production of complex machines such as railway vehicles, which determines relevant material handling needs due to the transport of delicate components and materials from suppliers to the main manufacturing and assembly site. Such activities can be defined as *transit transports* according to [2], and include the use many different solutions such as boxes, containers, crates, pallets and wrappers. The MORIMB project deals with the reduction of waste materials from the packaging used to handle the materials (namely, finished components and parts ready to be applied on the main vehicle assembly) supplied to Hitachi Rail STS site (HRS), located in Pistoia (PT, Italy). The motivation is based on the evidence that a relevant amount of packaging used by the Company is tailored on specific applications, since standardized formats cannot satisfy the requirements in terms of protection during transport and performances during warehouse storage. The consequence of such approach is that relevant amount of packaging devices are scrapped after a single use; therefore main target of the activity is the reduction of such events and the definition of devices enabling transport to the Company and take-back logistics to the suppliers. From a preliminary study we are noted that the waste produced in recent years is very high, and comprehends into wood, plastic, cardboard and metals, these latter especially for fasteners and similar parts. The present text illustrates the definition and the application of a methodology for the improvement of the sustainability profile of packaging applied on this sector and is organized as follows: section 2 describes the approach and main indicators and boundary conditions adopted for the study; section 3 is focused on the development of a relevant case study; section 4 provides final observation and remarks.

2. Methodology

An adapted DMAIC - Define, Measure, Analyze, Improve and Control - algorithm was used to find improvement solutions compared to the existing situation. The approach is aimed at enhancing the environmental profile of the HRS company according to the target of continuously improvement set by internal policy. According to the indications provided by the industrial partner, the project focused on a group of representative packaging units to be used as case study on which, through the application of the adapted DMAIC methodology, we defined a packaging waste reduction model that can be adopted and transferred to similar products. Main phases are:

- DEFINE: Two products were selected as reference case studies, namely the "Caravaggio" and "Masaccio". These railway vehicles are planned as long-term orders and they will be under production for a time frame compatible with the implementation of new practices. The possible case studies of components were selected due to the large quantity needed to set up a train and the volume occupied in the warehouse, and also on the basis internal know-how and operation needs.
- MEASURE AND ANALYZE: The materials involved for the various packaging and the quantities involved were analyzed, evaluating the costs of preparing the packaging unit, packing, transporting, unpacking, waste disposal, and warehouse storage; all these are part of the typical life. In addition, loss related to the damage of components due to non-compliances were considered, assuming that due to packaging re-design such events will be further reduced. The LCA (Life Cycle Assessment) approach was used to evaluate the current impact of packaging, as described below.
- IMPROVE: In this phase, the packaging was redesigned according to new requirements and specification; the output of such phase is the creation of a CAD model and a subsequent Finite Element Model (FEM) analysis to verify the coherence with design requirements; relevant solicitation can arise during transportation in road vehicles and warehouse stacking. In this phase, the costs for the creation of the new packaging were also evaluated, creating economic accounts and differential statements to evaluate the economic sustainability of these investments.
- CONTROL: After the selection of priority components which should adopt the newly designed components, implementation and control will follow. Such phase includes the inclusion of a number of prototypes to verify the effectiveness of the new packaging, the real-world cost and their functionality. Particular attention should be paid on the acceptance of the new approach by suppliers. Control phase therefore include Mid and Long term objectives aimed at continuously verify the effectiveness of the initiative and the correspondence to the target, eventually providing warnings in case of non-conformities and stimulation a reaction, which potentially starts back from Measure, Analyze and Improve phases. Particular attention is expected to be paid on the inverse logistics, which is not part of current practices and could imply iterative steps to be implemented and optimized over the time.

A few notes are necessary regarding the indicators adopted. For the assessment of the lifetime impact of the application, a simplified framework for the Environmental and Economic impact has been considered:

- Economic impact has been based on the measured costs (known by the Company) and other hypothesized costs derived for the design phase. Due to the relevant investment foreseen for the introduction of reusable packaging, the Internal Rate of Return IRR has been estimated on the basis of the expected product life.
- Environmental impact has been based on the assessment of GWP (Global Warming Potential) indicator of reference and new packaging unit, expressed as kg_{CO2_eq}.

The environmental impact has been estimated on the basis of two main sources. For existing and newly designed packages the use of CAD-integrated sustainability software such as the Solidworks Sustainability tool [3] has been adopted; due to the adoption of known materials (wood, metal, plastics etc.) and of conventional manufacturing technologies all main information were available. In order to provide information related to the direct and take-back transportation activities, data have also been extracted by the Sphera-Gabi software, considering proper transportation routes (e.g. by maritime or road transport) depending on the supplier, on its distance to the Company and on the expected occupancy of typical containers.

For both economic and environmental indicators, the identification of correct and up-to-date information required a strong interaction with the whole supply chain, as typical for improvement project.

During the design phase a relevant performance improvement target has been set: the reduction of total volume occupied in the warehouse for the supply of typical vehicle set (e.g. the stock of seats necessary for one carriage). The target has been achieved both providing resistance to multiple stacking and increasing the unused volume in a single package, designing tailored part housings and fastenings. Due to increase of the mass contained, the new packaging will have to comply to more restrictive requirements compared to the current situation, dealing with stability and resistance requirements, for example during road vehicle transportation [4]. Main design targets for the reusable packaging include:

- Life from 5 to 10 years.
- Significant volume reduction for the take-back phase, to be achieved using foldable parts but excluding part dismantling/separation.
- Stacking capability to improve warehouse utilization coefficient.
- Safety according to known best practices.
- Compatibility with conventional forklift handling.

3. Analysis of the proposed case studies.

3.1. The seats assembly case study

Considering the characteristics of the railway vehicles, various components were analyzed: the seats, interior trim, toilets, windows, panels and so on, focusing on body parts rather than on mechanical parts. See Figure 1 for a typical vehicle layout.

The priority for the adoption of the reusability criteria on these parts was unknown but, as a first case study, a component widely used was examined, namely the seats. Each carriage has on average around 60 seats, and each railway train of the Masaccio and Caravaggio, regional orders, is made up of 3 to 6 carriages. The quantity of seats fitted and the types of seats change based on the type of setup. Double seats, triple seats and benches are mounted and are divided into various types based on the position in which they must be installed.

Figure 2 shows two types of seats, which can be organized in double and triple assemblies; these two types are divided into right and left, with guides underneath and without guides. As visible in figure, such component is provided in mixed wood-cardboard boxes, which can be stacked on a maximum of three units. Such AS-IS condition has been analyzed considering the materials necessary to create each package, the quantities necessary to create a carriage; the costs and environmental impact deriving from them.

IOP Conf. Series: Materials Science and Engineering 1306 (2024) 012043 doi:10.1088/1757-899X/1306/1/012043

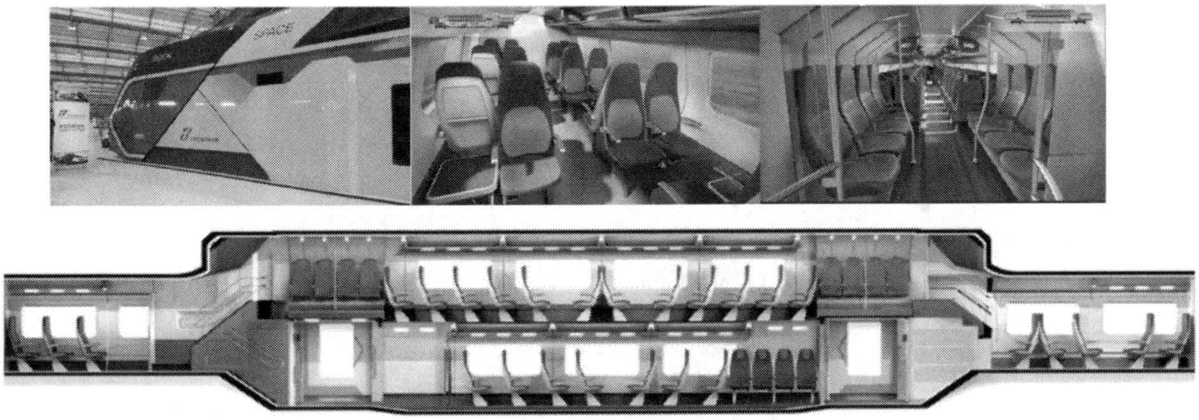

Figure 1. Seating layout of typical reference vehicle.

Figure 2. Seat packaging with waste materials

A pallet is made up of three boxes which contain a total of three seats, the dimensions of the boxes are 1120 mm wide, 1120 mm long and 530 mm high. In each pallet the packaging consists of 27 kg of cardboard, 16 kg of wood and 0.95 kg of plastic. This packaging is usually scrapped after one service. As said, the Global Warming Potential is adopted as an index to evaluate the environmental impact and the adopted data are reported in Table 1.

Table 1. Environmental impact data

Base material (1 kg)	Material and production	Transport $kg_{CO2}/km*kg$	End of life
GWP Wood	0,793	5,43E-05	0,506
GWP Card board	1,172	5,43E-05	0,035
GWP Plastic	2,859	5,43E-05	0,676
GWP Steel	0,756	5,43E-05	0,023

Figure 3 shows the impact in terms of CO2 equivalent for each material referring to a single seat. The analysis was conducted considering the various production phases from materials, to manufacturing, to transport and to end of life. Packaging for a seat has an impact of approximately 19.74 kg of CO_2

equivalent, the calculation also considering the transport from the manufacturer to HRS, in the city of Pistoia.

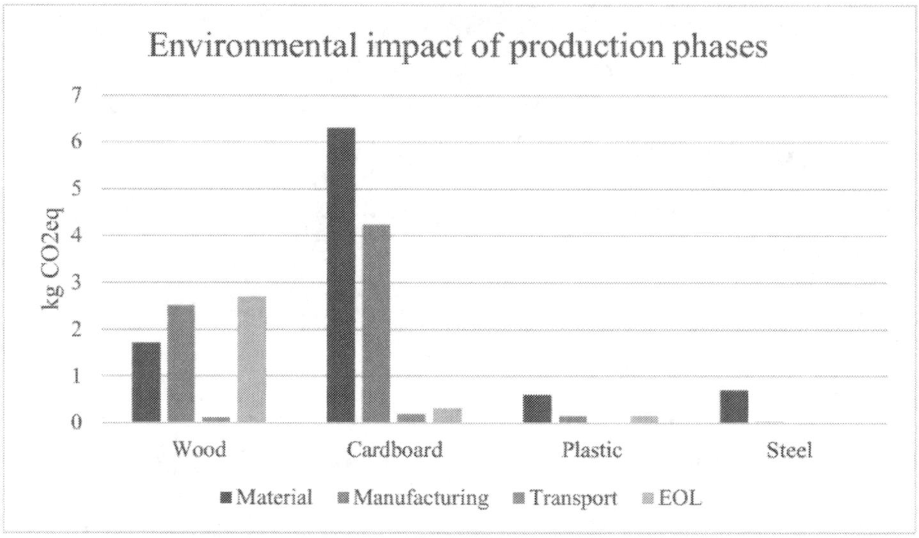

Figure 3. Environmental impact related to a seat

The costs deriving from the manufacturing of the packaging have been calculated, using data coming from market analysis and from interaction with supply chain. The cost of the wooden platform, the cardboard and the plastic present in each package were considered and subsequently normalized by referring them to each seat. Furthermore, the labor to assemble the packaging was also considered. The costs from the supplier to HRS were considered in the transport costs and the costs for internal handling, storage and unpacking were also considered. Instead, the costs of disposing of the various materials including internal movements were considered in the disposal costs.

The costs for the manufacturing of each package represent approximately 50% of the costs and another 40% is derived from the transport and storage phase, as shown in Figure 4.

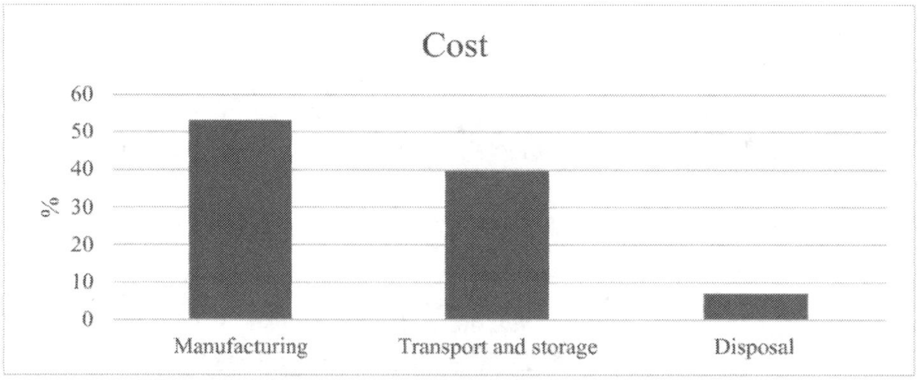

Figure 4. Normalized cost breakdown for seat assembly.

This type of packaging does not allow stacking on multiple levels, as the cardboard boxes deform and lead to non-conformity of the internal component, as parts of the internal seat coverings can be damaged. In the IMPROVE phase, where the packaging is redesigned, it was considered that the new packaging had to respond to various requirements: better use of the warehouse, better effectiveness of the new packaging with reduction of non-conformities, better compaction of the packaging in transport, reduction of waste materials by creating reusable and compactible packaging, better compaction factor of the material transported inside the packaging, increase in packaging safety. A packaging was created that could satisfy all the previously defined requirements, as shown in Figure 5.

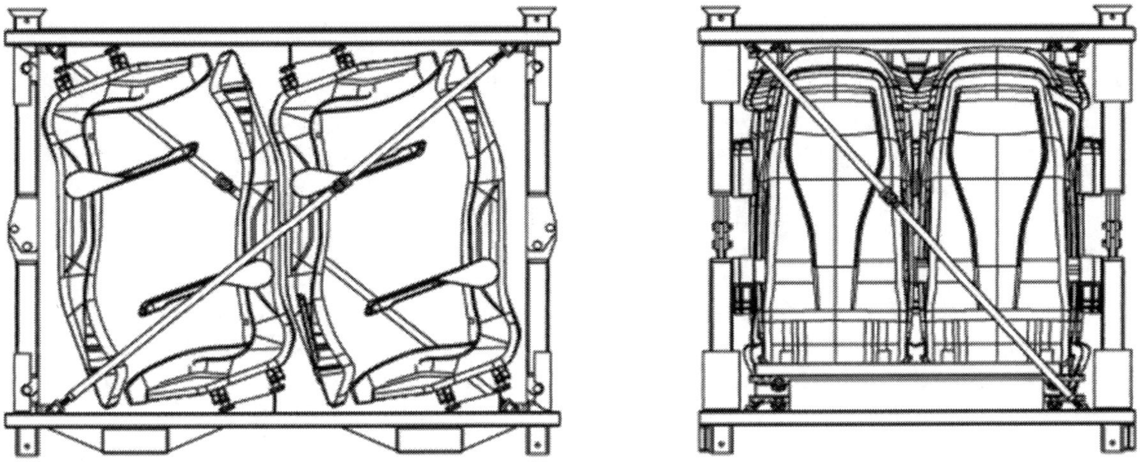

Figure 5. New reusable and foldable housing for seat assembly; 4 units of 2-seats version are contained.

A package was created that could contain four seats assembly (8 seats in total) to improve the packing factor. The packaging is foldable in order to improve the packaging factor when returning the packaging to the manufacturer. It was possible to achieve this by creating special hinges secured with pins that can be easily removed by any operator. The seats are fixed on slides, stopped with manual tightening systems, easily extractable by operators, as shown in Figure 6.

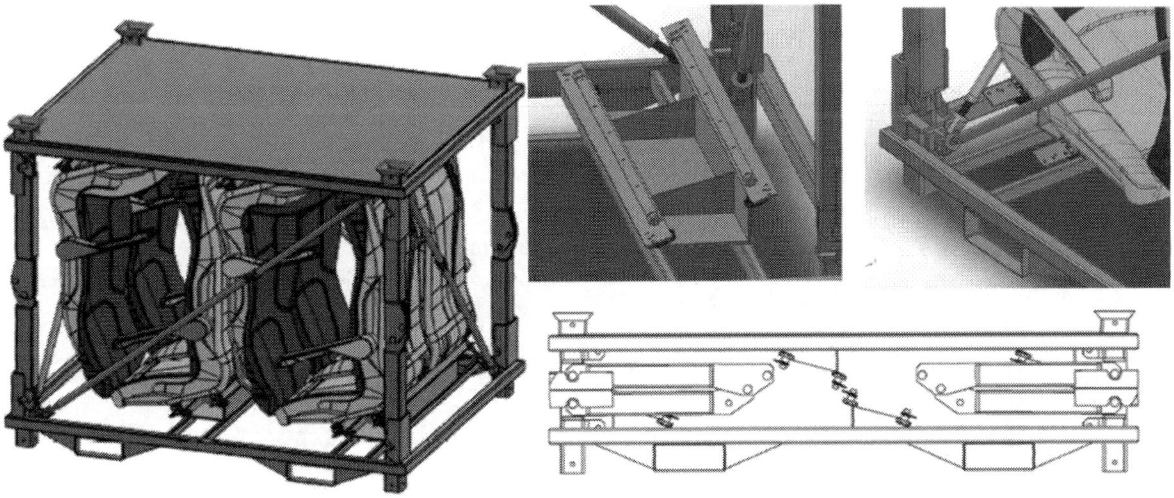

Figure 6. Details of the new assembly in transport and take-back status.

Furthermore, to avoid any lateral movement of the seats, locking pins have been provided. The EN 12195 standard [5] provides that during transport a package is subject to lateral accelerations equal to 0.8 g in the event of emergency braking, while accelerations equal to 0.5 g in the case of a curve, in which case the acceleration is directed at 90 degrees respect to the previous case; such boundary conditions were used to estimate the stresses and deformations acting on the packaging. The represented packaging unit is the solution obtained at the end of verification and correction phase, and is compliant to such solicitation.

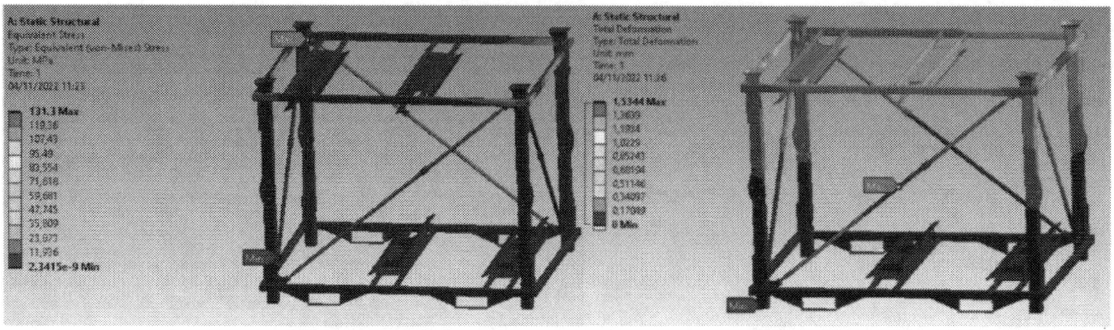

Figure 7. Finite element analysis, stress and strain.

Figure 7 shows that the maximum stresses never exceed 132 MPa, with maximum deformations. equal to 1.6 mm. It should be noted that each complete package can achieve a mass of approximately 250 kg, the metal foldable container accounting for approximately 120 kg. Such relevant mass, as demonstrated below, is not hindering the achievement of satisfactory environmental performances since for the reference case the saturation of the volume of container or of road vehicles loading platform typically occurs in terms of volume rather than in term of mass.

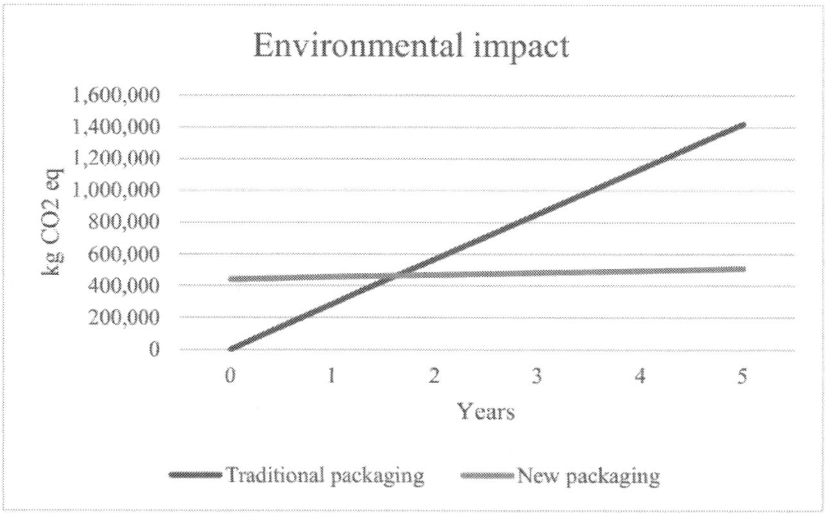

Figure 8. Environmental impact comparison.

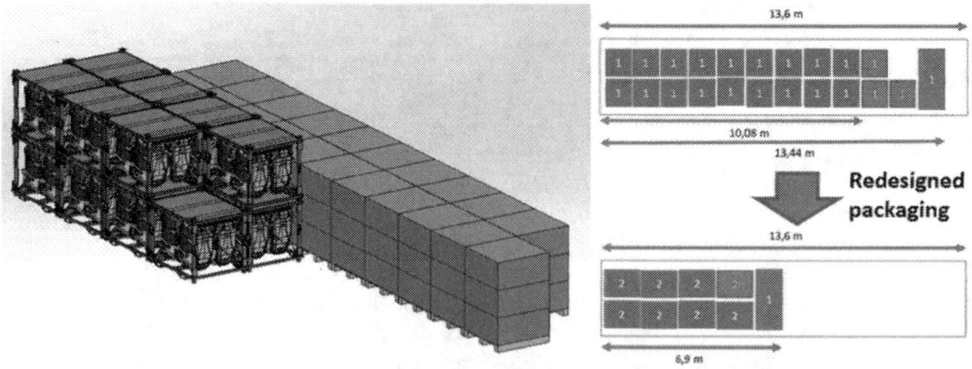

Figure 9. Comparison of the as-is and to-be packaging during transport. Right: Example of volume to be transported reduction for a same supply. Left: Occupancy of a transport container; blue, red and brown blocks correspond to different seat models, all being part of the same vehicle.

A comparative LCA analysis was performed between the current situation and with the redesigned packaging. The analysis is expressed as a parametric calculation over the years of adoption considering the environmental impact of the material, the manufacturing of the new packaging, the transport and its end of life both for reference and to-be case study.
Figure 8 shows that for the seat assembly case study the new packaging achieves environmental impact break-even-point in less than two years. Figure 9 shows that with the redesigned packaging it is possible to reduce the volume occupied during transport, approximately doubling the number of seats transported in comparison with the as-is situation. With the redesigned packaging it was also possible to optimize the space occupied in the warehouse, making better use of the spaces available and taking advantage of the entire height of the magazine. Figure 10 shows that a kit with the current packaging occupies approximately 18 m^2 while the new packaging will occupy approximately 7 m^2 for the same kit. For the comparison, as a conservative hypothesis it was assumed that conventional packaging would be stacked up to a 6 levels, which is mostly unfeasible due to resistance issues.

Figure 10. Comparison of the warehouse occupancy with baselined (brown) and redesigned packaging.

To evaluate whether this investment is also sustainable from an economic point of view, an economic analysis and a differential report were carried out between the current situation and the one with the redesigned packaging. Such result is shown in term of value over the initial investement.

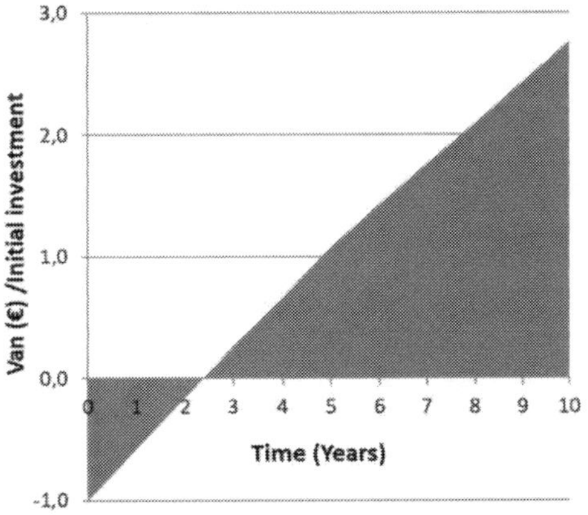

Figure 11. IRR of the proposed solution.

Figure 11 shows that the return on investment will take around 2 and a half years. The redesign of the packaging allows for a reduction of 2.8% in the annual waste produced by the company deriving from the wood used to make the packaging and by 35.5% from cardboard.

3.2. Railway vehicle components case studies.

The defined approach has been applied on other components such as interiors, windows, toilets, and other relevant vehicle body parts, in order to find out the priorities for the implementation of the reusable packaging, considering the need for an initial investment as demonstrated by the seats assembly case studies. Similarly, customized packaging were created, maintaining the requirements of foldability and introducing the possibility to provide reconfiguration for different components after the end of the order. In addition, a number of components to be provided as initial investment was suggested on the basis of the expected supply needs. Table 2 shows the number of packages to be created for each component analyzed, the breakeven point, the internal rate of return of each investment and the space saved in the warehouse. Considering the implementation of all solutions, the breakeven point of the analyzed components will be achieved approximately in two years and three months. The investments allow for a profitability of the investment equal to 37% and can reduce the space occupied in the warehouse by approximately 517 m^2.

Table 2. Economic summary for the implementation of the reusable packaging on other vehicle parts.

Description	Necessary reusable package units	BEP (Break Even Point)	IRR (Internal Rate of Return)	Volume saving
Seats	390	2 years, 8 months	34%	311 m^2
Interior 366	40	1 year, 2 months	81%	34 m^2
Interior 367	40	3 years	21%	30 m^2
Interior 500	40	1 year	96%	35 m^2
Interior 375	40	1 year	102%	14 m^2
Toilette HK	8	1 year, 4 months	74%	21 m^2
Toilette St	8	1 year, 7 months	69%	9 m^2
Windows	60	3 years, 1 month	21%	30 m^2
Veils	40	1 year, 11 months	49%	25 m^2
Vestibule	40	2 years, 9 months	27%	8 m^2
Total		2 years and 3 months	37%	517 m^2

Table 3 shows the reduction in waste that was possible to achieve for each component analyzed in relation to the total produced by the company and the reduction in terms of CO_2 equivalent. Furthermore, it can be observed in Figure 12 that the breakeven point from an environmental point of view is less than two years.

Table 3. Environmental summary. The baseline amount for the total percentage calculation is the total amount of that material inside the Company.

Description	Waste change in 5 years			GWP variation in 5 years ($t_{CO2\ eq}$)
	Wood (kg)	Cardboard (kg)	Plastic (kg)	
Seats	2.8%	35.5%		911
Interior 366	0.8%		5.8%	134
Interior 367	0.5%			14
Interior 500	1.1%			147
Interior 375	0.3%			44
Toilette HK	0.3%			16
Toilette St	0.1%	0.1%		0.8
Windows	0.6%		1.0%	37

Veils	0.5%		0.7%	36
Vestibule	0.5%		1.1%	64
Total	7.6%	35.6%	8.6%	1408

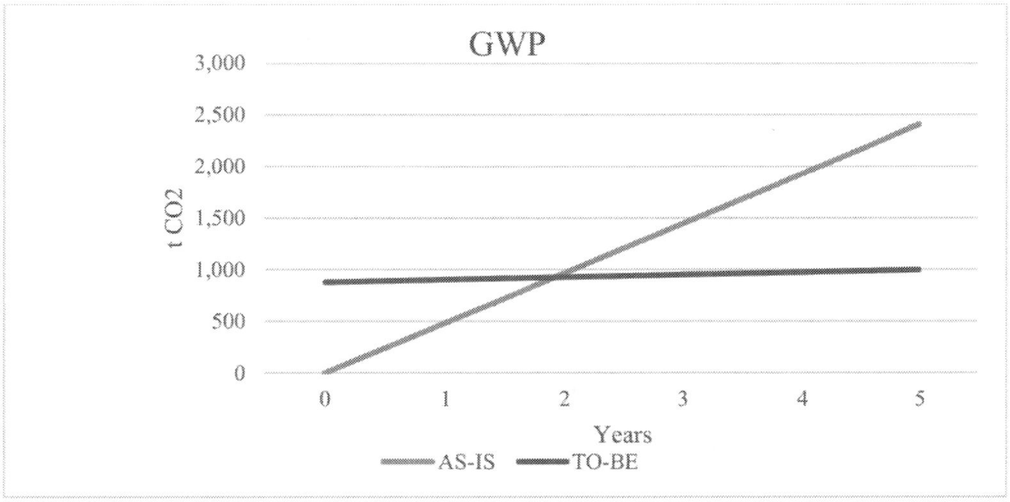

Figure 12. Environmental summary. Comparison of GWP impact for the "as is" and the "to be" solution considering the adoption of reusable packaging units for all the elements listed in Table 2 and Table 3.

4. Conclusions

The transit and storage of the components needed to produce a complex product such as a railway vehicles imply the manufacturing, the transport, the use of a remarkable amount of materials which, in the current practice, are frequently scrapped at the end of their service, best case being their recycling; reusability is hindered by the special format adopted for large and specific railway components. Due to the presence of repeatable supplies for the body of typical carriages, a re-design action has been described using the data for two different vehicles produced at HRS Company. Such activity implied the application of good design practices monitored by the adoption of environmental and economic indicators, used to assess the effective advantages of reusable components. Despite of the take-back costs, the results clearly indicates that for a large number of component categories relevant advantages are possible, main one being the volume reduction both in transportation and storage phase.

However, the need for a proper design of the packages highlights the need for accurate engineering design, proper management, and strong collaboration with supplier. It is therefore possible the adoption of such reusable packaging units has been hindered up to know by the initial barrier represented by the investment costs.

During MORIMB project various components were considered using primary and secondary data to demonstrate the potential, while the implementation is running in terms of detailed packaging design and manufacturing, which will be followed by implementation and supply chain modification. The approach defined within the activity is potentially able for implementation to other industrial case studies both in HRS and in other contexts.

Acknowledgements

The activity here presented is part of the MORIMB project (*Modello di riduzione degli scarti da imballaggi per il settore ferroviario*), funded by the Fondazione CARIPT (Cassa di Risparmio di Pistoia e Pescia) under the "Applied Research and industry innovation" call 14/2021. Authors gratefully acknowledge the Fondazione in the person of Beatrice Lombardi and the Hitachi Rail STS Company in the persons of Giacomo Marchini, Fabio Severi and Marco Rebbuzzi.

References

[1] Mahmoudi M and Parviziomran I 2020 Reusable packaging in supply chains: A review of environmental and economic impacts, logistics system designs, and operations management *International Journal of Production Economics* **228** 107730

[2] Coelho P M, Corona B, ten Klooster R and Worrell E 2020 Sustainability of reusable packaging–Current situation and trends *Resources, Conservation & Recycling: X* **6** 100037

[3] Torcătoru C and Săvescu D 2019 Analyzing the sustainability of an automotive component using SolidWorks CAD software *IOP Conf. Ser.: Mater. Sci. Eng.* **568** 012113

[4] European Commission 2014 *Directive 2014/47/EU of the European Parliament and of the Council of 3 April 2014 on the technical roadside inspection of the roadworthiness of commercial vehicles circulating in the Union and repealing Directive 2000/30/EC Text with EEA relevance* vol 127

[5] EN 2010 *EN 12195-1:2010 - Load restraining on road vehicles - Safety - Part 1: Calculation of securing forces*

AUTHOR INDEX